Modeling and Simulation Support for System of Systems Engineering Applications

Modeling and Simulation Support for System of Systems Engineering Applications

Edited by

Larry B. Rainey
Andreas Tolk

Copyright © 2015 by John Wiley & Sons, Inc. All rights reserved

Published by John Wiley & Sons, Inc., Hoboken, New Jersey
Published simultaneously in Canada

No part of this publication may be reproduced, stored in a retrieval system, or transmitted in any form or by any means, electronic, mechanical, photocopying, recording, scanning, or otherwise, except as permitted under Section 107 or 108 of the 1976 United States Copyright Act, without either the prior written permission of the Publisher, or authorization through payment of the appropriate per-copy fee to the Copyright Clearance Center, Inc., 222 Rosewood Drive, Danvers, MA 01923, (978) 750-8400, fax (978) 750-4470, or on the web at www.copyright.com. Requests to the Publisher for permission should be addressed to the Permissions Department, John Wiley & Sons, Inc., 111 River Street, Hoboken, NJ 07030, (201) 748-6011, fax (201) 748-6008, or online at http://www.wiley.com/go/permissions.

Limit of Liability/Disclaimer of Warranty: While the publisher and author have used their best efforts in preparing this book, they make no representations or warranties with respect to the accuracy or completeness of the contents of this book and specifically disclaim any implied warranties of merchantability or fitness for a particular purpose. No warranty may be created or extended by sales representatives or written sales materials. The advice and strategies contained herein may not be suitable for your situation. You should consult with a professional where appropriate. Neither the publisher nor author shall be liable for any loss of profit or any other commercial damages, including but not limited to special, incidental, consequential, or other damages.

For general information on our other products and services or for technical support, please contact our Customer Care Department within the United States at (800) 762-2974, outside the United States at (317) 572-3993 or fax (317) 572-4002.

Wiley also publishes its books in a variety of electronic formats. Some content that appears in print may not be available in electronic formats. For more information about Wiley products, visit our web site at www.wiley.com.

Library of Congress Cataloging-in-Publication Data:

Rainey, Larry B.
 Modeling and simulation support for system of systems engineering applications / Larry B. Rainey, Colorado Springs, CO, Andreas Tolk, SimIS, Inc., Portsmouth, VA.
   pages cm
 Includes bibliographical references and index.
 ISBN 978-1-118-46031-3 (hardback)
 1. Systems engineering–Data processing. 2. Engineering models.
 I. Tolk, Andreas. II. Title.
 TA168.R338 2015
 620.001′13–dc23

2014022284

Printed in the United States of America

10 9 8 7 6 5 4 3 2 1

This text is dedicated to **Mr. Kevin Hibbs** of the Missile Defense Agency (MDA), an avid supporter of embracing and exercising new modeling and simulation techniques to enhance the mission of MDA, in which he recognized many system of systems related challenges and was pivotal in utilizing modeling and simulation to address these challenges.

We also remember Colonel (US Air Force, ret.) **Thomas W. O'Brien** who contributed directly as an author to this book. His insights from practical experiences and academic reflections shaped many ideas that we are taking now for granted when we study system of systems.

Contents

Foreword xi

List of Contributors xiii

Notes on Contributors xvii

List of Acronyms xxxi

Part I Overview and Introduction

1. **Overview and Introduction to Modeling and Simulation Support for System of Systems Engineering Applications** 3
   Larry B. Rainey and Andreas Tolk

2. **The Role of Modeling and Simulation in System of Systems Development** 11
   Mark W. Maier

Part II Theoretical and Methodological Considerations

3. **Composability** 45
   Michael C. Jones

4. **An Approach for System of Systems Tradespace Exploration** 75
   Adam M. Ross and Donna H. Rhodes

5. **Data Policy Definition and Verification for System of Systems Governance** 99
   Daniele Gianni

6. System Health Management — 131
Stephen B. Johnson

7. Model Methodology for a Department of Defense Architecture Design — 145
R. William Maule

Part III Theoretical and Methodological Considerations with Applications and Lessons Learned

8. An Agent-Oriented Perspective on System of Systems for Multiple Domains — 187
Agostino G. Bruzzone, Alfredo Garro, Francesco Longo, and Marina Massei

9. Building Analytical Support for Homeland Security — 219
Sanjay Jain, Charles W. Hutchings, and Yung-Tsun Tina Lee

10. Air Transportation Systems — 249
William Crossley and Daniel DeLaurentis

11. Systemigram Modeling for Contextualizing Complexity in System of Systems — 273
Brian Sauser and John Boardman

12. Using Modeling and Simulation for System of Systems Engineering Applications in the European Space Agency — 303
Joachim Fuchs and Niklas Lindman

13. System of Systems Modeling and Simulation for Microgrids Using DDDAMS — 337
Aristotelis E. Thanos, DeLante E. Moore, Xiaoran Shi, and Nurcin Celik

14. Composition of Behavior Models for Systems Architecture — 361
Clifford A. Whitcomb, Mikhail Auguston, and Kristin Giammarco

15. Joint Training — 393
James Harrington, Laura Hinton, and Michael Wright

16. **Human in the Loop in System of Systems (SoS) Modeling and Simulation: Applications to Live, Virtual, and Constructive (LVC) Distributed Mission Operations (DMO) Training**   415

    Saurabh Mittal, Margery J. Doyle, and Antoinette M. Portrey

17. **On Analysis of Ballistic Missile Defense Architecture through Surrogate Modeling and Simulation**   453

    Tommer R. Ender, Philip D. West, William Dale Blair, and Paul A. Miceli

18. **Medical Enhancements to Sustain Life during Extreme Trauma Care**   479

    L. Drew Pihera, Nathan L. Adams, Tommer R. Ender, and Matthew L. Paden

19. **Utility: Problem-Focused, Effects-Based Analysis (aka Information Value Chain Analysis)**   515

    Thomas W. O'Brien and John F. Sarkesain

20. **A Framework for Achieving Dynamic Cyber Effects through Distributed Cyber Command and Control/Battle Management (C^2/BM)**   531

    John F. Sarkesain and Thomas W. O'Brien

21. **System of Systems Security**   565

    Bharat B. Madan

Part IV   Conclusions

22. **Toward a Research Agenda for M&S Support of System of Systems Engineering**   583

    Andreas Tolk and Larry B. Rainey

Index   593

Foreword

System of Systems (SoS) is a maturing field, which has been described and applied in many different ways. However, at the core of different SoS descriptions and approaches is the focus on integration and coordination of multiple complex systems to achieve levels of performance, offer capabilities, or achieve purposes that are beyond the grasp of individual constituent systems. In effect, getting a set of systems that were not initially formed as a unity to come together as a unity to support a higher level (SoS) purpose. It is exciting to see the incorporation of Modeling and Simulation (M&S) presented in this volume for support and maturation of the SoS field through engineering applications.

M&S presented in this volume assists in four critical challenges facing SoS field development. First, they enable a better systems-based understanding, appreciation, and representation in the landscape of the twenty-first century enterprise. This landscape is beset with:

- Proliferation of information intensive systems and technologies, where creating interoperability extends beyond the "hard system" technical dimensions to include "soft system" dimensions such as conceptual interoperability
- Divergence in stakeholders interests and perspectives that invoke considerations for exploring and representing multiple frames of reference and their potential incompatibility
- Expectations for immediate results without long view considerations, where consequences of decisions may be significantly separated in space and time from the point of decision/action
- Scarce and dynamically shifting resources that make traditional planning forums suspect and suggest the necessity for near real time analysis of SoS level impacts stemming from dynamic shifts
- Application environments fraught with high degrees of complexity, uncertainty, emergence, and ambiguity. These "hyperturbulent" environments are subject to rapid and unexpected shifts and are no longer aberrations but are the norm
- Increasing doubt concerning the efficacy of traditional approaches based on reductionism, cause and effect certainty, absolute understanding, and results repeatability to effectively engage complex system problems

Such conditions are inevitable in the twenty-first landscape and cannot be eliminated. However, they can be managed, that is, they invoke the SoS challenge to improve effectiveness through different thinking, approaches, and applications. It is compelling that this volume offers examples of a direct assault on the SoS problem domain through application of M&S.

Second, SoS problems are inherently multidisciplinary, and they require multidisciplinary thinking and approaches for effective resolution. The SoS problem domain crosses technology, human/social, organizational, managerial, policy, and political boundaries. By their very nature, M&S applications are multidisciplinary and the array of applications in this volume attest to the capability of M&S applications to address the multidisciplinary nature of the SoS problem domain.

Third, poor decisions related to integration and coordination of SoS can have dire consequences, ranging from significant wasted resources to poor system performance, or even death. I would suggest strengthening the ability to make better decisions is the strongest contribution that M&S makes to the maturation of the SoS field. As this volume demonstrates, M&S brings the following capabilities to support better analysis and decisions in SoS: (i) systemic formulation of multidisciplinary problems through M&S driven representations, (ii) examination of the SoS behavior and responses to potential system changes in a "failsafe" setting, (iii) understanding the consequences for "poor" decisions prior to implementation and examination of alternative courses of action, and (iv) compression of time to better understand the potential long-term impacts for decisions.

Fourth, although engineering has always been about the practicality of solving problems and fulfilling unmet needs, grounding of applications in theoretical and methodological foundations is critical to maturing the SoS field. It is this grounding that will ultimately provide coherent maturation of the SoS field as well as long-term sustainability. The focused concentration of this volume on the theoretical and methodological implications for applications is laudable and makes a significant contribution to the maturing SoS field.

I have confidence that the work of this volume will be embraced for the significant weigh point it provides in the continuing journey of both the SoS and M&S fields. The contributions of this volume represent an important and decisive step forward in demonstrating the theoretical and methodological contributions of M&S to SoS through the lenses of application. The multidisciplinary breadth of applications ranging from security to transportation to space demonstrates the wide ranging applicability of SoS as informed by M&S.

CHARLES B. KEATING
Norfolk, VA, USA

List of Contributors

Larry B. Rainey
Integrity Systems and Solutions, LLC, Colorado Springs, CO, USA

Andreas Tolk
SimIS Inc., Portsmouth, VA, USA

Nathan L. Adams
Georgia Tech Research Institute, Atlanta, GA, USA

Mikhail Auguston
Naval Postgraduate School, Monterey, CA, USA

William Dale Blair
Georgia Tech Research Institute, Atlanta, GA, USA

John Boardman
John Boardman Associates, Worcestershire, UK

Agostino G. Bruzzone
DIME, University of Genoa, Genoa, Italy

Nurcin Celik
Department of Industrial Engineering, University of Miami, Coral Gables, FL, USA

William Crossley
School of Aeronautics and Astronautics, Purdue University, West Lafayette, IN, USA

Daniel DeLaurentis
School of Aeronautics and Astronautics, Purdue University, West Lafayette, IN, USA

Margery J. Doyle
L-3 Communications, Link Simulation and Training, Air Force Research Lab, Wright-Patterson Air Force Base, OH, USA

Tommer R. Ender
Georgia Tech Research Institute, Atlanta, GA, USA

Joachim Fuchs
European Space Agency, ESTEC, Noordwijk, the Netherlands

Alfredo Garro
DIMES, University of Calabria, Calabria, Italy

Kristin Giammarco
Naval Postgraduate School, Monterey, CA, USA

Daniele Gianni
European Organization for the Exploitation of Meteorological Data, Darmstadt, Germany

James Harrington
The MITRE Corporation, McLean, VA, USA

Laura Hinton
The MITRE Corporation, McLean, VA, USA

Charles W. Hutchings
National Institute of Standards and Technology, Gaithersburg, MD, USA

Sanjay Jain
The George Washington University, Washington, DC, USA

Stephen B. Johnson
National Aeronautics and Space Administration, Marshall Space Flight Center, Huntsville, AL, USA
and
Dependable System Technologies, LLC and the University of Colorado, Colorado Springs, CO, USA

Michael C. Jones
The Johns Hopkins University Applied Physics Laboratory, Laurel, MD, USA

Yung-Tsun Tina Lee
National Institute of Standards and Technology, Gaithersburg, MD, USA

Niklas Lindman
European Space Agency, ESTEC, Noordwijk, the Netherlands

Francesco Longo
DIMES, University of Calabria, Calabria, Italy

Bharat B. Madan
Old Dominion University, Norfolk, VA, USA

Mark W. Maier
The Aerospace Corporation, Chantilly, VA, USA

Marina Massei
DIME, University of Genoa, Genoa, Italy

R. William Maule
Naval Postgraduate School, Monterey, CA, USA

Paul A. Miceli
Georgia Tech Research Institute, Atlanta, GA, USA

Saurabh Mittal
L-3 Communications, Link Simulation and Training,
Air Force Research Lab, Wright-Patterson Air Force Base, OH, USA

DeLante E. Moore
Department of Industrial Engineering, University of Miami,
Coral Gables, FL, USA

Thomas W. O'Brien
US Air Force, Colonel (Retired)

Matthew L. Paden
Children's Healthcare of Atlanta, Emory University, Atlanta, GA, USA

L. Drew Pihera
Georgia Tech Research Institute, Atlanta, GA, USA

Antoinette M. Portrey
L-3 Communications, Link Simulation and Training, Air Force Research Lab,
Wright-Patterson Air Force Base, OH, USA

Donna H. Rhodes
Massachusetts Institute of Technology, Cambridge, MA, USA

Adam M. Ross
Massachusetts Institute of Technology, Cambridge, MA, USA

John F. Sarkesain
SIOC Group, L.L.C. and The Aerospace Corporation, Ashburn, VA, USA

Brian Sauser
Department of Marketing and Logistics, University of North Texas, Denton, TX, USA

Xiaoran Shi
Department of Industrial Engineering, University of Miami, Coral Gables, FL, USA

Aristotelis E. Thanos
Department of Industrial Engineering, University of Miami, Coral Gables, FL, USA

Philip D. West
Georgia Tech Research Institute, Atlanta, GA, USA

Clifford A. Whitcomb
Naval Postgraduate School, Monterey, CA, USA

Michael Wright
US Army Project Executive Office for Simulation Training and Instrumentation, Orlando, FL, USA

Notes on Contributors

Nathan L. Adams is a Research Engineer II at the Georgia Tech Research Institute and has over 12 years of experience in systems engineering, embedded systems development, and analysis and simulation support tool development. His current area of focus involves systems engineering, integration, and test and evaluation of embedded aircraft systems. Prior to joining the Georgia Tech Research Institute, Mr. Adams worked for Lockheed Martin Aeronautics Corporation where he developed and integrated real-time simulation systems in support of various simulation laboratories. He holds a B.S. in Computer Engineering Technology from Southern Polytechnic State University and a Professional Master's in Applied Systems Engineering from the Georgia Institute of Technology.

Mikhail Auguston is an Associate Professor in the Computer Science Department, Naval Postgraduate School in Monterey, California. Dr. Auguston's research interests encompass programming language design and implementation, real-time and reactive software testing and debugging automation and safety assessment tools, software and system architecture formal specification and validation, and visual programming languages. He has more than 40 years of experience in these areas and has published more than 120 papers in refereed journals, conferences, and workshop proceedings. He was PI and Co-PI in projects funded by NSF, NASA, U.S. Army Research Office, U.S. Naval Research Office, U.S. Missile Defense Agency, USMC Technology Center, TARDEC, SPAWAR/SYSCOM, and U.S. Army/Monterey TRAC.

William Dale Blair is a Principal Research Engineer with Georgia Tech Research Institute (GTRI) and is a Fellow there. He recently completed a three year assignment as the Technical Director for the C2BMC Knowledge Center of the Missile Defense Agency. Since joining GTRI in 1997, Dr. Blair has led a multiorganizational team in the development of multiplatform–multisensor–multitarget benchmarks to both air defense and ballistic missile defense. His projects at GTRI focus mostly on the modeling and simulation and algorithm assessment associated with the sensor netting for the C2BMC. Dr. Blair's research is reported in over 200 articles that include 38 refereed journals. He served as the Editor for Radar Systems for IEEE Transactions on Aerospace and Electronic Systems (T-AES) 1996-99 and Editor-in-Chief (EIC) for IEEE T-AES from 1999 to 2005. He is a Fellow of the IEEE and recipient of the 2001 IEEE Nathanson Award for Outstanding Young Radar Engineer. He is Coeditor and coauthor of the book, *Multitarget-Multisensor Tracking: Advances and Applications III*, and the author of chapter 19 "Radar Tracking Algorithms" and coauthor of chapter 18 "Radar Measurements" of the new edition of *Principles of Modern Radar*. He has served on the Board of Governors for the IEEE Aerospace and Electronics Systems Society (AESS) for 1998–2003, 2005–2010, and 2012–2014.

John Boardman has been an engineer, consultant, researcher, teacher, and public speaker. He has held academic appointments in the United Kingdom and the United States. Most recently he was a Distinguished Service Professor at Stevens Institute of Technology in the School of Systems and Enterprises where he taught graduate classes on systems thinking and enterprise architecting. His specialty subjects have covered electrical engineering, computer engineering, software, and systems engineering. He has coauthored two books, with Brian Sauser, on systems thinking. He is a Fellow of the Institution of Engineering and Technology. He now concentrates on writing and has recently completed Part I of a trilogy: "Memories live longer than dreams."

Agostino G. Bruzzone is a Full Professor at DIME University of Genoa, Director of M&S Net (International Network involving 34 centers), Director of the MISS McLeod Institute of Simulation Science—Genoa Center (28 centers distributed worldwide), Founder and President of the Liophant Simulation, Member of the Simulation team, Vice President and Member of the Board of MIMOS (Movimento Italiano di Simulazione), and Member of the NATO MSG. He works on innovative Modeling & Simulation, AI techniques, application of Neural Networks, and Gas. He is Member of several International Technical and Organization Committees (i.e., AI Application of IASTED, AI Conference, ESS, AMS) and General Coordinator of Scientific Initiatives (i.e., I3M General Chair). He teaches "M&S" for the DIMS Ph.D. Program (Doctorship in Integrated Mathematical M&S). He is Director of the Master Program in Industrial Plants for the University of Genoa. He has a new appointment as Project Leader M&S at NATO STO CMRE—Centre for Maritime Research and Experimentation.

Nurcin Celik is an Assistant Professor in the Department of Industrial Engineering at the University of Miami. She received her M.S. and Ph.D. degrees in Systems and Industrial Engineering from the University of Arizona with magna cum laude. Her research interests are in the areas of integrated modeling and decision making for large-scale, complex, and dynamic systems focusing on electric utility resource planning and dynamic load dispatching in distributed power grids and microgrids. She is the recipient of several awards including 2013 AFOSR Young Investigator Research Award, 2011 WSC Best Paper Award, 2011 IAMOT (International Association for Management of Technology) Best Research Project Award, 2010 University of Miami Provost Award, 2009 IIE (Institute of Industrial Engineers) Best Graduate Research Award, and 2007 Diversity in Science and Engineering Award from Women in Science and Engineering Program.

William Crossley is a Professor of Aeronautics and Astronautics at Purdue University in West Lafayette, Indiana, USA. His teaching and research interests are in design optimization for aerospace systems and for system of system design problems. He initiated the "System of Systems" signature area in the College of Engineering in 2004 and is involved in the continuing development of systems and system of systems efforts at Purdue University. He earned a BSE (Aerospace) from the University of Michigan, then participated in the Arizona State University Industrial Fellows Program through which he worked in Advanced Concept Development at McDonnell Douglas Helicopter Systems in Mesa, AZ, while earning his M.S. and Ph.D. degrees. He is an Associate Fellow of the American Institute of Aeronautics and Astronautics and the chair of the Aircraft Design

Technical Committee, an Executive Committee Member for the Council of Engineering Systems Universities, a Member of the International Council on Systems Engineering, a Member of the Institute for Operations Research and Management Science, and a Member of the International Society for Structural and Multidisciplinary Optimization.

Daniel DeLaurentis is an Associate Professor in Purdue's School of Aeronautics & Astronautics and the Director of Purdue's Center for Integrated Systems in Aerospace (CISA), which is home to 20 faculty affiliates, three research staff, and numerous dedicated graduate students. His primary research interests are in the areas of problem formulation, modeling and robust system design, and control methods for aerospace systems and system of systems. This includes agent-based modeling, network theory, optimization, aerospace vehicle modeling, missile defense battle management architecting, and air transportation network analysis. Dr. DeLaurentis is an Associate Fellow of the American Institute of Aeronautics and Astronautics and served as Chairman of the AIAA's Air Transportation Systems (ATS) Technical Committee from 2008 to 2010. He was Co-Chair of the System of Systems Technical Committee in the IEEE System, Man, and Cybernetics Community and Associated Editor for the *IEEE Systems Journal* for several years.

Margery J. Doyle is a Research Consultant Cognitive Systems Research Scientist and Engineer with L-3 Communications Link Simulation and Training at the Air Force Research Lab 711 HPW/RHA Warfighter Readiness Research Division at Wright-Patterson Air Force Base, OH. Margery leads the Not-So-Grand-Challenge to support integration and use of cognitive- and behavior-based models, agents, and architectures in adaptive Live Virtual Constructive training environments. She earned her M.A. in Experimental Psychology with a certificate in Cognitive Psychology from the University of West Florida in 2007. While attending UWF, Margery made significant sustaining contributions to the field of Augmented Cognition through working on a seminal DARPA challenge entitled AUGCOG. In addition, Margery completed work toward a Ph.D. in Cognitive Science. Recently she coedited a special edition of Cognitive Systems Research focusing on stigmergic systems that display properties of emergence.

Tommer R. Ender is a Senior Research Engineer at the Electronic Systems Laboratory of the Georgia Tech Research Institute and serves as Chief of the Systems Technology & Analysis Division. His primary area of research includes development of collaborative systems engineering tools and methods as applied to complex system of systems, concerned with supporting decision making through a holistic treatment of various problems. His research focuses on the application of Model Based Systems Engineering, advanced design methods, uncertainty analysis, and multidisciplinary design optimization to defense related, hybrid energy, and other complex systems. Dr. Ender is an instructor and course developer for Georgia Tech's Professional Master's in Applied Systems Engineering, and the Georgia Tech Professional Education's Systems Engineering Certificate, teaching courses in systems engineering fundamentals, modeling and simulation, system of systems, architecting, and project management. He is an active member of and regularly publishes with IEEE, INCOSE, and NDIA, and is a certified Project Management Professional. He earned a B.S., M.S., and Ph.D. in Aerospace Engineering from the Georgia Institute of Technology.

Joachim Fuchs is the Head of the System Modeling and Functional Verification section at the European Space Agency in Noordwijk, The Netherlands. He is responsible to support most of ESA's space programs in the area of verification over the entire life-cycle, starting from the definition phase to the integration and testing. Model-based approaches are the focus of research and development activities of the group, with substantial effort in the definition and development of methods and tools to support the system engineering and architecting function. His group is leading in conceptual data modeling methods and model-based system engineering in the space system development. He was convener and member of standardization groups dealing with simulation standards at implementation level as well as life-cycle related in support of system engineering. He led the development of an engineering environment to support a "Virtual Spacecraft Design" process. He is Member of INCOSE, SCS, and SISO.

Alfredo Garro is an Associate Professor of Computing Systems at the Department of Informatics, Modeling, Electronics and Systems Engineering (DIMES) of the University of Calabria (Italy). From 1999 to 2001, he has been a researcher at CSELT, the Telecom Italia Group R&D Lab. From 2001 to 2003, he collaborates with the Institute of High Performance Computing and Networking of the Italian National Research Council (CNR). In February 2005, he received the Ph.D. degree in Systems and Computer Engineering from the University of Calabria. From January 2005 to December 2011, he has been an Assistant Professor of Computing Systems at the DIMES Department (formerly DEIS) of the University of Calabria. His main research interests include systems and software engineering, reliability engineering, modeling and simulation. His list of publications contains about 80 papers published in international journals, books, and proceedings of international and national conferences. He is a Member of the IEEE and IEEE Computer Society from 2005 and a Member of the IEEE Reliability Society and IEEE Aerospace and Electronic Systems Society. He is a Member of the International Council on Systems Engineering (INCOSE). He has been elected for the 2014 SPACE Forum Planning and Review Panel (PRP) of the Simulation Interoperability Standards Organization (SISO). He is Member of the Executive Committee of the MODRIO (Model Driven Physical Systems Operation) ITEA2 Project and the Technical Contact for his Institution in the Open Source Modelica Consortium (OSMC).

Kristin Giammarco is an Associate Professor in the Department of Systems Engineering at the Naval Postgraduate School, Monterey, California. Her research interests include the use and development of formal methods for the improvement of system behavior modeling and patterning system processes and pathologies at the architectural level. She teaches courses in systems architecture and design, systems integration and development, systems software engineering, and model-based systems engineering at NPS and also teaches system architecture and design as an adjunct at Stevens Institute of Technology. She is an Active Member of INCOSE, serving on both the Tools Database and Systems Science Working Groups, and is a Member of the Lifecycle Modeling Language Steering Committee. She has earned a Ph.D. in Software Engineering, an M.S. in Systems Engineering Management through the Joint Executive PD-21 Program, and a Certificate in Advanced Systems Engineering, all from NPS. She holds a B.E. in Electrical Engineering from Stevens Institute of Technology.

Daniele Gianni is Requirements Engineering Consultant at the European Organization for the Exploitation of Meteorological Data (EUMETSAT) (DE) where he supports the requirements management activities of future space programs with the definition of requirements processes and requirements models. Previously, he held research appointments at the European Space Agency (NL), the University of Oxford (UK), and the Imperial College (UK), where he introduced new M&S methods for the systems engineering activities in domains such as space, software architectures, software performance modeling, biomedical engineering, and emergency management, publishing over 50 papers in journals, conferences, and workshops. On the same topics, he also held visiting positions at the Auckland Bioengineering Institute (NZ) and the Georgia Institute of Technology. He is the co-organizer of several international workshops on the themes of Model-Driven Simulation Engineering (Mod4Sim in the Spring Simulation Multiconference) and Collaborative M&S (COMETS in IEEE WETICE), and he is currently leading the conclusion of a book on M&S-based Systems Engineering. Gianni holds a Ph.D. and an M.S. in Computer Engineering from the University of Rome TorVergata (IT).

James Harrington is a Principal Modeling and Simulation Engineer for the MITRE Corporation in McLean, Virginia. He received a B.S. and an M.S. in Computer Science from Virginia Polytechnic Institute and State University and the Johns Hopkins University, respectively. As a Member of MITRE's Modeling and Simulation (M&S) Engineering division, he serves as a Capability Lead for Simulation Interoperability and Live-Virtual-Constructive Architectures focus area. He has worked in direct support of the Army's Program Executive Office for Simulation Training and Instrumentation since 2004 with a particular focus on system of systems engineering for M&S-based training federations.

Laura Hinton is a Lead Simulation Engineer for the MITRE Corporation in McLean, Virginia, where she works with various Department of Defense distributed simulation programs. She holds a B.S and an M.S. in Engineering from Purdue University. She has been involved with Services Modeling & Simulation (M&S) for 25 years. As a Member of MITRE's M&S Engineering division, she serves as a Capability Lead for the Simulation Interoperability and Live-Virtual-Constructive Architectures focus area. Since 2006, she has been actively involved with constructive simulation environments used for training army unit commanders and staff.

Charles W. Hutchings currently serves in the Office of the Director, Operational Test and Evaluation in the U.S. Department of Defense (DoD) and is Guest Researcher at the National Institute of Standards and Technology (NIST). Prior to transferring to DoD, in 2012, he coordinated development of risk management methodology and tools—including models and simulations—to support systems development and program management in the Program Accountability and Risk Management Division of the Management Directorate, U.S. Department of Homeland Security (DHS). From 2007 to 2011, he was Deputy Director, Modeling and Simulation in the DHS Science & Technology Directorate. He is the author of four peer reviewed papers on the management and use of homeland security modeling and simulation. From 2001 to 2007, Dr. Hutchings served in the U.S. Navy both as an Engineering Duty Officer and as a Navy civilian. Prior to government service, Dr. Hutchings completed post-doctoral work and was appointed Visiting Professor in the Department of Physics at the University of Nebraska, Lincoln. He completed a

Ph.D. in Physical Chemistry at University of Heidelberg, Heidelberg, Germany in 1994. He completed M.S. and B.S. degrees in Physics and Mathematics at Syracuse University, Syracuse New York in 1990 and 1987, respectively.

Sanjay Jain is an Associate Industry Professor in the Department of Decision Sciences at the School of Business at George Washington University (GWU). His research interests are in the development and application of decision science techniques to complex systems. Prior to joining GWU, he was a Research Faculty Member at Grado Industrial & Systems Engineering Department at Virginia Tech. Before moving to academia, he accumulated over a dozen years of international R&D and consulting experience working at Accenture, Singapore Institute of Manufacturing Technology, and General Motors North American Operations Technical Center. He has over 80 publications including technical reports, and papers in technical journals, and refereed conference proceedings. His recent work has been published in the *European Journal of Operational Research, International Journal of Advanced Intelligence Paradigms,* and *International Journal of Production Economics*. He serves as an Associate Editor of the *International Journal of Simulation and Process Modeling* and has served as a Coeditor for the Proceedings of 2010 and 2011 Winter Simulation Conferences. He received a Bachelors of Engineering from Indian Institute of Technology (IIT), Roorkee, India, a Post-Graduate Diploma from National Institute for Industrial Engineering, Mumbai, India, and a Ph.D. in Engineering Science from Rensselaer Polytechnic Institute, Troy, New York.

Stephen B. Johnson is the analysis lead for Mission and Fault Management on the National Aeronautics and Space Administration's Space Launch System program, led by Marshall Space Flight Center. In this position, he oversees and performs qualitative and quantitative analysis of nominal and off-nominal system behaviors and of fault management effectiveness. He is the President of Dependable System Technologies, LLC and an Associate Research Professor with the Department of Mechanical and Aerospace Engineering at the University of Colorado, Colorado Springs. He is the General Editor for "System Health Management: with Aerospace Applications" (July 2011), the author of "The Secret of Apollo: Systems Management in American and European Space Programs," and many other articles and books in system health management, systems engineering, space economics, and space history. He has a Bachelor's degree in physics from Whitman College (1981) and a Ph.D. in the history of science and technology from the University of Minnesota, Twin Cities (1997). His current research focuses on the theory and application of system health management, the fundamental reform of systems engineering from a process into a product-based and model-based discipline, the philosophy of technology and engineering, and historical work in space history, cognitive psychology, and artificial intelligence. He is on the editorial board of the *International Journal of Prognostics and Health Management, Astropolitics,* and *Quest: The History of Spaceflight Quarterly*, and a Member of the IEEE PHM Standards committee.

Michael C. Jones is a Program Manager at the Johns Hopkins University Applied Physics Laboratory and a Lecturer in Systems Engineering at the Johns Hopkins University's Whiting School of Engineering. He earned a B.S. in Computer Science from the U.S. Naval Academy and completed 20 years as a submarine officer in the U.S. Navy. He holds an M.S. in Electrical Engineering and an MBA from the Naval Postgraduate School, and is a Ph.D. student in Modeling and Simulation at Old Dominion University. He is currently serving as the Science and Technology Advisor to the Commander, U.S. Pacific Fleet.

Yung-Tsun Tina Lee is a Computer Scientist of the Engineering Laboratory at the National Institute of Standards and Technology (NIST). Her major responsibility in recent years is to develop information models to support various manufacturing applications. She is the technical leader of the project of National Ambulance Specification. She holds a B.S. in Mathematics from the Providence University, Taiwan, ROC, and an M.S. in Applied Science from the College of William and Mary, Virginia, USA. She received the Department of Commerce/NIST Bronze Medal Award in 2011. She was the Coeditor of SISO-STD-008-2010 and SISO-STD-008-1-2012, standards for Core Manufacturing Simulation Data (CMSD) and is currently the chair of the CMSD Product Support Group of the Simulation Interoperability Standards Organization (SISO). She has authored over 80 technical papers relating to manufacturing systems integration and modeling and simulation.

Niklas Lindman is a System Engineer at the European Space Agency (ESA) in Noordwijk, The Netherlands, where he is currently working on the European GNSS program, Galileo. Apart from his various system engineering roles in several space projects he has also been responsible for R&D activities in the area of model-based system engineering, simulation, modeling, and system of systems engineering methods and tools at the technical and quality directorate (D/TEC) at ESA. He has also actively contributed to various standardization activities within a number of standardization bodies (ECSS, CCSDS and OMG) on the topics of system engineering, simulation, and modeling. He holds an M.S. in Space Engineering from the University of Umea in Umea, Sweden.

Francesco Longo took his degree in Mechanical Engineering, summa cum Laude, in October 2002 from the University of Calabria. He received his Ph.D. in Mechanical Engineering from University of Calabria in January 2006. He is currently Assistant Professor at the Department of Mechanical, Energy, and Management Engineering (DIMEG) of the University of Calabria. His research interests include Modeling & Simulation applied to Industry, Logistics, and Defense. He is Director of the Modeling & Simulation Center—Laboratory of Enterprise Solutions (MSC-LES), a laboratory operating at the DIMEG Department. He has published more than 150 scientific papers on international conferences and journals participating as speaker and chairman to different international conferences. He serves as Scientific Responsible and Principal Investigator of many research projects (including different stakeholders in the field of Logistics and Defense). He is Associate Editor of the *Simulation: Transaction of the society for Modeling & Simulation International* and Guest Editor of the *International Journal of Simulation and Process Modelling*. He has been Program Chair of the Summer Simulation Multi-Conference (2012 and 2013), Vice General Chair of the Summer Simulation Multi-Conference (SummerSim2011), General Co-Chair of the European Modeling & Simulation Symposium (EMSS 2008-2013), General Chair of the International Conference on Modeling & Applied Simulation (MAS 2009, 2010), and Program Chair of the International Mediterranean and Latin American Modeling Multi-Conference (2007 and 2008, 2011–2013). He is Member of the International Program Committees of the most important conferences in the simulation field.

Bharat B. Madan joined Old Dominion University in August 2012 as a Professor in the Department of Modeling, Simulation, and Visualization Engineering. He is the Director of the Advanced Engineering Graduate Certificate Program in Cyber Systems Security. Prior to joining ODU, he was with the Applied Research Laboratory, Penn State University, where he headed the Distributed Systems Department. He conducts research in the areas of cyber security, cyber-attack tolerance, cyber-attack attribution, hardware acceleration of crypto algorithm under power and space budget constraints, malware behavioral modeling, simulation and visualization of large distributed systems, modeling and analysis of financial time series data, and the application of web services to distributed autonomous systems. He conducted cyber security related research among others for the Air Force Office of Scientific Research (AFSOR), the Defense Threat Reduction Agency (DTRA), the Office of Naval Research under their Multi-disciplinary University Initiative (MURI), and the Defense Advanced Research Project Agency (DARPA). He has more than 25 years experience in academia at IIT Delhi (1976-1996), Naval Postgraduate School (1984-1985), University of Delaware (1988-1989), Duke University (2001-2003), and ARL Penn State University (2003-2012). He has also worked in the industry R&D for over five years, first at IBM (1996-1999) designing and implementing networking protocol stacks (for the OS2 and the main frame OS/390 operating systems) IPSec & Secure FTP and after that at Ericsson, where he was involved in R&D activities related to wireless voice and data networking, hand-held multimedia communication devices and mobile IP.

Mark W. Maier is an author and practitioner of systems architecting, the art and science of creating complex systems. He is coauthor, with Dr. Eberhardt Rechtin, of *The Art of Systems Architecting*, Third Edition, CRC Press, as well more than 50 papers on systems engineering, architecting, and sensor analysis. He is currently a Distinguished Engineer at The Aerospace Corporation, a nonprofit corporation that operates a Federally Funded Research and Development Center with oversight responsibility for the U.S. National Security Space Program. At Aerospace he founded the systems architecting training program (an internal and external training program) and applies architecting methods to government and commercial clients, particularly in portfolios-of-systems and research and development problems. He received the B.S. and M.S. degrees from the California Institute of Technology and the Engineer and Ph.D. degrees in Electrical Engineering from the University of Southern California. While at USC, he held a Hughes Aircraft Company Doctoral Fellowship, where he was also employed as a section head. Prior to coming to The Aerospace Corporation, he was an Associate Professor of Electrical and Computer Engineering at the University of Alabama at Huntsville.

Marina Massei studied at the University of Genoa, Humanistic Area on File Structures & Data Warehousing. She obtained 2-years diploma from the School of Archival Science at the National Archive of Genoa. She had attended Postgraduate Courses in Project Management, Modeling & Simulation, Operational Management, Total Quality Management, and Logistics. She obtained the certification as CAX Operator. She has operated with the team of Prof. Mosca and Prof. Bruzzone at DIME University of Genoa as project controller. She participated in the organization of SIREN Courses (i.e., HLA, VV&A, M&S), HMS2002/2003/2004 and IEPAL Meetings. She was involved in the organization of International Scientific Events (i.e., Summer Computer Simulation Conference—2003: Montreal, 2004: San Jose', 2005: Philadelphia) and in the coordination of Technical Council specialized in Advanced Techniques (i.e., SIMPLEST). She is Member of the International Program Committee of major M&S Conferences. She served as Chair for Applications in Management, Planning & Forecasting Workshop in 2004, 2005, and 2006 SummerSim in San Jose', Philadelphia, and Calgary. She is Associate Director of the McLeod Institute for Simulation Science located in Perugia University. She worked on special seminar on Problem Solving, Project Management, Data Analysis, and Team Working for undergraduates and postgraduates in DIME/DIPTEM Organized Courses. She was Finance Control & Administration Director of MAST. She conducted several projects on Mobile Simulation and Virtual Simulation. She is currently enrolled in DIME University of Genoa as Member of the Simulation Team of Prof. Agostino Bruzzone.

R. William Maule is Research Associate Professor at the Naval Postgraduate School where he is active in field research with the Navy fleet and joint forces. His research includes design, development, and deployment of service architectures and analysis of future command and control systems, sensor and fusion networks, intelligence, and surveillance and reconnaissance technologies. Research and development sponsors have included OPNAV, JFCOM, NAVNETWARCOM, USFFC, and SPAWAR. His team received the DoN CIO Award for IT Excellence. Before NPS, Dr. Maule worked for ten years in enterprise systems with a major telecommunications corporation and consulted with prominent Silicon Valley companies. Prior to this, he worked in research and development that included programming, knowledge management, artificial intelligence, and application security at a major federal supercomputer center. He has produced more than 150 technical reports and has published more than 65 refereed articles. His degrees are from Michigan State University, the University of Florida, and the University of California, Berkeley. He is a Member of ACM and IEEE.

Paul A. Miceli received the B.S. degree in electrical engineering from Manhattan College, Bronx, NY, and the M.S. degree in electrical engineering from the Georgia Institute of Technology, Atlanta. He is a Research Engineer for the Electronics Systems Branch of the Georgia Tech Research Institute. During his time in graduate school, his research focused on the development of signal processing algorithms for the suppression of urban noise in radar warning receivers. Since joining the Georgia Tech research faculty full time in May 2002, he has contributed to numerous programs in applications involving target tracking, radar, signal processing, and modeling and simulation. In addition to infrastructure and algorithm development, he has also long been involved with applied research studies in the fields of multisensor fusion and radar performance prediction.

Saurabh Mittal is a Research Scientist with L-3 Communications, Link Simulation, and Training at the Air Force Research Lab in the Warfighter Readiness Research Division at Wright-Patterson Air Force Base, OH. He earned a Ph.D. in 2007 and M.S. in 2003, both in Electrical and Computer Engineering from the University of Arizona, Tucson. In this capacity, he is working on integration and interoperability efforts in Live, Virtual and Constructive (LVC) Distributed Mission Operations (DMO) environments. He is currently Vice General Chair of Summer Simulation Multiconference 2014, sponsored by Society of Computer Simulation (SCS) International. He is also the founder and principal of Dunip Technologies, USA. He was the recipient of the U.S. DoD highest civilian contractor recognition 'Golden Eagle' in 2006 for the project GENETSCOPE for developing state-of-the-art component-based simulation model of High Frequency Global Communication System using DEVS formalism for U.S. Air Force and U.S. Navy. He has coauthored one book and over forty peer reviewed articles. He has over a decade of experience in engineering netcentric M&S-based solutions using formal systems modeling and systems/software engineering approaches. He is a Member of SCS, ACM, and IEEE.

DeLante E. Moore is a Master's student in the Department of Industrial Engineering at the University of Miami. He is also currently a Captain in the United States Army, serving as an Operations Research/Systems Analyst. Upon completion of his Master's degree, he will serve as an instructor in the Department of Systems Engineering at the United States Military Academy at West Point, NY, from where he also obtained his Bachelor's of Science in Electrical Engineering. His research interests include modeling and simulation of complex dynamic systems and utility resource planning.

Thomas W. O'Brien has been a former Air Force Colonel supporting an FFRDC and a not-for-profit Science Institute in Washington, DC. While working in Cyber Operations, his focus was on command and control for network, and kinetic and cyber operations. He held a five-year Aerospace Engineering degree with a focus on propulsion and supersonic flow dynamics from the University of Florida and an M.S. in Systems Management, with a focus on Multivariate Decision Analysis from the University of Southern California. His career included counterinsurgency training with Special Forces, and 26 years with the Air Force and Joint assignments: Combat Targeting, ICBM Crew Commander and Instructor; Program Manager, Atlas Agena and Atlas Centaur space launch programs; NASA Test Director for first Space Shuttle launch and landings; Director Space Support Division, USSPACECOM; Director in Joint Operations Defense Planning Staff for national missile defense–offense integration; and a Deputy Director in Collections, DIA. Accomplishments included creation of original "ISR" terminology, framework and analysis; creation of the space mission areas of Force Enhancement, Space Support, Space Control and Force Application; utility analysis leading to DARPA's Global Hawk, Predator, Dark Star UAV programs; JROC approval of Tactical Satellite MNS, and related small space booster program development; and invention of "kill chain" analysis and the broader analytical framework, on assessing value of decisionready information in terms of net-centric, mission outcomes. He was selected as J2 for USSTRATCOM, and he also led successful Red Flag flight demonstrations of "kill chain" tuning against Time Critical Targets—attributed by CJCS as USJFCOM's first major experimentation breakthrough. Colonel O'Brian passed away in September 2014.

Matthew L. Paden is an Assistant Professor of Pediatric Critical Care at Emory University and serves as the Director of Pediatric/Adult ECMO and Advanced Technologies at Children's Healthcare of Atlanta. His research focus is advanced technologies use, including extracorporeal membrane oxygenation, plasma exchange, and continuous renal replacement therapies in critically ill children. He is an inventor of KIDS-CRRT, a novel continuous renal replacement device specifically designed for use in the pediatric population. Additionally, he serves on the Extracorporeal Life Support

Organization's (ELSO) Steering Committee as the Co-Chair of the ELSO Registry, which contains clinical data on over 60,000 patients treated with ECMO. He received his M.D. at the University of Oklahoma, with his pediatric residency and critical care fellowship training at Emory University.

L. Drew Pihera is a Research Scientist at the Georgia Teach Research Institute and is the Head of the System Engineering Software Applications Branch. His professional experience encompasses over thirteen years' of developing software and leading software efforts including in-flight training applications, decision support systems, and systems engineering tools. His current focus is on web-based, collaborative applications for systems engineering purposes, decision support, and education. In addition, he teaches in GTRI short courses and advises students during their capstone projects in the Professional Master's of Applied Systems Engineering at the Georgia Institute of Technology. He earned his B.S. in Computer Science from Georgia State University and his Professional Master's in Applied Systems Engineering from the Georgia Institute of Technology.

Antoinette M. Portrey is a Program Manager with L-3 Communications Link Simulation and Training at the Air Force Research Lab, Warfighter Readiness Research Division at Wright-Patterson Air Force Base, OH. She leads the Warfighter Readiness Science and Technology Program in the development, integration, and evaluation of training research methodologies and the development of new technologies. She is also Coeditor for *Individual and Team Skill Decay: The Science and Implications for Practice* (2013). She completed her M.S. in Applied Psychology—Human Factors from Arizona State University in 2005 and an MBA in Project Management from Capella University in 2008.

Larry B. Rainey is the Founder and Senior Partner in the consulting firm Integrity Systems and Solutions, LLC, which specializes in modeling and simulation in both missile defense and space operations domains. He holds an M.S. and Ph.D. in Systems Engineering from the Ohio State University in Columbus, OH. He has been a Visiting Assistant Professor of Systems Engineering at the Air Force Institute of Technology at Wright-Patterson Air Force Base, OH, and an Assistant Professor of Systems Engineering at Colorado Technical University in Colorado Springs, CO. He has been the Lead Executive Editor for the three texts addressing modeling and simulation and applied operations research. He is a Consulting Editor for *The Journal of Defense Modeling and Simulation*. He is a Senior Member of SCS and MORS.

Donna H. Rhodes is a Senior Lecturer and Principal Research Scientist at the Massachusetts Institute of Technology. She is the Director of MIT's Systems Engineering Advancement Research Initiative (SEAri), a research group focused on advancing theories, methods, and practice of systems engineering applied to complex sociotechnical systems. Previously, she held senior management positions in systems engineering and enterprise practices at IBM Federal Systems, Lockheed Martin, and Lucent Technologies. She conducts research on innovative approaches and methods for architecting complex systems and enterprises, including predictive indicators, engineering systems thinking empirical studies, and designing for uncertain futures. Her research focuses on more predicative architecting of sociotechnical systems to address significant societal needs in a dynamic world, crossing multiple sectors including defense, aerospace, transportation, energy, healthcare, and commercial products. She has been very involved in the evolution of the systems engineering discipline and in several university programs. She has served corporate, government and academic boards, and study panels of national and international importance. Dr. Rhodes is a Past President and Fellow of the International Council on Systems Engineering (INCOSE) and an INCOSE Founders Award recipient. She received her Ph.D. in Systems Science from T.J. Watson School of Engineering at Binghamton University.

Adam M. Ross is a Research Scientist in the Engineering Systems Division at the Massachusetts Institute of Technology. He is cofounder and lead research scientist for MIT's Systems Engineering Advancement Research Initiative (SEAri), a research group focused on advancing the theories, methods, and effective practice of systems engineering applied to complex sociotechnical systems through collaborative research with industry and government. He has professional experience working with government, industry, and academia. He holds a dual Bachelor degree in Physics and Astrophysics from Harvard University, two Master degrees in Aerospace Engineering and Technology & Policy, as well as a Doctoral degree in Engineering Systems from MIT. He has research interests and advises students in ongoing research projects in advanced systems design and selection methods, tradespace exploration, managing unarticulated value, designing for changeability, value-based decision analysis, and system of systems engineering. He has received numerous paper awards, including the Systems Engineering 2008 Outstanding Journal Paper of the Year. He has published over 80 papers in the area of space systems design, systems engineering, and tradespace exploration. He serves on technical committees with both AIAA and IEEE and is recognized as a leading expert in system tradespace exploration and change-related "ilities."

John F. Sarkesain has over 30 years of engineering and senior level management and leadership experience in the Department of Defense. Today, he is a Senior Protect Leader for the Aerospace Corporation where he supports the Deputy Assistant Secretary of Defense for C3 and Cyber. He has led and managed significant work in information assurance (IA) and advanced Cyber Warfare technology development and operational definition; to include managing and co-architect for a "real-time" distributed *Cyber Command and Control/Battle Management* (C2/BM) system prototype effort. This work has been described in DoD as a *first of its kind.* In 2000, he and a colleague developed an Advanced Concept Technology Demonstration (ACTD) proposal for real-time distributed intrusion detection that was voted by the Department of Defense as the top ranked ACTD that year. While serving as the Deputy Director for IA at the Missile Defense Agency, he established and managed the first MDA *Cyber Defense Operations Center*. He holds Bachelor's degrees in Computer Science and History and Master's degrees in Computer Systems Management and National Security and Strategic Studies. He is also a graduate of the DoD's Defense Leadership and Management Program.

Brian Sauser is an Associate Professor in the Department of Marketing and Logistics at the University of North Texas (UNT). He currently serves as the Director of the Logistics and Supply Chain Simulation Laboratory and Associate Director of Research for the Center of Logistics Education and Research. Before joining UNT, he held positions as an Assistant Professor with the School of Systems and Enterprises at Stevens Institute of

Technology; Project Specialist with ASRC Aerospace at NASA Kennedy Space Center; Program Administrator with the New Jersey—NASA Specialized Center of Research and Training at Rutgers, The State University of New Jersey; and Laboratory Director with G.B. Tech Engineering at NASA Johnson Space Center. His research interest is in the governance of complex systems. This includes system and enterprise governance and the advancement of a foundational science of systems thinking. He teaches courses in Business and Logistics Analytics, Theory of Logistics Systems, Systems Thinking, and Systems Engineering and Management. In addition, he is a National Aeronautics and Space Administration Faculty Fellow, IEEE Senior Member, and an Associate Editor of the *IEEE Systems Journal*. He holds a B.S. from Texas A&M University in Agricultural Development with an emphasis in Horticulture Technology, an M.S. from Rutgers, The State University of New Jersey in Bioresource Engineering, and a Ph.D. from Stevens Institute of Technology in Project Management.

Xiaoran Shi is a Ph.D. candidate in the Department of Industrial Engineering at the University of Miami. Her research interests lie in the areas of simulation modeling and decision making for large-scale, complex, and dynamic systems such as distributed power networks in conjunction with the sequential Monte Carlo methods and their corresponding applications in these networks. She earned her Bachelor's degree from Department of Industrial Engineering at the Nankai University, China, in June 2011. She is the recipient of Second-grade Scholarship of Nankai University, Top-grade of Tianfu Company Scholarship, and Excellent Student Leader Award of Teda College and Excellent Student Award of Nankai University.

Aristotelis E. Thanos is a Research Assistant and a Ph.D. Candidate in the Department of Industrial Engineering at the University of Miami. He earned his Bachelor's and Master's degree from the Department of Applied Mathematical and Physical Sciences at the National Technical University of Athens, Greece, in February 2010. He also earned his Master's degree in Logics Algorithms and Computation at the National and Kapodistrian University of Athens, Greece, in July 2012. His research interests lie broadly in the simulation and optimization (discrete-event simulation, agent-based simulation, mathematical modeling, and algorithms) of dynamic large-scale systems under uncertainty with a focus on power networks and other energy applications. He is the recipient of the 2013 Graduate Committee Conference Travel Award at the University of Miami.

Andreas Tolk is the Chief Scientist for SIMIS Inc, Portsmouth, VA, where he is developing and evaluating long-term visions regarding developments in simulation. He is also Adjunct Professor of Engineering Management and Systems Engineering at Old Dominion University in Norfolk, Virginia, USA. He holds an M.S. and a Ph.D. in Computer Science from the University of the Federal Armed Forces in Munich, Germany. He has contributed more than 200 articles to journals, book chapters, and conference proceedings and edited several books on Modeling & Simulation and Systems Engineering. He received the Excellence in Research Award from the Frank Batten College of Engineering and Technology in 2008, the Technical Merit Award from the Simulation Interoperability Standards Organization (SISO) in 2010, the Outstanding Professional Contributions Award from the Society for Modeling and Simulation (SCS) in 2012, and the Distinguished Professional Achievements Award from SCS in 2014.

He is on the Board of Directors of the Society for Modeling and Simulation (SCS) as well as of the Association for Computing Machinery (ACM) Special Interest Group Simulation (SIGSIM). He is a Senior Member of IEEE and SCS.

Philip D. West received the BSEE from the University of Maryland in 1983 and the MSEE and Ph.D. degrees from Georgia Tech in 1984 and 1994. He has been employed at GTRI since 1984 and is currently a principal Research Engineer, a GTRI Technical Fellow, and GTRI's Technical Director for electronic warfare (EW) where he is responsible for capture, coordination, and execution of sponsored programs in EW across GTRI. He is a Senior Member of Institute Electrical and Electronics Engineers (IEEE), a Life Member of the Association of Old Crows (the international society for EW), and a Member of the AOC EW Technology Hall of Fame. He has held various leadership roles in the Ballistic Missile Defense Benchmark Modeling and Simulation programs and has led various related programs such as the Integrated Air and Missile Defense Benchmark, the ESM Benchmark, the Electronic Attack Benchmark and the Chem/Bio Agent Identification Algorithm Benchmark—where GTRI worked with DoD sponsors to formulate and solve algorithm development problems in a collaborative challenge-problem setting.

Clifford A. Whitcomb is Professor and Chair of the Systems Engineering Department at the Naval Postgraduate School in Monterey, California, USA. His research interests include model-based systems engineering for enterprise systems; defense system of systems; naval construction and engineering; and leadership, communication, and interpersonal skills development for engineers. He has more than 35 years of experience in defense systems engineering and related fields and has published several textbooks and textbook chapters as well as more than 30 publications. He is a principal investigator for research projects from the U.S. Navy Office of Naval Research, Office of the Secretary of Defense, Office of the Secretary of the Navy, and several naval system commands and naval warfare centers. He is a Certified Systems Engineering Professional (CSEP), has served on the INCOSE Board of Directors, and was a Lean Six Sigma Master Black Belt for Northrop Grumman Ship Systems.

Michael Wright is a Chief Engineer with the U.S. Army Program Executive Office for Simulation Training and Instrumentation (PEO STRI). He has worked for the past 19 years in various engineering leadership positions on Modeling and Simulation System of Systems projects and federations in support of Army Digital Command and Staff Training. His areas of expertise include M&S systems interoperability and integration for training. He holds a B.S. in Electrical Engineering from the University of Central Florida and an M.S. in Industrial Engineering Management from Texas A&M University.

List of Acronyms

2D	Two-dimensional
3D	Three-dimensional
A2C2	Autonomy type 2/connectivity type 2
AAR	After-action review
ABAC	Attribute-Based Access Control
ABM	Activity-based method
ABM	Agent-based modeling
ACM	Association for Computing Machinery
ACT	Attack Countermeasures Tree
ACTD	Advanced Concept Technology Demonstration
ACT-R	Adaptive Control of Thought—Rational
AcV	Acquisition Views
ADD	Architecture description documents
ADM	Architecture Development Method
ADS	Agent-directed simulation
ADS	Authoritative data source
ADT	Abstract data types
AES	Advanced Encryption Standard
AESS	Aerospace and Electronics Systems Society
AFOSR	Air Force Office of Scientific Research
AFRL	Air Force Research Lab
AG	Attack Graph
AgV	Agreements Views
AHRQ	Agency of Healthcare Research and Quality
AI	Artificial intelligence
ALSP	Aggregate Level Simulation Protocol
AML	Agent Modeling Language
AMT	Architecture Management Team
ANL	Argonne National Laboratory
ANOVA	Analysis of variance
AoAs	Assessment of SoS alternatives
AOI(s)	Area(s) of interest
AORs	Areas of responsibility
API	Application programming interface
AR	Aspect ratio

ARFORGEN	Army Force Generation
ARG	Attack Response Graphs
ATACKS	Advanced Tactical Architecture for Combat Knowledge System
ATC	Air traffic control
ATES	Automated Threat Engagement System
ATS	Air transportation system
ATSA	Aviation and Transportation Security Act
ATV	Automated Transfer Vehicle
AUML	Agent UML
AV	All View
AV-1	All View-1
AV-2	All View-2
BAM	Business Activity Monitoring
BDA	Battle damage assessment
BDD	Block Definition Diagram
BIRT	Business Intelligence and Reporting Tools
BLOS	Beyond Line-of-Sight
BM	Battle Management
BMDS	Ballistic Missile Defense System
BMS	Ballistic Missile Defense
BoK	Body of knowledge
BPEL	Business Process Execution Language
BPM	Business Process Management
BPMN	Business Process Modeling Notation
BSSM	Boardman–Sauser Systems Methodology
BTS	Bureau of Transportation Statistics
C/ACAMS	Constellation/Automated Critical Asset Management System
C^2	Command and control
C^2/BM	Cyber Command and Control/Battle Management
C2ISR	Command, Control, Intelligence, and Surveillance
C3IOS	Cyber Command and Control and Information Operations Systems
C4	Command, control, communications, and computer
C4ISR	Command, control, communications, computers, intelligence, surveillance, and reconnaissance
CA	Central authority
CAF	Combat Air Force
CAS	Complex adaptive systems
CB	Contingency basing
CBRN	Chemical, biological, radiological, and nuclear
CCD	Central Composite Design
CCS	Complex combat system
CCTT	Close Combat Tactical Trainer
CDC	Centers for Disease Control and Prevention
CEP	Circular error probable
CEP	Complex Event Processing
CERT	Computer Emergency Response Team
CFATS	Chemical Facility Anti-Terrorism Standards

CFD	Computational fluid dynamics
CGFs	Computer-generated forces
CHOA	Children's Healthcare of Atlanta
CI	Critical infrastructure
CIKR	Critical infrastructure and key resources
CIMIC	Civil-Military Cooperation
CIPDSS	Critical Infrastructure Protection Decision Support System
CIPR/sim	Critical Infrastructure Protection and Resiliency Simulator
CISA	Center for Integrated Systems in Aerospace
CMM	Capability Maturity Model
CMSD	Core Manufacturing Simulation Data
CND	Computer network defense
COA	Course of action
COC	Cyber operational commander
ConOps	Concept of operations
COTS	Commercial Off-The-Shelf
CPB	Cardiopulmonary bypass
CS	Constituent systems
CSDP	Conceptual Schema Design Procedure
CSEP	Certified Systems Engineering Professional
CSOs	Closely spaced objects
CSP	Communicating Sequential Processes
CSs	Cyber systems
CTIA	Common Training and Instrumentation Architecture
CV	Capability Viewpoint
CyberOps	Cyber operations
DAA	Designated approval authority
DAML	DARPA Agent Markup Language
DARPA	Defense Advanced Research Project Agency
DAU	Defense Acquisition University
DB	Database
DCC	Dynamic cell commander
DCO	Defensive cyber operations
DDDAS	Dynamic data-driven application system
DDMS	DoD Discovery Metadata Specification
DDoS	Distributed DoS
DDS	Data Distribution Services
DES	Digital Encryption Standard
DES	Discrete-event simulation
DEVS	Discrete-event system specification
DFD	Data flow diagrams
DHS	Department of Homeland Security
DIMEG	Department of Mechanical, Energy and Management Engineering
DIMES	Department of Informatics, Modeling, Electronics, and Systems Engineering
DinD	Defense in depth
DIS	Distributed Interactive Simulation

DISA	Defense Information Systems Agency
DIV	Data and Information Viewpoint
DM2	DoDAF Metamodel
DML	Domain Modeling Language
DMO	Distributed Mission Operations
DMON	DMO Network
DMSO	Defense Modeling and Simulation Office
DNP3	Distributed Network Protocol
DOC	Direct operating cost
DoD	Department of Defense
DoDAF	DoD Architectural Framework
DoDIN	DoD information network
DoE	Design of Experiments
DoS	Denial of Service
DOTMLPF	Doctrine, organization, training, materiel, leadership education, personnel, and facilities
DS	Distributed system
DSEEP	Distributed Simulation Engineering and Execution Process
DSS	Decision support system
DTC	Distributed Training Center
DTOC	Distributed Training Operations Center
DTRA	Defense Threat Reduction Agency
DVCs	Dispensing/Vaccination Centers
E&T	Evaluation and testing
EA	Enterprise architecture
EA	Environment abstraction
ECLS	Extracorporeal Life Support
ECMO	Extracorporeal membrane oxygenation
EDA	Event-driven architecture
EDS	Explosives detection systems
EEA	Epoch–era analysis
EEA	European Environmental Agency
EFFBD	Enhanced Functional Flow Block Diagram
EHCs	Enabling Homeland Security Capabilities
EIC	Editor-in-Chief
ELSO	Extracorporeal Life Support Organization
EM	Execution Management
EMA	Effective managerial authority
EMS	National Emergency Medical Services
EMSS	European Modeling and Simulation Symposium
EOC	Emergency Operations Center
EOH	Equivalent operating hours
EPF	Eclipse Process Framework
ERA	European Robotic Arm
ERF	Entity Resolution Federation
ESA	European Space Agency

ESA-AF	ESA Architecture Framework
ESB	Enterprise service bus
ESE	Early-stage evaluation
ESOC	European Space Operation Centre
ET	External tank
EU	European Union
EW	Electronic warfare
FCDR	Fragmentation, coding, dispersion, and reassembly
FCS	Future Combat Systems
FDA	Food and Drug Administration
FDIR	Failure detection, isolation, and response
FEDEP	Federation Development and Execution Process
FEMA	Federal Emergency Management Agency
FFBD	Functional Flow Block Diagram
FFRDC	Federally funded research and development corporation
FIPS	Federal Information Processing Standards
FiV	Financial Views
FLEET	Fleet-level environmental evaluation tool
FM	Fault management
FMC	Fractional management company
FMCLs	FM control loops
FMECA	Failure mode, effects, and criticality analysis
FOM	Federation object model
GAs	Genetic algorithms
GCCS	Global Command and Control System
GEF	Graphical Editing Framework
GEOSS	Global Earth Observation System of Systems
GFT	Goal-function tree
GIG	Global Information Grid
GIS	Geographic information systems
GMES	Global Monitoring for Environment and Security
GMF	Graphical Modeling Framework
GPS	Global Positioning System
GTRI	Georgia Tech Research Institute
GUI	Graphical user interface
GWU	George Washington University
HADR	Humanitarian Assistance Disaster Relief
HBA	Human Behavior Architecture
HBM	Human behavior models
HIL	Hardware-in-the-loop
HLA	High-Level Architecture
HM	Health management
HMR	Hazardous material release
HRSA	Health Resources and Services Administration
HSWG	Homeland Security Working Group
HVAC	Heating ventilation and air conditioning

I/O	Input/output
IA	Information assurance
IA	Intelligent agents
IA-CGF	Intelligent Agents Computer-Generated Forces
IAM	Integrated Architecture Modeling
IAP	Integrated Applications Program
IASD	Infrastructure Analysis and Strategy Division
IBD	Internal Block Diagram
IC	Intelligence Community
ICS	Industrial control system
ID	Intrusion detection
IDA	Institute of Defense Analysis
IDEF	Integration DEFinition
IEEE	Institute of Electrical and Electronic Engineers
IFTU	In-flight target updates
IIE	Institute of Industrial Engineers
IIT	Indian Institute of Technology
INCOSE	International Council on Systems Engineering
INL	Idaho National Labs
INT cell	Intelligence cell
IO	Inventory optimizer
IP	Infrastructure Protection
IP	Integer programming
IR	Intrusion response
ISO	International Organization for Standardization
ISR	Intelligence, surveillance, and reconnaissance
ISS	International Space Station
ISTS	Institute of Security Technology Studies
IT	Information technology
ITL	Integrated task list
IVT	Interface Verification Tool
Java EE	Java Enterprise Edition
JAXA	Japanese Space Agency
JCMS	Joint Capability Management System
JCSS	Joint Communication Simulation System
JESS	Joint Exercise Support System
JIE	Joint Information Environment
JLCCTC	Joint Land Component Constructive Training Capability
JLVC	Joint Live, Virtual, and Constructive
JNTC	Joint National Training Capability
JPO	Joint Program Office
JRVIO	Joint Reserve Component Virtual Information Operations Organization
JSIMS	Joint Simulation System
JTC	Joint Training Confederation
JTTI	Joint Training Transformation Initiative
JVM	Java virtual machine

List of Acronyms **xxxvii**

KML	Keyhole Markup Language
LAN	Local area network
LANL	Los Alamos National Laboratory
LAPIS	Lean Advanced Pooling Intelligent Optimizer and Simulator
LCIM	Levels of Conceptual Interoperability Model
LEO	Low Earth orbit
LHS	Latin Hypercube Sample
LOG cell	Logistics cell
LROM	Logical Range Object Model
LS	Lean simulation
LVC	Live, Virtual, and Constructive
LVCAR	LVC Architecture Roadmap
LVC-IA	Live, Virtual, and Constructive Integrating Architecture
M&S	Modeling and simulation
m2DIS	Model to DIS
M2M	Machine-to-machine
MAS	Modeling and Applied Simulation
MAS	Multiagent simulation
MAS	Multiagent system
MAS-ML	Multiagent System Modeling Language
MATE	Multiattribute tradespace exploration
MBSE	Model-Based Systems Engineering
MBT	Model-based testing
MC	Mission Command
McTrans	Microcomputers in Transportation
MDA	Missile Defense Agency
MDA	Model-driven architecture
MDD	Model-Driven Development
MDO	Multidisciplinary design optimization
MDSD	Model-Driven Software Development
MECs	Mission Essential Competencies
MIL	Man in the loop
MIMIC	Mixed-Initiative Machine for Instructed Computing
MINLP	Mixed-integer nonlinear programming
MODAF	Ministry of Defence Architecture Framework
MoE(s)	Measure(s) of effectiveness
MoM	Measure of Metrics
MoP(s)	Measure(s) of performance
MoP(s) for SoS	Measure(s) of SoS performance
MP	Monterey Phoenix
MREIS-PF	Minimum relative entropy-based importance-density selection rule
MRF	Multiresolution Federation
MSA	Modeling, simulation, and analysis
MSC-LES	Modeling and Simulation Center—Laboratory of Enterprise Solutions
MSpE	Mean square pure error
MTCs	Mission Training Centers

multi-INT	Multi-intelligence
MURI	Multidisciplinary University Initiative
N2C2M2	NATO Net-enabled Capability Command and Control Maturity Model
NAF	NATO Architecture Framework
NARAC	National Atmospheric Release Advisory Capability
NAS	National Airspace System
NASA	National Aeronautics and Space Administration
NATO	North Atlantic Treaty Organization
NCDS	Net-Centric Data Strategy
NCSS	Net-Centric Services Strategy
NCW	Net-centric warfare
NEC C2 M2	Net Centric Command and Control Maturity Models
NEMSIS	National Emergency Medical Services Information System
NERL	National Exposure Research Lab
NGIC	Next-generation intelligent combat
NHTSA	National Highway Traffic Safety Administration
NICs	Network interface cards
NIIRS	National Imagery Interpretability Rating Scale
NIMS	National Incident Management System
NIPP	National Infrastructure Protection Plan
NISAC	National Infrastructure Simulation and Analysis Center
NIST	National Institute of Standards and Technology
NLP	Nonlinear programming
NOC	Network operations center
NPPD	National Protection and Programs Directorate
NREL	National Renewable Energy Laboratory
NRL	Naval Research Labs
NS3	Network Simulator-3
NSC	National Simulation Center
NSGC	Not-So-Grand-Challenge
NSRDEC	Natick Soldier Center RD&E Center
OASIS-RM	Open architecture for Accessible Services Integration and Standardization Reference Model
OCBA	Optimal computing budget allocation
OCO	Offensive cyber operations
OMC	Object Model Compiler
OMG	Object management group
OMT	Object model template
OneSAF	One Semi-Automated Forces
ONISTT	Open Netcentric Interoperability Standards for Training and Testing
OODA	Observe, orient, decide, act
OPFOR	Opposing Force
OPS cell	Operations cell
OR	Operations research
ORDSS	Operationally Responsive Disaster Surveillance System
ORM	Object Role Modeling

OSD	Office of the Secretary of Defense
OSI	Open Systems Interconnection
OSMC	Open Source Modelica Consortium
OV	Operational View
OV-1	Operational Viewpoint 1
OWL	Web Ontology Language
PAC	Personality-Enabled Architecture for Cognition
PALM	Performance and Learning Models
PANOPEA	Piracy Asymmetric Naval Operation Patterns Modeling for Education and Analysis
PDUs	Protocol Data Units
PEO	Program Executive Offices
PEO STRI	Program Executive Office for Simulation, Training, and Instrumentation
PES	Physical Exchange Specification
PES	Pruned entity structure
PF	Particle filtering
PFEBA	Problem-Focused, Effects-Based Analysis
PI	Principal investigator
PKI	Public key infrastructure
PLCs	Programmable logic controllers
PM ConSim	Project Manager for Constructive Simulation
PM	Program manager
PM	Project manager
PMEC	Program Management Excellence Center
PNB	Perceived net benefit
PO	Planning optimizer
PR	Participation risk
PRP	Planning and Review Panel
PSM	Problem structuring methods
PSYOPS	Psychological Operations
PTRs	Problem tracking reports
PV	Photovoltaic
PV	Project Viewpoint
QoS	Quality of service
R&D	Research and development
RAD	Rapid application development
RAP	Ready Aircrew Program
RAS	Reliability, availability, and safety
RBAC	Role-Based Access Control
RDECOM	Research, Development, and Engineering Command
RFA	Red force agent
RMF	Risk management framework
RMI	Remote method invocation
ROEs	Rules of engagement
ROI	Return on investment
ROM	Rough order of magnitude

RPC	Remote procedure call
RSA	Rational Software Architect
RTDS	Real-time digital simulator
RWR	Radar warning receiver
S&T	Science and Technology
S2A2	System type 2/autonomy type 2
SA	Situation awareness
SA	Systems architect
SAIC	Science Applications International Corporation
SAMs	Surface-to-air missiles
SCC	Strategic cyber commander
SCS	Society of Computer Simulation
SDLC	Systems design life cycle or software development life cycle
SDR	Strategic Defense Review
SDR-NC	SDR New Chapter
SE	Systems Engineering
SEAri	Systems Engineering Advancement Research Initiative
SEE	Simulation Exploratory Experience
SERC	Systems Engineering Research Center
SES	System entity structure
SHM	Systems health management
SIGSIM	Special Interest Group Simulation
SIL	Software-in-the-loop
SISO	Simulation Interoperability Standards Organization
SLA	Service-level agreement
SMEs	Subject matter experts
SOA	Service-Oriented Architecture
SoaML	SOA Modeling Language
SOM	Simulation object model
SOPs	Standard operating procedures
SoS	System of Systems
SoSE	System of Systems Engineering
SoSs	System of Systems
SoS TSE	System of Systems tradespace exploration
SOV	Service-Oriented Views
SSA	Space Situational Awareness
SSM	Soft Systems Methodology
SSPs	Sector-Specific Plans
SST	Space Surveillance and Tracking
StdV	Standards Viewpoint
STK	Systems Took Kit
StV	Strategic Views
ST VM	Simulation Team Virtual Marine
SV(s)	Systems View(s)
SvcV	Services Viewpoint
SysML	Systems Modeling Language

T&E	Test and evaluation
T/W	Thrust to weight
T-AES	Transactions on Aerospace and Electronic Systems
TCC	Tactical cyber commander
TCTs	Time-critical targets
TDL	TENA Definition Language
TE	Tradespace Exploration
TENA	Test and Training Enabling Architecture
TIDE	TENA Integrated Development Environment
TOGAF	The Open Group Architecture Framework
TRADOC	Training, Analysis, and Doctrine Command
TRL	Technology readiness level
TSA	Transportation Security Administration
TSE	Tradespace exploration
TSIS	Traffic Software Integrated System
TTPs	Tactics, techniques, and procedures
TV	Technical Standards View
UAH	University of Alabama in Huntsville
UARC	University Affiliated Research Center
UAS	Unmanned aircraft system
UAV	Unmanned aerial vehicle
UC	Use cases
UDDI	Universal Description, Discovery, and Integration
UML	Unified Modeling Language
UNT	University of North Texas
UoD	Universe of Discourse
UPDM	Unified Profile for DoDAF and MODAF
UPIA	UML Profile-Based Integrated Architecture
USCC	U.S. Cyber Command
USMTF	U.S. Message Text Format
V&V	Validation and verification
VA	Vulnerability assessments
VBIED	Vehicle-borne improvised explosive device
VID-6	Visual Identification 6
VIOA	Virtual information operations agency
VM	Virtual machine
VMF	Variable Message Format
VO	Virtual organization
VV&A	Verification, Validation, and Accreditation
VV&T	Verification, Validation, and Testing
W/S	Wing loading
W3C	World Wide Web Consortium
WAN	Wide area network
WBM	WebSphere Business Modeler
WEZ	Weapon Engagement Zone
WRRD	Warfighter Readiness Research Division

WS	Web service
WSDL	Web Services Description Language
WTRD	Warfighter Training Research Division
XACML	eXtensible Access Control Markup Language
XCITE	eXperimental Common Immersive Theater Environment
XML	eXtensible Markup Language

Part I
Overview and Introduction

Chapter 1

Overview and Introduction to Modeling and Simulation Support for System of Systems Engineering Applications

Larry B. Rainey[1] **and Andreas Tolk**[2]

[1] *Integrity Systems and Solutions, LLC, Colorado Springs, CO, USA*
[2] *SimIS Inc., Portsmouth, VA, USA*

1.1 MOTIVATION

A little longer than a decade ago, the community of systems engineers started to seriously deal with a category of challenges: system of systems (SoS).

Systems Engineering (SE) is now better understood since its official birth as a discipline in the middle of the last century. It is widely understood to control the total systems life cycle process: definition, development, deployment, and retirement of a system. SE ensures that solutions are reliable, maintainable, and cost-effective. But its solutions focus on a system made up of components born from a common set of user requirements. Interactions with other systems was always possible, but these other systems were external and beyond the central systems boundaries.

Modeling and Simulation Support for System of Systems Engineering Applications, First Edition.
Edited by Larry B. Rainey and Andreas Tolk.
© 2015 John Wiley & Sons, Inc. Published 2015 by John Wiley & Sons, Inc.

In particular, the Internet showed the power of interconnectivity. Net-centric solutions were soon discussed that allowed the reuse of systems by loosely coupling them via information exchange. But this extended system was beyond traditional program, governance, and organization boundaries. New processes were needed to help synchronize activities, budgets, and schedules. Sage and Cuppan (2001) and Keating et al. (2003) contributed to the engineering managerial foundations to move toward System of Systems Engineering (SoSE) using the architecting principles identified by Maier (1998) only a couple of years earlier.

The technical challenges cannot be neglected. Designing interfaces and communications protocols to federate independently developed systems with each other without allowing for any significant redesign of the system is a huge challenge. However, the organizational challenges that had to be addressed by engineering managers were even bigger. For example, today's managers are not used to the task to bring systems together to fulfill a common task without having the power to establish common processes, a common administration, and a common governance.

In his work, which is actually summarized and continued in Chapter 2 of this handbook, Maier identified five criteria commonly recognized in SoSE literature:

1. *Operational independence of the individual systems*: An SoS is composed of systems that are independent and useful in their own right. If an SoS is disassembled into the component systems, these component systems are capable of independently performing useful operations independently of one another.

2. *Managerial independence of the systems*: The component systems not only can operate independently, but they generally do operate independently to achieve an intended purpose. The component systems are generally individually acquired and integrated, and they maintain a continuing operational existence that is independent of the SoS.

3. *Geographic distribution*: Geographic dispersion of component systems is often large. Often, these systems can readily exchange only information and knowledge with one another and not substantial quantities of physical mass or energy.

4. *Emergent behavior*: The SoS performs functions and carries out purposes that do not reside in any component system. These behaviors are emergent properties of the entire SoS and not the behavior of any component system. The principal purposes supporting engineering of these systems are fulfilled by these emergent behaviors.

5. *Evolutionary development*: An SoS is never fully formed or complete. Development of these systems is evolutionary over time and with structure, function, and purpose added, removed, and modified as experience with the system grows and evolves over time.

Modeling and simulation (M&S) in general and agent-directed simulation (ADS) in particular already support SE successfully. Yilmaz and Ören (2009) dedicated a whole book to the synergisms of ADS and SE. Simulated systems can be used to obtain, display, and evaluate operationally relevant data in agile contexts by executing models using operational data exploiting the full potential of M&S and producing numerical insight into the behavior of complex systems. ADS have been shown to have the ability to support the development of robust, fault-tolerant, adaptive, self-optimizing, learning, social-capable, autonomous, and agile solutions. ADS also expose emergent behavior similar to SoS, so that they can be used to better understand and utilize this criterion and enforce positive emergence while avoiding negative emergence.

The experts in the field invited to contribute chapters to this handbook were asked to utilize the five criteria provided above as a common foundation. However, the community is still very diverse when it comes to using M&S in support of SoSE applications so this book can merely provide a map of the landscape of approaches.

1.2 OBJECTIVE

Jamshidi's books provided an overview on SoSE from the methodological perspective (Jamshidi, 2010) as well as from the application perspective (Jamshidi, 2011). He also invited experts to contribute chapters, but the focus was clearly on the SE processes.

This book has the objective to focus more on M&S support, providing the foundations for a better research agenda. This research agenda, however, cannot only focus on M&S support questions, but it needs to understand the SoSE application cases as well. As such, the objective of this book is to contribute to find answers—or identify required research to provide answers—to questions such as the following:

- What are the processes of SoSE?
- What steps in the processes are supported by M&S?
- What steps in the processes can be supported in the future by M&S, and what are the necessary constraints?
- How can M&S be used to better understand SoS?

The chapter of this book will start to address them and hopefully initiate more research to fill the gap in the body of knowledge.

1.3 STRUCTURE OF THE HANDBOOK

Despite the challenges mentioned in the last section, this text has the objective to demonstrate how M&S can provide academic augmentation to the relatively new term of SoS. To facilitate this discussion, this text is divided into four major sections. Section I is the "Overview and Introduction." Its purpose is to provide the taxonomy and academic foundation for the rest of the text. Section II is the "Theoretical or Methodological Considerations." The purpose of this section is to address, as the title implies, more theoretical or more generalized approaches to the subject. Section III is the "Theoretical or Methodological Considerations with Applications and Lessons Learned." The purpose of this section is to identify specific cases where definitive applications can be abstracted and lessons learned drawn from the specific application. The final section of the text, Section IV, provides a review of what has been presented previously and draws major conclusions across both the theoretical and application spectrums.

1.3.1 Overview and Introduction

The main thrust of Section I is Chapter 2 "The Role of Modeling and Simulation in System of Systems Development." This chapter is the cornerstone to this text. In this chapter, Mark Maier defines SoS and identifies important systems and various categories of systems that the reader should be cognizant of. He also discusses M&S within the specific SoS category. He then moves ahead to address architecture, architecture description, and development. He closes his chapter with a summary and conclusions.

1.3.2 Theoretical or Methodological Considerations

Section II has five chapters that have been categorized as purely theoretical or methodological in nature.

In Chapter 3, Mike Jones addresses the M&S subject of "Composability." He has taken the attributes of an SoS as defined by Mark Maier to lay a foundation for his chapter. He then examines the M&S topics of conceptual modeling separately and then composability, interoperability, and integratability as a unit. The levels of the Conceptual Interoperability Model are then addressed. Current standards and current research are then addressed from Maier's SoS perspective. Mike finishes with his conclusions.

In Chapter 4, Adam Ross and Donna Rhodes discuss "An Approach for System of Systems Tradespace Exploration (TE)." First, they provide the reader a background on the subject of TE. Then they identify SoS-specific considerations for TE. Next, a specific approach is provided for TE. Then, an illustrative case is considered. They close their discussion with a chapter summary.

Daniele Gianni, in Chapter 5, pursues "Data Policy Definition and Verification for SoS Governance." First, in his background, he addresses a methodology based upon the terminology and concepts from the Unified Modeling Language, enterprise architecture frameworks, SoS governance, and conceptual data modeling. In his next section, he identifies the role of data policy methodology in the context of SoS governance. Next, he addresses the topic of the design of the data policy method methodology. Finally, he provides an example application for the European Space Agency (ESA) space situational awareness preparatory program. Daniele completes his chapter with conclusions.

In Chapter 6, Stephen Johnson addresses the subject of "System Health Management (SHM)." He starts off by laying a foundation of definitions from which to draw upon later in his chapter. Then SHM is addressed for a system. He follows this topic with another that discusses SHM from the perspective of models, simulations, and their applications. Next, he draws a distinction between a system and an SoS. Finally, he illustrates how SHM would pertain to an SoS. He closes his chapter with a conclusion section.

In the last chapter of Section II, Chapter 7, R. William Maule describes a Model Methodology for a Department of Defense Architecture Design. This chapter is an examination of the Department of Defense Architecture Framework (DoDAF) as looked at through the lens of an SoS. The first few sections address a reference architecture, model tooling, and model workflow. The rest of the chapter addresses each individual view as currently portrayed in DoDAF. The chapter closes with a chapter conclusion.

1.3.3 Theoretical or Methodological Considerations with Applications and Lessons Learned

Section III constitutes the bulk of this text. It has 14 chapters assigned to it. These chapters are those that have been categorized as truly methodological consideration in nature as they have definitive applications from which lessons learned can be drawn.

The first in this section is Chapter 8 by Agostino Bruzzone, Marina Massei, Alfredo Garro, and Francesco Longo. It is entitled "An Agent-Oriented Perspective on System of Systems for Multiple Domains." In this chapter, the authors first address the spectrum that ranges from large-scale systems to SoS. Then they address M&S approaches for SoS. Next is an agent-oriented perspective for SoS. Then they turn the discussion to exploiting the agent-oriented perspective with application examples in multiple domains. They end their chapter with conclusions and future works.

Chapter 9 by Sanjay Jain, Charles Hutchings, and Y. Tina Lee is entitled "Building Analytical Support for Homeland Security." The authors address the relationship between homeland security and SoS. Then they identify the need for M&S and analysis for homeland security. This discussion is followed by providing a knowledge sharing framework. The last consideration in their chapter is that of a prototype for an SoS application. Their chapter is completed with a chapter summary.

Chapter 10 by William Crossley and Daniel DeLaurentis makes the case that the "Air Transportation System" provides many examples of SoS. The authors address theoretical considerations, methodology, applications and is finalized with lessons learned.

Chapter 11 is entitled "Systemigram Modeling for Contextualizing Complexity in System of Systems" and is written by Brian Sauser and John Boardman. These authors first address what is meant by SoS thinking. They then develop the "Boardman–Sauser Systems Methodology: A System of Systems Thinking Tool." There are seven steps that are enumerated and defined that constitute their thinking tool. They then provide case examples to illustrate their tool. Their chapter closes with conclusions.

Chapter 12 by Joachim Fuchs and Niklas Lindman is entitled "Using Modeling and Simulation for System of Systems Engineering Applications in the European Space Agency (ESA)." These authors first introduce the reader to ESA's background and context. Then they explain the complexity of problems that exists within the ESA. Next, the authors present what they refer to as the ESA Architecture Framework. They explain their framework via an overview, identify the process involved, discuss the supporting modeling tool that supports their framework, discuss various application cases in general, and then identify two specific ESA programs to which their framework would be applicable to and what the results are of making this application. Overall lessons learned are then addressed. The authors finalize their chapter with conclusions.

Chapter 13 addresses "System of Systems Modeling and Simulation for Microgrids Using a Dynamic Data-Driven Application System (DDDAS)" authored by Aristotelis E. Tanos, DeLante Moore, Xiaoran Shi, and Nurcin Celik. In this chapter, the authors first define a microgrid in the context of an SoS. Next, they introduce the DDDAS framework for a particular microgrid design. In their third section, they provide detailed experiments and results of the DDDAS framework as it applies to microgrids. They finalize their chapter with conclusions about the simulation technique used.

In Chapter 14, Clifford Whitcomb, Mikhail Auguston, and Kristin Giammarco address the subject of "Composition of Behavior Models for Systems Architecture." Within the scope of their chapter, they discuss a lineage of topics: common characteristics for architecture descriptions, the Monterey Phoenix (MP) approach to behavior modeling, modeling component behavior, modeling component interaction, merging schemas, comparison of MP with common SE notations, assertions and queries, and implementation prototypes. The authors then provide a chapter summary of their work.

"Joint training" is the subject of consideration in Chapter 15 by James Harrington, Laura Hinton, and Michael Wright. This chapter is another lineage of topics to be addressed by the authors to support the development of their selected subject. They begin with an introduction to Joint training then continue building their discussion with the following topics: SoS characterization for Army and Joint training M&S, complexity of layered M&S of SoS, Joint Land Component Constructive Training Capability (JLCCTC) overview, SoS characterization for JLCCTC, JLCCTC SoSE functions, JLCCTC SE key themes, and JLCCTC SE processes. Their chapter is completed with their conclusions.

The next chapter is 16, which explores the subject of "Human in the Loop in System of Systems (SoS) Modeling and Simulation: Applications to Live, Virtual, and Constructive (LVC) Distributed Mission Operations (DMO) Training." The authorship of this chapter has been undertaken by Saurabh Mittal, Margery J. Doyle, and Antoinette M. Portrey. They approach their subject first by providing a background and scope for the reader and then transition into the detailed topics to unveil their subject. This process starts with a discussion of a Model-Based Systems Engineering (MBSE) process applicable to LVC DMO training. This is followed by the topic of "Not-So-Grand-Challenge (NSGC) Phase I: Integrating Various Modeling Architectures with Air Force Research Laboratory Systems." In the second phase, the authors address environment abstraction (EA) for advanced situation awareness and semantic interoperability. Applying EA to next-generation intelligent combat (NGIC) systems is the next topic, which is followed by a presentation of conclusions and discussion of future work.

Chapter 17 is the next chapter and is entitled "On Analysis of Ballistic Missile Defense Architecture through Surrogate Modeling and Simulation" and has been written by Tommer R. Ender, Phillip D. West, W. Dale Blair, and Paul A. Miceli. These authors start with their proposed approach, address the associated results, and make recommendations for future work and practical application. They finish their chapter with a conclusion.

Chapter 18 is "Medical Enhancements to Sustain Life during Extreme Trauma Care" as written by L. Drew Pihera, Nathan Adams, and Tommer Ender. This chapter starts off with the topic of taming the problem through MBSE. The authors build on this foundation by a discussion of "MBSE and Extracorporeal Membrane Oxygenation (ECMO) Phase 1: Understanding the Problem." In turn, the follow-on topic is "MBSE and ECMO Phase 2: Refactoring the Models to Better Support Trade Studies." The last topic addressed is future work and conclusions.

Chapter 19 is coauthored by Tom Obrien and John Sarkesain and is entitled "Utility: Problem-Focused, Effects-Based Analysis." They start off their subject with the topic of the need for a cybersecurity framework. The discussion transitions to a definition of the problem to be solved. In turn, the application of SE and other disciplines is addressed. The chapter closes with a summary. It also provides the operational foundations for the following chapter that goes into the details of building a framework to address the challenges describe here.

John Sarkesain and Tom Obrien have also coauthored Chapter 20 entitled "Framework for Cyber Command and Control/Battle Management Research and Development, Acquisition, and Operations." Based in the foundation provided in the foregoing chapter, they start their subject off by addressing the topic of "Information Assurance and Defense in Depth: A Failed Strategy." They then transition to a discussion of cyber command and control/battle management operational architecture. This topic is followed by addressing cyber command and control/battle management systems architecture. The last major topic covered in their chapter is the presentation of a cyber command and control/battle management and missile defense SoS use-case scenario. Their chapter is completed with their conclusions.

Chapter 21 is the last chapter of this text that supports methodological considerations that have definitive applications from which lessons learned can be drawn. This chapter is written by Bharat Madan and is entitled "System of Systems Security." The author begins his subject discussion with topics addressing SoS security requirements, SoS security challenges, and SoS security solutions. He then turns his attention to intrusion-tolerant SoS and modeling, simulation, and emulation SoS security. The chapter closes with his conclusions.

Chapter 22 summarizes the contributions and gives the conclusions of the editors for an ongoing research agenda for the M&S support for SoSE applications. Using examples of the book chapters as well as additional contributions in the public domain, the need for a better understanding of the interplay in SoS is shown.

A new category of engineers is needed to solve the challenges of SoSE. While old engineers build bridges over gap, such stationary solutions are no longer an option in the agile world we find ourselves in today. We have to admit that we do not provide too many answers in this book, but we hope that we at least raise the right questions to spawn new research in both communities—SE and M&S—that will collaboratively contribute to the scientific foundations for Modeling and Simulation Support for System of Systems Engineering Applications.

REFERENCES

Jamshidi, M. (Ed.) Systems of Systems Engineering: Principles and Applications. CRC Press Taylor and Francis Group, Boca Raton, FL (2010).

Jamshidi, M. (Ed.) System of Systems Engineering: Innovations for the Twenty-First Century. John Wiley & Sons, Inc., Hoboken, NJ (2011).

Keating, C., Rogers, R., Unal, D., Dryer, D., Sousa-Poza, A., Safford, R., Peterson, W., Rabadi, G. System of Systems Engineering. Engineering Management Journal 15(3), 35–44 (2003).

Maier, M. Architecting Principles for Systems-of-Systems. Systems Engineering 1(4), 267–284 (1998).

Sage, A., Cuppan, C. On the Systems Engineering and Management of Systems of Systems and Federations of Systems. Information, Knowledge, Systems Management 2(4), 325–345 (2001).

Yilmaz, L., Ören, T. (Eds.) Agent-Directed Simulation and Systems Engineering. Wiley-VCH, Berlin (2009).

Chapter 2

The Role of Modeling and Simulation in System of Systems Development

Mark W. Maier
The Aerospace Corporation, Chantilly, VA, USA

2.1 INTRODUCTION

This chapter provides a general introduction to the concept of system of systems (SoS), their definition, the position of the concept within related concepts, the relationship of modeling and simulation issues to SoS, and the relationship of modeling and simulation to the development of SoS. Section 2.2 of Chapter 2 defines SoS and a number of related concepts and pays particular attention to the concept of emergent properties and its role in understanding both SoS and models of systems. Section 2.3 reviews a number of important examples of SoS, which will be explored in depth with respect to modeling and simulation later in the book. Section 2.4 takes up modeling issues particularly associated with SoS. Section 2.5 relates the modeling issues to the structure of architecture description documents (ADD) and distinguishing architecture as a concept or as decision from architecture as modeling artifacts.

Five key lessons come from the discussion:

1. SoS are distinguished from other systems by formation from independently operated and managed components. While other factors also apply, it is the element of collaborative assembly and operation that is most significant.

Modeling and Simulation Support for System of Systems Engineering Applications, First Edition.
Edited by Larry B. Rainey and Andreas Tolk.
© 2015 John Wiley & Sons, Inc. Published 2015 by John Wiley & Sons, Inc.

2. Because of the relatively abstract nature of many SoS, modeling and simulation plays an even larger role in their development than in other systems categories.
3. "Emergence" and its relationship to SoS is best understood in the relationship of the properties in question as exhibited in physical systems to their exhibition in models of the system.
4. Important examples of SoS are widely arrayed along a spectrum of collaborative assemblage, from mostly centrally directed to essentially virtual.
5. Modeling pervades the SoS development process, regardless of the domain or type of SoS, but it is important to distinguish the role in setting out architectural invariants from detailed specification.

2.2 SoS AND CATEGORIES OF SYSTEM

Why declare one system an "SoS" and another just a "plain" or "monolithic" system? What specifically distinguishes those systems in the category of SoS from those in another category? Since this book is about SoS, it is clearly required that we address these distinctions. But a definition is essentially academic unless it can be justified on other than ad hoc grounds. In general, why identify some systems as members of a category (whether SoS or otherwise) and claim that they deserve some kind of special treatment different from systems identified as belonging to another category? After all, one can define an arbitrary number of "types" of system. What makes one categorization of a subset of systems more useful or important than another categorization? And what makes any particular definition of SoS most appropriate?

The original source for the concept of an SoS used in this book (Maier, 1996, 1998) justified the definition on the basis of a category distinguishing identifiable best practices. In the discussion to follow, we will review the definition used in this book, consider variations and extensions on that theme, and review how the definition captures significant best practices.

2.2.1 Systems Categories and SoS

In the primary source for the definition used throughout this book (Maier, 1998), it was argued that it makes sense to create a category of systems only if the members of that category share important and distinct best practices in development or operation that are different from systems not in the category. Membership in the category should provide significant guidelines on development. It should not merely be easy to identify the category; the category should matter. Failing to recognize the category should lead to doing things likely to make development or operation less successful.

The paper argues that a specific set of conditions, related to *collaborative* assemblage and operation, drive identifiable and distinct best practices. The collaborative assemblage is not the only such systems category; there are many others, but it is one such category and it is the one that best represents examples commonly identified an SoS. Other, related, categories include families of systems and portfolios of systems. While the definition concerns how the system is constituted, its use and importance is in development. By identifying a system of interest as an SoS, we are informed on development practices, some of which are modeling and simulation related.

Under this argument, an SoS is identified by five properties, that is, an SoS is an assemblage of systems, which are as follows:

1. *Operational independence* of the individual systems: An SoS is composed of systems that are independent and useful in their own right. If an SoS is disassembled into the component systems, these component systems are capable of independently performing useful operations independently of one another.
2. *Managerial independence* of the individual systems: The component systems not only can operate independently, they do operate independently to achieve their own purpose as well as the purpose of the integrated whole. The component systems are generally individually acquired and integrated, and they maintain a continuing operational existence that is independent of the SoS.
3. *Geographic distribution*: Geographic dispersion of component systems is often large. Often, these systems can readily exchange only information and knowledge with one another, and not substantial quantities of physical mass or energy.
4. *Emergent behavior*: The SoS performs functions and carries out purposes that do not reside in any component system. These behaviors are emergent properties of the entire SoS and not the behavior of any component system. The principal purposes supporting engineering of these systems are fulfilled by these emergent behaviors.
5. *Evolutionary development*: An SoS is never fully formed or complete. Development of these systems is evolutionary over time and with structure, function, and purpose added, removed, and modified as experience with the system grows and evolves over time.

Of these, 1, 2, and 4 are essential. Without them, as is discussed in the following, the category is not well defined. Properties 3 and 5 are typical, but are not absolutes. The original paper proposes the term "collaborative system" for systems meeting this definition, although the term "SoS" has remained more popular. Systems meeting this definition can be thought of as collections of independently owned and operated elements that do new and valuable things when together. In some ways, an SoS can be thought of as analogous to a franchise, where the components are independently owned and operated but subject to centralized rules and management in some degree.

If a composition of things does not possess any properties not possessed by the components, then the composition is not a proper system. Hence, property 4 is essential or the composition is not a system considered as a whole. In some discussions of SoS, the idea of emergent behavior is taken somewhat farther based on ideas of weak and strong emergence. We consider these in the context of modeling and simulation in a subsequent section.

Properties 1 and 2 distinguish the SoS category from other categories, such as family of systems or portfolio of systems. These properties define the nature of an SoS as interactive (it is about the operational interaction of the parts as opposed to their development) and that the interaction is at least partially voluntary. The components choose to collaborate rather than have no other choice.

The community is not wholly settled on a definition or even criteria for an SoS. Various authors have asserted either extensions to the definition in or alternatives. While there is general agreement on the three points discussed earlier (properties 1, 2, and 4), there is some diversity on the role or necessity of several other properties. Papers such as

Eisner (1993), Shenhar (1994), and Boardman and Sauser (2006) offer more detailed arguments on variations in definitions and alternatives. Some issues are as follows:

- *Geographic distribution*: Must an SoS be geographically distributed with the elements exchanging only information, or can it be physically compact?
- *Evolutionary development*: Does an SoS go through evolutionary development or can it be in a single configuration?
- *Network centricity*: Must an SoS be a networked system (related to geographic distribution previously)?
- *Heterogeneity*: Must an SoS be composed of dissimilar systems (varying properties, time or space scales) or can it be composed of similar or identical elements?
- *Transdomain*: Do the constituent systems necessarily span several technical or nontechnical domains?
- *Strong or unexpected emergent behavior*: Must an SoS exhibit unexpected or irreducible emergent behavior (beyond that intended by the integration of the constituent systems)?

The weight of the argument seems to be that these additional properties are common, even normal, but are not required and are not themselves distinguishing of development practices. Most SoS are geographically distributed and network linked. But one can imagine a physically compact SoS based on mass and energy exchange. These additional factors will have an impact on how modeling and simulation is brought to bear in the development process. For example, if an SoS involves mass and energy exchange as well as information exchange, it will have a strong impact on the kinds of modeling and simulation that are relevant. Most SoS undergo evolutionary development, but one could exist at a single point in time and in a single configuration. Hence, geographic distribution and evolutionary development are part of our basic definition. When a system is developed with an evolutionary process, it will affect architectural decisions, even though it might not affect the choice of modeling and simulation methods.

Heterogeneity and transdomain character are likewise typical, but there are important examples without them. Consider the many natural examples of composing very similar elements (e.g., social animals) and getting rich and complex properties as a result. These are commonly called swarms. Swarms exhibit the characteristics of SoS without heterogeneity. There doesn't appear to be anything unique about whether the components are heterogeneous or partially homogeneous in identifying best practices in engineering such combinations. Finally, unexpected emergent behaviors are commonly encountered in collaborative assemblages. Our definition requires only that simple or weakly emergent behaviors be present. As a practical matter, because strong emergent behaviors bring us out of the range of modeling and simulation in its usual sense, and because it is not of much practical significance, it is not part of the definition used here.

2.2.2 The Argument of Scale

Is it important to the concept of an SoS that it be "large?" The notion of what is "large" is somewhat ambiguous. Is a "large" number of components, number or requirements, or the count of something else? Is the Global Positioning System (GPS) with ten's of satellites larger than a space telescope that has one satellite, even though the GPS satellites are production line items while the space telescope is unique and itself quite complex in both parts count and performance objectives? Nothing in the foregoing discussion has implied that an assemblage regarded as an SoS be large in the sense of having more than a minimum number of elements or requirements or anything else.

A trivial way of thinking about SoS is that it is simply an aggregation of components sufficiently complex themselves to be regarded as systems. But there are numerous examples today of systems that are so constructed but have no properties that are distinguishing in development. Most complex electronic products (laptops, tablets) are composed of complex parts that are properly regarded as systems themselves. But those products do not require any distinguishing practices or models in development.

It is reasonable to believe that size is a complicating factor and that larger systems require larger engineering efforts than smaller ones. While it is reasonable, it is not necessarily even true. Some data (Honour, 2013) shows that system size is anticorrelated with Systems Engineering effort on a proportioned to size basis. One likely reason is that as systems grow, they grow by encapsulation of components into systems. But with absent delegated autonomy to those subsystems, there is little evidence of structural change in Systems Engineering practices. We can see systems with very large numbers of components (e.g., single company telephone networks) that can be centrally controlled because they exhibit great regularity. Deep central control of a highly irregular architecture would likely be very difficult, and so it is not implemented.

The point is that large scale, *by itself*, is insufficient to call for structurally different techniques. We would expect the use of greater automation in tools (Eisner, 1993) and careful attention to maintaining regularity but not a whole different set of heuristics and approaches. In contrast, when the elements are independently operated and managed, things change more dramatically.

2.2.3 The Argument of Collaborative Assemblage

Why are the principles of collaborative assemblage governing in defining this category? The paper Maier (1998) argues this on two points. First, collaboratively assembled systems exhibit distinct best practices in the form of heuristics. We can identify sets of heuristics that apply particularly to systems collaboratively assembled and operated. Second, when systems are misclassified with respect to being collaboratively assembled or not, a predictable set of problems arises. The classification clearly matters. Among the most important heuristics in SoS are the following.

2.2.3.1 Stable Intermediate Forms

Systems evolve more effectively when they contain stable intermediate forms than when they do not. Alternatively, in an evolving system, the architecture is in the invariants, the things that don't change, and proper choice of invariants is critical to allowing the system of interest to evolve. The poster-child example is the Internet, where the TCP/IP protocol has been nearly invariant from the 1990s until after 2010. The invariance of TCP/IP, particularly the Ipv4 protocol, has allowed massive but independent evolution of both physical layer and application layer systems. Physical layers, from coaxial cables to wireless, have come and gone over the time interval with minimal coupling to TCP/IP and applications. Similarly, huge numbers of applications have come and gone without coupling to the physical layers on which they run.

Conversely, the presence of invariants disables certain evolutionary pathways. Since Ipv4 does not pass quality of service guarantees, there is little incentive on the part of physical layer providers to develop or deploy physical systems with quality of service capabilities, and there is little incentive for application developers to exploit such guarantees (since they cannot access them through Ipv4 interfaces).

This heuristic is particularly important in the collaborative assemblage context since loose coupling among the components is essential. The presence of well-defined invariants greatly facilitates the independent development of the components and raises the likelihood successful interoperation.

2.2.3.2 Policy Triage

Triage means to concentrate resources on things that will not survive or thrive without attention and withhold resources from things that will thrive regardless or that are highly likely to fail regardless of the resources you are able to devote. In a monolithic system, the program authority is responsible for all parts of the system and their integration. The triage heuristic generally applies only at the level of technology, as where the program must decide whether to invest in desired new technologies or not. The heuristic suggests that only technologies whose development can be meaningfully affected by the program and which will not mature without the program's investment should be investment targets. The program should avoid investment where the technology is likely to be available without investment or where any investment by the program would be ineffective. While this seems obvious, it is not hard to find examples of where it has been violated in both directions during development, usually with bad effects.

In the SoS case, the heuristic generally applies directly at the component systems level. The program attempted to bring about an SoS does not and cannot control the development and provision of all of its component systems. If it could so control, it would no longer meet our baseline definition. While the development and provision of the component systems cannot be fully controlled, it can be influenced. Where to invest for that influence, and the scale of the investment, should be undertaken informed by the triage consideration.

2.2.3.3 Leverage at the Interfaces

In general, in a system, the greatest leverage is at the interfaces. This is because the essence of a system is the creation of new properties that appear only through integration of the parts. That is, the properties that define a system exist only through the system's interfaces. The situation in an SoS is even more extreme since the party with responsibility for the whole does not control the parts. At best, the party in control of the whole controls the interfaces through which the SoS comes together. In an SoS, the interfaces *are* the system.

Consider again a large intelligent transportation SoS. The components (cars, public transport systems, traffic information service providers, etc.) are owned and operated by individuals or different agencies (Maier, 1997). Their ability to collaboratively interoperate into a greater whole, say through information fusion or control for collective behavior, comes about only through interfaces that one of them own or control, and that may not be fully controlled by anybody. An authority that seeks to influence how the whole emerges will have to work primarily through influencing the interfaces.

2.2.3.4 Design in Cooperation

The essence of an SoS is collaborative interaction. Why do the owner/operators of the component systems choose to collaborate rather than operate independently to achieve their own ends? To the point, have mechanisms that drive owner/operators toward productive collaboration been designed into the SoS? This is of central interest in a closed SoS where there is a central authority who chooses how much authority to distribute. In this case, there is a central designer, and that designer can choose how to distribute authority or operation. In an open SoS, there is still a central designer, but the central designer lacks the straight authority to force the components to do things. There is still the notion of designing for collaboration here, but it can't be executed in the same way as it would be for a closed SoS.

2.2.3.5 Misclassification

Finally, a key observation is that misclassification, such as treating what is actually an SoS as a monolithic system causes problems in predictable patterns. The most obvious is that when engineering a monolithic system, one pursues efficiency by employing redundancy only where it is needed to enhance reliability. In an SoS, one allows redundancy anywhere it is used to maintain the pattern of independent operations and management of the components. Consider a system like the International Space Station (ISS). The ISS is an assemblage of systems provided by the space agencies of several countries. Each of those countries has independent funding, policy, and direction (from their respective government). Provision of an element to the space station is done largely for the interests of the providing nation. The whole has only a limited ability to count on the parts, because the parts are provided by the national space agencies to the extent they find it in their individual best interests. Now consider a module that is extensively dependent on other modules for support. From a whole systems perspective, this makes sense, since relying on shared infrastructure is a key approach to efficiency. But, from another perspective, this is a poor choice since the module provider has no authority over the services provided by other agencies, and the group responsible for the whole cannot completely count on provision of the module.

In practice, stakeholder relationship and commitments may be sufficient to make it all work, but the situation is precarious. A more robust approach would be to encourage independence of modules, at an admitted cost in overall efficiency. Whether one wants to live with the difficulties of force fitting an SoS management structure onto a more efficient technical structure, or would rather pay the cost of independence, has to be evaluated for each case.

2.2.4 Dimensions of Complexity and SoS

Another way to characterize SoS is by what complexity footprint they present. We can characterize any systems development situation by how complex its various aspects are. Which factors in the problem and its associated systems solution present relatively high complexity and which present low complexity? There is no fully standardized set of

Table 2.1 Problem and Solution Complexity Factors in System Developments.

Attribute	Simplest			Most complex
Sponsors	one sponsor with funding	many sponsors with funding	one sponsor without funding	many sponsors without funding
Users	Same as sponsors	Aligned with sponsors	Distinct from sponsors	Unknown
Technology	Low	Medium	High	Super high
Feasibility	Easy	Constrained	Barely	No
Control	Centralized	Directed distributed	Collaborative distributed	Virtual
Situation objectives	Tame	Discoverable	Ill structured	Wicked
Quality	Directly measurable	Indirectly measurable	Semimeasurable	One-shot and unstable

There is a characteristic set of values with respect to this table for system of systems.

complexity factors, but the table first appearing in Maier (2007), from which Table 2.1 is adapted, provides seven useful measures: sponsors, users, technology, feasibility, control, situation objectives, and quality. These are defined as follows:

- Sponsors: Who pays for the systems development? In the simplest case, there is a single sponsor. In the most complex, the developers must hold together a coalition of sponsors none of whom can take individual responsibility to fund development.
- Users: Who uses the system, and how distinct are their interests from the sponsors?
- Technology: How much component technology development must be included in the systems development in order to realize a system that meets basic expectations?
- Feasibility: How many systems concepts are there that meet basic expectation (from a great many to none at all)?
- Control: How will the built system be controlled? In simple cases, control is centralized with the sponsors or users. In the most complex cases, nobody controls the system.
- Situation objectives: To what extent do stakeholders know what they want, can they communicate what they want, and are they expected to change their minds during and after development?
- Quality: How do we recognize the presence, or absence, of desired quality factors in the system? At the simple end, quality factors can be directly measured during development and operation. At the most complex end, quality factors can be observed only after the system is operationally expended.

Table 2.1 shows the range of measures of complexity for each of the aspects. The "footprint" for an SoS is primarily its complexity with regard to sponsors, users, and control. It will typically also have a more complex position with regard to situation objectives, although this is not essential. An SoS problem is relatively complex, but is not complex in every dimension. An SoS carries its complexity with respect to how it is sponsored, how its users relate to the sponsors, and (especially) how it is controlled. In addition, the objectives are typically ill structured or even wicked. This means the objectives change with experience with the system, and the rate of change may be fast compared to anybody's ability to make changes to the system.

Sponsors are those who fund the development of a system. An SoS will have multiple sponsors. There will be no single sponsor with the money required who can be called upon to arbitrate development trade-offs. Users use the system. In the SoS case, they will be distinct from the sponsors and possibly unknown. The component systems may have users fully aligned with each component's sponsors, but the integrated whole cannot.

Most significantly, there is no simple, centralized control for an SoS. Control is, at best, distributed over the component system's users and sponsors. At worst, control may be virtual in the sense of a purely emergent collaboration among sponsors and users unknown at development time.

There are also identifiable subcategories of SoS. Among the more popular and recognized distinctions are:

- Closed SoS: The assemblage as a whole has a purpose and a manager. The manager of the assemblage has all of the funding and authority but deliberately devolves objectives, funding, and authority to the component systems in the service of the aims of the whole. Single-service air defense is typically a closed SoS.
- Acknowledged SoS: Both the assemblage as a whole and the component systems have their own purposes, managers, and authorities. The manager of the assemblage can use his or her funding and authority to make the SoS work better, but the funding and authority are limited. Many multiservice defense systems are of this type where the assemblage has a corresponding funded program office, but the component programs are likewise independent entities.
- Open SoS: In this case, the assemblage has an overall purpose, and there is a manager for the assemblage, but the manager has little or no funding and no authority to make the components do as he or she wishes. The manager of the whole can convince the components, but cannot compel the components. Collective action is voluntary. In practice, the manager is often an organization or process rather than a designated person. The modern Internet largely operates this way, as do many multinational defense systems.
- Virtual SoS: In the virtual case, there is no central purpose or management. The components of the assemblage interact with each other voluntarily as they see fit. Operation as an integrated whole is as an undirected emergent property.
- Organic SoS: This term is sometimes used to describe sociotechnical systems that evolve without central direction of any sort but exhibit clear order. These are virtual SoS with a very heavy social rather than technical flavor. Probably the most famous example is the "Tiffin Wallah" system of Mumbai.

2.2.5 Closely Related Categories and Carry-Over Lessons

There are two other categories of assemblages of systems that are quite distinct from the SoS concept but are similar enough to sometimes be confusing. They are the family of systems or product line and the portfolio of systems.

A family of systems or product line is a collection of systems related by common design and manufacturing. The elements of a family of systems share design features, parts, software, protocols, or other elements, but they do not generally interact to form a greater whole. They are built, deployed, and operated largely as stand-alone systems. Their organization into a family with shared elements is normally to realize economies of scale and reduce cost. For example, the product line of printers made by a printer manufacturer is a family of systems. There is extensive joint design and manufacturing to cut costs and facilitate marketing, but the printers do not communicate with each other to form a greater whole. The architects of the product line structure the line as a whole to find effective combinations of need and technologies (product "niches"). The architects of the product line as a whole deliberately defer much choice about the content of individual products to downstream product architects. The elements held in common and shared across most or all of the products will be those where economies of scale are maximized and the value of product-by-product customization is minimized.

Of course, there can be cases that cross categories. Categories are not mutually exclusive. Somebody may set out to create something that deliberately contains both SoS and family of systems elements. Consider the overall Android phone enterprise. There are elements of a family of systems (the shared open-source operating system and development environment elements, the chipsets within a phone producer). There are also SoS elements in the interaction between app developers, phone developers, and users. The pieces communicate and a greater whole (actually, many greater wholes) emerges. The interaction is not centrally controlled, except to a limited degree. Google has a very strong role, created by continuing to sponsor and develop the core system, but other major players are free to pursue their own purposes.

A portfolio of systems is a collection of systems that are jointly managed to a shared budget in service of some larger purpose or goals. The portfolio manager is responsible for allocating resources among the elements of the portfolio. The elements of the portfolio jointly contribute to the fulfillment of the portfolio manager's goals. However, the components of the portfolio may share no technology or development, have no design in common, and may never interact. Their only connection may be resource allocation, manager, and contribution to higher-level goals.

Good practices in portfolio-of-systems development share somewhat with other categories but also has an identifiable set of practices of its own. The situation is clearly distinct from the SoS case, as the portfolio manager has extensive central authority where the SoS sponsor does not. The presence of central authority is more like the family-of-systems case. However, unlike the family-of-systems case, there are likely no economies of scale to be had from commonality. The elements of a portfolio of systems are distinct and heterogeneous. They may share little or nothing in technology or design but be linked just by mission.

2.2.6 What Is Emergence, and Is It Important to SoS?

To be an SoS, the collective must possess properties or behaviors that are not possessed by any of the components. This is an "emergent property." The same term is also sometimes used to describe unexpected or unpredictable collective properties, which is something of an antithesis to deliberately designed and desired collective properties. Because the presence and nature of emergent properties plays a central role in understanding types of systems, we examine it in some depth here.

The concept of emergence is simultaneously very simple and enormously profound. The basic concept of emergence is quite simple. An emergent property is a property possessed by an assemblage of things that is not possessed by any members of the assemblage individually. Consider an assemblage of electronic parts and software. Weight is not an emergent property within this definition. The weight of the whole is simply the sum of the weight of the parts. Each component possesses weight and weight adds up to the whole. On the other hand, if the assembled parts are a radar system, the property of detecting and tracking airplanes is emergent since it is possessed only by the assemblage as a whole. The antenna cannot track an airplane, the transmitter cannot track an airplane, and the software cannot track an airplane; only the radar as a whole can track an airplane. The core definition of "system" is that of an assemblage of elements that possess emergent properties. The whole has to do things that the components do not.

In the definition of SoS, we insisted that the assemblage of independently owned and operated systems do something the component systems do not. That is, the assemblage must itself have the characteristic of being a system and not just a loose assemblage. Thus, an assemblage that is an SoS must have emergent properties.

In much of the literature, there are two specified types of emergence: weak and strong. Weak emergent properties are commonly defined as those that can be reduced to the interactions of the components; they come about from the interaction of the parts in a way that is well described by the interaction of component-level properties. Strong emergent properties are not reducible to the interactions of the components. In practical terms, these are not terribly useful definitions alone. There is no unqualified example of a true strong emergent property in the sense just given. To be an unqualified example of a strong emergent property, it would have to be known that the emergent property *cannot* be reduced to or explained by the interaction of the parts. The strongest possible case would be that the emergent property is actually inconsistent with the properties of the parts. If, for example, it were possible for a system collectively to violate conservation of energy when all of the components operating individually do not violate it, then we would have an example of a true strong emergent property.

For our purposes here, some more refined definitions that directly incorporate distinguishing between a system and models of the system are more useful. These definitions introduce the notion of models of lower complexity than the system itself. An "abstracted model" of a thing describes its behavior or properties without being that thing. A "simulation model" of a thing describes its behavior or properties by composing models of the things' components. Consider an FM radio transceiver. An abstracted model would describe the transformation of audio into radio waves and their reception and reverse transformation. It would do this mathematically at a level above direct simulation of the operation of the components. A simulation of the FM radio would be a collection of component-level mathematical models, where the component level could be chosen variably. It might model at the level of blocks like microphones, amplifiers, and transmitters or might model all the way down at the level of transistors and other electronic components.

With this notion in mind, we can reclassify types of emergence in terms of how they map to models of the system.

Simple Emergence: An emergent property readily predicted (not just explained) by simplified models of the system's components. The property can be readily and predictably produced in lower complexity, abstracted models of the actual system.

Weak Emergence: An emergent property that is readily and consistently reproduced in simulations of the system, but not in reduced complexity nonsimulation models. It can be understood through reduced complexity models of the system after observation, but not consistently predicted in advance.

Strong Emergence: An emergent property that is consistent with the known properties of the system's components but which is not reproduced in any simplified model of the system. Direct simulations of the system may reproduce the emergent property but do so only inconsistently and with little pattern to where they do so and where they fail. Reduced complexity models or even simulations do not reliably predict where the property will occur.

Spooky Emergence: An emergent property that is inconsistent with the known properties of the system's components. The property is not reproduced in any model of the system, even one with complexity equal to that of the system itself, even one that appears to be precisely simulating the system itself in all details.

Consider a variety of examples of various types of emergence. Two FM radios acting together as a communications system exhibit simple emergence. Neither radio alone allows voice communication, but composed they do. Their behavior is readily modeled, and the characteristics of the communications capability produced can be predicted. A component in the radio could be removed and replaced by a digital simulation of that component, or a mathematical equivalent and the communications property would predictably remain. The communications function of the whole is readily composed from component properties or decomposed to component properties during design.

Consider instead the operation of a large cell phone network with millions of users or the road network in a large city with millions of cars and other vehicles. Such networks are frequently observed to exhibit things like congestion waves or bottlenecking based on interactions across users or across protocol levels. These phenomena frequently qualify as weak emergent properties, given the definition earlier. These phenomena can be simulated with models of the components that are simplified relative to the full complexity of each component. Simulations will produce reasonably accurate predictions of collective phenomena, at least after such phenomena have been observed in real systems. It is usually possible to build reduced complexity, nonsimulation models of the phenomena, at least after the fact. For example, it has been possible to build continuum flow models for traffic flow that exhibit "shock wave"-like phenomena that map to traffic disruptions. Typically, the phenomena are empirically observed in operating networks, then modeled and understood, and then modified through interventions in the system (not always with predictable results). A reductionist perspective is sufficient to understand these phenomena once observed, but not to predict their appearance prior to observation "in the wild." While modeling and simulation works after the fact, it does not always work in prospect. Engineers on large networks are frequently surprised by the appearance of collective phenomena, although they can be explained after the fact.

As the examples show, engineering in immature fields, especially when dealing with complex systems, is working in the realm of weak emergence. In mature engineering disciplines, simple emergence will be more common, though weak emergence persists in many areas. The scientific disciplines, physics, chemistry, and biology provide familiar examples of strong, though not spooky, emergence. Of course, there are many examples of weak emergence in all of these. It is common for scientists to observe a new phenomenon, struggle with an explanation, and eventually discover relatively simple or low complexity models that reproduce the phenomena. However, there are also many good examples where strong emergence remains and may never be simplified.

Quantum mechanics is presumably a complete model for atoms. In principle (if not in practice), it should be possible to predict chemical properties from quantum physics. Likewise, in principle, it should be possible to predict the existence of self-perpetuating chemical machines (biology) from knowledge of atomic properties and thermodynamics. But, again in practice, it is much more difficult. Theoretical chemists can use quantum physics to predict only a very limited subset of observed chemical properties. This is not because more complex chemical properties are inconsistent with micro-level physics (no such inconsistency has ever been observed), it is presumably because our ability to predict collective properties is inadequate (computational and perhaps conceptual limits).

Similarly, we observe the existence of biochemistry based around persistent encoding of information in the DNA molecule and a wide range of affiliated energy transport and structure creating molecules and cycles. While we can analyze the chemistry and physics of the life processes we know, we do not know, nor do we have any conceptual way to discover, if the known laws of chemistry will admit of other equally complex life processes based on another chemistry. We have never discovered one, but we have no way to exclude one existing. Nothing in the chemistry of live collectives is known to be inconsistent with any understood laws of chemistry or physics, so chemistry and biology are not examples of spooky emergence. They are examples of strong emergence in that no reduced complexity model or simulation of them can be shown to reproduce the relevant phenomena.

Because something is regarded as a strong emergent property today does not mean it will be forever regarded as one. Today, human consciousness is clearly an example of strong emergence. We cannot produce it with any simulation or model. We cannot predict whether or not any given system will have it. We cannot design a system that will produce it. Some speculate that we may eventually accidentally produce consciousness in a system, but if we did, its appearance would be a surprise. Even in the biological space, we cannot distinguish why one brain exhibits consciousness in operation and another does not. However, it is possible that in the future consciousness might become well understood enough to fall to being a weak or even a simple emergent property. It might be the case that experiments with neural networks above some threshold size or within some currently unknown pattern will consistently yield behavior we regard as conscious. It might be the case that over time we will develop descriptions of the behavior of large neural blocks that can be "swapped" for the actual execution of the neural network or a detailed simulation of the network with equivalent collective behavior. If this happened, the phenomena would be moved downward from being strongly emergent.

Similarly, we might at some point in the future discover reduced complexity patterns in chemistry and biology that explain why some molecular combinations can form self-replicating machines and others cannot. Such patterns might allow prediction and synthesis of alternative molecular systems of life. That we have no idea today of what such patterns would look like does not exclude the possibility of discovering them in the future, but neither is the existence of such reduced complexity patterns assured. It is possible that biochemistry, consciousness, and other emergent phenomena are inherently complex and have no reduced models that can ever reproduce them.

Strong emergence is a challenge to a reductionist perspective, but is not inconsistent with it. Reductionism does not help us understand a strong emergent property. If we can understand one through reduction, then it almost automatically becomes a weak emergent property. In contrast, spooky emergence is inconsistent with a reductionist view. The presence of a spooky emergent property is direct evidence of some nonreductionist law of nature.

There are no clear examples of spooky emergence, and one philosophical position is that none will ever exist. Hypothetically, life or consciousness could be examples of spooky emergence, although neither is known to be. As a thought experiment, suppose it was possible to produce an electrochemical "clone" of a brain that preserves a snapshot in its instantaneous state. Suppose it were further possible to then let the clone resume operation in a vat while communicating through neural pathways. If that clone did not exhibit consciousness even while observed to be operating at a micro-level identically with a brain in a human, then we would be forced to consider consciousness as an example of spooky emergence. Similarly, if it could be shown that chemical reactions were thermodynamically different when occurring in a living system, then when occurring under identical conditions in a nonliving system, we would be witnessing a spooky emergent property.

All models and simulations are simplified versions of the actual system. We do not simulate the physics and chemistry of the components (normally). We model the components as having some understood (usually designed) properties. A model of the whole system models the interactions of the represented properties of the components through known interaction paths.

It is a common occurrence that we find in a system emergent properties that were not predicted in advance from models of that system, but that are readily understood after the fact through extension or elaboration of the models. This situation is commonly encountered in studies of systems failures. The models of the system did not predict the failure situation, but having observed the situation models can be readily extended to encompass it.

Consider a somewhat famous case of the Lockheed Electra failures (Boyne, 1998). In a short period of time, two Lockheed Electra aircraft lost their wings and crashed in flight with no apparent cause. After extensive study, it was found that there was coupling between the elastic structure of the wing, aerodynamic forces on the wing, and gyroscopic forces in the rotating mass of the engine and propeller. Under certain conditions, these could form an oscillating positive feedback loop that tore the wings off. To make things more complex, the destructive coupling was not generated by specific environmental conditions, but only by environmental conditions coupled to a degree of looseness or damage in the engine-to-wing mounting. Once the mechanism was perceived, it was readily modeled, and a solution for it engineered. However, the nature of the interaction had not been imagined in advance. This is an example of a weak emergent property, in that modeling was not sufficient to make a prediction in advance of the empirical data, but a simplified model was quite sufficient to explain the emergent property once it was observed. Further advances in understanding aeroelasticity and structural analysis might even make it a simple emergent property when modeling tools are sufficient to explore a wide space of off-nominal conditions.

The distinguishing characteristic of SoS is the independent nature of the components. Depending on the case, the components may be said to have volitional behavior. This has significant impact on how we do modeling and simulation of SoS. While the behavior of the whole is the "roll-up" of the behaviors of the parts, the behavior of the parts may be very complex. If a component has independent volition, then in a philosophical sense, it may be said to be following rules, but it may be impossible to fully specify those rules or how the rules interrelate the extended consequences of the interaction of the rules. Ideally, we might be able to find some small set of deterministic rules, an abstraction of the real situation, that yield a reasonably correct integrated description. But, even if this is possible, we could not have confidence that it would work robustly, if the whole were to exhibit strong emergent behavior.

2.3 IMPORTANT SYSTEMS AND CATEGORIES

Subsequent chapters of this book discuss a variety of examples of systems, usually describing them as SoS. Are they, given the discussion earlier, SoS in the collaborative systems sense? What is the role of modeling and simulation in their development? By and large the answer is they are collaborative systems in the sense defined earlier, but in some cases the issue is somewhat complex. Several bear some examination here before they are fully discussed later to facilitate some comparisons.

2.3.1 Energy Systems, Conventional, and Sustainable

On the global scale, it is obvious that energy systems meet all of the outlined tests. Their major components are large-scale systems themselves, they are independently owned and operated for owner/operator goals, and they are geographically distributed and evolutionary. As a collective, they exhibit emergent properties, often unintended (e.g., the presence of cascading failure modes shown in large-scale blackouts). One of the interesting aspects of energy systems is that the components do not only exchange information, they are also coupled through much more physical interfaces, such as electric power exchange and fuel exchange. Emergent behaviors come about not just from information exchange, but are also manifestations of the physical exchanges. The mechanisms are not just human logic, but underlying physics. This obviously has a large impact on how modeling and simulation is used. The modeling and simulation must incorporate the physics, not just the information flow. Modeling and simulation for energy systems crosses significant disciplinary boundaries.

On a smaller scale, energy system does not necessarily meet the criteria. Individual power generation companies may run a network that they wholly own and operate. Even here, the behavior of uncontrolled consumers may introduce an SoS element.

A further aspect that particularly affects modeling and simulation in this domain is the interaction of different time scales: here, an operational time scale and an investment time scale. At the operational time scale (seconds to days), we are interested in how an SoS with essentially fixed configuration operates with independent agents operating segments. On the scale of years, the human elements choose investments, in capacity, transmission, or consumption. On the scale of years, the human elements choose the technical configuration, though they do it through a largely uncoordinated process. Both time scales are ultimately of interest.

Energy systems make a good case study because people have evolved mechanisms for managing their SoS nature, although some of those mechanisms are coming under challenge. We have understood for some time that interconnecting power networks can yield valuable emergent properties (like the ability to do cross-region backup and net efficiencies) while also potentially creating undesirable emergent properties. We have evolved both technical and policy approaches to encourage the desired properties and discourage the undesired ones. By and large, those policy and technical approaches work well, as exhibited by energy being very reliable in the developed world, at least where they can be applied effectively. At the same time, there are new forces at work that are altering the relationship with effects that are not yet understood.

- Deregulation is changing the policy underpinnings on which the collaborative energy SoS is managed.
- Supplier monopolies, and then their breakup, have affected the integrated nature of the system.
- Making renewable, intermittent sources more than a small fraction of total supply is an enormous challenge to the integrated system.

Consider the issue of converting a large fraction of the energy supply system to intermittent renewable sources (e.g., solar and wind). This conversion is a dual challenge that goes directly to the nature of the collaborative system. The first challenge is handling intermittency. Given that some sources are intermittent, are other suppliers obligated (by policy) to make up for any lack of those intermittent sources? If they have to make up for any lack, then what happens when the intermittent sources overproduce? How are costs and responsibilities distributed? If all elements were owned and operated by a single utility (as could happen but probably won't), the lines of responsibility would be clear.

Consider the current issue of distributed generation of solar power by homeowners with solar panels. When the homeowner is both a consumer and supplier and chooses to become a supplier, the power company is not in an SoS relationship with its own customers. The customers are now suppliers, with a relationship that is largely governed by policies set by neither (the rules being set by government agency). Responsibility for intermittency is now distributed. If the power company is, by policy, responsible for maintaining the performance of the grid, but does not control important grid inputs and outputs, it will most likely have to substantially overbuild, at least relative to average demand. Some studies indicate that to maintain reliability at current levels (well above 99%), the degree of overbuilding will have to be considerable, perhaps to nearly 100% of the maximum demand ignoring intermittent sources. This imposes large costs. How are those costs to be distributed, and how will the cost distribution effect the individual incentives to be power suppliers? Since the players are independent actors, we have a classic SoS situation. Exactly which form it takes (closed or open) depends on political-level policy decisions.

2.3.2 Air Transportation Systems

Air transport is fairly self-evidently an SoS, by the definition earlier, at least in most countries of the world. The elements of the overall system are owned and operated by multiple organizations and interact through collaborative mechanisms. Except for a country where all airlines are state entities and comanaged with the airspace, there is no single entity "in charge" of the whole. Each major player, the airlines, airport authorities, national air traffic control authorities, and other private aviation entities, interacts with considerable autonomy. It is, however, more of an open collaborative system rather than virtual. Central authority does exist, but its coercive power is limited, although this situation is variable from country to country.

Why is it that a collaborative system exists even though there is a national air transportation authority? In general, the national authority has significant control over real-time operations but much less control over longer-term operational planning and still less control over investments. The major operators, for example, airlines, have to cede authority for real-time operations to the national authority but plan their own routes and schedules (within limits of airports) and make their own investment decisions on aircraft and operational structures. There are management mechanisms at the international level as well, though their direct authority is even less.

Air transportation systems, like energy systems, have a significant element of interface beyond just information. The elements of the system exchange people and cargo as well as information. The entities involved have strong interfaces to the physical world, for example, being affected by weather. The operation of the whole depends on physical exchange as well as information exchange. A network description that included only information flow and state and did not incorporate the physical limitations of people, cargo, fuel, and so forth would be very limited. Modeling and simulation of air transportation systems has to take into account both the information nature as well as the physical nature of component systems interactions. The physical interactions have significant limitations. For example, a passenger might transfer from one airline to another if a flight is canceled, but an aircraft would not be transferred between airlines, except at a long time scale, for policy reasons.

As the energy systems case, there is an operational and an investment time scale. The technical configuration is largely fixed at the operational time scale, but subject to extensive change on a longer time scale. The nature of operations, for example, if an operator is making a profit or not, clearly influences investment decisions. Decisions by operators and consumers on long time scales may couple and have significant operational consequences at a later time.

2.3.3 Command, Control, Communications, Computing, Intelligence, Surveillance, and Reconnaissance Systems

Military Command, Control, Communications, Computing, Intelligence, Surveillance, and Reconnaissance (C4ISR) systems can span the full range from monolithic system to SoS without any central control and many intermediate points. The variable nature of military C4ISR on the SoS scale comes both from environmental as well as design choices. Within a given military organization, there is an assumption of unity of command. In principle, a commander exists with authority to direct down to the lowest level. In practice, of course, military operations have to be conducted with somewhat distributed authority. The main commander provides general direction to subordinate commanders, who refine the direction they are given to lower-level commanders, in a process that continues to the actual operational units. Moving in the other direction, raw surveillance and reconnaissance data is processed and analyzed in an "upward flow" that ultimately reaches the commander in chief. The highest-level commander is very unlikely to receive much raw data and would be unable to make much effective use of it, except where it has been pruned to only the most significant pieces.

Where a clear central authority exists who has explicitly devolved authority, we have the classic definition of a closed SoS. The central authority directs operational and managerial independence for well-understood reasons. The devolution is a response to both simple and complex factors. Among the simple factors are:

- The overwhelming amount of information involved makes complete situation awareness and command authority impossible. Each commander can manage information overload only by aggregation.
- There are too many involved persons to permit complete person-to-person communication.
- Communications limits make it impossible to move all data upward or all command information downward.

The more complex factors have to do with using the distribution of independence to achieve particular attributes. Military C4ISR systems are subject to violent and unexpected disassembly in normal use. They are attacked, disrupted, jammed, and exploited. Their robustness in the face of unexpected enemy action is a primary desired attribute. Militaries known from long combat experience that allowing units substantial managerial and operational independence is an effective way of achieving robustness in the face of hostile action.

A second source of managerial and operational independence comes from political diversity. Many C4ISR systems are multinational. Consider C4ISR in a large coalition operation, such as a major peacekeeping or disaster response incident. There is no longer a clear central authority with the power to direct all players. Each player answers to its own political authority, and the political authorities interact through treaty or entirely ad hoc arrangement. If the various authorities have compatible views and long experience, they may act with considerable unity, but players carrying out contradictory actions have happened in large coalition operations.

Modeling and simulation of multinational C4ISR then carries many complex challenges.

1. The number of entities is quite large, they are arranged to interact in a layered fashion, and stakeholder concerns may be at any layer.
2. The entities interaction is mostly informational but can also be physical. To the extent that entities can carry out kinetic or nonkinetic attacks on each other, their interactions become physical as well as informational.
3. Each entity received central direction, but not reliably, and may or may not fully carry out that direction.
4. Commanders may have different goals and objectives, which will be reflected in direction to entities under their command. Commanders may collaborate with each other, sometimes on the basis of their goals and objectives, but possibly also on some other basis (such as shared experience or antipathy).

The challenge of modeling and simulation military C4ISR systems can run from purely technical (data flows and processing) to cyberphysical (including kinetic effects) and to strongly sociotechnical.

Here, also, there are different operational and investment time scales. A further complicating factor is that the participants may change their overall concepts of operation on a time scale in between day-to-day operations and long-term investments. If military actions are effective, those on the losing end will try to adapt and those on the winning side will try to exploit what works best. Either side may introduce a previously unknown operational concept if it appears fruitful. Configurations are only stable on the shortest time scales.

2.4 M&S IN THE SoS CATEGORY

What, if anything, is distinctive about modeling and simulation of SoS? Does the placement of a system in the SoS category imply that modeling and simulation is a different sort of challenge, that different tools are required, or that modeling and simulation plays a different role in development or operation? Many aspects of modeling and simulation for SoS are not unique; they share the disciplinary links to any type of system. If an SoS is built on a communications network (as it usually will be), the communications network is not conceptually different from a network in monolithic system. If the underlying technologies are the same, the models that capture the operation of those technologies will likewise be the same.

While many aspects are the same, we can identify some modeling and simulation aspects that are particular to SoS.

1. An SoS is normally unique; it exists in only a single copy. Except for direct, experimental work with the system of interest, developers and operators can only work with models of the system. Models become stronger surrogates for the actual system than in cases where the system exists in multiple copies and can be examined in controlled environments.
2. SoS are large scale, in the sense that the number of components is large enough that bottom-up simulation is normally impossible. Simulations of an SoS have to work with abstractions of the components, typically with abstractions of the component systems.
3. The components of an SoS operate together, at least in part, voluntarily, based on a volitional choice of the component system's owner and/or operator. Models of the SoS have to take into account the somewhat independent nature of the components.

It is also useful to compare and contrast SoS modeling issues with those in other categories such as families of systems and portfolio of systems.

2.4.1 An SoS Is Unique

Consider some classic examples of an SoS, a fused collection of intelligent transportation services in a metropolitan area or a collection of interconnected power grids. While there might be multiple metropolitan intelligent transportation systems or interconnected power networks, each one will have a unique configuration, users, owners, operators, and issues. When each system of interest is unique, it forces great responsibility on modeling and simulation. Not only can we only assess the system of interest via modeling and simulation during design (as is typical in any system requiring Systems Engineering), but once in operation, we have no example to work with other than the operational system itself and models of the operational system.

Again, consider a metropolitan-wide intelligent transportation system. Suppose we want to assess the effectiveness of current policies and changes to those policies. We can observe the effects of current policies on the operational system. To assess the effect of new policies, we either have to rely on modeling and simulation, or we have to do a "live experiment" where we change the actual operational system and observe the results. In most cases, we will be loathe to make alternations to the operational system unless we have strong confidence in the results. That confidence can only be obtained through modeling and simulation. In smaller scale, monolithic systems, we will often have multiple copies of the system of interest. While the copies may not be in the operational environment (e.g., "extra" satellites are not usually in orbit, they are on the ground), they do exist and we can do isolated, controlled experiments with them. This means that modeling and simulation plays an even more central role in the engineering, and operation, of SoS than it does monolithic systems.

2.4.2 M&S and Scale Challenges

A second challenge for modeling and simulation in SoS is scale. Again, this is not strictly unique to SoS, but it is characteristic of them. SoS are large, in the sense that they have many components and those components themselves are systems with many more components. An attempt to do true bottom-up modeling starting from a level where the components are simple is likely to be futile. The "natural" scale of components will still be large, and the components at this level will be significantly complex and cannot be modeled without significant abstraction.

Consider again metropolitan-level intelligent transportation. The natural "component" units are vehicles, people, road networks on a scale of a few meters (comparable to component sizes), public transportation elements at a similar level, and transportation control entities. In a metropolitan area, there will be several million such "components." Since all of the important ones incorporate a human operator, who follows his or her own objectives and is only loosely bound to higher-level rules, each component is potentially very behaviorally complex. Because of the behavioral complexity and diversity of the components, it is not possible to have assurance that one is modeling them "correctly." The component models will be behavioral abstractions of the real things based on the modelers beliefs about what factors are important.

As computer power advances, it is easier to deal with scale through brute force. It is not implausible to consider running an agent-based simulation with a million agents. The question of scale here becomes what confidence is there that simulation scale is a good descriptor of real SoS scale. If the system has a million components, is there good reason to believe that a simulation with a million agents is a better simulation than one with say 10,000? In fields, very large-scale simulations that work bottom-up are normal (e.g., fluid dynamics, weather); we have some reasons to believe that scaling up a simulation improves results. But, as a cautionary note, we know from experience that scaling up is not always an improvement, and we usually only learn this from careful cross-comparison with experiment, something particularly difficult in most SoS situations.

2.4.3 M&S and Collaborative Systems Assemblage Aspects

The essential characteristic of an SoS is its collaborative assemblage. This introduces an element into modeling and simulation that is normally absent, having to model what is essentially volitional behavior on the part of components. One does not normally think of the components of a system as "having a mind of their own." We do not assume that the axle of a car will behave differently because the axle operators will change their minds about integration of the whole car. There are no axle operators. The axle operates according to the laws of physics as applicable to its physical structure within the car.

Software is much more functionally complex, but software likewise, as a component, is not normally presumed to have independent volition. Software implements a behavioral model and executes that behavioral model. The behavioral model may be complex, but it is normally presumed to be deterministic and fixed.

In a power network, on the other hand, especially one that encompasses many independent operators and consumers, the consumers clearly do make uncoordinated decisions. Power consumers choose to activate, or not, their appliances. Each consumer is in control of his or her own level of consumption. Consumers may be responsive to incentives, like variable pricing, and perhaps to voluntary requests, but ultimately as long as power is available, the consumer can consume it or not. The consumer's consumption choices come from the consumer's view of his or her own best interests, possibly modified by some willingness to follow a collective. Likewise, if the operators are different companies, they will produce power in accordance with their own best interests.

Modeling and simulation comes into play in both descriptive and prescriptive senses. Descriptively, we want to understand the dynamics of such a power network where the dynamics include individual choices. We want to understand how production and loads balance or not based on the policies of the various participants. We want to understand how various individual policies will affect overall dynamics. Prescriptively, we want to understand how to design technical elements, and influence usage policies, to accomplish various collective goals. For example, we may want to know what mandatory load management rules are necessary to stabilize a renewable heavy power network, given likely behavior on the part of the players. We may want to know what mixture of production sources, if independently managed, will yield given stability criteria even in the face of independent management.

Probably the most powerful and appropriate modeling and simulation technique for SoS is agent-based simulation (Railsback and Volker, 2011). In agent-based simulation, we can encode rules we believe to be behaviorally accurate into each agent, where each agent represents an independent decision maker in the SoS. Assuming that the rules are a good surrogate representation for the real agent, the behavior of the agent-based simulation should represent well the behavior of the real thing. We can also code alternative rules into the agents to investigate, for example, the impact of different individual behaviors on achieving collective goals. This is most appropriate because it provides a mechanism for directly capturing the presumed volitional behavior of an element.

While agent-based simulation is a powerful tool for analysis and investigation, it is still weak with regard to synthesis. We do not yet have a rich set of techniques for synthesizing agent rules to yield desired collective behaviors, much less know how to make them correspond to real-world behaviors. This is essentially a problem in generating weak or even strong emergent properties. By stating it this way, the challenge is apparent. The properties we are interested in with an SoS will be within the definition of a weak or strong emergent property given earlier. But, by definition, such properties are between difficult and impossible to replicate in a reduced complexity model of the system of interest. Fortunately, the "impossible" end of the spectrum should be unusual. But the "difficult" end of the spectrum will be common.

Another approach is to look for ways to avoid the brute force of simulation to understand an emergent property and instead discover the insight that reduces it from weak to a simple emergent. This reduction can happen if we can discover some order or pattern that explains the emergence by means simpler than direct simulation. For example, in economics, we believe that the interaction of independent agents will robustly reduce to a general equilibrium condition and that the equilibrium may be computable by means other than brute force simulation.

Describing the behavior of an SoS by relating patterns across disciplines can work in more than economics. Examples of where models can be extended from one discipline to an environment where the components have complex behavior include:

1. Economic models where the behavior of a system of many complex components can be modeled as a supply-and-demand equilibrium process
2. Fluid dynamic models where cars or people are the components and the movement of aggregates is modeled through wave equations

2.4.4 M&S in Closely Related Categories

While the focus here is clearly on SoS, other closely related categories also have significant modeling-and-simulation-related issues. It is useful to compare some of those issues to those unique to SoS, since given systems often overlap.

A family of systems or product line is a group of systems related by joint design and/or manufacturing, not by interoperation, usually to achieve a favorable mix of customization to users while also gaining economies of scale. In this case, the problem of modeling and simulating the whole does not turn on interactive behaviors of the whole; those are presumed not to be the focus, but on properties of the joint development process. We aren't looking for emergent properties of the members of the family interaction, we are looking to evaluate economies of scale and user need match across a diversity of users.

In most families of systems, the primary goal is to achieve a mixture of effects in customer match and economies of scale that cannot be achieved with an individual system. Much of the modeling and simulation challenge is that at the family of systems level, the concerns are largely economic (market match and economies of scale), while the "design variables," the things the engineer can choose, are mostly technical. Models are of little use unless they can span these worlds. While a typical SoS has a very important information network component (calling for models of the same), a family of systems may have no such information network component (at least not at the family-of-systems level). Thus, the tools of greatest use in SoS modeling may be of little use in families-of-systems modeling, and vice versa.

Likewise in a portfolio of systems, the focus is on the properties of the joint development process, not the interacting systems. A portfolio of systems is defined by common resource management across many systems to achieve enterprise mission goals. What we seek to model is how our distribution of resources, or alternative distributions of resources, across systems and projects produces enterprise mission goals.

As in the family-of-systems case, the concerns in portfolios are largely economic. Again in contrast to SoS, the information network modeling element is likely to play a small and noncentral role. In portfolios of systems, we are often interested in collections of objectives. The notion of aggregation over multiple objectives to a single utility measure is likely to be weaker, especially if the portfolio has many stakeholders without a convergent point of authority.

The finance world offers a number of tools for portfolio study, some of which can transfer to portfolios of systems, but others that do not generalize. As an example, consider real-option theory and tools (Leslie and Michaels, 1997; Shishko et al., 2004). The concept of real options is well known in the business literature. As a concept and as an intellectual model, it works very well in portfolios of systems. On the other hand, quantitative option tools are commonly used in financial products that have very limited carry-over.

2.5 ARCHITECTURE, ARCHITECTURE DESCRIPTION, AND DEVELOPMENT

This chapter has, so far, discussed a particular category of system, the SoS, and issues associated with modeling and simulation of SoS. Certainly, some may study modeling and simulation in the abstract as an important topic on its own. But in the remainder of this chapter, we will direct our attention to narrower issues, such as modeling and simulation of SoS related to architecture, architecture description, and the development process for SoS. A particular concern is how we understand the architecture of an SoS and how modeling and simulation relates to that architecture. To address this, we need to understand the varying interpretations of the concept of architecture.

2.5.1 Interpretations of Architecture

Architecture is generally understood as fundamental or unifying structure of a thing (IEEE, 2000; ISO/IEC/IEEE, 2011). This picks up the analogy to the civil architecture profession where the architect is responsible for overall conceptual definition to the point the client (the building's future owner) can make an informed decision to proceed with construction and hire a general contractor to proceed. In the broader context of systems development and acquisition, architecture can be thought of within four broad perspectives: as conceptual structure, as a document, as a response to mandates, and as a bureaucratic exercise. For the remainder of this chapter, we will consider architecture as a conceptual or organizing structure, and not as any of the other three perspectives.

2.5.1.1 Architecture as Conceptual Structure

If we ask the question "What is its architecture?" about a building or other manufactured object, it is generally understood that the answer is about the object's overall structure or style. So, for example, if we discuss the architecture of a building we discuss its style, its layout, how the form embodies the functions of interest to the sponsor, and the esthetic effects its form produces. While we recognize that the architects produce concrete artifacts (like drawings), the drawings are not themselves the "architecture" of the building. Thus, we can think of both conceptual and physical artifacts of the architecture. The conceptual artifacts are the decisions to build a system one way, and not another. They are the decisions as to which stakeholder concerns to address and which to reject. While this architecture is conceptual, it is obviously central to the ultimate success of the system. The physical artifacts are the drawings and other documentary products.

2.5.1.2 Architecture as a Document

In the document perspective, architecture is a document or, more generally, a collection of models (many of which may not normally be printed on paper as a standard document). In this case, when we refer to "The Architecture," we are referring to the document (or collection of models). We should note both the relationship between and the differences with the conceptual structure perspective. In a design activity, we make decisions and document those decisions in documents. The documents are not the decisions; they are representations of them. We can consider the wisdom and quality of the decision and documents largely separate from each other. We might have made excellent decisions but documented them poorly, and we might have very high-quality documentation of very foolish decisions.

2.5.1.3 Architecture as a Mandated Response to Concerns

In this perspective, architecture is produced as a response to specifically mandated concerns. So, for example, a group in authority is concerned about the ability of systems to interoperate. They express those concerns by mandating the creation of a set of architectures for the systems, so that they might be compared. The architectures are created in response to the mandated concerns. Note, at this point, the mandate may require particular decisions (in analogy to building codes) or may require particular documents being produced. In the SoS world, this appears most often because of mandates to use a particular architecture framework, such as the DoD Architecture Framework (DoDAF) (DOD, 2010).

2.5.1.4 Architecture as a Buzzword, Bureaucracy Driven

This might be thought of as the toxic cousin of the foregoing. When we create architecture documents without being able to trace their contents to clearly identifiable purposes, but only because of mandates, we are engaged in bureaucracy-driven architecture.

In the subsequent sections, we will be using the term "architecture" to refer to the fundamental and organizing structure of a system, the invariants of that system as it evolves, or the common elements within a product line. The architecture is thus conceptual, and not itself an artifact. The architecture artifacts are documents or models that we develop to describe the architecture. Modeling and simulation is thus part of the world of architecture description but will often be used to inform, or even drive, the decision part of architecture and architecting.

2.5.2 Architecture and Development

The classic view of architecture is as a front-end activity (Maier and Rechtin, 2009). Architects work with sponsors to select the problem to be solved, the scope of the requirements, the overall systems concept, and the structure of the development program. Once the decision to develop has been made, the resources allocated the more detailed parts of the process take-over and execute the initial, hopefully well-chosen, systems concept.

SoS are not developed in a waterfall fashion; they are evolutionary. There is no fixed configuration for production, but rather a sequence of delivered systems, each building on and extending on the versions that came before. In this scenario, architecture (in the conceptual sense) is largely characterized by the invariants, or the elements or aspects that do not change as we go from one incremental release to another. So, in an SoS, we expect to find the architecture in the invariants, although not absolutely exclusively in the invariants, as we discussed in previous sections.

The classic example of architecture as invariants is considering the architecture of the Internet as the TCP/IP protocol suite. The TCP/IP protocol suite has been largely stable for 15 years, a period in which the Internet has undergone radical change, both in physical structure and technology and in applications.

2.5.3 What Are Architecture Descriptions?

Given that architecture itself is conceptual, the abstracted basic structure of a system, we as architects and engineers have to have something concrete to work with. We have to have descriptions of the architecture, and the vocabulary of descriptions of systems is composed of models and simulations.

An ADD is a document that describes the architecture of a system or SoS by organizing a set of description elements. As discussed earlier, we should be careful not to conflate the ADD with the decisions embedded within one. It is possible to have a system with very fit architectural decisions, but poor architectural documentation, and to have excellent documentation that documents very poor decisions. Obviously, we prefer high quality in both decisions and documentation, but forced to make a choice, we would be better advised to take quality decisions. Quality documentation can be backfitted in a program, but if the program has made decisions that doom it, no documentation work will rescue it.

A number of standards are available for ADD. In principle, standards in this area can separately deal with the description methods themselves, processes for developing the documents, methods for conducting architectural activities (and thus making architectural decisions) on programs, and normative requirements systems. In practice, the standards do not necessarily cleanly separate among these issues.

2.5.3.1 IEEE/ANSI 1471, ISO 42010

These standards (IEEE, 2000; ISO/IEC/IEEE, 2011), which are very similar and directly related to each other, are all referred to here as 42010. The first member of the series was the IEEE 1471. The IEEE standard was submitted to the ANSI and approved. It was later submitted to the ISO and approved in its original form, subject to its modification in an update cycle. The update cycle has extended and added to the original models but retains the same essential concepts.

42010 is more of a meta-standard for architecture descriptions than a direct standard. It is a meta-standard in that it defines how to construct a detailed standard for writing ADD. For an ADD to be conformant to 42010, it must meet a set of normative requirements given in Section 5 of the standard. The details of conformance are beyond the scope of this chapter, but the principle requirements are:

- The ADD must explicitly list all stakeholders for the system described and the concerns of those stakeholders.
- The stakeholder set must include a defined minimum set.
- The models of the system itself must be arranged into views of the system.
- The methods used in each view must be formally defined in a particular format called a "viewpoint."
- All concerns identified are traceable through the viewpoints to the models provided in the views.
- The results of completeness and consistency analyses must be recorded.

42010 is as applicable to an SoS as to a general monolithic system as to a software system. Of course, the implementation will differ in the details. Different views and associated models will be chosen, but the overall approach will be the same.

2.5.3.2 DoDAF

Probably the best known of the ADD standards is the U.S. DoDAF (DOD, 2010). The DoDAF is a description standard, not a systems standard. This means that a set of documents which describe a system or SoS can be assessed for conformance to the DoDAF, but the system itself cannot be assessed for conformance. Description documents conform, or don't, to the DoDAF; systems do not conform. In contrast, most communications protocol standards are systems standards, not document standards. For example, we can talk about whether or not a communications system conforms to the TCP/IP standards independent of whether or not it has a set of formal description documents.

The DoDAF requires that an ADD include models arranged into three primary views, along with some additional information. The three primary views are known as operational, system, and technical. The elements of an Operational View are to define entities and information flow that are operationally significant. That is, the entities and flows are things that are relevant to the conductors of military operations. In the systems view, the entities and flows correspond to physically distinct systems. The concerns in the technical view are data and information technology standards.

In the original conception of the DoDAF, the primary purpose of a compliant document was to assist with the conduct of interoperability analysis. A justification for building-compliant documents was (and is) to foster the development of SoS with robust interoperability. If we parse that purpose, we see that the information required to facilitate interoperability analysis overlaps with, but is not a proper subset of, the information needed to conduct the acquisition of an individual system. If we were to develop just the DoDAF specified information in an ADD, we would not have all the information needed to run an acquisition, but we would have some additional information beyond that needed in an acquisition.

2.5.3.3 Ministry of Defense Architecture Framework

In part as a response to issues in actual use of the DoDAF, the U.K. Ministry of Defense has defined some extensions to the DoDAF, known logically enough as the MODAF (Ministry of Defence, 2008). The MODAF extends the DoDAF with a limited set of additional views that are particularly oriented to supporting systems acquisitions. The primary differences between the MODAF and the DoDAF are:

1. Terminology has been adjusted and, in some cases, sharpened. The concepts and terminology are generally close to those in ANSI/IEEE 1471, now ISO 42010. The terminology associated with what is an operational versus a systems node has been sharpened.
2. Various models are broken out into more formal pieces. For example, the high-level operational depiction, OV-1 in DoDAF, is broken into a purely pictorial element and other tabular and even quantitative parts.
3. A "strategic viewpoint" has been added. This viewpoint specifies models of policy, capability deployment, and related trade-offs for larger-scale planning. Its intended audience is mainly higher-level planners and staffers.
4. An "acquisition viewpoint" has been added. This is largely in response to the practice of mandating framework compliant documents for acquisition programs. If compliant documents are going to be required for an acquisition to go forward, then it would be desirable that the standard incorporate acquisition concerns. Compared to the full range of models usually used by project managers, the specified set here is rather thin. However, the intent in the MODAF is mainly to support planning and visibility between projects, so the models focus on dependencies and the clustering of projects.

2.5.4 Why Are Architecture Descriptions Important?

The most important reason why architecture descriptions are important is that architectures are important. Architecture decisions (as opposed to the architecture documents) can by themselves make or doom a program. A poor set of architecture decisions, by definition, renders the system unfit for purpose. No set of artifacts (documents) can recover a development program from poor architecture decisions. We say that architecture decisions can doom or make a program be the definition of what is an architecture decision. Architecture decisions are those that largely determine the value, cost, and risk of developing a system. The philosophy of architecture as decisions is to identify the normally small set of decisions that dominate value, cost, and risk.

While no set of architecture documents can rescue a bad architecture, a good architecture can be fatally compromised by bad architecture documents. Architecture documents that fail to capture or convey key decisions that determine value, cost, or risk are severely flawed. Basing a development effort on such documents will be itself a high risk.

The ADD conveys architecture decisions into the development process. In a waterfall program, this happens through a flow-down to requirements, detailed design, and eventually a systems build. In an evolutionary program, the ADD has a much longer useful life. In an evolutionary program, the ADD conveys the invariants. While good documents cannot recover poor architecture decisions, good decisions can be lost through poor or absent documentation. Good and complete documentation assists in maintaining the integrity of the concept through development. Good documentation communicates the organizing decisions throughout subsequent developers. Good documentation facilitates testing and verification that the system implemented is consistent with the original concept. Of course, conversely, bad documentation greatly inhibits all of these benefits.

2.5.5 How Do You Assess Goodness?

Given a set of architectural decisions, or an ADD, how does one assess the "goodness" of either? The description document will contain a collection of models and possibly simulations. The architecture decisions stand on their own conceptually, but the only vehicle we may have to examine and compare them may be models or simulations. We have to deal with the question of how to assess goodness. Absent clear attention to this issue it is easy to get lost in modeling and simulation for its own sake with no clear connection to achieving the real purpose, which is presumably facilitating the design and assembly of a valuable SoS.

What measures can be applied to artifacts, whether abstract and conceptual as in decisions or concrete as in an ADD? The most desirable measures of goodness must reach outside the context of the thing being assessed to some larger standards of merit. We would prefer not to measure goodness only on internal measures of consistency, as these tend to lead to self-referential activities, sometimes derisively referred to as "self-licking ice cream cones." Unfortunately, an architecture effort by its nature tends strongly toward inward looking. The point of a classic architecture effort is to develop a concept that satisfies its sponsor. Since the sponsor is inside the scope of the effort, and the sponsor is the source of measures of success, the process is inherently inward looking. Since the sponsor owns his or her own values, there is no "higher" authority to which to appeal.

A sponsor may well ask how well his architect does compared to another, but he can learn that only by engaging several architects and cross-comparing. This helps assess the work of an individual architect, but provides no measures for the architecture as a whole, whose boundary circumscribes all the architects engaged by the sponsor.

This is especially dangerous in the SoS case where the sponsor does not own or control the system of interest. The system of interest is, by definition, beyond the control of a single individual or organization, even when the individual or organization is solely responsible for the architecture of the SoS. This places the architect in a quandary. Satisfying the sponsor solely may be the logical and personally secure thing to do. But satisfying the sponsor solely may be effectively at odds with the sponsor's own larger purpose of making the overall SoS happen.

2.5.5.1 Measures of the Architecture

While noting the inherently self-referential aspects of the problem, we can define some useful types of measures of the goodness of a set of architectural decisions. Here, we are examining a set of decisions about the configuration of a system (which may itself be an abstract system). The decisions have been developed in response to the needs of one or more sponsors.

- Are the decisions, taken as a whole, *sufficient* to accomplish the programmatic purposes of the sponsor, even when those purposes go beyond the sponsor's scope of control?
- Are the decisions *self-consistent*? Is it possible to construct a system consistently with all of the decisions together?
- Is the value of system proposed consistent with its estimated cost? Are there no other readily known courses of action that would yield similar value at much less cost? Do the values delivered and costs represent locally desirable courses of action for all independent stakeholders, not just the sponsors (since in an SoS many stakeholders must decide to proceed).
- Are the risks of proceeding with the selected course of action understood by, and acceptable to, the sponsor and all independent stakeholders?
- Is there evidence that no other course of action superior in value, cost, and risk can be readily found? It may be impossible to prove that the recommended course of action is optimal among all possibilities (including unenumerated possibilities), but can we provide evidence that no superior course can be readily found? Is the desired course of action a stable equilibrium for all independent stakeholders (to avoid the likelihood of defection in the face of locally, if not globally, superior courses of action)?

2.5.5.2 Measures on an ADD

The quality of an ADD can also be measured, understanding that measuring document quality is not the same as measuring the quality of the thing the document describes. In devising measures for an ADD, we must be careful to place those measures in an appropriate context. Is it desirable to place quality measures on a document that have nothing to do with the quality of the decisions it embodies? For example, we would get little real benefit from having perfectly drawn blueprints of a building if the building design was so structurally flawed that the building would fall down in the first major storm. Document quality metrics primarily concern their ability to reliably communicate, not the quality of the underlying represented object. We must be careful not to substitute document quality measures for missing measures of the quality of the underlying architecture.

2.5.5.3 ADD Content Measures versus Syntax Measures

A basic measure of the content of an ADD is to what extent does it allow the architecture measures given in the previous section to be assessed? For example, can we assess the sufficiency of the architecture decisions from what is in the ADD? Are all elements of a view in the ADD consistent with the elements in another view? Does the collection of elements contain an assessment of value, cost, and risk? Is there evidence that selections are the best available? Are all stakeholder concerns "covered" by elements of the architecture description?

In addition to measures that map back to the architecture content itself, we can define some purely internal measures. These are syntax measures, in that they mainly concern the syntax of the modeling methods used. We can assess the models presented, one by one, for syntactic correctness. Obviously, a set of syntactically correct models may well describe a system the sponsor has no interest in. This only tells us that syntactic checks in and of themselves cannot be sufficient. It may be argued that syntactic checks are at least necessary. This is partially true. Syntactically incorrect models should be a concern, but the architecture description is not typically a set of construction drawings; it is more analogous to a set of sketches. Sketches need not be syntactically perfect in their representations to adequately convey the key points that should be of interest.

2.6 SUMMARY AND CONCLUSIONS

We can summarize by reviewing these key points:

1. System of systems are distinguished from other systems by formation from independently operated and managed components. While other factors also apply, it is the element of collaborative assembly and operation that is most significant.

2. Because of the relatively abstract nature of many systems of system, modeling and simulation plays an even larger role in their development than in other systems categories. A system of systems is almost always unique. Aside from the thing itself, of which there are no copies, we have only models and simulations of it as tools for description, specification, and reasoning.

3. "Emergence" and its relationship to system of systems is best understood in the relationship of the properties in question as exhibited in physical systems to their exhibition in models of the system. We are primarily interested in properties that emerge in both the system itself and in reduced complexity models of the system. At times, we may be interested in properties that emerge in high-fidelity simulations but not in reduced complexity models. Beyond this, we are not doing engineering.

4. Important examples of system of systems are widely arrayed along a spectrum of collaborative assemblage, from mostly centrally directed to essentially virtual. In each case, we can see the central role played by collaborative assemblage and the mechanism, or lack thereof, for constraining the actions of the components.

5. Modeling pervades the system of systems development process, but it is important to distinguish the role in setting out architectural invariants from detailed specification. System of systems architecture is largely contained in invariants, and the key elements of development are in determining those invariants.

REFERENCES

Boardman, J., & Sauser, B. (2006), System of Systems-the meaning of. In *Proceedings of the Conference on System of Systems Engineering,* 2006 IEEE/SMC, Los Angeles, CA, pp. 6.

Boyne, W. J. (1998), *Beyond the Horizons—The Lockheed Story*, St. Martin's Press, New York.

DOD Architecture Framework version 2.02. (2010), Available at dodcio.defense.gov/Portals/0/Documents/DODAF/DoDAF_v2-02_web.pdf. Accessed July 23, 2014.

Eisner, H. (1993), RCASSE: Rapid Computer-Aided Systems of Systems (S2) Engineering. In *Proceedings of the 3rd International Symposium of the National Council on System Engineering, NCOSE*, Arlington, VA, Vol. 1, pp. 267–273.

Honour, E. C. (2013), Systems Engineering Return on Investment [dissertation], School of Electrical and Information Engineering, University of South Australia.

IEEE. (2000), Recommended Practice for Architectural Description of Software-Intensive Systems. IEEE Std 1471-2000, pp. i,23. doi:10.1109/IEEESTD.2000.91944

ISO/IEC/IEEE. (2011), Systems and Software Engineering—Architecture Sescription. ISO/IEC/IEEE 42010:2011(E) (Revision of ISO/IEC 42010:2007 and IEEE Std 1471-2000). doi: 10.1109/IEEESTD.2011.6129467

Leslie, K., & Michaels, P. (1997), The Real Power of Real Options. *McKinsey Quarterly*, Number 3, pp. 97–108.

Maier, M. W. (1996), Architecting Principles for Systems-of-Systems. In *Proceedings of the Sixth Annual Symposium of the International Council on System Engineering*, Boston, MA, pp. 567–574.

———. (1997), On Architecting and Intelligent Transport Systems. Joint Issue *IEEE Transactions on Aerospace and Electronic Systems/System Engineering*, AES33:2, pp. 610–625.

———. (1998), Architecting Principles for Systems-of-Systems. *Systems Engineering*, 1:4, pp. 267–284.

———. (2007), Training the National Security Space Workforce in Systems Architecting. *Crosslink*, 8:1, pp. 30–37.

Maier, M. W., & Rechtin, E. (2009), *The Art of Systems Architecting*, 3rd edition, CRC Press, Boca Raton, FL.

Ministry of Defence. (2008), *MODAF Ministry of Defence Architectural Framework, v1.2*. Available at www.modaf.org.uk. Accessed July 23, 2014.

Railsback, R., & Volker, G. (2011), *Agent-Based and Individual-Based Modeling: A Practical Introduction*, Princeton University Press, Princeton, NJ.

Shenhar, A. (1994), A New Systems Engineering Taxonomy. In *Proceedings of the 4th International Symposium of the National Council on System Engineering*, San Jose, CA, Vol. 2, pp. 261–276.

Shishko, R., Ebbeler, D. H., & Fox, G. (2004), NASA Technology Assessment Using Real Options Valuation. *Systems Engineering*, 7:1, pp. 1–13.

Part II
Theoretical and Methodological Considerations

Chapter 3

Composability

Michael C. Jones
The Johns Hopkins University Applied Physics Laboratory, Laurel, MD, USA

Amateurs study tactics; professionals study logistics

—unknown

3.1 INTRODUCTION

Composability can be defined as "the capability to select and assemble simulation components in various combinations into simulation systems" (Petty and Weisel, 2003a). It is at the heart of modeling and simulation (M&S) for System of Systems (SoS) Engineering (SoSE). In some sense, composability can be viewed as the M&S analogy to SoSE. SoSE focuses on combining heterogeneous independent systems to create a larger system capable of providing new, hopefully expanded capability. Composability focuses on combining heterogeneous independent simulation systems into one larger simulation system. Frequently, a model, or several models, exists for each of the systems that will be combined into the SoS (Jamshidi, 2009b). Just as simulation is a fundamental tool for the systems engineer, composability allows for the creation of federations of simulation systems that become a fundamental tool for the SoSE. Composability, when done correctly, allows the SoSE to combine the simulations of the constituent systems into a single

---

Modeling and Simulation Support for System of Systems Engineering Applications, First Edition.
Edited by Larry B. Rainey and Andreas Tolk.
© 2015 John Wiley & Sons, Inc. Published 2015 by John Wiley & Sons, Inc.

simulation or a federation of simulations that represents the SoS and allows the SoSE to gain insight into the SoS and its behavior. Composability, when done incorrectly, can lead to serious errors and improper inclusions.

3.1.1 Chapter Organization

An M&S engineer working on a traditional system may follow a traditional Systems Engineering process such as the one described in Institute of Electrical and Electronic Engineers (IEEE) standard 15288. The systems model, much like the system the model represents, moves through a life cycle (Kossiakoff and Sweet, 2003). A brief look at the defining attributes of SoS will reveal that this is not the case for the M&S engineer working at the SoS level. The task at this level is much more focused on combining, or composing, systems-level simulations to gain insight into the behavior of the SoS. The remainder of this chapter will cover some key concepts used to successfully combine simulations into an SoS simulation. Section 3.2 provides motivation for the remainder of the chapter by reviewing Maier's attributes of an SoS and the implication of these attributes on the work of the simulation engineers supporting the development of an SoS. Section 3.3 introduces conceptual modeling, which we will show to contain the key to successful simulation composition. Section 3.4 clearly defines what we mean by composability and contrasts it with similar concepts such as interoperability and integratability. Building on the foundation of conceptual modeling and our definition of composability, Section 3.5 introduces the Levels of Conceptual Interoperability Model (LCIM). LCIM provides a common lens to view composability as well as a metric for quantifying the success of composability efforts. Section 3.6 describes some of the more common tools and standards employed by practitioners today, including Distributed Interactive Simulation (DIS), Test and Training Enabling Architecture (TENA), and the High-Level Architecture (HLA). Section 3.7 explores some of the areas of current research, including recent work on ontologies and semantic modeling as well as agent-based simulation. Concluding remarks are contained in Section 3.8.

3.2 MAIER'S ATTRIBUTES OF AN SoS

SoSE, like M&S, is a relatively young field. There is still disagreement on exactly what constitutes an SoS. There is general agreement within the literature that any SoS has a majority of these five attributes that have been discussed in the previous chapters (Maier, 1998):

1. Operational independence
2. Managerial independence
3. Geographic distribution
4. Emergent behavior
5. Evolutionary development

These attributes have significant implications for the simulations used to study the systems as well as for the systems themselves. This chapter focuses on supporting SoSE by combining heterogeneous models, each representing a constituent system or portion of a constituent system, to form a model of the larger SoS under study. Combining models in this manner can yield valuable insight into the performance of the SoS, empower the operators or designers to improve the SoS, reveal opportunities for new applications of the SoS, and provide a powerful tool for training operators. Naively connecting models, even if each is valid and accredited, does not guarantee that the final product will be a valid model of the SoS. Even worse, the final product may produce results that are incorrect and lead to poor decisions.

3.2.1 Operational Independence

Operational independence means that the constituent systems are able to be disconnected from the larger SoS and remain capable of performing useful functions independent of the SoS (Maier, 1998). Operational independence may bring with it an additional complication for the simulation professional. Constituent systems frequently have goals and objectives separate from those of the SoS. Whether operating as a part of an SoS or operating independently, the system will strive to satisfy its own goals and objective. These goals and objectives may be complementary, orthogonal, or even opposed to those of the SoS. The constituent systems may choose to operate within the SoS when doing so will support their own goals. This often results in negotiations and the establishment of quid pro quo relationships. Divergent goals and objectives result in situations where optimization of individual systems may not be the same thing as optimization of the SoS. In practice, this is often the case. A model at the SoS level must recognize these goals and objectives and provide a capability for systems to operate independently to achieve them.

3.2.2 Managerial Independence

Managerial independence has two aspects. First, following on the operational independence attribute of constituent systems being capable of independent operations, managerial independence adds the criteria that the constituent systems frequently do operate independently (Maier, 1998). If the constituent systems are theoretically capable of independent operations, but never exercise that capability, then the issues of divergent goals and objectives discussed earlier are moot. It is only when the systems exercise the operational independence that the divergent goals and objectives become significant. Managerial independence enables operational independence.

The second aspect of managerial independence is potentially more significant to the M&S engineer. Managerial independence also means that the constituent systems are independently acquired and integrated into the SoS (Maier, 1998). In practice, this often means the constituent systems are also independently financed, designed, upgraded, and maintained. As a result, systems may have fundamentally different timelines and underlying technologies. The set of standards to which they choose to conform may be different as well as their degree of conformance to those standards. Some constituent systems may have well-documented and accredited models with open interfaces. Some systems may have legacy models that are widely accepted within their user community and yet have little documentation. Some systems may have models that reflect the previous, or even outdated, versions of the system. Some constituent systems may have no models at all. Composability of the models of these systems becomes a significant challenge when the models, as well as the systems themselves, are under the management of different organizations.

3.2.3 Geographic Distribution

The geographic distribution of SoS can be quite large. Some, such as the Global Earth Observation System of Systems (GEOSS), cover the entire Earth, including the atmosphere and satellites in geosynchronous orbit (Jamshidi, 2009a). The main result of this characteristic is that the constituent systems can exchange information, but often cannot exchange significant amounts of energy or material (Sage and Cuppan, 2001). Even the successful exchange of information cannot be assumed: some SoSs must operate in environments where communication is denied, intermittent, or of limited bandwidth. A simulation engineer modeling the SoS must accurately model the communications capability between the constituent systems, including any limitations.

3.2.4 Emergent Behavior

Fisher describes emergence as attributes or behaviors of an SoS that are not inherent in any constituent system (Fisher, 2006). These attributes or behaviors may be beneficial or detrimental to the SoS as well as to the constituent systems. Keating, among others, tells us that emergent properties come about through operation of the system and cannot be deduced or anticipated before the system is operated (Keating and Katina, 2011). Emergence is such a foundational characteristic of SoS that, in many cases, the SoS itself emerged from existing systems. Emergence may represent the biggest challenge for the M&S community in supporting SoSE.

While not a perfect analogy, the story of Kitty Genovese provides a powerful and dramatic illustration of emergence. The details of this story are incomplete and in dispute but motivated significant research in one aspect of emergent properties in groups of humans. Ms. Genovese was brutally robbed, raped, and murdered over a period of more than 30 min near her home. The attack, or portions of it, was heard or seen by at least a dozen people. None of them took adequate action to intervene in the attack or save her life. Most of these people, if acting alone, would likely have taken more significant action. Together, and each believing others would act, these observers exhibited a form of emergent behavior called dilution of responsibility or the bystander effect (Gladwell, 2000).

Emergent behavior, even in the case of a mature SoS, may result from the addition of new constituent systems or from seemingly minor changes in existing systems (Chen et al., 2004; Boardman and Sauser, 2006). Emergence may also result from changes in the operating environment, the objectives of the SoS, or even the development of new systems external to the SoS.

Emergent properties may result in changes to the objectives or concept of operations (ConOps) for the SoS (Chen and Clothier, 2003). As the end user recognizes the emergent behavior, he or she may recognize new applications for the SoS that may become part of a new ConOps. This new ConOps may, in turn, result in new emergent behavior.

This characteristic of SoS represents a particular challenge for the M&S community. A model for SoS can be built by composing models of the constituent systems. The validity of these individual models is a necessary, but not sufficient, condition for the validity of the larger model for the SoS. How does the simulationist account for the user recognizing a new application for SoS or for the SoS reorganizing itself in an unexpected manner or for the SoS being applied to an entirely new problem set?

The emergent properties of SoS also represent a significant opportunity. The SoS can be operated through M&S before the actual SoS is built, or even before the design is finalized. This may provide the SoS engineer an opportunity to mitigate the detrimental effects of the emergent properties as well as to facilitate beneficial emergent properties.

3.2.5 Evolutionary Development

Traditional systems have a systems life cycle that usually includes requirements analysis, design, development, test and evaluation, fielding, support, and retirement phases (ISO/IEC, 2008). This is often not the case with SoS. They tend to emerge from existing systems, add and remove systems as the SoS evolves, alter their structure, and alter their objective over time (Rechtin, 1991). The Internet is a classic example of an SoS that is undergoing evolutionary development. The original ARPANET evolved by adding new technology, changing the architecture, and adding new service objectives that were not found in the original concept. For example, the Internet now enables players around the globe to participate in massively multiplayer online role-playing games using smartphone technology. This was not a capability resident in, or even envisioned in, the original ARPANET.

The main implication of evolutionary development, from the M&S perspective, is that the M&S tasks will evolve along with the SoS. New systems will be added to the SoS, and the models of these new systems must be integrated with the existing SoS model. The systems themselves will evolve as they are upgraded to add new capability or incorporate new technology. This often results in the requirement to integrate several versions of the same model into the SoS model along with the ability to select from available versions during simulation execution (Under Secretary of Defense, 2006).

3.3 CONCEPTUAL MODELING

Composability for M&S is more complicated than interoperability among information technology (IT) systems or command and control (C2) systems (Tolk et al., 2012). This topic will be discussed in more detail later in this chapter. The difference between composability in M&S and interoperability of IT or C2 systems hinges on conceptual modeling. To understand the issue of composability, we must first explore conceptual modeling.

Conceptual modeling is recognized as one of the most important steps in most simulation studies, but it remains one of the least documented steps (Robinson, 2006). Systems Engineering teaches us the value of formally capturing the requirements of a system before we begin designing it (Maier and Rechtin, 2000; Kossiakoff and Sweet, 2003). Conceptual modeling is the process by which we capture the requirements for a simulation system. It is often said that "All models are wrong. Some are useful" (Box and

Draper, 1987). A model is a purposeful abstraction or simplification of reality. Conceptual modeling may be viewed as the art of abstracting from a real-world system to an executable model that meets the goal of being useful (Balci et al., 2011). The conceptual model exists in the space between the system being studied and the executable model.

3.3.1 The Reference Model

Zeigler (1976) provides a valuable framework for developing a conceptual model, which has been expanded and modified by other authors. Zeigler starts with the system in the real world. Some authors call the real-world system the referent, and we will adopt that practice for the remainder of this chapter. In many cases, the referent may not yet exist or may not be perfectly observable. Different stakeholders may have different perspectives of the referent or different interpretations of the referent.

Zeigler's first step in creating a model is to establish the experimental frame. This is the set of circumstances or conditions for which the model must be valid. In other words, it is the limited set of circumstances or conditions that can be observed from the simulation. The experimental frame ensures the final model will be capable of answering the particular questions the M&S study is attempting to answer. The experimental frame should include the types of inputs expected as well as the input/output relationships (Zeigler et al., 2000). Many authors now reserve this step for later, and we will consider it in the next section.

Zeigler's second step is the base model. Many authors now call this the reference model, and it will be referred to by this name for the remainder of this chapter (Davis and Anderson, 2004). *The reference model is defined as the collection of everything which fully describes the real-world system, including its attributes, capabilities, relations, assumptions, and constraints under all perceptions and interpretations* (Tolk et al., 2013). The reference model aims to be as comprehensive as possible, even if that means that it is inconsistent. The reference model becomes the reality for the remainder of the M&S study. In practice, the reference model is often nothing more than a collection of artifacts from the Systems Engineering process used to create the system, including systems drawings, functional decompositions, DoDAF views, the various concept documents (e.g., ConOps, concept of employment, and concept of maintenance), training materials, and maintenance documentation. This approach is inadequate since the disparate views of various stakeholders are not well represented, conflicts and inconsistencies are not highlighted, words may be interpreted in different ways by different documents, or multiple words may be used for a single concept. Current research, as described in Section 3.5.1, recommends creating a much more formal reference model using a tool such as the Web Ontology Language (OWL).

In an SoS environment, the reference model is never complete. No matter how much effort is put into documenting an SoS and its performance, some aspects will remain unknown and undocumented. If all models are wrong, but some are useful, it follows that all reference models are incomplete, but some are useful (Law and David Kelton, 1999). This is even truer in an SoSE environment. Since the SoS undergoes evolutionary development and emergence, the reference model will have new systems attributes to document and new emergent behavior to capture. The key is to be a complete as is reasonably possible, balancing the desire to capture every nuance and detail with the reality of limited time and resources. The reference model must be sufficient to fully describe the referent, including inconsistent views and perspectives, in adequate detail to ensure the resulting models adequately answer the questions that will be asked.

3.3.2 The Conceptual Model

Zeigler's third step is the lumped model. This may be the closest to what is now called the conceptual model. The conceptual model is a subset of the reference model. It includes only those aspects of the reference model that are significant in terms of the experimental framework and those aspects that are needed to answer the questions being asked in the simulation study (Zhu et al., 2008). The conceptual model is independent of the platform or computer language being used to create the executable model. The conceptual model follows the guidance that a model "should be as simple as possible, but no simpler" (Einstein). This is easier said than done. In addition to the attributes that are necessary to answer the question, additional aspects of the system are often included to aid in validation of the model or to help the end user accept the results of the simulation as representative of the real-world system (Kelton et al., 2007).

While there is not a widely accepted definition of the conceptual model, or even a list of the contents of a conceptual model, in practice, it now contains a subset of the reference model along with elements of the experimental framework such as the data the model must be able to accept as well as the outputs the model must be able to generate (McGinnis et al., 2011). One popular definition of *a conceptual model is "a non-software specific description of the simulation model that is to be developed, describing the objectives, inputs, outputs, content, assumptions, and simplifications of the models"* (Robinson, 2008).

While the reference model aims to be complete, the conceptual model aims to be consistent. It is a subset of the reference model but includes only portions of the reference model that are internally consistent. Inconsistencies can be removed by eliminating portions of the reference model that are causing the inconsistency or by introducing constraints that resolve the conflict. These constraints may not hold universally, but they will hold within the limited set of conditions relevant to the experimental frame or the modeling question to be answered.

According to Robinson (2008), a well-documented conceptual model:

- Minimizes the likelihood of incomplete, unclear, inconsistent, and wrong requirements
- Helps build the credibility of the model
- Guides the development of the computer model
- Forms the basis for model verification and guides model validation
- Guides experimentation by expressing the objectives, experimental factors, and responses
- Provides the basis of the model documentation
- Can act as an aid to independent verification and validation when it is required
- Helps determine the appropriateness of the model or its parts for model reuse and distributed simulation

In practice, the conceptual model is often captured using Unified Modeling Language (UML) diagrams or SySML diagrams.

3.3.3 The Simulation Model

Zeigler's fourth step is the computer model, which some authors call the executable model or the simulation model. We will use the term simulation model for the remainder of this chapter. This is the step where software code is written to implement the conceptual model. Compared to the previous steps, this step is fairly straightforward. In this author's opinion, this step is more science, while the previous steps are more art. There are many ways to create a reference model or conceptual model, the correctness of which is arguable and subject to interpretation within the context of the study's objective. While there are many ways to generate a computer model, the correctness of the simulation model can be (and in almost all cases is) rigorously assessed and quantifiable. From one perspective, the step of creating the simulation model from the conceptual model is in the realm of verification, while all of the previous steps are in the realm of validation. This may explain why there is a large body of literature on generating and validating simulation model, while there is comparatively little written about conceptual modeling.

Before leaving the topic of conceptual modeling, we will use a short example to illustrate the relationship between the various types of models. We will consider the task of an M&S engineer supporting the design of a new air superiority fighter aircraft being designed for use by the members of a coalition such as the NATO. This example is illustrated in Figure 3.1. This aircraft is intended to be employed, with minor variation, by all branches of the U.S. military and potentially several other NATO member nations. It is easy to see that the real-world system, or referent, is the aircraft. Similarly, the reference model is everything that is, or can be, known about this aircraft. The engineer has not assembled a reference model, but instead has access to an electronic repository where the program management office stores all of its documentation for the new aircraft. The engineer is currently supporting two independent projects. The first project is tasked with identifying required modifications to the naval variant's landing system to allow the fighter to land on aircraft carriers. The second project is developing a training system for the crew who will fly the aircraft, including the naval variant. You may find it useful to consider the conceptual models developed for these two studies. While a conceptual model for either of these studies would be beyond the scope of this chapter, and even this book, we can list enough of the attributes to highlight some of the differences. What subset of the reference model should be included in the conceptual model for each study?

By procedure, aircraft have specified operational limitations for landing. The conceptual model for the first study would include the limited range of speeds allowed when landing, possibly with some additional range for a margin of safety and to allow for pilot error. Similarly, a reduced range of weight for fuel and weapons may be considered. The design characteristics for physical shock would certainly be included. Since aircraft normally apply power upon landing on a carrier, in case the arresting cable breaks, the thrust characteristics of the engines would be included along with assumptions about the reaction times of the pilots. Aspects of the aircraft carrier such as speed, length of the deck, and the limitations of the cable would be essential elements of this study.

The second study would include many of these aspects. It is certainly important for the pilot to be able to land the aircraft. The trainer would not be limited to the subset of the speed and weight characteristics proscribed for landing. The pilot must also be able to take off, which would include a heavier weight limit than landing, as well as combat, which would allow the full range of speeds the aircraft is capable of. The design strength

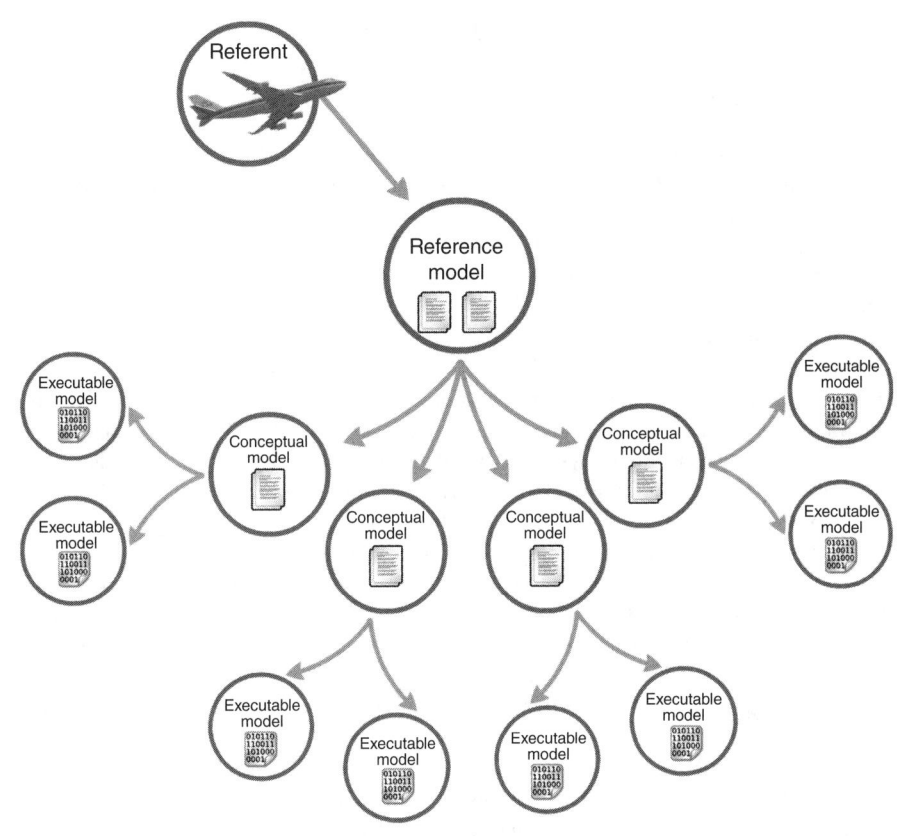

Figure 3.1 Relationship between referent, reference model, conceptual model, and simulation model.

of the arresting cable may not be required in the simulator, but the ability for the instructor to part the cable when desired (even when the pilot does everything correctly) may be essential to training the pilot to deal with this emergency.

It is important to notice that both simulations required aspects of the aircraft carrier as well as the aircraft. It is equally important to notice that the fidelity, the degree to which the model matches the real world, is different. In the first case, the fidelity of the arresting cable is critical. In the second case, a high-fidelity model of the cable may be detrimental by preventing the simulation from parting when this is desired for training.

3.4 COMPOSABILITY, INTEROPERABILITY, AND INTEGRATABILITY

Our world is becoming more interconnected each day. We can create a document on a Mac, edit it on a Windows-based PC, view it on an iPad, and even update it collaboratively with teammates around the world using Google Docs. Even inexpensive consumer electronics advertise that they are "plug and play." I recently purchased a new diving computer for scuba diving. This device measures the depth, water temperature, ascent/decent rate, and a myriad of other parameters throughout the dive and automatically calculates the amount of each gas absorbed in the body. It uses these calculations to let me know how long I can safely stay in the water. When I connected it to my laptop, the laptop immediately recognized the new device. The laptop identified and loaded the required drivers, identified and installed the compatible software from the manufacturer's Web page, matched the device to the software, downloaded my data from the device, and started the program. Within a few minutes, I had graphical and tabular data aesthetically presented in an easily digestible format to irrefutably prove that I am a poor diver. More importantly, before I ever connected the device, I was confident that this level of "plug-and-play" interoperability would be built into my new toy.

Why would the users of modern M&S products expect any less? Consider the case described in the previous section where a program office has a model for a new aircraft design. Further, assume that a second program office has a model for a missile, a third program office has a model for an electronic countermeasure, and a fourth office has a model for an adversary radar system. It would be logical to ask if these models can be combined to assess the military effectiveness of the electronic countermeasure by simulating the aircraft using the missile to attack the radar site while employing the electronic countermeasure. As in the case of my diving computer, many program managers expect their models to be "plug and play." In my experience, few managers are willing to pay the additional cost to achieve this capability. In this section, we will look at what it means to be "plug and play" and why that capability may be more challenging in M&S than in many other domains (Tolk et al., 2012).

We will start this discussion by examining a few definitions. The simulation engineer must understand composability, interoperability, and integratability. Composability may be defined as "the ability to rapidly assemble, initialize, test, and execute a system from members of a pool of reusable, interoperable elements" (Powell and Noseworthy, 2012). As presented in Section 3.1, *composability is "the capability to select and assemble simulation components in various combinations into simulation systems"* (Petty and Weisel, 2003a). Automatic composability is "the Holy Grail" of the M&S community. Automatic composability has not been achieved and may never be achieved, but the pursuit of improved techniques for composability is essential for providing the M&S community with a useful set of tools.

Interoperability is a term that is often used by systems engineers but one that is used without a strict definition. This lack of rigor can lead to confusion and costly mistakes. According to the IEEE, *interoperability is "the ability of two or more systems or components to exchange information and to use the information that has been exchanged"* (IEEE, 1990). Diallo et al. (2011) provide a more useful definition of interoperability. They provide two necessary and sufficient conditions for systems to be interoperable. Namely, the systems must be capable of information exchange, and the information exchanged must be usable. Since interoperability means the systems must be capable of exchanging information, and a particular system may be capable of exchanging information with one system but not another, interoperability may be viewed as a property that is not inherent in a system but rather shared between two or more systems. Furthermore, the first condition only addresses the technical aspect of communication between systems. Merely passing a message from one system to another, even if the message is trivial or meaningless, satisfies this requirement. The first condition does not address the semantic, syntactic, or pragmatic aspects of the information. The requirement that the information exchanged must be usable addresses these higher levels of communication between systems.

In practice, especially in complex systems, interoperability is a matter of degree. Systems specifications may list required data rates, protocols, message formats, or radiated power requirements. Message formats may include a small number of fixed format messages, or they may include semistructured and unstructured data fields. Even structured data fields that are well defined may leave room for ambiguity. Consider a simple message that is intended for a system to report its current location, course, and speed to another system. One method is to specify that the first field will contain the latitude to a certain accuracy, the second field will contain longitude to a certain accuracy, the third field will contain altitude in meters relative to a reference level such as sea level, the fourth field will include heading in degrees to a certain accuracy, and the fifth field will contain the speed in kilometers per hour to a certain accuracy. This may appear adequate and in the case of vehicle on the ground may be adequate, but what about aircraft and ships? An SoS intended for use by infantry systems may adopt this message format. When the SoS evolves to include aircraft, it must account for the fact that the air itself is moving. Does the fifth field contain the aircraft's speed over the ground or the speed through the air? These are different values. In some cases, especially in military systems, determination of speed relative to the ground may be unavailable.

Interoperation may be viewed as a necessary, but not sufficient, condition for composability. If several simulation components are to be combined into simulation systems, the individual models should be capable of exchanging information and using the information exchanged (Under Secretary of Defense, 2007). Otherwise, it is hard to see the value in combining the models.

Integratability is another term often used along with interoperability and composability, but it has a significant distinction from the other two (Tolk et al., 2009). Integratability is the ability of systems to be integrated. Fisher defines integration as "the process of composing or combining subsystems to form a unified system. Historically, both subsystems and the integrated system of which they are a part were viewed as monolithic" (Fisher, 2006). Integration is a fundamental part of the classic Systems Engineering process. It is listed as one of the 10 technical processes in the systems life cycle processes (ISO Press, 2008). This definition of integration has been expanded to its current use. Petty and Weisel (2003a) define integration as "the process of configuring and modifying a set of components to make them interoperable and possibly composable." Therefore, *integratability becomes the ability of systems to be configured or modified to make them interoperable and possibly composable*. This definition encompasses the portion of the Fisher definition referring to combining systems but specifically excludes the reference to a unified or monolithic system.

Page et al. (2004) present the concepts of composability, interoperability, and integratability as three orthogonal dimensions of simulation interconnection. All three dimensions are required for successful interconnection of simulation systems, but the dimensions are independent of each other. They propose that:

- Composability is in the realm of the modeler and is concerned with the objectives and underlying assumptions of a model.
- Interoperability, in the context of simulation models, is in the realm of the software designer and is concerned with issues such as data types.
- Integratability is in the realm of the site hosting the simulation and is concerned with issues such as running fiber and ensuring network interface cards (NICs) are working properly.

56 Theoretical and Methodological Considerations

A different view is taken by authors such as Tolk, Petty, and Weisel who view integratability as a necessary enabler for interoperability and interoperability as a necessary enabler for composability. This view is similar to the Capability Maturity Model (CMM) successfully applied to a wide variety of domains by the Carnegie Mellon Software Engineering Institute (Curtis, 2000). In this view, the degree of communication between two simulation systems is a spectrum. At the low end of the spectrum, the systems are independent and isolated. As the systems are integrated, they move up the spectrum. Interoperable systems are higher on the spectrum. Fully composable systems occupy the high end of the spectrum.

3.5 THE LCIM

The spectrum of connectivity of simulation systems from isolated systems up to full conceptual interoperability is expanded and visualized in the LCIM. LCIM was first presented by Tolk and Muguira (2003). It was embraced by the simulation community and improved (Tolk et al., 2007). The current version is shown in Figure 3.2.

LCIM now includes the following seven layers of interoperability:

1. Level 0: No interoperability. The simulation systems are independent and do not share data or services.
2. Level 1: Technical interoperability. At this level of interoperability, the systems have the ability to exchange containers that could convey information. This level consists of network protocols and all the important details reference in the earlier discussion of integration. Systems at this level can successfully exchange data.

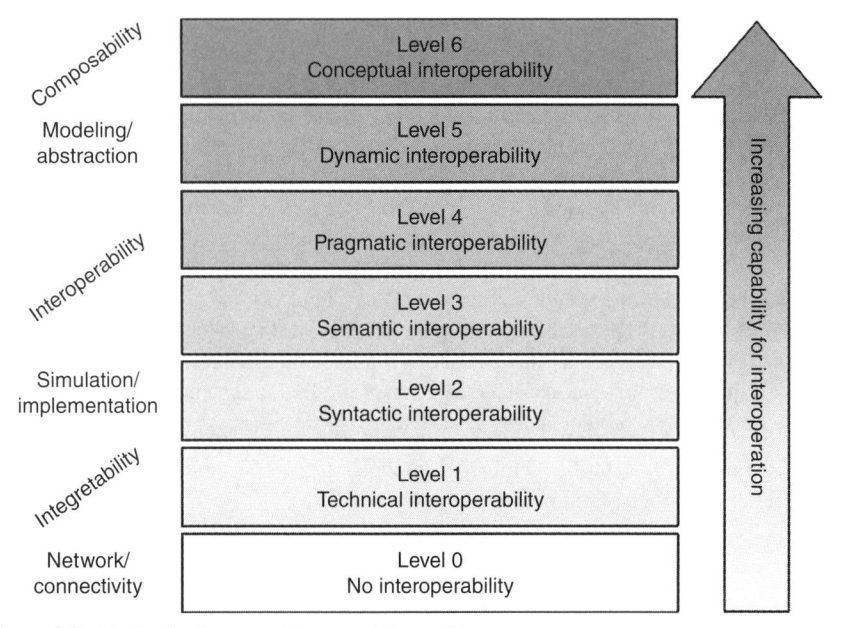

Figure 3.2 The levels of conceptual interoperability model.

3. Level 2: Syntactic interoperability. Since LCIM is concerned with communication, it seemed natural to borrow terms from the study of linguistics. Syntax is the study of how sentences are formed and structured within a language, but not the meaning of the sentences. This is the first level of interoperability that meets the definition of interoperability provided previously. Systems at the syntactic level can exchange symbols or containers that could contain information, as provided by level 1, but they also have the ability to structure information and place it into the containers in accordance with the protocols as well as to extract and parse information from a container received from other systems.
4. Level 3: Semantic interoperability. Semantics is the study of meaning. From a linguistics perspective, semantics is concerned not only with the definition of words but the connotation as well. Systems at the semantic level of interoperability can extract and parse information from a container, as required in level 2, and they can interpret the meaning of the information. This capability is sometimes referred to as aligning static data.
5. Level 4: Pragmatic interoperability. Pragmatics is the study of how the context contributes to, or changes, meaning. Systems at the pragmatic level of interoperability can not only recognize and interpret information, as required for level 3, but they can also recognize patterns of information. This includes recognizing the state of other systems and the processes used in other systems to fully understand the meaning of the information in the context of the situation. This capability is sometimes referred to as aligning dynamic data.
6. Level 5: Dynamic interoperability. Adaptive systems may react differently depending upon their current state. Systems at this level of interoperability recognize the state of other systems and how the state affects the underlying model assumptions and constraints of that system.
7. Level 6: Conceptual interoperability. This level represents the full alignment of the conceptual models of the individual systems. In other words, the assumptions, constraints, and simplifications of the underlying models, as well as the inputs and outputs, are fully documented and consistent with the other systems (Tolk, 2010).

3.6 CURRENT STANDARDS

This section will introduce three current standards or methods used to provide simulation interoperability. We will discuss the DIS, the HLA, and the TENA. This list is meant to introduce the reader to the variety of methods currently in practice and is not meant to be a complete list.

3.6.1 DIS (IEEE Standard 1278)

DIS was created to address a specific set of requirements within in the simulation community, namely, military training requirements. The U.S. Department of Defense (DoD) has been a major consumer of simulation products since the first simulators since Ed Link created the first mechanical flight simulator in 1929. His device allowed pilots to learn the basics of flying an airplane without the expense of a dedicated aircraft or the risk of the loss of the aircraft, personnel injury, or death from a pilot error. Link's simulator was so successful that it became the standard method for training pilots from most of the nations involved on both sides in World War II (Smith, 2010).

The role of simulation in military training, of course, predates Ed Link's flight simulator. The earliest warriors practiced fighting one another, which would be called live simulation in today's terminology. Simulation has not only been used to train individual members of the military, but it has been used to train at the unit level as far back as the sand tables of the Roman Empire (Smith, 2010). The role of simulation in training has continued to grow as the underlying technology improves. In today's military, simulation is used to train individual operators of systems, the people who must maintain the systems, and the units who collectively operate the systems (Jones, 2008). These units can range from a few soldiers, sailors, airmen, or marines practicing small unit tactics up to multinational exercises encompassing ground, air, naval, space, and cyber forces around the world.

As computer-based simulators improved in their usefulness for training individuals, DoD invested heavily connecting these simulators to train teams. The DARPA took the lead for DoD by creating the SIMNET program (Pimentel and Blau, 1994). One of the earliest successes was the Close Combat Tactical Trainer (CCTT) (Johnson et al., 1993). CCTT provides a means of networking several simulators used to train mechanized infantry and armor units. Each simulator provides the ability to train the crew of one combat vehicle. The simulator provides a place for each crew member to operate his or her equipment within the vehicle. The vehicle is placed within a synthetic environment with computer-generated forces playing the parts of allies, adversaries, and neutral actors upon the battlefield. By networking these simulators, CCTT provides the capability to train the crews of two or more vehicles together, exercising unit-level tactics as well as training unit commanders in decision making and C2 of combat units.

CCTT was the first program to successfully deploy a simulation system based on DIS. Since DIS was developed to address the very narrow problem of connecting existing military simulators, the developers of DIS chose to emphasize ease of use and standardization over robustness. DIS focuses on the syntactic and semantic levels of interoperability and relies upon widely accepted networking protocols such as the Ethernet, Token Rings, and Web services to provide the technical level of interoperability. Within a federation operating in accordance with DIS, each simulator is responsible for generating and displaying the shared terrain from the perspective of the individual simulator and publishing the information required for the other simulators to do the same. Information is shared using a well-defined set of messages called Protocol Data Units (PDUs). The PDUs are the only means of communicating between entities within the federation. Each entity has a state, which includes such information as the entities' current location and velocity in three dimensions, acceleration in three dimensions, and details required to allow the other federates to determine the entity's position between updates. One of the first things each simulator must do upon joining the federation is to send a PDU to establish its state. Any time a simulator must interact with its environment, or the entities within the environment, it must select one of the available PDUs, fill in the appropriate fields, and then transmit to PDU to the rest of the federation. For example, when the simulator changes speed or turns, it must send an updated state PDU. The enforcement of the use of these well-defined PDUs provides syntactic interoperability. Since the simulators used in a DIS federation are used for only the specific application of military training, semantic interoperability can be enforced.

We will use the term federation to discuss several simulations interoperating to provide a service or accomplish a task, in this case providing training to a military team. Similarly, we will refer to a single simulation within the group as a federate. Strictly speaking, these terms are defined by the HLA, which we will discuss in the next section. We introduce them here for ease of terminology and to highlight the similarities as well as the differences in DIS and HLA. Many simulation professionals use these terms for HLA, but not for DIS, because HLA emphasizes strict central control of the simulation systems, while DIS allows each simulation to maintain greater autonomy. We will return to this distinction in the next section.

Consider the case of two simulators, each representing armored vehicles from opposing forces, engaging one another. As the vehicles move toward each other, the exchange of state PDUs ensures that each simulator is aware of the other. (This does not mean that the crews being trained within the simulator are aware of each other. One crew may be using a tactic of approaching from behind a ridge or tree line to conceal its location. The crew in the other simulator may see only the ridge or tree line.) When one simulator maneuvers and sends the appropriate PDU, the network technology delivers the PDU, providing technical interoperability. The definition of the state PDU ensures each simulator understands exactly what data is contained in each field of the PDU, ensuring syntactic interoperability. The receiving simulator knows that the PDU should be interpreted to mean that the sending simulation has altered its state, and the receiving simulator must update its representation of the other armored vehicle within its own simulation in order to keep the two simulations synchronized. This ensures semantic interoperability.

While clearly defining each PDU, the designers of DIS provided a wide variety of PDUs to allow each simulator within the federation to meet its specific training objective. There are currently 72 PDUs (IEEE, 2012), up from 67 in the previous version of DIS. The PDUs are divided into 13 families, including:

1. Entity information/interaction
2. Warfare
3. Logistics
4. Distributed emission regeneration
5. Radio communications
6. Entity management
7. Live entity
8. Non-real time
9. Information operations
10. Minefield
11. Synthetic environment
12. Simulation management
13. Simulation management with reliability

DIS has proven to be very effective for its intended purpose, namely, creating federations of simulations for real-time training of military teams of all sizes. Antoine de Saint-Exupery is credited with saying, "Perfection is achieved, not when there is nothing left to add, but when there is nothing left to take away." In this sense, DIS achieved (near) perfection. The PDUs currently defined are all essential for training modern warriors. DIS does not provide any central management: each simulation is autonomous and merely reacts to each PDU as it sees fit. DIS does not attempt to reach beyond the military training domain by providing PDUs for operations analysis or optimization. It does not specify unnecessary requirements at the network level. DIS does not provide time management functions, and it does not provide the capability to run simulations at non-real time. Each simulation merely transmits a "heartbeat" every 5 s to let the rest of the federation know that it is still participating. As a result of its many strengths, DIS is now in use throughout the world.

The strengths of DIS can also be considered its weaknesses. DIS excels at its particular niche and has been successfully applied in a few other areas such as space systems (Kunz, 1993; Dewar et al., 1996) and medicine (Gorman et al., 1999). While PDUs could be defined for a wide variety of domains beyond military training, such as fantasy role-playing games, to date, they have not. Since PDUs would have to be defined for each domain, DIS is not considered appropriate for general-purpose simulation. The lack of time management makes it impractical for Monte Carlo simulations, which are usually run much faster than real time, or simulations involving computational fluid dynamics (CFD), which are usually run much slower than real time. The need for a more generally applicable interoperability standard leads us to our next area of discussion, the HLA.

3.6.2 HLA (IEEE Standard 1516)

While DIS aims to excel in a narrow domain and has been applied in a few areas outside the original application, HLA strives for more general solution to the interoperability problem that can be easily applied to a wide variety of application domains. The purpose of HLA is to allow simulation systems to work together, but it does not concern itself with the domain of the simulation as much as the interoperability of the simulation systems. The difference in philosophy between DIS and HLA becomes apparent from the very beginning. While DIS emphasizes autonomy of the simulation systems, HLA exercises strict control over each of the simulations within a federation. This strict control allows HLA to be applied in a wide variety of application domains with the same degree of confidence that DIS brings to the military training domain. In essence, HLA trades autonomy for flexibility in the application domain.

HLA has its own set of terms. We have introduced two of them already and used the terms with loose definitions. We will provide a more formal definition here and use the more formal definition for the remainder of this chapter. A federate is a simulation entity that is HLA compliant. This entity may be a simulation system, as we have been using the term when discussing DIS, or it may be a data recording device or an interface to a live system such as a radar. A federation is a collection of simulation entities working together to provide a service or solution, each connected to middleware called the run-time infrastructure or RTI, using a common object model template (OMT). An object is a collection of data exchanged within the federation. We will discuss the OMT in more detail, but for now, it is a framework for communication within the federation. An interaction is an event that is communicated within the federation.

Before a federation can be established using HLA, an agreement must be reached and documented. This agreement, called the federation agreement, details how federates will exchange service and will be different for different simulation domains. The federation agreement describes contents of the federation object model (FOM). The FOM describes the type of information that will be exchanged within the federation. It must include object classes, interaction classes, and data types that may be included within the federation. The IEEE standard includes an OMT that can be used as a guide in creating the FOM for a specific federation. The FOM is usually created as an XML file that can be referenced by the members of the federation during execution.

In addition to the FOM, the OMT also includes a document for creating a simulation object model (SOM). Where the FOM describes shared objects, attributes, and interactions for the entire federation, an SOM describes the shared objects, attributes, and interactions for a particular federate within the HLA federation. Each federate is required to publish its SOM when it participates in an HLA execution.

HLA federations are built on a bus topology. This bus is a component of the RTI. The RTI also includes software that provides HLA services to the federation. The RTI provides the ability for simulation systems to publish information they produce or alter and to subscribe to information that they require, along with essential management functions. The RTI hosts the XML file containing the FOM.

There are at least 15 vendors producing RTIs, including governments, universities, commercial companies, and open-source projects. As the foundation upon which the federation is built, the performance and cost of the RTI are critical to the success of the federation. Knight et al. (2002) provide a useful set of benchmarks for evaluating RITs. One critical concern is compatibility with other federations. It is common to build a successful federation only to discover a reason to connect it to another federation. Increased requirements for interoperability with larger federations are more the rule than the exception. Federations operating on RTIs from different vendors can work together, but this often requires the use of a bridge to connect the two federations. Besides the additional cost, the use of a bridge increases latency and reduces stability. Early consideration of the potential for eventually connecting with other federations, and adopting an RTI in use by those federations, can avoid the requirement for a bridge between RTIs.

While DIS provides a predefined set of PDUs that fully describe all allowed communications between simulation systems, HLA provides a structure or format for the communications, but does not attempt to define the content of the communication. HLA provides this structure as an interface specification. The interface specification allows the group of federates to agree upon the definition of objects and interactions that are required by the particular simulation domain. The interface specification is divided into the following seven service groups:

1. The federation management group provides instructions for creating the federation, joining or leaving the federation, saving federation states to save and restore functionality, keeping track of individual federates, and disestablishing the federation.
2. The declaration management group provides instructions for each federate to declare the information it will publish and the information to which it will subscribe.
3. The object management group (OMG) provides federates the ability to register new objects, discover objects that have been created and registered by other federates, update attributes of objects, and send and receive interactions.

4. The ownership management group provides the ability to transfer ownership of objects, and the responsibility for updating them, between federates.
5. The time management group provides the centralized management of time within the simulation. This includes delivering messages with time stamps and advancing simulation time.
6. The data distribution management group defines how object and interaction data is published to and received from the RTI, in accordance with intentions declared in accordance with the declaration management group.
7. The support service group provides various utility functions used within the federation.

In addition to the FOM, SOM, RTI, and interface specification, HLA also provides a recommended process for establishing HLA-compliant federations. This process is based on lessons learned and best practices from a wide variety of experiences in distributed simulation. Until 2007, this process was called the Federation Development and Execution Process (FEDEP) and included in the IEEE standard as IEEE 1516.3. In 2007, the process was updated, renamed the Distributed Simulation Engineering and Execution Process (DSEEP), and published as an independent standard as IEEE 1730. While still considered a part of any standard HLA instantiation, the creation of a separate IEEE standard provided recognition and community acceptance of the DSEEP process beyond HLA.

Finally, HLA requires each federate to agree to comply with a set of rules:

- Federations shall have an HLA FOM, documented in accordance with the HLA OMT.
- In a federation, all representation of objects in the FOM shall be in the federates, not in the RTI.
- During a federation execution, all exchange of FOM data among federates shall occur via the RTI.
- During a federation execution, federates shall interact with the RTI in accordance with the HLA interface specification.
- During a federation execution, an attribute of an instance of an object shall be owned by only one federate at any given time.
- Federates shall have an HLA SOM, documented in accordance with the HLA OMT.
- Federates shall be able to update and/or reflect any attributes of objects in their SOM and send and/or receive SOM object interactions externally, as specified in their SOM.
- Federates shall be able to transfer and/or accept ownership of an attribute dynamically during a federation execution, as specified in their SOM.
- Federates shall be able to vary the conditions under which they provide updates of attributes of objects, as specified in their SOM.
- Federates shall be able to manage local time in a way that will allow them to coordinate data exchange with other members of a federation.

The bus architecture, including the middleware provided by the RTI, ensures technical interoperability within the federation. Syntactic interoperability is provided by the OMT, which clearly defines the syntax used to exchange information. Unlike DIS, which enforces semantic interoperability by allowing only predefined PDUs, HLA does not strive to enforce interoperability at the levels above syntax. In order to allow HLA to be adopted by a wide range of simulation domains, HLA provides a means for ensuring interoperability up through the conceptual level through carefully crafted FOMs and SOMs. In practice, this level of interoperability is rarely achieved within the federation. More commonly, semantic and pragmatic interoperability is achieved within the federation.

HLA has demonstrated that, at least in some cases, less is more. HLA provides only the essentials upon which to build efficient and effective distributed simulations and leaves the domain-dependent details to be added as each federation is created. The structure provided by HLA has been successfully applied to a wide variety of distributed simulation federations in domains ranging from air traffic control, medicine, emergency management, and ship handling trainers. Even with this wide acceptance, HLA is not appropriate for every application. One application in particular the requirements of the live, virtual, and constructive (LVC) simulation community led developers to look for a solution beyond HLA. This search resulted in our next distributed simulation solution, the TENA.

3.6.3 TENA

While some practitioners view TENA as a general improvement over HLA, most view it as an alternative that is appropriate for certain applications. TENA was created to address the simulation needs of military ranges used by the U.S. DoD as well as those of other nations. DoD has a series of weapons ranges spread across the United States as well as the rest of the world that are used to conduct testing of weapons systems and training. They are instrumented with high-precision sensors that collect large quantities of data in real time to assess the performance of the weapons systems. In addition to the ranges owned and operated by the DoD, several key allies also have ranges that provide unique capabilities not found on U.S. ranges. These ranges work for different branches of the military, directly for the DoD staff, or the military of another country. They have, in SoS terms, managerial independence, operational independence, geographic dispersion, evolutionary development, and emergence. By their very nature, these ranges are a large-scale, distributed, real-time, and embedded system (Noseworthy, 2008). The designers of TENA had the benefit of the lessons learned in the development of DIS and HLA and chose aspects of each that enhanced their performance in this demanding application.

The customers who use these ranges are also interested in leveraging the power of this simulation technology to bring together real and simulated people operating real and simulated systems into a single environment for the event (Noseworthy, 2005). For example, an event may involve a real aircraft launching a missile. Instead of flying a second aircraft to act as the wingman to the first aircraft, another pilot may participate in a flight simulator at a base hundreds of miles from the test event. The adversary aircraft who will serve as targets may be synthetically generated forces. This event may be integrated into a larger exercise involving real aircraft at another range and a ship off the coast. All of these entities must be brought together into a single environment so that they each sense each other and interact as if they were on a single battlefield (Coolahan and Allen, 2012).

This type of simulation is called LVC simulation. Live simulation involves real people operating real equipment. In the aforementioned example, the firing aircraft as well as the aircraft on the second range and the ship off the coast are all live players. Virtual simulation involves real people operating simulated equipment. In the aforementioned example, the wingman in the flight simulator is a virtual simulation. Constructive simulation involves simulated people operating simulated systems. In the aforementioned example, the adversary aircraft are examples of constructive simulation. The term LVC is used when all three types of simulation are present in a single event (Powell and Noseworthy, 2012).

Test events on these ranges can involve tanks, aircraft, ships, and hundreds or even thousands of troops. In some cases, real missiles are fired at real targets. A single event can cost millions of dollars. Repeating an event due to a failure of the simulation may be prohibitively expensive or even impossible. Since these ranges are widely dispersed and staffed by a cadre of dedicated engineers and technicians at each site as well as supporting contractors around the world, distributed simulation techniques are not sufficient. In addition to distributed simulation, the DoD ranges require distributed development of the simulation as well. To meet the testing requirements, TENA was required to be resilient, have short latency, and accommodate high data rates (Powell and Noseworthy, 2012). While TENA was developed for DoD ranges, the intention from the very beginning was to create a system that could be applied beyond this narrow set of users. Therefore, the inherent flexibility of HLA was also desired for TENA.

TENA is built around the concept of a logical range. A logical range represents the space used for a specific exercise or event. All entities within the logical range are integrated together as if they were on a single physical range although they may be on physical ranges scattered around the world, in simulators in training facilities around the world, or generated forces being simulated in facilities around the world.

TENA requires the use of a tailored application to interface between equipment on a physical range, such as a radar system, or a virtual or constructive object and the logical range. Many pieces of equipment are TENA compliant and already have TENA applications. TENA Integrated Development Environment (TIDE), a tool provided with TENA, makes creating and updating these applications fairly easy. Equipment that is not TENA compliant can be integrated into the logical range through the use of a gateway application. TENA also comes with Gateway Builder. Gateway Builder is specifically designed to streamline the process of building gateways, reducing the time, effort, cost, and risk of integrating equipment into a TENA logical range. Gateway Builder has easy support for most versions of DIS and HLA. TENA also provides a series of tools that facilitate monitoring and managing the logical range. These tools come in the form of applications (Powell and Noseworthy, 2012).

The integration within the logical range is accomplished using any available network infrastructure. As with HLA, TENA relies on an existing network to provide technical interoperability and focuses on higher level of interoperability. TENA Middleware, which is analogous to the RTI introduced in HLA, is installed on each computer connected to the logical range. TENA Middleware manages the communications between all of the applications on the logical range in near real time. Latency requirements measured in milliseconds are common in live range events. Unlike RTI, TENA Middleware is owned by the U.S. government and available to any user at no cost. Unlike HLA, communications in TENA are peer to peer, but TENA retains HLA's publish/subscribe distribution of information. This is accomplished via TENA Middleware and the ability of each application to simultaneously be a producer of information, called a server, and a consumer of information, called a client. Each object is owned by only one server but may have a large number of subscribers. As a server, the application has a servant object that can publish the state of the object and send the updates to all subscribers of that object. As a client, each application that subscribes to a particular object has a proxy object that will receive updates on the object's state directly from the servant object on the server. This method of communications reduces the latency of updates by requiring only a single message to convey an update instead of one message to publish the state followed by a second to convey the change to the subscriber.

TENA contains a TENA object model, which provides syntactic and semantic interoperability. The TENA object model provides the language used for all communications within the logical range. Since TENA aims to remain flexible, it allows users to define objects for specific events. Each model must conform to the TENA object model, but it is flexible enough to allow a wide range of objects to be unambiguously defined. There are several standard objects that have been tested by several different ranges on several different events and approved by the Architecture Management Team (AMT). This is similar to the PDUs maintained for DIS. While not as mature as DIS and in no small part because of a result of the success of DIS, the list of standard objects in TENA is growing quickly. The collection of objects used on a specific logical range is called the Logical Range Object Model (LROM). The LROM may include both standard objects and user-defined objects.

In order to meet the stringent reliability requirements of live ranges, TENA places heavy emphasis on detecting errors before runtime. One way TENA does this is using automatically generated code. TENA leverages the success of the OMG in creating the model-driven architecture (MDA). A TENA object model may be created using either the UML or TENA Definition Language (TDL). If UML is chosen, MagicDraw with a TDL plug-in is used to convert the UML diagram into TDL. TDL is a text-based language that was created to provide a formal, easy to understand, unambiguous representation of relationships between concepts and constructs in TENA. The Object Model Compiler (OMC) automatically converts the TDL code into C++. This automatic code generation reduces the time required to create code for individual objects, therefore also reducing cost. More importantly, automatic code generation reduced errors, which could cause problems at runtime. The C++ code for each application, along with the TENA Middleware, is compiled and linked together. This allows the complier to identify and correct additional errors that would be found at runtime.

In order to further reduce the occurrence of runtime errors, developers are encouraged to use the Interface Verification Tool (IVT) to test new or modified TENA applications. IVT is a software tool that is provided with TENA. It has four basic functions. It is used to:

1. Verify application interfaces.
2. Test network functionality.
3. Generate scenarios/platforms.
4. Provide real-time monitoring and analysis of events.

The standards discussed so far address interoperability. As we have seen, interoperability is a necessary but not sufficient condition for composability. The technical driving requirements for TENA include interoperability, reuse, and composability. Powell and Noseworthy define interoperability as "the characteristic of an independently developed software element that enables it to work together with other elements toward a common goal." They define reuse as "the ability to use a software element in a context for which it was not originally intended." Reuse requires interoperability along with well-documented software with metadata that fully describes the interfaces. Finally, they define composability as "the ability to rapidly assemble, initialize, test, and execute a system from members of a pool of reusable, interoperable elements" (Powell and Noseworthy, 2012). Composability requires a repository of the reusable components along with metadata that describes the software in sufficient detail. To allow true composability, as defined by Petty and Wiesel, the metadata description of the software should include the objectives, inputs, outputs, content, assumptions, and simplifications of the models. In other words, it should contain the conceptual model from which the software was written.

TENA provides a repository as part of the TENA Common Infrastructure. This repository stores TENA Middleware, the TENA OMC, various TENA tools, applications, standard objects, user-defined objects, and documentation from logical ranges, which are made available for reuse in future ranges. When fully matured and equipped with tools to make searching the repository convenient, this repository will enable composability on a large scale. For now, the repository provides a large collection of reusable software. The ability to search through any repository to find a software element to fit a specific purpose is an area of current research.

3.7 CURRENT RESEARCH

The simulation community has made great strides in improving interoperability of simulation components, as exemplified by the examples of DIS, HLA, and TENA. While there is still much work to be done on interoperability, many experts believe the biggest challenges now are in the areas of conceptual modeling (Tolk and Miller, 2011). Specifically, there is no agreement on a method for standardizing conceptual models in a manner that facilitates easy discovery and assessment. Tools such as DIS, HLA, and TENA have proven effective at combining well-understood simulation components. The challenge now becomes determining which simulation components to combine. In order to allow automation of the tasks of identifying the most appropriate simulation components to address a specific requirement, unambiguous and machine-readable documentation of the conceptual model is required (Petty and Weisel, 2003b; Weisel, 2004). Semantic modeling is showing promise in providing this level of documentation. Section 3.6.1 presents some of the current research on the use of ontologies and ontology languages to support semantic modeling. Once the documentation is available in machine-readable format, simulation agents will be able to search through a repository of simulation components, such as the one provided by TENA, to identify candidate simulation components to create new federations to solve new problems or even to modify existing federations in real time in response to an unfolding scenario. Section 3.6.2 will review current research on agent-supported simulation.

3.7.1 Ontologies and Semantic Modeling

Ontology is the branch of philosophy dealing with the nature of being and the categories of being. It includes questions such as what exists, what properties of an object are essential to the object's identity and which properties are not essential, how things can be grouped, and how things can be related within a hierarchy. In computer science and artificial intelligence, an ontology is a formal representation of knowledge as a set of concepts and the relationships between those concepts (Yilmaz, 2007). Ontologies typically include classes of objects, individual objects that are instantiations of the classes, attributes that are properties of the objects, relations that are ways classes or objects are related to each other, and rules that describe logical inferences. Rules typically take the form of if-then statements. Several languages have been produced for creating ontologies, and several inference engine tool sets have been produced to read ontologies and reason about the information contained within them.

The Web, by its very nature, is composed of a wide variety of unstructured documents and services that are difficult to discover and reuse. There have been several attempts to create ontologies of the Web to make the task of discovery and reuse manageable. In 1999, the DARPA formed a project to create a machine-readable representation of the Web. The project produced the DARPA Agent Markup Language (DAML) that evolved into the OWL. In its current form, OWL is a family of languages that have been included in several semantic editors used for writing ontologies. OWL has been adopted by the World Wide Web Consortium (W3C) as a component of its Semantic Web effort.

The M&S community recognized the similarity between the W3C goal of making Web resources discoverable and reusable and the composability goal of making simulation components discoverable and reusable, and this recognition led to a highly productive line of research (Tolk et al. 2006). A special edition of the *Journal of Simulation* was released in 2011 titled "Enhancing Simulation Composability and Interoperability using Conceptual Modeling/Semantic/Ontological Models," which highlighted selected examples. From this journal, the Open Netcentric Interoperability Standards for Training and Testing (ONISTT) is of particular interest because of its potential to complement TENA (Ford et al., 2011).

ONISTT attempts to answer the question of which components to put into a federation to address a particular testing or training objective. It determines what capabilities are required to meet the objective and then searches through repositories, such as the one advocated by TENA, to find the components that might be combined to provide those capabilities. ONISTT uses OWL "and Semantic Web Rule Language to capture information about the roles and capabilities required to complete a task, and the detailed attributes of candidate resources." It operates in two phases: knowledge capture and operations. In the knowledge capture phase, ONSITT uses formal ontologies to create knowledge bases of both the resources available and the tasks to be completed. The resources may be simulation components or live systems such as aircraft or range sensors. Many of the ontologies required for this phase had to be created specifically for this project. The ONSITT developers plan to turn over responsibility for maintaining these ontologies to international standards organizations or communities of interest. Once the knowledge bases are complete, ONSITT moves to the operations phase where the resources are matched to the tasks to provide a candidate set of resources assigned to roles within the task plan. Examples of the application of ONSITT are available in Ford et al. (2011).

3.7.2 Agent-Supported Simulation

An agent is someone or something that is authorized to act on behalf of someone else or something else, such as a real estate agent or a talent agent. Software agents are merely computer programs that take action on behalf of humans or other programs in the role of an agent. Tolk and Uhrmacher (2009) propose a definition that is more useful for any discussion of software agents. They state that an agent requires the following capabilities in order to accomplish its objective:

- An agent must be able to perceive its environment and be able to act within the environment. Inherent in this requirement is the ability to make decisions about how actions will impact the environment and contribute to, or detract from, accomplishing their goal.
- If multiple agents are present, agents must be able to communicate with one another. This does not mean that every agent must be able to communicate with every other agent. Communication may be limited by range or line of sight between agents in the environment. Agents may compete with each other or they may collaborate. Occasionally, agents will compete in some areas while collaborating in others.
- Agents should be autonomous.

There are three main areas of study for the use of agents in simulation (Yilmaz and Oren, 2009):

1. Agent simulation: the simulation of agents for use in other applications
2. Agent-based simulation: the use of agents to generate model behavior within a simulation study
3. Agent-supported simulation: the use of agents to support or facilitate computer assistance in problem solving or experimentation

The use of agent-based simulation has been widely accepted by the simulation community, which has required extensive implementation of agent simulation to facilitate the agent-based simulation. Agent-supported simulation has not been as widely accepted in practice. However, agent-supported simulation has been the subject of several recent research efforts aimed at improving dynamic composability of heterogeneous simulation systems.

One such research effort is an agent-supported metalevel interoperation architecture (Yilmaz and Oren, 2009). This approach includes four types of agents acting together to select simulation components and integrate them into the simulation in a seamless fashion. In order to avoid interfering with the simulation that may be in progress, these agents work on an independent network but communicate with the simulation components within the federation, or which potentially could become a part of the federation.

Each simulation component, potentially including live systems, has a facilitator agent. The facilitator agent serves as the single point of interaction between the simulation component and the rest of the agents. As such, the facilitator agents are also the points of communication between the two networks. The facilitator agents reside on the agent network but communicate with the simulation components on the simulation network. The frequency of communication between the simulation component and the facilitator is low enough that it will not significantly impact the performance of the simulation network. The facilitator agent determines the capabilities as well as the needs of the simulation component and informs the component if and when it is required by the federation. Ideally, these capabilities and needs will be written based on a common ontology as described in Section 3.7.1, but that is often not the case.

Facilitator agents register their simulation component with the broker agent. The broker agent brokers between content producers and consumers and makes recommendations on which simulation components are capable of providing specific content. Each potential pair of provider and consumer of content is evaluated for suitability in the next phase.

Recommended pairs of providers and consumers of content are passed to a matchmaker agent. The matchmaker agent applies distance metrics to assess the capability of various pairs of consumers and producers to communicate and fulfill the requirement. The pair with the best capability is selected and directed, via their facilitator agent, to cooperate within the federation to accomplish the task.

Since heterogeneous simulations often do not share ontologies, a mediator agent is required. The mediator agent converts between units of measure and reference points to provide a common language for the simulation. In addition to simple unit conversion, mediator agents resolve four types of conflicts. Semantic conflicts exist when local scheme must be aggregated or disaggregated but fail to match. Descriptive conflict exists when a concept is described using synonyms, homonyms, or different attributes or values. Heterogeneous conflicts exist when concepts are described in terms of substantially different methodologies. Structural conflicts exist when concepts are defined in terms of different structures in the schemas (Tolk and Diallo, 2005).

The four types of agents work together to form the federation required to address a particular need such as executing a scenario for training or integrating geographically disparate systems to test a weapons system. As the scenario unfolds, and new requirements identified, the agents will identify new simulation components to integrate into the federation to address the new requirements.

3.8 CONCLUSION

Any SoSE project, almost by definition, deals with a constantly expanding set of systems. The individual systems are evolving or maturing at an increasing pace (INCOSE, 2010). New systems are constantly joining the SoS. The environment changes while the users either employ the system in new ways or in whole new environments. These changes result in new and emergent behavior. In addition to predicting the behavior of the SoS in the presence of these changes, the managers of the SoS must train new operators and maintainers of the SoS while training existing operators on the use of the system in the new environments or employment. In order to address these challenges, the role of M&S within SoSE is constantly growing. This means the size and scope of the simulation federations used in SoSE must keep pace with the growth and evolution of the SoS itself. Integrating new or modified simulations into existing federations, and building new federations, will be required at an increasing rate.

The M&S community has made great strides in the composability of models and the interoperability of simulations, but even greater strides are still required (Zeigler et al., 2008). DIS, HLA, and TENA provide a means for facilitating interoperability at the technical, syntactic, and potentially semantic level. Current federations, while successful, remain manpower intensive and brittle. The selection of simulation elements remains primarily a manual task performed by expert engineers. Interoperability on levels above the semantic level is not assured. The current solution is not scalable to the problem of maintaining a simulation federation in support of an evolving and expanding SoS. New tools, standards, and formalisms are required to provide the interoperability and composability required by the SoSs now being developed.

The chapter began with the quote, "Amateurs study tactics; professionals study logistics." It is often attributed to General Omar Bradley, but it is unclear if he was the first person to use the phrase. Tactics is arguably more fun to study than logistics and remains the cornerstone of the study of military science. However, it is often logistics that gives one side a significant advantage on the battlefield. General Bradley was trying to convince the military community to focus more of its attention on this critical, but often underappreciated, discipline within military science. Perhaps it is time for a similar paradigm shift in the M&S community. Perhaps one day soon, we will be saying "Amateurs study programming; Professionals study composability."

REFERENCES

Balci, Osman, James D. Arthur, and William F. Ormsby. 2011. Achieving reusability and composability with a simulation conceptual model. *Journal of Simulation* 5 (3): 157–165.

Boardman, John, and Brian Sauser. 2006. System of systems-the meaning of. Paper presented at 2006 IEEE/SMC International Conference on System of Systems Engineering, Los Angeles, CA.

Box, George E. P., and Norman R. Draper. 1987. *Empirical model building and response surfaces*. New York: John Wiley & Sons, Inc., p. 424.

Chen, Pin, and Jennie Clothier. 2003. Advancing systems engineering for systems-of-systems challenges. *Systems Engineering* 6 (3): 170–183.

Chen, Pin, Ronnie Gori, and Angela Pozgay. 2004. Systems and capability relation management in defence systems-of-system context. Paper presented at The 9th International Command and Control Research and Technology Symposium, Copenhagen, Denmark.

Coolahan, James E., and Gary W. Allen. 2012. LVC architecture roadmap implementation—Results of the first two years. Paper presented at Spring Simulation Interoperability Workshop (SIW), Orlando, FL.

Curtis, Bill. 2000. Guest editor's introduction: The global pursuit of process maturity. *IEEE SOFTWARE* 17 (4): 76–78.

Davis, Paul K., and Robert H. Anderson. 2004. Improving the composability of DoD models and simulations. *The Journal of Defense Modeling and Simulation: Applications, Methodology, Technology* 1 (1): 5–17.

Dewar, James A., Steven C. Bankes, James S. Hodges, Thomas Lucas, and Desmon K. Saunders-Newton. 1996. *Credible uses of the distributed interactive simulation (DIS) system*. Santa Monica, CA: Rand Corporation, RAND/MR-607-A.

Diallo, Saikou Y., Heber Herencia-Zapana, Jose J. Padilla, and Andreas Tolk. 2011. Understanding interoperability. Paper presented at the 2011 Emerging M&S Applications in Industry and Academia Symposium, Boston, MA.

Fisher, David. 2006. *An emergent perspective on interoperation in systems of systems*. Pittsburgh, PA: Carnegie Mellon University Software Engineering Institute, CMU/SEI-2006-TR-003.

Ford, Reginald, David Martin, Daniel Elenius, and Mark Johnson. 2011. Ontologies and tools for analysing and composing simulation confederations for the training and testing domains. *Journal of Simulation* 5 (3): 230–245.

Gladwell, Malcolm. 2000. *The tipping point: How little things can make a big difference*. New York: Little, Brown.

Gorman, Paul J., Andreas H. Meier, and Thomas M. Krummel. 1999. Simulation and virtual reality in surgical education: Real or unreal? *Archives of Surgery* 134 (11): 1203.

INCOSE. (ed.) 2010. *Systems engineering handbook: A guide for system life cycle processes and activities*. 3.2nd ed. San Diego, CA: International Council on Systems Engineering (INCOSE).

Institute of Electrical and Electronics Engineers. 1990. *IEEE standard computer dictionary: A compilation of IEEE standard computer glossaries*. New York: IEEE Publication.

Institute of Electrical and Electronic Engineers (IEEE). 2012. *IEEE standard 1278, distributed interactive simulation*. New York: IEEE Publication.

International Standards Organization/International Electrotechnical Commission (ISO/IEC). 2008. *ISO/IEC standard 15288:2008 system lifecycle processes*. Geneva, Switzerland: ISO/IEC.

Jamshidi, Mohammad (ed.) 2009a. *System of systems engineering: Innovations for the 21st century*. Wiley series in systems engineering and management. Hoboken, NJ: John Wiley & Sons, Inc.

———. 2009b. *Systems of systems engineering: Principles and applications*. Boca Raton, FL: Taylor & Francis Group.

Johnson, William R., Thomas W. Mastaglio, and Paul D. Peterson. 1993. The close combat tactical trainer program. Paper presented at the 25th conference on Winter Simulation, Los Angeles, CA.

Jones, Michael C. 2008. Simulation across the spectrum of submarine training. Paper presented at Proceedings of the 2008 Spring Simulation Multiconference, Ottawa, Canada.

Keating, Charles B., and Polinpapilinho F. Katina. 2011. Systems of systems engineering: Prospects and challenges for the emerging field. *International Journal of System of Systems Engineering* 2 (2): 234–256.

Keller-McNulty, Sallie, Kirstie L. Bellman, and Kathleen M. Carley. 2006. *Defense modeling, simulation, and analysis: Meeting the challenge*. Washington, DC: National Academy Press.

Kelton, W. David, Randal P. Sadowski, and David T. Sturrock. 2007. *Simulation with arena*, 4th ed. New York: McGraw Hill.

Knight, Pamella, Aaron Corder, Ron Liedel, Jessica Giddens, Ray Drake, Carol Jenkins, and Paul Agarwal. 2002. Evaluation of run time infrastructure (RTI) implementations. Paper presented at Huntsville Simulation Conference, Huntsville, AL.

Kossiakoff, Alexander, and William N. Sweet. 2003. *Systems engineering principles and practices*. Hoboken, NJ: John Wiley & Sons, Inc.

Kunz, Andrea A. 1993. *A virtual environment for satellite modeling and orbital analysis in a distributed interactive simulation*. Wright-Patterson AFB, OH: Air Force Institute of Technology.

Law, Averill M., and W. David Kelton. 1999. *Simulation modeling and analysis*, 3rd ed. McGraw-hill series in industrial engineering and management science. Boston, MA: McGraw-Hill Higher Education.

Maier, Mark W. 1998. Architecting principles for systems-of-systems. *Systems Engineering* 1 (4): 267–284.

Maier, Mark W., and Eberhardt Rechtin. 2000. *The art of systems architecting*. Boca Raton, FL: CRC Press.

McGinnis, Leon, Edward Huang, Ky Sang Kwon, and Volkan Ustun. 2011. Ontologies and simulation: A practical approach. *Journal of Simulation* 5 (3): 190–201.

Noseworthy, J. Russell. 2005. Developing distributed applications rapidly and reliably using the TENA middleware Paper presented at the Military Communications Conference (MILCOM), Atlantic City, NJ.

———. 2008. The test and training enabling architecture (TENA) supporting the decentralized development of distributed applications and LVC simulations. Paper presented at International Symposium on Distributed Simulation and Real-Time Applications. Vancouver, BC.

Page, Ernest H., Richard Briggs, and John A. Tufarolo. 2004. Toward a family of maturity models for the simulation interconnection problem. Paper presented at the Spring 2004 Simulation Interoperability Workshop, Arlington, VA.

Petty, Mikel D., and Eric W. Weisel. 2003a. A composability lexicon. Paper presented at the Spring 2003 Simulation Interoperability Workshop, Kissimmee, FL.

———. 2003b. A formal basis for a theory of semantic composability. Paper presented at the Spring 2003 Simulation Interoperability Workshop, Kissimmee, FL.

Pimentel, Ken, and Brian Blau. 1994. Teaching your system to share. *Computer Graphics and Applications, IEEE* 14 (1): 60–65.

Powell, Edward, and J. Russell Noseworthy. 2012. The test and training enabling architecture (TENA). In *Engineering principles of combat modeling and distributed simulation,* ed. A. Tolk, pp. 449-477. Hoboken, NJ: John Wiley & Sons, Inc.

Rechtin, Eberhardt. 1991. *Systems architecting: Creating and building complex systems*, Vol. 199. Upper Saddle River, NJ: Prentice Hall.

Robinson, Stewart. 2006. Conceptual modeling for simulation: Issues and research requirements. Paper presented at the 38th conference on Winter Simulation, Monterey, CA.

Robinson, Stewart. 2008. Conceptual modelling for simulation Part I: Definition and requirements. *Journal of the Operational Research Society* 59 (3): 278–290.

Sage, Andrew P., and Christopher D. Cuppan. 2001. On the systems engineering and management of systems of systems and federations of systems. *Information Knowledge Systems Management* 2 (4): 325–345.

Smith, Roger. 2010. The long history of gaming in military training. *Simulation & Gaming* 41 (1): 6–19.

Tolk, Andreas. 2010. Interoperability and composability. In *Modeling and simulation fundamentals: Theoretical underpinnings and practical domains*, eds. J. A. Sokolowski and C. M. Banks, pp. 403–433. Hoboken, NJ: John Wiley & Sons, Inc.

Tolk, Andreas, and Saikou Y. Diallo. 2005. Model-based data engineering for web services. *Internet Computing, IEEE* 9 (4): 65–70.

Tolk, Andreas, Saikou Y. Diallo, Robert King, and Charles Turnitsa. 2009. A layered approach to composition and interoperation in complex systems. In *Complex systems in knowledge-based environments: Theory, models and applications,* eds. A. Tolk and L. C. Jain, pp. 41–74. Heidelberg: Springer.

Tolk, Andreas, Saikou Y. Diallo, and Jose J. Padilla. 2012. Semiotics, entropy, and interoperability of simulation systems: Mathematical foundations of M&S standardization. Paper presented at 44th conference on Winter Simulation, Berlin, Germany.

Tolk, Andreas, Saikou Y. Diallo, Jose J. Padilla, and Heber Herencia-Zapana. 2013. Reference modelling in support of M&S—Foundations and applications. *Journal of Simulation* 7 (2): 69–82.

Tolk, Andreas, Saikou Y. Diallo, and Charles D. Turnitsa. 2006. Ontology driven interoperability–M&S applications. Paper presented at The Interservice/Industry Training, Simulation & Education Conference (I/ITSEC), Orlando, FL.

———. 2007. Applying the levels of conceptual interoperability model in support of integratability, interoperability, and composability for system-of-systems engineering. *International Journal Systemics, Cybernetics and Informatics* 5: 65–74.

Tolk, Andreas, and John A. Miller. 2011. Enhancing simulation composability and interoperability using conceptual/semantic/ontological models. *Journal of Simulation* 5 (3): 133–134.

———. 2004. M&S within the model driven architecture. Paper presented at The Interservice/Industry Training, Simulation & Education Conference (I/ITSEC), Orlando, FL.

Tolk, Andreas, and Adelinde M. Uhrmacher. 2009. Agents: Agenthood, agent architectures, and agent taxonomies. In *Agent-directed simulation and systems engineering,* eds. L. Yilmaz and T. Oren, pp. 75–109. Hoboken, NJ: John Wiley & Sons, Inc.

Under Secretary of Defense (Acquisition, Technology and Logistics). 2006. *Department of defense acquisition modeling and simulation master plan.* Office of the Under Secretary of Defense (Acquisition, Technology and Logistics) Defense Systems.

———. 2007. *DoD modeling and simulation (M&S) management.* Office of the Under Secretary of Defense (Acquisition, Technology and Logistics), DoDD 5000.59.

Weisel, Eric W. 2004. Models, composability, and validity. PhD., Old Dominion University.

Yilmaz, Levent. 2007. A strategy for improving dynamic composability: Ontology-driven introspective agent architectures. *Journal of Systemics, Cybernetics and Informatics* 5 (5): 1–9.

Yilmaz, Levent, and Tuncer Oren (eds.) 2009. *Agent-directed simulation and systems engineering*. Agent-Directed Simulation and Systems Engineering. Weinheim: Wiley-VCH Verlag GmbH & Co. KGaA.

Zeigler, Bernard P. 1976. *Theory of modeling and simulation.* New York: John Wiley & Sons, Inc.

Zeigler, Bernard P., Saurabh Mittal, and Xiaolin Hu. 2008. Towards a formal standard for interoperability in M&S/system of systems integration. Paper presented at GMU-AFCEA Symposium on Critical Issues in C4I, Fairfax, VA.

Zeigler, Bernard P., Herbert Praehofer, and Tag Gon Kim. 2000. *Theory of modeling and simulation*, 2nd ed. Amsterdam: Academic Press.

Zhu, Yifan, Weiping Wang, and Dongxiang Zhou. 2008. Conceptual framework of composable simulation using multilevel model specification for complex systems. Paper presented at The second IEEE International Conference on Semantic Computing, Santa Clara, CA.

Chapter 4

An Approach for System of Systems Tradespace Exploration

Adam M. Ross and Donna H. Rhodes
Massachusetts Institute of Technology, Cambridge, MA, USA

4.1 INTRODUCTION

Tradespace exploration (TSE) is a technique for evaluating a large number of alternatives in order to generate knowledge and insights into trade-offs, including costs and benefits. For a traditional system, key architectural trade-offs are made in the initial phase of the life cycle. While also true for system of systems (SoS), architecting continues throughout its operational lifespan. As a result, TSE can be a useful tool to support System of Systems Engineering (SoSE) for the ever-evolving SoS.

Architecting SoS is more complex than traditional systems. Managerial and operational independence of constituent systems, as well as the dynamic, time-dependent composition of the SoS, presents significant decision challenges. Several heuristics and guidelines proposed in the literature support architectural decision making (Maier, 1998; Keating et al., 2003). However, the ability to quantitatively compare SoS architecture alternatives, taking into account specific SoS considerations, largely remains a challenge.

Since SoS change over time, in order to quantify the particular strengths and weaknesses of an SoS, you should consider many possible alternative instances of that SoS. The approach to system of systems tradespace exploration (SoS TSE) we describe in this chapter is a promising model-based quantitative approach for enabling the ongoing

---

Modeling and Simulation Support for System of Systems Engineering Applications, First Edition.
Edited by Larry B. Rainey and Andreas Tolk.
© 2015 John Wiley & Sons, Inc. Published 2015 by John Wiley & Sons, Inc.

exploration of the SoS tradespace. We illustrate the SoS TSE approach through application to an Operationally Responsive Disaster Surveillance System (ORDSS).

4.1.1 M&S Challenges in SoS TSE

Model-based systems TSE has increasingly become part of engineering practice. The exploration of an SoS tradespace, however, presents multifaceted challenges as compared with exploring traditional systems tradespaces.

One unique challenge in SoS TSE is comparing diverse, multiconcept SoS on the same basis. As with systems-level TSE, the objective is to select architectures that will be *value robust* over a number of changing context scenarios. Value robustness is the ability of a system to deliver stakeholder value in the face of the dynamic world in which the system operates over its lifespan. As compared to traditional systems, this is a daunting challenge for SoS, given greater, more diverse sets of stakeholders, long lifespans, potentially reduced degree of control, and emergent constituent systems interactions.

The ability to explore SoS tradespaces, with quantitative comparison of alternative architectures, is less mature than at the systems level. A quantitative method needs to address the SoS-specific characteristics that distinguish SoS engineering from traditional Systems Engineering. You must be able to compare a large number of SoS architectures consisting of heterogeneous constituents, on a common basis. Models are essential to performing such a comparison. However, the potential size of an SoS tradespace presents a very practical challenge in performing a model-based analysis.

In TSE, each architectural concept involves design choices that satisfy the particular *attributes* valued by a given stakeholder. Systems attributes are stakeholder-perceived metrics of value that determine the overall benefit of the system to that stakeholder. The emergent value of combined constituent systems determines the benefit provided by an SoS. We have found that we need model-based approaches to effectively combine attributes. While progress has been made, this remains an open area of inquiry. This chapter introduces one technique for approximating attribute combination value.

4.2 BACKGROUND ON TSE

Researchers have demonstrated TSE methods as effective for gaining knowledge of the design space (Wolf, 2009; Ross et al., 2010b). These methods enable enumeration of alternative concepts and evaluation and selection of promising architectures that you can further investigate. Model-based TSE enables comparison of a much larger number of architectural designs than is possible with heuristics and qualitative approaches. By exploring the tradespace, in full or in part, you gain knowledge of possible systems performance, relative to stakeholder value or benefit. Exploring tradespaces can reveal key trade-offs as well as multistakeholder preference agreements and conflicts.

The SoS TSE approach enables multiconcept architecture comparison on a common performance and cost basis. Using the approach, an SoS architect can distinguish between constituent systems having high likelihood of participation in the SoS and those with lower likelihood of participation, based on the level of *EMA* that the architect has over the constituent.

4.2.1 Multi-attribute Tradespace Exploration

The SoS TSE approach extends from a prior MIT-developed model-based method, multi-attribute tradespace exploration (MATE) (Ross, 2003; Ross et al., 2004). Numerous case studies over the past decade have demonstrated the usefulness of MATE as a formal method for TSE (Ross and Hastings, 2005; Ross et al., 2010a). MATE was initially conceived for developing and exploring a static tradespace, that is, for fixed context and needs. Subsequent development of the method extended it to Dynamic MATE to enable consideration of dynamic issues such as changing stakeholder preferences and changing systems context (Ross, 2006). The MATE method, extended for SoS TSE, enables value-driven TSE of the SoS design space.

Figure 4.1 illustrates the MATE method.

A *set of attributes* represent value to a given stakeholder, which together are decision criteria elicited from that stakeholder. These are defined by the *stakeholder preferences*. When presented with two or more alternatives, the attributes are the criteria by which a given stakeholder decides which of the alternatives is better than the others (Keeney, 1992). For example, if deciding between two satellite constellations, a given stakeholder might care about image resolution and revisit time. A 1 m image resolution constellation with a 90 min revisit time might be more attractive than a 2 m image resolution constellation with a 120 min revisit time. More generally, the set of attributes can be

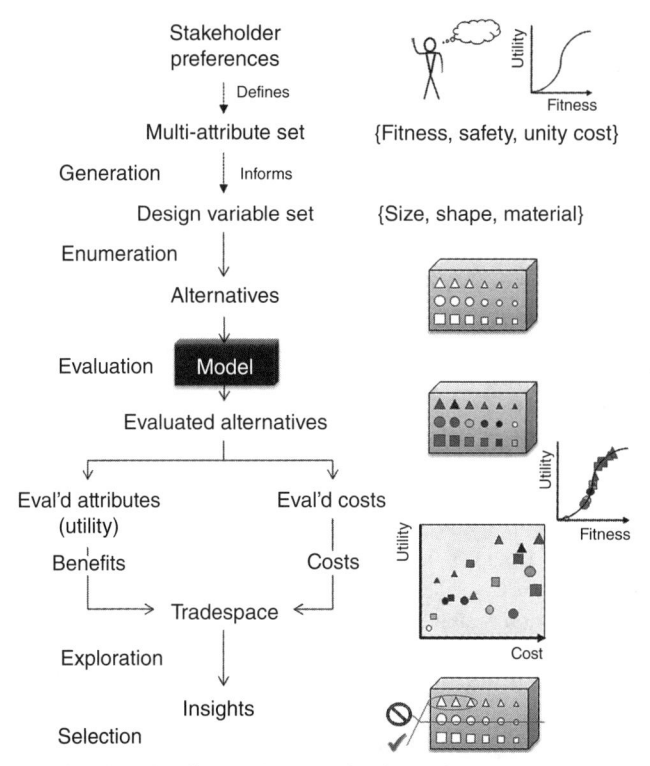

Figure 4.1 Illustration of multi-attribute tradespace exploration method.

elicited from a decision maker and then aggregated using techniques, such as multi-attribute utility theory (Keeney and Raiffa, 1993). MATE typically uses multi-attribute utility as the aggregate of benefit for each stakeholder, although alternative aggregation metrics can be used.

Each of the key stakeholders might have the same or different set of attributes. Ideally, attributes relate more to performance rather than form, since good performance is typically what a stakeholder desires. One of the attributes can be cost, but it is often left as a separate type of decision criterion in order to facilitate later exploration of "what can I get for different costs?" When following MATE, it is your job to identify the critical stakeholders (those that have influence over the resources or driving needs for the system) in order to reduce the chance of surprise emergent needs later in the study.

The elicited set of attributes forms the potential value space within which you will evaluate potential alternative systems. The parameterization of value using stakeholder-defined attributes frees you up to explore a large set of potential systems alternatives on a common basis. Informed by the attributes, you next *generate* alternative systems concepts that will potentially do well in terms of the attributes (and cost). The *design variable set* is a parameterization of (each) considered concept. These variables represent design factors within your control as SoS architect. For example, you might control the number of satellites for a constellation design, as well as the intended orbit.

Once design variable set(s) have been defined, you next *enumerate* particular choices for each of the variables. Together, the enumerated choices define the set of *alternatives* that you may consider in your MATE study. Continuing the satellite constellation example, choices for number of satellites in the constellation could include 12, 24, and 48, and intended orbits might include low inclination, 600 km circular; mid inclination, 800 km circular; and high inclination, 1200 km circular. The definition of choices for the design variables is intended to increase the chance of finding a good alternative in terms of the attributes. It is conceivable that you could enumerate hundreds, thousands, or millions of potential alternatives.

From the enumerated set of alternatives, you next select a subset for *evaluation* in a model. The *model* could take the form of anything that can evaluate a given set of design variable choices in terms of attributes. The more automated the model and the shorter the evaluation time, the larger the set of alternatives that you can evaluate in the time allocated. Model types include expert judgment, regression based, parametric, or simulation based, among others. Typically, the higher the fidelity of the model, the longer it will take to execute for evaluation.

After you evaluate each of the alternatives, you can combine the evaluated attributes for each alternative into utility. The set of utility and cost information for each systems alternative represents the *tradespace* of cost–benefit trade-offs. Using a scatterplot representation, you can *explore* patterns of relationships between design choices and feasible cost–utility pairs. The most efficient designs are those that are on the Pareto frontier of highest utility at a given cost or lowest cost for a given utility. "Pareto optimality" is achieved when a solution is nondominated, that is, a solution cannot be improved in a particular objective without making others worse off than before (de Weck, 2004). You can gain *insights*, such as design drivers for attributes, utility, and cost, through interaction with the data. Ultimately, the exploration activity concludes with the *selection* of one or more alternatives for further investigation. Such further investigation could include higher-fidelity evaluation or more detailed design work. It is essential to interpret the tradespace results in light of the fidelity of the evaluation models used, as well as your particular parameterization of design and value.

4.2.2 Epoch–Era Analysis in TSE

Epoch–era analysis (EEA) moves analysis into the time domain and is a useful approach for enhancing TSE (Ross and Rhodes, 2008b). In EEA, you define periods of time, *epochs*, where stakeholder preferences and systems operating environment are fixed. Within such epochs, a traditional MATE study can be conducted. But what if stakeholders change their mind? What if constraints and assumptions of the operating environment change? If either the needs or context for the system changes, then a new epoch has begun. In EEA, key uncertainty factors that could impact the success of the system are parameterized as *epoch variables*. For example, suppose the satellite constellation design example has uncertainty in the radiation environment (high radiation or low radiation) and the mission (particular locations in the globe as imaging targets). For various combinations of these epoch variables, the performance of a given design alternative will change.

Comparing the relative cost and utility of given alternatives across various epochs indicates how sensitive the alternatives are to the considered uncertainty. A value robust alternative is one that delivers efficient utility for cost across a large number of potential epochs. The *normalized Pareto trace* of a given alternative is the fraction of evaluated epochs in which the alternative is Pareto efficient in utility and cost (Ross et al., 2009). This is a useful metric for identifying value robust alternatives. If you desire insight into the impact of time dependence, a sequence of epochs can be used to describe an *era*. You can use various eras to analyze how alternatives perform over time across such epoch sequences. Such considerations may be particularly useful for SoS TSE.

4.3 SoS-SPECIFIC CONSIDERATIONS FOR TSE

In order to extend beyond traditional systems, you must take into account five specific considerations for SoS TSE. The first relates to having multilevel stakeholders, who derive and contribute value at a local (constituent system) level, at the global level, or both. As such, their respective value propositions may align or may be in conflict. The second consideration is the dynamic composition of the SoS, with an architecture that evolves over time, as legacy and new constituents exit and join the SoS over its lifespan. The third consideration is that SoS are typically a mix of legacy and new systems elements, with operational and other constraints. Interfaces may be where you have leverage and can implement architecting strategies to accommodate these limitations. The fourth consideration relates to managerial and operational independence that leads to participation risk (PR). Architects need to have strategies to address the potential situation where elements of the SoS may, or may not, be present or operate as expected. Finally, there may be other extenuating factors to consider for a given SoS. These may include type of SoS, geographic considerations, sociopolitical factors, stakeholder saliency, and other factors stemming from the architecting of large-scale sociotechnical systems.

80 Theoretical and Methodological Considerations

4.3.1 Multilevel Stakeholder Value

A specific consideration for SoS TSE relates to stakeholders who derive and contribute value (benefits) to the system. Constituent systems stakeholders give and receive value at the local level for their respective constituent. SoS are comprised of constituent systems with some degree of independence. A constituent having managerial independence makes decisions about the system and its operation, both when it participates within and outside of the SoS. Some, but not all, constituent stakeholders are global stakeholders who give and receive value directly from the SoS. Figure 4.2 illustrates this concept of local and global stakeholders.

Given both global and local stakeholders, architects are confronted with a multilevel stakeholder value proposition. Sometimes, there is alignment, and sometimes, there is conflict. Multistakeholder negotiations may require aggregating and trading preferences of decision makers, depending on relationships between the local and global stakeholders.

Architects need to incorporate local and global distributions of costs and benefits into a multilevel value proposition for the SoS. Architects must acknowledge the independent decision-making capability of each constituent system. You need to account for the risk of any key constituent failing to participate in the SoS as part of any SoS TSE activity.

4.3.2 Dynamic SoS Composition

SoS exist in a dynamic world and accordingly evolve over time. Constituent systems join or exit the SoS for various reasons. The constituents available to participate in the SoS at any given time can vary. This may be influenced by stakeholder preferences changing. The viability and desire to participate in the SoS can change as well, driven by changing political, economic, or resource-related factors.

Figure 4.3 notionally illustrates the potential changing SoS composition, as well as legacy and new constituent systems within the SoS. The architecture of the SoS evolves with significant changes in the SoS or its context, such as a constituent system joining or leaving the SoS. Architects face the challenge of architecting the SoS with the understanding that capability may be evolutionary, with different configurations of

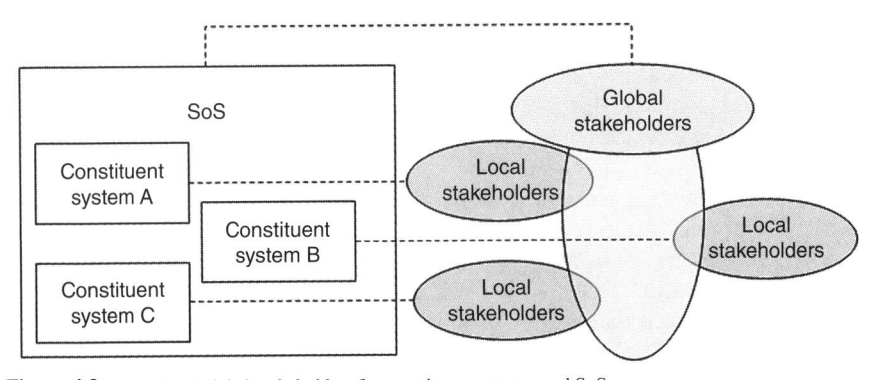

Figure 4.2 Local and global stakeholders for constituent systems and SoS.

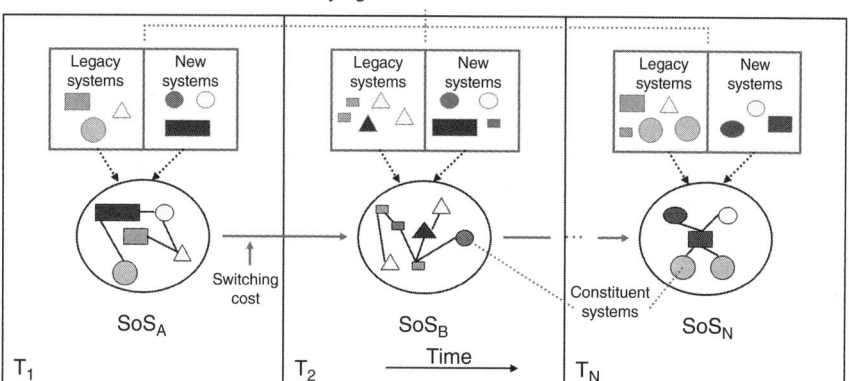

Time-varying available constituent sets

Figure 4.3 Legacy and new constituent systems in time-varying SoS composition.

constituent systems within the SoS. SoS TSE needs to accommodate decision making across multiple time increments with unique SoS configurations.

4.3.3 Strong Inheritance

Typically, an SoS is composed of both legacy and new constituent systems, as well as existing and newly designed interfaces between constituents. Architects do not always have the ability to make all design decisions regarding enhancements and upgrades to legacy systems, interfaces, and systems operations. In some cases, the legacy system will have constraints that prohibit this. In other cases, the managerial or operational independence of a constituent may restrict certain decisions. Strategies and design principles can be employed to overcome limitations. The "systems shell" is a useful construct when the architect cannot alter a constituent (Ross and Rhodes, 2007). Designing a wrapper or shell around the legacy constituent enables integration into the SoS without adversely affecting the legacy operations of the constituent. Similarly, it enables switching constituents in and out of an SoS with minimum impact on the SoS operation. Similarly, architects need to recognize differences in their control of design decisions in legacy and new constituent systems. It is often the design of the interfaces where architects can gain the most leverage.

4.3.4 Incomplete Control over SoS

The degree of design control that the architect effectively has over each constituent impacts the SoS architecture. Figure 4.4 illustrates a framework developed by Chattopadhyay et al. (2008) to characterize this challenge. The relationship between the SoS architect making SoS-level design choices and the constituent decision makers can be represented with two quantities—*managerial control* and *influence*. Managerial control is the amount of hierarchical control that the SoS architect has over the constituent system. Where participation of constituent systems is voluntary, the SoS architect may employ various methods, collectively referred to as influence, to persuade constituent

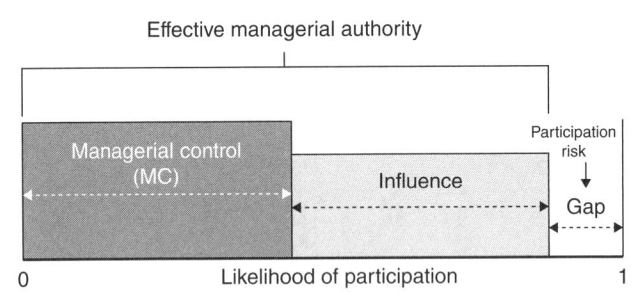

Figure 4.4 Managerial control, influence, and participation risk.

systems to engage in a manner that is beneficial to the SoS. Such influence is intended to sway the attractiveness of constituents participating in the SoS and could include incentives, information, infrastructure, integration, and institutions (Shah, 2013).

The constituent decision makers determine participation in the SoS based on their local perception of benefits and costs of participating. The change in *perceived net benefit* (PNB), which is the difference between local benefit and local cost as perceived by the constituent, is a metric that can be used by the constituent leadership to determine whether to join the SoS. The SoS architect may be able to change the PNB for a constituent system by providing incentives or additional nonmonetary value to influence the constituent system.

The *effective managerial authority* (EMA) that an architect has over the constituent systems is composed of the aforementioned managerial control described and the influence exerted to change the PNB of the constituent system. A greater EMA over a constituent system leads to greater likelihood of participation of the constituent system in the SoS. The risk of nonparticipation of a constituent system in the SoS is defined as the *PR* of the constituent system from the perspective of the SoS architect. PR represents the uncertainty involved in attaining a particular desired SoS design configuration. The concept of PR allows the SoS architect to differentiate between easy-to-achieve SoS designs and relatively difficult-to-achieve SoS designs during TSE.

In the SoS tradespace, PR can be used as a way to filter out difficult-to-achieve SoS designs. This information allows you to compare designs on the basis of performance and cost, as well as the risk of not achieving those performance and cost values. Thus, PR can be considered a measure of the uncertainty associated with attaining a particular desired SoS design configuration. In the SoS tradespace, this can be represented as a third dimension, along with utility and cost. The PR of the constituent systems can be estimated based on approximations for the SoS architect's managerial control over the constituent and the estimated influence, either monetary or otherwise, that you expect. In this approach, the effective managerial control (and the PR derived from it) is a variable for a given particular constituent. Effective managerial control can be varied over a range, generating many possible SoS designs that you can plot on the tradespace. The architect can identify high-performance, low-cost designs as in traditional TSE. The added PR dimension also enables you to differentiate between configurations that are easy to achieve versus those that are more risky. This enables you to compare high-performance, high-risk designs with relatively lower-performance, lower-risk designs that may be a more attractive solution, depending on risk tolerance.

4.3.5 Extenuating Factors

The SoS architect is faced with any number of extenuating factors relevant to designing and evolving large-scale sociotechnical systems. There may be geographic considerations that impact the architectural configuration as well as operational choices. Sociopolitical factors may impose choices of constituent systems in the architecture or impact operational choices. Given the SoS mission, certain stakeholders may be more important or have more authority in driving architectural decision choices. The overall impact is that the number of factors you may want to consider could outweigh what you can practically include in TSE. The architect's role is to decide which, if any, of these factors must be brought into the activity of TSE because they are critical for differentiating and valuing SoS alternatives.

4.4 AN APPROACH FOR SoS TSE

This section describes an approach for SoS TSE that begins to address the SoS-specific considerations to help gain value-centric insights into SoS alternatives.

We have leveraged MATE and EEA to formulate a ten-step approach for evaluating alternative SoS (Chattopadhyay, 2009).

The ten-step approach for SoS TSE includes:

1. Determine the SoS mission.
2. Generate a list of (potential) constituent systems.
3. Identify stakeholders and decision makers for SoS and constituent systems.
4. Classify constituent systems by managerial control and PR.
5. Define SoS attributes and utility expectations.
6. Define potential SoS context changes.
7. Model SoS performance and cost.
8. Conduct tradespace analysis.
9. Conduct EEA.
10. Select value robust SoS designs.

As a model-based approach, SoS TSE enables quantitative evaluation of alternatives while addressing the SoS-specific considerations described earlier as challenges to M&S of SoS.

4.4.1 Steps of Approach

In applying SoS TSE, it is likely that you may not have accurate or complete information for characterizing potential alternative SoS. The approach is iterative by design, and the key output of the approach is generation of **insights** rather than specific predictive outcomes. TSE seeks to help you identify patterns and relationships among factors within and not within your control, especially as it relates to stakeholder-perceived value.

An example application of this approach can be found in Section 4.5. What follows is a brief description of the ten-step approach:

1. Determine the SoS mission: The SoS mission is the problem that the SoS must solve or capability gap that the SoS must fulfill, generated through interaction with the SoS stakeholders.
2. Generate a list of (potential) constituent systems: A list of legacy systems that may fully or partially fulfill the mission need is generated, and new systems concepts can be proposed to fill any capability gaps. The output of this step is a list of potential constituent systems for the SoS.

   Identify stakeholders and decision makers for SoS and constituent systems: Local constituent systems stakeholders and SoS-level stakeholders are identified. In this approach, it is assumed that the SoS and each legacy constituent system have a single decision maker each—this may require an aggregation of stakeholder preferences by defining a hypothetical benevolent leader. The benevolent leader is a primary decision maker who takes actions to allocate resources in such a way as to satisfy the stakeholder network he/she represents.
3. Classify constituent systems by managerial control and PR: The identification of the primary decision maker for the SoS and each of the potential constituent systems enables the SoS designer to estimate the amount of managerial control that the SoS has over each of the constituent systems. This is necessarily an iterative step, as the estimation of managerial control may not be accurate at this stage, due to limited available information.
4. Define SoS attributes and utility expectations: Attributes (performance metrics used to measure systems value delivery to each of the stakeholders) and utility (used to represent aggregate stakeholder satisfaction based on systems performance on the tradespace) information is obtained through interviews of the SoS stakeholders.
5. Define potential SoS context changes: A number of possible future context changes, such as SoS stakeholder preference changes, changes in control and PR, and changes in availability of constituent systems for inclusion in the SoS, are identified in order to study the changes in SoS value delivery over time in EEA.
6. Model SoS performance and cost: This step includes modeling both legacy systems and new systems, as well as appropriately modeling the SoS to allow for the calculation of expected performance and cost. This allows for the generation of tradespaces where each design alternative is represented in terms of utility and cost.
7. Conduct tradespace analysis: Tradespaces including both single-system designs and multiconcept SoS are analyzed, and predefined criteria are used to select SoS designs of interest for further study. One example criterion is using the Pareto set of designs, which are the most utility-cost-efficient alternatives in a given tradespace.
8. Conduct EEA: The models created for the SoS performance calculation can be repeatedly executed for each future epoch described by the context changes defined in step 6. Thus, many tradespaces are obtained, and tradespace statistics can be generated over sets and sequences of epochs. Sequences of epochs are called eras, which each represent a potential future scenario for the SoS.

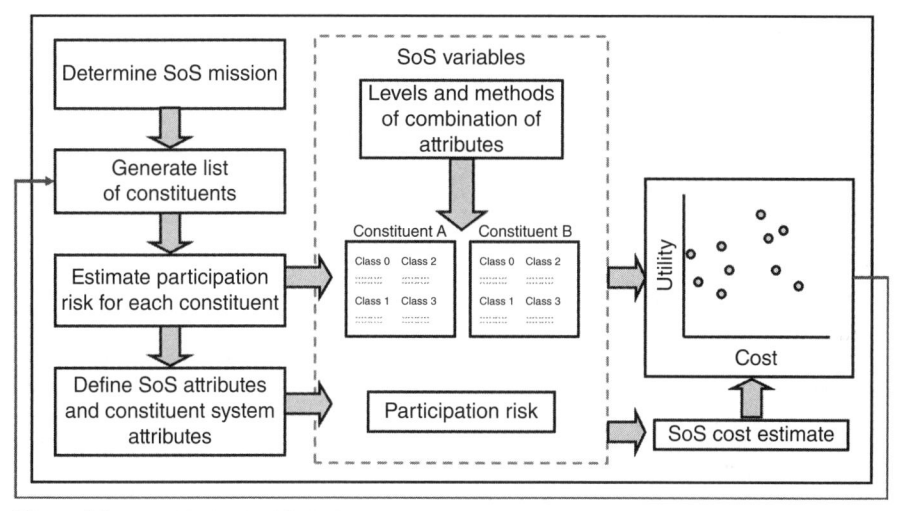

An Approach for System of Systems Tradespace Exploration   **85**

Figure 4.5 Activities involved for SoS tradespace analysis.

9. Select value robust SoS designs: Alternatives, or sets of alternatives, that consistently deliver adequate stakeholder value are considered robust in value, or value robust. Tradespace metrics such as normalized Pareto trace can be used to select designs that are value robust over multiple epochs or eras.

This approach can be conducted iteratively as you gain experience in the various steps and uncover insights about what is important (or not) in the tradespace. A subset of steps in the approach is illustrated in Figure 4.5, showing how to generate SoS tradespaces.

4.4.2 Addressing the SoS Considerations Using the Approach

Table 4.1 shows where in the approach the various SoS-specific considerations are addressed. In particular, the approach is intended to guide the SoS designer to explicitly consider these extra factors on top of what might be considered in a more traditional systems tradespace study.

4.4.3 Using Attribute Abstractions for Concept-Level Modeling of SoS Value

Two abstractions are described in the following text to aid in quantifying value for SoS. These include attribute classes and attribute combination complexity. The first helps to categorize potential benefits of the SoS in terms of cost to accomplish. The second helps to quantify how constituent systems attributes can be combined to provide SoS-level attributes.

Table 4.1 SoS TSE Approach Steps with SoS Considerations

Step/consideration	Multilevel stakeholder value	Dynamic SoS composition	Strong inheritance	Incomplete control over SoS	Extenuating factors
Determine the SoS mission	X		X	X	
Generate a list of (potential) constituent systems		X	X		
Identify stakeholders and decision makers for SoS and constituent systems	X		X		
Classify constituent systems by managerial control and participation risk				X	X
Define SoS attributes and utility expectations	X				
Define potential SoS context changes		X		X	X
Model SoS performance and cost		X	X	X	X
Conduct tradespace analysis	X			X	
Conduct epoch–era analysis		X		X	X
Select value robust SoS designs	X	X	X	X	X

4.4.3.1 Attribute Classes to Categorize Type of Desired Value

Attributes can be classified on the basis of whether or not they are articulated by the SoS decision maker, as well as the cost to "display" the attribute. Attributes that a decision maker communicates to a systems designer are *articulated attributes*. Systems performance is assessed on the basis of these articulated attributes. Attributes that the system exhibits by design are referred to as *existing attributes*. Table 4.2 shows a classification scheme for attributes using five classes (Ross and Rhodes, 2008a).

Class 0 attributes are always considered in systems design, given that these are the *articulated value* attributes. Value delivery to constituent systems stakeholders is measured using Class 0 attributes. This is, however, a subset of the possible attributes displayed by the system. The system may also display attributes that have not been articulated by any systems stakeholder. Given these already exist, they represent *free latent value* in that system. These are the Class 1 attributes that can be potentially utilized at no cost, as they are already displayed by the system.

Combining Class 0 and 1 attributes at some small cost can provide *combinatorial latent value*, or Class 2 attributes. Respecifying displayed attributes is often all that is required to provide the additional value that is already available from the system. The system may or may not need to be modified to obtain Class 2 attributes.

Class 3 attributes, *accessible value*, are achievable by modifying the original system at some cost to obtain new performance characteristics. Cost may vary from small to

Table 4.2 Attribute Classification Scheme from Ross and Rhodes (2008a)

Class	Name	Property of class	Cost to display
0	Articulated value	Exist, performance is assessed	None
1	Free latent value	Exist, but performance is not assessed	None
2	Combinatorial latent value	Can exist through combining Class 0 and 1 attributes	Small
3	Accessible value	Can be added through changing the design variable set (scale or modify system)	Small to large
4	Inaccessible value	Cannot be added through changing design variable set given constraints or prohibitive cost	Large to infinite

large, so trade-offs of value for cost will be necessary. Class 4 attributes are those that would only be available through drastic changes to the original system at large to infinite cost. These attributes have *inaccessible value*, due to the prohibitive cost or physical constraints of the systems design.

SoS constituents have some degree of managerial and operational independence and accordingly have local stakeholders. Local stakeholders may or may not derive value from the participation of the constituent in an SoS. Class 0 attributes of each constituent system derive from its local stakeholders. These are unlikely to coincide with all required SoS attributes articulated by the SoS stakeholder set. SoS attributes, therefore, are a subset of the combined set of Class 0, 1, 2, and 3 attributes of the constituent sets. Class 0 and 1 attributes are more readily available than Class 3 attributes. The most efficient way to achieve SoS attributes is to directly utilize Class 0 and 1 attributes of the constituent systems. Additional value at a small cost can be attained through combining Class 0 and 1 attributes of constituents. Accessible SoS value can be achieved through changing constituent systems design, yielding Class 3 attributes. The necessary systems design change will require additional costs that may be small to large. Trade-offs will be needed to determine if the cost expenditure is worthwhile, given the additional value that would be accessible at the SoS level. This additional value, added specifically for the SoS, may or may not be desired by the constituent system itself. Generating the SoS attribute values can require some combination of the constituent attributes. The method by which the constituent attributes are aggregated is determined by classifying the "level of combination complexity" selected for the particular attribute.

SoS value emerges through linking of two or more constituent systems together into a higher-order system. Evaluation of emergent benefits is not trivial. As more interactions are anticipated, the potential complexity of evaluation can rapidly grow beyond resources of a particular study. In order to allow for the use of the SoS TSE approach described in this chapter, appropriate simplifications may be needed. This comes with the caveat that simplifications often reduce the fidelity of the output. However, the trade-off of fidelity with modeling effort is not a new one. In applying the approach for SoS TSE, simplification is implemented in combining constituent-level attributes to SoS-level attributes.

4.4.3.2 Attribute Combination Complexity for Aggregating SoS Emergent Value

The way in which SoS constituents interact during operations is an integral factor for determining interfaces between constituents. This also helps you determine how you can combine constituent attributes to achieve the desired SoS attributes. In concept exploration, where there are few constraints on the design space, the SoS architect needs to explore attribute combination possibilities. This space of possibilities is large. As a first-order estimate, we define three types of attribute combination: low complexity, moderate complexity, and high complexity. We assume that the higher the complexity of attribute combination, the higher the cost for creating an interface capable of this kind of combination.

Low-complexity attribute combination involves taking the best performance from the set of constituents in the SoS to achieve the SoS value. There may be cases where constituent systems operate in parallel, but only the best performance is chosen. This would be the case when an SoS constituent with high performance is only available to deliver that attribute for part of the operating time. An interface between the constituents that generates an attribute through this level of combination will have relatively low complexity and therefore lower cost.

Moderate-complexity attribute combination corresponds with added complexity in SoS concepts of operation. For example, when there is a handoff between different assets in the SoS, such that multiple constituents are involved in delivering a single-attribute performance, the resulting SoS attribute performance is a combination of the two constituent attribute performances. SoS attribute combination with moderate complexity may involve techniques such as time-weighted averaging, such as the concept of time-weighted average utility (Richards et al., 2008). Due to the complexity of operations in this case, systems designers need to consider carefully the additional costs required to create interfaces that will enable operations.

High-complexity attribute combination is required when multiple SoS constituents deliver performance relating to the same SoS attribute simultaneously. In this situation, fusion of attributes at a more detailed level, rather than just averaging, is required. A technique for high-complexity attribute combination is data fusion. Data fusion is a well-developed field with methods available for combining data-related attributes, such as image resolution. Data fusion combines data from multiple sources along with database information to draw inferences beyond that obtainable through a single data source alone (Hall and Llinas, 1997; Llinas et al., 2004). Chattopadhyay (2009) describes data fusion for attribute combination in further detail.

An illustrative example describes the SoS attribute resulting from multiple constituent systems sensors. Consider the attribute combination required in a multimodal surveillance SoS consisting of a satellite radar asset and an aircraft radar asset, illustrated in Figure 4.6. If there is no overlap between the SoS relevant attribute sets required from each asset—the satellite performs imaging and the aircraft performs target-tracking activities—the attribute combination method simply consists of taking the attribute performance of each constituent and representing it as the SoS attribute. There may be additional SoS-level attributes, such as responsiveness of the SoS (i.e., "how quickly after a request coverage can be obtained of a particular area of interest?"). In the case of the SoS responsiveness attribute, the best performance between the satellite and the aircraft is an appropriate SoS performance measure. Here, low-complexity attribute combination is sufficient to model the SoS value delivery.

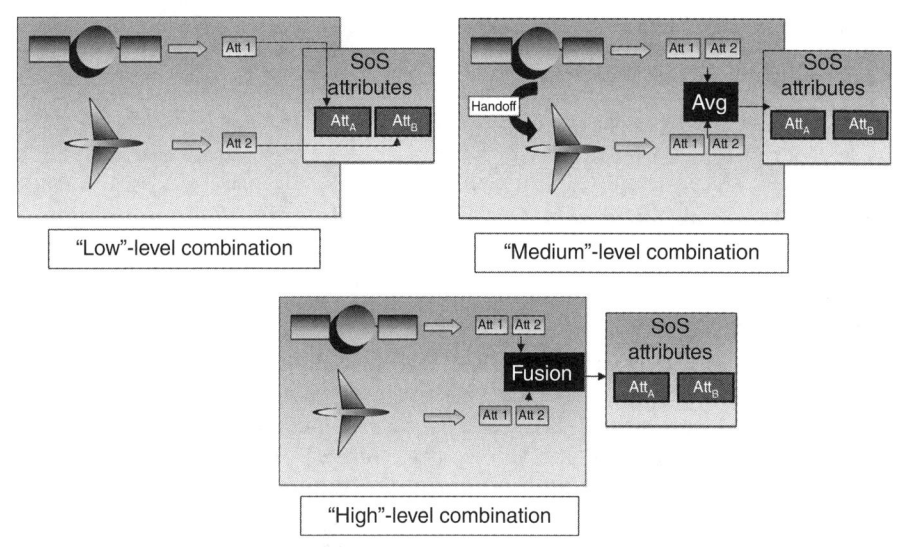

Figure 4.6 Levels of attribute combination complexity example (Chattopadhyay, 2009).

Continuing with this example, suppose the concept of operations was changed such that the satellite would identify a target in its field of view, notify the aircraft, and then track the target until the air asset arrived and the track could be handed off. An SoS attribute such as track life (i.e., the length of time for which the SoS can track an object of interest after successful identification) would be a combination of the two track life attribute performances of the constituent systems. Similarly, an attribute-like image resolution could be represented as a time-weighted average of the constituent systems resolutions during the aforementioned operation. Here, moderate-complexity attribute combination will be necessary, having an associated cost for the implementation of the interface. If the satellite and aircraft operate simultaneously (e.g., imaging overlapping fields in different wavelengths), high-complexity data fusion may be required. In this case, we may require complex interfaces and additional constituent systems in order to achieve the data fusion.

4.4.3.3 Effects of Attribute Classes and Combination Complexity on the Tradespace

SoS attributes and costs are influenced both by the attribute class of the attributes combined and by the attribute combination complexity used to merge those attributes. The attribute combination complexity required to attain a particular SoS attribute drives the technique for how you combine constituent attributes. Once you select the technique, you can obtain the SoS attribute value as a function of the constituent systems attributes. Based on this, you calculate SoS single-attribute utility in the TSE approach, using defined stakeholder utility preferences. You can then aggregate the single-attribute utilities calculated for each attribute using multi-attribute utility theory to obtain an SoS multi-attribute utility. You can use this utility to compare various SoS concept designs in a tradespace.

In order to estimate the additional cost required, you use the attribute-class information for each constituent attribute involved in the SoS attribute calculation. In concept

Table 4.3 SoS and Constituent System Class 0 Attributes

Class 0 attributes: Elicited from stakeholders		
Aircraft	Satellite	SoS
Acquisition cost	Acquisition cost	Acquisition cost
Resolution	Imaging capability (NIIRS)	Imaging capability (NIIRS)
Target track life	Percentage of area of interest covered	Field of regard
		Target track life

exploration, the SoS designer is primarily concerned with determining a set of suitable constituent systems. The interfaces are not yet defined, because the composition of the SoS may still be undetermined. The additional cost required to bring together the independent constituent systems into an SoS must be estimated. Together, the class of a constituent attribute and required attribute combination complexity form an indicator of "integration cost." Combining two constituent attributes that are Class 0 requires less effort and cost than combining two attributes that are Class 3. The SoS designer, perhaps in consultation with the constituent leadership, determines the additional cost for a Class 3 attribute versus a Class 0 or Class 1 attribute.

Considering each attribute in the SoS attribute list, if high-complexity attribute combination is required, this results in additional cost added to the total SoS cost estimate. In addition to the cost of the constituent systems participating in the SoS, the total estimate for cost encompasses the difficulty and expense required to integrate these constituents. This method for estimating SoS cost enables the comparison on the same basis of different SoS designs with different constituent compositions and includes the consideration of difficulty of achieving the final configuration.

Chattopadhyay (2009) provides a simple qualitative example of SoS attribute combination. The SoS is comprised of two constituent systems, an aircraft and a satellite, with a mission to conduct surveillance. This example is an extraction from the illustrative case of an ORDSS case study discussed later in this chapter.

Table 4.3 lists the elicited Class 0 (articulated) attributes for the aircraft, satellite, and SoS.

SoS attributes are individually considered to determine the classes of attributes as well as the level of combination required to achieve that SoS-level attribute. Class 0 and 1 attributes are identified using information about the performance of the system. Class 2 and 3 attributes are generated by recombination of Class 0 and 1 attributes and by enumerating possible modifications to the systems design variables, respectively. What follows is a brief description of how we assessed the cost implications for each of the SoS attributes.

Acquisition Cost Acquisition cost is an explicitly requested attribute for both constituent systems and the SoS. This means it is a Class 0 attribute for the SoS itself and the two constituent systems. The combination of two Class 0 attributes is relatively easy to accomplish and does not add significantly to the cost of the SoS. With low-complexity combination of constituent attributes, a simple combining method such as addition can be used to obtain the SoS acquisition cost attribute value.

Imaging Capability National Imagery Interpretability Rating Scale (NIIRS) levels are used to measure imaging quality (IRARS, 1996). This is done using imaging resolution as well as additional information about image interpretation. For the satellite, imaging capability is a Class 0 attribute, while it is a Class 2 attribute for the aircraft. This is because calculating the NIIRS level for the aircraft requires a combination of the Class 0 attribute of resolution, along with other existing imaging information relating to "interpretability" to generate NIIRS levels. The cost incurred to combine a Class 0 and 2 attribute is approximated at 10% of the constituent systems cost. This additional cost needs to be considered for this imaging capability attribute. A high level of combination is selected in this case, since this is a highly important attribute for the surveillance system. The SoS imaging capability attribute is generated through data fusion algorithms.

Field of Regard For this attribute, appropriate Class 0 attributes were not articulated. However, the constituent systems each had appropriate latent value (Class 1) attributes. These are available for use since they already exist, despite the fact that they were not explicitly asked for by stakeholders. The field of regard SoS attribute is then obtained by combining a Class 1 (latent value) attribute for the aircraft with a Class 1 attribute for the satellite. As Class 1 attributes, cost contribution is not a significant factor. A moderate attribute combination complexity is selected for this SoS attribute, obtained with simple averaging of the two constituent latent value attributes.

Target Track Life This attribute is the continuous length of time that the system can track a particular target. While this is a Class 0 attribute for the aircraft, it is a Class 3 attribute for the satellite. The primary mission of the satellite is imaging, and so displaying this attribute would require significant resources to equip the satellite to measure track life. Here, the cost for this Class 3 attribute to contribute to the SoS-level attribute is approximated as 30% of the satellite constituent cost and added to the SoS cost. The attribute combination chosen is high complexity as it requires tailored algorithms for track life combination. The concept of operations for the SoS factors into this attribute heavily. Coordination between assets, where the satellite identifies a target and passes off the track to the aircraft, results in a complex track life calculation.

Once the single-attribute values are obtained, the single- and multi-attribute utility values for the SoS design can be calculated. Despite the constraints on concept of operation possibilities for constituent systems in the SoS, the SoS designer may have several levels of attribute combination complexity to choose from when designing the SoS. Additional designs can be generated and added to the SoS tradespace by varying attribute combination and methods utilized to combine attributes.

4.5 ILLUSTRATIVE CASE: ORDSS

The following is an illustrative case to demonstrate the SoS TSE approach at a simple level. Severe natural disasters can have substantial impact on human life, such as Hurricane Katrina in New Orleans in 2005, the California wildfires of 2007 and 2008, and the Sichuan earthquake in China in 2008. There is a clear need for effective and

Table 4.4 Participation Risk (PR) Changes for a Range of Managerial Control and Influence Values

	Sensors	Aircraft	Satellite	Sat + aircraft	Sat + sensors	Sat + aircraft + sensors
MC	0.8–1	0.4–0.7	1	–	–	–
Influence	0–0.1	0–0.2	0	–	–	–
PR	0–0.2	0.1–0.6	0	0.1–0.6	0–0.2	0.1–0.7

timely observations of disaster location areas of interest (AOIs) in order to aid first responders:

1. **Determine the SoS mission.**
   For disaster surveillance information to be of use to first-response and disaster relief efforts, it must be provided as soon as possible after the disaster. In other words, an operationally responsive system with short response time is necessary to generate disaster observing data that is useful for time-critical disaster relief (Chattopadhyay et al., 2009).

2. **Generate a list of (potential) constituent systems.**
   For this case, heterogeneous SoS comprised of multiple single-system concepts, both legacy and new, are compared on the same performance and cost basis in a tradespace, alongside single-system concepts. Three different constituent systems concepts are considered in the case: satellites, aircraft, and sensor swarms.

3. **Identify stakeholders and decision makers for SoS and constituent systems.**
   Two SoS stakeholders—a firefighter and the ORDSS owner—are simultaneously considered.

4. **Classify constituent systems by managerial control and PR.**
   For this case, it is assumed that the satellite will be under the control of the SoS, while sensor swarms and aircraft assets may be partly independent (either operational or managerial, or both). Table 4.4 illustrates an estimated degree of managerial control over the three asset types, along with estimation of range of influence to shift the likelihood of participation of each asset. The PR range is calculated for each individual asset, as well as the range for various combinations of constituent systems for the SoS. Further consideration of these assets is dependent on the willingness to accept the chance that constituent systems with higher PR may not be available when needed during SoS operations. Modeling of the SoS can include these estimates in expected value calculations of performance of the SoS.

5. **Define SoS attributes and utility expectations.**
   The attributes for quantifying value to each of the stakeholders include the following:
   - Acquisition cost ($M)
   - Price/day or cost/day ($K/day)
   - Time to initial operating capability (days)
   - Responsiveness (hours)
   - Max percent AOI covered (%)
   - Time to max coverage (minutes)

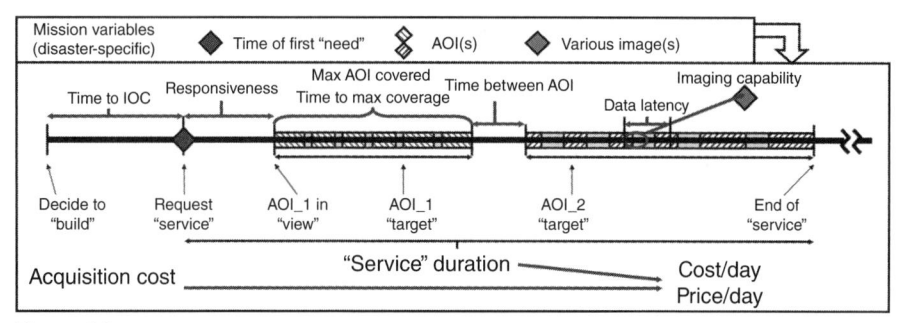

An Approach for System of Systems Tradespace Exploration   93

Figure 4.7 Attributes of interest indicated along the mission timeline for an Operationally Responsive Disaster Surveillance SoS.

- Time between AOI (minutes)
- Imaging capability (NIIRS* level)
- Data latency (minutes)

Figure 4.7 illustrates the attributes against a notional mission timeline.

For each of these attributes, the stakeholders each have acceptability ranges, with larger scores perceived as better. Figure 4.8 illustrates the single-attribute utility functions showing perceived benefit to the firefighter for different levels of each attribute. The utility evaluation of each potential alternative is scored using an aggregation function that combines the attribute scores for each evaluated alternative. The aggregation function used in this case is a linear weighted sum of the single-attribute utility scores interpolated using the functions in Figure 4.8, as well as the indicated weights (k).

6. **Define potential SoS context changes.**

   The system will need to observe a large number of different types of disasters at different locations during its operational lifetime. Three scenarios are considered: (1) hurricane disaster area, modeled on the Hurricane Katrina disaster area; (2) a forest fire disaster, based on the Witch Creek Fire in California; and (3) a cyclone disaster area, associated with the Myanmar cyclone in 2007.

7. **Model SoS performance and cost.**

   The constituent systems performance and cost are modeled individually, and then the SoS are evaluated through combining attributes. SoS consisting of multiple single systems (such as an aircraft and a satellite together) are also modeled. The model flow is briefly depicted in Figure 4.9. For this case, the aggregation of performance among assets for a particular SoS is done by selecting the best attribute score of each asset. This corresponds to low-complexity attribute combination. Additional coordination and interface costs are estimated by multiplying constituent systems costs by a fixed factor (in this case multiplying costs by 110%).

8. **Conduct tradespace analysis.**

   Using the evaluated data in the prior step, we generate tradespaces that we use to compare both single assets and SoS consisting of sets of one or more types of assets (including aircraft, satellite, and sensor swarms). Figure 4.10 illustrates the

---

*NIIRS is a standardized scale used for rating imagery obtained from imaging systems (IRARS, 1996).

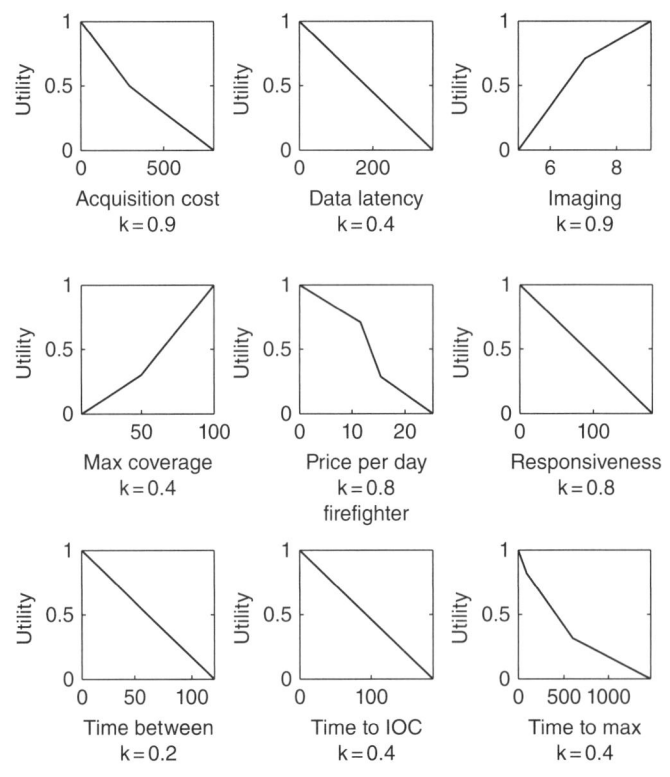

Figure 4.8 Single-attribute utility curves for firefighter stakeholder indicating the most desirable (utility = 1) and the least desirable (utility = 0) levels for each attribute.

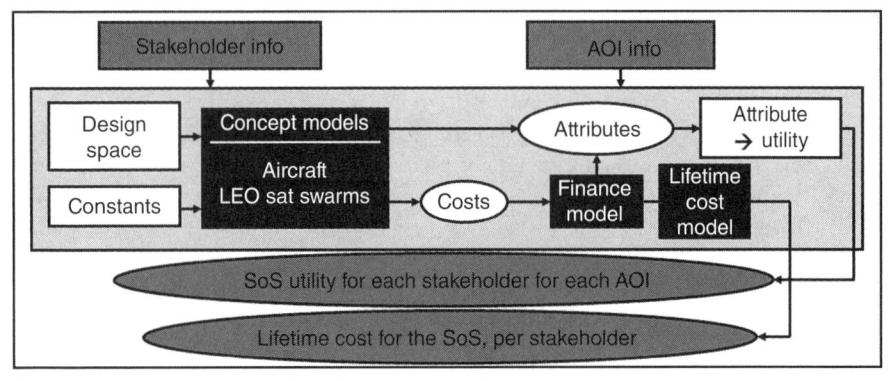

Figure 4.9 Description of modeling effort to evaluate utility and cost for each SoS alternative.

An Approach for System of Systems Tradespace Exploration 95

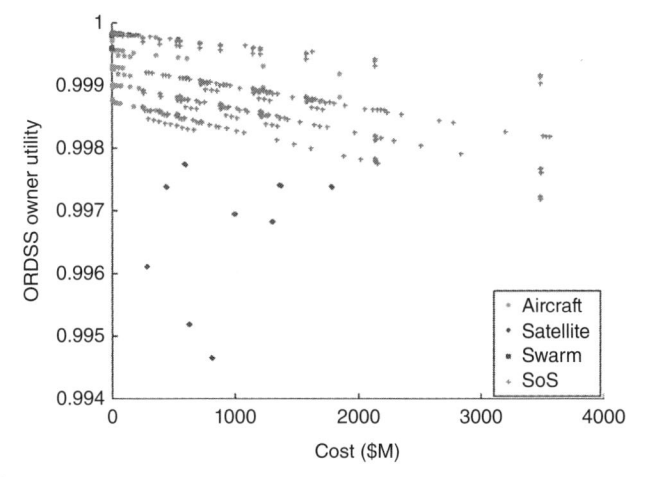

Figure 4.10 Example tradespace for ORDSS owner with utility and cost for various single-concept and SoS alternatives (Chattopadhyay, 2009).

three constituent systems individually as well as SoS in terms of utility to the ORDSS owner and their cost. Sensor swarms alone are clear low-cost alternatives, with SoS providing both higher utility and most cost.

9. **Conduct EEA.**
   Figure 4.11 shows three epochs in which the target surveillance area is the context change under consideration. The change in AOI location clearly impacts the possible utility to the ORDSS owner.

10. **Select value robust SoS designs.**
    Tradespace metrics such as normalized Pareto trace (Ross et al., 2009) are used to identify passively value robust designs among the single-system and SoS design concepts and then used as the set of designs to consider in further detail. Table 4.5 shows the four concepts with high normalized Pareto trace across the three considered epochs. Two of the concepts are single constituent system, while two are SoS with aircraft and satellites.

    This set of Pareto efficient designs are those that provide high value to both stakeholders, across each of the three AOI considered. The first three designs listed in Table 4.5 are designs that are high performance in all three selected AOI, while the fourth is a valid option if the system were only used for the continental United States.

    Table 4.5 illustrates that the approach for SoS TSE enables the consideration of a diverse set of concepts on the same basis. This allows you to consider many different efficient options that would have not been possible if the study had been limited to a single-concept or single-mission context. You can find further details of the ORDSS illustrative case in Chattopadhyay (2009).

4.6 CHAPTER SUMMARY

This chapter described an approach for SoS TSE. The benefits of TSE for traditional systems are multiplying with the growing complexity of modern systems. Using TSE, you can quickly identify the core cost and benefit trade-offs as determined by proposed systems

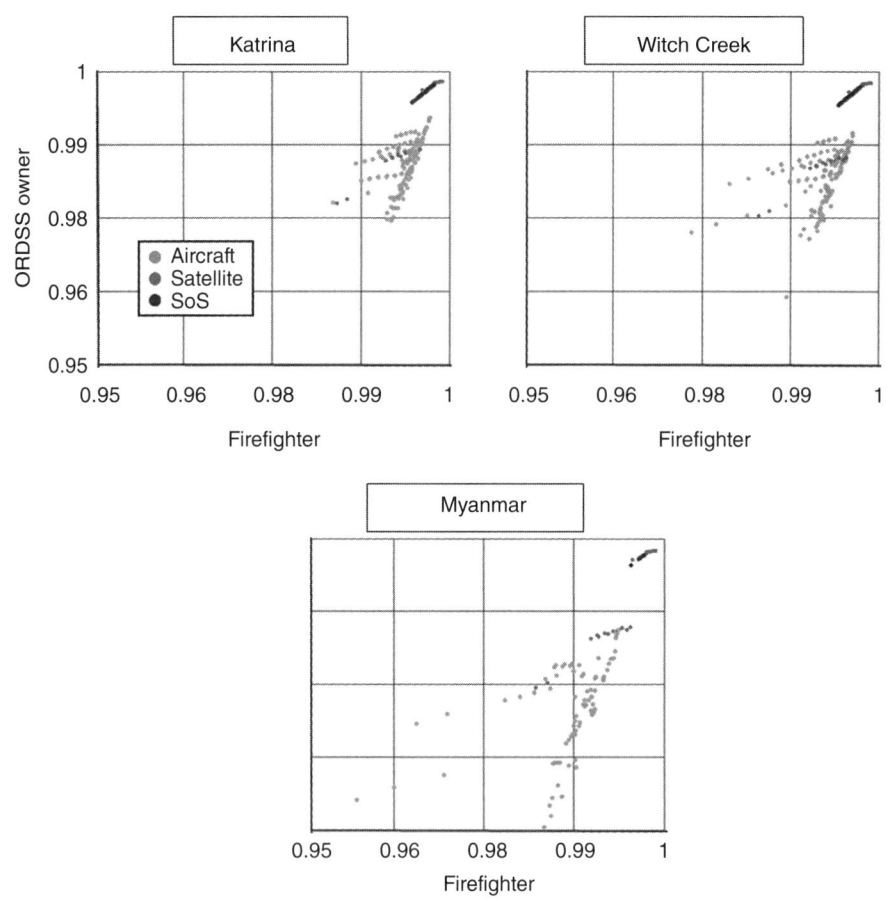

Figure 4.11 Epoch–era analysis with varying AOI locations, showing multiple single concepts along with SoS on the same tradespace (Chattopadhyay, 2009).

Table 4.5 Selected Pareto Designs for Three Epochs with Change in AOI

				Pareto efficient for		
Concept	Description	Lifetime cost ($M)	Normalized Pareto trace ($N=3$)	Katrina disaster	Witch Creek Fire	Myanmar cyclone
Aircraft	Existing ScanEagle	0.70	1.00	Yes	Yes	Yes
Satellite	120 km, sun-synch orbit, IR payload	12.40	1.00	Yes	Yes	Yes
SoS	Aircraft (small UAV w/ piston) + satellite (800 km, sun-synch orbit, IR payload)	3.86	1.00	Yes	Yes	Yes
SoS	Aircraft (existing Cessna 206) + satellite (120 km, 23 deg inclination orbit, IR payload)	3.22	0.67	Yes	Yes	No

alternatives and valued by stakeholder preferences. Applying TSE to SoS requires you to consider additional factors beyond that of a traditional system. These considerations include multilevel stakeholders (both constituent level and SoS level), dynamic SoS composition (the SoS may change over time), strong inheritance (often constituents and legacy constraints limit future options), incomplete control (constituents often retain a degree of independence that results in PR to the SoS), and other extenuating factors (constituents may be geographically dispersed, among other concerns). These considerations multiply the complexity of an SoS M&S effort considerably. We intend for this approach to ensure that you can begin to address the SoS-specific considerations in a quantitative manner:

The ten-step approach for SoS TSE includes:

1. Determine the SoS mission.
2. Generate a list of (potential) constituent systems.
3. Identify stakeholders and decision makers for SoS and constituent systems.
4. Classify constituent systems by managerial control and PR.
5. Define SoS attributes and utility expectations.
6. Define potential SoS context changes.
7. Model SoS performance and cost.
8. Conduct tradespace analysis.
9. Conduct EEA.
10. Select value robust SoS designs.

The SoS TSE approach enables a designer with a large design space to consider many possible options with a relatively small amount of modeling and analysis effort. The approach can be used to identify a set of value robust designs suitable for further detailed study. The quantitative comparison of a variety of single- and multiconcept designs is in contrast to narrow concept design as sometimes done in traditional systems design. As a result of using a tradespace approach, you are more likely to be able to identify a larger number of value robust solutions than is possible with qualitative methods or traditional concept exploration methods alone. While the illustrative case used in this chapter is a simplified one, the approach-based insights are generalizable.

Quantitative modeling of SoS adds another layer of computational complexity to that of traditional systems modeling. Using a tradespace approach increases the breadth of evaluated alternatives. This allows you to focus on identifying patterns and relationships between factors within your control and factors outside of your control. You can see the impact of changing preferences and contexts and how PR of constituent systems relates to the value they deliver. Since the goal of TSE is not generating precise cost or performance estimates for a particular SoS design, the pressure to develop accurate models decreases. Finding the appropriate fidelity models for SoS TSE is an open area of research, and the reader is encouraged to iterate when possible, as the primary outcome of SoS TSE is knowledge creation.

REFERENCES

Chattopadhyay D. (2009). A method for tradespace exploration of systems of systems [master's thesis]. Cambridge (MA): Massachusetts Institute of Technology. Available from MIT libraries.

Chattopadhyay D, Ross AM, and Rhodes DH. (2008). A Framework for Tradespace Exploration of Systems of Systems. Proceedings of the 6th Conference on Systems Engineering Research, Redondo Beach, CA, April 4–5, 2008.

———. (2009). Demonstration of System of Systems Multi-Attribute Tradespace Exploration on a Multi-Concept Surveillance Architecture. Proceedings of the 7th Conference on Systems Engineering Research, Loughborough University, Loughborough, UK, April 20–23, 2009.

de Weck O. (2004). Multiobjective Optimization: History and Promise. Proceedings of the Third China-Japan-Korea Joint Symposium on Optimization of Structural and Mechanical Systems, Kanazawa, Japan, October 30–November 2, 2004.

Hall DL and Llinas J. (1997). An Introduction to Multisensor Data Fusion. *Proceedings of IEEE* 85(1): 6–23.

Imagery Resolution Assessments and Reporting Standards (IRARS) Committee. (1996). Civil NIIRS Reference Guide. Available at http://www.fas.org/irp/imint/niirs_c/guide.htm (accessed July 24, 2014).

Keating C, Rogers R, Unal R, Dryer D, Sousa-Poza A, Safford R, Peterson W, and Rabadi G. (2003). System of Systems Engineering. *Engineering Management Journal* 15 (3): 36–45.

Keeney RL. (1992). *Value-Focused Thinking—A Path to Creative Decisionmaking*. Cambridge, MA: Harvard University Press.

Keeney RL and Raiffa H. (1993). *Decisions with Multiple Objectives—Preferences and Value Tradeoffs*. 2nd ed. Cambridge, UK: Cambridge University Press.

Llinas J, Bowman C, Rogova G, Steinberg A, Waltz E, and White F. (2004). Revisiting the JDL Data Fusion Model II. Proceedings of the 7th International Conference on Information Fusion (FUSION 2004), Stockholm, Sweden, June 28–July 1, 2004.

Maier MW. (1998). Architecting Principles for Systems-of-Systems. *Systems Engineering* 1 (4): 267–284.

Richards MG, Ross AM, Shah NB, and Hastings DE. (2008). Metrics for Evaluating Survivability in Dynamic Multi-Attribute Tradespace Exploration. Proceedings of the AIAA Space 2008, San Diego, CA, September 9–11, 2008.

Ross, AM. (2003). Multi-attribute tradespace exploration with concurrent design as a value-centric framework for space system architecture and design [master's thesis]. Cambridge (MA): Massachusetts Institute of Technology. Available from MIT libraries.

———. (2006). Managing unarticulated value: Changeability in multi-attribute tradespace exploration [doctoral dissertation]. Cambridge (MA): Massachusetts Institute of Technology. Available from MIT libraries.

Ross AM and Hastings DE. (2005). The Tradespace Exploration Paradigm. Proceedings of the INCOSE International Symposium 2005, Rochester, NY, July 10–15, 2005.

Ross AM, Hastings DE, Warmkessel JM, and Diller NP. (2004). Multi-Attribute Tradespace Exploration as a Front-End for Effective Space System Design. *Journal of Spacecraft and Rockets* 41 (1): 20–28.

Ross AM and Rhodes DH. (2007). The System Shell as a Construct for Mitigating the Impact of Changing Contexts by Creating Opportunities for Value Robustness. Proceedings of the 1st IEEE International Systems Conference, Honolulu, HI, April 9–13, 2007.

———. (2008a). Using Attribute Classes to Uncover Latent Value During Conceptual Systems Design. Proceedings of the 2nd IEEE International Systems Conference, Montreal, Canada, April 7–10, 2008.

———. (2008b). Using Natural Value-Centric Time Scales for Conceptualizing System Timelines through Epoch-Era Analysis. Proceedings of the INCOSE International Symposium 2008, Utrecht, the Netherlands, June 15–19, 2008.

Ross AM, Rhodes DH, and Hastings DE. (2009). Using Pareto Trace to Determine System Passive Value Robustness. Proceedings of the 3rd IEEE International Systems Conference, Vancouver, Canada, March 23–26, 2009.

Ross AM, McManus HL, Rhodes DH, and Hastings DE. (2010a). Revisiting the Tradespace Exploration Paradigm: Structuring the Exploration Process. Proceedings of the AIAA Space 2010, Anaheim, CA, August 30–September 2, 2010.

———. (2010b). A Role for Interactive Tradespace Exploration in Multi-Stakeholder Negotiations. Proceedings of the AIAA Space 2010, Anaheim, CA, August 30–September 2, 2010.

Shah NB. (2013). Influence strategies for systems of systems [doctoral dissertation]. Cambridge (MA): Massachusetts Institute of Technology. Available from MIT libraries.

Wolf DR. (2009). An assessment of novice and expert users' decision-making strategies during visual trade space exploration [masters thesis]. University Park (PA): The Pennsylvania State University. Available from Penn State libraries.

Chapter 5

Data Policy Definition and Verification for System of Systems Governance

Daniele Gianni
European Organization for the Exploitation of Meteorological Data, Darmstadt, Germany

5.1 INTRODUCTION

The concept of *governance* was originally introduced in the corporate world, in which mechanisms were needed to assign responsibilities to the corporate roles (e.g., board of directors, managers, shareholders) and to provide the structures for setting and pursuing the corporate objectives (Goergen, 2012). The overall *governance* aim was to align the investment strategy, and more generally the decision-making processes, to the stakeholders' best interests, while ensuring the conformity with the societal and market regulations. The concept was subsequently specialized to the IT infrastructure, upon which modern enterprises are highly dependent, with the objective of ensuring that the IT systems could most effectively support the core corporate objectives and the stakeholders' interests (Weill and Ross, 2004). In the years, direct experience and common sense have driven the definition of a number of enterprise IT *governance* frameworks for establishing and administrating the IT infrastructure. However, these frameworks are based on two implicit key assumptions (Grembergen and De Haes, 2009): (i) the enterprise IT systems are under full and centralized control; (ii) the IT system architecture does not vary autonomously. Consequently, these frameworks cannot be immediately applied to information-intensive

---

Modeling and Simulation Support for System of Systems Engineering Applications, First Edition.
Edited by Larry B. Rainey and Andreas Tolk.
© 2015 John Wiley & Sons, Inc. Published 2015 by John Wiley & Sons, Inc.

SoS as the respective assumptions do not hold for SoS: a SoS is not subjected to centralized control and the SoS architecture can vary autonomously, depending on the participating systems (Maier, 1998). Therefore, SoS *governance* requires newly thought approaches that overcome the above assumptions and that can contribute to address the new challenges introduced by the inherent characteristics of SoS. In particular, in information-intensive SoS, the issues related to the data distribution and use are central to the SoS *governance* as these issues may hinder the delivery of the expected SoS services and the fulfillment of the planned capabilities. For example, systems managers may withdraw their systems from the SoS configuration, if their data distribution requirements are not satisfied and the trust on the requirement implementation is not gained.

In this chapter, we specifically focus on the *governance* aspects related to the data distribution and use policies—shortly data policies, with the definition and application of the *Data Policy Methodology* (Gianni et al., 2011a), a *model-based* methodology for supporting the SoS engineering activities related to data policies for information-intensive SoS. The methodology consists in three pillars: *data policy* definition, SoS functional design guidance, and SoS physical design verification. The methodology was defined in two phases, one related to the problem definition and one related to the problem solution.

The chapter is structured as follows. The section "Background" provides the foundational and relevant terminology for the methodology motivation and definition. The section "The Role of Data Policy Methodology in SoS" identifies the questions that a *data policy* must address. The section "Design of the Data Policy Methodology" describes the design of the *model-based* methodology. Finally, the section "Example Application" illustrates how the methodology can support Space Situational Awareness (SSA) programs, in which data sharing can be a critical, yet enabling, factor.

5.2 BACKGROUND

The methodology has been defined using the terminology and concepts from the UML language, *enterprise architecture (EA) frameworks*, SoS, and *governance*, and conceptual data modeling. However, the UML language and *EA frameworks* are of wide dissemination and therefore not discussed below.

5.2.1 IT Governance and SoS

The IT Governance Institute states that "overall objective of IT *governance* ... is to understand the issues and the strategic importance of IT, so that the enterprise can sustain its operations and implement the strategies required to extend its activities into the future. IT *governance* aims at ensuring that expectations for IT are met and IT risks are mitigated" (IT Governance Institute, 2003). Although other different definitions are available for IT *governance*, all the definitions share the common point that IT *governance* involves policies for the alignment of IT resources to business goals, control and coordination of IT resources, enforcement of those policies, and measurement of the outcome (Morris et al., 2006).

Currently, several frameworks are available for enterprise IT *governance* (Grembergen, and De Haes, 2009). These frameworks are inherently calibrated for enterprise contexts in which the control and the *governance* actuation are centralized—though stakeholders' influence may likely cross the enterprise boundaries. These frameworks generally consist of responsibility assignment, processes for eliciting the stakeholder needs and maintain/evolve the IT infrastructure, and maturity models for evaluating the alignment of the IT systems with the business goals.

However, existing frameworks are deemed to be unsuitable for SoS as SoS presents the following unique characteristics that distinguish these systems from conventional systems (Jamshidi, 2009):

- Operational Independence of Elements: if the SoS is disassembled into its component systems, the component systems must be able to usefully operate independently.
- Managerial Independence of Elements: the component systems not only can operate independently; they do operate independently.
- Evolutionary Development: the SoS does not appear fully formed; its development and existence are evolutionary with functions and purposes added, removed, or modified.
- Emergent Behavior: the system performs and carries out purposes that do not reside in any component system. The principal purposes of the SoS are fulfilled by these behaviors.
- Geographic Distribution: the geographic extent of the component systems is large. The components can readily exchange only information.

The above characteristics implicitly invalidate the assumptions upon which the existing enterprise IT *governance* frameworks are defined. Consequently, new approaches are needed to address the new challenges introduced by the above characteristics in the *governance* of SoS. In their previous work, Morris et al. (2006) identify the key characteristics that should be met by "good" *governance* models:

- Collaboration and authority: no individual participating organization has full authority on the entire SoS *governance*, and also if formal authority may exist in large programs, the authority might hardly understand the nuances involved in effective control.
- Motivation and accountability: the decentralized and collaborative nature of the SoS *governance* requires that the motivation of individual organization must be considered. Similarly, each organization must be held accountable for the agreed *governance* decisions to ensure an effective control.
- Multiple models: different level of *governance* control may need to be defined, depending on the life cycle phase. Consequently, models at different resolution and scope need to be provided.
- Expectation of evolution: the SoS *governance* must consider the evolution of individual systems and of the SoS configuration as a whole, defining rules and guidelines to ensure effective SoS *governance* while supporting the SoS evolution.
- Highly fluid processes: *governance* processes must deliver prompt responses to the SoS variations to ensure an effective SoS *governance*.
- Minimal centrality: the SoS *governance* should rely on minimal centrality, except for cases of dominant system and SoS infrastructure *governance*.

Furthermore, in information-intensive SoS, the SoS *governance* does not only cover the definition of communication and notification processes for governing the HW and SW resources. Indeed, a prominent role is taken by the policies for the distribution and use of the data, namely, *data policy*, as the data constitute the primary form of product of the services provided in and by the SoS configuration.

5.2.2 Conceptual Modeling

An information system can be represented at four distinct levels: conceptual, logical, physical, and external (Halpin and Morgan, 2008). The conceptual level provides the capabilities for information representation in the most natural format to humans, who can draw the business model and validate against their understanding. Differently, the logical level concerns the representation of the conceptual model in a modeling paradigm, such as entity relationships, object oriented, or hierarchical. The physical level provides the technology-specific capabilities for the actual information system implementation (e.g., MySQL, Oracle, IBM DB2, UML, and XML). Finally, the external level is a variation of the conceptual level in which the view on the entire business domain is restricted, depending on the privileges assigned to the respective actor.

The most prominent conceptual data modeling language is the Object Role Modeling (ORM) language, which is built on formal logical foundation and is used for modeling and querying information systems at conceptual level (Halpin and Morgan, 2008). The ORM name originates by two of the fundamental language primitives: Object and Roles. Objects represent entities or value; and roles represent the involvement of Objects in relationships. Compared to logical modeling languages, such as ER or UML, the ORM language has been proved to bring considerable advantages, for its formal definition, graphical modeling, and verbalization capability. Moreover, the implementation independence contributes to raise the modeling level by hiding irrelevant implementation details for a system data specification phase (Melli and McQuinn, 2008), therefore offering semantic stability to ORM models. Moreover, the verbalization capabilities can be used to potentially identify the conceptual model by listing all the true natural language statements concerning the Universe of Discourse (UoD), which identifies the business domain to be modeled. With the objective of supporting the modeling activity, the ORM language is also provided with the Conceptual Schema Design Procedure (CSDP), a modeling process consisting of seven steps: from the transformation of familiar UoD examples into elementary true statements, to the drawing of the Fact Types and all the constraints, to finally reach the final validation checks.

Currently, several tools support the development of ORM modeling, for example, the NORMA plug in for Visual Studio or the FAMOUS environment (Lemmens et al., 2009), each offering diverse sets of functions for model verbalization, model verification, and ER, UML, and XSD derivations from ORM schemas.

The ORM language defines the following primitives:

- Object Type, which classify business objects in Value Type and Entity Type.
- Value Type, which represents a simple Object that can be expressed with a lexical expression; Entity Type, which represent an Object that cannot be expressed with a lexical expression.

- Role, which represent an Object Type involvement in a relationship.
- Fact Type, which classifies facts and involves Objects and Roles through a predicate.
- Constraints, which limit the possible instances of Object Types and Fact Types. Constraints can be either internal or external.
  - Internal constraints are Mandatory Constraint, Internal Uniqueness Constraint, and Value Constraint.
  - External constraints are External Uniqueness, Equality Constraint, Exclusion Constraint, Inclusive Or Constraint, Exclusive Or Constraint, Subset Constraint, Frequency Constraint, and a variety of Ring Constraints.

As general interpretation key, each of the above constraints applies to one or more Roles, imposing logical conditions. For example, the Exclusive Or Constraints identifies XOR relationships between the Objects participating in the Roles linked through the respective constraint symbol.

All these concepts are auto-explicative, and interested readers are invited to refer to Halpin's book for a thorough presentation. A brief summary of the graphical notation for the symbols used in this chapter is available from the URL (http://www.orm.net/pdf/ORM2_TechReport1.pdf).

5.3 ROLE OF A DATA POLICY METHODOLOGY IN THE SOS GOVERNANCE

A Data Policy Methodology must support the definition of individual *data policy* models that meet the above Morris' recommendations for SoS *governance* models:

- Collaboration and authority: the model is inherently collaborative as it requires the sharing of data distribution and use requirements as well as the design models of the SoS infrastructure and constituting systems.
- Motivation and accountability: the model aims to strengthen the motivation in the participation in the SoS configuration and relies on the implementation of the architectural design models.
- Expectation of evolution: the model inherently supports the evolutions of the *data policy* requirements, introducing traceability links from the *data policy* to the functional and physical architectural models.
- Highly fluid processes: the model can support prompt response to SoS variations, though human intervention and data monitoring infrastructure need to be implemented beyond the model definition, on the actual SoS.
- Minimal centrality: the model does not rely on any centrality as the model provides the means for expressing the data distribution and use requirements and to reassure partners on their consideration in the SoS design.

Differently, the Morris' multimodel recommendation (i.e., the availability of models in different scope and resolution) does not necessarily need to be addressed in a first release of a Data Policy Methodology, and this recommendation can be therefore temporarily omitted.

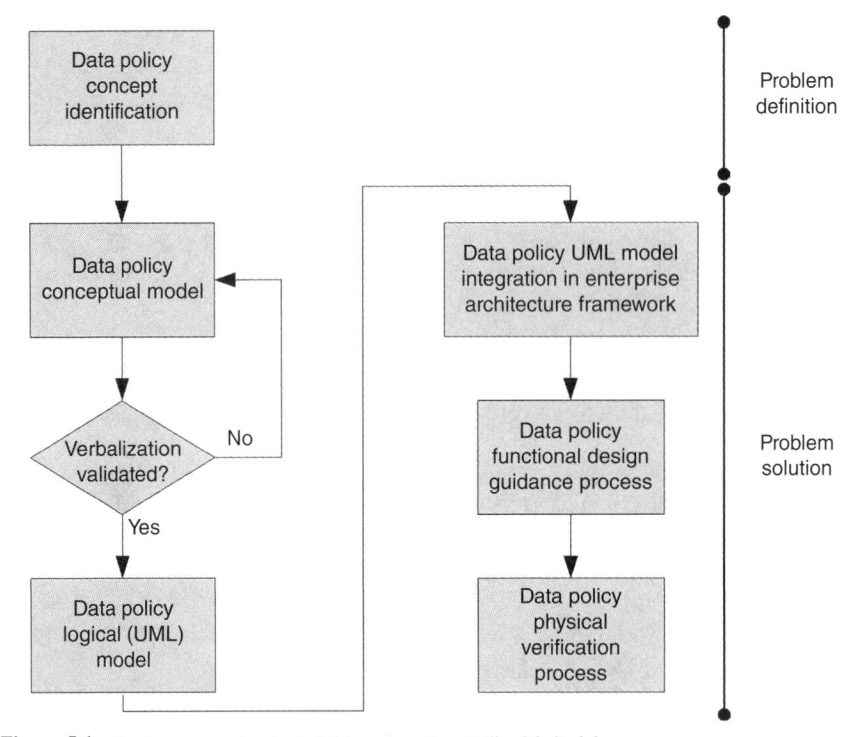

Figure 5.1 Design process for the definition of our Data Policy Methodology.

A Data Policy Methodology can be defined in two phases: problem definition and problem solution. The problem definition phase concerns the role of the *data policy* in the SoS *governance*, addressing the identification of the *data policy* concept. Differently, the problem solution phase concerns the design of the *Data Policy Methodology*. Particularly, Figure 5.1 shows the flow-chart-like diagram representing all the steps undertaken in these two phases. As part of the problem definition phase, the data policy concept identification is described in this background section to lay the problem understanding for the methodology definition.

5.3.1 Data Policy Concept Identification

A *data policy* can be defined as "*an agreed set of rules that regulates the production, use and dissemination of data*" (Gianni et al., 2011a). As such, a *data policy* must accurately define the details of the actors involved in the respective scenarios, the data types, the purpose, the modalities with which the purpose can be achieved, and the data provisioning modalities, for example. Using the five Ws, the concept of *data policy* can be more accurately defined by identifying the set of questions that a *data policy* must answer. This can be obtained by associating the possible concerns to each of the five Ws, as shown in Figure 5.2.

For an accurate and unambiguous *data policy* specification, the above questions must be formulated so to address only one concern. Consequently, the answers can be

Data Policy Definition and Verification for System of Systems Governance   105

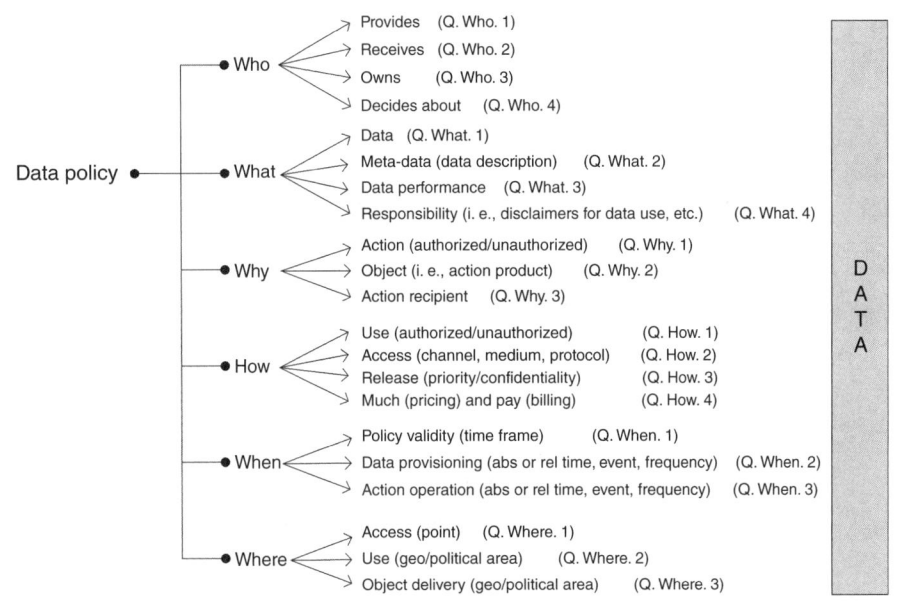

```
                        ┌→ Provides    (Q. Who. 1)
              ┌─ Who ───┼→ Receives    (Q. Who. 2)
              │         ├→ Owns        (Q. Who. 3)
              │         └→ Decides about  (Q. Who. 4)
              │
              │         ┌→ Data        (Q. What. 1)
              ├─ What ──┼→ Meta-data (data description)  (Q. What. 2)
Data policy ──┤         ├→ Data performance  (Q. What. 3)
              │         └→ Responsibility (i. e., disclaimers for data use, etc.)  (Q. What. 4)
              │
              │         ┌→ Action (authorized/unauthorized)  (Q. Why. 1)
              ├─ Why ───┼→ Object (i. e., action product)    (Q. Why. 2)
              │         └→ Action recipient   (Q. Why. 3)
              │
              │         ┌→ Use (authorized/unauthorized)     (Q. How. 1)
              ├─ How ───┼→ Access (channel, medium, protocol) (Q. How. 2)
              │         ├→ Release (priority/confidentiality) (Q. How. 3)
              │         └→ Much (pricing) and pay (billing)   (Q. How. 4)
              │
              │         ┌→ Policy validity (time frame)   (Q. When. 1)
              ├─ When ──┼→ Data provisioning (abs or rel time, event, frequency)   (Q. When. 2)
              │         └→ Action operation (abs or rel time, event, frequency)    (Q. When. 3)
              │
              │         ┌→ Access (point)   (Q. Where. 1)
              └─ Where ─┼→ Use (geo/political area)   (Q. Where. 2)
                        └→ Object delivery (geo/political area)   (Q. Where. 3)
```

DATA

Figure 5.2 Questions for the identification of the data policy concept (Gianni et al., 2011a).

expressed in a simple form consisting of a logical formula of predicates concerning combinations of objects and/or object properties, which regard all the same subject. As a result, the above wh-questions might be further refined to more closely guide *data policy* makers in the *data policy* definition.

Each group of questions addresses a common concern. The purpose of the who group is to identify all the relevant agreement stakeholders. This group defines the following questions:

- Who provides the data?
- Who owns the data?
- Who decides about the data? (and what can it decide?)

Analogously, the purpose of the what group is to specify what the agreement is about. This group defines the following questions:

- What data is the *data policy* about?
- What metadata will be provided within this *data policy*?
- What are the data performance (i.e., quality)?
- What are the data provider responsibilities (i.e., claimers and disclaimers)?

Following the why group, whose purpose is to identify the objective of the agreement, specifies the final purpose of the data provisioning. This group concerns the following question:

- Why is the agreement being defined?

However, this question can be answered only by a complex answer consisting of an action (which the agreement recipient is authorized to perform) and an action recipient (which is the actor to whom the agreement recipient can deliver the action output). As a consequence, the following two simple questions are defined in this group:

- Which are the authorized actions on the data? (n.b.: for exclusion, all the unspecified authorized actions will be unauthorized.)
- Which are the authorized outputs that can be obtained by the data, by means of the above actions? (n.b.: for exclusion, all the unspecified authorized output will be unauthorized.)
- Which actors can be the recipient of authorized action output? (n.b.: for exclusions, all the unspecified recipient will be unauthorized.)

Next, the how group aims to identify the delivery modalities. These include the data provisioning modalities, which concern agreement holder and agreement recipient, and the data use modalities, which concern agreement recipient and action recipient (with the action being defined by the above question group).

Data provisioning modalities can be defined by answering the following questions:

- How can the data be accessed? (i.e., which are the requirements for the access point, the communication channel, the communication protocol?)
- How will data be released? (i.e., at what confidential and priority level?)
- How much will the data provisioning cost?

Data use modalities can be defined by answering the following questions:

- How can the data be used to achieve the authorized action?
- How cannot the data be used to achieve the authorized action?
- How often (when) the action can be operated?

Next, the when group has the purpose of identifying time aspects related to the time in a *data policy* definition. These aspects concern the *data policy*, the data flow between agreement holder and recipient, and the data flow between agreement recipient and action recipient.

Concerning the *data policy*, the following question is defined:

1. When is the *data policy* agreement valid?

Concerning agreement holder-recipient data flow, the following question is defined:

2. When are the data to be delivered to the recipient? (Absolute time/relative time/frequency/event.)

Concerning agreement recipient-action recipient data flow, the following question is defined:

3. When can the action be operated and the result delivered to the recipient? (Absolute time/relative time/frequency/event.)

Finally, the where group can be used to specify physical/geographical aspects of the *data policy*. In this group, aspects concerning data access, data use, and action output (i.e., product) are addressed.

Concerning the data access, the following question is defined:
- Where can the data be accessed?

Concerning the data use, the following question is defined:
- Where can the data be used?

Concerning the action product delivery, the following question is defined:
- Where can the action product be delivered?

A last observation regards the ways each of the above questions can be answered: (i) explicitly mentioning the actual objects or individuals and (ii) describing the properties that objects or individuals must satisfy. For example, the *data policy* question "Who receives data<X>?" can be answered with, for example:
- "Italian Ministry of Defense (MoD)"

or
- A public military institution of a nation of the European Union

The *data policy* definition is only the most evident of the stakeholders' concerns, which pertain to the assurance to the actual implementation of their *data policy* specification. Consequently, the SoS design should also provide the means to ensure that the *data policy* can be fully incorporated into the SoS engineering activities. Intuitively, the reader may immediately observe the analogy between the above question groups and the dimensions in the Zachman framework (Zachman, 1987). From this observation, it becomes possible to imagine the relationship between a *data policy* model, as part of the stakeholders' requirements, and the EA model, as part of the SoS design output.

5.4 DESIGN OF THE DATA POLICY METHODOLOGY

The methodology objective is to provide a *model-based* mean to specify *data policy* requirements and to integrate these requirements in *model-based* SoS engineering activities. The *Data Policy Methodology* achieves this in three points. Firstly, the methodology defines the modeling structures to support *data policy* definitions. Secondly, the methodology defines process models to enable SoS engineers to verify *data policy* against the architectural model. Thirdly, with the objective of reducing rework efforts in the SoS engineering activities, the *Data Policy Methodology* defines process models to inject the *data policy* requirements into the SoS functional design. For communicating the widest insight on the methodology idea and definition, we have decided to present the data methodology through all the steps undertaken in its design. Although the methodology objectives are distinct and well identified, the methodology design has not similarly proceeded linearly as the above three points are highly coupled and the entire methodology definition was an explorative activity. Consequently, three concurrent and interdependent activities were undertaken after the initial identification of the *data policy* concept, and several iterations were performed to identify possible "answer values" that could be used in preliminary (and unstructured) answers for the *data policy* questions. Particularly, the activities were Requirements Classification, Relevant Environment Classification, and *data policy* definition—as shown in Figure 5.3.

The Requirements Classification aimed at identifying the main actors, their properties, and their expected operations. Similarly, the Relevant Environment Classification

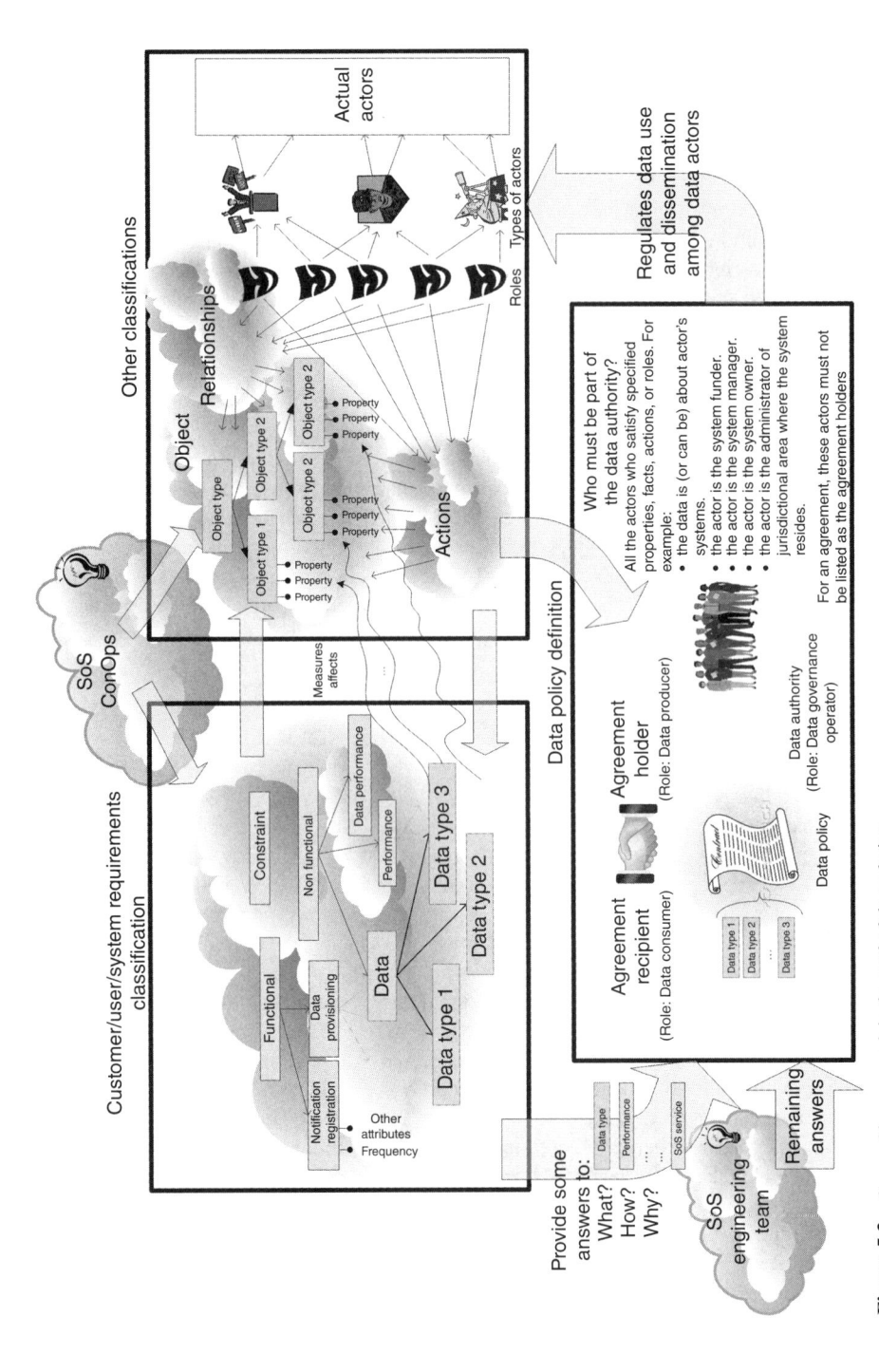

Figure 5.3 Overarching approach in the methodology design.

aimed at a similar objective but in the scope of the external environment in which the SoS is supposed to live and operate. Both activities can receive input from the SoS concept of operations (ConOps) documents and the preliminary requirements documents, when available. Finally, using the above question schema, the third activity aimed at identifying extensional and intentional responses that could be provided in a *data policy* definition. While most of the activity was based on the iteratively refinement of the output from the former two activities, this activity can also require direct feedback from the SoS engineering team, when the above documents are insufficient to support accurate and reliable analysis.

Although apparently different, the above activities share a fundamental analysis approach that was based on the following tasks:

- Roles Classification: identification and categorization of the possible roles within the expected scenarios described in the concepts of operations document.
- Systems Classification: identification and categorization of the possible systems within the expected scenarios described in the concepts of operations document.
- Actions Definition and Classification: allocation of the above roles on the actors identified in the ConOps document. Classification of the actors depending on nonfunctional properties.
- Types of Actors Definition: abstraction of the identified actors in groups characterizes by relevant shared values of the nonfunctional properties.
- Types of Scenario Classification: grouping of the scenarios identified in the ConOps document by the involvement of the identified Types of Actors.
- Types of Scenario Definition: analysis and review of the scenarios identified in the ConOps document.

Differently from the activities, the tasks were undertaken iteratively, in the above shown order. Moreover, intrinsic synchronizations and data exchanges were implemented during the tasks iteration within the three activities, as partially shown by the arrows in the above diagram.

5.4.1 Model-Based Data Policy Methodology

The *model-based Data Policy Methodology* has been designed in three steps: (1) structuring the above identified set of answers according to a controlled natural language form—standard approach that is purposely omitted from this manuscript, (2) conceptual data model definition from controlled natural language of the identified answers, and (3) derivation of a UML logical model from the defined *data policy* conceptual model.

5.4.1.1 Data Policy Conceptual Model

The *data policy* model has been initially drafted applying iteratively the CSDP process of the ORM methodology to formalize the natural language model encoded in the questions in Figure 5.2. In particular, the data population for the *data policy* schema was identified by listing the possible responses that could be provided for the questions in the section

"Data Policy Concept Identification." However, these responses were iteratively improved so to provide a general structure that could most closely conform to the one obtained for the verbalization of ORM schemas.

Using the capabilities of the existing software tools, the conceptual data model could be easily inferred by the listed set of data policies, and a part of the model is shown in Figures 5.4 and 5.5—these diagrams purposely omit all the constraints related to Object Types participation in Roles, for presentation sake.

After defining the conceptual data model, ORM-controlled natural language description (a.k.a. textual verbalization) can be derived to validate the *data policy* conceptual schema against the above questions and the general understanding of the *data policy* concept. Specifically, Figures 5.4 and 5.5 can be verbalized using the standard ORM/NORMA coloring schema,[1] as shown below:

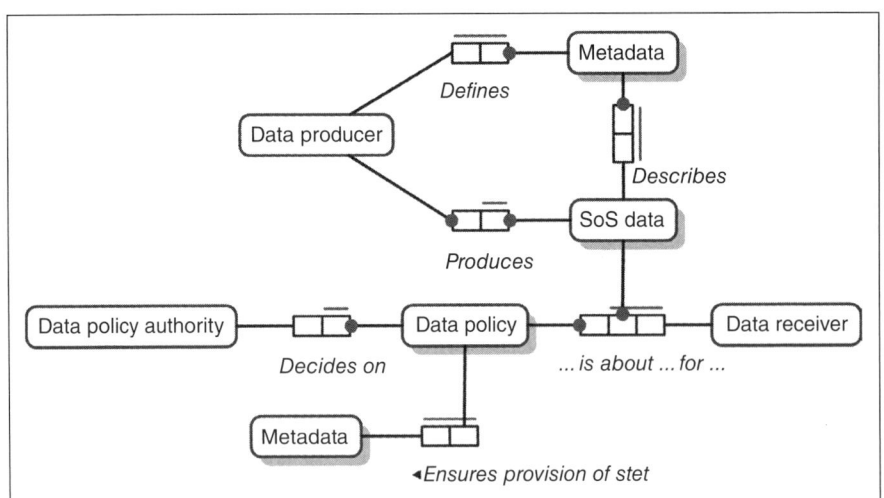

Figure 5.4 Excerpt of the data policy conceptual model—core elements.

Data Policy *is about* SoS Data *for* Data Receiver.
Each Data Policy *is about* **some** SoS Data *for* **some** Data Receiver.
For each SoS Data,
some Data Policy *is about* **that** SoS Data *for* **some** Data Receiver.
For each Data Receiver **and** SoS Data,
at most one Data Policy *is about* **that** SoS Data *for* **that** Data Receiver.
This association with Data Receiver, SoS Data **provides the preferred identification scheme for** DataPolicyIsAboutSoSDataForDataReceiver.
Data Policy *ensures provision of* Metadata.
It is possible that more than one Data Policy *ensures provision of* **the same** Metadata
and that the same Data Policy *ensures provision of* **more than one** Metadata.

---

[1] The coloring schema is **<ORM reserved Word>**; <Object Type>; and <*Predicate constituent*>

In each population of Data Policy *ensures provision of* Metadata, **each** Data Policy, Metadata **combination occurs at most once.**
This association with Data Policy, Metadata **provides the preferred identification scheme for** DataPolicyEnsuresProvisionOfMetadata.
Data Policy Authority *decides on* Data Policy.
For each Data Policy, **exactly one** Data Policy Authority *decides on* **that** Data Policy.
It is possible that the same Data Policy Authority *decides on* **more than one** Data Policy.
Data Producer *defines* Metadata.
It is possible that more than one Data Producer *defines* **the same** Metadata **and that the same** Data Producer *defines* **more than one** Metadata.
In each population of Data Producer *defines* Metadata, **each** Data Producer, Metadata **combination occurs at most once.**
This association with Data Producer, Metadata **provides the preferred identification scheme for** DataProducerDefinesMetadata.
For each Metadata, **some** Data Producer *defines* **that** Metadata.
Data Producer *produces* SoS Data.
Each Data Producer *produces* **some** SoS Data.
For each SoS Data, **exactly one** Data Producer *produces* **that** SoS Data.
It is possible that the same Data Producer *produces* **more than one** SoS Data.
Metadata *describes* SoS Data.
It is possible that more than one Metadata *describes* **the same** SoS Data **and that the same** Metadata *describes* **more than one** SoS Data.
In each population of Metadata *describes* SoS Data, **each** SoS Data, Metadata **combination occurs at most once.**
This association with SoS Data, Metadata **provides the preferred identification scheme for** MetadataDescribesSoSData.
Each Metadata *describes* **some** SoS Data.

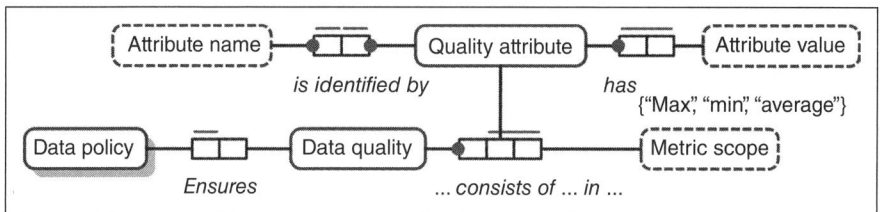

Figure 5.5 Excerpt of the data policy conceptual model—data quality.

Data Policy *ensures* Data Quality.
Each Data Policy *ensures* **at most one** Data Quality.
It is possible that more than one Data Policy *ensures* **the same** Data Quality.
Data Quality *consists of* Quality Attribute *in* Metric Scope.
Each Data Quality *consists of* **some** Quality Attribute *in* **some** Metric Scope.
For each Quality Attribute **and** Metric Scope,

> **at most one** Data Quality *consists of* **that** Quality Attribute *in* **that** Metric Scope.
> **This association with** Quality Attribute, Metric Scope **provides the preferred identification scheme for** DataQualityConsistsOfQualityAttributeInMetricScope.
> Attribute Name *is identified by* Quality Attribute.
> **Each** Attribute Name *is identified by* **exactly one** Quality Attribute.
> **For each** Quality Attribute, **exactly one** Attribute Name *is identified by* **that** Quality Attribute.
> Quality Attribute *has* Attribute Value.
> **It is possible that more than one** Quality Attribute *has* **the same** Attribute Value **and that the same** Quality Attribute *has* **more than one** Attribute Value.
> **In each population of** Quality Attribute *has* Attribute Value, **each** Attribute Value, Quality Attribute **combination occurs at most once**.
> **This association with** Attribute Value, Quality Attribute **provides the preferred identification scheme for** QualityAttributeHasAttributeValue.
> **Each** Quality Attribute *has* **some** Attribute Value.

5.4.1.2 Data Policy UML Logical Model

Once validated, the conceptual data model has been converted into a logical model based on the UML language. Although the conceptual data model offers business stakeholders a viable way for bridging the gap between the *data policy* natural language model and the model validation, the conceptual data model is not immediately suitable for integration in *model-based* SoS engineering activities. These activities are primarily developed using *EA frameworks*—such as DoDAF, MODAF (Bailey, 2008), or ESA-AF (Gianni et al., 2011c)—which are based on UML- and XML-based technologies. Consequently, the conceptual model was translated into a logical and physical model (UML-based object oriented), which was iteratively refined to better integrate the logical model into the standard design conventions adopted in common enterprise architectural frameworks. However, the refinement process did only rely on standard model-driven and information engineering practices, such as the mapping between concepts to classes or class attributes, and therefore, it is not further discussed in this chapter. Nevertheless, we provide an insight of the resulting UML *data policy* model with the excerpt in Figure 5.6—which represents the respective UML part for the conceptual model in Figures 5.4 and 5.5.

5.4.1.3 Model-Based Data Policy Verification

The *data policy* verification is inherently established by the link between by above questions and the possible answers. The possible answers are exemplified among the possible values that could be experienced on the actual systems. As such, these values reflect the properties that the systems must present. Consequently, the *model-based data policy* verification simply becomes the process of checking whether the responses—in the form of object type or object instance—are met by the actual instances that handle the data in the physical design.

The verification methodology has been defined using the BPMN standard (http://www.bpmn.org). In particular, Figure 5.7 shows the overarching process that, given a data item, guides the SoS engineers to verify the SoS physical design against the data policies.

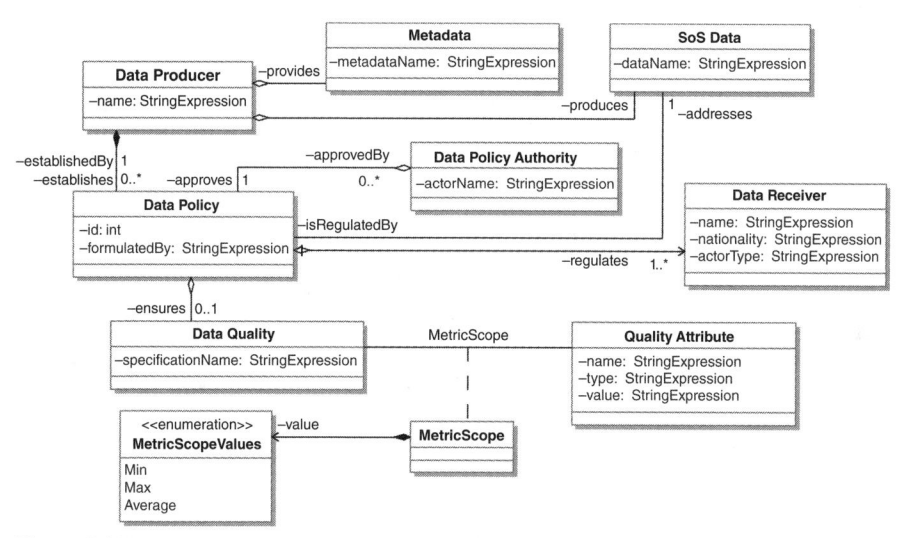

Data Policy Definition and Verification for System of Systems Governance   113

Figure 5.6 Logical–physical data policy model.

After retrieving all the data policies associated to the given data item, the overarching process continues in two distinct branches: one for the verification of the individual systems design and one for the verification of the network design. In turn, each branch is structured into two macro tasks: (1) the identification of all the elements (systems or networks) that become in contact with the data item and (2) the verification subprocesses for the identified elements. The first macro task needs no further detailing as it can be immediately carried out either manually—visual inspection on the EA model—or automatically—model processing using business intelligence or reporting software. Differently, the second task has been detailed as it encapsulates the central verification know-how. Figure 5.8 shows the verification subprocesses for the system and network elements.

However, these two verification subprocesses are to be considered exemplificative models as they only address the verification of two key properties—receiving system (Q.Who.2) and authorized action (Q.Why.1)—and leave undefined the verification of the remaining *data policy* requirements related to the other questions. Nevertheless, more comprehensive processes can be easily defined by further detailing the task "Verify Remaining Questions" for the remaining *data policy* requirements.

Although the above *model-based* approach can already bring considerable benefits to SoS engineering activities, the approach is inherently based on the static properties, which are explicitly embedded into the design models. These properties regard SoS characteristics that do not vary in the specific course of development of a scenario, such as system classification or network type. Differently, dynamic properties—for example, response time in a particular scenario, data distribution delay—might involve aspects related to the specific course of the scenario development. As these properties are not embedded in EA models, the design and verification of these properties cannot be supported by extension of the above processes. However, different models and tools are available to support behavioral and statistical representations of the relevant phenomenon and to perform simulation-based analysis and verification. Nevertheless, these models

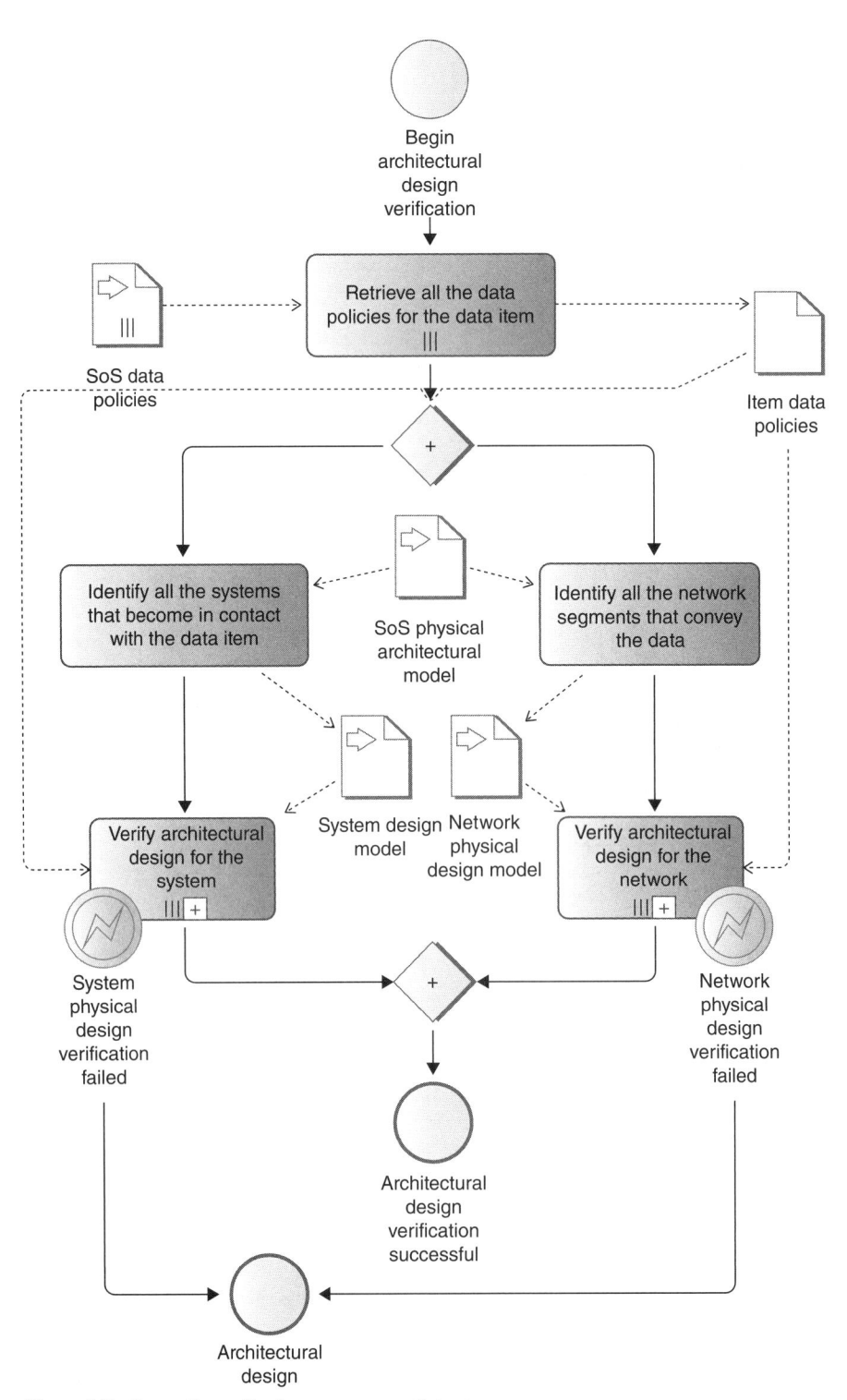

Figure 5.7 Data policy verification process—overall structure.

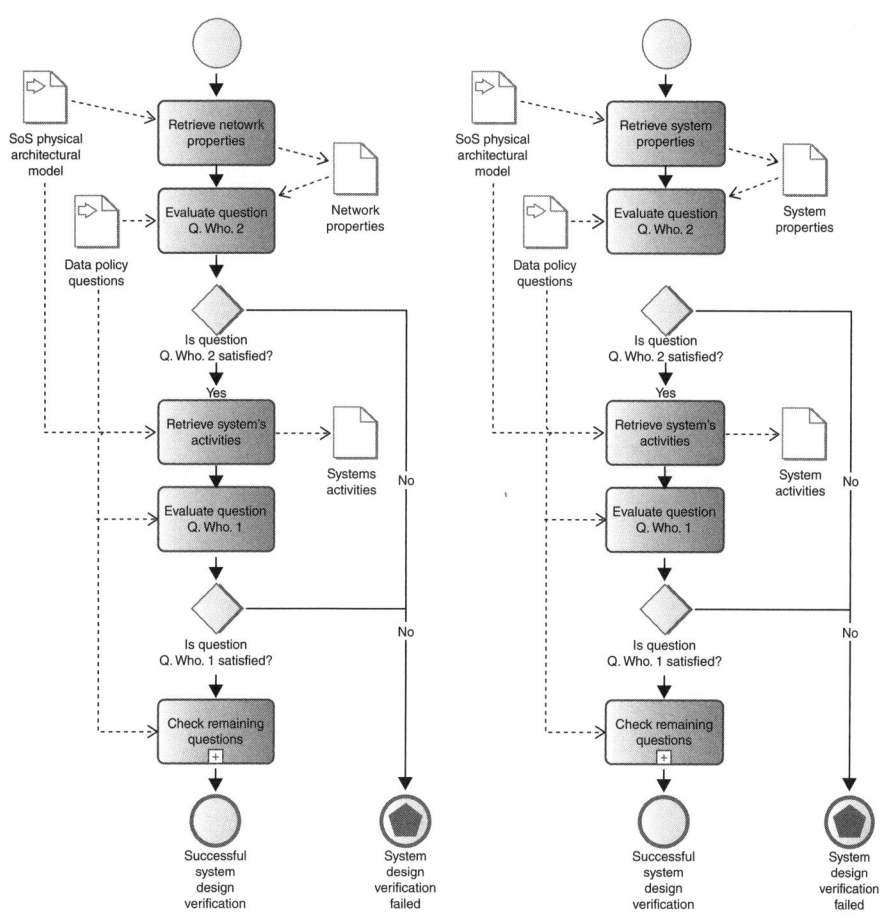

Figure 5.8 System and network segments verification subprocesses.

and tools are often integrated or linked to the architectural ones for providing a comprehensive and integrated approach of the entire SoS design and verification. In contemporary literature, two relevant key areas for the dynamic properties are system performance engineering—for the quality of data from sensing systems or for the performance of telecommunication networks (Smith, 1990)—and executable architectures (Mercer, 2008). In these areas, *model-based* approaches are widely available in both the M&S and Systems Engineering communities.

5.4.2 Integration in SoS Engineering Activities

A *data policy* encapsulates a critical part of the requirements for the *governance* of an SoS, along with front-end user requirements and customer requirements (if available). Consequently, the integration of the *data policy* definition within the SoS engineering

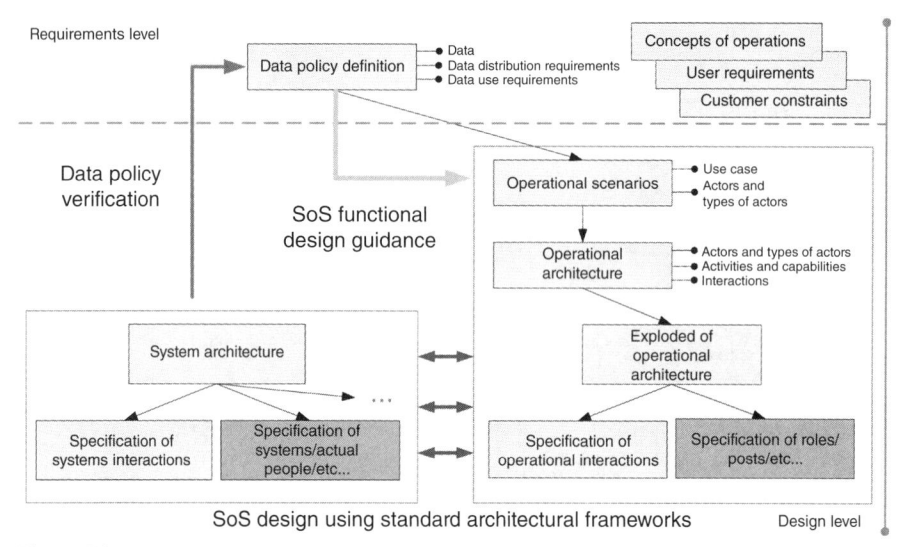

116 Theoretical and Methodological Considerations

Figure 5.9 Integration of *DPM* in SoS engineering based on standard architectural frameworks (Gianni et al., 2011a, 2011b).

process can bring both political and technical strategic advantages to the SoS design and operation. The political strategic advantages can be gained by ensuring the involvement of the stakeholders' needs and concerns. In particular, these advantages can be summarized in four points: (1) their requirements are fully and transparently considered from the beginning; (2) the SoS engineering proceeds effectively and efficaciously since the beginning; (3) integrated and prompt SoS design adaptation upon *data policy* evolution; and (4) preventive impact analysis on the variation of *data policy* requirements. Differently, the technical strategic advantages originate from a reduced number of rework activities and a reduced effort in the project with respect to a blind omission of the *data policy* requirements from the design, in the first place. As the *data policy* requirements can contribute to drive the SoS engineering activities, the probability of unsuccessful verification of the physical architectural design can be sensibly lower than the one when the *data policy* requirements are not fully integrated into the SoS engineering cycle.

To address this need, *the DPM* methodology can be fully integrated with architectural frameworks as shown in Figure 5.9. Specifically, the diagram shows that the *data policy* definitions become integral part of the requirements guiding the SoS functional design. The functional design will produce the functional architecture, which considers the *data policy requirements*, the customer and user requirements, and the ConOps. The functional architecture becomes the base for the SoS physical design, using standard Systems Engineering and SoS methods (also presented in this book). Moreover, the physical design will map the functional elements to the available (or procured) systems, considering also the programmatic aspects and other constraints (Forsberg et al., 2010). Finally, the SoS engineering loop is closed by the *DPM* verification methods that enable SoS engineers to verify that the physical design satisfies the *data policy* requirements.

Table 5.1 Mapping Outline Between the Data Policy Concepts and the UPDM-Related Concepts

Data policy concept	UPDM-related concept	Type of relationship (DP UML model→UPDM)
System	System/operational node	Equality
System property (any)	Class attribute (a.k.a. tagged value) in system/operational activity	Equality
Activity	Operational activity (for design guiding)	Equality
Activity	Function (for design verification)	Implements
Network requirements (any)	Class attribute in operational flow	Equality
Data item	Information asset (for design guiding)/ message (for verification)	Equality
Data item property	Relationship to EA element, for properties that can be immediately related to elements in the EA framework (e.g., system, operational node) Information item property (for design guiding)/message property (for verification), for all the other properties	Equality

To achieve an integration of the methodology in SoS engineering, the following points must be addressed:

- Mapping between *data policy* UML model and the adopted Enterprise Framework UML Models (e.g., *UPDM* UML Profile; Hause 2010)
- Definition of a *data policy* functional design guidance process to inject *data policy* requirements into the functional architecture

5.4.2.1 Data Policy Mapping to EA Frameworks

Using the *data policy* identification and conceptual data model, the *data policy* UML model can be easily mapped on the elements of your custom *EA framework*. For *UPDM* based, a general mapping is outlined in Table 5.1.

5.4.2.2 Data Policy Functional Guidance Process

Figure 5.10 shows the process for guiding the design of the SoS functional architecture. Similarly to the verification process, the functional guidance process only considers a subset of the *data policy* questions identified in the section "Data Policy Concept Identification." However, the functional guidance process has a different input scope with respect to the verification process one: the functional guidance process must be applied individually to all the data items in the defined *data policy* models.

The process model is defined in terms of three subprocesses:

1. Provide guidance on the node design, which is shown in Figure 5.11.
2. Provide guidance on the operational flow design, which is shown in Figure 5.12.

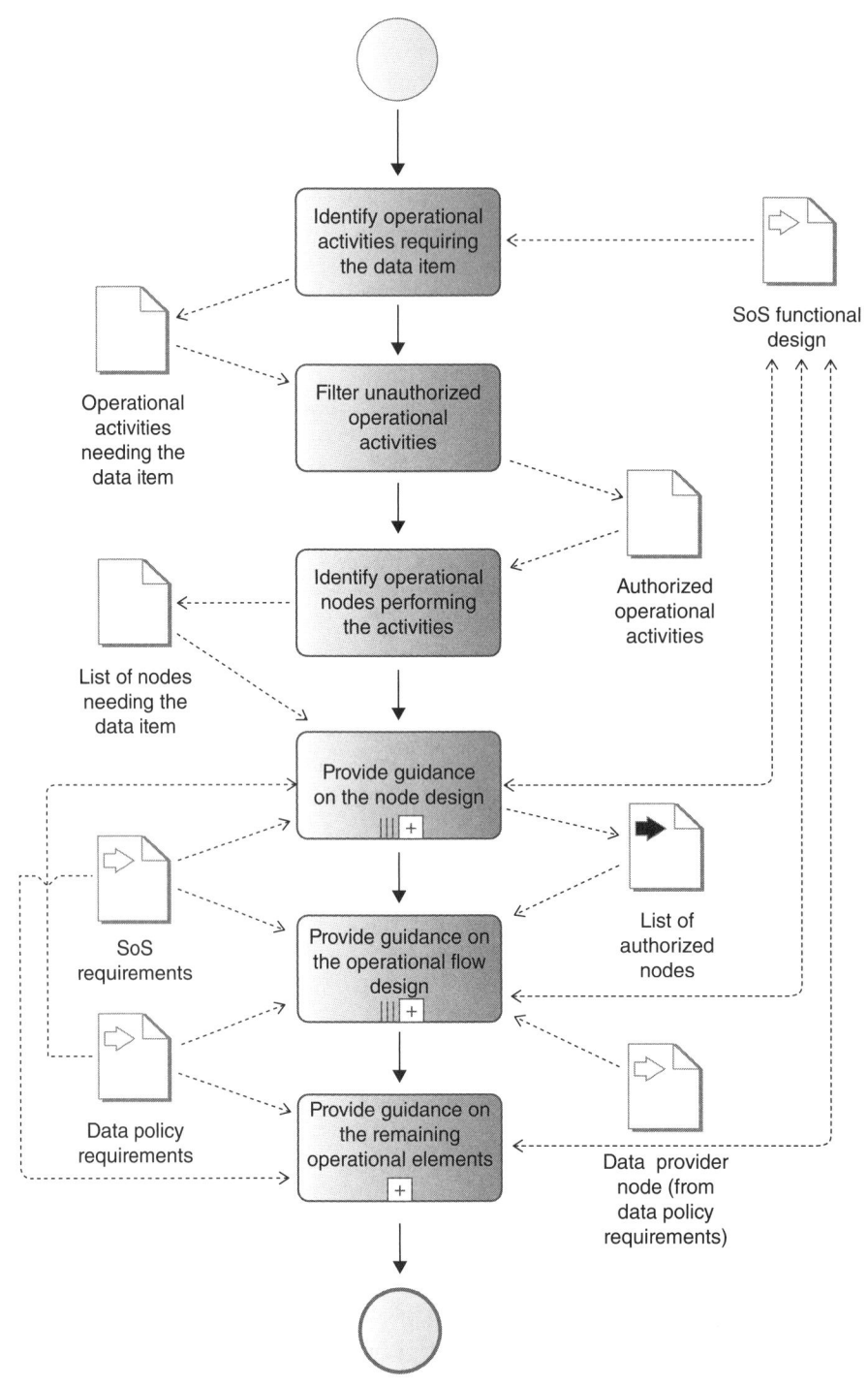

118 Theoretical and Methodological Considerations

Figure 5.10 Data policy functional guidance process.

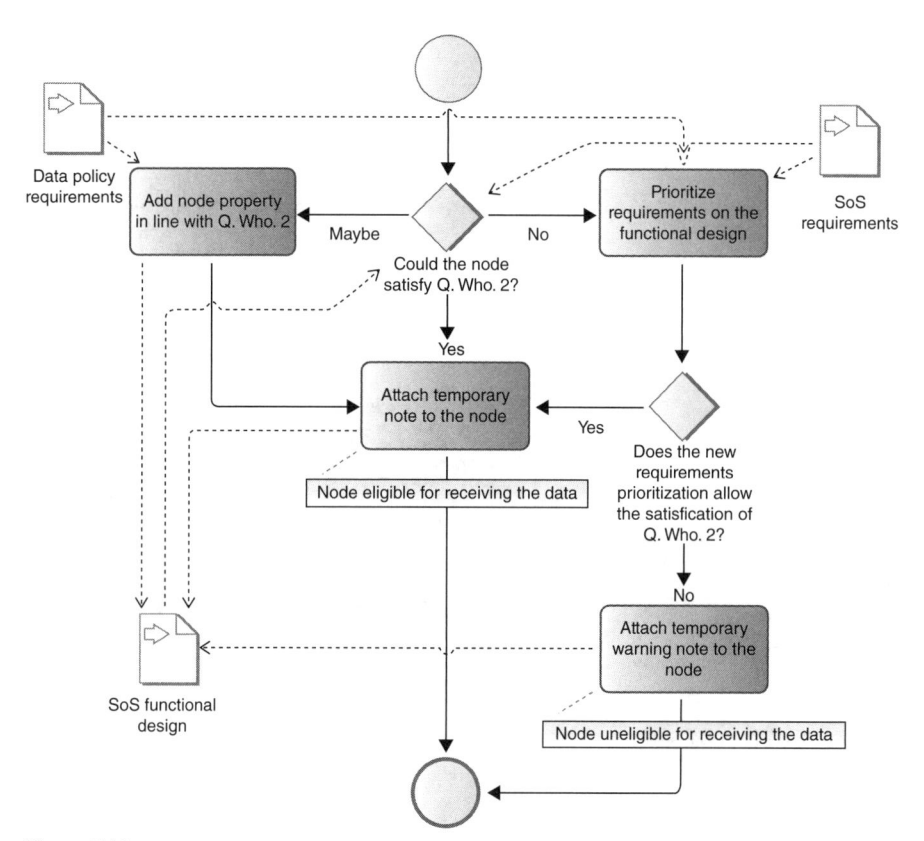

Data Policy Definition and Verification for System of Systems Governance   119

Figure 5.11   Provide guidance on the operational node design.

3. Provide guidance on the remaining operational elements, which can be defined to cover all the *EA framework*-specific elements and the remaining *data policy* questions.

5.5 EXAMPLE APPLICATION

Part of the presented *Data Policy Methodology* was originally designed in the context of an experimental activity (Gianni, 2010), for the European Space Agency (ESA)'s SSA preparatory program (http://www.esa.int/Our_Activities/Operations/Space_Situational_Awareness) (Bobrinsky and Del Monte, 2010), for which other homonymous projects exist (e.g., NASA SSA). These programs aim to establish sensing and forecasting capabilities that can provide information on the current and future status of the space environment surrounding the Earth. These programs typically cover three distinct areas:

1. Near-Earth Objects: this area concerns asteroids and objects that may fall on the Earth.
2. Space Weather: this area concerns the electromagnetic storms originating from solar eruptions and affecting space- and ground-based critical infrastructures, such as navigation systems, power grids, and telecommunication systems.

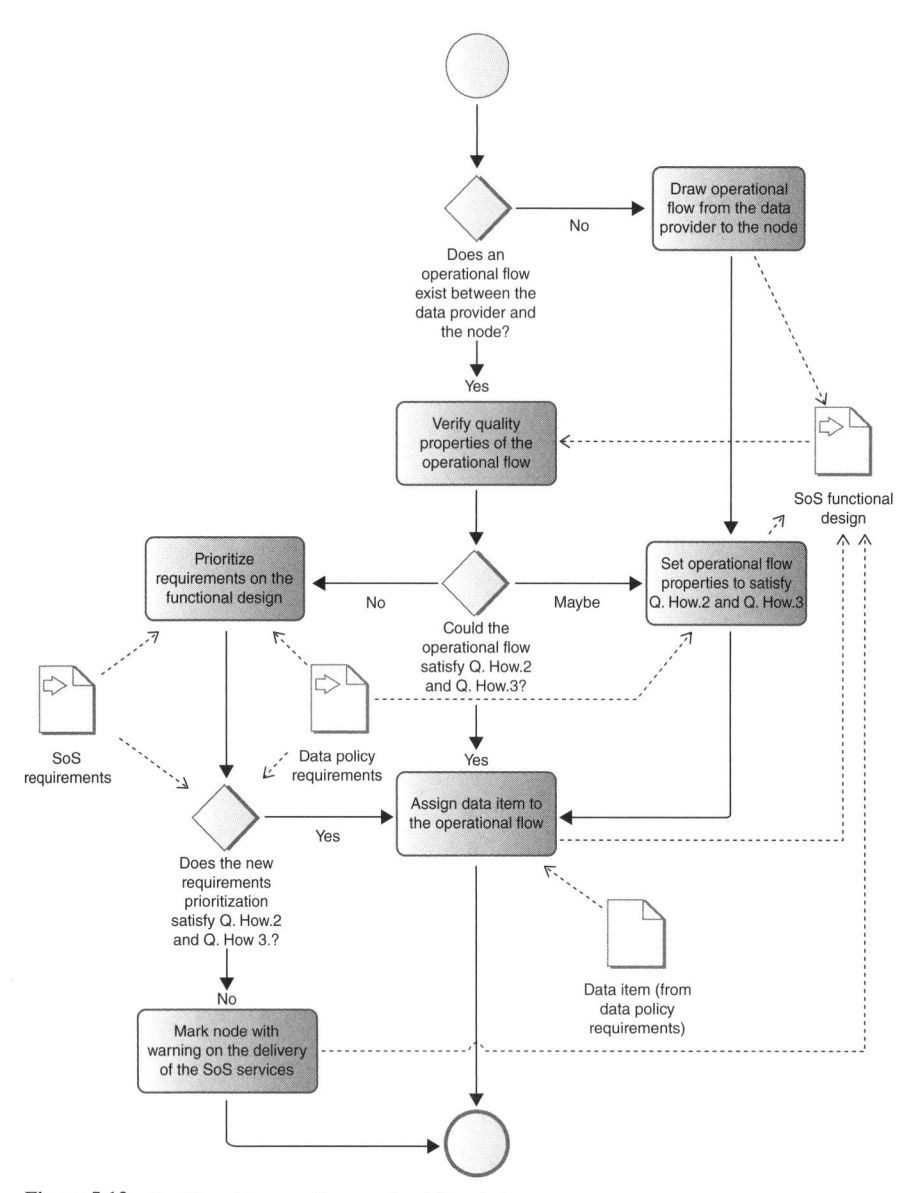

Figure 5.12 Provide guidance on the operational flow design.

3. Space Surveillance and Tracking: this area concerns the monitoring of the objects (satellites and debris) that are orbiting around the Earth. This area is critical to avoid collision among satellite and space debris. However, satellite owners might be rather reluctant to share or authorize the distribution of Satellite Orbital Data regarding satellite that support activities of confidential or secret nature, such as the military ones.

SSA programs would inherently benefit from collaboration of countries and institutions, for two reasons mainly: (1) the magnitude of investment and know-how required for designing, implementing, and operating SSA systems and (2) the intrinsic field-of-view limitation deriving from the geographical position of an SSA sensing systems. Consequently, data sharing becomes an enabler for delivering SSA services, though new political and technical challenges arise for the confidentiality of data and for the possible commercial exploitations.

This context has motivated the initial activity on the definition of the methodology, in which the UML structures were further calibrated to the specific technical needs of the agency and implemented as part of the ESA-AF (Gianni et al., 2011c) by a European company (Stoitsev et al., 2011). An example application was developed as proof of concept, for the overall methodology internal validation and for presentation at the First European Conference on Space Situational Surveillance (Gianni et al., 2011a). The example model has been developed using general understanding of SSA programs and general input from the program activities, though the names used in the models are only examples and any reference or association is fictitious and purely coincidental.

The example application has been developed on the three foundation pillars of the *Data Policy Methodology*: *data policy* model, functional architecture guidance, and physical architecture verification.

5.5.1 Example Data Policy Model

Figure 5.13 shows an example *model-based data policy* specified using the *DPM* methodology. The diagram can be read as follows: "Example SSA SST Satellite Orbital Data Policy" is a *data policy* regulating the production, use, and dissemination of Satellite Orbital Data, in the SST area. For example, this *data policy* could be promulgated by the "European SSA" and concerns the provision of Satellite Orbital Data and Orbit Accuracy to European Military Institutions. The *data policy* authorizes the use of these data for the purposes of (P1) Supporting the Identification of Unknown Natural or Manmade Objects and of (P2) Supporting Satellite Launches. The *data policy* also defines constraints on when, where, and how of these purposes, using the provided data. Specifically, purpose P1 can be achieved only once a month and only in the premises of European MoDs. Similarly, purpose P2 can be achieved only at European Space Operation Centre (ESOC) premises. In addition, this purpose is attached to a legal claim stating that SSA assumes no responsibilities on the consequences deriving from the use of the data. This example *data policy* also specifies the provisioning modalities with which Satellite Orbital Data and Orbit Accuracy will be provided to European Military Institutions.

Figure 5.14 shows the specification of the provisioning modalities. Similarly to the above diagram, this diagram can be read as follows. The provision modality, which we have conventionally identified with the name Level 2 Standard Provisioning Modality, requires that the data is provided upon request of the recipient within 1 min. The provision modality also indicates requirements for the communication medium, the channel, and the protocol.

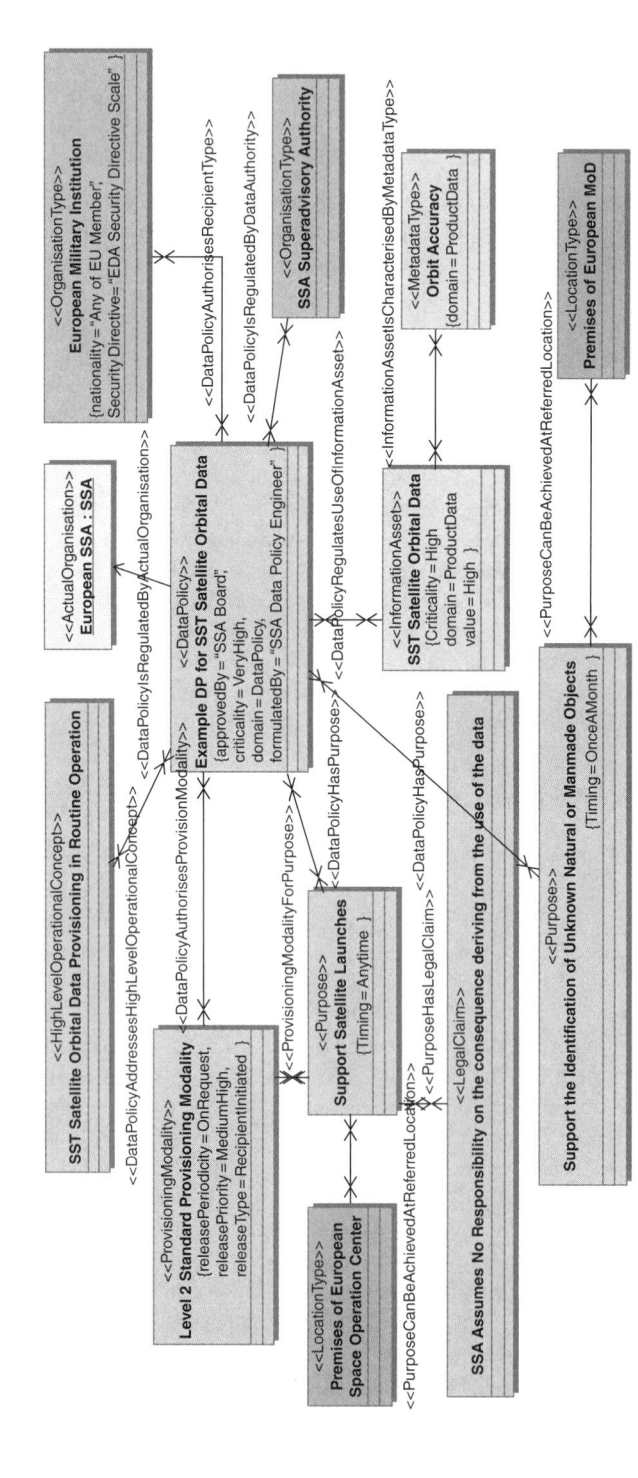

Figure 5.13 Example data policy for SST Satellite Orbital Data—policy context definition (Gianni et al., 2011a).

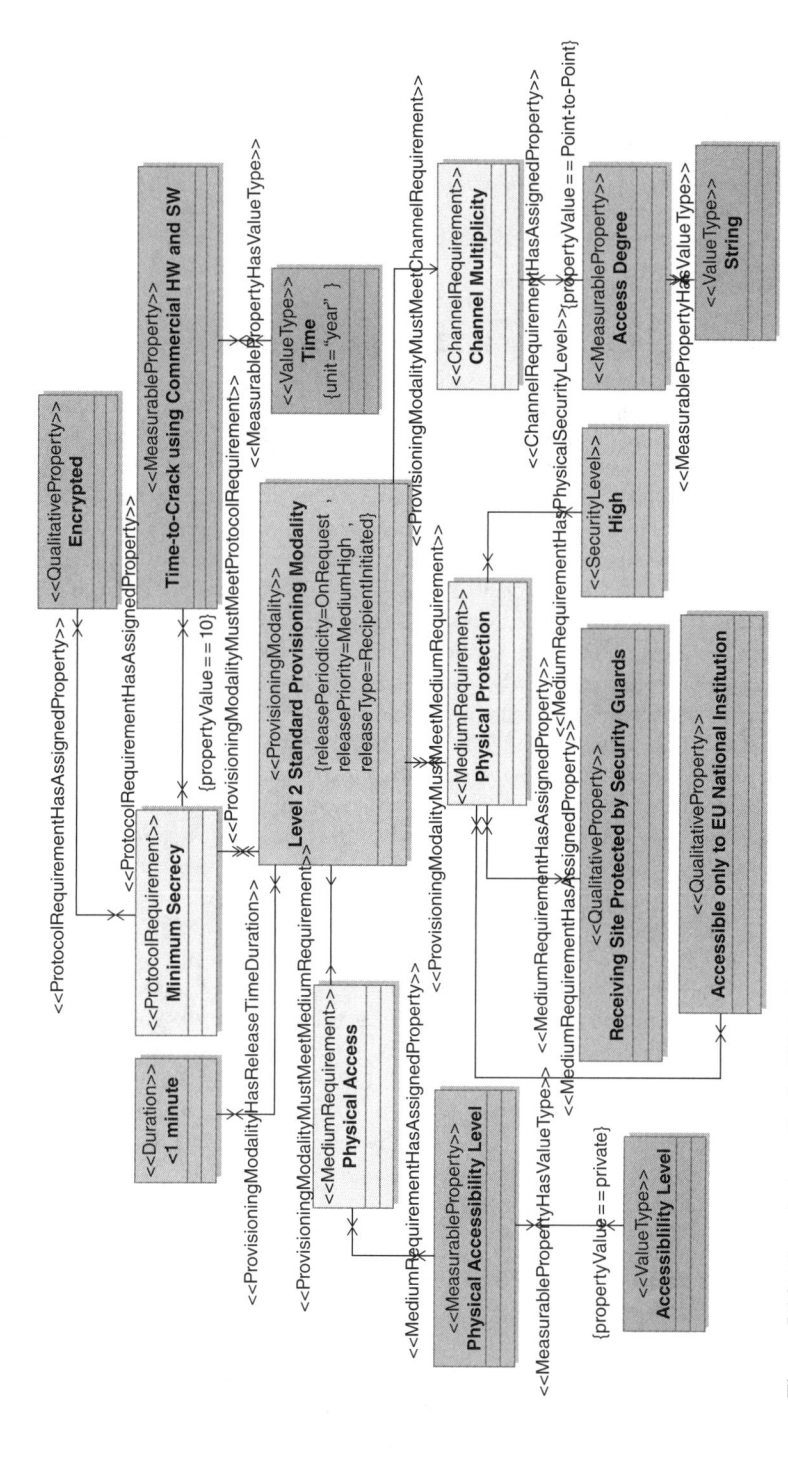

Figure 5.14 Example data policy for SST Satellite Orbital Data—specification of the provisioning modalities (Gianni et al., 2011a).

This *data policy* specifies that the medium must be physically protected, ensuring a high level of security. Specifically, the medium must be accessible only to EU National Institutions and the receiving site must be protected by security guards. The channel is required to have an access degree of point-to-point (i.e., multicasting or broadcasting channels are forbidden). Finally, the protocol must satisfy a minimum secrecy, which is quantified by the 10 years needed to crack the protocol encryption using commercial HW and SW.

For the sake of conciseness, we have shown only a relevant excerpt of the entire *data policy* example model. Other aspects can be easily inferred from the definition of all the questions in Figure 5.2.

5.5.2 Example Functional Design Guidance

Figure 5.15 shows a simplified and high-level diagram of the SoS functional architecture, which concerns roles involved in the collection and distribution of SST Orbital Data. The vertical chain illustrates the interactions for the request and collection of SST Orbital Data. In the diagram, each interaction is represented with a labeled arrow from the sender to the recipient of the data items. At the bottom of the diagram, the end users are displayed.

In particular, we focus on the information exchange "Classified Satellite Orbital Data Delivery from SSA," which occurs between SoS Front-End and European Military Institution. In a *UPDM*-based framework, this information exchange can be detailed by another type of diagram, for which Figure 5.16 shows a possible example. This information exchange can be characterized by a set of properties either qualitative or measurable. Customer and user requirements can determine which properties must be associated to this information exchange. However, the *data policy* requirements specifically concern the data distribution and they can be more restricting than customer or user requirements, for a given information exchange. Other times, these requirements do not at all concern information exchanges. In all these cases, feeding the *data policy* requirements into the functional design is necessary to minimize the verification effort of the SoS physical design. Aside from which properties must be considered, *data policy* can require minimum values for such properties. For example, the above example of SST Satellite Orbital Data Policy requires that SST Orbital Data are transmitted using secure protocols that can guarantee a value of 10 years to be cracked using commercial HW and SW.

Once the entire SoS functional architecture has been defined, SoS engineers can take this architecture in input to design the physical architecture. Particularly, using SoS engineering methods, the design can proceed through the integration of existing assets and with the identification of the new systems matching the uncovered properties and operations of the functional architecture. Once the physical architecture is finalized in the *EA framework*, the *DPM* methodology enables the *data policy* verification on the design, as shown by the example below.

5.5.3 Example Data Policy Verification on the Physical Architectural Model

Figure 5.17 (left-hand side and lower-right diagram) represents an example physical design concerning the integration of the SSA SoS with the Italian MoD, thus implementing the subpart of the functional design concerning the SoS Front-End and

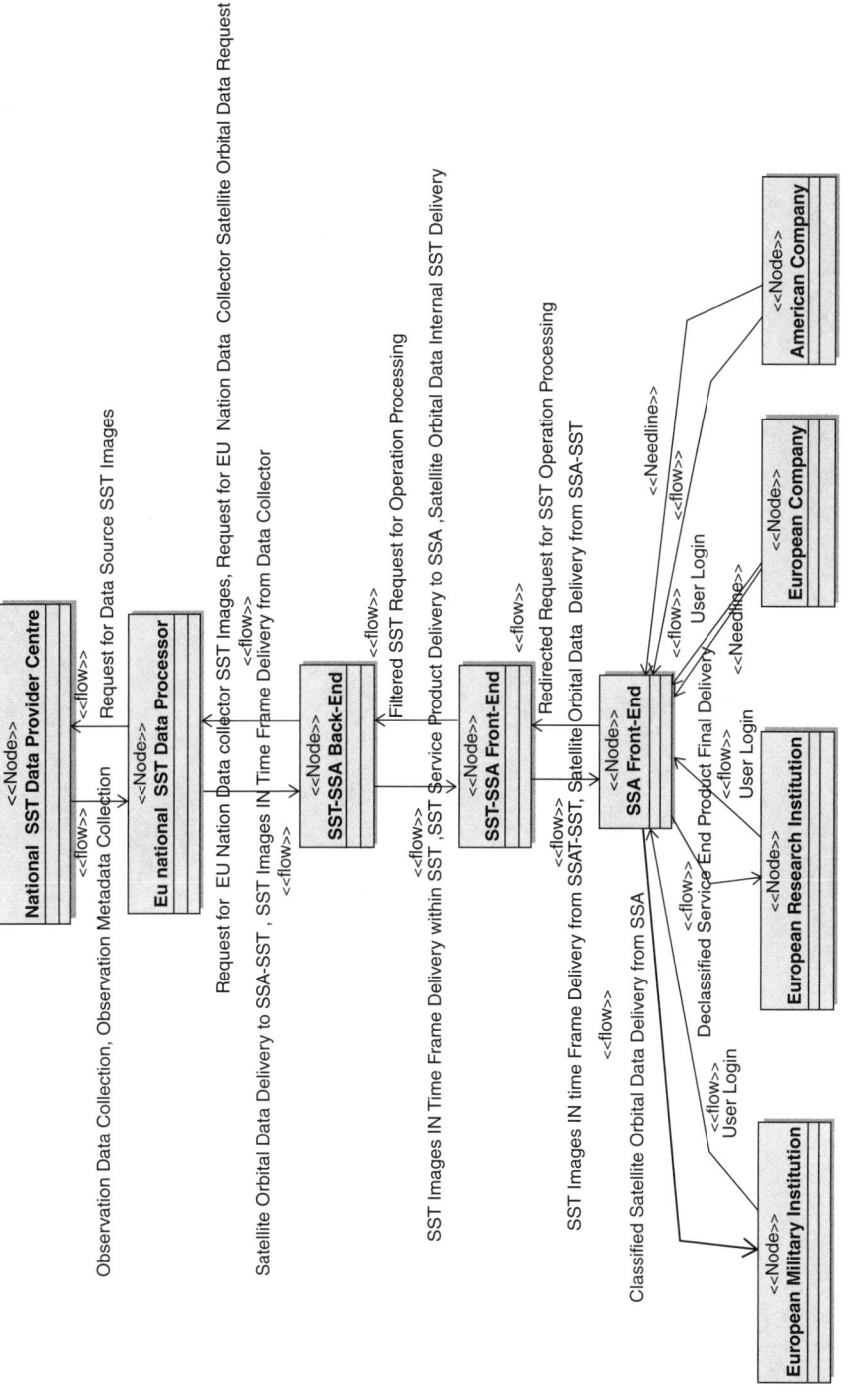

Figure 5.15 Simplified example of a possible functional architecture for an SSA SoS (Gianni et al., 2011a).

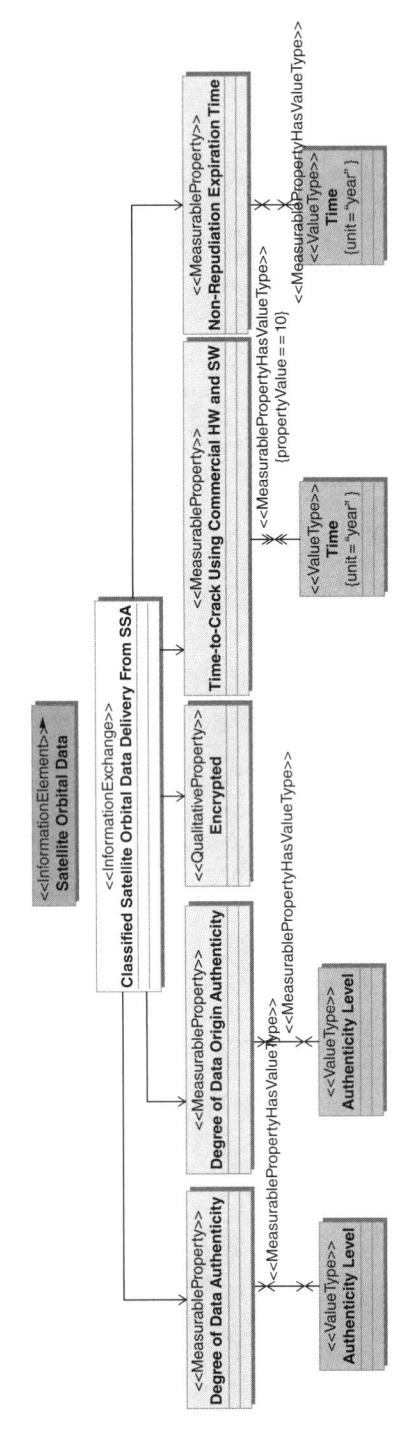

Figure 5.16 Example data policy requirements input to the functional architecture (Gianni et al., 2011a).

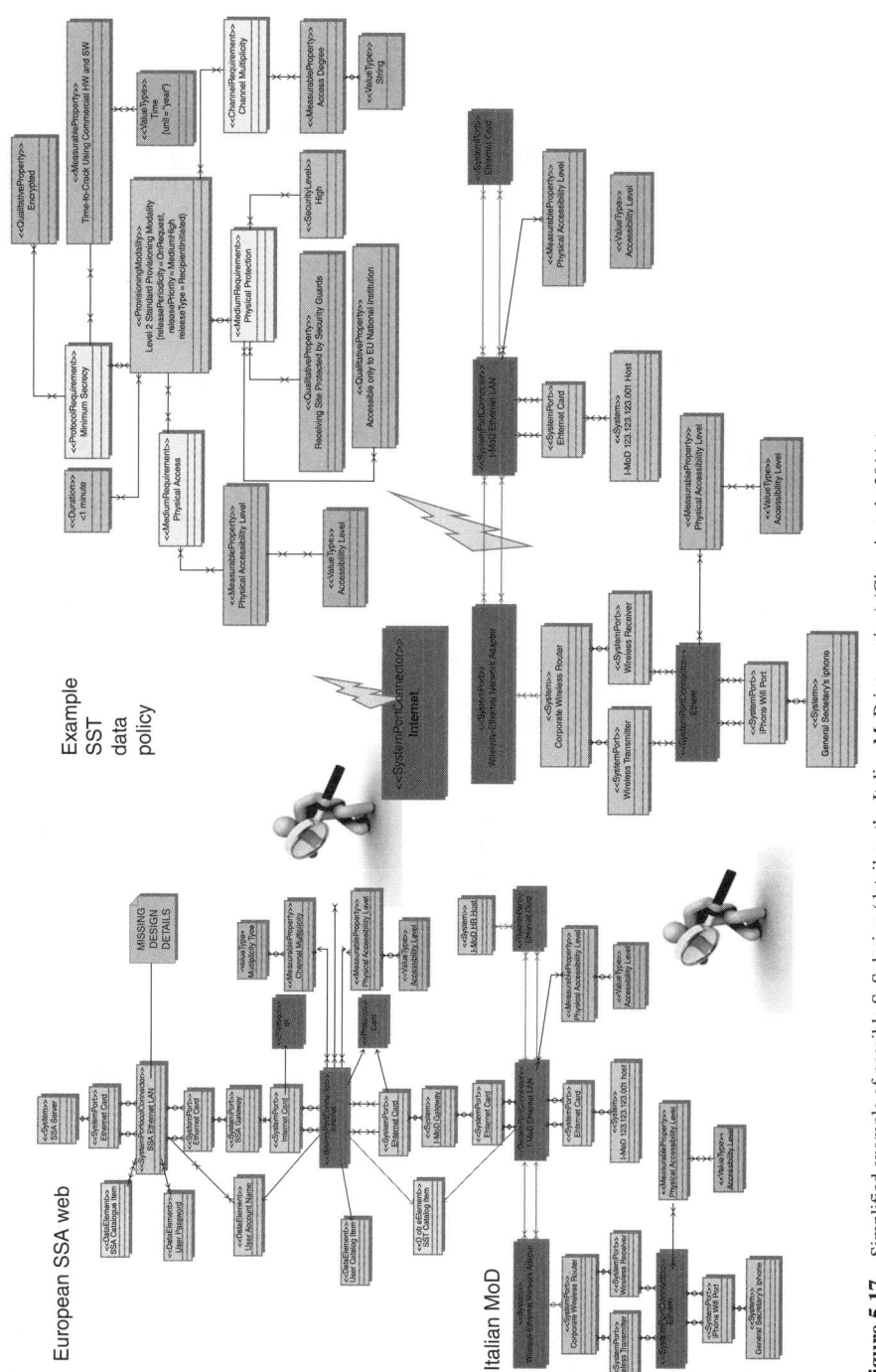

Figure 5.17 Simplified example of possible SoS design (detail on the Italian MoD integration) (Gianni et al., 2011a).

127

European Military Institution (Figure 5.15). The diagram defines the physical systems involved in the data distribution, their interconnections, and the data item transmitted. Part of the *DPM* verification can be developed by identifying the systems that become in contact with SST Satellite Orbital Data and by verifying that these systems satisfy the requirements defined in the above defined *data policy*. Differently, Figure 5.17 (upper-right diagram) shows a detail of this verification for a part of the two diagrams in Figure 5.17, as indicated by the lens icons. In this example, the Internet connector is matched against the defined *data policy* specification and shown not to satisfy the property of "Physical Accessibility == Private" for SST Orbital Data. More intuitively, the internal design of the Italian MoD can be immediately matched to the defined *data policy* specification, by verifying the presence of all required properties and verifying the conformance of the values of these properties. The diagram is purposely represented at this zooming level to show the relationships from a bird-fly perspective. The interested reader is furthermore encouraged to refer to Gianni et al. (2011a) regarding the additional use of colors supporting visual matching and verification as it is support by respective data management tools.

5.6 CONCLUSIONS

Conventional enterprise *governance* frameworks cannot be applied to the SoS *governance* as these frameworks are inherently designed on the assumptions of static enterprise structure and centralized enterprise control, which are not met by SoS. As such, SoS *governance* requires a fundamentally newly thought *governance* approach that considers the dynamic evolution of an SoS—depending on the goals of participating organization—and the decentralized control, which however may not necessarily be a fully democratic one. In information-intensive SoS, a central *governance* issue is related to the regulation for the data dissemination and use, briefly named *data policy*. Data can be of various nature (from secret to commercial), and their distribution and use should be therefore restricted to the conditions dictated by the relevant stakeholders, which participation may be pending. As such, this issue does not only concern the definition of a *data policy*. Indeed, each SoS participant must also be reassured that the SoS physical design satisfies the defined data distribution and use conditions.

In this chapter, we have presented the definition of the *Data Policy Methodology*, a *model-based* methodology that can support SoS engineering teams in the definition of data policies and in their subsequent verification against the SoS architectural design. The methodology is based on three pillars: *data policy model-based* definition, functional design guidance process, and physical verification process. As such, the methodology can be used to specify the *data policy* requirements and also to gain the trust of SoS stakeholders on the requirements design. Consequently, the methodology can contribute to the formation and evolution of SoS, ensuring the traceability of the *data policy* requirements on the functional and physical design, for the initial SoS design and for supporting the SoS evolution. The chapter also shows that the methodology can be integrated in standard *EA frameworks*, such as those based on *UPDM*, and therefore can naturally become part of the SoS engineering activities. A simple example is also discussed to show the methodology application and its potential exploitation.

5.7 ACKNOWLEDGMENTS

The author would like to thank the members of the ESA's Software Systems Division and the members of the ESA's SSA Team, for providing the initial input that gave origin to part of the results presented in this chapter. The activity was partially supported by the ESA Internal Research Fellowship schema.

5.8 DISCLAIMER

The views and opinions expressed in this article are those of the author and do not necessarily reflect the official policy or position of any European or national agency or the ESA. Example models presented within this chapter are only examples and any reference or association is fictitious and purely coincidental.

REFERENCES

Bailey, I. (2008). "Brief Introduction to MODAF with v1.2 Updates." *IET Seminar on Enterprise Architecture Frameworks,* September 2008, London, pp. 1–18.

Bobrinsky, N. and Del Monte, L. (2010). "The Space Situational Awareness Program of the European Space Agency." *Cosmic Research*, vol. 48, no. 5, pp. 392–398.

Forsberg, K., et al. (2010). *INCOSE Systems Engineering Handbook,* v. 3.2, INCOSE.

Gianni, D. (2010). "SSA Data Policy Definition Document." Internal Report, European Space Agency, Software Systems Division, December.

Gianni, D., et al. (2011a). "SSA-DPM: A Model-Based Methodology for the Definition and Verification of European Space Situational Awareness Data Policy." *Proceedings of the 1st European Space Surveillance Conference*, June, Madrid.

Gianni, D., et al. (2011b). "A Model-Based Approach to Support Systems of Systems Security Engineering for Data Policies." *INCOSE Insight*, vol. 14, no. 2, pp. 18–22.

Gianni, D., et al. (2011c). "Introducing the European Space Agency Architectural Framework for Space-Based Systems of Systems Engineering." In O. Hammami, D. Krob, and J.-L. Voirin, *Complex Systems Design & Management*, Berlin: Springer, pp. 335–346.

Goergen, M. (2012). *International Corporate Governance*. New York: Prentice Hall.

Grembergen, W.V. and De Haes, S. (2009). *Enterprise Governance of Information Technology*. Berlin: Springer Verlag.

Halpin, T. and Morgan, T. (2008). *Information Modeling and Relational Databases*, 2nd ed. Burlington, VT: Morgan Kaufmann.

Hause, M. (2010). "The Unified Profile for DoDAF/MODAF (UPDM) Enabling Systems of Systems on Many Levels." *Systems Conference, 2010 4th Annual IEEE*, IEEE, San Diego, CA, pp. 426–431.

IT Governance Institute (2003). *Board Briefing on IT Governance*, 2nd ed. Rolling Meadows: IT Governance Institute.

Jamshidi, M. (2009). *Systems of Systems Engineering: Innovation for the 21th Century*. Hoboken: John Wiley & Sons, Inc.

Lemmens, I., Sgaramella, F., and Valera, S. (2009). "Development of Tooling to Support Fact-Oriented Modeling at ESA." *On the Move to Meaningful Internet Systems, OTM Workshops, 2009*, Springer Verlag, pp. 714–722.

Maier, M.W. (1998). "Architecting Principles for Systems-of-Systems." *System Engineering*, vol. 1, pp. 267–284.

Melli, G. and McQuinn, J. (2008). "Requirements Specification Using Fact-Oriented Modeling: A Case Study and Generalization." *On the Move to Meaningful Internet Systems, OTM Workshops*, LNCS, Springer, pp. 738–749.

Mercer, B. (2008). "Enabling Executable Architecture by Improving the Foundations of DoD Architecting." *Collaborative Technologies and Systems, 2008. CTS 2008*, Irvine, CA, IEEE, pp. 558–559.

Morris, E., et al. (2006). "System-of-Systems Governance: New Patterns of Thought, Technical Note (2006-TN-036)." Carnegie Mellon University/Software Engineering Institute.

Smith, C.U. (1990). *Performance Engineering of Software Systems*. Pittsburgh: Addison-Wesley.

Stoitsev, T., et al. (2011). "System of Systems Engineering with the ESA Architectural Framework." *International Astronautical Congress*. Cape Town, South Africa.

Weill, P. and Ross, J.W. (2004). *IT Governance: How Top Performers Manage IT Decision Rights for Superior Results*. Boston: Harvard Business School Press.

Zachman, J.A. (1987). "A Framework for Information Systems Architecture." *IBM Systems Journal*, vol. 26, no. 3, pp. 276–292.

Chapter 6

System Health Management

Stephen B. Johnson[1,2]
[1]*Dependable System Technologies, LLC and the University of Colorado, Colorado Springs, CO, USA*
[2]*National Aeronautics and Space Administration, Marshall Space Flight Center, Huntsville, AL, USA*

6.1 INTRODUCTION

Systems health management (SHM) addresses what might be considered the "dark side" of Systems Engineering. That is, for every goal, objective, or requirement (see the definition of these terms in the following), there is the possibility that this goal will not be achieved. SHM refers to the collection of methods, processes, procedures, designs, and design attributes of a system that ensure that the system can achieve all or some of its goals despite potential or actual failures. As such, it encompasses aspects of many historic, existing subfields. These include safety; reliability; availability; maintainability; failure (or fault) detection, isolation, and response (or recovery) (FDIR); fault or failure tolerance; vehicle health management (HM); prognostics; diagnostics; and dependability. SHM is the aspect of "resilience engineering" that deals with failures of the system itself, whether from internal or external causes, as opposed to aspects of the environment that an otherwise "healthy" system cannot successfully accommodate or address (Hollnagel et al., 2006).

Since a system of systems (SoS) has goals that it intends to achieve, the possibility of failure to achieve these goals exists for SoSs as much as for systems. Thus, SHM, or perhaps an extended variant of SHM such as "SoS health management," is as necessary for SoSs to achieve their goals as it is for systems.

The purposes of this chapter are to describe the concepts and theory of SHM and the differences between systems and SoSs that we must understand to determine how SHM

Modeling and Simulation Support for System of Systems Engineering Applications, First Edition.
Edited by Larry B. Rainey and Andreas Tolk.
© 2015 John Wiley & Sons, Inc. Published 2015 by John Wiley & Sons, Inc.

must be modified to accommodate SoSs and to discuss how systems and SoSs can be modeled to address SHM attributes and concepts. To set the stage to achieve these chapter goals, we must first define a number of terms to provide a common language in which these issues can be addressed.

6.2 DEFINITIONS

According to the International Council on Systems Engineering (INCOSE), a *system* is "a construct or collection of different elements that together produce results not obtainable by the elements alone" (INCOSE, 2010). This definition describes systems of all kinds, including both natural and engineered systems. For our purposes, natural, biological systems are of no interest except for their use as part of engineered systems, in which case they have a specific intended purpose. The kinds of systems and SoSs being considered in this book all have specific purposes or goals. Thus, the "results" described in the INCOSE definition are not arbitrary. The results to be achieved are based on the goals set for the system. Put simply, a system is designed, built, and operated to achieve one or more goals. A more useful definition is of an *engineered system*, which we will define here as "a construct or collection of different elements that together produce results that achieve one or more intended purposes not obtainable by the elements alone." As we will see later in this paper, the fact that systems exist to achieve intended purposes is critical to modeling and understanding SoSs.

We will follow Jamshidi's preferred definition of *SoSs*: "Systems of systems are large-scale integrated systems which are heterogeneous and independently operable on their own, but are networked together for a common goal" (Jamshidi, 2009). For the purposes of engineering, this definition significantly improves on the INCOSE definition of system, because of its explicit reference to a "common goal," though we could quibble with this by noting that there could be more than one goal for an SoS.

SHM is defined as "the capabilities of a system that preserve the system's ability to function as intended" (Johnson et al., 2011). SHM is a set of capabilities, as opposed to being a subsystem or a specific technology.

A *function* is defined as it is in mathematics as "a mathematical correspondence that assigns exactly one element of one set to each element of the same or another set" (Merriam-Webster, 1991). Or more casually, a function maps the states (values) of the input state vector $\mathbf{x}$ to the output state vector $\mathbf{y}$, as in $\mathbf{y} = f(\mathbf{x})$.

A *goal* is defined as "the end toward which effort is directed" (Merriam-Webster, 1991). In this chapter, the terms *goal* and *objective* are considered synonyms.

Intention is defined as "what one intends to do or bring about" (Merriam-Webster, 1991). A synonym for intention is *purpose*. When humans create an engineered system or SoS, they have an intention, for which they set a goal (or goals) and document these as "requirements."

A *requirement* is defined as a formal statement of a goal. Requirements might or might not capture all of the original intentions and goals for the system. They frequently do not.

A *failure* is defined as the unacceptable performance of intended function (Johnson et al., 2011). Failures can exist at any level or part of the system or the system as a whole. While failures are usually unexpected events, sometimes, they are perfectly predictable, such as the power of the Voyager 2 nuclear power source degrading to the point when no further communication to Earth is possible. This is an unacceptable performance of the power system, but based on the design, it is inevitable and predictable.

A *fault* is defined as a physical or logical cause internal to the system, which explains a failure (Johnson et al., 2011). Faults are explanations of failures. Failures are unacceptable behaviors. Faults are the explanation of why the unacceptable behaviors occurred, if the event was internal to the system. A typical failure example is that a car tire slowly goes flat, which is the failure to be explained. The explanation is that all tires have some amount of pressure leakage and the owner did not maintain the tire properly by refilling the tire with air. In this case, "the system" inherently involves the owner of the car, and hence, the owner is "at fault" for causing the failure. Failures can also be caused by events outside the system, which are not considered faults. The same tire running over a nail in the road is no fault of the automobile or the owner ("the system"); rather, an external event caused the failure.

Faults and failures are interdependent, recursive concepts of cause and effect. Seen from one perspective, a fault explains a given failure, but from another, that same fault is seen as a failure that needs explanation. For example, in the Columbia space shuttle tragedy of 2003, the hole in the leading edge of the wing is the failure that needs explaining, and its explanation (fault—cause) is a chunk of insulation foam falling off the external tank (ET) hitting the wing during ascent. However, from the perspective of the ET designers, the failure to be explained is the foam falling off the ET, and the explanation or fault is the expansion of air bubbles in the foam during ascent. In turn, the air bubbles can be considered the failure, with flaws in the foam insulating material or application procedures being defective.

An *anomaly* is defined as the unexpected performance of intended function (Johnson et al., 2011). Anomalies can be good or bad, but they are always unexpected. For example, when the power level of the Mars Exploration Rovers unexpectedly increased, this was an anomaly, later explained as a dust devil blowing dust off of the solar panels. Anomalies signify a lack of understanding of an event, and the usual response is to investigate. Once the behavior is understood, it is no longer an anomaly.

A *model* is defined as a representation of the attributes of some subset of reality. To understand a system or SoS, those humans trying to understand their behaviors rely on models, which can be formal or informal, documented in computer code, in mathematical equations, or merely in the humans' minds. When an anomaly is investigated and understood, this understanding occurs because the models upon which the prior understanding was based are updated to incorporate or explain the new, formerly anomalous phenomena. If the behavior occurs again, it will not be anomalous for long, as it will be quickly classified as a failure or normal behavior based on the predictions of the new models.

A *state* is defined as the value(s) of a (set of) state variable(s). A behavior is the time evolution of states over time. Since states are just sets of values, a set can be arranged from values over time, and hence, this behavior itself can be considered a state as well. However, it is nonetheless useful to generally consider states as values at a point in time and behaviors as the values of states over time. In the equation $F = m \times a$, all three variables are state variables. Their values at a point in time are the states of those state variables at that point in time.

6.3 SHM FOR A SYSTEM

SHM exists to ensure that a system performs its functions correctly. To do so, two basic strategies exist: to prevent a failure from occurring and to detect and respond to failure, usually called failure (or fault) tolerance.

Preventing failures in turn has two substrategies: to prevent faults (internal causes of failures) from occurring and to predict that failure will occur in the future and take action to prevent the predicted failure from occurring. The fault prevention strategy typically involves increasing the reliability of components through quality assurance methods and testing. Predicting that a failure will occur in the future is called prognostics, and taking preventive action is often implemented through schedule-based or condition-based maintenance for components that can be repaired. Other actions can include operating the system differently to reduce component wear and tear, which delays failure.

Failure tolerance requires detecting and responding to failures and is often called FDIR. Failure tolerance can be subdivided into three major kinds of substrategies: failure masking, failure recovery, and goal change. Failure masking means that the detection and response operate so quickly that failure effects are "masked" so that they do not affect the function in question. A typical example is a redundant computer voting scheme in which two computers can outvote a single third computer by running the same identical data and calculations through the redundant computing suite. Failure recovery means that the FDIR operates more slowly, such that the function is temporarily compromised, but the FDIR eventually recovers the full capability of the function. The last type of failure tolerance strategy is goal change, which means that functionality cannot be fully restored when the failure occurs. In the goal change strategy, some new goal is selected when the failure occurs, because the original goal(s) cannot be achieved. Usually, this new goal is a subset of the original goal(s). In a crewed launch vehicle case, an "abort," which returns the crew back to Earth, is a typical goal change action. The launcher cannot achieve orbit and may in fact be exploding, so the crew is removed from the hazardous situation. In this case, the goal of achieving orbit is abandoned, but the goal of keeping the crew safe is retained.

Among these strategies and substrategies, only the "fault prevention" (technically called "design time fault avoidance") strategy is a passive strategy that requires no actions during systems operations. This can be thought of as increasing the reliability of a single string system. The remaining four substrategies all require operational actions of some kind. The division between passive and active strategies is highlighted in SHM by classifying all active SHM strategies as *fault management* (FM). By definition, FM is the operational subset of SHM.

All FM mechanisms inherently detect current or future failures, or current anomalies, and then take some action to respond. For current or future failures, which signify that the nominal system is unable to control the system's behavior within acceptable bounds, FM acts as a means to bring the system back to a state in which the system's behaviors can once again be maintained by "classical" control methods. In other words, FM acts as a "metacontrol system" that takes over when the system's nominal (passive or active) control mechanisms are unable to keep the system's behaviors under control. For anomalies, FM mechanisms detect if the system's behavior is unexpected, different from that predicted from models of the system and from prior behaviors. As noted in the previous section, the usual response to an anomaly is to investigate the event so as to determine its cause. In other words, an anomaly indicates a problem with knowledge, whereas a failure indicates a problem with control. This correlates to the distinction in control theory between state estimation, which relates to knowledge, and system control based on that knowledge.

All FM mechanisms inherently require redundancy. This explains the frequent use of the phrase "redundancy management" to describe FM functions and mechanisms. Redundancy is necessary to compare information about how the system should behave to how it is actually behaving (detection) and to provide mechanisms to respond if off-nominal behaviors are detected. There are several types of redundancy: hardware identical redundancy, functional (analytic/dissimilar) redundancy, information redundancy, temporal redundancy, and knowledge redundancy. Hardware identical redundancy is just as its name describes, whereas functional redundancy uses different mechanisms for its redundant strings. Information redundancy is exemplified by error detection and correction codes, which add extra bits of information so as to reconstruct the original message in case of bit flips. Temporal redundancy means rerunning a process or calculation again to determine if an original calculation is correct or in case of an invalid calculation. Knowledge redundancy uses sources of knowledge different from the operational system to cross-check to determine if current systems behaviors are valid.

SHM also utilizes key assumptions from social science to understand the causes of failures and from this understanding to influence SHM designs. History clearly shows that humans can and do build systems that generate behaviors that the designers and operators do not expect or (at least initially) understand. The theory of SHM postulates that the root causes of the vast majority of failures are ultimately due to human cognitive, performance, and communications faults (or mistakes). Accident, mishap, and failure investigations typically find that most failures appear "technical" on the surface. However, as the investigation proceeds, it is almost invariably true that the investigators find that humans made one or more mistakes in communicating with each other and performing simple manual (such as a poor solder.) or cognitive (such as making a sign error in a mathematical calculation.) tasks. Most failures, in retrospect, are due to very simple or even trivial mistakes. The problem is that there are millions of opportunities for mistakes, and humans will inevitably make some.

Finally, different people and groups will have different views as to the goals of the system and how to interpret its performance. This implies that people can and will judge the events differently; some may assess an event as a failure, while others may judge the same event as a success. Examples of this abound, such as the O-ring charring problems in the early space shuttle program that were interpreted as "normal maintenance" issues by some personnel and as failure by others. Interpretations by different individuals and groups can change over time. The behaviors of the system in practice are far more complex and detailed than any set of specifications can fully describe. The theory of SHM assumes that these are normal and expected situations. There is no set of specifications or single agreed interpretation that covers all systems behaviors and requirements for all personnel that design, build, and operate the system.

Overall, SHM provides mechanisms to ensure that the system functions properly and achieves its intended goals. Failures can be prevented or detected and responded to. In the latter case, FM uses redundancy in a "metacontrol system" to reestablish a controllable state when the normal control system cannot keep the system within acceptable performance. Humans frequently build systems that generate behaviors that the designers and operators did not expect. The causes of failure are normally simple human-caused mistakes, which cause systems behaviors that different humans designing, building, and operating the system can and do interpret differently.

6.4 SHM MODELS, SIMULATIONS, AND USES

SHM is an emerging discipline in the sense that there are efforts ongoing to replace the various ad hoc practices and designs that have characterized its operational FM capabilities. As such, modeling representations and simulation practices specific to SHM are in their infancy.

Early in the design process, when design specifics do not yet exist, Systems Engineering often uses a functional decomposition to define systems functions. Making the traditional functional decomposition far more precise, the goal-function tree (GFT) representation enables SHM analysis of failure detection coverage assessed against system's goals. As illustrated in Figure 6.1, the GFT enables this because of its use of the basic relationship $y = f(x)$ to define functions and the inputs and outputs to those functions. Goals are defined as constraints on the range of the output state vector y and functions as the transformation from input state vectors x to the output state vector y.

Off-nominal goals are inserted into the GFT to apply when the state variables associated with a goal are monitored and go out of range. When the states go out of range and are detected, this activates an off-nominal FM response. If this response is able to maintain the system's goals, then this acts like an OR gate in the GFT (which normally uses AND logic), indicating it is a redundant mechanism to achieve the goal. If the response changes the system's goals, then the FM activates a new set of GFT branches corresponding to the system's new goals. Because of the use of state variables that define how the system intends to control the physics, the GFT representation is physically accurate, enabling the GFT to be used for analysis, as well as providing a near-complete representation of goals and functions (Johnson, 2013).

Since a systems architecture is not a tree structure, how is it that a tree structure can usefully model a system? What does a tree structure actually model? It represents *intentionality*. Any engineered system is created to achieve some intended purpose, and this purpose and associated function are represented as the top node of the tree (the function) and its constrained output state variable (the goal). All other functions and their

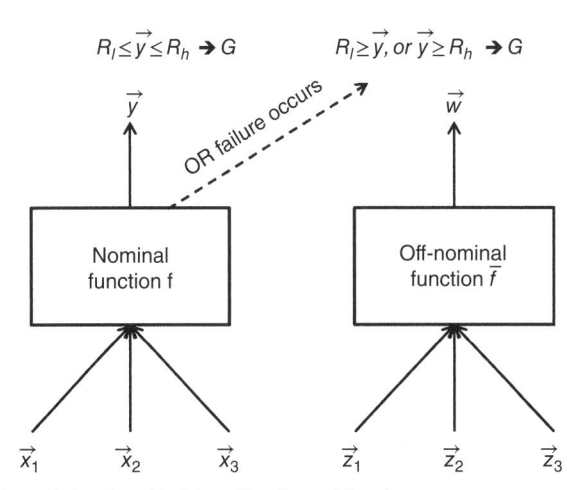

Figure 6.1 Mathematical and graphical formalism for goal-function trees.

associated goals are subsidiary in the tree structure because they support the top-level goal. Because intentionality is one of the key distinctions between systems and SoS, the GFT will be one of the key representations to differentiate the two.

The GFT enables several useful SHM analyses. The first is an assessment of failure detection coverage. SHM exists to protect the system's top-level goal(s). While some goals may be protected by providing improved single string reliability, it is often true that designers will want to detect and respond to any failures that can threaten the top-level goals. In a GFT representation, failures that compromise low-level goals will compromise goals further up the tree, unless there is some FM with its inherent redundancy that provides another way for that higher-level goal to be achieved. In other words, unless the failure is detected and responded to, its effects will propagate up the tree and cause systems failure; that is, the top-level goal(s) will not be achieved. With some subtle exceptions that we will not discuss here, the assessment of failure detection coverage is merely a scan of all paths up the tree to determine if there is at least one failure detection and response along every path leading up to the top-level goal.

The GFT also provides inputs to the definition of failure scenarios. Every path from the bottom to the top of the tree is a unique failure scenario, because failures at any location in the tree then propagate up the tree until a node with redundancy is met. Put another way, when a goal in the tree cannot be achieved, higher-level goals will also fail to be achieved, and with the GFT ensuring that the tree structure matches the systems physics, the progression of goal failures from bottom to top correctly describes the progression of failure effects. The GFT does not describe all possible failure scenarios, because some failures create new connections between components and functions that do not exist nominally. For example, an electrical short circuit creates a new electrical path that does not normally exist. Because the GFT is a success space representation, only nominal connections will be represented.

Because FM is a control mechanism, its performance is assessed using classical concepts from control theory for state estimation and state control. Failure detection is assessed using truth (or confusion) tables of false positive, false negative, true positive, and true negative. Fault isolation, which determines the location of the cause of a failure, is also a state estimation function and is assessed with similar kinds of truth tables. State control is assessed by comparison of the race of the failure responses versus the failure effects, in which the failure response must be faster than the failure effects it is responding to. The quantitative models to assess performance are all structured by this theory.

Another implication of FM being a set of control mechanisms is that system-threatening interactions of FM mechanisms with each other or with other parts of the system, sometimes called "the deadly embrace," are usually interactions of two or more control loops trying to control the same state variables. By developing models that represent the system's many control loops and functions, the intersection points of these control loops can be identified. These are the danger points for system-threatening interactions. Once identified, typical discrete-event simulations and state machine models can be used to assess the interactions that can occur at these danger points.

Another typical model used in SHM is the directed graph representation used for diagnostic analysis and operations. Directed graphs are built to model the system's component connectivity, the failure modes of these components, and the failure effect propagation paths resulting from these failure modes. These enable forward- and backward-chaining logic to determine the system's ability to detect failures (in this case from the bottom-up design model) and to determine the location of the causes of failures from observable symptoms.

Prognostic models are typically physics-based models that aim to determine how systems wear and tear leads to systems failure and how long this takes. A typical example of a prognostic model is of the structural physics of aircraft wings to determine how flaws in the metal alloys or composites lead to microfractures and then the growth of micro- into macrofractures that can lead to wing failure after many flights that cycle structural and thermal loads and stresses.

Other typical failure models include fault trees, which are failure space hierarchical representations. These are at least in part logical complements of the GFT success space representation. Fault trees can contain the new off-nominal paths that success space tree models cannot. However, fault tree models have not to date used the state-variable approach of the GFT. This hampers their ability to match the logical rigor and completeness of the GFT. Fault tree models are used to determine the ways in which a system may fail to achieve its top-level goals, though these goals are not explicitly and rigorously represented as is done in the GFT.

The ideal design process for the HM of a system proceeds in the following manner. First, a GFT model of the system's goals, functions, and state variables is built. Next, for every goal in the tree, the SHM designer queries what, if anything, should be done if that goal cannot be achieved. This leads to the determination of whether that goal needs to be protected and, if so, whether it should be done by improving component reliability or by adding FM mechanisms to detect and respond to failure of that function. These FM detections and responses are control loops and hence can be called "FM control loops" (FMCLs). The GFT model is modified by adding the FMCL detection goals and functions at the locations of the monitored state variables.

The GFT model, with the detection goals and functions in it, can be used to assess the completeness of the proposed FM detections. The GFT model can also be used to allocate the proposed reliability needed to achieve each goal, including the proposed reliability of the FM mechanisms needed to achieve the overall intended systems reliability, availability, and safety (RAS) for the system.

Finally, once a proposed set of FM functions and reliability improvements have been agreed, the design can proceed, and models of the SHM design can be built and used to determine the effectiveness of that design against the SHM metrics. These models include the suite described earlier, including fault trees, discrete-event and state machine models, quantitative (usually probabilistic) models of state estimation and control response effectiveness, directed graphs for diagnostics, and physics-based models to support the models noted earlier and for prognostics. The design and the models are then progressively refined in the typical design spiral, with changes occurring to each to meet the performance and cost goals of the program. Testing proceeds through a variety of fault injection tests to stimulate the FM designs, at progressive levels of systems integration. The results of these tests are fed back into the SHM analytic models, and the systems validation is achieved analytically, since in general it is impossible to completely test the extremely large off-nominal state space.

6.5 DIFFERENCES BETWEEN SYSTEMS AND SoS

A topic of considerable debate in the SoS community is determining and defining the difference between a system and an SoS. In this chapter, we assume that the definitions in Section 6.2 encapsulate the results of many of these debates. The key portions of these definitions indicate that an engineered system is designed to achieve one or more intended purposes and that SoSs integrate a number of these "independently operable" systems to achieve a "common goal." In the SoS definition, these "independently operable" systems each have their own goals, which might or might not be the same as the goal(s) of the SoS. Thus, the distinction between systems and SoSs is that the constituent systems of SoSs each have goals that can differ from the SoS and that they can continue to operate as independent systems to achieve those original goals. In other words, the difference is *intentionality*.

The difference in the intentionality of systems as opposed to SoSs can be observed clearly through GFT models. A system's goals (the intention for the system) and its supporting subgoals are represented in a specific GFT hierarchical model. Each system of an SoS can be modeled in this way, leading to a unique GFT for each system. The SoS has its own goals, and hence, it can be modeled using a GFT as well. However, the SoS uses its constituent systems to achieve its goals. This means that the SoS GFT inherently contains within it the GFTs of its constituent systems.

Because the GFTs represent the goals and intentions of a system, with the top node of the GFT representing the system's primary purpose and function and lower nodes and branches representing supporting goals and functions, each of the systems' GFTs has different goals at the top (representing their different purposes) and differing subgoals and branches further down. The SoS that uses these systems might not be using the top-level goals of these systems, but may instead be using lower-level goals within these systems. This is represented in the GFTs by the fact that the SoS GFT will only incorporate some branches of those systems' GFTs, not the entire GFT for those systems.

The SoS will only use a system as part of its network of systems if the goals and intentions of that system are compatible between the SoS and at least some part of the system. This translates in the GFT representation to compatibility of a fragment of the systems' GFT with the goals of the SoS GFT. In other words, at least one of the goals of the SoS GFT must utilize the same state variable as the systems' GFT and with at least partially overlapping ranges of the output state variables of that goal, as shown in Figure 6.2. Whether it is cost-effective to use the system to achieve the SoS purpose depends in part on having compatible goals and functions, which represent its semantics. Cost-effectiveness also depends on the syntax, that is, the details of the interface design needed to connect the SoS goal to the system's goal. The syntax and semantics are key attributes of the spaces and interfaces between constituent systems of an SoS, what Garrett et al. refer to as the SoS "interstitials" (Hollnagel et al., 2006; Garrett et al., 2010). The GFT represents the semantics (the purpose and functions) correctly, but does not provide the design details that define the syntax.

As seen from the GFT perspective, the SoS is in key respects not significantly different from a system. For both SoSs and systems have top-level goals, with supporting goals, functions, and designs that provide the capability to achieve them. The SoS differs from the system in its use of entire systems or parts of systems for its own purposes, while those systems can still be "independently operated" for their original purposes without reference to the SoS.

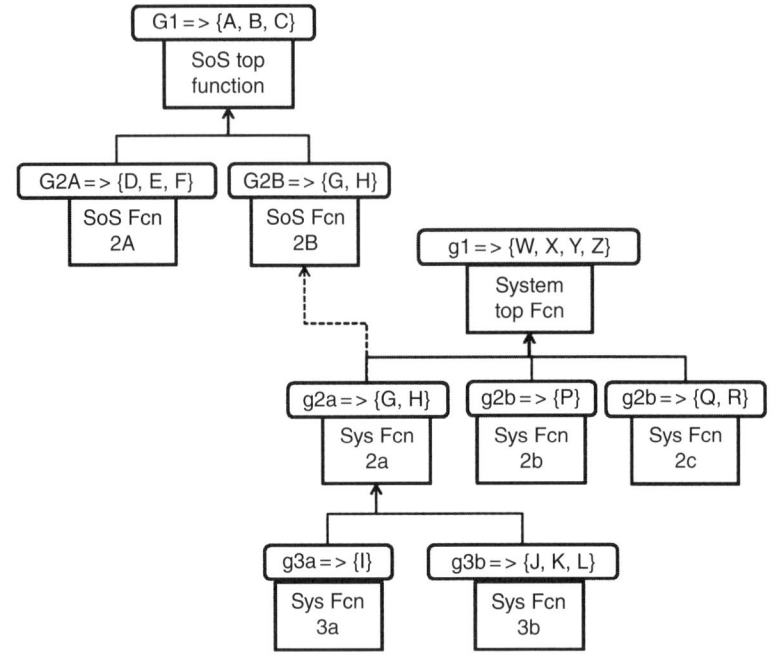

Figure 6.2 An SoS GFT showing use of lower-level instead of top-level system's goals.

6.6 SHM OF SoS

As an SoS has fundamental similarities to a system, the HM of SoSs has fundamental similarities to the HM of systems. The purpose of the HM is the same: to ensure that the SoS can achieve all or some of its goals. HM for an SoS still deploys either improved component reliability or adds components that provide systems redundancy, which operates as metacontrol loops. The systems or portions of systems that are incorporated into the SoS bring their own existing HM techniques. These heritage designs therefore become part of the SoS, just as the nominal designs are incorporated into the SoS.

To the extent that the SoS has new goals that differ from the goals of its constituent systems, it may also need new FM mechanisms to protect these new goals. These FMCLs could be incorporated inside the boundaries of its constituent systems, or they could cross systems boundaries. Determining whether these are needed and are cost-effective is evaluated the same way as they are for systems, by determining the improvements to SoS RAS and comparing that to the life cycle cost of the FMCL.

In Chapter 2 of this book, Maier states that one of the fundamental properties of SoSs is that they perform functions that do not exist in any of their constituent systems. This is true, but it can be true in the nonobvious way of providing redundancy to ensure performance of functions that already exist in at least one of the constituent systems. Thus, the new functionality is providing an alternate means of performing an already existing function. It is "new functionality" only in the sense of providing an alternate way to achieve an already existing capability if that existing capability fails. This would be clear from the GFT representation, as failure recovery would show identical goals, and goal change would identify new subgoals that are activated only if the original goals cannot be met. This type of "redundancy adding SoS" differs from more typical SoSs, which might or might not add redundant functions, but do add functions that are new and nonredundant.

Maier also notes in Chapter 2 that constituent systems of an SoS can have their own redundancy to ensure their operational and managerial independence. This could provide an oversupply of redundancy from an SoS perspective, insofar as some of the systems could be providing redundancy for each other as described in the previous paragraph. However, each constituent system can and often does maintains its own systems redundancy to enable its independent management and operation.

One of the frequently discussed attributes of SoSs is their propensity to generate the so-called emergent phenomena. McCarter and White define emergence as "something unexpected in the collective behavior of an entity within its environment, not attributable to any subset of its parts, that is present (and observed) in a given view and not present (or observed) in any other view." As noted in this same chapter, this definition notes that "the description of a phenomenon as emergent is contingent, then, on the existence of an observer; ... Clearly the existence of an observer is a sine qua non for the issue of emergence to arise at all." Identifying a phenomenon as emergent "... centers on an observer's avowed incapacity (amazement) to reconcile his perception of an experiment in terms of a global world view with his awareness of the atomic nature of the elementary interactions" (McCarter and White, 2009).

In Chapter 2 of this book, Maier defines and elaborates four subcategories of emergence: simple, weak, strong, and spooky. He then notes the relationship of knowledge about an SoS at any given time and the classification of any given behavior as emergent and the category type of that emergence. In all cases, there is "something unexpected in the collective behavior of the SoS" (McCarter and White, 2009).

Note the similarity to the SHM definition of an anomaly, which is "unexpected performance of intended behavior." We hypothesize here that there is no significant difference between the concept of anomaly and the concept of emergence, except in the relatively large amounts of literature devoted to emergence and its various subclassifications. Emergent behavior, just like anomalous behavior, can be good or bad. Both signify differences between actual and expected behavior and are thus knowledge problems. In the parlance of control theory, emergence and anomalies signify large errors in state estimation and hence are knowledge errors. Knowledge error is defined as a difference between the true state of a system and the systems estimate of that state.

Inherent in the definitions of emergence and anomaly is the necessity of an observer. There can be no emergence or anomaly unless there is an expectation from observers and measurements or observations by an observer. In essence, emergence and anomaly are measures or indicators of our ignorance. More concisely, they refer to differences between reality and our predictions of reality, which are always based on our models and simulations of reality.

Within the theory of SHM, there has been to date no particular defined need for subclassifications of the concept of anomaly. Hence, there does not seem to be any particular need for detailed subclassifications of emergence (i.e., simple, weak, strong, spooky) either. All are simply knowledge errors. Regardless of any subtypes that could potentially be defined, the response to anomalies is the same in all cases: investigate the anomaly until it is understood. Once understood, the behavior will be considered nominal or failed behavior, and the relevant models and simulations are updated so that the formerly anomalous behavior will no longer be considered anomalous in the future. This presumes that for the purposes of SHM for systems or SoSs, "spooky emergence" and hence "spooky anomalies" do not exist.

6.7 CONCLUSION

The issues and hence the models and simulations involved with designing SHM for an SoS are not fundamentally different than the issues involved with design or redesign of a system. SoSs can be built either to accomplish new goals that none of its constituent systems can achieve or to provide improved dependability to achieve some existing system's goals through the addition of redundant functions.

Cost is perhaps the primary driver to decide whether to build an SoS or to create a new specialized system. If it is less expensive to build a new system to achieve goals than to use other existing systems in an SoS, this would be done, as a custom-built system can be optimized to achieve the new goals. Cost is also a key metric to decide which new HM should be implemented in the SoS. The designer must always compare the benefits of the new HM with RAS goals and compare those with the costs of implementing the new HM functions or improvements.

For modeling and simulation, the GFT appears to be the primary tool to understand the differences between systems and SoSs, because they primarily differ in the intentionality of the SoS compared to the intentionality of its constituent systems. Whether a system can be used in the new SoS depends on the goals that the systems can support, compared to the goals that the SoS is designed to achieve. The GFT clearly indicates the differences in the goals, functions, and state variables being used by the systems and by the SoS. In combination, these define the "semantics" of the SoS and its constituent systems, which is the first step in determining their compatibility.

Because the SoS GFT model incorporates all of the goals and functions of the constituent systems in a single tree structure, HM design can proceed in the same way as in systems design by asking what the designer intends to do if a goal in the GFT cannot be achieved. If something must be done, then the state variables related to that goal (or to lower-level goals that support it) can be monitored, and if a failure occurs, it will be detected, leading to response functions. Once the off-nominal requirements for detection and response are specified after being modeled in the GFT, then a suite of HM models and simulations can be developed and used to assess the proposed HM designs. These include fault trees, discrete-event and state machine models, quantitative control metric models, diagnostic directed graph models, and physics-based models to support all of the others and for prognostics. Finally, fault injection test is used to verify the HM design as off-nominal behaviors are necessary to stimulate the FMCLs. Their results are fed into the analytic validation of the HM design of the SoS.

REFERENCES

Garrett, Robert K., Jr., Steve Anderson, Neil T. Baron, and James D. Moreland, Jr. (2010) Managing the Interstitials, a System of Systems Framework Suited for the Ballistic Missile Defense System. *Systems Engineering*, 14, 87–109.

Hollnagel, Erik, David D. Woods, and Nancy Leveson, eds. (2006) *Resilience Engineering: Concepts and Precepts*. Burlington: Ashgate.

International Council on Systems Engineering (INCOSE) (2010) *A Consensus of the INCOSE Fellows*, http://incose.org/practice/fellowsconsensus.aspx (accessed September 1, 2013).

Jamshidi, Mo, ed. (2009) *Systems of Systems Engineering: Principles and Applications*. Boca Raton: CRC Press.

Johnson, Stephen B. (2013) Goal-Function Tree Modeling for Systems Engineering and Fault Management. AIAA Infotech@Aerospace (I@A) Conference, August 19–22, 2013, Boston, AIAA Paper 2013-4576.

Johnson, Stephen B., Thomas J. Gormley, Seth S. Kessler, Charles D. Mott, Ann Patterson-Hine, Karl M. Reichard, and Philip A. Scandura, eds. (2011) *System Health Management: With Aerospace Applications*. Chichester: John Wiley & Sons, Ltd.

McCarter, Beverly Gay and Brian E. White (2009). "Emergence of SoS, Sociocognitive Aspects." In Mo Jamshidi (ed.), *Systems of Systems Engineering: Principles and Applications*. Boca Raton: CRC Press.

Merriam-Webster (1991) *Webster's Ninth New Collegiate Dictionary*. Springfield: Merriam-Webster.

Chapter 7

Model Methodology for a Department of Defense Architecture Design

R. William Maule
Naval Postgraduate School, Monterey, CA, USA

7.1 INTRODUCTION

In this chapter, we briefly discuss common modeling practices as applied in the Department of Defense (DoD) for system of systems (SoS) architecture design and development. SoS architectures are prevalent throughout the DoD, and the Department of Defense Architecture Framework (DoDAF) models provide the reference for systems integration. Formal processes and models found throughout DoDAF serve as guides for architecture development and SoS integration.

DoDAF can be used to satisfy the principal characteristics of a SoS, as described by Maier in Chapter 2 of this text. DoD systems are designed to operate independently so that a failure will not cascade—yet at the same time be highly interoperable to enable data integration across wide geographic regions. Collectively, services of these independent yet highly integrated systems support decision makers with information beyond what is available from the individual systems—the whole is truly greater than the sum of the parts. Additionally, this approach enables the enterprise SoS to evolve with the addition of each new independent and interoperable system component.

While overwhelming to most of us at first glance, the rationale for DoDAF for SoS engineering is straightforward. First is the complexity of extremely large systems

---

Modeling and Simulation Support for System of Systems Engineering Applications, First Edition.
Edited by Larry B. Rainey and Andreas Tolk.
© 2015 John Wiley & Sons, Inc. Published 2015 by John Wiley & Sons, Inc.

integration projects and the need for a broad and deep model base. Second is the need for a model framework that can serve as the basis for communication from top to bottom in an organization and within and between units in different organizations. Finally, DoDAF is the required modeling technique for those in the DoD and for organizations supporting the DoD: "DoD Components are expected to conform to DoDAF to the maximum extent possible in development of architectures within the Department—Conformance ensures that reuse of information, architecture artifacts, models, and viewpoints can be shared with common understanding" (DoD CIO, 2010).

Commercial tools that support DoDAF models are often employed to help code and enact the architecture, and some of the more popular commercial tools will be discussed. To more fully understand this process, we will begin with a brief overview of DoDAF, and this will serve as the foundation for examples throughout this chapter.

7.2 DoD ARCHITECTURE

The DoD uses models for a range of activities—from concept representation to network design, to interface specification, and to code generation. DoDAF is perhaps the most comprehensive modeling architecture available. It is not a tool but rather a series of specifications for model development.

The impetus to develop a comprehensive modeling framework across the U.S. government received momentum with the Clinger–Cohen Act in 1996, which mandated that federal agencies develop and maintain enterprise information technology architecture (GAO, 2002). Enterprise architecture (EA) is defined as SoS architecture that clearly illustrates functionality and precisely models interoperability.

Additional federal support for EA integration came from the Office of Management and Budget Circular A-130, which established policy for the management of federal information resources and called for the use of EA, and the E-Government Act of 2002, which called for the development of EA to enhance the management and promotion of electronic government services and processes (DoD CIO, 2010).

DoDAF is focused on interoperation of disparate systems used in support of military missions. The framework is intended to support command, control, communications, computers, intelligence, surveillance, and reconnaissance (C4ISR) mission systems development with a common modeling language to enable large groups of stakeholders to work effectively on large complex systems (Rational, 2002).

To accomplish this task, DoDAF addresses six core processes: (1) Joint Capability Integration and Development; (2) Defense Acquisition; (3) Systems Engineering; (4) Planning, Programming, Budgeting, and Execution; (5) Portfolio Management; and (6) Operations (DoD CIO, 2010).

In military modeling, we characterize DoDAF models by their intended function—which tends to be dependent on their intended audience. Models for conceptual understanding are at the highest level of the modeling hierarchy. In these models, we seek to convey technical ideas at an operational level. Our audience would be leadership in the organization. An example would be a new technical innovation for which we seek adoption by leadership. Our intent is to persuade or inform decision makers through our models.

In DoDAF, these are referred to as Operational View (OV) models. In OV models, we depict the potential impact of an innovation on everyday routines or processes.

Innovations may range from minor software upgrades to major changes for personnel, processes, or systems.

As we move down the "complexity curve" from the conceptual at one extreme to the highly technical at the other extreme, we find models that present architectural context in increasing degrees of specificity. Models at the more technical levels will typically show active systems or networks. Into these models, we inject our proposed new additions.

The audience for this level of model would tend to be more technical than for the previous OV models, and their role would be to evaluate the technical feasibility of the innovation. In DoDAF, we use these Systems View (SV) models to convey detailed processes, for example, networks, servers, or applications. SV diagrams play a critical role in systems integration projects. We also use them to gain security accreditation for our innovation.

How do we know what all these systems really do? How exactly do they share data? What happens to the network and other systems when our innovation is active? Are there potential security concerns? These are the issues we seek to resolve through models and simulations at the systems level.

In the DoD and for corporations that support the DoD, the DoDAF reference models are typically the starting point. Through DoDAF, we model down to specific events in operational, systems, or service views that show the execution of our application, step-by-step, component by component, location by location, over time.

We address intervening systems, human-in-the-loop operations, and decision processes—both human and automated. Once fully modeled, a simulation may then be run with our system to approximate actual operations over time. We examine the output of the simulation for inconsistencies or problem areas.

While there are overlap and repetition within the broad classifications and their respective models, the broad premise that OV models present organizational perspectives and SV models present engineering specifications generally holds.

In the remainder of this chapter, we will look deeper at model development and practices within the DoD. We will include some examples of commercial tools that are commonly used to develop DoDAF models and simulations. First, perspective is important, so we begin with a quick look at DoD architecture within the broader architecture community.

7.3 REFERENCE ARCHITECTURE

Why do we need reference architecture and frameworks? Simply put, to communicate. The DoD is characterized by very large SoS projects. My large and complex system needs to interface with your large and complex system—which in turn interface with other large systems, different types of networks, various security devices, and so on. Reference architecture helps us make sense of it all.

Important for the reader to note is that DoDAF was not developed independent of other modeling approaches. The DoDAF developers based their design on the best practices at the time, and they upgrade DoDAF when new software practices or systems capabilities become available. DoDAF is therein a cumulative process in which the framework is updated to include new innovations from other reference architectures as feasible and appropriate to meet the specific needs of DoD clientele.

Some of the prominent reference architectures and supporting documentation that provides a basis for DoD modeling efforts include:

- Federal Enterprise Architecture Framework
- Federal Enterprise Architecture Reference Models
- Enterprise Architecture Assessment Framework
- Enterprise Architecture Management Maturity Framework
- DoD Global Information Grid Architectural Vision: Vision for a Net-Centric, Service-Oriented DoD Enterprise
- DoD Information Enterprise Strategic Plan 2010–2012
- DoD Architecture Registry System
- DoD Information Enterprise Architecture
- Open architecture for Accessible Services Integration and Standardization Reference Model (OASIS-RM) for Service-Oriented Architecture (SOA)
- Office of the Assistant Secretary of Defense, Networks and Information Integration, Reference Architecture Description
- UML Profile-Based Integrated Architecture (UPIA)
- SOA Modeling Language (SoaML)

As we discussed earlier, a very important use of modeling within the DoD is for modeling SoS integration. Reference architectures provide guidance for costly and time-consuming systems integration efforts. EA documents address standards, agreements, security, and communications protocols, Web service specifications, XML namespaces, measures for data quality, etc. Collectively, these reference documents and the supporting federal frameworks provide a conceptual basis for our models.

Reference models or documentation can also be used to justify models. If we are questioned on the direction we are heading, it helps to have a high-level reference handy—especially one approved by a higher authority.

DoDAF has evolved through several generations, and as the broader architecture community creates new reference architecture, the DoDAF community quickly responds—adopting important new modeling concepts and integrating these into the framework. A recent and very significant upgrade to DoDAF occurred with the programming evolution into SOA. Some background information may help to understand the importance of this evolution.

The reader will recall that architectural descriptions include views that contain information about a system from a particular perspective. These are often referred to as "products" and classified according to their specific architectural attributes.

DoDAF V1.5 included the All View (AV), OV, SV, and Technical Standards View (TV). In telling our "story" about the system that we are developing through our models, we develop models in each of these categories—representing our innovation from the different perspectives of those viewpoints. Again, the audience is the key.

The model librarian would want an AV to establish the innovation and all its model products. Leadership would want an OV to provide the capability in context. The engineering community would insist on SV models to evaluate technical feasibility. A TV diagram could address security and appropriate operational standards.

Thus, aspects of systems architecture are often best described through multiple views or models. Relationships between significant architectural elements of an enterprise will quite often require many views.

In expressing views, the core elements of the models are "nodes," "needlines," "services," and "information exchanges." Nodes can be physical or logical and include locations, systems, components, and humans. Needlines show relationships and dependencies between nodes and are the primary means to characterize complex SoS relationships. Services are a newer approach, popularized through SOA, and represent information exchanges and the data types that traverse the service architecture. The service element is expanded in DoDAF Version 2.0 (V2.0) (DoD CIO, 2009).

DoDAF V2.0 shifted the underlying modeling paradigm to data-driven "viewpoints" with an extensible data model (Hughes and Tolk, 2013). Additional perspectives were added to assist in the transition to service architecture. The Capability Viewpoint (CV) helps us articulate requirements and functions to be achieved to address capability requirements. In the Data and Information Viewpoint (DIV), we specify relationships between data elements and structures, including databases. The Project Viewpoint (PV) helps us describe the association between operational elements and dependencies. Finally, through the Services Viewpoint (SvcV), the most specific to SOA, we communicate service-based solutions, complete with "performers," "activities," "services," and their "exchanges" (DoD CIO, 2010). In V2.0, the TV has been renamed the Standards Viewpoint (StdV). The remaining V1.5 "views" are reclassified in V2.0 to "viewpoints" with the same acronym.

To help us realize all of the models required to fully document an EA, we rely on commercial modeling tools and practices. Some of these support our DoDAF requirements; some even output their models in DoDAF formats.

Unified Modeling Language (UML) is overseen by the Object Management Group (OMG), a not-for-profit industry standards consortium (Object Management Group, 2011). UML and the tools that support UML model development are instrumental at the lower, more technical levels of model design. It is at these lower levels that DoDAF begins to waiver and some of the commercial modeling approaches begin to shine. Specifically, when we evolve to the far end of the modeling perspective and need to generate operational computer code from our models, execute and monitor the code from the models, revise the code through the models, and/or deploy the code to operational enterprise systems. For this, we need robust commercial tools, and many of these are based on UML.

UPIA builds upon UML to present design viewpoints that can model data-level structures for SOA—complete with service specifications, service ports, and service consumers and providers at both operational and system levels of abstraction. UPIA is a DoDAF V2.0 and eXtensible Markup Language (XML)-complaint metalanguage that can import and export DoDAF V2.0 Physical Exchange Specification (PES) architectural data (IBM, 2010).

PES and the DoDAF Metamodel (DM2) are two of the more significant aspects of DoDAF V2.0. DM2 provides a high-level view of data elements to better support reuse of architectural information among DoD agencies and partners (Mittal and Martin, 2013). Collectively, PES and DM2 provide the capability to describe a model in XML vice a physical/visual model and can therein speed model exchange—assuming one has a compatible software package to use the PES. Additionally, UPIA includes utilities for model validation. Most of the concepts defined in DM2 can be modeled in UPIA.

UPIA modeling provides a standard approach for modeling systems and EA and supports both DoDAF V2.0 and the U.K. Ministry of Defence Architecture Framework

(MODAF) (IBM, 2010). Additionally, UPIA can be integrated with Model-Driven Software Development (MDSD) methodologies to enable the generation of operational code from the models.

Technically, UPIA models are a type of UML model that uses specialized elements to depict enterprise and SoS architectures. Fortunately, we have some pretty robust modeling tools to help us through this complexity. The reader will benefit from additional background information to place these advanced architecture concepts into context.

SoaML is an open-source specification project, also from OMG, which describes a UML profile in a metamodel format for services within a SOA. SoaML proponents argue that existing models and metamodels—such as The Open Group Architecture Framework (TOGAF) for describing systems architectures—were insufficient to describe SOA in a precise and standardized way. UML at the time was considered too general and needed clarification and standardization specific to SOA.

Additionally, a means was required to operationalize SOA as advanced in the OASIS-RM. SoaML provides SOA-specific tooling, and many consider it a viable instantiation of OASIS-RM.

The models in the following sections were developed using DoDAF and UPIA frameworks, along with SoaML tooling. We use UML use cases (UC) to model architectural context and conceptual architecture as discussed previously and to provide environmental perspective. Models specific to SOA were designed to convey real-time, high transaction operations as would be found in many DoD architectures.

Prior to our discussion of these detailed models, we present some examples of commercial tools that are commonly used in the DoD to model architecture and/or to simulate operations. Before we leave this section, we must also note similar and associated military-specific modeling systems that somewhat align with DoDAF but have different implementations specific to their organizational requirements. These include the NATO Architecture Framework (NAF) and the previously mentioned MODAF.

Integration DEFinition (IDEF) models are also significant in the DoD. Similar to DoDAF, the IDEF models range from high-level functional models to low-level object-oriented design and simulation models. IDEF provides useful operational representations in addition to precise data/information models (IDEF, undated).

7.4 MODEL TOOLING

Selection of the correct tool for design and development of our models is critically important. The sections that follow present a few of the more popular modeling and simulation (M&S) tools in the DoD. We will note those that directly output DoDAF models.

7.4.1 Joint Communication Simulation System

The Joint Communication Simulation System (JCSS) uses DoD-specific systems nomenclature. JCSS is a Defense Information Systems Agency (DISA)-endorsed Commercial Off-The-Shelf (COTS) M&S tool based on the OPNET Modeler product line. The software is free for DoD civilians, DoD contractors, and military personnel.

The intent in providing this software to the DoD community is to develop a common modeling base across the command, control, communications, and computer (C4)

Model Methodology for a Department of Defense Architecture Design   151

systems communities (JCSS, undated). As the name implies, this software is focused on communications planning and on simulation of communications effects on networks.

In operation, we begin with construction of a network topology. Then, we add information systems, databases, and communication-specific devices—such as routers, satellite antennas, multiplexors, etc. We note expected transmission capacity, including the bandwidth that will be available and the expected demands on that bandwidth.

Next, we run simulations using "what-if" analysis to help determine if the communications capacity is sufficient to accommodate the expected network traffic. Of course, all of this implies that the modeler has access to accurate data—such as past network performance for similar environmental constraints and transmission conduit or frequencies, the load generated by the applications, and the degree of latency tolerated by the systems.

JCSS can generate reports to help us analyze our models, including utilization statistics, failure reports, and network optimization analysis. We can run discrete event simulation as a "state machine" in which change in the state of a machine, network connection, or transmission capacity can help predict impact on systems and applications—with metrics for jitter, latency, and queuing delay for different traffic.

We can render JCSS reports into a DoDAF OV-3 Operational Information Exchange Matrix or a DoDAF SV-6 Systems/Services Data Exchange Matrix. The higher-level views generated in JCSS can serve as DoDAF OV-1 High-Level Operational Concept Description models, and the more detailed views can present DoDAF SV-2 Systems Communications Description models—if appropriate for our situation (see Figure 7.1).

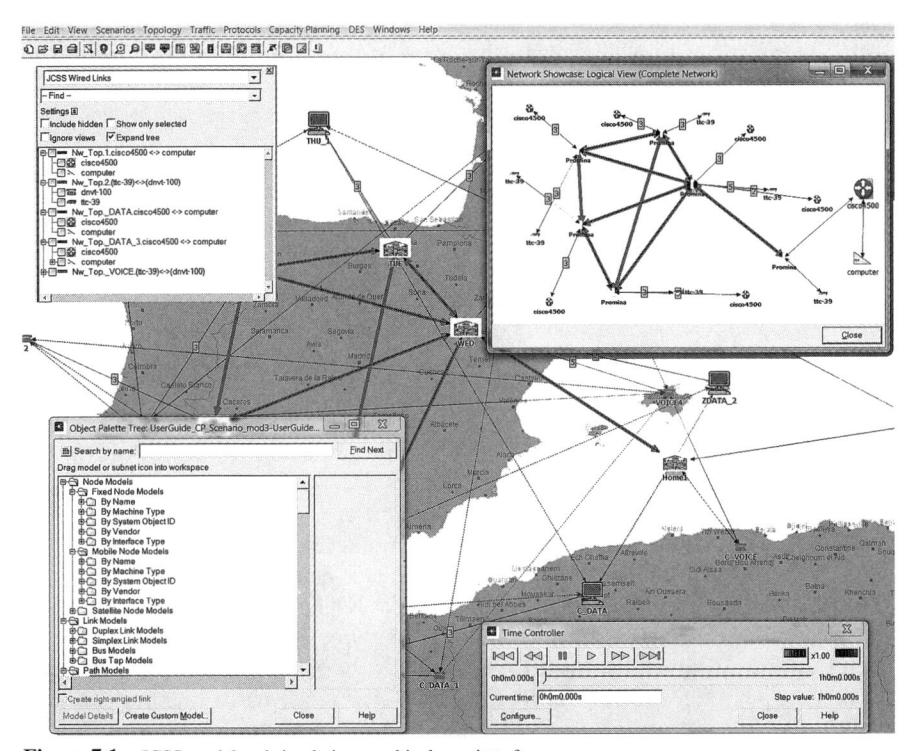

Figure 7.1   JCSS model and simulation graphical user interface.

7.4.2 Systems Tool Kit

Systems Took Kit (STK) M&S software from AGI Corporation excels in model development for space defense industries. STK is widely used in the DoD for satellite and aircraft M&S. Recent enhancements to the product have introduced opportunities to include air, land, and sea assets in the models, complete with high-resolution event simulations (AGI, undated).

STK orbit and ephemeris data can be exported to JCSS. However, STK is in a distinctly different area of modeling from JCSS. We would use JCSS in situations where detailed communications modeling and analysis was required, and we have access to data points with defined characteristics, such as networks, systems, and associated data/media.

Another difference is that JCSS models and simulations are 2-dimensional (2D), the reports highly detailed, and as discussed previously, the matrix views in DoDAF are supported. By contrast, STK is for 3-dimensional (3D) representation of moving objects—on land, sea, air, or space. Visuals in the simulations and animations can be nearly lifelike and simulated with radar patterns, satellite footprints, and aircraft.

If properly constructed, through STK, we can create precise, realistic events. The reports that we can generate are more in the area of resource allocation vice the JCSS communications performance reports.

In a perfect world, with time permitting, we would use both tools. We would model and then simulate the scenario in STK with primary resources and assets for 3D visual analysis, and we would simultaneously model communication specifics within the JCSS.

We would use STK for asset evaluation—for example, unmanned aircraft system (UAS) sensor packages within geographic areas or area coverage of unmanned aerial vehicle (UAV) flight patterns for search and rescue operations or analysis of communications coverage for mobile users in isolated terrain with geostationary satellite constellations supplemented by Beyond Line-of-Sight (BLOS) radios.

Prior to execution, we would model the communications in JCSS and during execution capture the traffic data, import that data to JCSS, and use that data to refine our STK event simulations and resource allocation models.

While there is overlap with JCSS in some of the communications and systems areas, the focus of each software package is distinctly different, and the user audience for the generated models can be similarly differentiated. Thus, we need to not only evaluate the project carefully before selecting a modeling tool but also the audience for the models. We would consider the output capability of our modeling tool—whether DoDAF models, 3D simulations, or formal reports.

STK does not directly render to DoDAF, but the visual representations can be captured for OV-1 models. The reports can be used as data points for the DoDAF resource matrix models. Figure 7.2 shows the STK user interface with rendering in 2D and 3D windows—which are active for both modeling and simulation.

A bonus is that we can integrate STK with a large number of complementary software packages for an easy exchange of ideas, such as Keyhole Markup Language (KML) for visualization to Google Earth. Or we can export our model specifications to the open-source Systems Modeling Language (SysML) for Systems Engineering specification and model validation. We can use QualNet from SCALABLE Network Technologies to capture and import live communications data into our STK simulations.

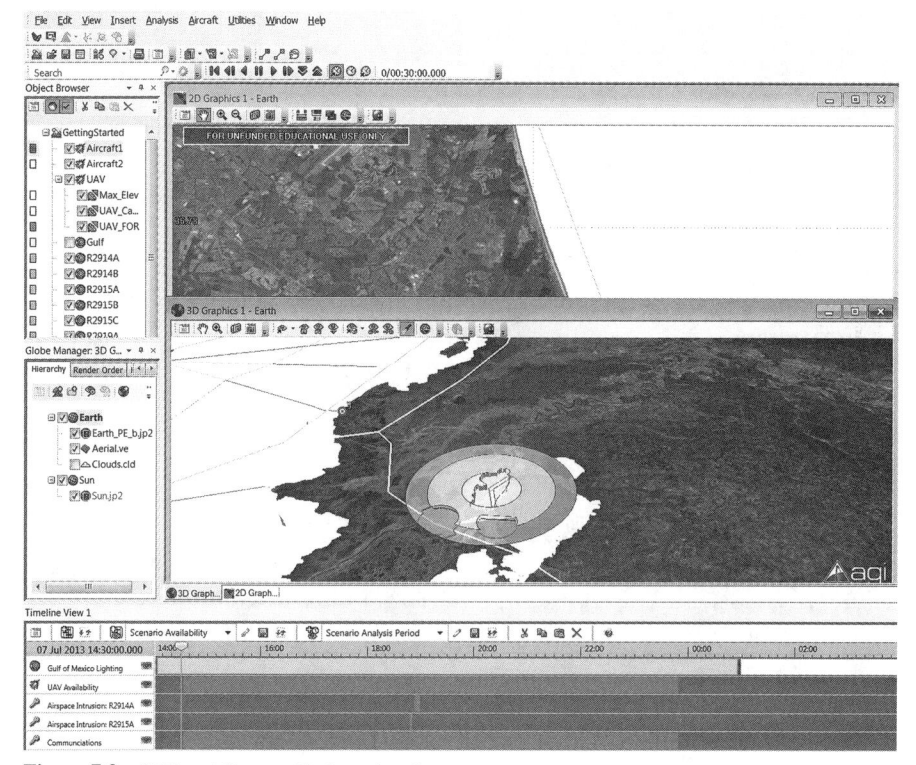

Figure 7.2 STK modeling graphical user interface.

7.4.3 IBM Rational

UML originated as an integration of modeling theories from Grady Booch, Ivar Jacobson, and James Rumbaugh who formed Rational Software in the 1990s to advance UML methodology and develop UML-based model tooling (Hamilton, 1999). UML was adopted by OMG in 1997 who manages its development and by the International Organization for Standardization (ISO) in 2000 as an industry standard for modeling software-intensive systems. IBM acquired Rational in 2002 to help integrate software across all of their computing segments (IBM, 2003).

As originators of UML, Rational was the prominent producer of modeling tools based on UML. Under IBM, the Rational UML approach was integrated across the IBM EA for systems and software engineering.

IBM acquired the Swedish company Telelogic for its application life cycle management software in 2007, which it also integrated into the Rational suite (Network World, 2007). Pertinent to our discussions is that Telelogic System Architect [now IBM Rational System Architect] has the most significant support for DoDAF of any of the modeling tools on the market.

Figure 7.3 shows the DoDAF model options built into the systems architect (SA) graphical user interface. There are two SA versions optimized for DoDAF architectures:

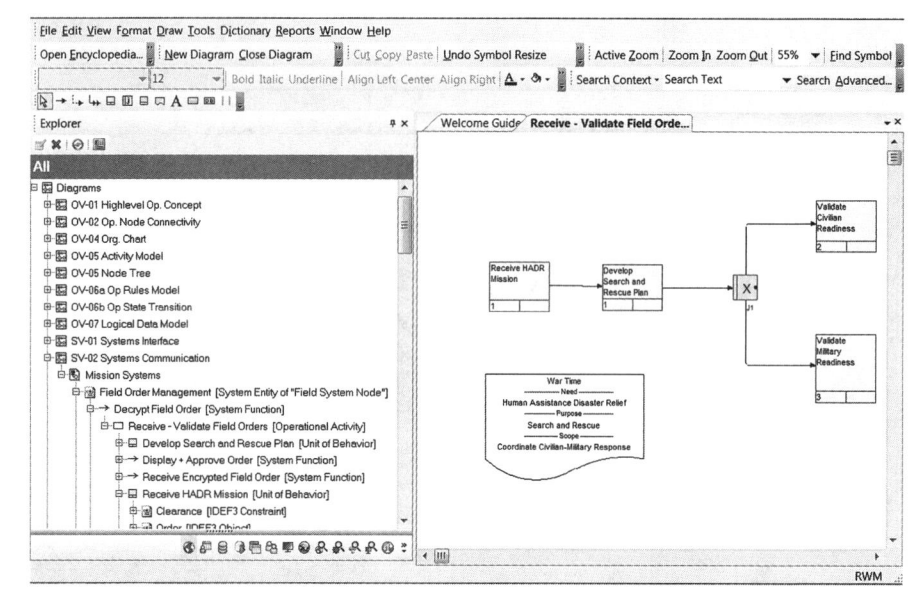

Figure 7.3 Systems architect graphical user interface with native DoDAF models.

SA for DoDAF with the MITRE Activity-Based Method (ABM) option and SA C4ISR—which has been renamed to SA for DoDAF—to build open architecture models around structured IDEF techniques.

Rational Software Architect (RSA) moves us to the next step in the systems design life cycle (SDLC) [or software development life cycle (SDLC)]—software engineering. RSA is built on UML and provides DoDAF-type models but without predefined DoDAF output modes. In other words, we need to understand the DoDAF model that we are designing and then use RSA to develop that model vice the SA process where we would select the DoDAF model and then automatically receive the necessary tooling.

RSA can be used to model and simulate enterprise applications in C++ and Java Enterprise Edition (Java EE) including SOA and Web services (Rational Software Architect, undated). RSA is built on the Eclipse open-source software framework, which is the industry standard for Model-Driven Development (MDD) in the DoD and private sector.

Figure 7.4 shows the RSA user interface in the traditional Eclipse design pattern—with the model in the center and the development tooling around the model with tabs and drop-down menus for the tooling options. In the figure, we have selected the Integrated Architecture Modeling (IAM) with UPIA SOA design options to develop SoaML models—as indicated in the selection in the left column.

We use simulation within RSA on several fronts. In the requirements specification phase, our model is still informal. A simulation can help us communicate to sponsors our understanding of the dynamics of the system and also to bring front and center any potential misunderstandings from incorrect assumptions or interpretations.

In the development phase—where we spend most of our time—we use simulation to debug our code. We initially run our code module by module in stand-alone mode and

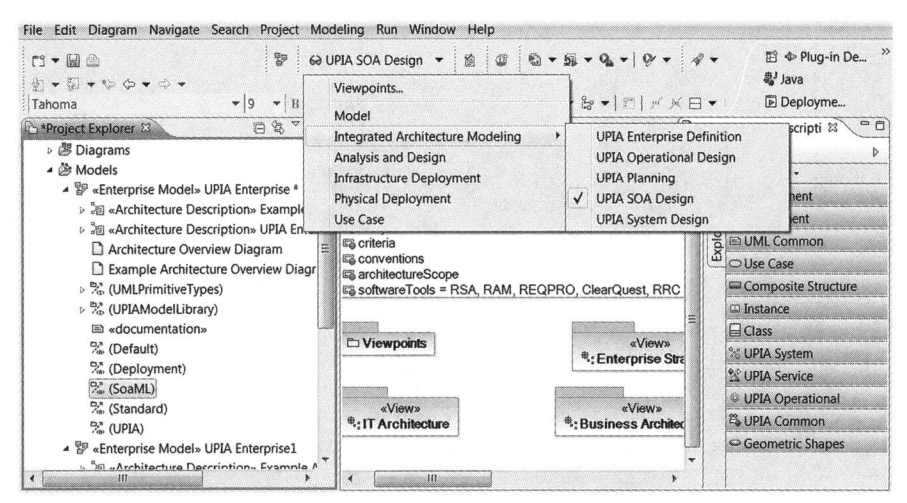

Model Methodology for a Department of Defense Architecture Design   155

Figure 7.4 Rational Software Architect UPIA and SoaML modeling options.

later collectively in our anticipated deployment environment. In the deployment phase, we use simulation to monitor our software when implemented against different network topologies (Mohlin, 2010). It is here we gain an understanding of potential communications bottlenecks or systems incompatibilities.

Next, and last in our tour of modeling tools with DoDAF references, is a derivative of the previously discussed software architecture modeling in which we use a specific tool to achieve increased fidelity in a specific type of model. There are numerous such instances where a specialized approach works best, and there are typically tools optimized for each instance. We have selected one in the Business Process Management (BPM) area specific to SOA processes. In this instance, for tooling, we will use WebSphere Business Modeler (WBM) (IBM, undated).

We continue the Humanitarian Assistance Disaster Relief (HADR) scenario we have been modeling across the SDLC and through the different levels of tooling. Figure 7.5 shows the cross-domain security components for our HADR SOA model. Note that the simulation component includes a module that calculates cost for return on investment (ROI) analysis of our proposed innovation—in this instance, for a complex role- and attribute-based security process.

Here, we are showing a means to semiautomate a largely manual process in need-to-know content clearance across DoD, coalition, and civilian authorities, basically, how to get the information to those who need it to process emergency supplies for victims of a natural disaster—without compromising security for the systems that host the content.

The figure shows only a small portion of a larger model but is sufficient to illustrate some of the different options for content clearance in this scenario. When the simulation is run, tokens animate through the model, and our cost calculator at the bottom of the screen dynamically updates. We can see where security bottlenecks occur under load—such as is typical in an HADR event.

Table 7.1 expands the simulation data with a report on the cumulative runs to show total cost of the new capability for 100 processes.

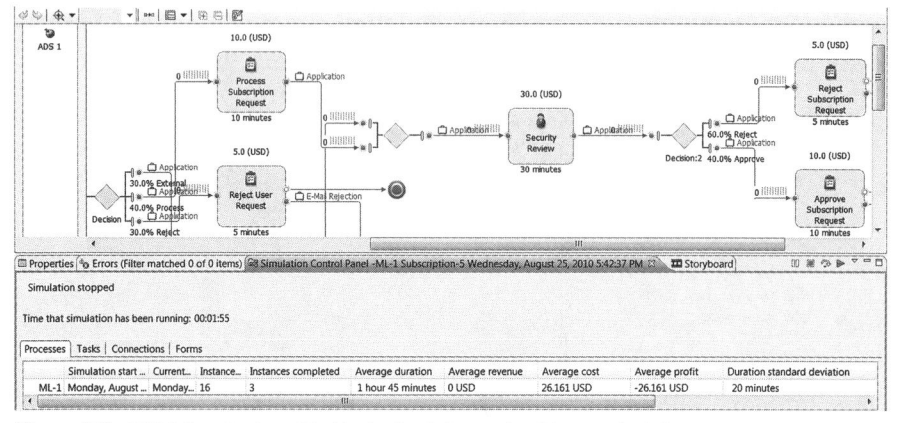

Figure 7.5 WebSphere Business Modeler in simulation mode with cost calculation.

Name	Total cost	Total revenue	Total run cost	Resource	Total profit	Total elapsed duration	Runs
Case 1	USD2,486.75	USD0.00	USD2,465.00	USD21.75	(USD2,486.75)	23 days 6 h 5 min	29
Case 2	USD816.75	USD900.00	USD810.00	USD6.75	USD83.25	7 days 6 h 50 min	9
Case 3	USD1,452.00	USD1,600.00	USD1,440.00	USD12.00	USD148.00	9 days 23 h	16
Case 4	USD1,135.42	USD0.00	USD1,125.00	USD10.42	(USD1,135.42)	18 days 1 h 25 min	25
Case 5	USD1,800.75	USD0.00	USD1,785.00	USD15.75	(USD1,800.75)	15 days 11 h 45 min	21
All	USD7,691.67	USD2,500.00	USD7,625.00	USD66.67	(USD5,191.67)	74 days 1 h 5 min	100

Table 7.1 WebSphere Business Modeler simulation data with cost calculation.

7.5 MODEL WORKFLOW

As we discussed earlier, DoDAF is the primary modeling system in the DoD. In support of DoDAF, some commercial M&S tools support one or more of the DoDAF models. Some provide specialized tooling for specific military needs. In our applications, DoDAF models provide the basis for communication between forces requiring capabilities and assets providing systems to meet operational requirements.

Simulation helps us display the capability conveyed in the model in a situational context. For specialized modelers at the technical end of the SDLC, there are modeling tools with code generation capabilities to enable software engineers to build and iteratively refine software directly from models. We have models with monitor and control capabilities for systems and network engineers to watch the system as it is executed and test and measurement models with data capture capabilities for software quality assurance.

DoDAF model "views" or "viewpoints" characterize conceptual, operational, systems, standards, and management perspectives. Ideally, in the development of a system, the DoDAF models will flow from the conceptual to the operational and then to systems development and management—with standards referenced as appropriate. High-level concept diagrams such as the AV will provide organizational structure and the OV conceptual understanding of the innovation, within context. The SV provides interfaces between systems and components, complete with data flows across interfaces. Under each of these "views" or "viewpoints" are up to 10 additional diagrams that provide increasing levels of specificity.

7.5.1 Unified Modeling Language

For engineers, the DoDAF suite provides necessary conceptual and operational understanding, plus the system and network context for development. At runtime, we use specialized modeling tools to monitor the applications and systems as they execute.

If fully implemented, we have an end-to-end life cycle model that can assist with test, measurement, and the debugging of our computer code. In practice, due to the extreme complexity of the SoS architecture active in the DoD, different teams of individuals tend to work on different models. So, full life cycle, end-to-end model integration is rarely achieved.

When modeling to the code level and modeling for SoS integration at a service level, the choice of most developers is UML-based systems. UML derivatives include the Unified Profile for DoDAF and MODAF (UPDM) and the SoaML referenced earlier.

SOA is perhaps the prominent methodology for DoD SoS integration—which we typically see as service-to-service integration across systems. Rather than hardwired models, we use a broker and publish/subscribe communications for data exchange across diverse systems and their data models.

Most DoD systems have already evolved to SOA or are very close. SOA is seen as a means to significantly increase overall information technology effectiveness and performance while substantially reducing hardware and software costs (DoD Business Transformation Agency, 2009). In enterprise SOA, the primary operations are shared and typically modeled "in the cloud" or within cloud computing architectures.

Services can be modeled cumulatively, building upon one another to form composite applications. In a composite service model, we capture and parse data, add logic, and publish for others to use. SOA modeling environments are thereby rich and interesting. Since SOA services are increasingly the norm in the DoD, we will reference SOA in our example models.

However, while SOA composite services provide users with a level of systems integration that had been previously cost prohibitive, the SOA composite application approach also introduces complexity and therein risk. This risk can be significantly lessened through the introduction of robust models, which serve as frameworks for the programmers.

SOA services are independent and loosely coupled—but tightly controlled through security and policies—"if" programmed to be truly secure. So, we must carefully integrate both security and policy at the service level into our models and across the entire SDLC. The devil is in the details—any lapse in security, in any model, at any phase of the SDLC, and we have a potential security issue. Complicating our models is that services are self-describing to users and may be assembled "ad hoc" to "orchestrate" processes. [Modeling of security for user-defined composite services is beyond the scope of our discussion.]

In the SDLC, a suggested approach is to first model enterprise systems that implement collaborating services as a high-level UC. Each UC model then provides a well-defined set of functions—in our case, available over a SOA network. All of this will become a little clearer as we evolve into the examples.

We will integrate UML models and UCs with DoDAF models throughout this chapter. UC models can provide a user perspective. UML-based modeling tools can provide a finer level of granularity than DoDAF and can be actionable to simulate operations with metrics and SOA-specific measurements.

158 Theoretical and Methodological Considerations

Through UML modelers, we can invoke operational code from models, measure network and systems performance and integrate those readings into models, simulate operational systems from models, and inject modeled/simulated components into live operational systems. So, the benefits of moving to code-based models can be significant.

In sum, we can use UML-based tools to model operational code as end-to-end processes in a SoS architecture. Models may evolve from high-level requirements to specific designs and be used as the basis for code development, implementation, monitoring, and maintenance.

7.5.2 DoD Discovery Metadata Specification

The DoD Discovery Metadata Specification (DDMS) is another model framework developed by the DoD. We can use DDMS for its common set of descriptive metadata elements to assist in modeling and systems development. DDMS metadata elements enable users and systems across an EA to discover a database exists and then access that data. This can be a critical element when modeling user-defined composite services.

As a DDMS UC example, we have modeled a Web service implementation of DDMS metadata elements within the DISA core enterprise service model. The UC is illustrated in Figure 7.6. An example would be a Web service provided as an XML-based geolocation capability for a geographically bound entity. This information is modeled as a capability for field HADR forces to receive alerts on portable devices. We have modeled Web services as a receive capability from a mobile node with synchronization to peer nodes in a distributed architecture. In this instance, the national asset provides imagery of devastation caused by the HADR event.

7.5.3 Business Process Execution Language

Another important model, often used to represent a DoDAF OV-5, is the Business Process Execution Language (BPEL) model. We use BPEL models to present human or process flows. BPEL models can be used to simulate a concept, develop computer code, and monitor code during real-world execution.

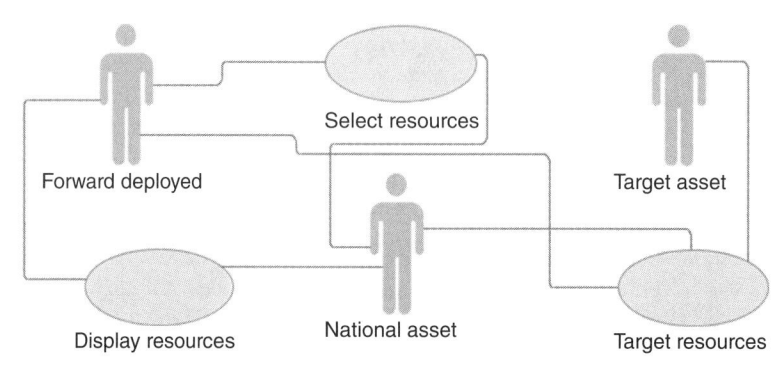

Figure 7.6 Use case for subscription service to HADR event area video.

As an example, in a publish/subscribe model, the developer may complete a BPEL workflow (or equivalent), deploy the service into a "container," and register the service through a Web Services Description Language (WSDL) document into a Universal Description, Discovery, and Integration (UDDI) directory. A subscriber locates the desired service through this UDDI and uses the document's description—from the WSDL—to get a copy of the document by machine-to-machine (M2M) connection with the Web service publisher.

During this process, a user will note the available methods [data objects] that are exposed and select those that provide or update the desired data. At this point, a fully automated M2M process can be modeled and implemented to receive and/or update data in the service. Such is the beauty [and complexity] of SOA.

7.5.4 Business Process Management

As referenced earlier, BPM is often used to model "governance" and is one of the more rapidly evolving components in SOA. We use BPM models to simulate workflow automation and operationally as a component to achieve cost and operational efficiencies. As with other governance tools, BPM models interface with operational components—such as the enterprise service bus (ESB) for SOA communications and BPEL for process execution.

BPM is another of the primary tools we can use to model "agile" or rapid application development (RAD) environments. Most interesting in BPM is the ease with which "patterns"—or preconfigured models with operational code—can be imported and orchestrated with activities that include people.

A pattern is typically modeled in layers, with the:

- *User interface layer*—modeled through Web-based tools integrated with BPM software or by custom programs with external interfaces to BPM models
- *BPM tools layer*—modeled within the core BPM functional areas
- *Storage layer*—modeled as the shared repository for user-defined processes
- *Interface layer*—to model user interaction with BPM components and workflows

We use the term "orchestration" to refer to the automated arrangement, coordination, and management of systems and services in models. Through orchestration, we can model end-to-end processes across the SDLC in a SoS architecture.

As discussed earlier, through BPEL models, we can introduce quality of service (QoS), end-to-end automation, and human-in-the-loop security processes. Orchestration models, in this context, can involve high-level governance—such as the previously discussed BPM and BPEL workflows.

7.5.5 Business Activity Monitoring

Another option for the modeler to consider is Business Activity Monitoring (BAM) software in which we model the monitor components of our SoS. In an SOA, BAM models are considered a prominent component to achieve full SDLC modeling. In this instance, BAM is the final component at the technical end of our SDLC, handling deployment, execution, real-time monitor and control, and maintenance of our SoS.

BAM components and their associated models enable precise monitor and control of processes throughout a SOA cloud—from the backbone to the various nodes in a distributed SoS architecture and down to the individual user. BAM can monitor hardware, software, processes, services, and applications—a capability that historically required multiple expensive hardware tools. We implement BAM in software.

Of interest to those with an eye on the bottom line is that analytic capabilities are typically built into a BAM suite. BAM suites also typically serve as a gateway for the integration of specialized assessment components—such as real-time stream analytics. Obviously, this can all be very complex.

We note the argument against governance systems is that they can introduce significant complexity and therein overhead. This can be true. They require highly skilled modelers. This is true. Another argument is that they can become additional processes to run and manage and may not be universally useful. This can also be true.

However, BAM tools and their real-time models are some of the best available means to secure and manage comprehensive, global, distributed systems—such as those found in the DoD SoS architecture. The gain is worth the pain. Without governance tools, it will be very difficult to manage a modern EA. It will be difficult to understand dynamic processes in a globally distributed SoS environment. Without the understanding provided by our governance models, our security is likely compromised. This trade-off is not well understood.

The reader should also note that systems modeled without governance from the onset of the project may be cost prohibitive to retrofit or convert at a later date. In other words, if your system is not modeled and designed from end to end and does not include the monitor and governance capabilities, then this is opportunity lost. The cost-efficiencies of service architecture will not be realized and perhaps just the opposite—costs may spin out of control—and especially so with each new systems integration project.

You may not have achieved the cost-efficiencies that you expected when you evolved from client–server to SOA. You may even be unaware that you have brought your old client–server infrastructure into your SOA and therein your old models and problems—and this too will prevent your expected cost-efficiencies.

Overall, without governance models, your system may evolve without the safeguards envisioned for SOA. Your system may not be effectively monitored end to end. Here, we are using the term "system" lightly. With service architecture, the number of processes explodes—an exponential increase over client–server. Accurate models are critical to understanding this complexity.

A final note is that without active BAM, BPM, BPEL, or equivalent models built into your projects from the start then end-to-end service monitor and control capabilities may not be feasible. In other words, we cannot cost-effectively secure and monitor complex SoS architectures that have been built without governance models. Usually, due to the complexity of SoS service architecture, a retrofit to include governance may be cost prohibitive so you will in essence be reengineering/remodeling your EA.

Thus, lax modeling leads to costly, difficult-to-understand systems—which results in systems that are overly expensive and easy to compromise. Promises of end-to-end workflow automation, comprehensive security, and the very significant cost-efficiencies that can be realized through service architecture will likely never be realized. But, enough the cautionary—we now have sufficient rationale for full, comprehensive modeling in our SoS architecture and across the entire SDLC. Now, into the models!

7.6 DoDAF ALL VIEWS

DoDAF AV products provide a high-level overview of our system. The AV is typically a text product that provides summary information about our models. The AV can therein provide a conceptual starting point for subsequent lower-level models—which will be presented in increasing detail as this section progresses.

7.6.1 AV-1: Overview

The All View-1 (AV-1) is typically a table of all systems components. An example would be the primary nodes or systems to be modeled in a distributed SoS architecture. Diagrams may support the AV-1 and illustrate basic concepts for the systems to be represented. Subsequent views break the high-level AV-1 attributes into operational, systems, and service components.

7.6.2 AV-2: Integrated Dictionary

In our All View-2 (AV-2), we provide the taxonomy or dictionary for terms used in our models. In this context, taxonomies are most easily characterized as common terminology with common definitions for the development of our models. We use taxonomies as building blocks for DoDAF-described models within the architectural description. We use the AV-2 to model metadata used in the architecture.

7.7 DoDAF OPERATIONAL VIEWS

We use DoDAF models in the operational viewpoints to describe tasks and activities, operational elements, and resource flow exchanges required to conduct operations. DoDAF OV models focus on implementation contexts and provide the basis for the technical systems viewpoints that model service, component, or systems integration.

DoDAF V2.0 adds CV models that provide higher-level constructs that we can map to capabilities and requirements (DoD CIO, 2010). DoDAF V2.0 model extensions are presented throughout this section, often integrated with DoDAF V1.5 model types. We supplement DoDAF with UC models and UML to increase model fidelity and reader understanding.

7.7.1 OV-1: Operational Viewpoint

We use the OV-1 to model systems at a high level, illustrating core participants and operations. We use the OV-1 to model interactions between architecture and the operational environment. In Figure 7.7, we have used the OV-1 to model SOA and authoritative data source (ADS) nodes along the Global Information Grid (GIG). We would develop models of increasing detail for this OV-1 in the systems viewpoints.

Our supporting UC would provide a high-level OV of primary participants and operations, with a finer granularity than the DoDAF OV-1. The core infrastructure through which forces access assets through the network operations center (NOC) and backbone GIG cloud is modeled in Figure 7.8.

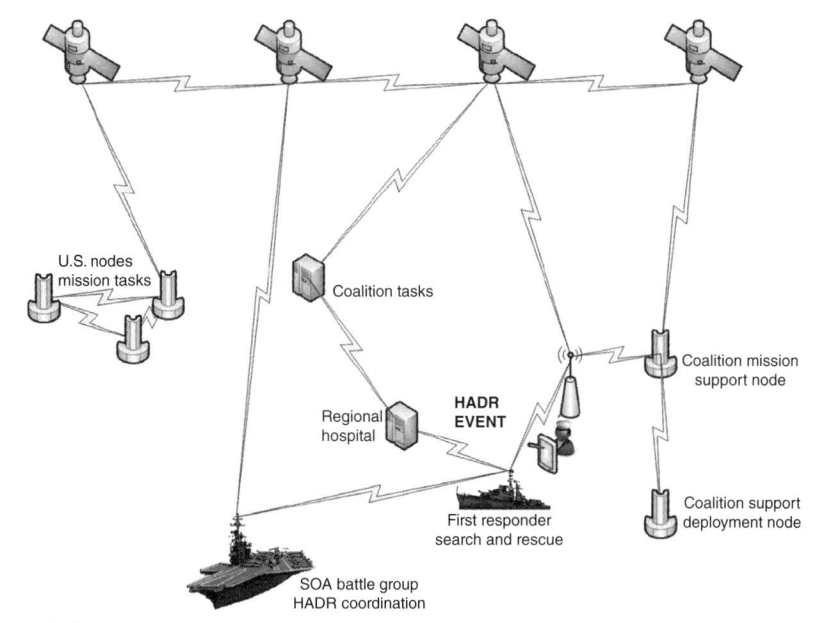

Figure 7.7 OV-1 for secure distributed service architecture.

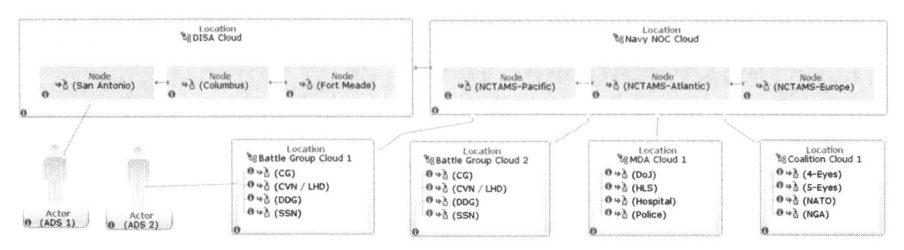

Figure 7.8 Use case for secure distributed service architecture.

7.7.2 OV-2: Operational Node Connectivity Description

In the OV-2, we define capability requirements within an operational context and depict operational "needlines" that indicate the exchange of resources (DoD CIO, 2010). Essentially, we illustrate resource flows but without prescribing the way specific resource flows are handled or specific solutions.

The OV-2 model we present in Figure 7.9 is another UML derivative with a more compact representation than a traditional OV-2 to enable a more direct focus on services. In this model, the nodes provide synchronous updates [publish/subscribe] to support online operations over a global communications network.

Synchronization of applications, services, and content in a SOA is a complex process, which we will address sparingly in the remaining models of this section. Figure 7.10 advances the concept as we assume a national cloud aggregates and synchronizes

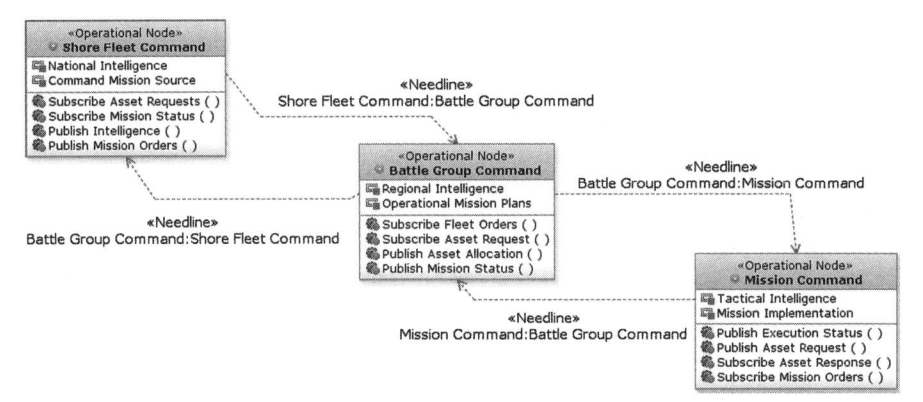

Model Methodology for a Department of Defense Architecture Design **163**

Figure 7.9 OV-2 use case for SOA nodes and content synchronization.

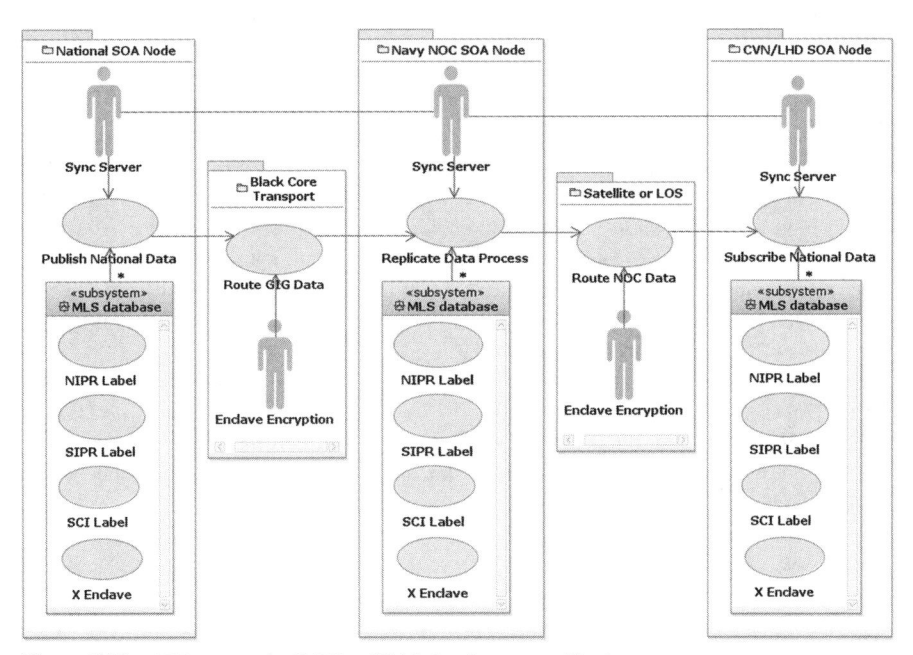

Figure 7.10 OV-2 use case for HADR publish/subscribe event notifications.

databases and publishes to intermediate nodes/clouds. In turn, intermediate nodes publish to end nodes, and subscribers therein receive their HADR event updates.

7.7.3 OV-3: Operational Information Exchange Matrix

The Operational Information Exchange Matrix (OV-3) we typically model as a table that shows information exchanged between nodes and relevant attributes of that exchange—such as media, quality, quantity, and interoperability.

7.7.4 OV-5b: Operational Activity Model

We use the Operational Activity Model to model operational activities (or tasks) and information flows between activities. The OV-5b is typically one of our more active models, often employed to show systems operations before and after our innovation.

We model inputs and outputs, constraints, and mechanisms that perform activities. In the current context, the activity model will continue our HADR example to model operations conducted to achieve a humanitarian assistance mission.

Historically, activity models in the DoD originated from IDEF activity modeling along with Federal Information Processing Standards (FIPS) published by the National Institute of Standards and Technology (NIST). More recently, activity model development has expanded to include Business Process Modeling Notation (BPMN) developed by the Business Process Management Initiative with specifications from the OMG.

BPMN provides a graphical notation based on flowcharting techniques, similar to UML activity diagrams. Our objective in BPMN is to model process management for both technical users and business users by providing a modeling notation that is intuitive to business users yet able to represent complex engineering processes.

Of interest to software engineers, BPMN provides a mapping between the graphics in the model and the underlying execution language. In other words, BPMN models can output operational computer code. As we discussed earlier, BPMN supports the full SDLC—processes can be modeled, code executed, a simulation run, and code deployed.

In Figure 7.11, we show a BPMN model with process flows across nodes. Nodes are modeled in "swim lanes." Operational activities are modeled, with both systems and services represented as resources or capabilities. Our BPMN activity model therein expands the information exchange in previous models wherein an ADS produces mission-critical information that is consumed by users.

In our model, we consider security to be an enabling process since it is a requirement for all DoD communications. At this stage in our model evolution, the security UC is a subset of the more comprehensive BPMN activity model described earlier. In that diagram, we showed behaviors and identified actors and functions of the system—but we did not show temporal information specific to security.

Security is invoked when a user requests that a resource be displayed. Processes validate the user to enable the query and to request assignment of an asset. At the data layer, discussed later in this section, access controls filter users for specific pieces of data.

In the instance of service-based architecture, contracts help ensure that services are kept in compliance with appropriate standards. Service contracts may be modeled formally as a service-level agreement (SLA) or QoS agreement or documented within our WSDL document.

In an activity diagram for a subscription contract, modeled in Figure 7.12, our process begins with a request for a specific information service. The request is sent simultaneously to the publisher and to the appropriate personnel for a security review. If the request is approved by the publisher and security officer, then a service contract is sent to the subscriber.

The contract may specify the terms of usage, provide options for QoS, or simply provide access information. Once the contract is received at the publisher, then an account is created, and the user is added to the registry for that service—in this example, the

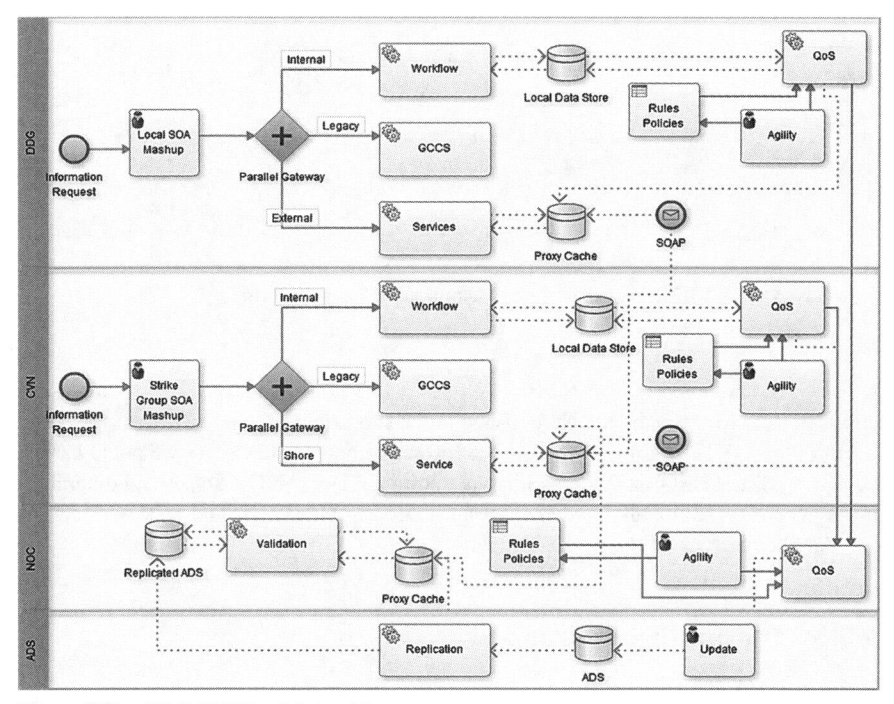

Model Methodology for a Department of Defense Architecture Design   165

Figure 7.11   OV-5b BPMN activity model.

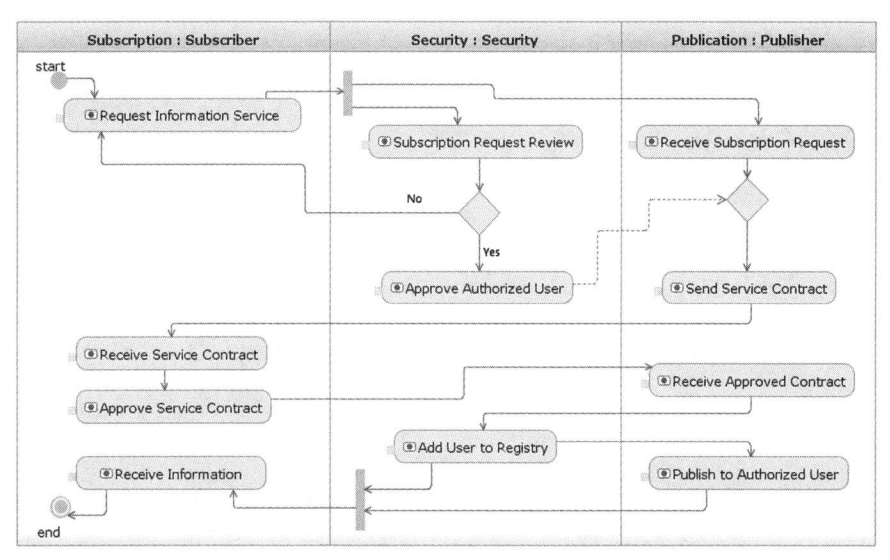

Figure 7.12   OV-5b activity diagram for security approval in a subscription contract.

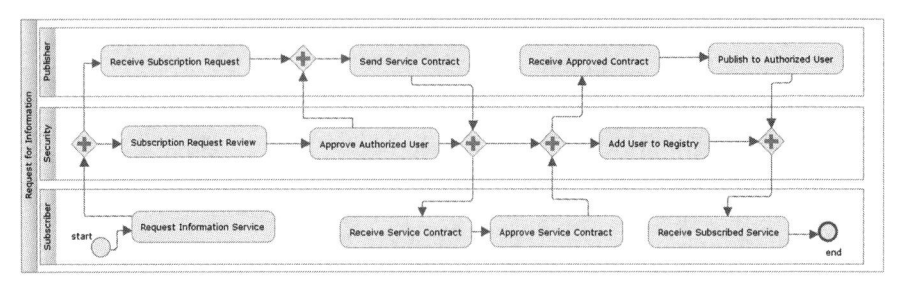

Figure 7.13 OV-5b BPMN model for security approval in a subscription contract.

registry is maintained by the security office. At this point, the subscription process is complete, and the information is published to the subscriber.

In Figure 7.13, we model the same information but in BPMN. As we noted earlier, BPMN tends to be SOA specific, while the traditional DoDAF OV-5b activity diagram is more generic. Additionally, code can be generated from BPMN, and processes can be simulated directly from the BPMN model.

7.7.5 OV-6c: Event-Trace

If properly modeled, the OV-6c and its partner the SV-10c event sequence models are probably two of our most important models for understanding how a system really works. In these diagrams, we model resource flows over time for each event. At the operational level, this typically includes locations, assets, people, and places. Often referred to as a sequence diagram, our event-trace model details data flows within a specific system or process.

The OV-6c is valuable for moving to the next level of detail from the initial operational concepts by defining interactions with the environment. In this instance, by "environment," we mean the experimental context that includes operational nodes, systems, or components. The OV-6c can help ensure that each participating operational activity and location has the necessary information it needs and at the right time to perform its assigned activity.

In Figure 7.14, we model a simple event sequence for publish/subscribe operations in the HADR UC. In this process, the node updates its databases by subscribing to the Web service published by the authoritative server, which in turn publishes to the end-user system. Our model shows an exchange wherein the tactical unit that has been assigned a specific task—in this case search and rescue—publishes into an asset allocation system a request for resources. In this instance, the tactical unit is both a publisher into the system and a subscriber of the system. The dashed lines illustrate asynchronous communications.

Our previously discussed BAM and BPEL models would monitor the event in real time, with human intervention able to override automated processes if required through the Web service policy manager. This level of automation will speed asset assignment to tactical units; real-time sensor updates from field units will improve overall situational awareness to help our responders rescue disaster victims.

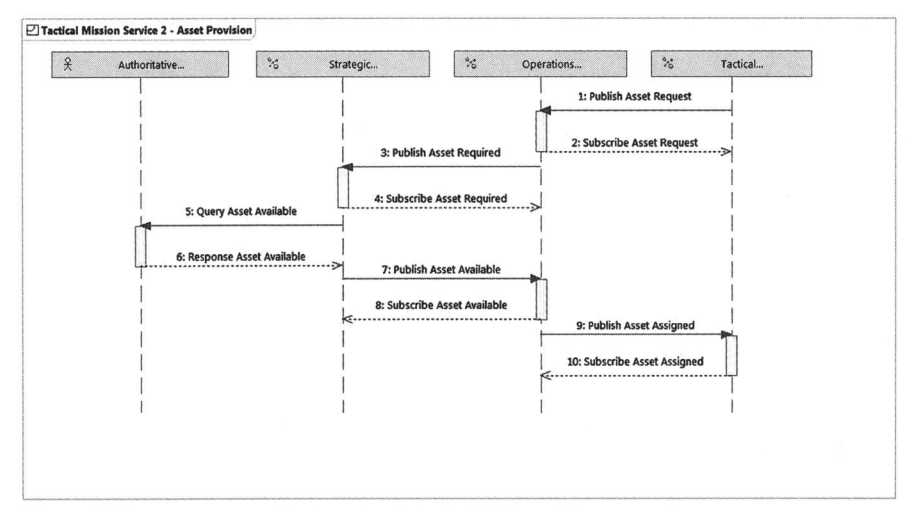

Figure 7.14 OV-6C asset publication/subscription in HADR workflow automation.

7.8 DoDAF SYSTEMS VIEWS

In the SV models, we are deeper into the technical operations. Specifically, we are looking at systems and their interconnections. Technical specifications for data flows and processes are modeled within the SVs. We use models to associate systems resources to operational and capability requirements—which in turn facilitate the exchange of information that drives our technical requirements and specifications.

7.8.1 SV-1: Interface Description

We use the SV-1 to model the composition of systems and the interaction of systems—linking together the operational and systems architecture models. Through the SV-1, we model how resources are structured and how they interact. We therein instantiate the logic of our OV-2 operational resource flows.

Through a systems resource flow, we can represent, for example, an information flow through a network—which we usually model with annotated "connector" lines. We use the SV-1 to model all systems resource flows between systems that are pertinent to our event.

An example of an SV-1 for the HADR UC that we have been evolving is presented in Figure 7.15. In this model, the shore command SOA node high-level processes include a Java EE application server that hosts publication/subscription (pub/sub) services and draws content from the primary ADS for the HADR mission. The application server and the ADS are replicated. The movement of Web services is through a message queue and broker, with the topic the means for content synchronization.

In Figure 7.16, we have modeled a set of interfaces for our publication service, including the subscription process, user authentication and security, and process monitor. An initial user request for information or access to a topic spawns a security audit and

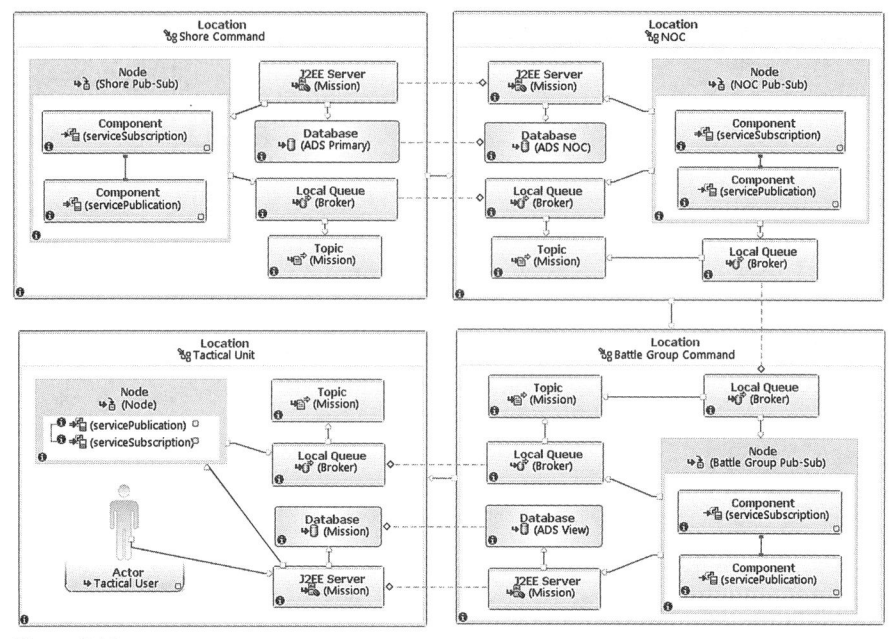

Figure 7.15 SV-1 interfaces for integrated operational and systems architecture.

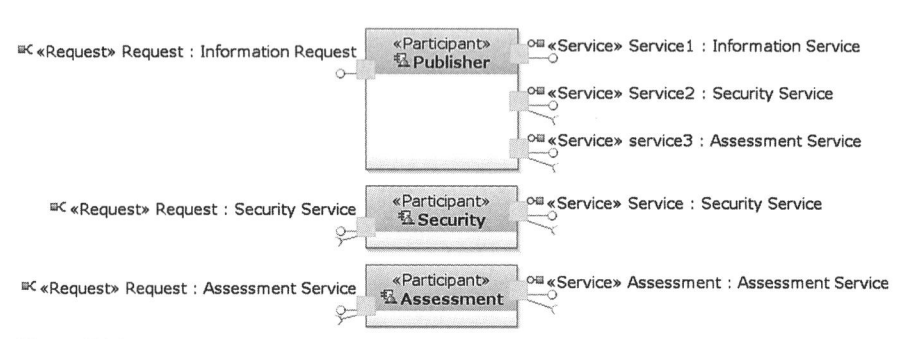

Figure 7.16 SV-1 SoaML interfaces for publish, subscribe, security, and analytics.

generation of a user record in our BAM monitor—which we use for analytics. This assumes a semiautomated workflow. For more extensive human-in-the-loop operations, we would initiate a BPM workflow.

7.8.2 SV-2: Resource Flow

The SV-2 diagram is another one of our mission-critical diagrams. The SV-2 is a necessity for understanding how a system operates and its interfaces. At a practical level, we can also use the SV-2 for our DoD security accreditation. In our SV-2, we will model data flows between systems, identify the protocols that will be traversing our system and the networks to which it connects, and specify the systems ports we will use for those protocols.

A model may be created for each resource flow, or we can model all resource flows on one diagram. Networks connect systems, and systems are components of networks—for example, parts of the SoS infrastructure. For example, Figure 7.17 models a Navy cloud that consists of telecommunications centers and user systems. Within the cloud are the nodes that provide our services. Connections are considered to be multilevel, but the same principles apply if we are serving only a single level or enclave. A source node hosts ADS and physical servers host virtual machines.

A server dedicated to HTTP/S operations hosts the UDDI directory services. A Java EE server dedicated to security provides user authentication and authorization as well as attribute-based security—which in this model references data labels for precise filtering of content for each viewer and for each piece of data.

In the model, we have a messaging manager that provides gateways, queuing services, subscription processing, and service translation through our ESB. A third server holds the portal for remote access and distributed operations. The portal is our primary interface for users, containing access to operating procedures, subscription topics, WSDLs, service-to-service interoperability specifications, and pub/sub guidelines. A fourth server is dedicated to content and security replication and synchronization across nodes—with federation as appropriate.

7.8.3 SV-3: Systems Matrix

In the SV-3 table or matrix, we provide an overview of all systems resource interactions specified in one or more of our SV-1 Systems Interface Description models. As noted earlier, many of the popular modeling tools output OV and SV matrices. The matrix provides a tabular summary of systems interactions and helps us identify potential commonalities, conflicts, and redundancies.

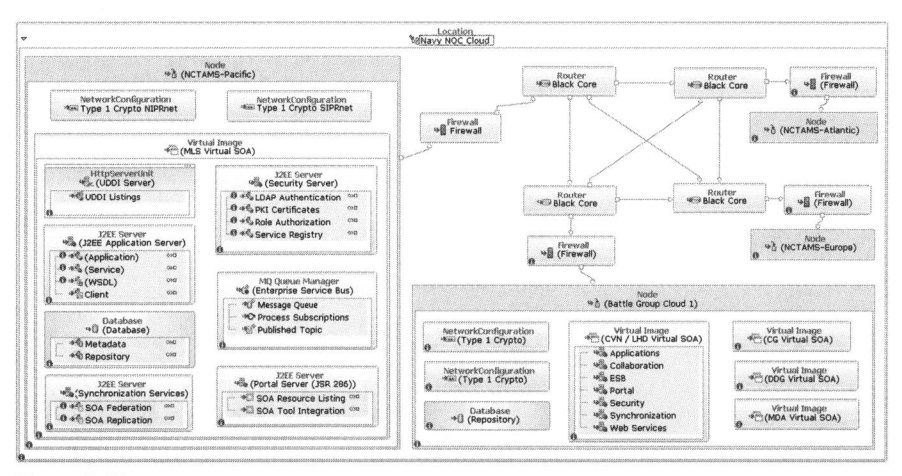

Figure 7.17 SV-2 for backbone cloud with subordinate fleet clouds.

In the matrix or table, the systems resources are typically listed in the rows and columns of the matrix. In the cells of the matrix, we indicate if there is an interaction or potential interaction between resources.

7.8.4 SV-4: Functionality Model

We use the SV-4 to specify the function of resources in the architecture. We typically model an SV-4 as (a) Taxonomic Functional Hierarchy that decomposes functions in a tree structure—typically used where tasks are concurrent but dependent—or (b) Data Flow Model that shows functions connected by arrows and data stores.

The Taxonomic Functional Hierarchy is useful in capability-based procurement processes where it is helpful to model functions that are associated with a particular capability. By contrast, the Data Flow Model is typically graphical and may feature BPMN-style "swim lanes." A function in a swim lane is associated with a system, resource, or performer executing an activity.

In Figure 7.18, we implement a modified data flow diagram to show functional resources, systems, and capabilities. Rather than a traditional data flow diagram, a class diagram is presented to better illustrate the expanded number of operations in our SOA.

A class diagram in UML is a static structure that models our system by showing classes, their attributes, their operations or methods, and their relationships. The class diagram is the main building block in object-oriented modeling and programming—used for general conceptual modeling of an application and for translation of models into code. In other words, the classes represent the objects, and the interactions represent the application functionality. Class diagrams thereby provide a means for us to model the structure and capability of systems active in our event sequences.

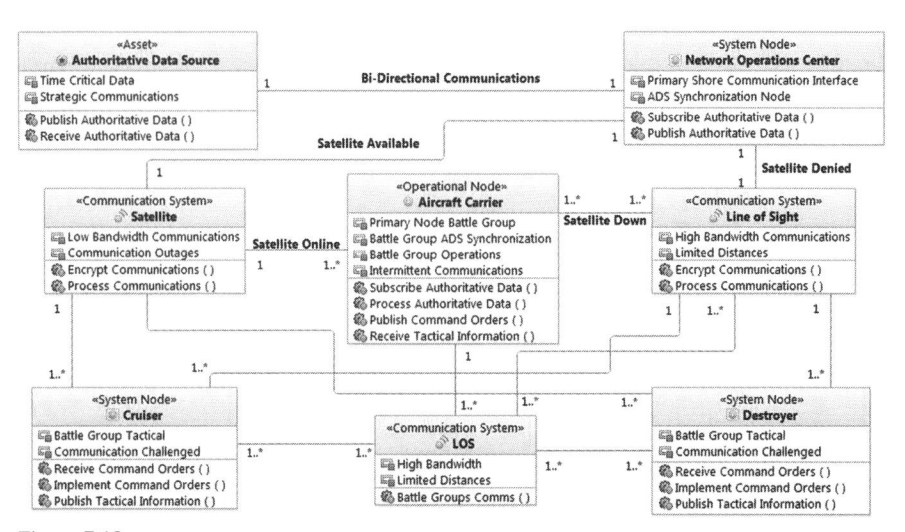

Figure 7.18 SV-4 systems function and communications class diagram.

7.8.5 SV-5a: Function Traceability

We use the SV-5a to trace activities from operational tasks to systems functions. Specifically, in the SV-5a matrix, we map human/systems functions to operational activities. Through the SV-5a, we can identify operational needs and how those are translated into systems operations and in an easily readable table or matrix format.

While not as pleasing to the eye as our graphic models and their simulations, these matrix-based resources support our requirements, specifications, and capabilities analysis. Through the matrix, we can associate needs with costs or human requirements with systems development tasks.

During the requirements definition phase of modeling, our SV-5a can play a particularly important role—helping to directly associate architectural elements with systems functions and user requirements.

7.8.6 SV-8: Systems Evolution

The SV-8 tends to be a different type of model. Through the SV-8, we try to model a life cycle view of resources (systems) and how they change over time. The modeler will typically structure resources over a timeline so the reader understands where we are today and where we will be headed in the future with the development of our system.

When we link the SV-8 with our other evolution models—such as the CV-3 Capability Phasing model and the StdV-2 Standards Forecast model—we have a means to understand how the enterprise system and its capabilities will evolve. In this manner, our SV-8 notional model can be used to support project transition plans.

Figure 7.19 models our service architecture progression over 10 years, beginning with XML extensions to SQL and object-relational databases. Adoption of XML led to

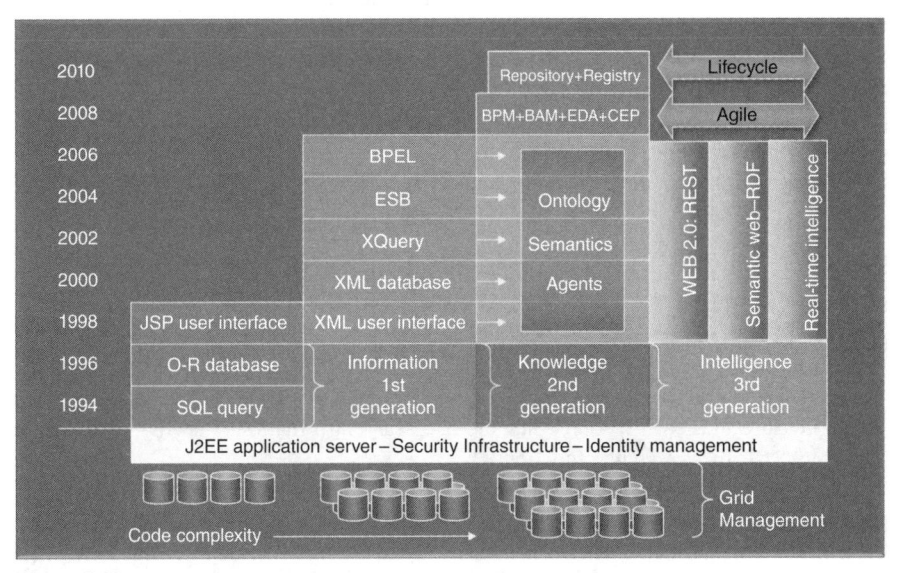

Figure 7.19 SV-8 notional model for Grid–SOA cloud functional evolution.

new capabilities at the application layer, which included service components such as BPEL and ESB and service extensions for XML data types.

In the mid-2000s, our SOA stack was upgraded to include BPM, BAM, Event-Driven Architecture (EDA), and Complex Event Processing (CEP). Metadata repositories and registries became a primary means to develop and manage our pub/sub services.

The next generation introduced semantic capabilities and real-time analytics of event streams. Our tools evolved significantly over this time period to enable full SDLC modeling with agile methods for code generation, monitor, and revision directly from our models.

7.8.7 SV-10a: Systems Rules

With the SV-10 series, we evolve into models that are especially useful for net-centric and service-oriented processes, including orchestration of services. We use the SV-10a to model or define functional and nonfunctional constraints on an architecture including both structural and behavioral elements. Constraints can include resources, functions, and data that make up our SV physical architecture.

7.8.8 SV-10b: State Transition

The state chart or transition diagram has a rich history in systems and software engineering. In its simplest form, we use the state transition to show before and after scenarios—what the capability was before our change and what the capability will be after our innovation is introduced into the system.

Technically, we use the SV-10b to model a resource or systems response to an event by illustrating a "change of state." The model presents an event to which a resource responds by taking an action to move to a new state. Each state transition specifies an event and an action.

We can use the SV-10b to describe sequences of functions, actions internal to a single function, or systems functions with respect to a specific resource. Behavior is modeled as a graph of specific states interconnected by transition arcs that are triggered by events.

7.8.9 SV-10c: Event-Trace

Similar to an OV-6c, but with more systems-level specifics, in our SV-10c, we model time-ordered interactions between functional resources and systems data interfaces. We seek to ensure that each resource has the necessary information it needs—at the right time—to perform its assigned function.

In the SV-10c, we identify functional resources, the owner of those resources, and their technical enablers—such as the systems port that receives the protocol that triggers the event. Similar to the OV-6c we modeled earlier, we use "lifelines" to illustrate event flow and sequencing. Specific points in time are identified. Resource flows from one resource/port to another are labeled events, and the timing of those events is modeled through the lifelines.

7.9 DoDAF DATA AND INFORMATION VIEWS

DoDAF V2.0 is a very significant advance over DoDAF V1.5. In DoDAF V2.0, we find support for DIV to model information requirements and rules to constrain activities. We accomplish this through abstraction. DoDAF V2.0 incorporates three levels of abstraction for DIV models: Conceptual, Logical, and Physical (DoD CIO, 2009).

7.9.1 DIV-1: Conceptual Data

In the DIV-1, we model information concepts at a high level of abstraction, focusing on information requirements and hierarchies. This would include structural process rules for our architecture. Model elements include information items, their attributes or characteristics, and their interrelationships. Most important to DoD modelers is that through the DIV-1 information products, we can directly support doctrine, standard operating procedures, concept of operations (CONOPS), etc.

7.9.2 DIV-2: Logical Data

We model data definitions through the DIV-2. We keep the definitions independent of any implementation-specific products. The DIV-2 is basically a common dictionary of data definitions. If everyone uses the dictionary, then model interpretation problems will be eliminated or significantly reduced. Ideally, we map the concepts of the DIV-1 to the logic of the DIV-2. Note that the DIV-2 replaces the OV-7 from DoDAF V1.5.

7.9.3 DIV-3: Physical Data

In the DIV-3, the data from the architectural description (DIV-2) is implemented in a physical schema. As such, our DIV-3 is the DIV model that most closely represents an actual system. Associations between the logical and physical model elements are relatively straightforward—such as entity types in a logical model versus relational tables in a physical model. Note that the DIV-3 replaces the SV-11 from DoDAF V1.5.

As an example, we have modeled the class diagram for the HADR UC systems domain in Figure 7.20. The model shows some of the classes in the system and some of the primary data types. In the system, our user can display resources, make asset assignments, and enact other capabilities as specified in our UC.

We use "realizations" to present collaboration elements for our HADR UC. A realization may model how a system implements a specific capability at the data and class levels. A realization may also illustrate a UC model for a specific participant relationship or for a workflow sequence through a use case.

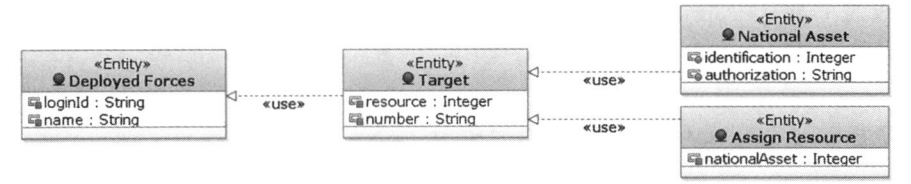

Figure 7.20 DIV-3 relationships between classes and data types.

7.10 DoDAF STANDARDS VIEWS

StdV models help us articulate rules for the arrangement and interaction of elements. The models can depict interdependence of elements in an architectural description. Rules help us ensure that a given solution satisfies a specific set of operational or capability requirements.

Similar to the DIV models discussed previously, our StdV models can represent doctrine, technical procedures, industry guidelines, or engineering specifications. Different from the DIV is that we use the StdV to represent written, formal government or industry policy or standards that tend to directly impact our model development.

7.10.1 StdV-1: Standards Profile

The DoDAF V2.0 StdV-1 model serves as the replacement for the TV-1 model in DoDAF V1.5. As the name implies, we use StdV-1 models to convey technical and operational standards, guidance, and policy. We can also use this model to convey context—the policies and standards applicable to the business or operational environment for our innovation. This can include rules for the implementation of our innovation, and constraints that we recommend are placed on modeling choices during the design and implementation of our architectural description. Standards for citation may include ISO standards, national standards, or organization-specific standards.

7.10.2 StdV-2: Standards Forecast

The StdV-2 Standards Forecast model helps us brief expected changes in the standards or conventions that we documented in our StdV-1 model. Forecasts for evolutionary change in standards can be correlated against the time periods we designed in our SV-8 Systems Evolution model, SvcV-8 Services Evolution model, SV-9 Systems Technology and Skills Forecast model, or SvcV-9 Services Technology and Skills Forecast model.

One of the primary purposes of the StdV-2 is to identify standards upon which we have relied throughout our architectural description and that if changed would impact our model designs. So, in the StdV-2, we address the standards upon which any of our models have been based or which we have assumed in our model design, including the life expectancy of those standards and the impact any changes to these standards would have on future development and maintainability of our innovation.

7.11 DoDAF SERVICES VIEWS

SvcV are new with DoDAF V2.0 and offer us a more precise means to model SOA. At a high level, DoDAF V2.0 supports DoD initiatives to define and institutionalize Net-Centric Data Strategy (NCDS) and Net-Centric Services Strategy (NCSS). In SvcV, we provide for the definition, description, development, and execution of services (DoD CIO, 2009).

We can employ SvcV models to articulate UC performers, systems activities, and data exchanges in support of operational capabilities. Through SvcV models, we help stakeholders visualize how structural and behavioral factors impact our architecture. We explicitly convey which functions are carried out entirely by M2M processes and which functions have a human in the loop.

Through SvcV models, we now have a means to directly model risk variables and address possible costs based on elements in those risk variables.

7.11.1 SvcV-1: Context Description

Due to the extreme complexity of the SoS architectures in the DoD, we have an implied assumption that "context is king," meaning that our models are assumed accurate only within the environmental context that we have defined. Our innovation will behave differently if we are on the moon, deep undersea, or finding a rescue route. We use the SvcV-1 to model resource flows and the provisioning of services—within context.

The SvcV-1 differs from our SV-1 Systems Interface Description that focuses on system-to-system, point-to-point communications for which source and target systems have an agreed-upon interface. The beauty of SOA is that predefined or hardwired interfaces are no longer required. We can potentially have dynamic, just-in-time information exchange—although the details for such capabilities are still evolving in the DoD due to the significant security ramifications.

We see this regularly in experimentation where systems integration projects that once required millions of dollars can now be accomplished over lunch when the software engineers agree to experiment with data exchange. Service architecture offers exponentially enhanced functionality for the users and exponentially reduced costs for the sponsors—if correctly implemented.

If not implemented correctly, then we have tremendous cost overruns and resultant confusion due to the exponential increase in complexity at the back end of service architecture. Of course, if we do not fully document and model what we have done, then we have lost a good deal of our efficiencies, cost reductions, and security.

Through the SvcV-1 model, we provide the context for each new capability. In our SvcV-1 model, the focus is on net-centric data strategies appropriate for publish and subscribe patterns. Specific to DoD implementations are additional requirements for security, including contracts with publishers for user authorization, role-based information access, and data-level authentication.

Access to specific data types by user role or information attribute is typically modeled in the SvcV-2. In the associated SvcV-3 matrix, we can document security functionality for auditors—for example, as an analytics capability available for subscription.

In Figure 7.21, we model publisher, subscriber, and security services. Publisher services include capabilities to receive subscription requests, send service contracts to

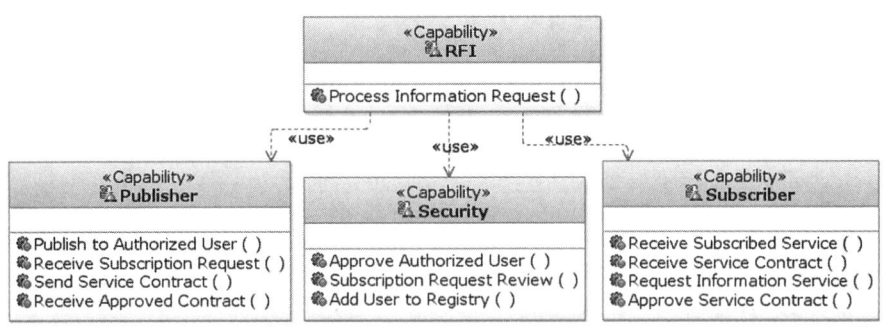

176 Theoretical and Methodological Considerations

Figure 7.21 SvcV-1 SoaML capability class diagram for RFI service contract.

subscribers, receive approved contracts, and publish to authorized users. This UC is at a high level of granularity to accommodate topic-based publish/subscribe patterns.

7.11.2 SvcV-2: Resource Flow

In our SvcV-2, we specify resource flows between services. We may also list the protocol stacks used in connections and therein give a very precise specification for a connection between services—which may be an existing connection or a specification of a connection that is to be made in the future.

A complementary capability in UML is the subsystem that we use to represent independent behavioral components of our system. In this context, our system can be characterized as a hierarchy of subsystems—represented in class, component, and UC models. We indicate behaviors of the subsystems through interfaces, services between interfaces, and resource flows that support our interfaces. We model attributes, operations, interfaces, and realizations in the internal structure of the subsystem.

In Figure 7.22, we again model our HADR natural disaster search and rescue efforts—as if our operations had the support of a comprehensive SOA. In the live operational tests upon which our models are based, there was not a comprehensive SOA overseeing the rescue operations. The experimental intent was to approximate an actual natural disaster, in this case an earthquake, tsunami, and tidal wave that destroyed major parts of an island.

This very large experiment featured numerous advanced new technologies to help in disaster recovery, ranging from search and rescue services, to hospital and medical systems coordination, to news coverage. The military teams, resources, and systems helped manage the rescue and reconstruction efforts.

Our model conveys a SOA that "could" have supported the efforts—if construction of such a SOA had been a focus of the experiment. Still, we came close to achieving much of this functionality as a series of independent architectures, many service oriented. Ideally, our HADR SOA would be an enabler for service integration—with the fleet node able to maintain an operational network in the disaster area and thereby support civilian and interagency communications.

The primary subsystems are the data repository, application servers, ESB to connect the servers, various database servers to coordinate emergency medical requirements, our user registry, and the security servers through which we retain cross-domain and need-to-know privileges.

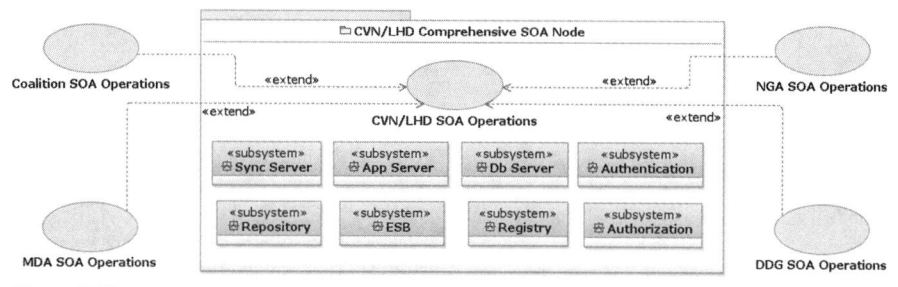

Figure 7.22 SvcV-2 SoaML representation of strategic subsystems and resource flows.

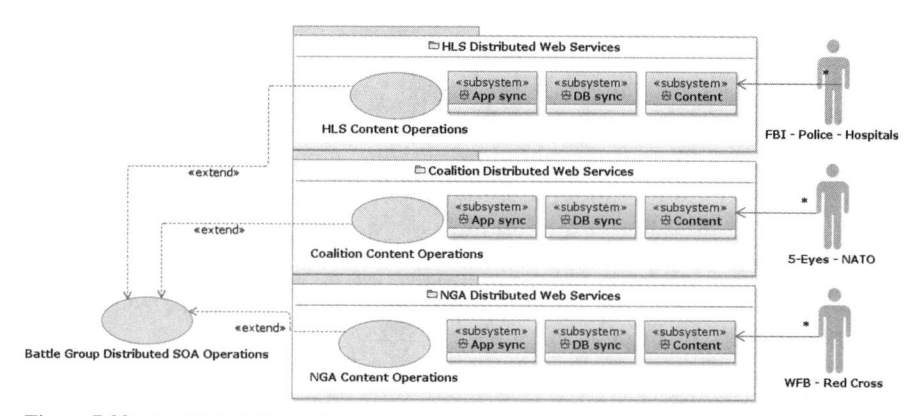

Figure 7.23 SvcV-2 SoaML participants for tactical Web services to managed clients.

Specific resource flows are not indicated at this high level since each component communicates with each of the others and for most transactions. Modeling of specific resource flows, for specific transactions, would be a logical extension of this model.

Figure 7.23 represents the tactical layer for our mobile users—ranging from medical personnel to rescue and construction crews. Interfaces in this UC included actors in the DoD as well as local emergency officials and medical personnel. In an actual systems model, our interfaces would be more detailed to illustrate support for authentication, authorization, and role-based services.

Local interfaces to police, fire, and medical organizations to support emergency services present a unique challenge for our SOA. In an ideal world, the DoD SOA, and the SOA of each emergency response organization, would be flexible enough to seamlessly share services. In reality, we are not there yet. Still, we can be optimistic and design our model to dynamically segment communications by communities of interest, with attribute-level controls on our resource flows.

7.11.3 SvcV-4: Functionality

We use the SvcV-4 Services Functionality Description to model service allocation, specifically the flow of resources between service functions. Modeling techniques for the SvcV-4 include the traditional data flow diagram with service functions connected by

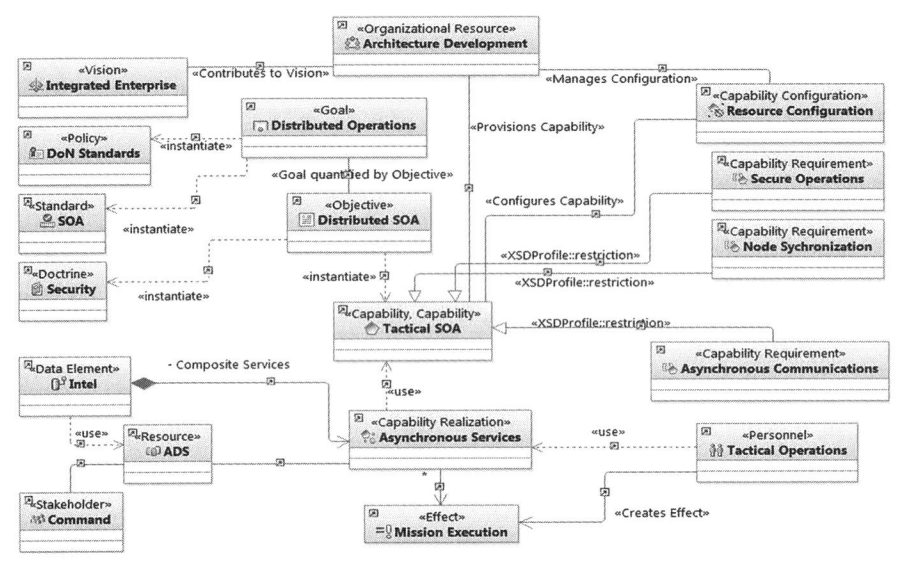

178 Theoretical and Methodological Considerations

Figure 7.24 SvcV-4 service functions, resources, and flows to provide capabilities.

data flow arrows to data stores and the newer taxonomic service functional hierarchy that models the decomposition of service functions in a tree structure.

In SvcV-4 models, we start with concepts, which flow to service functions and then to resources—which we model as capabilities between service functions. Figure 7.24 models the high-level concept or "vision" of an integrated enterprise and the goal or "objective" of distributed operations. To achieve the vision, we model overall architecture, and then we realize the functionality objectives through service policies.

7.11.4   SvcV-10a: Rules

Through the SvcV-10a, we model the rules for services and the structure and behavior of services. While services are typically automated M2M flows, we also include human elements as UC performers, organizations, and personnel types.

In the SvcV-10a, we link together the operational and service architecture models to illustrate how resources are structured and function to realize the logical architecture—as specified in our OV-2 Operational Resource Flow Description. We can use the SvcV-10a to model flows that are exchanged between resources and to present a solution in terms of capabilities and their physical instantiation on platforms.

For example, a candidate service may be represented as a capability in a service model. The service interface provides the application programming interface—typically a WSDL or other means to access the capability—from which we render our publish/subscribe processes. In this instance, the WSDL serves as the primary interface to contract for services that collectively achieve the capability and satisfy our mission.

This process can be compared to the legacy process of an RFP and lengthy and expensive systems integration. Instead, read the WSDL and have your capability in a few minutes. Such is the benefit of service-based SoS architecture—remembering that the

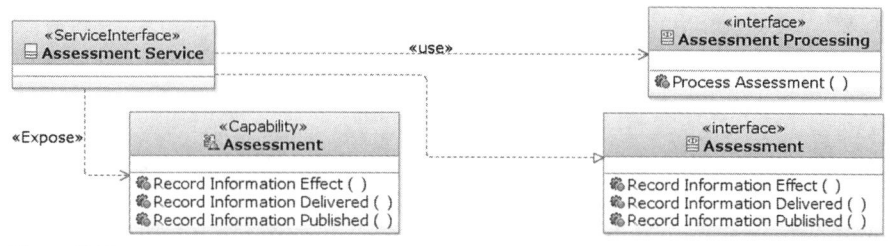

Figure 7.25 SvcV-10a SoaML analytics service interface.

exponential increase in complexity at the back end necessitates comprehensive, end-to-end modeling across the SDLC; else, we slip into huge cost overruns and a truly massive security problem.

Next is to model a service interface for each candidate service. The "expose" dependency relationships are implemented to maintain traceability from the service functions to the capabilities. In Figure 7.25, we model our analytics routines as a facet of an assessment capability in which we use attribute tags to provide content pedigree.

Data labels or tags enable each specific piece of data to be monitored as it traverses our SoS architecture. We can record the effect of data as it is aggregated in composite services. Our analytics services can help us extract metrics for the use of specific pieces of information and note any constraints on those services.

7.11.5 SvcV-10b: State Transition

Similar to the SV-10b but specific to services, the SvcV-10b is a graphical method for modeling a resource (or function) and its response to various events. The model represents a sets of events for which our resources respond—taking action to move to a new state as a function of a current state. We model each transition as an event and an action.

Our SvcV-10b also uses a state chart approach to model change in a sequence of service functions. Again, for reference, a state machine is a specification that describes all possible behaviors of some dynamic element. We model behavior as the traversal of a graph of specific states, all interconnected by transition arcs that are triggered by events.

In UML, we model two kinds of state machines: behavioral state machines and protocol state machines. Behavioral state machines model the behavior of individual entities (e.g., class instances). We use protocol state machines to model specific protocols, classifiers, interfaces, and ports.

In Figure 7.26, we employ an SoaML model to represent an SvcV-10b state machine. State change in this instance refers to change in our SOA due to changes in data elements or attributes—such as publication topics or subscriptions—or changes in security for any process.

The ovals in the diagrams are UML collaboration UCs for participant interaction—which is where our state change occurs. We use the connection boxes with handles to identify service interface ports and protocols. The service contract is the means to ensure QoS and enforce contracts.

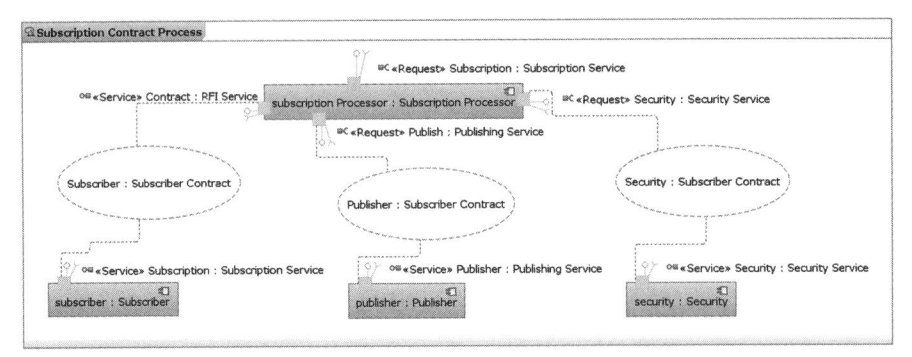

Figure 7.26 SvcV-10b SoaML service state diagram for publication/subscription.

The collaboration shows how our participants work together to provide services. Each participant plays a role in the service contract to help verify and ensure that underlying constraints are honored.

7.11.6 SvcV-10c: Event Trace

Visually similar to the OV-6c—but for services vice organization-level processes—the SvcV-10c provides a time-ordered examination of interactions between functions and resources. The SvcV-10c is valuable for moving to the next level of detail from the initial service design. Here, we model a sequence of service functions, service data interfaces, and service data flows.

Through our SvcV-10c model, we help ensure that each participating resource or service role, and associated port, has the necessary information it needs and at the right time to perform our assigned functions.

7.12 DoDAF CAPABILITY VIEWS

The CV models were introduced in DoDAF V2.0 to help manage the risks of complex procurements through visualization of capability increments across a portfolio of projects (DoD CIO, 2010). In these models, we capture complex relationships between interdependent projects and their associated capabilities.

For example, in our CV-1 Vision model, we define strategic context; in the CV-2 Capability Taxonomy, we model a hierarchy of capabilities across a timeline; and in the CV-3 Capability Phasing, we document the activities, conditions, desired effect, rules, and measures.

Our CV-4 Capability Dependencies model, as the name implies, maps dependencies between planned capabilities and logical groupings of capabilities. Similarly, our CV-5 Capability to Organizational Development Mapping models interconnections for capability phases, such as performers and locations.

Continuing this flow, in our CV-6 Capability to Operational Activities Mapping, we associate operational activities with capabilities, and in our CV-7 Capability to Services Mapping, we specify services supplied by or which support capabilities.

7.13 DoDAF PROJECT VIEWS

RELATED to the CV, our PV models designate capabilities within the context of the larger, governing project (DoD CIO, 2010). Specifically, we model how programs, projects, and portfolios deliver capabilities. The PV-1 Project Portfolio Relationships model illustrates dependency relationships between organizations and projects and helps us understand the structures required for effective portfolio management.

In the PV-2 Project Timelines model, as the name implies, we develop timelines for projects with milestones, deliverables, and interdependencies. A Gantt chart, as modeled in Microsoft Project or a similar project management tool, is the most common means to generate a PV-2.

Through the PV-3 Project to Capability Mapping model, we can show how specific projects, and/or the aggregation of projects, achieve a capability. This includes specific elements within projects and illustrates/documents how elements build cumulatively to provide a capability.

7.14 CONCLUSION

In this chapter, we introduced some of the modeling methods commonly used in the DoD and framed the discussion using the DoDAF. We looked at DoDAF models that we use to represent and visualize SoS architecture.

We examined M&S tools that provide lifelike representation at one extreme and executable code at the other extreme. In the former category are tools to assess overall communications, complete with aircraft, ship, and cyber simulations. In the latter category are tools that support UML and generate live operational computer code from models—with interfaces to DoDAF to ease understanding.

In combination, this modeling approach lets us see in 3D visual simulation the event as we expect it to unfold and then actually build and run the software for that event as a fully functioning system suitable for live operations. When complete, the modeling cycle evolves from high-level DoDAF views or viewpoints to low-level DoDAF architecture—from simulation with operational context to machine code.

If properly selected, the tools can generate low-level models suitable for the generation of computer code and intermediate-level simulations that execute the application in context. However, we must note that in practice today this ideal scenario of end-to-end modeling across the SDLC for complex, global SoS architecture is still evolving. In this chapter, we have presented an ideal state—what could be achieved—and a possible methodology for comprehensive end-to-end SoS modeling.

In practice, the projects and transitions from high-level concepts to low-level computer code are often not structured as comprehensive end-to-end processes. Contracting, specialization, the need to multisource and bid projects, and the extreme complexity of DoD systems integration tend to result in different personnel for different modeling and development tasks. Due to the manner in which contracts are awarded for extremely large projects, there may be only loose lines of communication between peer [and often competing] organizations. As such, the models hold the projects together. The models become the "glue" for SoS integration.

So, for large SoS projects, the SDLC is often not the smooth end-to-end modeling flow that one might expect or that we have presented. Still, this does not downplay the need for robust modeling practices at each level because the models are often the sole line of communication between different groups and even between different organizations and branches of the military—each with very different specializations and areas of responsibility.

Finally, and also ideally, once the models have been used to build the SoS architecture, we then have a direct interface between the model and the represented systems. We can use the models for real-time monitoring of the operations and directly modify the systems and their representative computer code as required and appropriate.

These real-time monitor models—or rather models that support this function—can oversee the designed system or application to provide a tight synergy between the design and runtime systems. Code debugging is improved as network and systems performance data is fed back into the models, and we achieve iterative model refinement and improvement. In this manner, we keep the model in sync with the operational code.

Ideally, performance results would feed back up to the high-level models. In reality, this may not occur—the previously mentioned issues of complexity, specialization, contracting, and organizational boundaries come into play. Time is a factor, as are funding priorities—and perhaps even politics. People come and go, contracts change, priorities shift, and code documentation may not be adequately enforced.

Still, despite the potential absence of comprehensive end-to-end SDLC SoS modeling—to code execution and real-time monitor with automated model updates from live operations—the technology advances over the years have been very significant and extremely impressive. This author believes that DoDAF provides the most comprehensive modeling framework available, and when coupled with supporting commercial modeling and code generation products—some of which were presented in this chapter—we have the best process currently available for global SoS architecture design, development, implementation and maintenance.

This chapter was written with the hope that the concepts and processes herein might help with our next generation of SoS infrastructure design and development.

REFERENCES

AGI. (n.d.). *Systems Tool Kit*. Available at http://www.agi.com.

DoD Business Transformation Agency. (2009). *Business Transformation Agency's Service-Oriented Architecture Implementation Strategy*. Available at http://www.bta.mil/directorates/dbsae/index.html.

DoD CIO. (2009). *DoD Architecture Framework Version 2.0: Volume 1: Introduction, Overview, and Concepts*. Available at http://jitc.fhu.disa.mil/jitc_dri/pdfs/dodaf_v2v1.pdf.

DoD CIO. (2010). *DoDAF Architecture Framework Version 2.02*. Available at http://dodcio.defense.gov/TodayinCIO/DoDArchitectureFramework.aspx.

GAO. (2002). *Enterprise Architecture Use across the Federal Government Can Be Improved, GAO-02-6*. Washington, DC: General Accounting Office.

Hamilton, M. (1999). *Software Development: Building Reliable Systems*. New York: Prentice Hall.

Hughes, T. and Tolk, A. (2013). Orchestrating Systems Engineering Processes and System Architectures Within DoD: A Discussion of the Potentials of DoDAF. *Journal of Reliability, Maintainability, & Supportability in Systems Engineering*, Winter: 13–19.

IBM. (2003). *IBM Completes Acquisition of Rational Software.* NYSE:IBM – News. Available at http://www-03.ibm.com/press/us/en/pressrelease/314.wss.

IBM. (2010). *UML Profile-Based Integrated Architecture (UPIA) DoDAF 2.01 Mapping Reference.* Available at https://www-304.ibm.com/support/docview.wss?uid=swg27019099&aid=1.

IBM. (undated). *IBM WebSphere Business Process Management Version 7.0 Information Center.* Available at http://pic.dhe.ibm.com/infocenter/dmndhelp/v7r0mx/index.jsp

IDEF. (undated). *IDEF Family of Methods: A Structured Approach to Enterprise Modeling & Analysis.* Available at http://www.idef.com/.

JCSS. (undated). *Joint Communication Simulation System.* Available at http://disa.mil/Services/Enterprise-Engineering/JCSS.

Mittal, S. and Martin, J. (2013). *Netcentric System of Systems Engineering with DEVS Unified Process.* Boca Raton: CRC Press.

Mohlin, M. (2010). *Model Simulation in Rational Software Architect: Simulating UML Models.* Cupertino: IBM.

Network World. (2007). *IBM to Acquire Telelogic.* Available at http://www.networkworld.com/news/2007/061107-ibm-offers-745-million-for.html.

Object Management Group. (2011). *Unified Modeling Language: Infrastructure Version 2.4.1.* Available at http://www.omg.org/spec/UML/2.4.1/Infrastructure.

Rational. (2002). *From Inception to Implementation: A Life-cycle Approach to Enterprise Architectures, TP901.* Cupertino: Rational Software.

Rational Software Architect. (undated). *Simplify Architectural Modeling with an Integrated Design and Development Platform.* Available at https://www.ibm.com/developerworks/rational/products/rsa/.

Part III

Theoretical and Methodological Considerations with Applications and Lessons Learned

Chapter 8

An Agent-Oriented Perspective on System of Systems for Multiple Domains

Agostino G. Bruzzone[1], Alfredo Garro[2], Francesco Longo[2], and Marina Massei[1]
[1]*DIME, University of Genoa, Genoa, Italy*
[2]*DIMES, University of Calabria, Calabria, Italy*

8.1 INTRODUCTION

System of systems (SoS) is an expression that indicates a class of systems constituted by a set of interconnected, and often geographically distributed, systems that interact for achieving common or conflicting goals and are capable of autonomous and independent behaviors (Maier, 1999). At this class of systems belong a wide range of systems in several application domains ranging from commercial and financial to military ones (Crossley, 2004). As a consequence, several research efforts are currently devoted to provide methods, models, and techniques for the analysis of SoSs. In this context, Modeling and Simulation (M&S) plays a central role as a primary tool for representing a SoS and for the study of its evolution, behaviors, and performances. However, with respect to large-scale systems (SoSs are often confused with large-scale systems that have been extensively experimented through M&S approaches), SoSs present different and peculiar aspects that need to be accurately taken into account and are essential to address for the definition of effective approaches for M&S of SoSs.

In this context, the chapter proposes a new perspective on SoS M&S based on the adoption of the agent-based paradigm (Jennings, 2001). Specifically, after the identification of the main issues for the M&S of SoSs (Section 8.2), some of the main approaches

---

Modeling and Simulation Support for System of Systems Engineering Applications, First Edition.
Edited by Larry B. Rainey and Andreas Tolk.
© 2015 John Wiley & Sons, Inc. Published 2015 by John Wiley & Sons, Inc.

for the M&S of SoSs are evaluated and compared (Section 8.3). Then, an agent-oriented perspective for SoS M&S, able to fully address the above identified issues and overcoming the limitations of the currently available solutions, is proposed (Section 8.4) and exemplified by application examples in multiple domains (Section 8.5). Finally, in Section 8.6, conclusions are drawn and future works delineated.

8.2 FROM LARGE-SCALE SYSTEMS TO SoS: MAIN M&S ISSUES

In this section, the main aspects that characterize SoS are discussed especially with respect to those of large-scale systems, with which they are often confused. The correct and complete identification and analysis of such features leads to the identification of several issues that are essential to address for the definition of effective approaches for M&S of SoSs.

Large-scale systems are constituted by a multitude of components organized so to form a whole with clearly defined boundaries (Crossley, 2004). Examples of this kind of systems are military and commercial aircrafts, spacecraft, satellites, power plant automobiles, etc. However, although the structure of these large-scale systems is rather complex (or better complicated), it remains quite the same during the systems life cycle; moreover, a great part of the systems components manifest a reactive behavior, and proactiveness is limited to a narrow subset of components. Moving from large-scale system to the SoS context, these assumptions typically do not hold (Maier, 1999; Crossley, 2004). Indeed, an SoS (e.g., a Coast Guard Integrated Deepwater System or an Air Traffic Management System) is constituted by a set of interconnected, and often geographically distributed, systems that interact for achieving common goals and are capable of autonomous and independent behaviors; moreover, the set of involved systems typically changes during the SoS life as new systems join the SoS and others dynamically leave it.

As a consequence, an SoS presents different and peculiar aspects with respect to that of a large-scale system that need to be accurately taken into account for its M&S, which thus requires for specific approaches and techniques.

A first issue concerns the M&S of the SoS structure. Indeed, differently to what typically happens for a system (think again to an aircraft), the structure of an SoS changes dynamically (new entities enter into the SoS and others leave it) so as to its configuration; moreover, the boundaries of an SoS are often not well identifiable as it is not clear when to consider an entity as part of the SoS and as belonging to the environment in which the SoS is situated (Siemieniuch and Sinclair, 2012). For the aforementioned reasons, the range of influence of an entity action is not clearly circumscribable.

A second issue concerns the M&S of the organizational structure of an SoS and of the norms, rules, and protocols that govern the SoS operation and evolution. Indeed, SoSs are typically characterized by distributed organizational structures (organization charts, command chains, etc.) that envisage roles, permissions, responsibilities, rights, and so on (Pavon et al., 2008; Darabi et al., 2012). Moreover, norms, rules, policies, and protocols can be defined so to organize and coordinate the SoS entities. Often, these norms are the most stable part of an SoS as the entities can dynamically enter and leave the SoS as long as they respect these rules.

A third issue concerns the M&S of the dynamic and goal-driven behaviors of the entities of an SoS. Indeed, in an SOS, while not lacking entities essentially characterized by reactive and passive behaviors, most of the component entities are characterized by autonomous and goal-oriented behaviors that in some cases change over time due to learning or adaption processes. This higher degree of autonomy in the behavior of the entities of an SoS makes its analysis and design even more challenging.

A fourth issue concerns the evaluation through M&S techniques of SoS properties and/or performances. Indeed, even the very definition of important systems properties is challenging for SoSs. As an example, let us consider reliability, which is a key property to guarantee especially for mission-critical systems where systems failures could cause even human losses; in the case of a system (even complicated such an aircraft), it is normally clear when the system fails to perform its mission and thus the systems failure modes and consequent effects could be clearly identified (e.g., by a failure mode, effects, and criticality analysis (FMECA)), whereas in the case of an SoS, the concept of failure is not so easily identifiable. As an example, if a Coast Guard Integrated Deepwater System is considered, it is not so immediate to identify when the SoS fails. In fact, if during the rescue of a man overboard, a helicopter fails due to an unexpected breakdown, the SoS can exploit a ship (by recruiting it also on the moment) for performing the same activity but with degraded performances (the time required for reaching the man). In a similar way, in the case of an Integrated Logistic System, if a damaged aircraft is no more available for the transportation of goods from the city A to the city B, a train can be used to substitute it, and again, the SoS as a whole did not failed but was able to perform its mission (transporting the goods from A to B) with degraded performances. Obviously, in both the described cases, below a certain threshold, the performances become so degraded as to be unacceptable (the ship is not able to reach the man in a given time, the train is not able to deliver the goods in a time useful for their distribution), thus leading to an SoS failure. These examples show that the concept of failure for an SoS is related but more articulated with respect to that for a system. As a consequence, the definition of reliability for an SoS is related to that for a system but, in a certain way, is a more wide and flexible concept that should take into account the more flexible and dynamic nature of an SoS. Same considerations hold for other main properties.

On the basis of how the identified features for SoSs are taken into account and how the consequent M&S issues are addressed (see Table 8.1), in the following section, some of the main approaches for the M&S of SoSs are evaluated and compared in the light of the five principles highlighted by Maier in the second chapter of this book. Then, in Section 8.4, an agent-oriented perspective for SoS M&S, able to fully address the aforementioned issues and overcome the limitations of the currently available solutions, is proposed with specific reference to the capacity of agent-based modeling to address the key aspect of "emergence."

8.3 M&S APPROACHES FOR SoS

Considering the complexity of SoS and the large number of elements and interactions involved, the potential to use M&S emerges evident; from this point of view, there are multiple classification criteria that are possible to adopt in analyzing the simulation paradigms in this context. For instance, by considering as classification criteria the purpose of the simulation, it is possible to identify a simulator addressing SoS with different uses.

Table 8.1 System of Systems and Large-Scale Systems: Main Features

Feature	Large-scale systems	System of systems
Physical structure	• Stable during the systems life cycle • Systems boundaries clearly identifiable • Range of influence of systems entity actions clearly circumscribable	• Changing during the systems life cycle • Systems boundaries not well identifiable • Range of influence of systems entity actions not clearly circumscribable
Organizational structure	Typically stable and clearly identifiable	Typically distributed and dynamic
Behaviors of systems entities	Essentially reactive or passive	Often autonomous and goal oriented, can change over time due to learning or adaption processes
Properties and performances	Clearly defined along with criteria for their evaluation	Clear definition for main properties and related evaluation criteria not available

This table summarizes the previously discussed main aspects that characterize SoSs and then their M&S also in comparison with those of large-scale systems.

As application example, the sector related to the SoS evaluation and testing (E&T) is proposed: in this sector, M&S is very interesting for the possibility to anticipate testing before all the systems are completed and/or commissioned. By this approach, it is possible to introduce hardware-in-the-loop (HIL) and software-in-the-loop (SIL) concepts, so it becomes possible to complete the tests of several systems interoperating with virtual simulators. The adoption of emulation approach is another common use of applying M&S for automation systems; in this case, it becomes possible to conduct test on the different real systems, while the reality and operational environment is reproduced by the simulation. In the aforementioned case, the real-time requirement becomes usually very challenging considering that automation systems are characterized by very demanding requirements in this sector. In most of SoS, the different systems are based on proprietary protocols and solutions from different companies, so for both commercial reasons and competitive issues, it is usually pretty challenging to be able to integrate the system together within a virtual framework. Indeed, the advantage of using simulation for the related tests represents at the same time the major challenge for its application, considering the understandable concern of the different companies in having a common virtual test context where the system could be evaluated extensively before the development is completed. Therefore, the advantage of this approach is also evident, so usually in order to succeed, the main contractor of customer (depending on the application sectors) needs to push the different players to work together to get the opportunity to successfully use M&S in this activity. Hence, the use of M&S for SoS testing and anticipatory commissioning of crucial systems is often very useful to speed up the setup and ramp up the new systems and the possibility to quick detection and solution of problems.

This approach is very important also in engineering as well as during early development phases for new SoS; indeed, simulation allows to model and evaluate many alternative solutions in order to anticipate identification of critical elements affecting the overall

SoS performance. Usually, these elements deal with interoperability issue among the different systems. Obviously, simulation in these sectors supports also validation of the original SoS requirements and systems specification as well as an opportunity to virtually evaluate the different Measure of Metrics (MoM) introduced by users. Indeed, in the development of new SoS, often, the customers/users have a limited knowledge of the configuration and scenario, being that these systems are devoted usually to face new challenges, so M&S represents an opportunity to create a common preliminary benchmark for evaluating alternatives between engineers and decision makers. Indeed, M&S for System of Systems Engineering (SoSE) allows to identify problems during the early phases of SoS development, reducing drastically the costs of errors and changes. In addition, the integration of simulation in E&T is a very important aid for guaranteeing that the final SoS objectives could be achieved. Indeed, the possibility to conduct early test allows to change engineering and functional solutions in order to guarantee SoS performance.

Another interesting area of application is related to the training and education; indeed, the SoS due to their complexity requires often to develop specific programs for addressing this issue, and obviously, simulation represents a major aid in this area. This kind of application is usually very important even when considering the high degree of obsolescence often related to SoS, where the new integration technologies and functions require specific training over complex scenarios; obviously, for these applications, the use of real-time simulation is often common, considering that the involvement of humans, man in the loop (MIL), which often deals with psychomotor, collective and operational training, is usually real time. Therefore, more recently and due to the high degree of autonomy of modern SoS, the supervision becomes a new demand for this system, introducing a shift in training requirements; based on this concept, it seems to assist to an evolution from an MIL to a man-on-the-loop approach. In these cases, the user of the simulation is required to supervise the SoS for different purposes, for instance, service management but also operations supervision; for instance, in intelligence, surveillance, and reconnaissance (ISR), the new SoSs based on integrating multiple systems and drones are becoming more and more demanding of developments under this approach, providing an evident opportunity to M&S to support the new supervision systems as well as an effective training equipment for these new operational roles. In these cases, the use of fast-time simulation could become interesting to train users in a wide spectrum of applications; therefore, real-time modes are expected also to be required even if the requirement could be a little bit relaxed with respect to automation context.

Another critical sector where M&S could provide a drastic improvement is the doctrine development; this element is sometimes not too directly related to SoS. However, as already anticipated, it is important to outline that often SoS requirements are driven by technological consideration that miss real operational issues. This is due to the fact that by using traditional techniques the development of a new SoS (i.e., a new autonomous swarm devoted to air supremacy) cannot be tested on the field until a very advanced development phase; due to these reasons the requirements are based on broad and generic concepts directed by the feelings of experts.

Therefore, today, it is possible to develop quickly and effectively, by adopting lean simulation (LS) approach, virtual experimentation frameworks able to create these operational frameworks and to investigate if the doctrinal approach and function required are really useful in future scenarios; obviously, these models are expected to be characterized by low fidelity, being based on preliminary information. Therefore, the integration of the different systems in the simulation allows to consider many interactions and interoperability issues within the operational framework providing immediately highlights about critical issue. Considering these boundary conditions, the application of LS methodology has usually to be identified as best solution for developing models for this specific case dealing with the early stage evaluation (ESE).

So the common phrase "devil is in details" for SoS could be partially mitigated by introducing simulation as early as possible in new systems definition and design.

Therefore, it is evident that in order to succeed in applying M&S to SoS, it is necessary to deal with several challenges:

- Innovative design of the SoS: It reduces the capacity to estimate the systems performances and characteristics during early phases.
- Very small quantities of SoS: SoS is usually produced in small numbers. This fact stresses the necessity to deal with a completely new SoS with mostly no support of statistical examples, also from similar units.
- Confidential data: In defense, the fact that most of the data are classified due to their military implications introduces additional difficulties in modeling. In addition to the military classification, it is important to state that also in industrial application, most of the data have also an industrial strategic relevance (they involve costs/performances/reliability characteristics) that restrict access and availability of such information.
- Multiple industrial actors: An SoS usually involves many different actors and companies active in different sectors (i.e., ICT, automation, power, engineering, etc.) in its development phases such as design, engineering, purchasing, constructions, commissioning, and service. All these entities have often a limited knowledge of the interactions among the components and operate often in a competitive environment each other.
- Multiple disciplines: The components to be integrated in the SoS are often related to completely different aspects, so the model needs to face many different issues.
- Horizontal features such as reliability, service, and logistics: These concepts are very well known in terms of importance from centuries; therefore, they are affected by almost all the SoS components. So the historical data about these aspects are usually not collected in comparable, nor consistent, format, mostly due to the very high number of hypotheses related to their measures (i.e., cost accounting system). So, also when few data are available, it is required a very detailed analysis to check their validity and real meaning.

In all the proposed applications, the benefit to use simulation due to its capability to create complex scenarios able to stress the SoS emerges evident; this element is probably the major benefit, so the importance to focus the models on these issues is evident with special attention to taking care of several aspects such as:

- Stochastic components: These affect reliability of systems and components but also operational performance and behavior of the element characterizing the scenario.
- Autonomy of the objects: The reproducing of actors within the scenario is a crucial element; most of the SoSs have to deal with mission environments involving many entities. From this point of view, the introduction of intelligent agents (IA)

represents a strategic advantage in modeling these aspects. A very good example in this case is the researches carried out in applying agents in simulating complex joint operations combining Psychological Operations (PSYOPS), Civil-Military Cooperation (CIMIC), ISR, and Region Control by using Intelligent Agents Computer-Generated Forces (IA-CGF) for controlling both Coalition and Opposing Force (OPFOR) in an asymmetric scenario (Bruzzone, 2013a).

- Interoperability issues: The capability to integrate different models as well as simulators with real systems is a major issue in M&S applied to SoS, so the necessity to guarantee this approach is evident. Usually, the adoption of state-of-the-art standards in the sector avoiding the introduction of proprietary solutions that usually don't guarantee higher performance and create a high risk of obsolescence of the models is strongly recommended. From this point of view, there is a very interesting experience for promoting the use of simulation in a complex SoS—a Moon Base: Simulation Exploratory Experience (SEE)/Smackdown initiative led by NASA personnel since 2011 in cooperation with scientific organizations and societies such as the Society for Modeling & Simulation International (SCS), Simulation Interoperability Standards Organization (SISO), Liophant, and Simulation Team with support of many institutions (i.e., MIT, Genoa University, Penn State, Calabria University, University of Munchen, University of Bordeaux, University of Alabama in Huntsville (UAH), Ajou University, etc.) and organizations (i.e., AEgis Technologies, Pitch, VT Mak, and Forwardsim). In this case, the teams composed of students in universities and in internship in companies and institutions develop the different simulators to be integrated into a Federation based on High-Level Architecture (HLA) Evolved. During Smackdown 2013, the Federation included a Lunar Satellite Constellation, a spaceguard system, and several moon vehicles and equipment, and the simulation involved a successful interception deviating in real time a small asteroid (Elfrey, 2013).

- Optimization capabilities: Considering the huge amount of variables and entities affecting SoS performance, the benefit to develop hybrid simulations able to combine the models with smart optimizers is evident; these elements could be effectively used in order to support users in evaluating the most effective solutions and configurations. Sometimes, it could make sense to adopt a nested simulation paradigm introducing these elements directly in the simulation loop to provide self-organization capabilities. In order to succeed in this area, the use of artificial intelligence (AI) techniques, often combined together, could result as very interesting with respect to traditional analytical approaches that are usually limited by the stochastic nature of the systems and by the high number of nonlinear constraints usually present in this sector of applications. An example is related to development of a smart management system for a wide regional network of stores combining human resource management, sales control, store operations planning, and logistics (Bruzzone and Longo, 2010).

- Human behavior models (HBM): The processes are strongly affected by human factors and represent critical elements in most of the SoS (i.e., the crew of a ship); indeed, the HBM are present in these aspects and have sometimes a very strong impact, so it becomes more and more important to properly consider these aspects. As an example, Figure 8.1 shows an example of simulation model able to

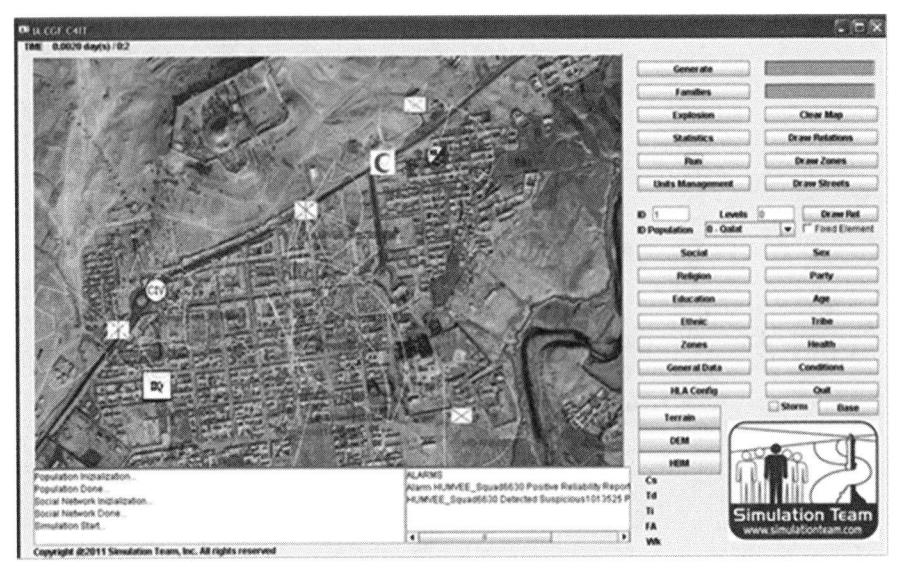

Figure 8.1 IA-CGF directing Coalition and Local Forces, Insurgents and village population in an HLA federation for analyzing C2 architectures within an Urban Operation in South Asia.

consider HBM in terms of IA-CGF directing Coalition and Local Forces, Insurgents, and village population in an HLA federation for analyzing C2 architectures within an Urban Operation in South Asia. As an additional example, the authors were involved in the past in evaluating the impact of human factors on the efficiency of a port terminal considering container handling operations as interoperability among different handling systems and yard control; for this case, an interoperable simulator was developed (see Figure 8.2), called Simulation Team Virtual Marine (ST_VM), combining use of biomedical devices for measuring human factors (i.e., fatigue and stress) dynamically federated in an HLA real-time Live, Virtual, and Constructive (LVC) simulation (Bruzzone et al., 2010; Bruzzone and Longo, 2013).

Obviously, these are just some of the critical elements in developing M&S for this purpose, and many other ones are important; for instance, reusability is a very crucial element in M&S, even if it is fundamental to tailor reusability with respect to the application area: in early phase evaluation, the model should be reused probably for other future similar SoS projects during the same phase of project life cycle and not in the future development of the SoS, considering that the resolution and granularity are completely different in a systems development process and it could be very hard to adapt the same simulator to these changes (Amico et al., 2000). Vice versa, considering the obsolescence in innovative systems, it could be not too much convenient to dedicate too many efforts to create reusable models of detailed functional aspects that could be limited in their reuse due to the change on the physical technologies used in the real system voiding the previous model validity.

In general sense for SoS, another critical issue, related to the aforementioned aspects, is related to legacy simulators; it is common that a new SoS could involve the previous

Figure 8.2 Mobile virtual simulator cave embedded in a 40′ high cube container integrating ST_VP in a real time HLA simulation for container handling operations.

systems where sometimes legacy models or simulators are available. In general, the reuse of the elements could be interesting; therefore, it is critical to check if the previous models were addressing objectives consistent with the new SoS: for instance, a simulator of a previous radar could be available. Therefore, if it is developed for evaluating this coverage operating in stand-alone way, it will be not usable for an interoperable simulation addressing operational use of the radar concurrently with other systems (i.e., fire direction) that could generate electromagnetic interference. In the next future, it is expected to get benefits of new technological trends; for instance, in SoS area, the authors are conducting researches related to the development of virtual experimental frameworks to be accessible remotely by applying Simulation as a Service paradigm for allowing multiple developers to connect and test their systems (Bruzzone, 2013b).

The most critical elements in using M&S for SoS are common to the general application of simulation, with specific declination within this area:

- Data: Data and knowledge of the system to simulate are probably one of the major efforts in M&S; for SoS, these aspects are reinforced by the presence of multiple systems. This element is even stressed by the usual challenge related to sharing information and data related to different companies/organizations. In addition, the SoS is usually referring to complex mission environments, and it could be pretty difficult to acquire data covering multiple cases; often, there are just few situations that were extensively captured, and the proposed scenario is subdivided in a simple set of basic cases to be addressed by SoS. This obviously introduces big risk in development of the new systems; for this reason, simulation represents a major aid, and the benefits introduced by creating scenarios by introducing IA and stochastic factors are evident.

- Conceptual modeling: The conceptual model is the opportunity to create the virtual SoS as well as the virtual mission environment for testing it; at the same time, this represents a challenge considering the complexity to create a model addressing different systems, usually operating within different domains with their specific control and architectures that need to interoperate for a complex common mission/goal. From this point of view, the authors stress the conceptual interoperability aspect already addressed by the radar example where the evidence that a good model for a purpose could result as being completely not useful for other applications emerges; in addition, SoSs are often involving strong impact of multidisciplinary requirements for the simulators even considering that the different systems are dealing often with different domains. The crucial element to be stressed in this context is the one related to the connection among different simulation developers and the subject matter experts of the different systems: someone should be able to get the whole picture to address properly the systems interoperability issues to be modeled as well as the proper understanding of the requirements for the virtual mission environment. The introduction of IA and AI within hybrid simulation solution has a great potential; therefore, it introduces the necessity to adopt a proper approach for describing and analyzing the different entity behaviors in order to properly design these elements as well as to support effective verification and validation.

- Verification, Validation, and Accreditation (VV&A): In general, as always in M&S, the main aspect to succeed is VV&A, which is pretty challenging within this context considering the multiplicity of elements of SoS and the common presence of interoperability requirements; it is necessary to complete VV&A along all the life cycle phase of the simulator. Another aspect the simulationist needs to address is the continuous and full alignment of the simulation models with the real SoS configuration that usually is subjected to changes and evolution during its development phase and sometimes even during deployment and service. Last but not least, it should be stressed in an SoS that validating and verifying a system is useful; therefore, Verification, Validation, and Testing (VV&T) should be conducted on the overall SoS including all systems. The high influence of stochastic components suggests to adopt extensive dynamical testing by applying analysis of variance (ANOVA) and Design of Experiments (DoE). The success in VV&A is strongly related to the involvement of users and subject matter experts both in simulation domain and in specific sectors to be addressed by each model and system.

- Implementation: The implementation of M&S for SoS often deals with interoperable simulation distributed over different teams; so even if this issue is traditionally overestimated with respect to other aspects, in SoS, the implementation aspect requires more capabilities in distributed team working with respect to other application fields. Another important aspect in this sector is the need to create transdisciplinary teams able to deal with the different aspects covered by SoS, as already mentioned for modeling and for VV&A. In implementation, the critical connection is between simulation developers of the different systems being able to communicate and collaborate considering critical issues such as approximation, time management, walk-around proprietary constraints, computational constraints, etc.

The authors have long experience in applying M&S to SoS for different specific applications; therefore, along last years, the diffusion of IA and their integration with simulation represented a real leap. By this approach, it was possible to create new complex scenarios quickly, extending the potential of simulation in terms of use and experimentation; the agents are allowed to simulate fast-time complex scenarios playing the role of MIL. This allows to speed up development and extend testing; for instance, in the case of the simulation of a new-generation vessel, usually, a single run requires weeks to get the availability of all console operators (i.e., naval officers), to bring them together, and to prepare them for the exercise. Often, few cases could be simulated in the available time frame over the simple available scenarios; vice versa, by IA, it is possible to substitute the operators and to create dynamically very challenging scenarios in a few minutes, simulation could run also fast time (i.e., if not HIL or SIL is present in the architecture in use for the experimentation), and it is possible to complete an extensive testing campaign within a few days (sometimes hours) and to create a massive virtual experimental data related to the new SoS that could support decisions, performance evaluation, reengineering, etc.

In fact, the authors have introduced a new generation of agents, defined IA-CGF, that are able to deal with multiple applications; IA-CGF was developed by Simulation Team originally for defense applications. Therefore, they are addressing the different critical aspects related to:

- Rational decision making
  - Intelligent individual behavior
  - Organization and hierarchies
- Emotions and attributes
  - Psychology, culture, and social
  - Crowd behavior
  - Social networks

The IA-CGF are devoted to simulate complex joint scenarios and include the following modules:

- IA-CGF units
- IA-CGF behavior library
- IA-CGF nonconventional frameworks

By these elements, it is possible to develop simulators for SoS addressing the aforementioned challenges and critical issues, and the authors will provide several application examples in the following sections.

8.4 AN AGENT-ORIENTED PERSPECTIVE FOR SoS

In this section, the effectiveness of the agent-based paradigm for the M&S of SoSs is discussed by highlighting how this paradigm is able to address the issues identified in Section 8.2 and thus to provide a full-fledged approach for the M&S of SoSs.

Agents and multiagent systems (MASs) (Jennings, 2001) represent a powerful and popular paradigm for modeling and analyzing natural and artificial systems that allows overcoming the limitations of more classical analytical-deductive modeling approaches, which are often characterized by strong simplification assumptions making their resulting models not powerful enough from the descriptive and predictive perspectives (North and Macal, 2007). Indeed, a system can be modeled in terms of entities capable to (i) perceive and act in an environment, (ii) exhibit autonomous and goal-driven behaviors, and (iii) interact with other agents and cooperate or compete to achieve specific objectives. According to the agent-based paradigm, these entities are called agents.

The previously depicted mind-set allows one to focus on the modeling of the single systems entities and their behaviors and interactions rather than on the representation of the system as a whole. The so obtained agent-based model of the system can be then simulated so that the knowledge of the behaviors of the single entities and their interactions (microlevel) can produce an understanding of the overall outcome at the systems level (macrolevel). In this way, simulation of agent-based models allows to observe and analyze emergent phenomena at macrolevel that are unpredictable and hard to catch with analytical-deductive techniques.

The abstractions offered by the agent paradigm allow to effectively address the issues identified in Section 8.2 concerning the M&S of SoSs (see Table 8.1); indeed, (i) agents are capable of autonomous and proactive behaviors (as well as reactive and possibly passive ones) that can change over time due to learning or adaption processes (thus, agents can be used to model the entities that typically are involved in an SoS) and (ii) agents can be organized in societies characterized by (dynamically changing) members, rules, norms, and organizational structures, thus allowing to represent the dynamic and articulated structure of an SoS.

The application of the agent-based mind-set to the modeling of SoSs thus naturally leads to conceive an SoS as a collection of *agent societies* that represent the systems, or better the *organizations*, involved in the SoS and are constituted by *agents* that can play different *roles* in the society and are characterized by different degree of autonomy and proactiveness as well as capabilities and social abilities. An *agent society* can be characterized by *rules*, *norms*, and *organizational structures* that govern its evolution and the relationships among its members. The relationships among the agent societies that dynamically are involved in an SoS can be regulated by *agreements*, *protocols*, or simple *interaction patterns*.

Starting from this agent-based representation of an SoS, in the following, a framework for M&S of SoSs is proposed.

An agent-based modeling approach for SoSs should be able to fully represent the different aspects of the SoS under consideration by grouping them in different views and at different abstraction levels.

In particular, the modeling of SoSs should be based on the principle of layering and exploit the well-known techniques of decomposition and abstraction thus to produce several views of the SoS, each of which is aimed to represent a specific aspect of the SoS (architectural, behavioral, organizational, etc.). Each aspect and the related views can be analyzed at various abstraction levels. In particular, the definition of a specific view starts from a desired abstraction level and then moves toward higher or lower levels of abstraction respectively by applying the zooming-out/zooming-in mechanisms to obtain the desired level of detail. As an example, in order to model the aspect concerning the architecture of the SoS under analysis, it is necessary to first identify a starting level of abstraction that leads to detect the SoS *entities* "observable" at this level of abstraction (e.g., *organizations*). Then, by zooming in, it is possible to "look inside" these entities so to individuate their (sub)entities; by zooming out, it is possible to group entities into a single one of a higher abstraction level.

With reference to the different possible SoS views, the proposed agent-based approach for SoSs supports the following ones: *environmental view*, *organizational view*, *goals view*, *interactions view*, *behavioral view*, and *properties views*; in the following, each of these views and related modeling constructs are discussed.

Environmental view: This view highlights links/interactions between the SoS under consideration and stakeholders, actors, and/or systems that are currently outside the scope of the SoS but that can affect its behavior. This view allows modeling the environment effects/factors that could influence the SoS without being directly involved in it.

Organizational view: This view is devoted to represent the *entities* that constitute the considered SoS (at a given time and abstraction level) and their *relationships*. In particular, this view provides constructs to model the following concepts: (i) *organizations* in terms of *agent societies*, (ii) interrelationships that describe the relationships among the *organizations* involved in the SoS (and "observable" at the considered level of abstraction), (iii) intrarelationships that represent the relationships among the members of an *organization* and that are observable by zooming in the organization, (iv) *norms* and *rules* that govern the evolution of an *organization* influencing the interactions and behaviors of its members, and (v) *agreements* and *protocols* that govern the interaction among interrelated *organizations*.

Goals view: This view provides the representation of the objectives that are pursued by the entities involved in the SoS and that are essential to guide their behaviors and support their decisions; in particular, this view allows to (i) organize the objectives hierarchically, (ii) establish links between the objectives and priorities and then an order of their achievement, and (iii) associate a specific (sub)goal to a specific system/organization.

Interactions view: This view provides a model for describing how the entities involved in an SoS interact among them by using specific communications mechanisms and interaction protocols.

Behavioral view: This view provides a representation of the behaviors of the entities involved in an SoS; specifically, each entity can be characterized by different degree of autonomy and proactiveness as well as capabilities and social abilities. The behavior of each entity can be described in several ways depending on the aforementioned features. For passive and almost reactive entities, it can be sufficient to specify the tasks associated to the services that the entity provides and/or to its reactions to external stimuli. For proactive entities, it could be useful to specify the decision rules which, on the basis of the goals to reach and both the internal and external perceptions, allow choosing how to act for achieving the goals.

Properties views: This view allows the representation of properties that should be useful to observe during the SoS evolution and/or performance indices that should be monitored and measured during the SoS. Properties can be associated to *assertions* that allow to evaluate them against desiderated (range of) values; similarly, performance indices can be associated with metrics for their evaluation. Properties whose assertions should be satisfied by the SoS during its evolution represent requirements for the SoS as well as performance indices that have to reach required values.

The previously described views allow to fully represent an SoS by using an agent-based perspective. Each view highlights specific aspects of the SoS; the exploited modeling constructs can be easily represented by using a popular UML-based modeling language such as Agent UML (AUML) (Bauer et al., 2001), Agent Modeling Language (AML) (Cervenka and Trencansky, 2007), and Multiagent System Modeling Language (MAS-ML) (Gulyas et al., 1999). The so obtained SoS model can be then simulated by one of the available agent-based simulation platforms (see Garro and Russo (2010) for a complete list). Although this requires to represent and codify the agent-based SoS model in terms of the constructs provided by the framework that has been chosen for the simulation, by using suitable model-driven techniques, the simulation code can be (semi)automatically generated with a significant reduction of programming and implementation efforts (Garro and Russo, 2010; Garro et al., 2013). Different scenarios can be set and simulated by suitably tuning SoS parameters and by interacting during the simulation with the SoS and its entities; finally, simulation results can be analyzed so to obtain not only insight on the SoS behaviors and evolution but also on emergent phenomena that are unpredictable and hard to catch with classical analytical-deductive techniques.

8.5 EXPLOITING THE AGENT-ORIENTED PERSPECTIVE: APPLICATION EXAMPLES IN MULTIPLE DOMAINS

In the following examples, the authors propose three different case studies addressing SoS simulation based on agents in defense and industrial applications; obviously, the results proposed in this section have been modified due to their confidential nature. Therefore, the proposed modeling approach and concepts are valid and represent an important opportunity to address SoS around their life cycle.

8.5.1 Aircraft Carrier

The first proposed example is related to a research conducted in reference to a new aircraft carrier where the authors were involved in addressing the SoSE by using M&S during the development phase (Bruzzone and Bocca, 2006).

The study was focused on the development of ad hoc models in order to analyze the life cycle of a new aircraft carrier and to cross-check the systems and subsystems considering operational capabilities, availability, operative costs, and overall performances; obviously, such a complex system deals with a complex identification process of data and variables. Therefore, the application of the simulation to such an issue allows obtaining an effective configuration management, providing support to shipbuilding design and to

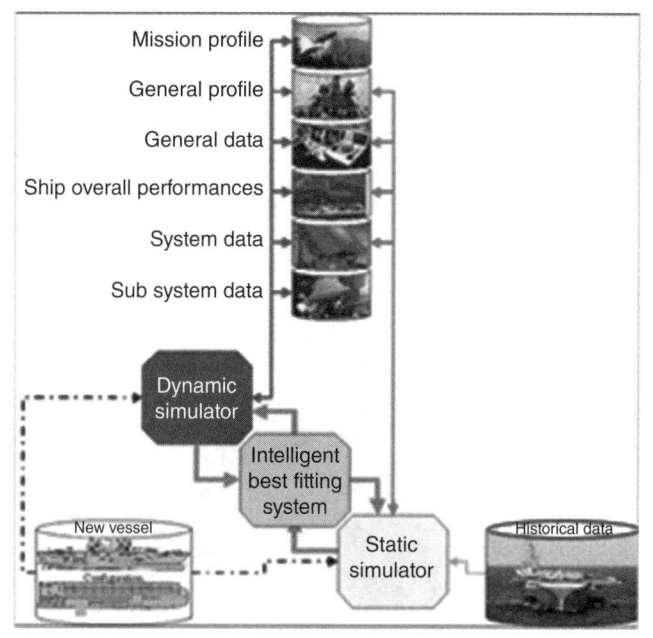

Figure 8.3 General architecture for new aircraft carrier integrated system for SoSE.

configuration analysis. Stochastic components are involved in life cycle analysis of a vessel as main aspects.

The simulation model could be useful to estimate not only the final results but also the relative confidence band and related risk, taking over from the traditional approaches based on average values (Figures 8.3, 8.4, and 8.5 show the main architecture and the mainframe of the simulation model). In fact, the authors reached the goal of creating an advanced life cycle analysis system, which could be able to support decision making in a brand-new vessel design; the crucial element of this development was the use of IA to create the mission environments and scenarios to be addressed by the carrier along its life cycle. This allowed to simulate the entire life cycle including port operations, regular sailing, exercises, real operations, middle life maintenance, etc.

In this case, the use of simulation was driven by several critical issues such as:

- Research and development: Most of the ship plants and systems are based on innovation and developments, often including critical systems to be concurrently developed ad hoc for the new vessel (i.e., a new 3D radar developed for a new destroyer).
- Mission profile: The mission profile is often the motivation for launching a new vessel; however, this profile is rarely detailed, and in several cases, it continuously evolves, sometimes including radical changes.
- Overall performance expectations: These performance requirements are usually vague from a quantitative point of view during early phase of the program.

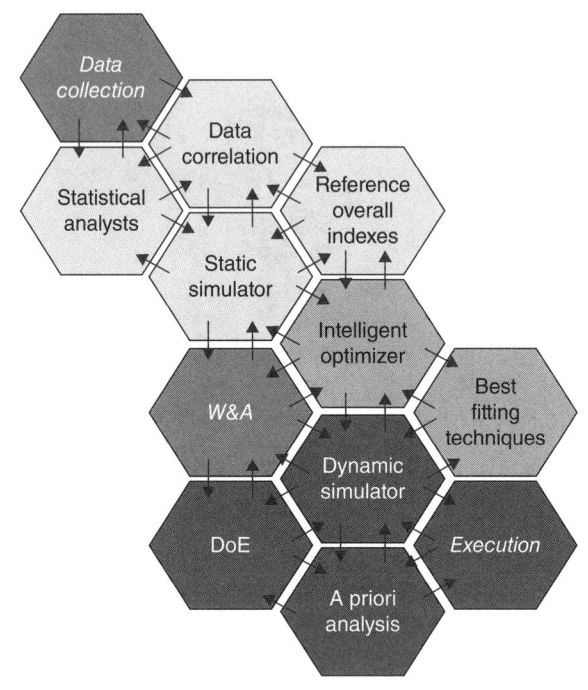

Figure 8.4 Methodological approach for addressing the new aircraft carrier life cycle simulation.

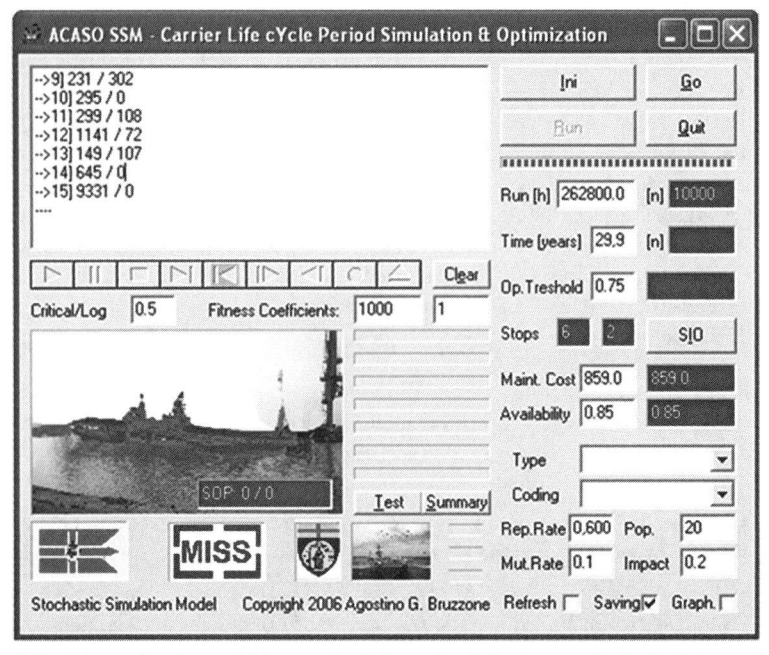

Figure 8.5 Advanced carrier acquisition cost simulation and optimization—stochastic simulation model GUI.

- Systems performances: The performances of the different ship systems are pretty unknown in advance, due to the fact that even if the system is not new, usually each new vessel involves strong tailoring and special solutions. In addition, normally due to the technology advance, all the strategic components (i.e., engine, radar, weapon systems) are pretty new.
- Systems interaction: The overall efficiency and effectiveness strongly depend on the combination of different ship plants, systems, and subsystems.
- Systems reliability and maintenance programs: These aspects strongly affect the ship availability and operative efficiency as well as costs.
- Historical data: Usually, the data are not easily available for many reasons, including their reliability and not homogeneous measurement procedures. Also, the suppliers don't share most of the related information about their system for commercial competitiveness reasons.

After the evaluations of such aspects, the authors decided to develop an architecture combining different simulators able to conduct analysis on similar case and to extract a correlation for parameters related to a new aircraft carrier design to be used in an advanced stochastic discrete-event simulator integrating agents for scenario definition and crew behavior analysis.

The very important issue analyzed in this study was the possibility to evaluate aspects in relation with such a such complex system such as a middle-sized aircraft carrier (length 240 m, 27,000 DWT, 1,200 people, 30 aircrafts) before producing it; in these cases, unfortunately starting from a total lack of critical parameters, it was necessary to establish an approach able to define an effective model and to optimize its tuning to support decision making. The adopted approach involved the combination of dynamic simulation with a smart optimizer able to direct the identification process of the independent variables and the self-tuning on the model by a correlation of a data set, elaborated by a static simulator correlating different historical aircraft carrier models, with a dynamic stochastic discrete-event simulation reproducing all systems and subsystems. By this approach, it was possible to get reference data to be used with hypotheses on new systems by reverse engineering techniques directed by genetic algorithms (GAs) driving directly the dynamic simulator. This approach allowed to properly estimate the new ship component characteristics and the consequent effect of different configurations on the overall performance; indeed, the dynamic simulator was including also the detailed model of crew operations based on IA.

So by this approach, it was possible to evaluate the different vessel configurations with respect to the automation solutions and crew use also conspiring human factors such as harmony, stress, fatigue, and turnover (see Figure 8.6); this allowed to estimate the overall efficiency of the new vessel with respect to systems and operations as well as mission environments considering different crew solutions in terms of spaces, polyfunctional assignments, automation solutions, and operative processes.

The simulator was used dynamically during this process and for evaluating especially new scenarios in terms of cost and benefits of different design solutions, operative modes, or general policies (Figure 8.7 shows parameter settings before running the simulation, while Figures 8.8 and 8.9 show an example of simulation results concerning reliability and availability aspects).

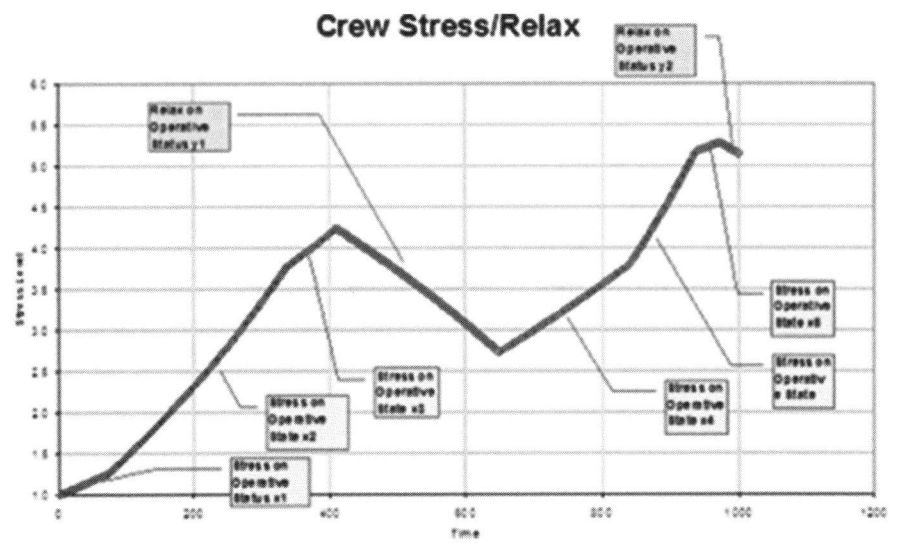

Figure 8.6 Psychological modifier evolves based on their status, trend and ship operative condition.

Figure 8.7 Synthetic settings about the life cycle a priori hypotheses.

Trend on bath tube effect sources

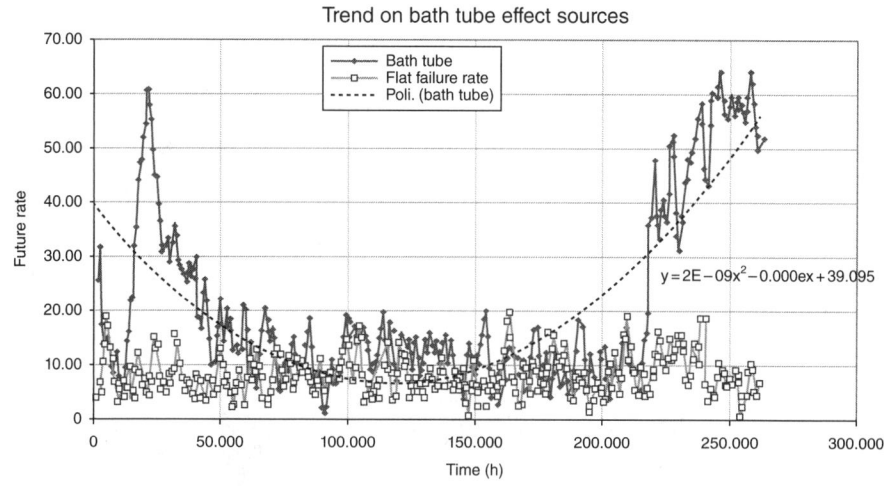

$y = 2E-09x^2 - 0.000ex + 39.095$

Figure 8.8 Instantaneous failures rates during simulation along carried life cycle.

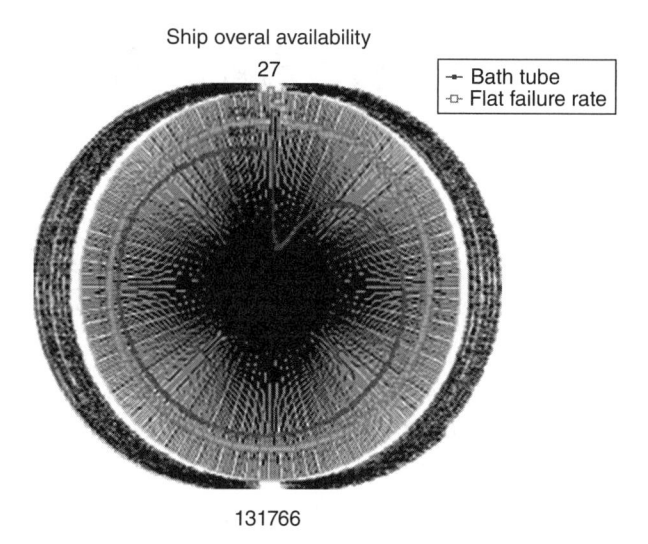

Ship overal availability

Figure 8.9 Ship availability along its life cycle with different hypotheses on bath-tube effect related to sub system failures.

8.5.2 Agile Multicoalition C2

This research was focusing on the creation of an innovative solution able to identify implications of different vessel characteristics on its performance along the whole life cycle.

This second proposed example is still related to marine domain, but in reference to a complex command and control (C2) able to demonstrate agility over different maturity levels (Bruzzone et al., 2011b). Indeed, this research was devoted to experimentation, by

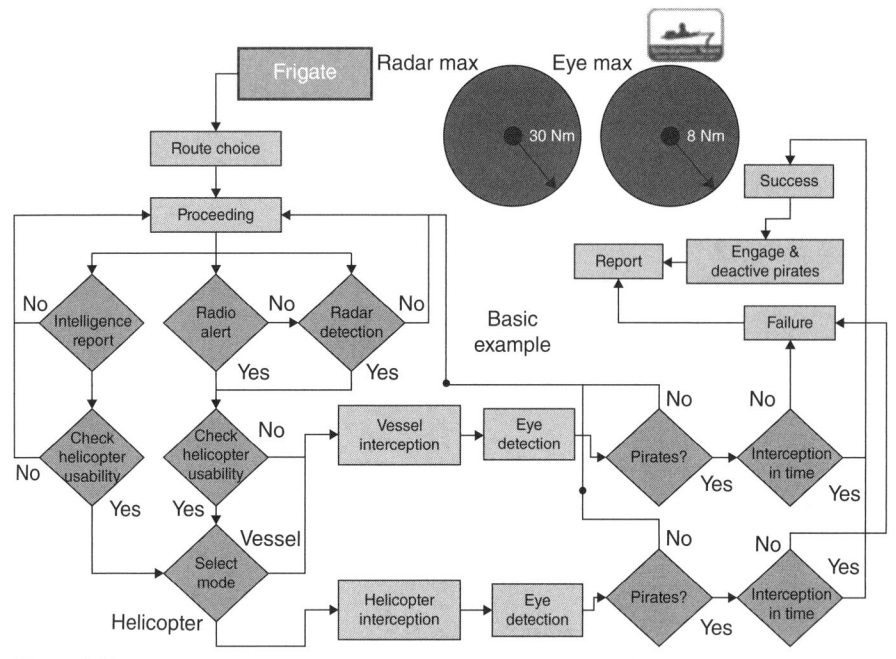

Figure 8.10 Basic example of agent rules for ship patrolling.

using simulation, on a complex maritime scenario where it was possible to evaluate different Net Centric Command and Control Maturity Models (NEC C2 M2) (Alberts et al., 2000; Bruzzone et al., 2009). In this case, the authors propose an experimentation based on a simulation model related to an asymmetric scenario in maritime domain with special attention to piracy by exploiting the simulator titled Piracy Asymmetric Naval Operation Patterns modeling for Education & Analysis (PANOPEA). Figure 8.10 shows an example of a conceptual model related to the implementation of the PANOPEA simulator. Figures 8.11 and 8.12 show the mainframe of the PANOPEA simulator and the animation during a simulation run, respectively. PANOPEA was developed by Simulation Team to analyze new asymmetrical war theaters focusing on scenarios of marine warfare versus pirates; the model was used by different Navies as well as the North Atlantic Treaty Organization (NATO) to conduct not-classified researches on C2 agility (Bruzzone et al., 2011c).

The case study here proposed is related to Aden Gulf mission environment; indeed, PANOPEA reproduces a piracy scenario in the Horn of Africa, a very critical area in terms of pirates' attacks against cargo ships. This scenario involves a very complex SoS involving navy vessels and helicopters, satellites, intelligence assets, and ground bases; obviously, in order to test this SoS, it is necessary to reproduce a large complex scenario over thousands of square miles including cargos as well as other boats (i.e., fisherman and yachts) and pirates hiding in the general traffic. All these entities are directed by IA-CGF and apply strategies for succeeding based on their scenario awareness. In addition, PANOPEA simulator allows to adopt different strategies in operating C2 over different architectures, including conflicted, deconflicted, collaborative, and edge solutions (see Figures 8.13 and 8.14). PANOPEA tool supports the authors in making experimental

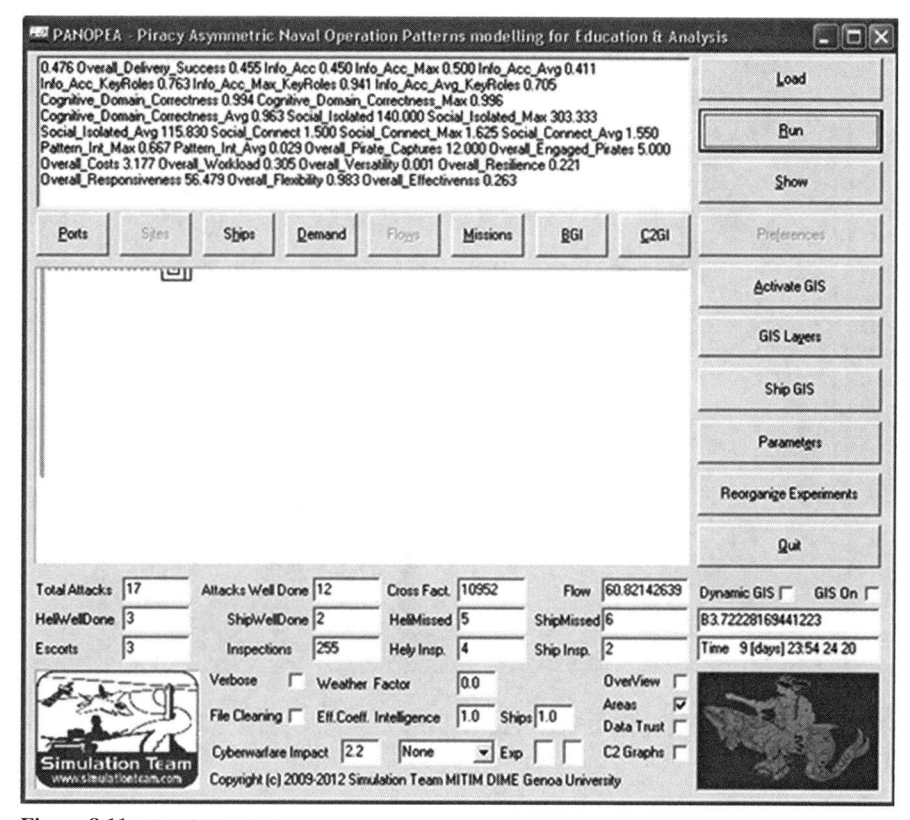

Figure 8.11 PANOPEA GUI a simulator integrating IA-CGF for complex asymmetric maritime scenarios.

analysis by modeling different C2 maturity levels and measuring the effectiveness and the efficiency of the proposed scenarios in order to investigate the agility of the C2 solutions and their influence in preventing attacks by implementing different policies and different organizational models.

Today, this scenario is quite interesting: in fact, maritime security is a very critical aspect of the marine framework and extends the concept of asymmetric warfare within marine environment with new threats such as piracy, conventional terrorism, and chemical, biological, radiological, and nuclear (CBRN). Therefore, the case proposed involves around 2000 boats and ships directed by IA, so M&S is critical to evaluate strategies in terms of efficiency to prevent and mitigate threats by improving policies, sensors, equipment, as well as C2 solutions that obviously affect detection, identification, decision making, and scenario evolution.

As target functions, the model allows to monitor, among the others:

- Actual, max, and average information access level by coalition vessels
- Actual, max, and average information by key role players
- Actual, max, and average cognitive domain correctness
- Actual, max, and average social isolated entities

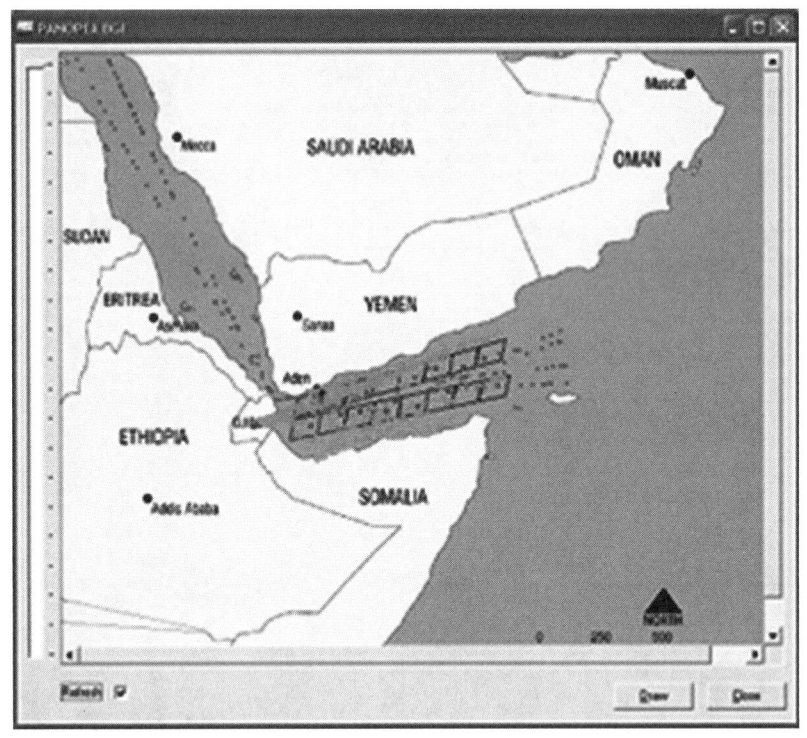

208 Theoretical and Methodological Considerations with Applications and Lessons Learned

Figure 8.12 Animation of the overall area patrolling areas for the different forces during a simulation run.

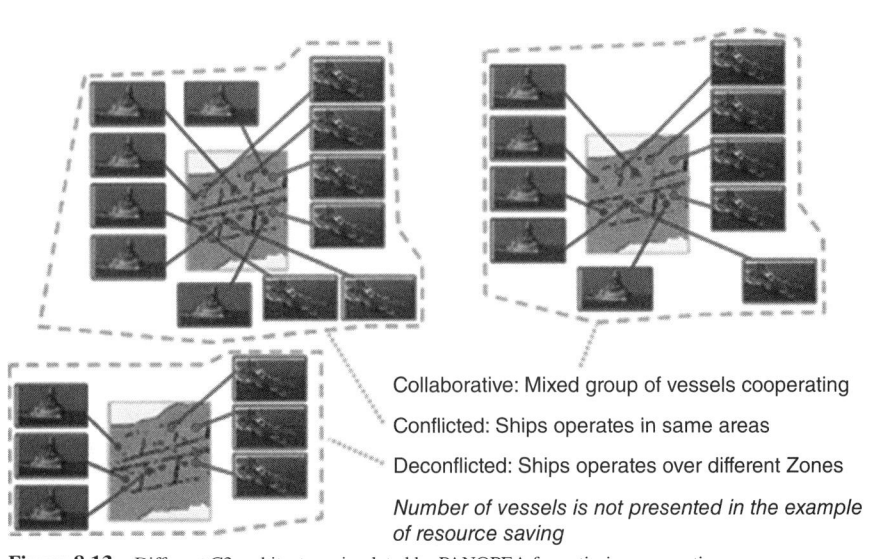

Collaborative: Mixed group of vessels cooperating

Conflicted: Ships operates in same areas

Deconflicted: Ships operates over different Zones

Number of vessels is not presented in the example of resource saving

Figure 8.13 Different C2 architecture simulated by PANOPEA for anti-piracy operation.

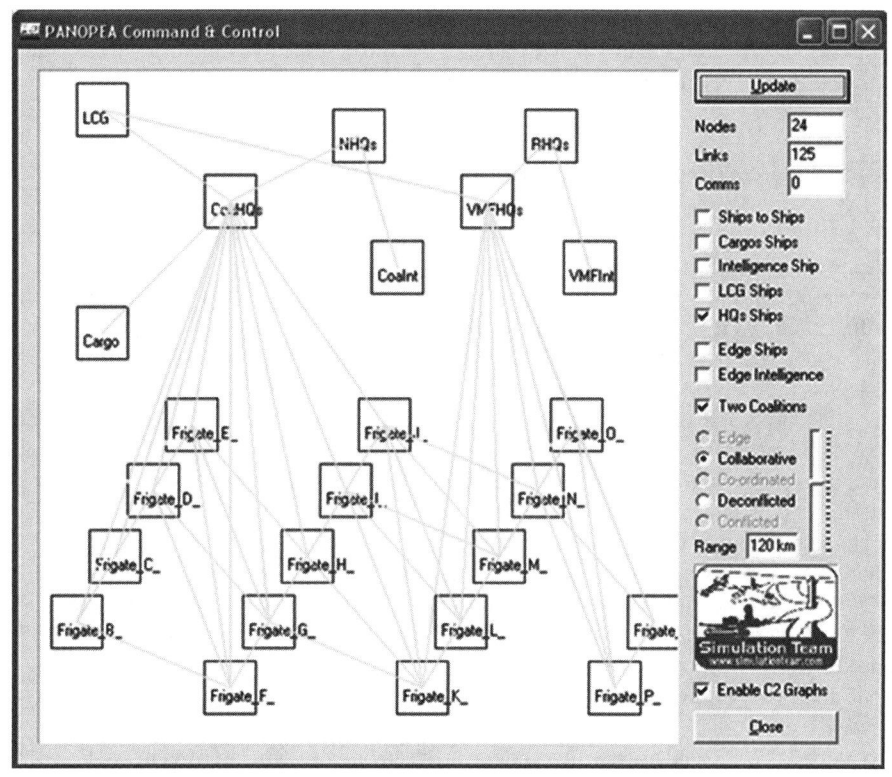

An Agent-Oriented Perspective on System of Systems for Multiple Domains 209

Figure 8.14 Example of dynamic collaborative C2 architecture between two different coalition fleet simulated by PANOPEA.

- Actual, max, and average social connected entities
- Actual, max, and average pattern interaction
- Actual, max, and average workload on vessel crew for different roles
- Overall success rate
- Overall pirate captures
- Overall engaged pirates
- Overall costs
- Overall versatility, overall resilience, overall responsiveness, overall flexibility, and overall effectiveness

The authors analyzed the possibility to cooperate with local coast guards and among different coalitions adapting the maturity level dynamically based on the specific situation and set of assets; this research included the combination of different coalition fleets over the area and the possibility to share different information.

This is very good example where the use of IA-CGF allowed to investigate both technological and operational issues related to a new complex SoS over a very articulated mission environment.

The experimental results based on PANOPEA simulation were able to evaluate the importance of systems agility and its requirements with respect to organizational models, C2 maturity levels, assets, and systems characteristics.

8.5.3 Power Plant Pool

The last example is related to an industrial application dealing with a power plant pool representing our SoS (Bruzzone et al., 2011a); the complexity to manage and guarantee most effective and profitable use of power plants during their operations is evident considering commercial, operative, technical, and market aspects. This example proposes a methodology based on integration of optimization and simulation driven by agents for supporting this SoS including several combined cycle power plants (steam and gas turbines).

The proposed approach is addressing pool service simulation; in this case, the agents are in charge of directing each single plant as well as centered maintenance operations and demand behavior. Indeed, the simulation allows to identify best configuration from design and management point of view as well to evaluate alternative strategies; a smart optimizer based on GAs is integrated with this simulator to identify most effective configuration and possible results starting from specific boundary conditions.

The proposal introduces a new metrics for properly estimating the real feasible best service performances in pool service management as reference for optimization processes.

Indeed, a pooling of complex systems such as power plants involving combined cycles (gas turbine, steam turbine, and related generators) is a very good example of SoS; to properly design and operate this system, the possibility to simulate systems and subsystems of each plant and to consider alternative scenarios as well as innovative approaches for serving the sites by clustering the machine in subsets able to guarantee time frames compatible with time cycle for each item for optimizing availability, costs, and technical commercial constraints, is required.

Therefore, in these real problems, the stochastic factors (i.e., spare part lead times and consumption, failures, inspection duration, etc.) as well as the complex processes (i.e., component refurbishment, supply expediting procedures) (Bruzzone and Simeoni, 2002; Longo et al., 2012) require decision support systems (DSS), and the use of stochastic simulation in joint combination with intelligent optimization represents a critical component as well as an innovative approach to address this issue.

This proposed approach allows to identify optimal configuration for spare part set definition, warehouse setup, and inventory management systems as well as to develop smart service schedulers considering both quality (i.e., availability, service time) and costs (new spare parts and refurbishment activities); in the proposed case study, the different contract elements (i.e., plant stop penalties, contract duration, terms for inventory reuse) in addition to technical constraints (i.e., intervals for inspections and revisions) require a complex model for the optimization.

The authors developed Lean Advanced Pooling Intelligent optimizer & Simulator (LAPIS) as a discrete-event dynamic simulation model driven by agents for the joint optimization/simulation.

Indeed, power plant service is a pretty complex framework: in fact, it includes high order of interactions among different entities, multiple objects, and multiple target functions as well as stochastic factors; modern service strategies deal with pooling the power plants. This means that many machines (i.e., generators, gas turbines, steam turbines, boilers) in multiple sites need to be maintained concurrently both in terms of preventive maintenance and in terms of failure recovery; each machine has many components that could be subjected to review, substitution, and/or refurbishment. In fact, for instance, it is possible, for many types of turbine blades, to proceed with hi-tech processes for refurbishing the blades after dismounting and to reuse in future for one, two, or even three times; the refurbishment of used component is usually costing about 1/10 of the acquisition of new elements, while at the same time some percentage of the refurbishment elements need to be substituted with new ones due to the too deep damages on the material (this is usually defined as scraping percentage). The failures are typical stochastic phenomena; in addition, even quantities to be substituted and scraping percentage, the duration of inspections, and minor and major revisions are affected by stochastic behaviors. The intervals among preventive events depend on consumption of equivalent operating hours (EOH); these are regulated both by the intensity of the use of the power plant (use to satisfy constant demand or peaks) and by the mode (many start-up or shutdown increase EOH with respect to solar time). Therefore, both have stochastic components and not so regular predictable behavior. Just mentioning these elements, the complexity of the service is evident; it is even important to identify that simulating this reality deals with at least two major target functions: availability and cost. These are obviously competing functions and require combined optimization; usually, the power plant service degree of freedom is related to the following major elements:

- Power plant preventive maintenance scheduling
- Power plant component inventory management
- Refurbishment component sequencing

The main purpose of this research was to provide power plant managers with a flexible simulator and intelligent decision support solution, combining different modules to define criteria and configurations for specific customers (i.e., a pool of 12 power plants for a private customer); to support service; to define inventory management systems, scheduling systems, and resource allocation criteria; and to measure plant performances. As mentioned in Curcio and Longo (2009), inventory management problem involving multiple plants (even along a supply chain) is a complex problem; the proposed solution allows users to test possible alternatives in terms of service and maintenance of power plants able to consider market changes (see also Bruzzone et al., 1998).

The concept of pooling power plant maintenance is strongly related with the possibility to optimize the reuse of items subject to refurbishment; in fact, by effective planning, it could be possible to obtain sets of kits for service from refurbishment of elements dismounted by other plants in the pool and to properly sequence the inspections and revisions.

In fact, as anticipated, many kinds of component kits could be refurbished one or more times, scrapping a percentage of components; the way to manage correctly inventory is to schedule major and minor revision intervals taking in consideration not only the technical constraints related to component life cycle but also in order to minimize costs, refurbish and kits for a subset of power plants. Obviously, this approach is more efficient and effective if combined with the maintenance plan of other plants so to share as many kits as possible. In this case, by creating a common power plant pool, the inventory levels are

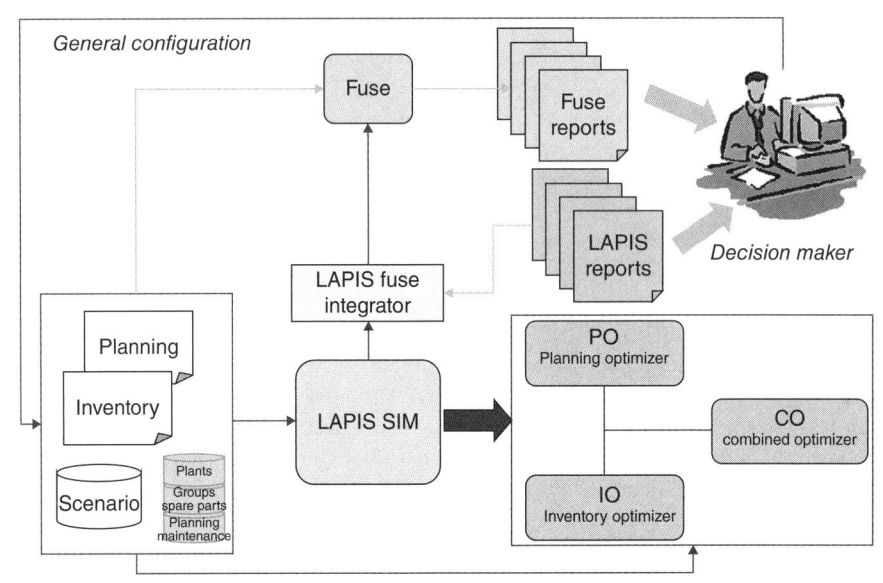

Figure 8.15 LAPIS architecture.

reduced, as well as safety stocks for many maintenance components; this allows to reduce the costs among different plants and to improve the effectiveness of maintenance resources (Bruzzone and Bocca, 2008).

The service quality was modeled in simple cases for different management strategies, and the result is that pooling always improves the service levels at all locations (Tagaras, 1989); demand pattern identifier based on statistical methods support has been demonstrated as reliable support for managers of service part inventories (Cerda and Espinosa de los Monteros, 1997; Paschalidis et al., 2004; Muckstadt, 2005; Sugita and Fujimoto 2005; Beardslee and Trafalis, 2006). The methodologies in inventory management have been investigated for similar case by several authors (Cohen et al., 1990; Silver, 1991; Nahmias, 1994; Harris, 1997). The proposed LAPIS simulation is based on following architecture (see Figure 8.15).

LAPIS combines dynamically simulation and optimization; in fact, simulator is connected to each part, because it is the head of the software. The second entity, one of the most important, is the box entitled "FUSE." Fuzzy techniques implemented in FUSE allow a correct evaluation of interaction among technical (machine life cycle), operational (interference among inspections), contractual (periods preferable for maintenance), and commercial factors (energy request from the market) (Cox, 1994). Considering the uncertainty of constraints, it is necessary to use fuzzy logic (Bruzzone et al., 2004). There are three types of optimization supported by this approach:

- Planning optimizer (PO)
- Inventory optimizer (IO)
- Combination of PO and IO

This model is an event-driven, stochastic combined simulator. Several events are taken in consideration and correspond to failures, planned maintenance events, and critical time

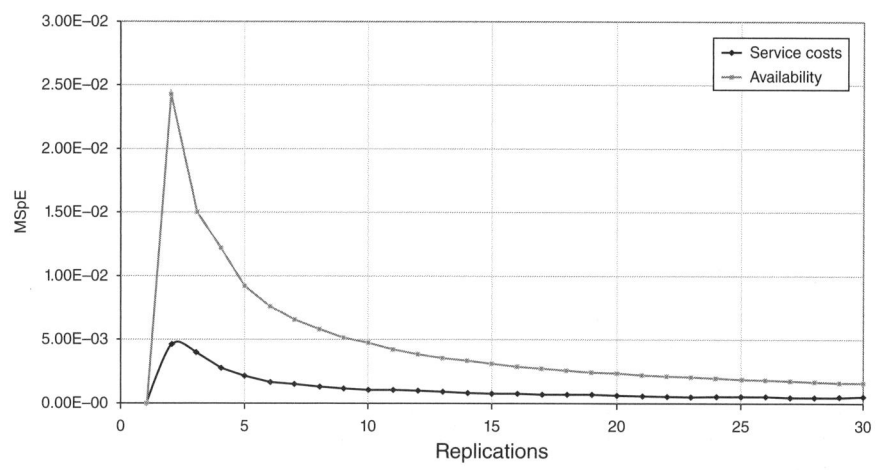

Figure 8.16 LAPIS verification and validation by total cost mean square pure error.

points such as shutdowns, start-up, contract closure, and item delivery, while the agents direct the countermeasures to face the contingencies and the impact of stochastic factors (i.e., expediting).

The way to compute power demand and unit EOH is to integrate the expected profiles between two consecutive events.

Based on the Monte Carlo technique, the simulator extracts values of the variables from statistical distributions in each time frame for each run. The statistical distributions have been built by historical data analysis conducted by statistical techniques such as chi-squared test and by subject matter experts in order to find the best fitting distribution. Historical data available were very few, due to short history, errors in records, and confidential nature of the information, so at the beginning, the authors used extensively beta distribution that optimizes the combination of historical data with expert estimations.

The optimizers use GAs in order to find robust and cost-effective solutions (Bruzzone, 1995).

The simulation model has been subjected to a VV&A process in several scenarios. The validation involved several months of coordinated desktop review and dynamic testing with user and experts; for the statistical test, ANOVA, mean square pure error (MSpE; see Figure 8.16), confidence band, statistical comparison, and sensitivity analysis were applied.

This approach allows estimating the proper criteria to be applied to GAs in order to get robust solutions considering inventory costs, stop costs, availability, contractual term respect and constraints respect. Considering the computational workload, the model was implemented in C++ and integrated in Office Suite for decision-maker output. The system that is developed is in addition integrated with fuzzy logic performance evaluator, carried out in the previous research and devoted to complete a priori planning evaluation (Bruzzone and Williams, 2005).

The database (DB) uses the existing planning and inventory, as well as established scenario, data about industries, spare parts associated, and maintenance planning. The user can decide if the solution is good or not, thanks to the report delivered by LAPIS and the fuse report. The number of data to input is huge and subject to errors due to manipulation of a lot of data.

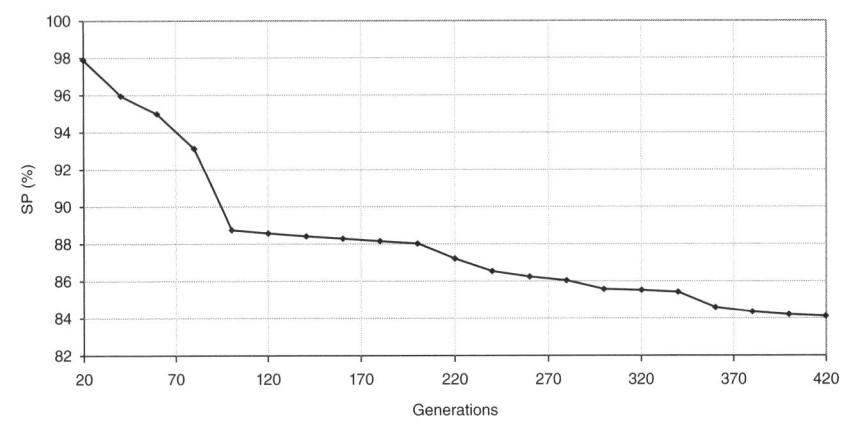

Figure 8.17 Lapis optimization evolution. Service combined optimization process on scheduling and inventory management.

The authors worked on optimization and simulation of pooling strategies for managing several power plants with different spare parts; therefore, due to complexity of the system, it is critical to create an effective interface to make the use of the program easier and to support all the functions.

In Figure 8.17, the results of the optimization process over a realistic scenario, involving a pool with nine power units that shows a significant reduction of costs corresponding to several millions USD over a period of 12 years, are proposed.

The DB industry contains data about machines and park machines. The DB scenarios are used to save scenario settings envisaged by the user; in order to be user-friendly, LAPIS generates three kinds of reports: a customer report (it is possible to send it directly to the customer), a precustomer report (this report needs to be checked by the user before sending them), and a user report (to control if the results are consistent with the related scenario).

The proposed approach successfully demonstrated the effectiveness of this architecture and the robustness versus this SoS and the related complex scenarios.

8.6 CONCLUSIONS AND FUTURE WORKS

M&S is gaining a key role as a fundamental tool for representing SoS and for the study of its evolution, behaviors, and performances. However, to deal with the different and peculiar aspects that characterized SoSs, suitable M&S techniques and methods should be provided. In this context, this chapter has proposed a new perspective on SoS M&S based on the adoption of the agent-oriented paradigm. Specifically, in the proposed perspective, an SoS is represented as a collection of agent societies that represent the systems, or better the organization, involved in the SoS and are constituted by agents that can play different roles and are characterized by different degree of autonomy and proactiveness as well as capabilities and social abilities. Starting from this agent-based representation of an SoS, six *views*, each of which aimed to represent a specific aspect of the SoS that can be analyzed at various abstraction levels, have been introduced: *environmental view*, *organizational view*, *goals view*, *interactions view*, *behavioral view*, and *properties view*. The specification of these views, whose modeling constructs can be easily represented by using popular UML-based modeling language, allows to obtain a full-fledged model of an SoS that can be then simulated by one of the available agent-based simulation platforms. From the analysis of the simulation results, it is possible to acquire not only insight on the SoS behaviors and evolution but also on emergent phenomena that are unpredictable and hard to catch with classical analytical-deductive techniques. The proposed agent-based perspective for the M&S of SoSs has been exemplified by application examples in multiple domains that proved its effectiveness. In particular, three different application examples have been proposed that provide evidence on the relevance of agent-based perspective for the M&S of SoSs: the first application example has regarded the design of a new aircraft carrier, the second application example has been related to an Agile Multicoalition Command and Control system, while the last example has referred to a power plant pool. In all the application examples, the proposed approach shows its robustness versus the SoSs considered and the related complex scenarios.

Future research efforts will be devoted to (i) specify a metamodel for the agent-based modeling of SoSs able to embrace all the introduced views so to provide a formal reference for the definition of SoS models that adhere to the proposed perspective, (ii) specify the UML-based constructs to exploit for the definition of the diagrams required by each introduced view, (iii) specify a method able to guide in the definition of SoS models that adhere the proposed perspective, and (iv) experiment the proposed framework in a wide range of application domains.

REFERENCES

Alberts David S, Gartska JJ, and Stein FP (2000). *Net Centric Warfare*, CCRP, Washington, DC.

Amico V, Guha R, and Bruzzone AG (2000). *Critical Issues in Simulation*. Proceedings of SCSC, Vancouver, BC.

Bauer B, Muller JP, and Odell J (2001). *Agent UML: A Formalism for Specifying Multiagent Interaction. Agent-Oriented Software Engineering*, pp. 91–103. Springer-Verlag, Berlin.

Beardslee EA and Trafalis TB (2006). *Discovering Service Inventory Demand Patterns from Archetypal Demand Training Data*. University of Oklahoma, Oklahoma.

Bruzzone AG (1995). *Adaptive Decision Support Techniques and Integrated Simulation as a Tool for Industrial Reorganisation*. Proceedings of ESS95, Erlangen, Germany.

Bruzzone AG (2013a). Intelligent agent-based simulation for supporting operational planning in country reconstruction. *International Journal of Simulation and Process Modeling* 8: 145–159.

Bruzzone AG (2013b). *Experimentation, Science & Technology Organization: Simulation as Strategic Technology for Coalition Force Initiative*. Keynote Speech at NATO CAX Forum, Rome, Italy.

Bruzzone AG and Bocca E (2006). *Vessel Management Simulation*. Proceedings of HMS2006, Barcelona, Spain.

Bruzzone AG and Bocca E (2008). *Introducing Pooling by Using Artificial Intelligence Supported by Simulation*. Proceedings of SCSC2008, Edinburgh, Scotland.

Bruzzone AG, Briano C, and Simeoni S (2004). *Power Plant Service Evaluation Based on Advanced Fuzzy Logic Architecture*. Proceedings of SCSC2004, San Jose, CA.

Bruzzone AG, Cantice G, Morabito G, Mursia A, Sebastiani M, and Tremori A (2009). *CGF for NATO NEC C2 Maturity Model Evaluation*. Proceedings of I/ITSEC2009, Orlando, FL.

Bruzzone AG, Fancello G, Fadda P, Bocca E, D'Errico G, and Massei M (2010). Virtual world and biometrics as strongholds for the development of innovative port interoperable simulators for supporting both training and R&D. *International Journal of Simulation and Process Modelling* 6: 89–102.

Bruzzone AG, Giribone P, Revetria R, Solinas F, and Schena F (1998). *Artificial Neural Networks as a Support for the Forecasts in the Maintenance Planning*. Proceedings of Neurap98, Marseilles, France.

Bruzzone AG and Longo F (2010). An advanced system for supporting the decision process within large-scale retail stores. *Simulation* 86: 742–762.

Bruzzone AG and Longo F (2013). 3D simulation as training tool in container terminals: The TRAINPORTS simulator. *Journal of Manufacturing Systems* 32: 85–98.

Bruzzone AG, Madeo F, and Tarone F (2011a). *Pool Based Scheduling and Inventory Optimisation for Service in Complex System*. Proceedings of EMSS, Rome, Italy.

Bruzzone AG, Massei M, Madeo F, Tarone F, and Gunal M (2011b). *Simulating Marine Asymmetric Scenarios for Testing Different C2 Maturity Levels*. Proceedings of ICCRTS2011, Quebec, Canada.

Bruzzone AG and Simeoni S (2002). *Cougar Concept and New Approach to Service Management by Using Simulation*. Proceedings of ESM2002, Darmstadt, Germany.

Bruzzone AG, Tremori A, and Merkuryev Y (2011c). *Asymmetric Marine Warfare: PANOPEA a Piracy Simulator for Investigating New C2 Solutions*. Proceedings of WAMS, Saint Petersburg, Russia.

Bruzzone AG and Williams E (2005). *Summer Computer Simulation Conference*, SCS, San Diego.

Cerda CBR and Espinosa de los Monteros FAJ (1997). *Evaluation of a (R,s,Q,c) Multi-Item Inventory Replenishment Policy Through Simulation*. Proceedings of the 1997 Winter Simulation Conference, Atlanta, GA.

Cervenka R and Trencansky I (2007). *The Agent Modeling Language—AML*. Whitestein Series in Software Agent Technology, Birkhäuser.

Cohen M, Kamesam PV, Kleindorfer P, Lee H, and Tekerian A (1990). Optimizer: IBM's multi-Echelon inventory system for managing service logistics. *Interfaces* 20: 65–82.

Cox E (1994). *The Fuzzy System Handbook*. AP Professional, Chestnut Hill, MA.

Crossley WA (2004). *System of Systems: An Introduction of Purdue University Schools of Engineering's Signature Area*. Proceedings of the Engineering Systems Symposium, Cambridge, MA.

Curcio D and Longo F (2009). Inventory and internal logistics management as critical factors affecting the supply chain performances. *International Journal of Simulation & Process Modelling* 5: 278–288.

Darabi HR, Gorod A, and Mansouri M (2012). *Governance Mechanism Pillars for Systems of Systems*. Proceedings of the 7th International Conference on System of Systems Engineering (SoSE), Genoa, Italy.

Elfrey P (2013). *SimSmackdown*. Key Note Presentation at Springsim 2013, San Diego, CA.

Garro A, Parisi F, and Russo W (2013). A process based on the model-driven architecture to enable the definition of platform-independent simulation models. *Simulation and Modeling Methodologies, Technologies and Applications*, Springer-Verlag, Advances in Intelligent Systems and Computing, Vol. 197, pp. 113–129.

Garro A and Russo W (2010). easyABMS: A domain-expert oriented methodology for agent based modeling and simulation. *Simulation Modelling Practice and Theory* 18: 1453–1467.

Gulyas L, Kozsik T, and Corliss JB (1999). The multi-agent modelling language and the model design interface. *Journal of Artificial Societies and Social Simulation* 2(3): http://jasss.soc.surrey.ac.uk/2/3/contents.html (accessed October 15, 2014).

Harris T (1997). Optimized inventory management. *Production and Inventory Management Journal* 38: 22–25.

Jennings NR (2001). An agent-based approach for building complex software systems. *Communications of the ACM* 44(4): 35–41.

Longo F, Massei M, and Nicoletti L. (2012). An application of modeling and simulation to support industrial plants design. *International Journal of Modeling, Simulation, and Scientific Computing* 3: 1240001-1-1240001-26.

Maier MW (1999). Architecting principles for systems of systems. *Systems Engineering* 1: 267–284.

Muckstadt JA (2005). *Analysis and Algorithms for Service Parts Supply Chains*. Springer, New York.

Nahmias S (1994). Demand estimation in lost sales inventory systems. *Naval Research Logistics* 41: 739–757.

North MJ and Macal CM (2007). *Managing Business Complexity: Discovering Strategic Solutions with Agent-Based Modeling and Simulation*. Oxford University Press, Oxford.

Paschalidis IC, Liu Y, Cassandras CG, and Panayiotou C (2004). Inventory control for supply chains with service level constraints: A synergy between large deviations and perturbation analysis. *Annals of Operations Research* 126: 231–258.

Pavon J, Sansores C, and Gómez-Sanz JJ (2008). Modelling and simulation of social systems with INGENIAS. *International Journal of Agent-Oriented Software Engineering* 2: 196–221.

Siemieniuch CE and Sinclair MA (2012). *Socio-Technical Considerations for Enterprise System Interfaces in Systems of Systems*. Proceedings of the 7th International Conference on System of Systems Engineering (SoSE), Genoa, Italy.

Silver EA (1991). A graphical implementation aid for inventory management of slow-moving items. *Journal of the Operational Research Society* 42: 605–608.

Sugita K and Fujimoto Y (2005). *An Optimal Inventory Management for Supply Chain Considering Demand Distortion*. Industrial Informatics, 3rd IEEE International Conference on (INDIN), pp. 425–430.

Tagaras G (1989). Effects of pooling on the optimization and service levels of two-location inventory systems. *IIE Transactions* 21(3): 250–257.

Chapter 9

Building Analytical Support for Homeland Security

Sanjay Jain[1], Charles W. Hutchings[2], and Yung-Tsun Tina Lee[2]
[1]*The George Washington University, Washington, DC, USA*
[2]*National Institute of Standards and Technology, Gaithersburg, MD, USA*

9.1 INTRODUCTION

This chapter discusses modeling and simulation (M&S) as a significant means for providing analytical support for homeland security problem solving and decision making that involves what can be considered a complex and dynamic system of systems—the current, human, domestic environment considered from a national perspective. Homeland security, in general, aims to understand the domestic environment and mitigate risks from both natural and human threats and hazards. Homeland security analyses and decisions occur at multiple levels, from individuals, communities, and regions to national governments and even international organizations. Sufficient, available knowledge of the domestic environment and likely threats helps decision makers at multiple levels with mitigating risks or in responding to and recovering from adverse events that occur.

Most homeland security risks involve significant unknowns and varying levels of uncertainty. An example of a natural hazard is a large hurricane approaching a coastal city, which is likely to occur each hurricane season. Coastal regions in the United States affected by hurricanes prepare for them and have implemented approaches such as building codes and evacuation procedures to mitigate risks to property and lives.

Modeling and Simulation Support for System of Systems Engineering Applications, First Edition.
Edited by Larry B. Rainey and Andreas Tolk.
© 2015 John Wiley & Sons, Inc. Published 2015 by John Wiley & Sons, Inc.

At the national level, organizations like the National Infrastructure Simulation and Analysis Center (NISAC) (Sandia, 2013a) operated by the U.S. Department of Homeland Security (DHS) provide analyses of the potential impacts of hurricane damage to an array of critical infrastructure (CI) such as the electric power grid, telecommunications, transportation systems, water supply systems, and healthcare systems. NISAC analyses aid decision making at the regional or national levels in planning for, responding to, and recovering from hurricane damage (Sandia, 2013b). Hurricane tracks and associated damage are not exactly predictable; however, weather patterns occur each year that lead to hurricane formation, and these storms are expected to threaten many areas. Other types of disasters due to unusual natural events, accidents, or terrorism are often unexpected and surprise decision makers and homeland security officials.

Incident management requires significant knowledge of the domestic environment, operating under normal conditions, as a basis for understanding each adverse event and its resulting interactions and impacts on multiple independent, interconnected systems. First responders and government officials often use exercises based on anticipated disaster scenarios to train for incident management, explore incident management techniques, and gain insights into organizational, logistical, and other incident management issues. The benefits of exercises include:

- Valuable experience to participants on relevant issues and responsibilities
- Testing of communications channels between first responders and higher-level incident management organizations
- Improved working relationships between participants to improve responses to and management of real disasters
- Identification of gaps in capabilities and/or procedures

Exercises provide a means to explore the unknowns and uncertainties associated with various threats and hazards but have some significant limitations. Depending on the scope and complexity of an exercise, these include:

- Contrived exercise scenarios based on known prior events
- Lack of analytical capabilities to assess incident management decisions and explore different courses of action to improve learning outcomes
- Difficulties in planning and successful orchestration
- Significant time and funding for exercise development and execution

M&S capabilities provide an alternative means to explore unknowns and uncertainties associated with various threats and hazards, especially when coupled with exercises or serious games. Understanding the functions of one type of domestic system under abnormal operating conditions imposed by an adverse event is an analytical challenge by itself. Understanding the functions of a system of systems under both normal and abnormal operating conditions requires advanced analytical techniques that include M&S capabilities since aspects of a domestic system of systems can respond in unexpected ways due to cascading effects of a disruption. Since a domestic system of systems cannot be recreated and experimentally studied under either normal or adverse conditions, M&S capabilities are the only available means to rigorously explore this domain and provide useful insights to analysts and decision makers.

The required M&S capabilities to support homeland security analyses are diverse, and the set of these capabilities used to analyze and manage homeland security could be considered a system of systems itself from the perspective of a government agency like the DHS. For example, each CI sector has its own analytical approaches—critical data, analysis methods, key performance indicators, experts, models, etc.—to understand and support internal management and decision making in each sector. A homeland security agency needs to engineer an analytical system of systems that integrates diverse analytical capabilities and data to support homeland security analyses and problem solving. A framework for homeland security knowledge sharing is essential to enable this analytical system of systems and the application of models and simulations as analytical components.

This chapter proposes a knowledge sharing framework to improve M&S as an analytical capability for homeland security. The next section identifies the characteristics of the homeland security domain that identify it as a system of systems as defined in Chapter 2. Section 9.3 establishes the need for an organized approach and framework to address the homeland security domain. Section 9.4 proposes a knowledge sharing framework. Section 9.5 describes a prototype application of M&S for system of systems to analyze an emergency incident. Section 9.6 summarizes the chapter. The last section provides the list of references.

9.2 HOMELAND SECURITY AND SYSTEM OF SYSTEMS

Homeland security self-evidently involves a system of systems according to the criteria described in Chapter 2. Homeland security aims to preserve the current, human, domestic environment from disruption or damage caused by various kinds of natural phenomena or human activities. The domestic environment is an assemblage of systems that exhibits properties of operational independence, managerial independence, geographic distribution, emergent behavior, and evolutionary development:

- Operational independence of the individual systems: The domestic environment is composed of multiple types of collaborative systems that are independent and useful in their own right. CI sectors such as energy, communications, water and waste water systems, and healthcare and public health sectors are examples. Although many systems are interdependent, for example, most systems are connected to the electric power grid, other component systems of the domestic environment are capable of performing useful operations independently of one another. Domestic environments existed well before development of the electric power grid.
- Managerial independence of the individual systems: The component systems of the domestic environment not only can operate independently; they operate independently to achieve their own purposes as well as the purpose of the integrated whole. The CI sectors such as the financial services and communications are examples.
- Geographic distribution: The component systems of the domestic environment such as the CI sectors are distributed and networked throughout geographic boundaries. Components of the communications system, for example, extend across the globe and into space.

- Emergent behavior: The domestic environment performs functions and carries out purposes that do not reside in any component system. For example, nations and distinctive national cultures emerge from domestic environments. In many nations, well-developed CI enhances the domestic environment and improves standards of living that in turn support national development and growth that can occur in unpredictable ways.
- Evolutionary behavior: Domestic environments are never fully formed or complete. They evolve over time. Technological innovations, such as development of the internal combustion engine, the digital computer, and telecommunications networks, alter domestic environments in profound ways.

Homeland security functions involve analyzing and understanding various components of the domestic system of systems to assess potential operational risks and related impacts on the domestic environment. Understanding the domestic environment and managing risks associated with natural and man-made threats and hazards requires the collaborative assemblage and use of vast amounts of information of varying types and quality from multiple, independent sources. For example, homeland security-related knowledge is generated by government organizations at multiple levels (e.g., town, city, county, state, federal), commercial enterprises, and private sector organizations and from physical systems.

The DHS is the primary U.S. government organization assigned to coordinate and lead the enterprise effort to secure the nation. The DHS is primarily a law enforcement organization that has a unique and multifaceted role compared to all other U.S. government organizations. The DHS functions include:

- Enforcing laws to ensure public safety and health
- Supporting or managing large-scale incidents and catastrophes
- Securing CI, national borders, and cyberspace
- Screening people and cargo for potential hazards
- Administration of benefits to disaster victims

In the DHS, the National Protection and Programs Directorate (NPPD), the Federal Emergency Management Agency (FEMA), and the Transportation Security Administration (TSA) focus on different aspects of homeland security. The NPPD gathers information on and analyzes risks to the 16 CI sectors currently identified (DHS, 2013). The NPPD also leads cybersecurity efforts for the U.S. government information technology infrastructure. The FEMA gathers information, assesses risks, and plans response to a variety of emergencies and disasters such as hurricanes, tornadoes, and major floods. The FEMA plans and executes high-level exercises each year involving multiple federal, state, and local organizations to improve incident management. The FEMA stages materiel to support responses to areas overwhelmed by a catastrophe and administers benefits to victims. The TSA coordinates security for transportations systems, such as the air transportation system. The TSA assesses risks to these systems and implements means to mitigate identified risks.

Air transportation systems are an example of a system of systems as identified in Section 2.3.2 of Chapter 2. Securing air transportation systems against terrorist threats illustrates the system of systems nature and some of the challenges involved with homeland security. Commercial aviation has encountered a number of man-made threats since the

1970s as shown in Table 9.1. For explosive threats, a watershed event occurred with the Pan Am Flight 103 Lockerbie bombing in 1988, which resulted in catastrophic loss of the aircraft due to an explosive device inserted in checked baggage. This resulted in increased focus on explosives as an aviation threat including U.S. legislation to improve aviation security. The Aviation Security Improvement Act of 1990 (Public Law 101-604) (Congress, 1990) defined and specified requirements for explosives detection systems (EDS) including:

- Performance criteria
- Explosive type, configuration, and amounts
- Detection rates
- False-alarm rate
- Throughput rate

The U.S. government also began to utilize live fire explosive testing to study the effects of explosions on aircraft structures to serve as a basis for aircraft hardening measures and explosive detection standards. After implementation of new measures, there was a decrease in the frequency of aircraft bomb blasts worldwide after 1990.

The U.S. enhanced aviation security again as a result of the September 11, 2001, attacks with enactment of the Aviation Transportation Security Act (Public Law 107-71), passed in November 2001 (Congress, 2001). This identified a comprehensive set of security requirements for all modes of transportation. Suicide bombing attempts by Reid in December 2001 and Abdulmutallab in December 2009 using improvised explosive devices created new security challenges that require improved approaches and technologies for screening passengers and carry-on baggage.

Table 9.1 Aviation Security Threats and Responses

Time period	Event/threat	Vulnerability	Response
1970s	Traditional hostage/hijacking	Guns, weapons	Magnetometers
December 1988	Pan Am 103, Lockerbie	Bomb in baggage	Baggage scans
September 2001	World Trade Center, Pentagon attacks	Boxcutters, small knives, etc. on person or in carry-ons	Aviation and Transportation Security Act (ATSA)
December 2001	Reid hijacking attempt	Shoe bomb	Shoes removed
August 2004	Chechen suicide attacks	Vests	Pat-downs, backscatter
August 2006	Heathrow liquids plot	Novel liquid bomb	Liquids ban
June 2007	Glasgow airport attack	Vehicle-borne improvised explosive device (VBIED) to soft target	Increased awareness and focus on emerging threats to physical transportation infrastructures
December 2009	Nonmetallic body bomb	Body bomb in sensitive area	Explosive trace detection, sensitive pat-down, whole-body imaging

U.S. stakeholders in aviation security include the TSA, DHS Science and Technology (S&T) Directorate, airports, airlines, and airline passengers/users of commercial aviation. Roles for each of these stakeholders include:

- TSA
  - Establish policy
  - Identify and assess threats
  - Issue and enforce regulations
  - Approve security plans and programs
  - Inspect and monitor operations for compliance
  - Provide operational direction
  - Initiate necessary changes
  - Screen passengers and baggage

- DHS S&T Directorate
  - Implement research and development (R&D) to enhance aviation security
  - Provide paths to commercialization of technology for explosives detection and mitigation

- Airlines
  - Secure baggage and cargo
  - Protect aircraft

- Airports
  - Protect air operations area
  - Provide automated access control
  - Provide law enforcement support

- Airline passengers/users
  - Use commercial aviation as necessary
  - Comply with laws and regulations

Potential attackers will attempt to evade security procedures and explosive detection technologies to carry out an attack. Several means or vectors exist for introducing explosives onto an aircraft. These threat vectors include:

- Flight crew carry-on bags
- Flight crew
- Catering and cleaning services
- Cargo
- Mail
- Ground crew services
- Passenger introduction through carry-on, checked, gate-checked, or transfer baggage or personnel-borne devices

Since security procedures limit the number of threat vectors available for introduction of explosives on aircraft and are continually evolving based on new threats, adversaries must continually modify their tactics and techniques. For example, this includes use of suicide bombing devices using homemade explosives and/or improvised explosive devices, which evade screening and detection technologies. Homeland security organizations, such as the DHS, must understand the air transportation system, continually assess the risks, and implement actions to mitigate potential dangers.

Homeland security concerns extend well beyond air transportation systems and encompass multiple security concerns in the domestic environment. Government and other organizations that lead and coordinate homeland security need a framework or architecture to understand the domestic environments in normal operating conditions, evaluate and assess threats, and mitigate risks for disruptions. If a significant disruption does occur, they need the appropriate situational awareness and understanding to intervene and provide the necessary capabilities to cope with the damage and restore the domestic environment to its normal operating condition as quickly as possible, ideally with minimal losses. Integral to the homeland security analytical architecture is an ability to share knowledge and support the development and use of analytical tools and capabilities such as all types of models, simulations, and associated data.

9.3 MODELING, SIMULATION, AND ANALYSIS FOR HOMELAND SECURITY

Modeling, simulation, and analysis (MSA) techniques and capabilities are extensively developed and used for strategic planning, operations analysis, and systems development for national defense. MSA techniques and capabilities can potentially support problem solving across many homeland security domains by providing insights to analysts and decision makers in many complex areas of importance such as social behavior, natural phenomena, environments, economy and finance, organizational performance, CI, and other systems (McLean et al., 2008). Science-based MSA capabilities, judiciously applied to homeland security domains, can enhance risk analyses and support response planning for natural and man-made threats and hazards. Currently, M&S capabilities are being developed in an ad hoc manner within the DHS to address specific problems identified by a few homeland security "customers" such as first responders (Hutchings, 2009). For example, national laboratories, industry partners, and universities are working on a number of DHS-sponsored MSA developments. Laboratories, private firms, and academic researchers are developing MSA capabilities to support needs and provide analytical capabilities and management tools for various homeland security sponsors in other organizations, for example, at regional, state, or city levels, independently. MSA developments should be coordinated to efficiently identify gaps to advance capabilities and avoid duplication of effort. A DHS Critical Infrastructure Modeling and Simulation Workshop (Adam, 2008) recognized a variety of MSA activities and recommended:

- Developing and regularly updating a master compendium of available models and related research from labs, academia, and industry
- Conducting research to define the model attributes and characteristics that must be included for the entries in the compendium
- Developing methods for communicating the scope and limitations of models in a manner as transparent as possible

A knowledge sharing framework for homeland security MSA can promote a common understanding of current analytical tools and techniques and support development of a model compendium. A suitable knowledge sharing framework provides a classification schema that defines a set of categories into which various concepts or artifacts can be arranged, understood, and evaluated. Homeland security analysts can use such a framework to logically organize knowledge assets, including MSA assets and capabilities, and identify R&D gaps and needs. A knowledge sharing framework, as described in the next section, provides structure for MSA development and use. This framework differs from enterprise architecture schemes developed by Zachman (2008) or Department of Defense (DoD) Architectural Framework (DoD CIO, 2010).

Homeland security stakeholders—domain experts, capability sponsors, developers, managers, and analytical and operational communities—should collaborate to create a common body of knowledge and reach consensus on homeland security issues, concepts, analytical methodology, and existing and needed MSA capabilities. A sufficient knowledge sharing framework should provide the necessary structure to support stakeholder collaboration and address significant homeland security problems and decision making. A purpose-built framework supporting homeland security decision makers to address problems can guide framework development. For example, rational problem-solving and decision support typically proceed in distinct phases. These might include identification of a problem situation, problem definition and formulation, creation of an evaluation model, problem analysis, and creation of results or recommendations for action. An appropriate MSA development approach might be analogous to the rational problem-solving methodology and include the following major phases:

- Problem identification
- Analytical requirements definition and specification
- Capability development
- Capability evaluation
- Capability application

This approach provides a common means to guide MSA development and is applicable across homeland security domains of interest.

9.4 KNOWLEDGE SHARING FRAMEWORK

A knowledge sharing framework that integrates MSA R&D efforts is shown in Figure 9.1. The framework identifies knowledge asset needs and requirements based on input from key stakeholders—the subject matter experts, researchers, and users—and provides a number of benefits:

- Users capture and share known research, development, and implementation issues.
- Developers identify currently available capabilities, ongoing projects, and facilities to avoid duplication of efforts, which ensures the best use of constrained resources for developing useful MSA tools.

Building Analytical Support for Homeland Security   227

- Best practices are shared to allow dissemination of lessons learned from experiences of others.
- Current and needed standards are identified to ensure interoperability of developed MSA tools.

MSA developers should be able to access the knowledge and insights captured in the framework in a report or online using mechanisms such as a secure portal or Wiki. The following subsections describe the elements of the knowledge sharing framework shown in Figure 9.1.

9.4.1 Scope of the Framework

A knowledge sharing framework for a coherent area of homeland security MSA needs to include the subareas that are considered relevant by each corresponding community or technical interest group. For CI, for instance, the scope should include crosscutting concerns with MSA in all 16 CI sectors currently identified by the DHS (2013). The framework needs to incorporate information across all phases of infrastructure life cycle including planning, designing, building, operating, and decommissioning. As CI MSA capabilities evolve,

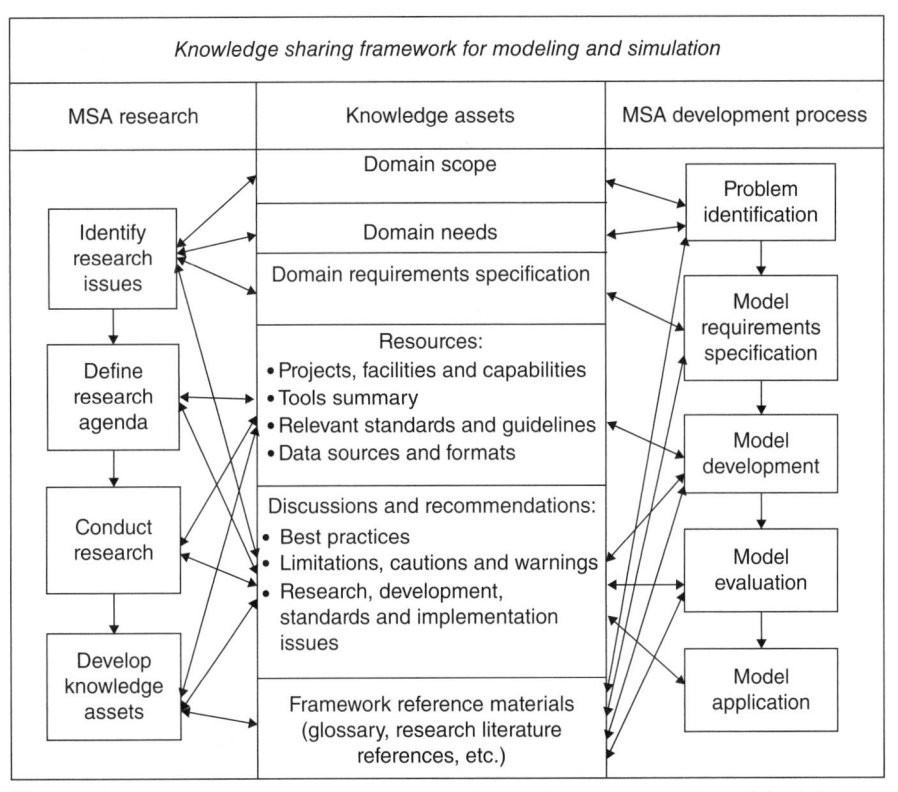

Figure 9.1 Overview of knowledge sharing framework for homeland security modeling and simulation technical areas.

separate knowledge sharing frameworks may be needed for each CI sector to reflect sector-specific needs and concerns. Similarly, for the hazardous material release (HMR) area, the scope may include M&S for release of chemical, biological, nuclear, and radiological agents, as well as natural occurring releases such as volcanoes and wildfires. Associated models and simulations may involve the release of materials into the atmosphere, within buildings and other structures including heating ventilation and air conditioning (HVAC) systems, bodies of water and watershed systems, as well as ground contamination.

The scope for the healthcare systems technical area would include simulation and modeling activities that support analysis, planning, and training needs for the healthcare institutions, epidemics, and other healthcare-related emergencies. Simulation models may be used to understand healthcare systems, interdependencies with other systems, their vulnerabilities, and the impact of emergency incidents on the population and healthcare community.

For all the technical areas identified, the MSA tools may be used for all the application types defined in McLean et al. (2008) to inform analysis and decision support, planning and operations, Systems Engineering and acquisition, and training, exercises, and performance measurement.

9.4.2 Needs Analysis Overview

Stakeholders at all levels need analytical tools and methods to address problems and support decision making. The purpose of this section of the framework would be to capture problem situations that are commonly encountered by stakeholders across each identified technical area and how M&S capabilities are used to support the problem-solving/decision-making processes at more than one level. For example in the case of CI, given a major hurricane and potential impacts, the FEMA will have a set of concerns, state and local officials in potentially affected areas another set of concerns, and power generating facilities still other concerns. This will promote a common understanding of analytical issues from multiple perspectives and support the development and sharing of analytical tools and M&S capabilities across stakeholder organizations.

This section of the framework will identify, document, and catalog problems that are most likely to be faced in the particular technical area. It will help define the relevant analytical needs and determine the right decision-relevant questions that need to be addressed, establishing the purpose and objectives for MSA capabilities. Examples of questions that may be captured as needs for using MSA for CI include:

- How will the impact of a natural disaster such as hurricane, tornado, or violent lightning storms striking parts of the power grid affect outage management?
- What will be the impact of disruption in one CI on other colocated or connected CI systems? For example, how will disruption in the power grid affect the water delivery infrastructure for a city?

Brase and Brown (2009) identified research questions for complex network questions. Since CI systems are networked systems, the identified questions contribute to needs analysis as indicated by examples below:

- Can we use, understand, and quantify the efficacy of new security approaches for computer networks?

- Can we improve the design of computer or communications networks to be more robust against partial failures or intentional attacks?
- Can we understand how populations will respond to the availability of new energy sources or to changes in energy policy?

Similarly for the HMR area, on the occurrence of a release, officials want to know—What is the hazard? Where is it going? Who is at risk? How do we respond? Specific questions that may be answered using simulation include the following:

- What is the forecasted transport direction of the plume and what areas may be under hazard?
- What are the estimated potential damages, casualties, illnesses, and fatalities?
- What are the estimated emergency assistance requirements?
- What are the areas where buildings, land, agricultural crops, bodies of water, and other man-made or natural resources are or will be contaminated?

9.4.3 MSA Requirements Specifications

The high-level needs analysis should form the basis for requirements specifications for the technical area. To address the analytical needs, each MSA capability needs clearly defined requirements. This section of the framework will include:

- Intended use
- Data and metadata requirements
  - Quality
  - Provenance
  - Timeliness
  - Management
  - Interoperability
  - Security
- Functional requirements
- Interactions with other MSA capabilities
- User interface requirements
- Performance requirements
- Credibility and evaluation requirements
  - Theoretical corroboration
  - Model components verification
  - Corroboration (independent data)
  - Sensitivity analysis
  - Uncertainty analysis
  - Robustness determination
  - Comparison to evaluation criteria

Examples of functional requirements for HMR are provided below:

- Predict the initial direction, travel, and dispersion of a plume over time from a single or multiple sources taking into account the type of source, material/chemical properties, release location, weather conditions, terrain, urban areas, and other man-made structures.
- Predict the concentration of the chemical or biological agent within the plume and flow-through drainage areas over time.
- Estimate deposition and contamination levels for air, water, ground, and building surfaces.
- Identify exposed population and predict exposure levels over time.

9.4.4 Identification of Existing MSA Resources

This section of the framework will capture MSA resources available for systems analyses for each technical area. These are categorized into the following subsections.

9.4.4.1 Projects, Facilities, and Capabilities

The subsection will identify the ongoing projects, facilities, and capabilities focused on the relevant technical areas. Examples for the CI sector include the NISAC (Sandia, 2013a) and Chemical Sector Supply Chain and Economics Project and are discussed in the following paragraphs.

NISAC is an MSA program within the DHS comprising personnel in the Washington, DC, area as well as from Sandia National Laboratories (Sandia) and Los Alamos National Laboratory (LANL). A facility dedicated to NISAC is located at Sandia National Laboratory, Albuquerque, NM. Congress mandated that NISAC serve as a "source of national expertise to address CI protection research and analysis." NISAC prepares and shares analyses of critical infrastructure and key resources (CIKR), including their interdependencies, vulnerabilities, consequences, and other complexities, under the direction of the Office of Infrastructure Protection (IP), Infrastructure Analysis and Strategy Division (IASD). As of September 2008, NISAC had conducted and published 11 analyses of hurricanes covering the entire U.S. Gulf and Atlantic coast (DHS, 2008).

The Chemical Sector Supply Chain and Economics Project is a key component of a larger effort to deliver Enabling Homeland Security Capabilities (EHCs) for the modeling, mapping, and simulation program. The first goal of this project is to populate a detailed data set of the chemical and petrochemical manufacturing, supply and distribution components that comprise the chemical infrastructure supply chain. The second goal is to develop a means to mathematically analyze not only the consequence of significant threats but also the resiliency of the supply chain to recover from these impacts. This project was part of the Critical Infrastructure Protection Thrust Area and the MSA program of the Infrastructure and Geophysical Division of the DHS.

Ongoing projects will also include efforts in academia as described in research literature. For example, recent reported efforts on infrastructure simulations in academia include energy distribution (Baxevanos and Labridis, 2007) and water supply (Qiao et al., 2007).

Examples for the HMR area include the National Atmospheric Release Advisory Capability (NARAC) and the National Exposure Research Lab (NERL) and are briefly described below.

NARAC—The NARAC facility is located at Lawrence Livermore National Laboratory in Livermore, CA. It provides tools and expert services to map the spread of hazardous material accidentally or intentionally released into the atmosphere (LLNL, 2012).

EPA's NERL—Located in Research Triangle Park, North Carolina, the NERL provides scientific understanding, information, and assessment tools to reduce and quantify the uncertainty in the Agency's exposure and risk assessments for all environmental stressors. The Atmospheric Sciences Modeling Division provides numerical and physical modeling support to the homeland security mission in protecting against the environmental and health effects of terrorist acts (EPA, 2013). This involves numerical modeling complemented by physical modeling in the Division's wind tunnel. For example, a 1:600 scale model of lower Manhattan was built and the dispersion of material from the collapse of the World Trade Center towers was studied under various meteorological conditions. Also, dispersion of airborne material around the Pentagon was simulated in the wind tunnel.

9.4.4.2 MSA Tools Summary

This subsection will provide a collection of available MSA tools (the tools under development are identified in the preceding subsection as projects). The tools will be briefly described to provide an overview of their capabilities. In the long term, this information should be enhanced to grow the collection to the compendium of models called for in the workshop focused on CI (Adam, 2008). The brief tool descriptions will provide a useful resource until the research issues related to the proposed compendium are addressed. An interested reader can use the description to quickly develop a short list of tools that may be applicable for their purpose and can then follow up to find more details. Examples of such descriptions of MSA tools for CI area are provided in the following paragraphs.

The Critical Infrastructure Protection Decision Support System (CIPDSS) (LANL, 2013a) has been developed jointly by the LANL, Sandia National Laboratories, and Argonne National Laboratory (ANL); the set of tools under the CIPDSS program models the impact of CI on the economy, government, and population. LANL developed the city level models, Sandia developed the national level models, while ANL provided the decision support part. The set of tools is intended to provide "orders of magnitude" results quickly. It was used for the analysis underlying NISAC's report on potential impact of pandemic influenza (DHS, 2007).

The Critical Infrastructure Protection and Resiliency Simulator (CIPR/sim) has been developed by Idaho National Labs (INL). The CIPR/sim allows emergency planners to visualize the real-time cascading effects of multiple infrastructure failures before an actual emergency occurs. It uses a common operating framework that allows the tool to import real-time data from numerous existing analysis modules, including real-time digital simulator (RTDS) for electric grid analysis, QualNet for telecommunications analysis, and PC Tide for wind speed and flood surge analysis (INL, 2013).

Examples of MSA tools for healthcare systems area include EpiSimS (LANL, 2013b) and MedModel (ProModel, 2013). EpiSimS is an epidemic simulation engine. It is a C++ application that runs on high-performance computing clusters. It is a stochastic agent-based discrete-event model that explicitly represents every person in a city and every place within the city where people interact. A city or region is represented physically by a set of road segment locations and a set of business locations. EpiSimS can simulate various pharmaceutical and nonpharmaceutical interventions, including panic-based stay-home behavior, therapeutic and prophylactic use of antivirals, contact tracing, vaccination, wearing of masks, social distancing behaviors (increased interpersonal separation, hand washing, cough etiquette, etc.), household quarantine, and closures of schools. More information is available in Stroud et al. (2007).

MedModel is a simulation tool designed specifically for the healthcare industry. MedModel is used in the evaluation, planning, and redesign of hospitals, clinics, and other healthcare systems. In the hands of a trained and experienced analyst, MedModel models can be used to identify inefficiencies in an existing process and test a variety of scenarios. The animation and graphic output results show the behavior of the system under any set of circumstances (ProModel, 2013).

9.4.4.3 Relevant Standards and Guidelines

This subsection will identify the known applicable standards and guidelines. A few examples for this subsection for CI area are presented next.

National Infrastructure Protection Plan (NIPP) is a coordinated strategy that defines CIKR protection roles and responsibilities for federal, state, local, tribal, and private sector security partners. The NIPP sets national priorities, goals, and requirements for effective distribution of resources, which will enable the government, economy, and public services to continue in the event of a terrorist attack or other disasters. Sector-Specific Plans (SSPs) have been developed for each of the identified CI sectors supporting the NIPP.

Chemical Facility Anti-Terrorism Standards (CFATS) is a regulatory program to secure national high-risk chemical facilities. At the outset of the program, the DHS expected that roughly 30,000 facilities would require registration and regulatory compliance to monitoring standards of which approximately 6000 facilities would fall into one of the four high-risk categories that require further regulation. As of March 2011, almost 40,000 chemical facilities had registered with the DHS and completed the Top-Screen process. Of these facilities, the DHS considered more than 8,064 as high risk and required them to complete and submit site vulnerability assessments (Congress, 2011).

Examples of standards and guidelines for healthcare systems include those for smallpox response and mass prophylaxis. The "Smallpox Response Plan and Guidelines" document from the Centers for Disease Control and Prevention (CDC) outlines the public health strategies that would guide the public health response to a smallpox emergency and many of the federal, state, and local public health activities that must be undertaken in a smallpox outbreak.

The "Community-Based Mass Prophylaxis: A Planning Guide for Public Health Preparedness" has been developed by the Agency of Healthcare Research and Quality (AHRQ) to help state, county, and local officials meet federal requirements to prepare for public health emergencies (AHRQ, 2013). It outlines five components of mass prophylaxis response to epidemic outbreaks and addresses dispensing operations using a comprehensive operational structure for Dispensing/Vaccination Centers (DVCs) based on the National Incident Management System (NIMS) (FEMA, 2013).

9.4.4.4 Data Sources and Formats

The sources of data relevant to MSA technical areas and their identified formats will be provided in this subsection.

An example for CI sector is the Constellation/Automated Critical Asset Management System (C/ACAMS) (CAL EMA, 2013). It is a web-enabled information service portal that helps state and local governments build critical protection programs in their local jurisdictions. ACAMS is a secure online database and database management platform that allows for the collection and management of CI asset data; the cataloging, screening, and sorting of this data; the production of tailored infrastructure reports; and the development of a variety of pre- and postincident response plans useful to strategic and operational planners and tactical commanders.

An example for healthcare systems area is provided by the National Emergency Medical Services (EMS) Information System (NEMSIS) (NEMSIS TAC, 2013). The system is the national repository that will be used to potentially store EMS data from every state in the nation. Since the 1970s, the need for EMS information systems and databases has been well established, and many statewide data systems have been created. The involved organizations include National Highway Traffic Safety Administration (NHTSA), Health Resources and Services Administration (HRSA), CDC, University of Utah, and University of North Carolina.

9.4.5 Discussions and Recommendations

The discussions and recommendations section in the framework will include identified best practices, limitations, cautions, and warnings and research, development, standards, and implementation issues. Government or other sponsors can then consider recommendations for funding and execution.

9.4.5.1 Best Practices for Development and Use of MSA Tools

The development and use of MSA tools can benefit through development and use of lessons learned and sharing of best practices. This subsection would document and capture lessons learned and best practices acquired in development, evaluation, and use of MSA tools for the technical area. Some examples of best practices include:

- Conceptual modeling practice
- Innovative approaches
- Software engineering practice

9.4.5.2 Limitations, Cautions, and Warnings

Models provide results with varying levels of error and uncertainty. This subsection is intended to highlight and document the limitations associated with MSA applications to minimize improper use and highlight potential areas for further development.

9.4.5.3 Research, Development, Standards, and Implementation Issues

This section of the framework is intended to capture challenges and issues related to MSA tools. The challenges and issues should be prioritized for maximum value.

A number of R&D challenges need to be addressed in modeling potential threats for the technical area and impact of disruptions due to involved phenomena. Common challenges across the technical areas should be identified such as the development of simulation application architectures and integration of models and simulations for the technical area. In addition, challenges need to be identified and prioritized for each technical area. For example, for CI area, computational modeling to understand the vulnerability of a dam to explosions would be useful in mitigating this threat. Physical models are necessary to collect data on impact of explosions to support evaluation of the modeling capability to ensure credibility of result. However, a physical model of a dam cannot be scaled down to laboratory settings since the physics changes based on the scale. Hence, expensive large-scale experiments may be required for the purpose. An initial straw-man list of such challenges for the healthcare systems area may include increasing reality in healthcare M&S training exercises and devices and access to and usage of healthcare M&S applications by systems personnel.

The success of a coordinated approach will largely depend on the ability to apply data-driven MSA tools across a range of scenarios. This ability in turn depends on use of standards. Gaps in available standards should be identified and prioritized to guide standardization efforts. At times, the issue may be to identify a common standard from among multiple ones available. For example, issues with multiple geographic information systems (GIS) standards should be identified, and an approach to identify one common GIS standard to be followed for outputs by all MSA tools for a particular area should be identified or developed.

Deployment of new MSA tools for homeland security has to follow a carefully organized approach. The approach may vary across the technical areas depending on the state of technology, the familiarity of the end users with MSA, and the interfaces with the end users. Common implementation challenges across the technical areas include return on investment to stakeholders and sponsors for research projects and ownership and usage of publicly versus privately developed tools. For example, the 18 identified CI domains involve a myriad of stakeholders including government, quasigovernment, and private sector operators; a number of oversight agencies across federal, state, and local levels; a range of jurisdictions involved in normal operation; and another range of jurisdiction that may be affected under disrupted or interrupted operations. Some sectors may require use of "reachback" centers, that is, entities with expertise available to guide the users in interpretation of results. Other sectors may have sophisticated users who can be trained to use the MSA tools independently.

9.4.6 Framework Reference Materials

The knowledge sharing framework will need to be supported by a number of reference materials provided via appendices. While every attempt should be made to use standard terminology, more often than not, this may not be possible due to its absence. A technical interest group should guide the generation of standard terminology that should be defined in a glossary.

The framework is expected to include technical discussions and build on existing current literature. A reference section should be included to capture the relevant publications.

Potential and existing users of MSA tools may need guidance from experts in respective areas. A list of identified experts for each major aspect of each technical area should be captured. The list may be restricted to authorized personnel and the names on the list may be included only with permission. The experts may need support for the time they may have to spend fielding questions.

This section described a knowledge sharing framework for MSA for homeland security. It is suggested that the wide area of homeland security be divided into several technical areas based on existing communities of researchers. The knowledge sharing framework can then be used to pull together and create an information source for each technical area. The collection and maintenance of information, identification of research challenges, and development of the research agenda should be coordinated by technical interest groups. Establishing such a framework for MSA capabilities is one approach to encourage collaboration and coordination for systems analysis, problem solving, and decision making for each of the technical areas. The framework should be documented and made available to stakeholders as an online resource such as a secure portal or Wiki to enable development and progress of MSA for homeland security purposes.

9.5 A PROTOTYPE SYSTEM OF SYSTEMS APPLICATION

The utility of a knowledge sharing framework for developing and using MSA capabilities for analyzing incident management is illustrated by a homeland security training scenario, which uses a variety of virtual reality, simulation, and gaming capabilities. A conceptual architecture for integrating these capabilities is shown in Figure 9.2.

The integrated simulation modules are intended to provide technically correct solutions. The gaming modules are included to provide the interaction required for training. The simulation and gaming modules should be integrated together through a data synchronization and transfer processor. The integrated capability will allow joint training of first responders and the management-level personnel in the preparedness phase. The integrated simulation modules by themselves can be used throughout the incident management life cycle including the phases of prevention, preparedness, response, recovery, and mitigation.

A prototype was developed to demonstrate the concept to potential users. A prototype helps explain MSA concepts to those who are not familiar with simulation and gaming and demonstrates MSA applicability to incident management. A hypothetical scenario involving a dirty bomb explosion in Washington, DC, was created, and selected aspects of the incident and the response were modeled to demonstrate the capabilities of simulation and gaming. Integration of various modules highlights the advantages of this approach.

The next subsection briefly describes the hypothetical scenario. A number of simulation and gaming modules were developed to help understand the issues involved in modeling and integration. The following subsections discuss the simulation and gaming modules that have been included in the concept demonstration. Each of the implemented module is discussed in the following with the full capability desired and the subset implemented

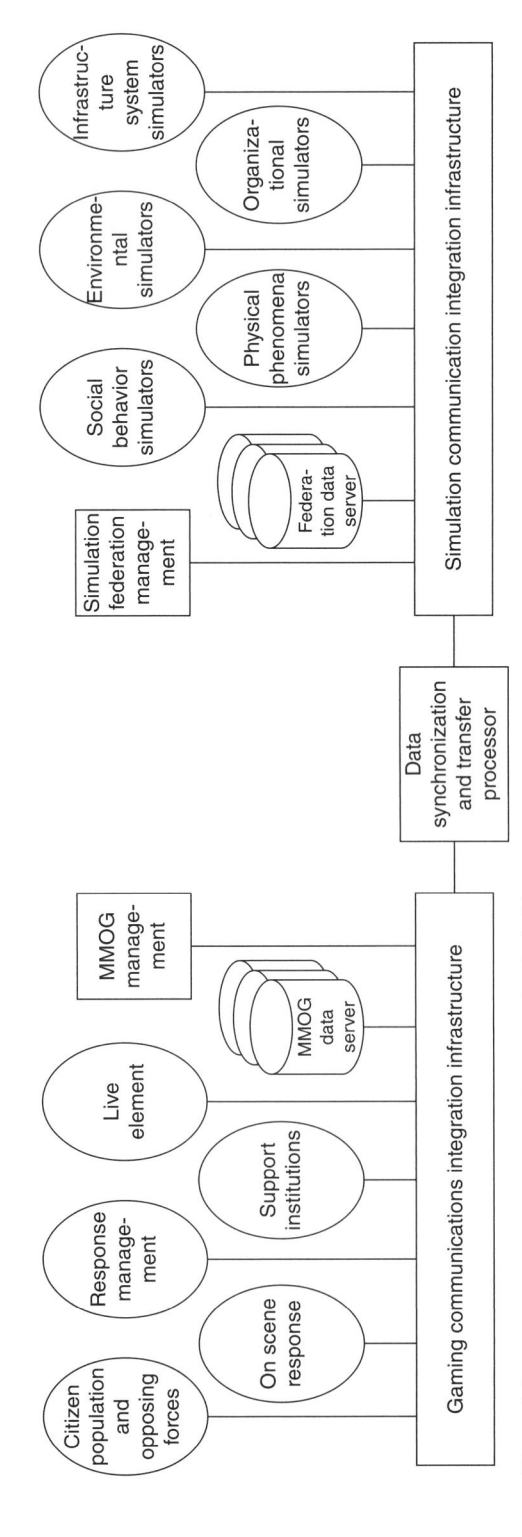

Figure 9.2 Architecture concept for simulation and gaming for incident management.

for the demonstration. The proposed approach for integration including a test implementation for two of the modules is discussed next. The data needs for building and executing the simulations are briefly discussed. The issues identified from this experience of developing the prototype are discussed.

9.5.1 Hypothetical Scenario

The scenario for the concept prototype is based on a dirty bomb attack in Washington, DC, on the evening of July 4. The fireworks on the National Mall on July 4 attract a large crowd. A large number of people utilize the metro rail system to get to the National Mall. The metro rail authorities actually close the Smithsonian metro station that is nearest to the National Mall to allow better management of the crowd flow on July 4. It does not take much imagination to identify streets nearby metro station entrances as potential targets for terrorists. The selection of public places like a street as the incident location also avoids any concerns that may be raised on selecting privately owned location such as stadiums for such a study.

The scenario uses the area outside the Federal Triangle metro station, which is the second closest station to the National Mall, as the target for detonation of a dirty bomb by terrorists. The scenario did not consider the feasibility or means of getting a dirty bomb to the identified location. The probability of such an occurrence is expected to be very low with the typical high security surrounding such an event. The focus of the scenario was on the consequences if such an incident occurs.

The near-term consequences of a dirty bomb explosion include the casualties and radiation exposure among the crowd in the immediate vicinity and in the area covered by the plume and response by police, fire department, and emergency medical technicians. The major consequences of the incident and the response need to be modeled for incident management purposes.

9.5.2 Simulation Modules

The simulation modules included in this effort are as follows.

9.5.2.1 Plume Simulation

This module falls in the category of physical phenomena simulators shown in Figure 9.2. It should model the dispersion of plumes of various kinds including chemical, biological, and radiological agents. Inputs may include the characteristics of the agent released, release mechanism used, the location of release point, terrain and structures around the release point, and weather conditions. Inputs may alternately be based on the sensor readings over time in the area of interest indicating the presence of an agent and the direction(s) of the spreading plume. Outputs may include time profile of the plume, and exposure profile for the population in the region affected by the plume over time.

This module was implemented using CT-Analyst software from Naval Research Labs (NRL, 2013). This tool provides the desired capabilities for modeling plume dispersion as described in the preceding paragraph. It models the spread of the plume from the identified location taking into account the weather and the geometry of the buildings in the surrounding areas.

9.5.2.2 Crowd Simulation

The capability of modeling crowd behavior is a part of the social behavior simulators group in Figure 9.2. It should model crowd status and movement at locations of interest under different event scenarios, crowd behavior, and crowd management strategies. The locations of interest may include areas around actual and potential emergency incident sites, major business, commercial and residential areas that may be affected by evacuation directives, and major public transportation points such as bus and train stations, local rail transport stations, and airports. Different event scenarios may include normal, rush hour, terrorist attack, accidental fire, natural disaster, etc. The models may predict crowd movement and crowd density variations along movement directions, predict occurrence of stampede and casualties, perform route planning through the crowd for selected individuals (such as first responders), determine location of individuals as a function of time, and predict individual movement times between selected points. Inputs may include street layouts including pedestrian areas, layouts within public buildings such as train stations and public parks, crowd volumes and density data, probabilities for stampede and casualties, weather conditions, location of emergency incidents, behavioral models of individuals, sensor data, and communications. Outputs may include location and status of specific individuals in the crowd, crowd volumes and density by city block and passages within public buildings and parks, crowd movement times between selected points, and crowd management systems data.

The crowd simulation module has been implemented by researchers from University of Arizona using AnyLogic software using the agent-based simulation paradigm. Individuals and small groups are defined as agents with each of them having parameters, such as age, mobility, and knowledge of the area, that determine their reaction to the incident and the behavior (Shendarkar et al., 2008).

9.5.2.3 Traffic Simulation

Traffic simulation is another module that falls in the group of social behavior simulators. It should provide models of general traffic flow and specific vehicle movements for a given region under different event scenarios (normal, rush hours, off-peak hours, terrorist attack, natural disaster, evacuation, etc.), driver behavioral models, and traffic management strategies. The model may perform automatic route planning for selected vehicles, generate random events that disrupt traffic flow (vehicle breakdowns, accidents, traffic management systems failures), determine vehicle locations as a function of time, predict travel times between locations, etc. Inputs may include road network layout and characteristics, traffic management systems description and status, individual vehicle locations and status, driver moods, historical traffic volume and vehicle density data, pedestrian data, probabilities for accidents, incidents, weather conditions, location of emergency incidents, behavioral models of vehicle operators, sensor data, and communications. Outputs may include locations and status of specific vehicles, traffic volume and densities by area or road segment, travel times between selected locations, accident data, and traffic management systems data.

This capability has been implemented using two modules that simulate traffic at different levels of detail. The Emergency Response Vehicle Simulator models the traffic at a macrolevel. It has been developed by NIST researchers using Java and GeoTools, an open-source GIS toolkit (GeoTools, 2013). It mimics the movement of the response vehicles from their initial locations to the site of the incident. While individual response vehicles are modeled, the effect of the rest of the traffic is modeled using congestion factors for each road segment that they go through. The travel route is determined using Dijkstra's algorithm.

The microlevel traffic simulation capability has been implemented using the Traffic Software Integrated System (TSIS) developed at the University of Florida and available through the Center for Microcomputers in Transportation (McTrans) (UOF TRC, 2012). The model simulates movement of individual vehicles in the immediate area around the National Mall before and after the incident. Following the incident, a number of vehicles come out of parking garages in the area and attempt to leave resulting in traffic jams. The software allows capturing the congestion factors for each road segment defined. The congestion factors determined through microlevel simulation of one area can be used to estimate congestion factors for the wider area modeled in the macrolevel traffic simulation.

9.5.2.4 Healthcare Simulation

Healthcare simulation module is part of the organizational simulator category in Figure 9.2. It should model the actions of the healthcare organizations (including emergency medical technicians and hospitals) in response to an emergency incident including the deployment of resources and actions for triage and treatment of injured at the incident site, movement of casualties to hospitals, and treatment at the hospitals. The model logic will include relevant policies and procedures for emergency situations including calling in medical staff, using temporary accommodations for the injured, acquiring needed supplies and equipment. Inputs may include the number, location, and type of casualties from an emergency incident, the availability of staff at work and off (on-call), the availability of resources (own and those that can be acquired quickly from surrounding jurisdictions), the time and resources required for attending to each casualty type, and the probabilities of death from different casualty types over time. Outputs may include the operation of the healthcare system over time including the number of people treated and released, admitted, dead, and waiting for treatment and the state of the staff and facilities (to determine their capability to deal with another incident).

The concept demonstration includes a model for only one part of the healthcare system, namely, the emergency department. The operation of a hypothetical emergency department for handling the casualties from the incident is simulated. It was developed by NIST researchers using ProModel. Casualties arriving at the emergency department include serious cases of trauma and cardiac cases brought in by ambulances and walk-ins with minor injuries and the worried well. The model indicates the buildup of queues for the walk-ins. The ambulances carrying serious cases are occasionally diverted to other hospitals based on the status at the hospital modeled.

9.5.2.5 Transportation Simulation

Transportation simulation is a part of the infrastructure systems simulators group in Figure 9.2. It should mimic the transportation systems infrastructure including highways and road network, rail network, waterways, and marine and air transport. It should model the impact of man-made or natural disasters on the transportation infrastructure components. Inputs may include the description of the transportation systems infrastructure together with its network, characteristics of node points, traffic volumes across arcs and through the nodes, traffic control mechanisms, failure characteristics of major control mechanism and equipment, operation and maintenance resources, multimodal links, and links to other CI. Outputs may include the impact of modeled emergency events on the operation of the transportation infrastructure over time.

A metro rail simulation model was developed for the purpose of demonstrating the concept of transportation simulation. The model was developed using AutoMod by NIST researchers with active support from the vendor, Brooks Software. It models the evacuation of people from the incident area using metro rail system. The metro systems lines passing through the incident area are modeled. The model helps determine the rate at which the crowd can disperse using the metro system.

9.5.3 Gaming Modules

The gaming modules would be especially useful for incident management training applications. Two of the modules were implemented, one at responder level and the other at the management level.

9.5.3.1 Triage Application

Triage is part of the On-Scene Response group of gaming applications in Figure 9.2. This application allows trainees to play the role of emergency medical technicians conducting triage following a dirty bomb explosion. This module was developed through collaboration between researchers from the NIST and the Institute of Security Technology Studies (ISTS) at Dartmouth College. The ISTS researchers had previously developed the triage application for an airplane crash scenario (McGrath and Hill, 2004; McGrath and Carella, 2005; ISTS, 2013). The NIST researchers created the 3D geometry of the incident location and worked with ISTS researchers to set up the application.

The gaming application allows a user to move around in the 3D space representing the incident site. The user can see the fire caused by the explosion, the casualties lying on the ground, the fire trucks, other responders, objects and structures on the street, and the surrounding buildings. They can go to each victim and perform triage by looking for the vital signs and asking specific questions if possible. The victims requiring immediate attention can be carried away on stretchers through a gross decontamination station created using hoses from two fire trucks. The application includes audio effects to make the experience closer to reality. A user has to contend with sounds of sirens, victims, and limited lighting conditions in performing his/her responsibilities for conducting triage.

9.5.3.2 Incident Management Strategy Gaming

The strategy gaming application would fall under the Response Management category of gaming applications shown in Figure 9.2 and is targeted at the management-level personnel for the responding agencies. This may be used by personnel at the Emergency Operations Center (EOC) to plan out the response resource deployments. The module was developed by NIST researchers using C#.

The module shows a map of the incident site together with the locations of response resource providers including police stations, fire stations, and hospitals. The map also shows the important buildings around the incident site. The interface provides the capability to place icons representing response resources on the map thus making and visualizing the deployments. The map can be updated based on reports from the incident site. All the icons used are based on standards defined by the Homeland Security Working Group (HSWG, 2012).

The application allows decision makers to develop an awareness of the situation and make decisions for resource deployment. These decisions can then be communicated to the responding teams. The strategy board can be updated with locations and damage information as reports are received from the incident. The board can be used with a real incident or with a simulated incident modeled using the concept demonstration prototype.

9.5.4 Integration of Simulation Modules

The benefit of the individual simulation and gaming modules can be synergistically increased through integration. In the absence of integration, each simulation will have to either make assumptions about the phenomenon modeled by other simulations or utilize summary statistics from the other simulations. For example, the emergency department simulation will have to utilize the arrival rate of ambulances determined by the emergency vehicle response simulation model. Utilizing a distribution based on results from other simulations will result in piecewise simulations of the overall system. Integrating the simulations together such that they can exchange entity information will allow modeling the whole system together. The integrated set will thus allow increased accuracy.

The integration of simulation modules can be accomplished using the High-Level Architecture (HLA) (Kuhl et al., 1999; IEEE, 2010). However, the traditional approach of integrating the simulation using the HLA is quite resource intensive and requires major coordination among the developers of the simulations being integrated. A modified approach involving adapters developed at the NIST (McLean et al., 2000) was used for the integration. Two of the simulation modules, the emergency department simulation and the emergency response vehicle simulation, were integrated together using the adapters. Figure 9.3 shows a screenshot of the integrated execution of the two modules. As the simulated ambulance arrives at the hospital location in the emergency vehicle response simulation window shown in the right half of the screen, it enters the emergency department simulation window shown in the left half of the screen. The ambulance discharges casualties at the hospital and leaves. Again, the movement between the two simulations is coordinated. The two simulations are integrated using a conservative approach with a simulation time step of 30 s.

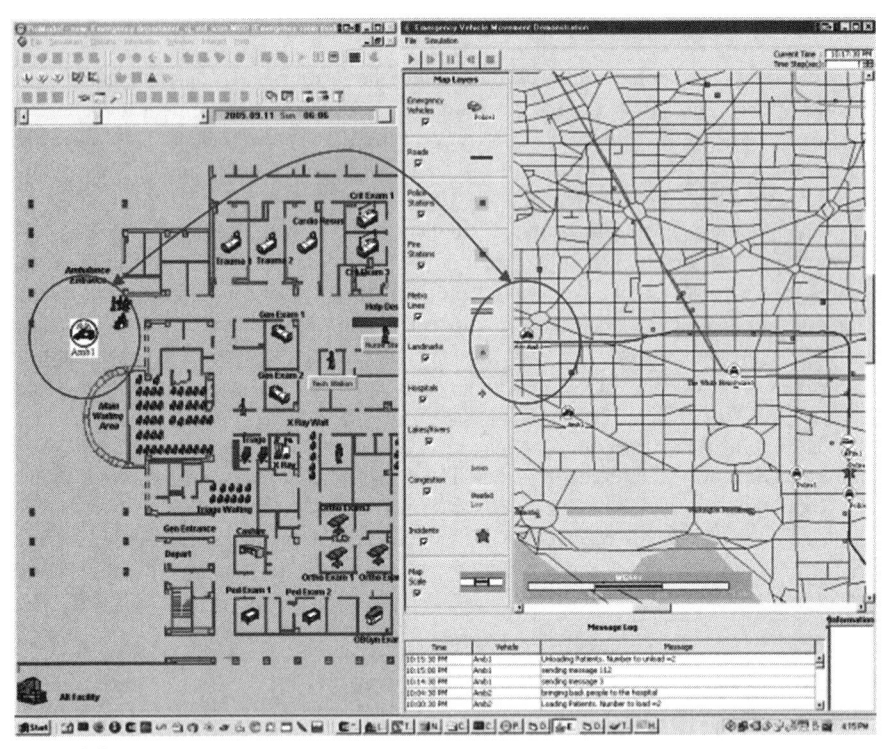

Figure 9.3 Screenshot of integrated execution of emergency department simulation (left half) and the emergency vehicle response simulation (right half). Added annotations show the corresponding ambulances in the two simulations.

The integration of gaming and simulation modules allows joint training of personnel from multiple levels of organizations. The first responders and management-level personnel can be trained together on the same simulated incident. This allows them to share experiences leading to better teamwork and performance. Strategies for addressing the significant technical challenges in the integration of the gaming and simulation modules are topics for future research. The challenges include:

- Synchronization of time between gaming modules that should execute in real time and simulation modules that have the capability to execute in accelerated time
- Resynchronization of gaming and simulation modules after a fast forward or "jump" in simulated time
- Establishing communications with gaming software since they have generally been developed with proprietary architecture
- Affecting the event flow in simulation based on actions taken in gaming clients, and vice versa
- Achieving agreement between results of simulations executed at an abstract level and games executed at a detailed level

9.5.5 Simulation Data Needs

Development and execution of simulations such as those discussed in this chapter require a large and varied set of data. Two databases have been developed to indicate the kind of data that will be needed to support the simulations. One database includes reference information that may be useful for any incident, while the other is specific to the incident. The example databases are developed as two units for illustrative purposes only. It is recognized that the real data may reside in a number of databases spread at different organizations. It may be worthwhile to read some of the data directly from the existing databases, while other parts of the data may be read in and stored locally depending on the communication and storage infrastructure available to a community.

The reference database may include information such as natural disaster intensity scales for earthquakes and hurricanes, characteristics of biological and chemical agents, triage categories and tags, and capabilities of emergency response equipment such as police cars, fire trucks, and personal protective devices. Such a database will provide the information for modeling the behaviors of the various entities and physical phenomena. It can also serve as the reference for decision makers.

The incident database may include information specific to the incident. Such information would comprise of the map of the area; locations of nearby responding agencies; sites of local, state, or national importance in the area; current status; traffic densities on the streets in the area; etc. The information in the incident databases will be utilized to populate and update the simulation models. For example, the incident database may indicate that two fire trucks of type A are close to the scene. The corresponding simulation model will be initialized with that information. The information on capabilities and features of fire trucks of type A will be retrieved from the reference database to correctly model their actions.

9.5.6 Issues Identified

The development of the concept demonstration prototype helped identify the issues involved in building a system with integrated simulation and gaming modules. The issues are summarized below:

- The numerical data inputs for the simulators were generally in proprietary formats. Some of the data had to be entered using input screens of the simulators. The tools developed internally at the NIST allowed XML inputs.
- The GIS data inputs could be in a number of several GIS "standard formats." There are multiple file formats defined with different versions, multiple earth models, and multiple projections. There is no best combination of these factors, and translation errors between the formats are common.
- The imported graphics required varied formats also. For an earlier version of the strategy board, the map had to be downloaded as a bitmap and then converted into Targa (.tga) format. Another tool required the bitmap to be converted to Jpeg (.jpg) format. Yet another required conversion to AutoCAD format.

- The open-source software used was not found to be as robust or well documented as commercial software. The lack of documentation such as UML diagrams made it difficult to comprehend the code.
- Communication with some of the proprietary tools was hard to implement as they were not designed to interact with external programs.
- One of the tools was restricted to use by government personnel only.

There were some instances of default standards that helped the process. For example, inputs of 3D models for the two different gaming software used was possible using 3D Studio Max format.

The experience underlined the need for standards for data inputs and outputs for the simulation and gaming modules. Standards are also needed to enable plug and play interfacing of the modules.

The demonstration prototype helped illustrate and explain concepts for integrated gaming and simulation for incident management to the user community since the responder community associates simulations with live exercises and tabletop models. The concept demonstration prototype helped elucidate the value computer simulation and gaming may provide to them. The demonstration prototype development process also helped verify some of the major issues that were anticipated in realizing the use of simulation and gaming in incident management. Further work is needed to identify and build the infrastructure required for enabling integrated gaming and simulation using independently developed modules. Such infrastructure will include defined standard architecture, interfaces, and data formats that allow bringing together desired modules for incident management for different scenarios.

9.6 SUMMARY

This chapter presents an approach for enabling M&S capabilities to analyze and understand the current, human, domestic environment and support homeland security concerns using a knowledge sharing framework. Homeland security concerns encompass a large number of diverse, interacting systems, organizations, and individuals, which form complex and interwoven system of systems. A large incident causes significant disruption of the domestic system of systems and can damage multiple components, such as CI, that adversely impact a significant population. Effective management of a large incident requires appropriate actions from a range of government agencies, nongovernmental organizations, private sector organizations, and individuals. Homeland security analysts and decision makers need to understand the nature of an incident on the domestic environment, the cascading damage resulting from the incident, the related impacts of cascading damage, and carefully coordinate actions to address the damage and restore the domestic environment.

Analyzing the domestic system of systems requires the use of advanced M&S capabilities and techniques, which need development. A suitable knowledge sharing framework can enable a coherent body of knowledge and the use of a wide variety of data and M&S capabilities to understand the functions and performance of individual systems and the integrated components of a system of systems. An appropriate framework would support organized development of M&S capabilities for a wide range of incidents over large geographical areas by coordinating knowledge and guiding the work of personnel involved in research, development, and practice of M&S for homeland security applications.

This chapter presents a knowledge sharing framework to support homeland security M&S development and use. The framework provides for necessary knowledge assets to set up and maintain the ability to develop and deploy homeland security M&S capabilities. Framework development involves stakeholders, subject matter experts, and M&S researchers in building and maintaining essential knowledge assets.

A prototype system of systems example shows how such an analytical capability can be implemented to understand and plan for a large incident such as a bomb explosion during the Fourth of July celebrations at the National Mall in Washington, DC. Models of crowd behavior, vehicular traffic, emergency response deployment, and emergency healthcare system are all integrated together to create a system of systems model for study and analysis. Such a capability is useful for identifying poorly understood aspects of an incident or incident response and can serve as a basis for further study, data acquisition, or experimentation to improve knowledge. The demonstration prototype also highlighted some of the technical challenges that need to be addressed to integrate a diverse set of M&S capabilities and data.

Homeland security issues are especially challenging because they are grounded in lack of knowledge or uncertainty about physical phenomena or human intentions related to potential threats and hazards. Risk analysis and management focus on adverse effects of known unknowns; however, homeland security organizations are often unprepared for unforeseen disasters or catastrophes such as the 9/11 attacks, the impact of Hurricane Katrina, and the Deepwater Horizon disaster in the United States over the past decade. Homeland security analysis needs to employ strategies and capabilities to deal with known unknowns and unknown unknowns. A sufficiently rich body of knowledge augmented by MSA capabilities in an appropriate knowledge sharing framework may help analysts to better explore uncertainties and discover unknowns to address this problem.

ACKNOWLEDGMENTS

The Science and Technology Directorate of the U.S. Department of Homeland Security (DHS) sponsored the production of part of this material under Interagency Agreement HSHQDC-08-X-00418 with the National Institute of Standards and Technology (NIST). The NIST sponsored a part of this work to the George Washington University under grant number 707NANB8H8167. The work described was funded by the U.S. government and is not subject to copyright.

The work reported here gained from substantial technical and leadership contributions by the late Charles R. McLean. The concept prototype was developed through contribution of several people including the following NIST researchers: Damien Bertot, Swee Leong, Yan Luo, Guillaume Radde, Benjamin Raverdy, Frank Riddick, and Guodong Shao. External collaborators include Dr. Dennis McGrath, Douglas Hill, and Jenny Bodwell of the Institute of Security Technology Studies at Dartmouth College, Dr. Gopal Patnaik of Naval Research Labs, Dr. Young Jun Son and his graduate students from University of Arizona, and personnel from Brooks Software.

DISCLAIMER

The material discussed and views expressed in this chapter are the authors' personal perspectives and do not reflect official views or policies of the U.S. Department of Commerce or DHS.

Some software products or services may have been identified in the context in this paper. This does not imply a recommendation or endorsement of such products or services by the authors, NIST or DHS; nor does it imply that such products or services are necessarily the best available for the purpose.

REFERENCES

Adam, N. 2008. *Workshop on Future Directions in Critical Infrastructure Modeling and Simulation, Final Report*. U.S. Department of Homeland Security, Infrastructure & Geophysical Division, Science and Technology Directorate, Washington, DC.

AHRQ (Agency for Healthcare Research and Quality of U.S. Department of Health & Human Services). 2013. *Community-Based Mass Prophylaxis*. Available at http://archive.ahrq.gov/research/cbmprophyl/ [accessed December 22, 2013].

Baxevanos, L.S., and D.P. Labridis. 2007. Implementing Multiagent Systems Technology for Power Distribution Network Control and Protection Management. *IEEE Transactions on Power Delivery*, 22(1): 433–443.

Brase, J.M., and D.L. Brown. 2009. *Modeling, Simulation and Analysis of Complex Networked Systems: A Program Plan*. Livermore, CA: Lawrence Livermore National Laboratory. Available at http://science.energy.gov/~/media/ascr/pdf/program-documents/docs/Complex_networked_systems_program_final.pdf [accessed September 19, 2014].

CAL EMA (California Emergency Management Agency). 2013. *Constellation/Automated Critical Asset Management System (C/ACAMS)*. Available at http://develop.oes.ca.gov/WebPage/oeswebsite.nsf/Content/D290C3544ECABEF788257561006A1812?OpenDocument [accessed December 22, 2013].

Congress (U.S. Congress, Senate and House of Representatives). 1990. Public Law 101-604—Nov 16, 1990. Available at http://www.gpo.gov/fdsys/pkg/STATUTE-104/pdf/STATUTE-104-Pg3066.pdf [accessed December 22, 2013].

Congress (U.S. Congress, Senate and House of Representatives). 2001. Public Law 107-71—Nov. 19, 2001. Available at http://www.gpo.gov/fdsys/pkg/PLAW-107publ71/content-detail.html [accessed September 19, 2014].

Congress (U.S. Congress, House of Representatives). 2011. *Full Implementation of the Chemical Facility Anti-Terrorism Standards Act*. House Report 112-211. U.S. Government Printing Office. Available at http://www.gpo.gov/fdsys/pkg/CRPT-112hrpt211/html/CRPT-112hrpt211.htm [accessed December 22, 2013].

DHS (U.S. Department of Homeland Security). 2007. *National Population, Economic, and Infrastructure Impacts of Pandemic Influenza with Strategic Recommendations*. Available at http://info.publicintelligence.net/PI%20FINAL%20-%2012-21-07.pdf [accessed December 21, 2013].

DHS (U.S. Department of Homeland Security). 2008. *Fact Sheet: Critical Infrastructure and Homeland Security Protection Accomplishments*. Available at https://www.hsdl.org/?view&did=235174 [accessed December 21, 2013].

DHS (U.S. Department of Homeland Security). 2013. *Critical Infrastructure Sectors*. Available at http://www.dhs.gov/critical-infrastructure-sectors [accessed December 22, 2013].

DoD CIO (U.S. Department of Defense Chief Information Officer). 2010. *The DoDAF Architecture Framework Version 2.02*. Available at http://dodcio.defense.gov/Portals/0/Documents/DODAF/DoDAF_v2-02_web.pdf [accessed December 21, 2013].

EPA (U.S. Environmental Protection Agency). 2013. *Atmospheric Modeling and Analysis Research*. Available at http://www.epa.gov/AMD/ [accessed December 21, 2013].

FEMA (Federal Emergency Management Agency of U.S. Department of Homeland Security). 2013. *National Incident Management System (NIMS)*. Available at http://www.fema.gov/national-incident-management-system [accessed December 22, 2013].

GeoTools. 2013. *GeoTools: The Open Source Java GIS Toolkit*. Available at http://www.geotools.org/ [accessed December 2013].

HSWG. 2012. *Homeland Security Working Group: Symbology Reference*. Available at http://www.fgdc.gov/HSWG/index.html [accessed December 19, 2013].

Hutchings, C.W. 2009. *Enabling Homeland Security with Modeling & Simulation (M&S)*. Interservice/Industry Training, Simulation, and Education Conference (I/ITSEC), Paper ID# 9512, Orlando, FL, November 30–December 3.

IEEE. 2010. *IEEE 1516-2010—IEEE Standard for Modeling and Simulation (M&S) High Level Architecture (HLA)—Framework and Rules*. Available at http://standards.ieee.org/findstds/standard/1516-2010.html [accessed December 19, 2013].

INL (Idaho National Laboratory). 2013. *Modeling and Simulation: Critical Infrastructure Protection and Resiliency Simulator (CIPR/sim)*. Available at https://inlportal.inl.gov/portal/server.pt/community/national_and_homeland_security/273/modeling_and_simulation/1707 [accessed December 21, 2013].

ISTS (Institute for Security, Technology, and Society). 2013. *Synthetic Environment for Emergency Response Simulation*. Available at http://www.ists.dartmouth.edu/projects/archives/synth.html [accessed December 19, 2013].

Kuhl, F., R. Weatherly, and J. Dahmann. 1999. *Creating Computer Simulations: An Introduction to the High Level Architecture*. Upper Saddle River, NJ: Prentice Hall.

LANL (Los Alamos National Laboratory). 2013a. *CIPDSS: Critical Infrastructure Protection Decision Support System*. Available at http://www.lanl.gov/programs/nisac/cipdss.shtml [accessed December 22, 2013].

LANL (Los Alamos National Laboratory). *EpiSimS: Epidemic Simulation System*. Available at http://www.lanl.gov/programs/nisac/episims.shtml [accessed December 22, 2013].

LLNL (Lawrence Livermore National Laboratory). 2012. *National Atmospheric Release Advisory Center*. Available at https://narac.llnl.gov/ [accessed December 21, 2013].

McGrath, D., and C. Carella. 2005. *Synthetic Environments for Emergency Response Simulation*. Proceedings of the 2005 Game Developers Conference, San Jose, CA. Available at http://www.ists.dartmouth.edu/library/118.pdf [accessed December 19, 2013].

McGrath, D., and D. Hill. 2004. *Unreal Triage: A Game-Based Simulation for Emergency Response*. Proceedings of the Huntsville Simulation Conference, Huntsville, AL. Available at http://www.ists.dartmouth.edu/library/58.pdf [accessed December 19, 2013].

McLean, C.R., S. Jain, and Y.T. Lee. 2008. *A Taxonomy of Homeland Security Modeling, Simulation, and Analysis Applications*. Simulation Interoperability Workshop (SIW), Paper No. 08S-SIW-098, Providence, RI.

McLean, C.R., S. Leong, and F. Riddick. 2000. *Integration of Manufacturing Simulations using High Level Architecture (HLA)*. Proceedings of the 2000 Advanced Simulation Technologies Conference, April 16–20, Washington, DC.

NEMSIS TAC (NEMSIS Technical Assistance Center, University of Utah School of Medicine). 2013. *National EMS Information System (NEMSIS)*, Version 3. Available at http://www.nemsis.org/v3/ [accessed December 22, 2013].

NRL (Naval Research Laboratory). 2013. *CT-Analyst*. Available at http://www.lcp.nrl.navy.mil/ct-analyst/Home.html [accessed December 19, 2013].

ProModel Corporation. 2013. *MedModel—The Industry Standard for Healthcare Simulations*. Available at http://www.promodel.com/products/medmodel/ [accessed December 22, 2013].

Qiao, J., D. Jeong, M. Lawley, J.P. Richard, D.M. Abraham, and Y. Yih. 2007. Allocating Security Resources to a Water Supply Network. *IIE Transactions*, 39(1): 95–109.

Sandia (Sandia National Laboratories). 2013a. *National Infrastructure Simulation and Analysis Center* (NISAC). Available at http://www.sandia.gov/nisac/ [accessed December 21, 2013].

Sandia (Sandia National Laboratories). 2013b. Sandia Labs New Release: Hurricane Season: Predicting in Advance What Could Happen. Available at https://share.sandia.gov/news/resources/news_releases/hurricane_nisac/#.UraEEvRDs1M [accessed December 22, 2013].

Shendarkar, A., K. Vasudevan, S. Lee, and Y.J. Son. 2008. Crowd Simulation for Emergency Response Using BDI Agents Based on Immersive Virtual Reality. *Simulation Modelling Practice and Theory*, 16: 1415–1429.

Stroud, P.D., S.Y. Del Valle, S.J. Sydoriak, J.M. Riese, and S.M. Mniszewski. 2007. Spatial Dynamics of Pandemic Influenza in a Massive Artificial Society. *Journal of Artificial Societies and Social Simulation*, 10(4): 9.

UOF TRC (University of Florida Transportation Research Center). 2012. *Traffic Software Integrated System—Corridor Simulation (TSIS-CORSIM)*. Available at http://mctrans.ce.ufl.edu/featured/tsis/ [accessed December 22, 2013].

Zachman, J.A. 2008. *John Zachman's Concise Defintion of The Zachman FrameworkTM*. Available at http://www.zachman.com/about-the-zachman-framework [accessed November 20, 2014].

Chapter 10

Air Transportation Systems

William Crossley and Daniel DeLaurentis
School of Aeronautics and Astronautics, Purdue University, West Lafayette, IN, USA

10.1 INTRODUCTION

The air transportation system (ATS) provides many examples of system of systems (SoS). Airlines use multiple aircraft and coordinated crews in an effort to make profit while meeting passenger travel demand. Multiple competing airlines use numerous airports and share the same airspace as they also compete in the marketplace. Air navigation service providers coordinate multiple pieces of geographically dispersed technological and human infrastructure to provide efficient (and fair) service while maintaining an incredibly high level of safety. The airports themselves are also geographically separate, typically independently managed entities. Finally, in the midst of these multilevel, multiorganization dynamics, the passengers seek to meet their own needs for high-speed, long-distance transportation. By most definitions and descriptions of SoS, these examples from the domain of air transportation indeed are examples of SoS. Further, there are more than those identified here, there are strong interactions between and among many of them, and they are all subject to radical transformation from new technologies and/or policies.

Modeling and simulation (M&S) of the different SoSs (and, where appropriate, multiple interacting ones) in the air transportation domain should support decisions about design, planning, and operation to provide the desired capabilities (safe, rapid,

---

Modeling and Simulation Support for System of Systems Engineering Applications, First Edition.
Edited by Larry B. Rainey and Andreas Tolk.
© 2015 John Wiley & Sons, Inc. Published 2015 by John Wiley & Sons, Inc.

long-distance travel) with an efficient use of available resources. A high-profile example from the late 1990s and the early 2000s highlights the critical nature of decision making in the midst of the ATS SoSs. The two major commercial aircraft manufacturers—Airbus and Boeing—both launched major aircraft design and development programs that would lead to operational aircraft; these were the A380 for Airbus and the 787 for Boeing. Both of these aircraft were intended to enter the market at nearly the same time, but the marketing and design teams at the two companies took a rather different view of how the commercial airline market and ATS would evolve over the time these two aircraft would enter and operate in service. The A380 is a large-capacity aircraft, capable of carrying roughly 525 passengers; this concept would allow increased capacity on hub-and-spoke service networks by bringing more people into and out of hub airports while using roughly the same number of aircraft flights. The Boeing 787 is a medium-capacity, long-range aircraft that carries about 225 passengers that would enable more point-to-point operations, skipping stops at hub airports for long routes with moderate passenger demand. Both aircraft development programs required significant investment from the companies; the phrase "you bet your company" is often associated with the launch of a new large aircraft development effort. Obviously, both companies relied upon best-practice decision making and likely used M&S to determine what they thought was the best new aircraft design for the commercial airline market and that could be supported by next-generation airports and airspace management. At the time of this chapter in 2013, it is still too early to make a definitive statement about whether one, both, or neither of these decisions was correct. However, the point of this anecdotal example is to emphasize that making decisions about changes (here, the design of new aircraft for introduction to the airlines) to ATS of systems has significant cost and risk and that the better the use of M&S to support these decisions, the better the outcome.

The intent of this chapter is to provide the authors' perspective of SoS, how this translates to the air transportation domain, and a framework for M&S in this context. Some theoretical considerations about M&S of SoSs appear in the following, including lexicon and taxonomy to describe or categorize different types of SoSs. A discussion of some methodologies used for air transportation SoS follows the theoretical considerations. Highlights of work related to SoSs in air transportation show the potential to use M&S methodologies in this domain. A reflection on lessons learned from this work concludes the chapter.

10.2 THEORETICAL CONSIDERATIONS

To provide a basis for the discussion of SoSs in air transportation and the associated M&S methodologies employed for these SoSs, discussions of several theoretical considerations are helpful. SoSs from numerous domains require similar considerations; the intent here is to highlight these similarities and make clear how these apply in the air transportation domain.

10.2.1 Behavior Traits and Time Scales

While there have been alternate proposals both before and after, the most cogent distinguishing traits of SoSs have been presented by Maier (1998) and elaborated in Chapter 2 of this volume. He identifies operational and managerial independence of constituent systems as the key distinction for SoSs; otherwise, a collection of systems is either a monolithic system or a geographically distributed, but centrally controlled, system. Therefore, SoSs are collaborative and rely on some form of communication/coordination to operate as an SoS. Emergent behaviors (changes in capability and/or performance, both beneficial and detrimental, that would not be anticipated by assessing only the performance of each individual entity in the SoS) and evolutionary behaviors (the capability and/or performance of the SoS changes over time) are likely outcomes due to this nature of SoSs. This view appears to have fairly wide acceptance, because it captures many of the important characteristics actually observed in SoSs. It is noted here that when referring to emergent behavior, we mean the weak form of emergence as described in Chapter 2. DeLaurentis proposed three implications on M&S for SoS from this view: networks as an underlying representation, heterogeneity of systems as an expected challenge for many SoSs, and "transdomain" nature of the considerations needed in models, including policy, economic, and operations variables (DeLaurentis, 2005).

Like many other systems, SoSs may exhibit different levels of the aforementioned traits when they are considered using differing time scales. The authors of this chapter proposed that there are three distinct time scales that lead to different emphases in the decision-making process for SoSs (DeLaurentis et al., 2011). As a result, M&S approaches should be able to accommodate these time scales appropriately. These decision time scales are:

1. The *design time scale*, which is long (years or potentially decades) and implies that an enduring SoS will exist for a long time horizon and the design development of new constituent systems for the SoS will require significant time. The open nature of SoS, with constituent systems entering and leaving the SoS faster than the development and entry of new systems, complicates the M&S here. Additionally, the M&S approaches face the general challenge of making predictions for long time horizons that appropriately recognize the uncertainty of these predictions.

2. The *planning and implementation time scale* is shorter than the design time scale but may still be measured in months or years. In the planning and implementation time scale, decisions are made to develop a strategic approach to providing desired capabilities. This time scale would typically cover the introduction of new systems or the removal of current systems from the SoS. This time scale also covers general trend changes in the operating environment of the SoS.

3. The operational time scale is real time, or near real time, so the decision windows are short. With many different constituent entities managed by separate entities, real-time decision making can be difficult. Weather and unscheduled maintenance events greatly affect the daily (short) operations time scale of the air transportation SoS; the SoS should be able to handle short-term impacts. The M&S here needs much finer time fidelity and must be fast to support near real-time decision making.

Over its existence, an SoS generally touches all of these time scales. Because of this, improved understanding of SoS features and characteristics relevant to the design, planning, integration, and operations time scales is needed. For example, the design of an ATS of systems must account for the multiyear time scales required for the development of new aircraft, new air traffic control (ATC) technology, or for the construction of a new runway that will impact the performance of the SoS. This long "design" time scale must also account for the development of new technology that could upgrade existing assets as frequently as every year. The planning and implementation cycles for an air transportation SoS account for transient seasonal variation in passenger demand and the more steady overall increase in passenger demand over time; the design of the SoS should also reflect this shorter time scale. Weather and unscheduled maintenance events greatly affect the daily (short) operations time scale of the air transportation SoS; similarly, the design of the SoS should consider the ability to respond to short-term impacts. The literature in disruption recovery models for airline operations (Bratu and Barnhart, 2006; Kohl et al., 2007; Clausen, 2010) is one example along these lines.

10.2.2 Lexicon and Taxonomy

A taxonomy is a means of classifying entities according to their natural relationships. Our contention is that an important use of a taxonomy for SoS is to inform designers about which methods are appropriate to assist with decision making in an SoS context. The taxonomy here uses three characteristic dimensions or axes: (1) type of systems, (2) control of systems, and (3) connectivity of systems. Other concepts and taxonomies exist to help describe interacting systems problems; these date back many years. We believe that the three-axis taxonomy covers the most relevant aspects of SoSs, but we also wish to recognize that others have developed similar ideas.

An early example appears in Jordan (1968); a dimension-based taxonomy of systems is proposed in which he identified three dimensions along which systems could be distinguished: rate of change (structural static vs. functional dynamic), purpose (purposive or not), and connectivity (mechanistic or organismic). Although not a taxonomy, the concept of sociotechnical systems recognizes that systems heterogeneity (human–technical systems) is a crucial aspect of modern endeavors. Large-scale sociotechnical systems is a term used in Thomas Hughes' systems theory (Bijker et al., 1987). To avoid the lengthy acronym, Nikolic et al. have proposed the term λ-systems (Nikolic et al., 2009). "The term indicates a class of systems that span technical artifacts embedded in a social network, by which a large-scale, complex socio-technical artifact emerges. λ-systems include, for example, organizations, companies and institutions that develop around and sustain a particular industrial system, be it a single plant, industrial complex or set of interconnected supply-chains."

Krygiel proposed a form of taxonomy around the dimensions of autonomy, heterogeneity, and distribution to delineate an SoS from a "federation" of systems, arguing that a federation has greater degree of these three characteristics, implying more distributed, and less centralized, control (Krygiel, 1999). The work of Dennis et al. proposes multidisciplinary design optimization (MDO) formulations of SoS problems based on the degree of control exerted by a central authority (CA): inactive, guiding, mediating, and omnipotent (Dennis et al., 2005). Dennis and colleagues postulate that decreasing the activity of the CA, or increasing the autonomy of the subsystems, would allow larger instances of SoS to be posed and solved in an optimization problem. Essentially, this relates to the "control of systems" axis in our taxonomy.

Beyond engineering methods familiar in design of aircraft and similar large, complex, monolithic systems, methods from areas such as operations research (OR), complexity science, artificial intelligence, systems theory, competitive games, etc. can apply as M&S approaches for some SoS problems. The application problems presented later in this paper document how our three-axis taxonomy of system type, control, and connectivity describes and guides effective application of some of these to air transportation problems.

While taxonomy organizes features and behaviors as a prerequisite for appropriate modeling, a lexicon that articulates hierarchic structure and organization is needed to properly interpret outcomes of using taxonomy. A simple lexicon to frame the SoS taxonomy was put forth by DeLaurentis and Callaway (2004). The lexicon consists of two major structures: categories of systems and levels of organization. Table 10.1 shows four levels and four categories; however, there can be any number of levels and categories depending on the circumstance. Each category has a hierarchy of levels. To avoid confusion from ambiguous labels (e.g., is this a "system," "SoS," or "architecture"?), the lexicon employs Greek letters to establish the hierarchy: alpha (α), beta (β), gamma (γ), and delta (δ) indicate relative position within each category.

A β-level network comprises a collection of α entities and their connectivity. Likewise, a γ-level network is an organized set of β networks; this continues through the chain of levels. For many applications, there will be orders of magnitude difference between the numbers of α-level systems and β-level systems. This intuits an important point about SoS problems: **the physical manifestation may be most obvious at the α-level, but the behavior of the SoS is dominated by the structure and organization at higher levels.** Relevant questions for SoS design might be: "How does the preferred or observed behavior at the upper levels (e.g., γ-level) affect the possibilities for alternatives

Table 10.1 Lexicon for Describing System of Systems

Category	Description
Resources	The entities (systems) that give physical manifestation to the system of systems (SoS)
Stakeholders/values	The nonphysical entities that give intent to the SoS operation through values
Operations	The application of intent to direct the activity of physical and nonphysical entities
Policies	The external forcing functions that impact the operation of physical and nonphysical entities

Level	Description
Alpha (α)	Base level of entities in each category, further decomposition will not take place
Beta (β)	Collections of α-level systems (across categories), organized in a network
Gamma (γ)	Collections of β-level systems (across categories), organized in a network
Delta (δ)	Collections of γ-level systems (across categories), organized in a network

at the lower levels (α and β)?" and "How can one make design decisions at the lower levels (α and β) to improve a performance metric or meet constraints at the upper levels (γ)?" Not coincidentally, the most consequential decisions usually arise at the upper levels. For instance, describing desired capabilities at the γ-level can have great consequences on the design space at the α- and β-level, but these consequences are usually not obvious to one making a decision at the γ-level.

Employing this lexicon and with an eye toward understanding methodological needs for modeling SoS, there are three dimensions that serve as a starting point to characterize an SoS problem taxonomy. In a view similar to the "spanning issues" proposed by Rouse (2003), we believe that (at least) three major dimensions of SoS problem characteristics exist. These are (i) type of systems, (ii) control of systems, and (iii) connectivity of systems. In each of these dimensions, a given SoS problem will have a (qualitative) "value" or "measure." The "location" of an SoS problem in this three-axis space indicates how the problem might be cast and which method(s) might be best suited for use.

10.2.2.1 Type of Systems

SoSs are comprised of numerous independent systems. Following Table 10.1, "resources" are the physical entities representing independent systems; "stakeholders," the nonphysical entities. The design of an SoS requires that analysis methods be appropriate to the type of entities that constitute the SoS. Some SoSs consist predominantly of technological systems—independently operable mechanical (hardware) or computational (software) artifacts. Technological systems have no purposeful intent; that is, these resources require operation by, programming by, and activation by a human or organization. Other SoSs consist predominantly of humans and human enterprise systems—a person or a collection of people with a definitive set of values/skills. While these systems are physical entities, they primarily act as operators of the technological systems, as service providers (both with and without the support of the technological systems), and/or as consumers of services.

Each SoS lies on a spectrum between wholly technological and wholly human enterprise. For example, the Army's initial vision for the Future Combat Systems (FCS) was a mixture of 18 + 1 + 1 systems. The 18 systems are technological—stationary sensors, ground vehicles, and air vehicles. The first "+1" is the network linking the entities and allowing information exchange; this, too, is a technological system. The last "+1" is the solider—clearly a human system. A healthcare SoS largely consists of doctors, nurses, lab technicians, and other healthcare practitioners, all of whom are humans or human enterprise systems. However, diagnostic equipment and information systems are also included, which are technological. The ATS SoS embraces large numbers of both types of systems. The aircraft, airports, airways, information systems, etc. constitute the technological systems, while the aircraft designers, air traffic controllers, maintenance technicians, pilots, etc. contribute the human or human enterprise systems.

10.2.2.2 Control of Systems

The second SoS dimension is the degree of control of authorities over the entities or the autonomy granted to the entities. This relates to Maier's discussion of operational independence and managerial independence of systems within an SoS (Maier, 1998).

Managerial independence means that decisions governing the control of an entity within the SoS have no dependence on control decisions made for other entities in the SoS. Operational independence means that an action taken by an entity within the SoS does not require actions of other entities within the SoS. These are not absolute; there is a spectrum of these characteristics, and this relates to the control imposed or autonomy granted within the SoS. Emphasizing the importance of control/autonomy, Sage et al. refer to a collection of systems with operational, but limited managerial, independence as an "SoS" and a collection of systems with little CA as a "federation of systems" (Sage and Cuppan, 2001). This also follows from Krygiel's classification (Krygiel, 1999).

In principle, the military has a chain of command, and there is a single high-level set of objectives/capabilities/needs described by some high-level decision maker for an SoS. The constituent systems in a defense SoS have independence—an air vehicle and a ground vehicle in the FCS would operate without direct linkage to each other or without requiring explicit instructions for every move—but strategic SoS decisions are made at a high level. Ultimately, someone is responsible for directing the military SoS to provide the capabilities. The DoD SoS Engineering Guide helps here by identifying the four types of SoSs based on the degree of centralized management (in order from the most to the least centrally directed): directed, acknowledged, collaborative, and virtual. Most military or defense SoS problems are acknowledged or collaborative (U.S. Department of Defense, 2011).

In air transportation, each airline is seeking to make profit by providing air transportation while following the requirements of safety imposed by regulations and policy; however, one airline cannot directly control or make decisions for another airline. The Internet, however, contrasts with the CA exhibited by a defense SoS. Information exchange uses a set of protocols that most users agree upon, but no entity enforces adherence to these protocols. Individual computers connect into local area networks that have administrators, but the individual systems that comprise the Internet are very loosely connected, and there is no clear chain of command or central controlling authority. Arguably, this provision of autonomy (or lack of control) allows the Internet to successfully provide the services requested of it. Finally, the ATS application lies in between the extremes of the defense and Internet examples. A subset of resources, mostly at the α-level, is centrally controlled—an airline controls the operations (scheduling, maintenance, crew assignment) of the aircraft in its fleet—while other systems, typically stakeholders at the β-level and above, operate with a high degree of autonomy. Importantly, however, these degrees of control at any level may shift with either time or context, or both, presenting a challenge to M&S.

10.2.2.3 Connectivity of Systems

The level descriptors in the lexicon of Table 10.1 highlight the importance of connectivity. Systems involved in an SoS are interrelated within a level and through the hierarchy. Operationally, these interrelationships often, but not always, are manifest as communications links and thus form a communications network. In terms of development, intended connectivity of the SoS implies the necessity for understanding programmatic interdependencies that exist in addition to the individual systems issues. Thus, a key focus for design methods research in an SoS context lies in analysis and exploitation of interdependencies, and this focus often takes shape via analytical means of analyzing resulting networks in addition to the attributes of systems in isolation (Mane and DeLaurentis, 2010; Mane et al., 2011). Two overarching implications of connectivity are the potential emergent behaviors that result (hopefully producing a beneficial impact on capability and performance) and the fact that the connectivity itself may evolve over time. An example of emergence due to connectivity—albeit this example shows a detrimental impact on capability and performance—is the reduced total capacity of adjacent airports in the New York City area: JFK, LaGuardia, and Newark. Because of airspace interdependencies between the three airports, their combined capacity is less than the sum of their individual capacities assessed if they were isolated (Ayyalasomayajula and DeLaurentis, 2009). This became evident as air travel demands and airline operations at these airports increased over time. However, emergent behavior can provide robustness (as one independent system becomes incapacitated, another system alters its operations to accommodate), which is a beneficial impact on capability. The evolution of connectivity characteristics can fundamentally change operational behavior of the SoS; thus, SoS models must possess evolutionary capability, so that independent systems can be removed, replaced, or upgraded over time. Additionally, appropriate metrics must be generated, because time-varying behavior and the tracing of underlying drivers are crucial in making design choices. In large measure, all of the SoS applications addressed in this volume, not just air transportation, exhibit some form of significant connectivity; it is incumbent on the analyst to choose modeling approaches wisely in seeking to address it.

In the ATS, for example, the impact of changes in connectivity can be profound. For example, shifting operations at one airport can have a great impact on the operations at another, especially if they are "proximal" in the network sense (i.e., there are many flights between them). Flight delays at New York JFK can hinder operations at Chicago ORD because of this type of strong interdependency. The March 2010 closure of one of the runways at JFK (Schlangenstein, 2010) led airlines like Delta and JetBlue to reduce the number of departures from JFK, stating that they "prefer a four-month closure over the alternative," which was delays in the operations at connecting airports. As actors in the SoS who wield managerial independence, airlines indeed can change their network of operations; they create new routes between some cities or discontinue service between others. SoS models must capture these evolutionary connectivity characteristics of component systems.

Figure 10.1 offers a representation of the taxonomy that illuminates variations in system type, control, and connectivity within an SoS; notionally, the position on each axis depicts where an SoS analysis or design problem lies on the spectrum of these three features. Subregions in the three spaces can be delineated (as shown) and used to describe and communicate a particular problem type. For instance, a problem at the origin consists of only technical systems that are independent of each other and under centralized control (i.e., not an SoS problem at all). An SoS consisting of mostly human/organizational systems that are independent of each other and decentralized in their control would present problems in the "system type 2/autonomy type 2 ($S2A2$)" region. An "autonomy type 2/ connectivity type 2 ($A2C2$)" problem has all technical systems whose interdependence is strong and whose control is distributed. Examples of SoSs discussed in this paper are located in these three regions. A coupling exists between the location in this taxonomy and appropriate modeling to support decisions about the SoS design and/or evolution. Additionally, if consideration of multiple SoS contexts is important to the problem at hand (e.g., developmental and operational contexts), then they must be included and represented in using the taxonomy.

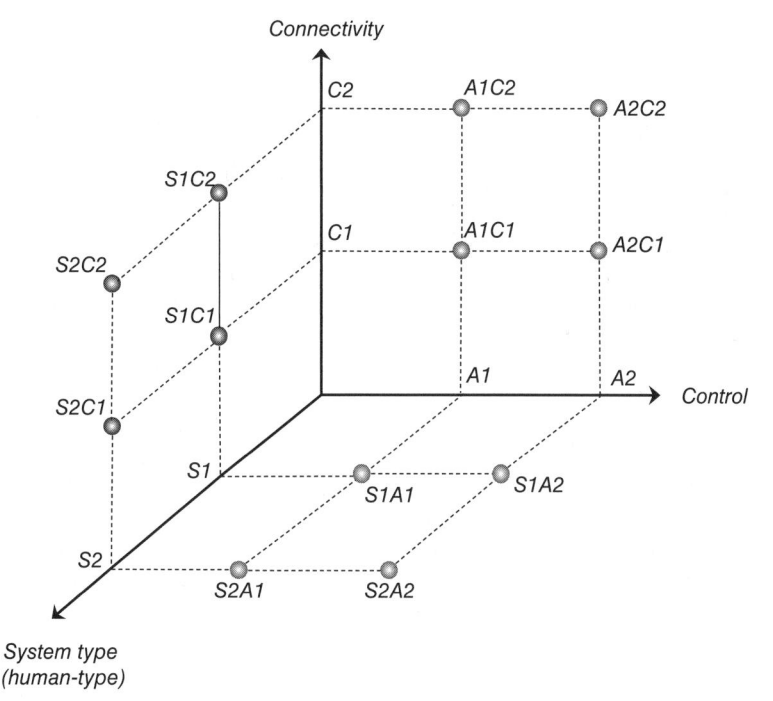

Figure 10.1 Dimensions for a partial SoS taxonomy: relative location of SoS application problems guide model and simulation development.

10.3 METHODOLOGY

We hypothesize that the location of an SoS problem within the space represented in Figure 10.1 indicates which M&S methods are appropriate for analysis and design—determining the appropriate mix of entities with associated architecture and including yet-to-be-designed entities in an SoS to provide a set of desired capabilities within a set of constraints. Subsequently, an assessment can be made concerning whether (i) advances are required within a method domain to handle size and complexity of the SoS application or (ii) advances are required to facilitate/enable interoperability between methods. Especially in air transportation SoS settings, important areas of foundational methods include optimization, resource allocation, agent-based modeling (ABM), and network topology analysis.

10.3.1 Optimization to Model Airline Behavior

In the context of air transportation, optimization commonly supports decision making by the airlines; in particular, various forms of integer programming (IP) problem formulations help determine schedules, assign aircraft to routes, assign crews to aircraft, etc. In this sense, optimization models an airline's decision-making process about how to use its array of aircraft, personnel, etc. while providing a capability (satisfying passenger demand)

and meeting a high-level objective of maximizing profit. References like Clark and Smith (2004) and Barnhart et al. (2003) describe many of these applications of optimization to support decision making. Thus, because airlines use optimization to support their decisions, the premise here is that optimization problems can model or simulate the decision-making behavior of an airline.

Purposefully considering an airline as an SoS within the air transportation domain, the airline has centralized control. The optimization problems solved to determine schedules, fleet assignments, and crew assignments describe many of the strategic aspects of the airline's operations. However, these decisions do not specifically tell individual pilots how to fly the routes between airports, but they do describe which aircraft the pilots will use on that route. The operation of one specific aircraft within the airline's route network does not significantly affect operations of other aircraft in the airline, as long as the instructions of ATC are followed.

Extending the idea that the airline is an SoS within a larger ATS of systems, these optimization problems used by airlines for strategic decision making can also serve as an M&S tool that represents how a profit-seeking airline would use its multiple assets to maximize profit while satisfying demand. In some of the following examples, these airline-level optimization problems take the form of a resource allocation problem, the results of which help provide M&S support for decisions about what kinds of new aircraft or technologies might best improve SoS-level performance.

10.3.2 ABM

ABM has emerged as an approach of choice for M&S of a wide variety of situations involving sociotechnical systems in ATS SoS settings. ABM employs a collection of autonomous decision-making entities called agents imbued with simple rules of behavior that direct their interaction with each other and their environment. Agent functionality is quite flexible, with behavior types ranging from simply reactive (change state or take action based on fixed rules) to learning/adaptive (change state or take action after updating internal logic schema via learning). If a given environment has multiple, diverse agent types, it is described as a multiagent simulation (MAS). However, it is good modeling practice to limit the complexity of a MAS to only that required to answer specific questions; as modeling complexity increases, so too does the effort of verification and validation.

Employing ABM for SoS problems in which distinct decision-making entities exert control has a challenge—how to validate that the agent models properly reflect real human/organizational behavior. This is a critical question aimed at the trustworthiness of simulation results. The literature on this subject within the ABM domain is growing. The most common approach uses as much historical data as possible to validate the individual agent behavior models and then to trust that the emergent behavior from agent interactions will be realistic. Taken together, model validation for SoS problems that have especially high degrees of human/organizational behavior components must make best use of state-of-the-art methods. Further, rather than a predictive mentality, the outcomes of ABM-type models should be exploratory in nature, seeking to understand distinct regimes of likely behavior rather than identifying locally specific solutions with high predictive accuracy.

Table 10.2 Characterization of Several Networks in the ATS

Network	Node	Link	Time scale of change
Demand	Homes/businesses	Demand for trips	Months/years
Mobility	Origin/destination locations	Actual passenger trips	Days/weeks
Transport	Airports	Service routes	Days/weeks
Operator	Aircraft, crew	Missions	Hours
Infrastructure	Waypoints and airports	Air routes	Months

10.3.3 Networks for Analysis of Connectivity

There are multiple interacting networks embedded in the ATS. One characterization of these networks has been presented by Holmes (2004) using the analogy of an ISO stack in communications network design. A modified version of this characterization, adopting a more clearly top-down construct, is presented in Table 10.2. In light of this characterization, a network-centric abstraction and analysis approach for study of the ATS seems appropriate (Lewe et al., 2006). In such an approach, some networks employ the same resources but in a different operational context. For example, nodes in the transport network are airports and the links represent service routes between nodes. The mobility network also contains airports in its nodal structure, in this setting as passenger points of entry into the ATS. Consequently, these networks are coupled in a physical manner (e.g., node resource sharing) and in an operational manner (e.g., overlapping time scales).

Another advantage of network-centric abstraction of the ATS is its applicability at different hierarchical levels. Network analysis can address interactions among networks across these hierarchical levels with unaltered methodology—demonstrating extensibility of the approach. The transport network provides one example of hierarchy. The overall transport network is composed of the scheduled and nonscheduled service subnetworks. The scheduled service subnetwork is composed of the individual airline subnetworks.

The availability of data is a key consideration in network-oriented M&S. Much of the recent advancement in modern network science has been motivated by the ability to query, analyze, and visualize real, large-scale networks such as those in computing, biology, and social interactions. Thus, the ability to validate theoretical models is increased, though the potential to extrapolate conclusions beyond what the data shows (absent a validated underlying theoretical model) is also a danger to be faced.

10.4 APPLICATIONS

In this section, we focus on M&S specifically for air transportation. Two research projects undertaken by the authors provide experiments for the use of the taxonomy and associated methods to address SoS problems. The two projects provide an opportunity to describe how the type, connectivity, and control of systems are taken into consideration and how appropriate methodologies are used.

10.4.1 How to Design New Aircraft

The introduction to this chapter anecdotally uses the decisions of Airbus and Boeing to pursue different new aircraft concepts at the close of the 1990s and the start of the 2000s. The motivation behind that anecdote was to illustrate how M&S approaches might help improve decisions about a new system that needs to operate along with existing systems in an SoS context. For this discussion, the new system is a new passenger aircraft type that will serve along with an existing fleet operating in and out of airports in an operator's route network. In this sense, the operator is viewed as the SoS of interest, a β-level network of α-level resources. Its goal is to assemble and operate a fleet that satisfies the demand for passenger transportation while maximizing the profit as a higher-level objective.

The references Mane et al. (2007) and Mane (2008) present two experiments to commingle tools from OR and MDO into an SoS design method. The motivation to look for OR and MDO tools in these applications follows the preceding discussions of SoS characteristics. The first experiment, concurrent aircraft design and resource allocation with new aircraft design for airline operations, seeks to determine the characteristics of a new aircraft for allocation along with an airline's existing fleet to meet passenger demand (providing transportation is a capability). The problem must consider the utilization of both new and existing/legacy aircraft along with the implications that this entails. Reported historical information about the number of passengers carried on the routes in an airline's network provides the demand for the airline problem. This provides an appropriate size and structure to the passenger demand; it does not predict future demand, in this formulation.

The second experiment, concurrent aircraft design and resource allocation with new aircraft design for operations of a fractional management company (FMC), also seeks to determine the characteristics of a new aircraft for allocation, but demand is uncertain and is expressed as a probability distribution of trips between city pairs. Further, the fractional operator has the option to hire charter aircraft to complement its fleet capability to satisfy demand. FMCs utilize this option to cover demand when the owned aircraft are unavailable due to unscheduled maintenance events or when demand is so large that owned assets are insufficient. In fractional operations, aircraft "owners" purchase shares of business jets starting at shares as small as 1/16th from FMCs based on the aircraft type and yearly number of hours the owner perceives to best fit his/her needs. The FMCs manage and operate the aircraft to satisfy the flight/trip requests of the shareowners. A fractional aircraft ownership operation is unlike airline operations in many ways. Airlines decide the flight schedule several months in advance, while FMCs must schedule their aircraft on a few hours' notice. While airlines must decide which routes and frequency of flights on those routes are most profitable, fractional operators must respond to the demanded flights and frequency of its customers. To book a flight, a shareowner makes a request, with as little as 4 h notice, to indicate the desired departure time, departure location, destination location, and number of passengers traveling with him/her. This makes the FMC allocation problem highly uncertain. Ensuring that all owner requests are satisfied, and doing so while maximizing profits or minimizing costs, is a complex problem of resource allocation and scheduling. Randomly generated trips that are informed by—but are not actual—FMC operations provide the demand structure for this problem.

The solution, for both problems, seeks to minimize direct operating cost (DOC) as a surrogate for maximizing profit (providing the operator with profit is also a capability, or profit could be a measure of the effectiveness for providing the transportation capability). These problems reflect the intent of the airline's and FMC's management. Both problems consider only one β-level entity (the airline or the FMC) that consists of several α-level entities (airplanes, both existing and yet-to-be-designed, and routes).

Solutions to these two problems give insight into the implications of approaching this type of problem as an SoS problem. In the airline example, the mixture of new and legacy aircraft captures the implications of how existing assets along with allocation impact the aircraft design. In the FMC problem, uncertainty of demand and the option to utilize charter aircraft accounts for the possibility of unexpected behavior (i.e., the charter aircraft are utilized differently from current practices).

By posing these as a simultaneous allocation and new aircraft design problems, optimizing the β-level metrics (here, minimizing fleet DOC) directly drives decisions about the new α-level entity (here, the design variables describing the new aircraft). While aircraft designers do seek to improve the operators' key metrics, they traditionally do so indirectly, by optimizing an aircraft-specific performance metric that relates to the operator's key metric, rather than optimizing the key metric directly.

The decision time scales for an SoS include the three components previously mentioned: design, planning, and operations. Time scale is not one of the axes on the taxonomy diagram (Figure 10.1), but it might influence the interpretation of the problem and/or the choice of methods.

As posed in Mane et al. (2007) and Mane (2008), the combined allocation/aircraft design problems focus upon the longer time scales—predominated by the design and life cycle of the new aircraft. The airline currently operates (and allocates) a fleet of aircraft, and it plans to continue to do so with the addition of the newly designed aircraft; the same is true for the FMC. The problems, therefore, must consider the utilization of legacy and new systems along with its implications on the choice of methods for the airline problem and the utilization of charter aircraft and newly designed aircraft under uncertain trip demand for the FMC.

This formulation allows a look at how the new aircraft design problem may be tied directly with an allocation problem. For the airline problem, while fleet allocation is often a planning time-scale activity, the problem formulation does not address features of growing average demand over time. At an even finer time scale, the daily schedule of arrival and departure times represents an operations time scale. With its strategic focus, the airline problem does not consider the operations time scale. For the FMC problem, on the other hand, the problem formulation addresses the variability of demand by expressing the trip demand as a random selection of city pairs. While not directly including the operations time scale and growth in average demand, the FMC problem captures variations in demand and the implications that this has on the planning time scale. Decisions regarding the characteristics of a new aircraft must consider that demand is uncertain and that solution to the allocation problem must ensure that all demanded trips are served, with the yet-to-be-designed aircraft and/or charter aircraft.

10.4.1.1 System Type

In the airline problem, the airline can utilize up to seven (7) existing aircraft types as well as the yet-to-be-designed aircraft, and it must provide service on 31 different city pairs or routes. To provide service, aircraft and routes are not sufficient. The airline must engage in planning to coordinate the operations of its aircraft on the available routes.

"Operations"—the plan by which the airline allocates aircraft to routes, schedules aircraft departures, and schedules aircraft maintenance—is a highly technological system. This represents the aircraft and crew scheduling, revenue management, etc. arm of the airline. With the strategic focus of this problem, this is abstracted here to be just the allocation problem; and along with the aircraft and routes, it represents the technological systems of this SoS.

The FMC problem formulation is somewhat different. The FMC must decide the number of yet-to-be-designed aircraft that it will own and the number of charter aircraft that it will utilize to satisfy demand between uncertain city pairs. As with the airline problem, the allocation problem, the aircraft, and city pairs represent the technological systems of the SoS.

In the airline problem, by treating the passenger demand on predetermined routes as static, the passengers do not have a direct impact on the airline. This is an assumption, because passenger demand does fluctuate; but with this assumption, the airline SoS consists solely of technological systems. In the FMC problem, however, trip demand is uncertain. The city pairs (and therefore the trip distances) are randomly distributed. Therefore, the FMC problem also includes human systems that introduce uncertainty to the planning process. While increasing the complexity of the allocation problem, this aspect makes possible observation of unexpected outcomes that result from the uncertain nature of demand.

10.4.1.2 Control

The problems view the airline and the FMC as the sole "top-level" entities. Because of this, centralized control exists; the airline's management can make decisions about which aircraft types and how many of each type to assign to each route and the FMC's management can decide how many aircraft to own, how much charter to utilize, and how these aircraft will fly the demanded trips. Centralized control implies that the management has objectives that it wishes to minimize or maximize. There may be many objectives, but the location of the airline and FMC SoS on the control axis implies that optimization is applicable to formulate the problems (as opposed to when only satisfaction or equilibrium is applicable when there is high autonomy).

10.4.1.3 Connectivity

While the airline is not establishing new routes and the existing route structure is fixed, the FMC faces different routes each day. For the airline, this means that there is little coupling between the routes and the individual aircraft; aircraft must have sufficient range to be assigned to a given route. Coupling does exist, however, between the passenger capacity of the new aircraft and the allocation problem. Passenger capacity of the aircraft is a variable in the aircraft design problem and is influenced by the allocation problem, which assigns aircraft to routes based on their cost as well as passenger capacity. Furthermore, between various aircraft, the only coupling is that the total capacity deployed on a given route meets or exceeds demand; for example, if one aircraft does not provide sufficient capacity, a second aircraft is needed on the same route.

For the FMC, on the other hand, coupling is stronger. Uncertainty in demand means that the lengths of demanded trips greatly impact aircraft design, namely, design range.

The distance between any two city pairs will dictate the design range of the aircraft, for example, or force the FMC to utilize charter aircraft with longer design range. Because the FMC uses only one type of aircraft, the coupling between aircraft is that the design range of the newly designed aircraft and the range of chartered aircraft must be large enough to ensure that all demanded trips can be served.

In both problems, however, the operator does not have the option to turn down demand; it must serve all passenger demand. While airlines turn down demand through overbooking, this generally ensures high load factors and does not drive the selection of which aircraft type to utilize on a given route, nor does it keep the airline from providing service on the route. Hence, while the allocation problem serves as an evaluator of a given aircraft, the aircraft characteristics cannot, in themselves, influence demand. This limited (unidirectional) connectivity between the constituent α-level systems and its basis on the flow of information (i.e., the aircraft design problem needs to know the length of the longest route and the allocation problem needs to know the aircraft passenger capacity, for the airline problem, and the design range and velocity of the aircraft, for the FMC problem) implies that a decomposition strategy may be employed.

10.4.1.4 Problem Solution Strategy

Given that an optimization approach seems applicable for the concurrent aircraft design and aircraft allocation problems and the nature of the connectivity, the appropriate model is to pose these as mixed-integer nonlinear programming (MINLP) problems. The optimization seeks to minimize the expected daily operating cost of the FMC while determining the optimal aircraft design (e.g., design requirements and aircraft characteristics) and the optimal operations (e.g., aircraft assignment to routes and number of aircraft to be owned and operated). However, the size of the MINLP problems is such that solving all but the most simplistic versions is impractical. Because connections between the constituent systems are few, a decomposition approach allows solution to these problems. For the airline problem, the aircraft design and aircraft allocation problems are solved as a nonlinear programming (NLP) problem and an IP problem, respectively (Figure 10.2a); the results are used as function evaluations in a much smaller top-level IP problem (IP because passenger capacity is the only variable). A similar approach solves the FMC aircraft design and aircraft allocation problem. However, because demand is uncertain and is expressed as a distribution, a Monte Carlo simulation is performed to allocate the aircraft to the different sampled demand scenarios for every function evaluation of the top-level NLP problem. The top-level problem is an NLP problem, because the multidisciplinary design variables are continuous variables for the FMC problem, namely, aircraft design range and cruise velocity (Figure 10.2b).

In one example application for the airline problem, using a 31-route structure and an existing aircraft fleet consisting of seven different aircraft types, the optimization strategy via decomposition resulted in a new aircraft design and its allocation along with existing aircraft that reduced the airline's daily operating costs by nearly 13%. The new aircraft had a capacity of 250 passengers, wing aspect ratio (AR) of 9.5, wing loading (W/S) of 135 lbs/ft^2, and thrust-to-weight ratio (T/W) of 0.31. The values of AR, W/S, and T/W not only support the reduction of the fleet operating cost via the top-level problem, but they also meet specific performance requirements for the new aircraft. In this example, the

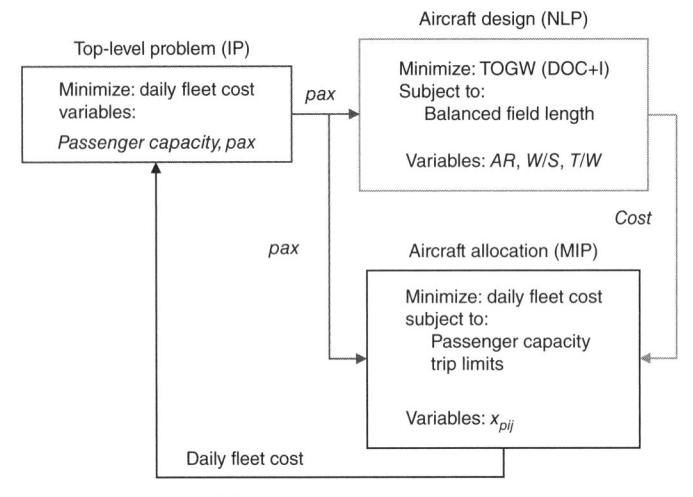

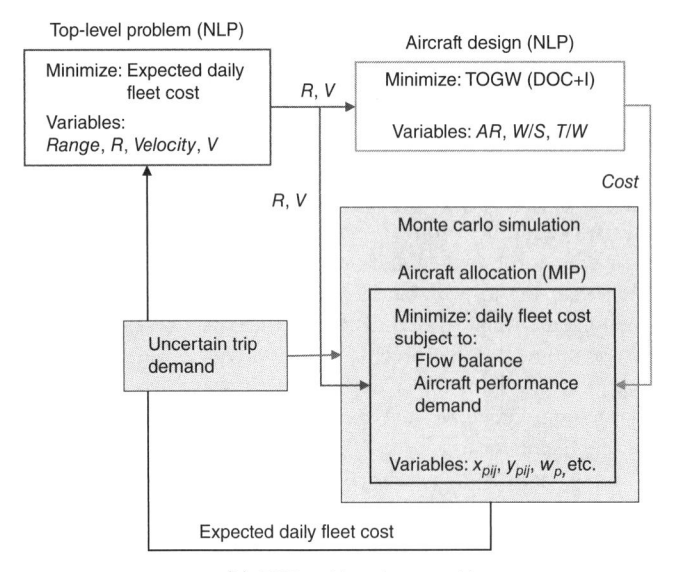

Figure 10.2 Decomposition strategy in (a) airline and (b) FMC application problems.

new aircraft must meet a constraint on maximum allowable takeoff distance, and the aircraft design NLP problem ensures this is met.

Similarly, in an example application of the FMC problem, serving demand between ten (uniformly) randomly selected cities, the optimization resulted in the design of a new aircraft that, when allocated in concert with chartered aircraft, reduced the daily expected cost of operations by nearly 1% with respect to operating an existing aircraft on the routes and allowing charter. The new aircraft had a design range of 1549 nmi, a cruise velocity of 438 knots, AR of 7.51, W/S of 44.8 lbs/ft^2, and T/W of 0.464; as with the airline-based problem, this aircraft not only reduces the FMC's operating cost, but it also meets performance constraints on the aircraft. The design range of the new aircraft was shorter than the longest demand trip (1600 nmi). As a result, the newly designed aircraft would not be able to serve all demanded trips, but the FMC must rely on charter aircraft that have longer range to satisfy all demand.

FMCs currently use charter on a regular basis generally because owned aircraft are unavailable or because a spike occurs in the daily demand; this is the only way to satisfy demand, so the lowest cost solution includes these aircraft. Here, allowing chartered aircraft in the problem formulation leads to a result that suggests a change in operations, where the FMC plans for the utilization of charter aircraft on the long, but infrequent, trips and uses the owned aircraft for the shorter, more frequent trips. The taxonomy guided an optimization problem formulation, and the formulation tried not to dictate α-level decisions (e.g., ensure that the new aircraft must fly all possible routes and only use charter to address unavailability or abnormally high demand). While not exactly emergent behavior, this is an unexpected result that points to the ability of this problem formulation as an SoS to capture potential emergent behavior and changes in the component systems of the SoS.

The decomposition approach to allocate a yet-to-be-designed resource is not limited to the airline or FMC application. An analog may be seen in a supply chain application in which a company is seeking to design new workstation tooling to improve not only the performance of the workstation but that of the entire manufacturing plant. If the role of humans in the plant is largely supervision of automated tooling, the SoS here is mostly technological systems. With one β-level entity, the manufacturing plant's management exerts centralized control, and the problem may be posed as an optimization problem. With fixed production demand, the new workstation or tool is loosely connected to other tools, and decomposition may be possible. A method using a small top-level problem to coordinate the decomposed tool design and job allocation problems may have promise, because this application shares many features of the airline problem.

While these two examples focused on the design of a new aircraft to enable an optimal fleet utilization, the emphasis can "shift higher and broader" in the SoS lexicon of Table 10.1 with appropriate shift in model setup and simulation parameters. For example, the authors of this chapter created the fleet-level environmental evaluation tool (FLEET) to assess the environmental impact of aviation under different scenarios of aircraft technology availability, market demand, and policy implementation (Moolchandani et al., 2013). FLEET enables an assessment of the environmental impact of aviation by modeling airline operations, again, as a resource allocation problem with subsequent calculation of emissions in the form of CO_2, NOx (oxides of nitrogen), and noise from aircraft performance lookup tables. Additionally, the tool has a systems dynamics-inspired approach that mimics the economics of airline operations, models the airlines' decisions regarding retirement and acquisition of aircraft, as well as represents passenger demand growth in response to economic conditions. These additional features begin to address how current conditions can lead to future conditions in which airlines would operate new aircraft.

10.4.2 How to Design New Operations/Network

DeLaurentis and Ayyalasomayajula (2009) present a second air transportation design study that was guided via the lexicon and taxonomy. This problem concerns a means to assist decision makers attempting to transform the ATS to a state that can satisfy growing demand with greater levels of efficiency. The methodology objective of this problem is to generate a design solution space at a higher level of aggregation than the previous problem and considering two distinct time scales. Further, as indicated from the lexicon (Table 10.1), a comprehensive SoS M&S approach for ATS considers not only the technical aspects, but it also incorporates policy, socioeconomic, and alternative transportation systems considerations. However, because such an objective is overwhelmingly complex if pursued at the lowest levels of detail, judgment is necessary in order to model alternative air transportation architectures at appropriate levels of abstraction in a hierarchy that follows the previously described lexicon. For problems focused on higher levels in the hierarchy (say, at the γ- or δ-level in the lexicon), the individual systems models (at the α- or β-level in the lexicon) need only be detailed enough to enable the higher-level analysis that seeks to identify regions of good solutions. For example, simple models of airports are all that is needed to determine the best network topology or configuration of nodes and links. The final product of such an analysis is a concept consisting of a set of rules of behavior and network structure that satisfies the transportation goals. Further, the high-impact rules (policies) are identified that accomplish those goals by allowing agents in the system to "do the right thing" naturally.

10.4.2.1 System Types

The problem combines human/organizational and technological systems with high degrees of connectivity and autonomy. The primary organizations modeled are an aggregate airline and an infrastructure/air navigation service provider. As with the first problem presented in this paper, the passengers are not modeled explicitly, but implicitly through the infrastructure provider's desire to minimize saturation in the network. These features place it in the most challenging region of the taxonomic space in Figure 10.1.

10.4.2.2 Time Scales

As currently posed, the problem involves time scales that are between the *design time scale* and the *planning time scale*. For this problem, a conceptual transportation model encompassing all four entity categories, ranging from the α- to γ-levels, enables the generation, study, and analysis of alternative futures for the ATS. Included are models of the resources and the economic and regulatory drivers (via actions of stakeholder agents) that comprise the ATS, placing it near the middle of the system-type axis. In particular, each layer or network topology is unique in its makeup and time scale (see Table 10.2). A simulation-based approach using a variety of methods, including ABM, presents a better opportunity to obtain a solution.

Stakeholder agents at the γ-level making choices based on simple rules of self-interest determine the evolution of a transport network. In general, these include such choices as adding capacity at airport nodes and route network growth and reconfiguration (e.g., spreading demand more evenly via point-to-point travel instead of hub-and-spoke). In the current study, we implemented only two stakeholder agent classes: service providers (airlines) and infrastructure providers (the FAA and airports).

10.4.2.3 Control

In this example, the air transportation provider (airlines) and the infrastructure provider (airports) operate with some level of autonomy; neither has full authority over all of the resources within the SoS. A decision made by the airline will have an impact on airports, and vice versa; however, neither party has the ability to make a unilateral decision about how the other party will behave. The simulation and modeling must address this level of autonomy among this SoS; therefore, it needs to account for the behavior of entities that control different resources within the SoS. The simulation results should reflect both cooperation and competition, with the intent of establishing conditions where the various entities do cooperate in a way that provides a higher-level capability.

ABM has emerged as an approach of choice in this setting where there is no single centralized controlling authority. However, employing ABM for SoS problems in which distinct decision-making entities exert control has a challenge—how to validate that the agent models properly reflect real human/organizational behavior. This is a critical question aimed at the trustworthiness of simulation results. The literature on this subject within the ABM domain is growing (Barreteau et al., 2003; Fagiolo et al., 2007; Marks, 2007; Windrum et al., 2007). The most common approach uses as much historical data as possible to validate the individual agent behavior models and to then trust that the emergent behavior from agent interactions will be realistic. In this air transportation example, calibration of the agent behavior rules relied upon historical airline route selection data and the ability to increase airport and airspace capacity. The particulars of this calibration appear in DeLaurentis and Ayyalasomayajula (2009). Taken together, model validation for SoS problems that have especially high degrees of human/organizational behavior components must make best use of state-of-the-art methods. Further, rather than a predictive mentality, the outcomes of ABM-type models should be exploratory in nature, seeking to understand distinct regimes of likely behavior rather than identifying locally specific solutions with high predictive accuracy.

10.4.2.4 Connectivity

The connectivity aspect enters when examining how agents configure the transportation nodes (e.g., airports) and links (e.g., routes and aircraft). Among other approaches, the connectivity generated by interacting agents can be compared to "preferred" network topologies (determined via network theory methods) that display desirable traits such as scalability and robustness. Such an approach can determine the possible "rules of behavior" that lead naturally to desired performance of the ATS. Network science examines connectivity (links) between the entities (nodes) in a topology through various statistical properties. For instance, the degree of a node is the sum of all links associated with a node, and the degree distribution represents the topology of a network. Further, average shortest path reflects the efficiency of propagating information across the network, whereas average clustering coefficient measures a network's cohesiveness. These properties (and others) are highly relevant to understanding the robustness, vulnerability, and overall efficiency of the network of systems. A study of the properties of the networks can help determine preferred patterns of the connectivity in an SoS. The long-term intent is to build upon this knowledge base to create new, integrated network topological measures and apply them as design objectives to manage and improve the performance of the SoS.

10.4.2.5 Problem Solution Strategy

The combination of ABM and network theory provides the core of the air transportation SoS simulation. The method constitutes a nondeterministic approach, which means that, fundamentally, it asks and answers different questions than deterministic models. The nondeterministic method is necessary primarily due to the marriage of human systems with technological ones in a partially unknown set of future worlds. The goal is to simulate how the SoSs, human and technological components combined, evolve. Observing these simulations allows understanding of this process. The simulation makes significant use of actual data from today's transportation system obtained from the Bureau of Transportation Statistics (BTS). Once initialized, a validation exercise confirmed that the simulation could represent the reality of today's system, within the bounds on fidelity.

The overall framework for the integrated simulation is as follows: stakeholder agents (e.g., service providers, infrastructure providers) act to evolve an initialized air transportation capacity network under various scenarios (Figure 10.3). Each agent employs its logic to guide its decisions and actions. In subsequent time steps, the agent sees consequences from the environment and updates its behaviors. As this process unfolds, the magnitude and shape of the mobility network (demand) also changes, and the actions of agents must respond by manipulating the capacity network topology. Thus, a family of new network topologies is created over time, and their structure and network-theoretic analysis tracks their behavior. The key question is as follows: Do the evolved networks exhibit good performance in terms of capacity? To address this question, a network evaluator compares the evolved networks to topologies that do exhibit preferred behaviors. Using this method, the evaluator can function as the search direction generator for a design/optimization problem.

An example outcome from this simulation approach, presented in DeLaurentis and Ayyalasomayajula (2009), appears in Figure 10.4.

This result indicates the average saturation of airport nodes in the network (indicated via color) as both the amount of nodal capacity (x-axis) and time needed to implement nodal capacity increase (y-axis) are varied. For this study, the intent was to provide a visualization of the decision space so that a decision maker(s) can consider options given the boundaries of behavior change. Selecting a singular optimum from this decision-space analysis was not intended here. If subsequently desired, the behavioral rules, connectivity structure, and engineered systems capability can act as design variables to explore the generation of preferred outcomes over ensemble of plausible scenarios. This particular result from the ABM showed that a distinct regime of desired behavior emerged from the interactions between the airline agents growing and restructuring their routes and the varying capability of the infrastructure agent to "keep up" via capacity addition.

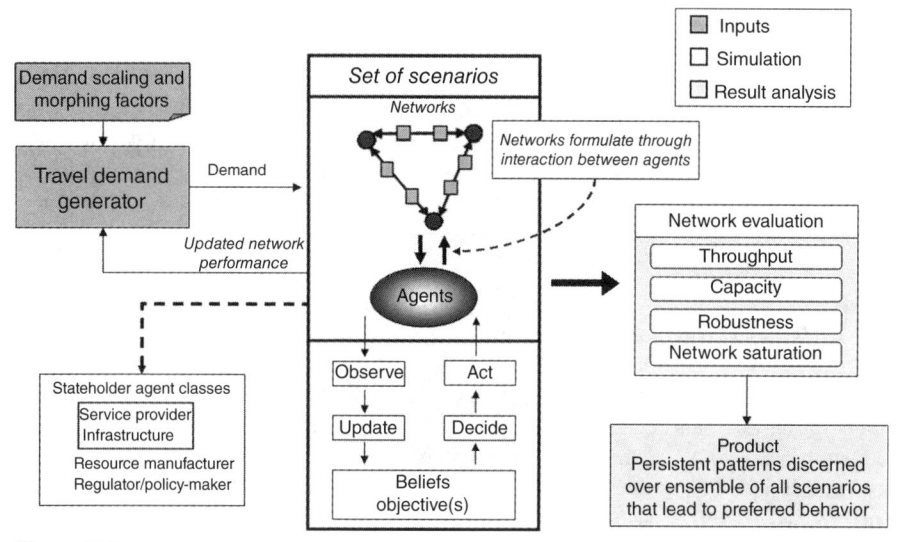

Figure 10.3 Simulation environment for example problem.

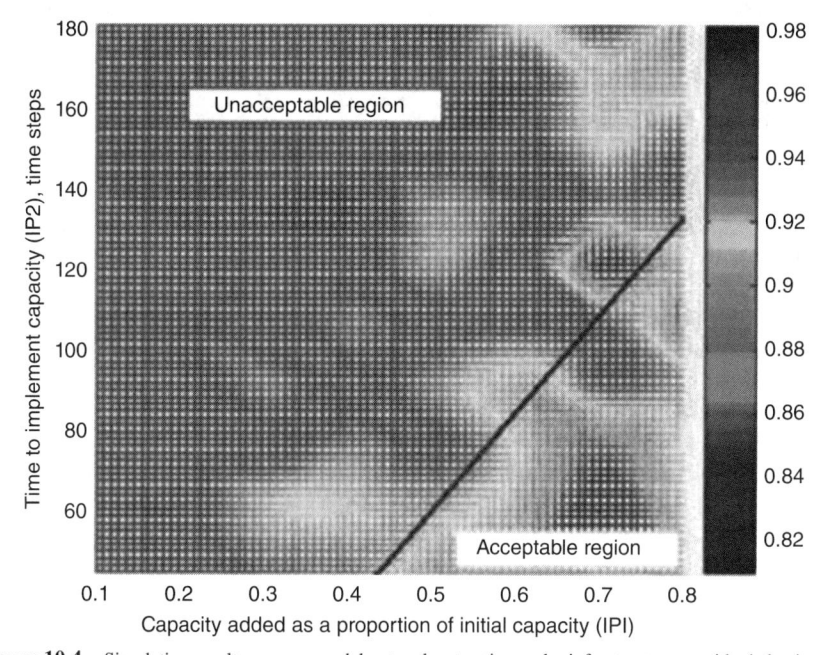

Figure 10.4 Simulation results: average nodal network saturation under infrastructure provider behaviors.

10.4.3 Linkage between the Two Examples

Beyond the obvious domain overlap in air transportation, more specific links or connections exist between the new aircraft design example and the new network example. Given that an operator would seek to maximize profit and/or minimize operating costs in the first example, the transportation provider in the second example would have similar aims. One could see how additional improvements in either example could result from working to combine some of the M&S concepts, within reasonable computational limits. For instance, the rules governing the behavior of the airline agent perhaps could use the results of some form of optimization. The quality of the future aircraft design decision may improve, if the infrastructure service provider opens new routes for the airline operator. Clearly, there are several overlaps in the basic parameters here as well as in the questions to be answered.

10.5 LESSONS LEARNED

Only recently is an understanding of SoSs developing; hence, the M&S approaches to support analysis and design are not very well developed. While SoS analysis can be undertaken using conventional M&S and systems engineering tools, albeit at an increased cost and complexity, efforts to develop an SoS with significantly incomplete understanding can lead to future service failure or otherwise undesirable results. A further complication is that though various taxonomies do exist, all SoSs do not exhibit all traits at all times. To overcome these issues, the authors suggest the use of three dimensions to categorize SoS—type, control, and connectivity. While not a complete set, use of these substantially improves the modeling choices for an SoS problem and the determination of whether new methods or modes of communication between methods need to be developed. The authors present two examples to demonstrate these ideas. The first example is related to aircraft allocation to different routes first in a scheduled airline and later in an FMC. The difference between the two lies in the time scale for decision making and the inherent uncertainty of operations. The second example, the evolution of an airline network, is simulated using ABM to arrive at the optimal network for the satisfaction of systems requirements. This example admits highly distributed networks with high connectivity and long dynamic time scales as compared to the previous one, which uses a more centralized and less connected SoS on a long but static time scale. These examples demonstrate that for SoS problems, especially in air transportation, there are likely to be multiple subproblems involved and that, on some occasions, they may give nonintuitive results. Thus, modeling some situations as SoS would lead to results that are not possible otherwise. In any event, for ATS application domain, a spectrum of M&S tools will almost always be needed and their proper integration is a challenge (in addition to the individual tool development). This mirrors the fact that ATS problems involve not "an SoS problem" but a series of embedded ones.

Therefore, key questions to consider looking to the future center on this challenge of SoS M&S tool integration and "insight" integration. Ultimately, M&S supports human decision makers. If the M&S tools cannot provide reasonably accessible insight to the decision makers, then they have failed. The wide breadth of SoS problems, exemplified by the ATS domain, also calls for continued work toward a common model-based representations methods and applicable across all multiple domains (e.g., healthcare, business, etc. or a single SoS comprised of systems from across these). Any taxonomy should effectively reflect this diversity. Further, one or more of the taxonomy dimensions presented in this chapter might need further subdivision to properly distinguish effects appropriate to the application domain and problem.

REFERENCES

Ayyalasomayajula, S., and DeLaurentis, D. 2009. "Developing strategies for improved management of airport metroplex resources." *Proceedings of the AIAA 9th Aviation, Technology, Integration, and Operations Conference (ATIO)*. AIAA 2009-7036, Hilton Head, SC. DOI: 10.2514/6.2009-7036.

Barnhart, C., Cohn, A. M., Johnson, E. L., Klabjan, D., Nemhauser, G., and Vance, P. 2003. "Airline crew scheduling." In *Handbook of Transportation Science*, edited by R. Hall, Vol. 56, pp. 517–560. Norwell, MA: Kluwer Academic Publishers.

Barreteau, O. and others. 2003. "Our companion modelling approach." *Journal of Artificial Societies and Social Simulation (JASSS)* 6(2): 1.

Bijker, W. E., Hughes, T. P., and Pinch, T. (Eds.) 1987. "The evolution of large technical systems." In *The Social Construction of Technological Systems*, pp. 51–82. Cambridge, MA: MIT Press.

Bratu, S., and Barnhart, C. 2006. "Flight operations recovery: New approaches considering passenger recovery." *Journal of Scheduling* 9(3): 279–298. DOI: 10.1007/s10951-006-6781-0.

Clarke, M., and Smith, B. 2004. "Impact of operations research on the evolution of the airline industry." *Journal of Aircraft* 41(1): 62–72. DOI: 10.2514/1.900.

Clausen, J., Larsen, A., Larsen, J., and Rezanova, N. J. 2010. "Disruption management in the airline industry—concepts, models and methods." *Computers & Operations Research* 37(5): 809–821. DOI: 10.1016/j.cor.2009.03.027.

DeLaurentis, D. 2005. Understanding transportation as a system-of-systems design problem. *Proceedings of the 43rd AIAA Aerospace Sciences Meeting and Exhibit*. AIAA-2005-0123, Reno, NV. DOI: 10.2514/6.2005-123.

DeLaurentis, D., and Ayyalasomayajula, S. 2009. "Exploring the synergy between industrial ecology and system-of-systems to understand complexity: A case study in air transportation." *Journal of Industrial Ecology* 13(2): 247–263. DOI: 10.1111/j.1530-9290.2009.00121.x.

DeLaurentis, D., and Callaway, R. K. 2004. "A system-of systems perspective for future public policy." *Review of Policy Research* 21(6): 829–837. DOI: 10.1111/j.1541-1338.2004.00111.x.

DeLaurentis, D., Crossley, W., and Mane, M. 2011. "Taxonomy to guide systems-of-systems decision-making in air transportation problems." *Journal of Aircraft* 48(3): 760–770. DOI:10.2514/1.C031008.

Dennis, J. E., Arroyo, S. F., Cramer, E., and Frank, P. D. 2005. "Problem formulations for systems of systems." *Proceedings of the IEEE International Conference on Systems, Man and Cybernetics* 1: 64–71. DOI: 10.1109/ICSMC.2005.1571123.

Fagiolo, G., Birchenhall, C., and Windrum, P. 2007. "Empirical validation in agent-based models." *Computational Economics* 30(3): 189–194. DOI: 10.1007/s10614-007-9109-z.

Holmes, B. 2004. "Transformation in air transportation systems for the 21st century." *Proceedings of the 24th Congress of the International Council on the Aeronautical Sciences (ICAS)*. Plenary paper, Yokohama, Japan. Accessed February 27, 2014. Available at: http://www.icas.org/media/pdf/ICAS Congress General Lectures/2004/F. ICAS 2004-Holmes.pdf.

Jordan, N. 1968. *Themes in Speculative Psychology*. London, UK: Tavistock.

Kohl, N., Larsen, A., Larsen, J., Ross, A., and Tiourine, S. 2007. "Airline disruption management—perspectives, experiences and outlook." *Journal of Air Transportation Management* 13(3): 49–162. DOI: 10.1016/j.jairtraman.2007.01.001.

Krygiel, A. J. 1999. *Behind the Wizards Curtain: An Integration Environment for a System of Systems*. Washington, DC: National Defense University Press.

Lewe, J., DeLaurentis, D., Mavris, D., and Schrage, D. 2006. "Entity-centric abstraction and modeling framework for transportation architectures." *Journal of Air Transportation*, 11(3): 3–33.

Maier, M. W. 1998. "Architecting principles for system-of-systems." *Systems Engineering* 1(4): 267–284. DOI: 10.1002/(SICI)1520-6858(1998)1:4<267::AID-SYS3>3.0.CO;2-D.

Mane, M. 2008. "Concurrent aircraft design and trip assignment under uncertainty as a system of systems problem," PhD Dissertation, Purdue University School of Aeronautics and Astronautics, West Lafayette, IN.

Mane, M., Crossley, W., and Nusawardhana, A. 2007. "System-of-systems inspired aircraft sizing and airline resource allocation via decomposition." *Journal of Aircraft* 44(4): 1222–1235. DOI: 10.2514/1.26333.

Mane, M., and DeLaurentis, D. 2010. "Network-level metric measuring delay propagation in networks of interdependent systems." *Proceedings of the 5th IEEE International Conference on System of Systems Engineering*, Loughborough, UK. DOI: 10.1109/SYSOSE.2010.5544080.

Mane, M., DeLaurentis, D., and Frazho, A. 2011. "A Markov perspective on development interdependencies in networks of systems." *ASME Journal of Mechanical Design* 133(10): 101009. DOI:10.1115/1.4004975.

Marks, R. 2007. "Validating simulation models: A general framework and four applied examples." *Computational Economics* 30(3): 265–290. DOI: 10.1007/s10614-007-9101-7.

Moolchandani, K., Agusdinata, D., DeLaurentis, D., and Crossley, W. 2013. "Assessment of the effect of aircraft technological advancement on aviation environmental impacts." *Proceedings of the 51st AIAA Aerospace Sciences Meeting*. AIAA 2013-0652 Dallas, TX. DOI: 10.2514/6.2013-652.

Nikolic, I., Dijkema, G., and van Dam, K. H. 2009. "Understanding and shaping the evolution of sustainable large-scale socio-technical systems: Towards a framework of action oriented industrial ecology." In *The Dynamics of Regions and Networks in Industrial Ecosystems*, edited by M. Ruth and B. Davidsdottir, pp. 156–178. Cheltenham, UK: Edward Elgar.

Rouse, W. 2003. "Engineering complex systems: Implications for research in systems engineering." *Transactions on Systems, Man, and Cybernetics- Part C Applications and Reviews* 33(2): 154–156. DOI: 10.1109/TSMCC.2003.813335.

Sage, A., and Cuppan, C. 2001. "On the systems engineering and management of systems of systems and federations of systems." *Information, Knowledge, Systems Management* 2(4): 325–345.

Schlangenstein, M. 2010. "Fliers face fewer choices on kennedy runway work (update 2)." *Bloomberg Business Week*. Accessed October 15, 2014. Available at: http://www.bloomberg.com/apps/news?pid=newsarchive&sid=a0hIK6HJ6u14.

U.S. Department of Defense. 2011. "Systems engineering guide for system-of-systems." Accessed January 12, 2011. Available at: http://www.acq.osd.mil/se/docs/SE-Guide-for-SoS.pdf.

Windrum, P., Fagiolo, G., and Moneta, A. 2007. "Empirical validation of agent-based models: Alternatives and prospect." *Journal of Artificial Societies and Social Simulation* 10(2): 8.

Chapter 11

Systemigram Modeling for Contextualizing Complexity in System of Systems

Brian Sauser[1] and John Boardman[2]
[1] Department of Marketing and Logistics, University of North Texas, Denton, TX, USA
[2] John Boardman Associates, Worcestershire, UK

11.1 INTRODUCTION

Pictures play a key role in our understanding of even the simplest problems, and often, we use pictures, even if within our mind, when describing something with words. We have what some describe as an "inner eye" that allows us to see the scenes that words educe. We will often use these pictures to understand form, function, and structure based on impressions from words. These pictures allow us to make graphical formulations of the links between words and create a representation that we can better understand—a model of the words if you will. With the balance of words and pictures, we can tell a more comprehensive story that can be more universally understood. In solving problems, we use words and pictures to convey meaning. But, when understanding problems, we are confronted by the challenge of unifying multiple perspectives. We need to be able to understand the unstructured problem, with the "big picture" in mind, and appreciate that there are multiple perspectives (with conflicting objectives (Jenkins, 1969)). The inability to effectively do this has been described as a key point of failure in the practice of many complex problems (Ryschkewitsch et al., 2009). Doing this right is what Senge (1990) would define as a "shared vision."

Modeling and Simulation Support for System of Systems Engineering Applications, First Edition.
Edited by Larry B. Rainey and Andreas Tolk.
© 2015 John Wiley & Sons, Inc. Published 2015 by John Wiley & Sons, Inc.

In solving problems, attention and understanding become key. The role of attention in problem solving must be given to relevant information in the problem and what necessary information is missing. Why? Because, in our current culture, attention is limited and attention given to irrelevant information distracts the individual from thinking about the problem. Once the essential information in the problem is identified, the problem needs to be represented. Representation means how you state the problem, the clarification and representation of rules and regulations, and the role of context in solving the problem. The ability to solve problems is tied to specific context and is termed situated cognition.

As a fundamental theory, method, and/or tool, systems thinking has often been described as a way to balance multiple perspectives for understanding and guiding to a problem resolution. Systems thinking as a practice is intended to provide an ability to see the big picture, grasp how things come together into a single system, understand the environment in which it should perform, identify the synergy and emergent properties of combined systems, and describe the system from all relevant perspectives (Boardman and Sauser, 2008).

The marriage of systems thinking with pictures and words is not new for solving simple or complex problems. There have been numerous instantiations of diagrammatic arrangements that were founded upon principles of systems thinking, for example, casual loops (Senge, 1990), problematique diagrams (Warfield and Perino, 1999), and problem structuring methods (PSM) (Mingers and Rosenhead, 2004). Of these, PSM have focused predominately on what a problem is and identifying an agreed-upon framework for the problem. Within PSM, Soft Systems Methodology (SSM) has been used in the framing and definition of the issues constituting the problem. Checkland (2000a), the father of SSM, intended SSM to make this tension an opportunity in terms of both problem definition and synthesizing feasible changes that addresses the defined problem. SSM, also defined as a method within Action Research, can reveal emergent theory from initial theory (Checkland and Holwell, 1998).

Munro and Mingers (2002) showed SSM to be the prevalent approach used with a combination of other methods in multimethodological practice for addressing this question. In addition, SSM has not been limited in its domain application nor its effectiveness in expressing solutions to problems with a systems thinking approach. For example, Molineux and Haslett (2007) demonstrated how it could enhance group creativity; Duczynski (2004) proposed a new economic model for development of indigenous people in Australia; Brocklesby (1995) identified competency requirement in human resource management; Taylor et al. (2007) implemented it in computer game design; Bustard et al. (2006) integrated SSM with visual systems methodology to understand a way of designing autonomic systems; Shalhoub and Qasimi (2005) were able to effectively apply SSM to the analysis of nonprofit organizations; Jacobs (2004) was able to use SSM to improve performance in the English National Health Service; Neves et al. (2004) demonstrated a new way to think about energy efficiency initiatives; Lehaney and Paul (1996) developed a simulation of outpatient services at Watford General Hospital. SSM has been shown to be an effective tool for well over thirty years at identifying solutions to systemic problems (Reisman, 2005).

While the application and variation of SSM has been shown to be an effective diagrammatic expression of conceptual thinking, it may be limited in its linkage of diagram to text as an expression of a problem. Diagrams are essentially networks having two elements—nodes and links. A third element we reason is text, which after "decomposition" into parts and relationships can allow the assemblage in diagrammatic form. This is a foundational motivation of a derivative of SSM entitled the Boardman Soft Systems Methodology (BSSM), which uses a systemic diagramming method called systemigrams. In this chapter, we will describe BSSM and the method and technique for creating systemigrams and demonstrate how systemigrams have been used to bring context to some complex problems. The objective is to present a method for better contextualizing a system of systems (SoS) from a multistakeholder perspective that will allow those stakeholders to achieve a unified perspective and guide them to the start of an SoS resolution.

11.2 SYSTEMS THINKING AND SoS

In an evolutionary state, the rates at which change will occur, the order of change, the legacy information upon which changes are imposed, and the environmental challenges during the periods of the change are all unknown. The error in the design of an evolutionary system is that change may be seeking a specific goal, and basic to the assumption of evolution is that no goal exists. The lack of a goal limits the analysis and ability to calculate a likelihood or failure; thus, most of the assumptions in a system are not verifiable. This becomes compounded as it is readily accepted that SoS can be characterized as having an evolutionary development. That is, an SoS is never finished; it continually evolves as needs change and newer technologies become available (Maier, 1998).

Our approach to managing or engineering the evolution of systems, that is, their developmental life cycle, has been focused on a well-bounded and firmly established set of requirements and procedures. For SoS, we have learned to leverage and satisfice (Gorod et al., 2009) the independent development of multiple systems to implement operational flexibility with interoperability (Gorod et al., 2008). This has forced us to think as much from systems, as we do about systems, and find a resurgence in the value of systems thinking (Boardman and Sauser, 2008). Keating et al. (2003) also described this in SoS as not being narrowly bound in scope, complexity, time, or geography. They state that this has created a deficiency of whole systems analysis for SoS or what some would describe as the application of systems thinking (Jackson, 2003). Systems thinking has been clearly articulated as a core competency in the practice of good Systems Engineering (Bahill and Gissing, 1998; Derro and Williams, 2007; Jansma and Derro, 2007; Haskins et al., 2008). We believe that systems thinking becomes a foundational paradigm to build our understanding of SoS. Based on our appreciation for systems thinking, we have defined four principles that underlie SoS thinking.

First, there is the notion of assessing the preexisting systems that give the backdrop to what might be envisioned. These in effect become legacy or hereditary systems and reputedly constituent systems of the SoS (Sage, 2003; Boardman and Sauser, 2006). Such an assessment at first is necessarily informal and perhaps intuitive, but as thinking develops, it may need to become more formal. The importance of legacy as described by Henderson and Clark (1990) in architectural innovation has an impact on a systems technical evolution, organizational experience, recurrent task, and technical knowledge as they relate to the component linkages, product architecture, communication channels, and problem-solving strategies. An SoS has a higher purpose than any of its constituent systems, independently or additively. Legacy systems may need to undergo change, even radical change, in order to serve in an SoS.

Secondly, there is the envisioning of a superior system (e.g., SoS) that can conceivably achieve not only a valiant goal but one that no system that is not an SoS could ever achieve (Baldwin et al., 2012). This latter point is important because nobody should consider developing an SoS if the goal can be achieved by "merely" a system. Thirdly, there is the idea of the pressing demand of the elements that are a network of enigmas that interfere with one another (DiMario et al., 2009). Separating them out is not feasible, and the associated urgency is that it be treated for its systematic solution rather than its systemic resolution. Finally, there is the idea of a framework and the need for such. The fundamental value in a framework is being able to correctly associate a definition and arrangement of a system (Sauser, 2006). A framework draws the legacy systems together along with new technology systems, thereby making up the SoS with its envisioned capability. With such a framework, the legacy systems are in effect migrated to serve a new higher purpose in addition to fulfilling their original terms of reference. This is possible because of new ways in which they can interoperate with systems, both legacy and new. The preexisting systems may act as a source of inspiration to the SoS envisioning, which in turn calls for an integrating framework. This begins to be used to revector the legacy systems in addition to identifying gaps to be filled by new systems.

So, we argue, the essence of the rationale for an SoS is legacy assessment, state-space solutioning, problem demystification, and integration framework. These components, in the spirit of systems thinking, need to be handled interdependently knowing that small changes in any one may hugely impact the other three, which may easily lead to gross instability in the evolution of the SoS itself. In this context, we want to present a methodology that can allow for these SoS thinking constructs to be contextualized that we may better understand the SoS's problem space before we attempt to understand the solution space.

11.3 BSSM: AN SoS THINKING TOOL

In the early practice of Systems Engineering, the direct application to the broader conditions of management was not well understood. Peter Checkland was one of the forefathers for understanding and defining a combination of systems thinking with real-world practice to build a bridge between what he would define as "hard" (e.g., engineered) and "soft" (e.g., management) systems. Checkland defined this bridge through SSM (Checkland, 1999, 2000a, b; Checkland and Scholes, 1999). One of Checkland's notable contributions from SSM is that the first step is to understand the unstructured problem or situation. It is unstructured because the problem or situation has not been clearly defined or bounded. In Systems Engineering, the step of defining the problem is too often already started with a structured problem. This becomes the same starting point for BSSM. Likewise, BSSM follows seven steps that can be viewed as an iterative process for defining an ill-defined problem (or system of interest). The seven steps of BSSM are depicted in Figure 11.1. While Checkland describes the concept of rich pictures to model systems thinking, BSSM creates systemigrams. In the following sections, we will describe the seven steps of BSSM both in text and with various scene of a systemigram of BSSM (Figure 11.2). We will then follow this section with four case examples where systemigrams have been used to better understand complex systems.

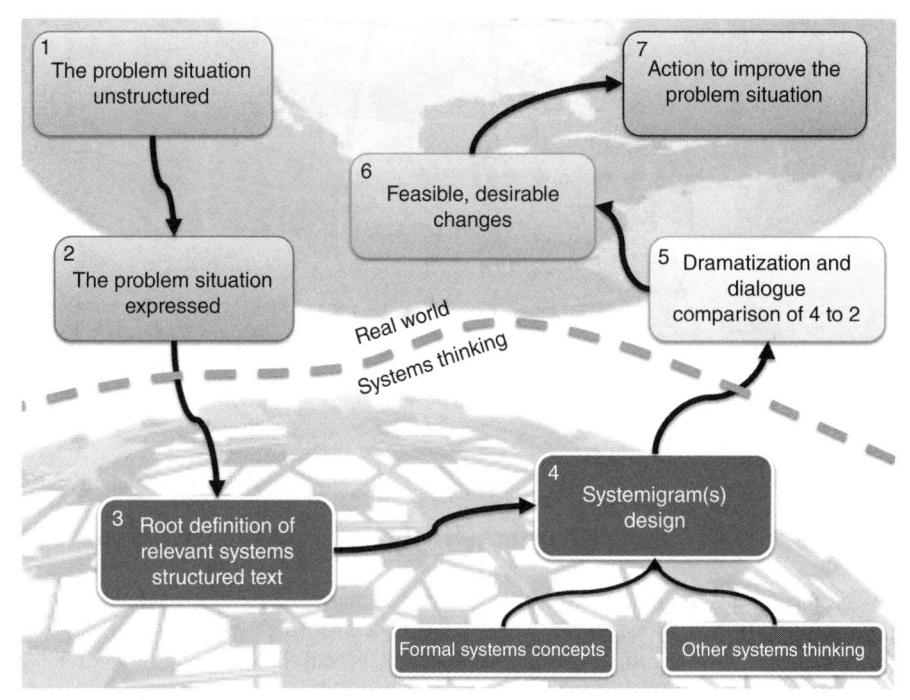

Systemigram Modeling for Contextualizing Complexity in System of Systems 277

Figure 11.1 Boardman soft systems methodology (Boardman and Sauser, 2008).

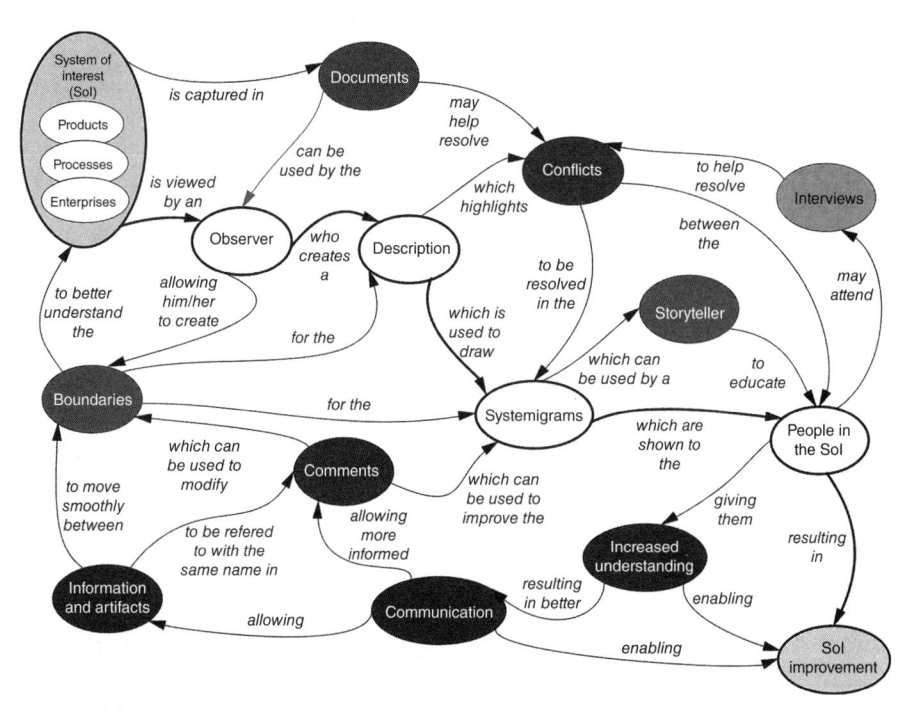

Figure 11.2 Systemigram of the Boardman soft systems methodology.

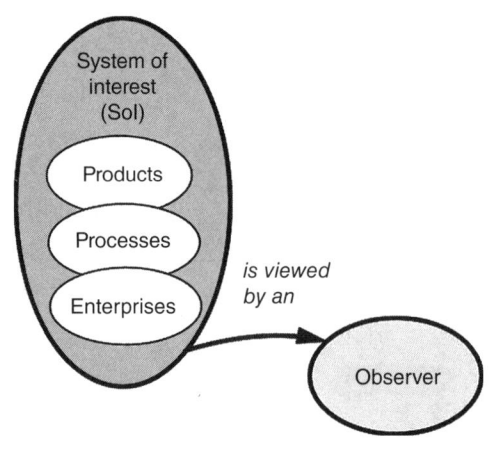

Figure 11.3 The problem situation observed.

11.3.1 Step 1: The Problem Situation—Unstructured

The problem situation is first expressed textually (e.g., presentations, reports, meeting notes, publications), verbally (e.g., interviews, observations of oral discussions), and graphically (e.g., presentations, pictures), as it is by the stakeholders (Figure 11.3) without considering a resolution or solution to the problem. This step may seem at first reverse of traditional approaches that often start with some end in mind. The challenge is that this step can be based on many presumptions though every attempt should be made not to extrapolate about the nature of the situation without:

- Confirmation bias—seeking information that confirms early beliefs and ideas
- Ease of recall bias—relying too much on information that is easy to recall from memory
- Anchoring bias—emphasizing too much, the first piece of information encountered (Hitt et al., 2005)

At this point, the objective is to understand the underlying structure of a phenomenon.

11.3.2 Step 2: The Problem Situation—Expressed

A description of the situation within which the problem occurs is formulated (Figure 11.4). Both the logic and the culture of the situation are taken into account at this point. The key to a good problem definition is ensuring that you deal with the real problem—not its symptoms. At this stage, it is also important to ensure that you look at the issue from a variety of perspectives. If you commit yourself too early, you can end up with a problem statement that's really a solution. A group of stakeholders can overemphasize information held by a majority, failing to be mindful of information held by one or a few stakeholders (Hitt et al., 2005). A good practice to use at this point is triangulation, which is largely used in case study research (Yin, 2008). Triangulation uses multiple sources of data collection and information to provide stronger substantiation of constructs.

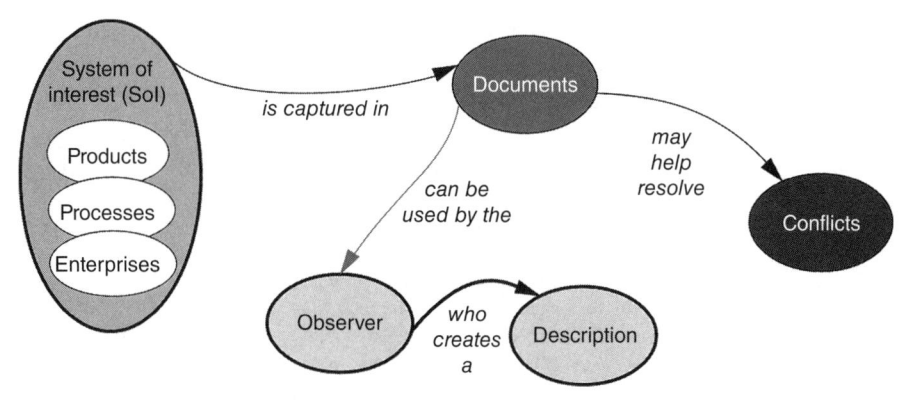

Systemigram Modeling for Contextualizing Complexity in System of Systems 279

Figure 11.4 The problem situation expressed.

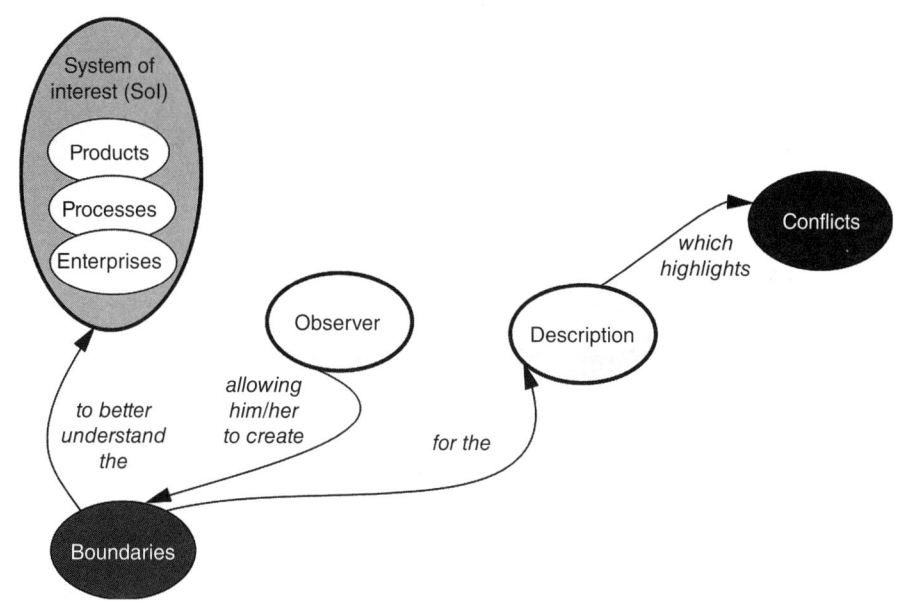

Figure 11.5 Creating the system description.

11.3.3 Step 3: Structured Text

The problem situation or system of interest is conceptualized in structured text (Figure 11.5). The structured text identifies the key elements with attention to systems thinking modeling and analysis requirements, that is, systemigrams. This structured text must be excessively "intelligent." It must be about strategic intent, not procedural tactics. The text is not a checklist but rather a well-crafted piece that searches out the minds of its readers and stretches the mind of the author. It need not be lengthy, for example, 2000 words is sufficient; however, it can be a large document, but this would need to become an executive summary if the scope of the to-be-created systemigram were meant to address the scope of the document.

Below is a sample of a structured text that was created for the systemigram presented in case Example 11.4.1 (The Network-Enabled Challenge). This was first published in Blair et al. (2007) and is being reprinted with permission from Wiley and Sons and the *Systems Engineering* journal:

> *UK Defense policy directs the development of future UK Operational Concepts for Land, Air, Space, Maritime, and Logistics. These operational concepts are guided by the UK Joint Vision, Joint High-Level Operational Concept, Effects-Based Operations Concept, and the Defense White Paper. The operational concepts are informed by emerging concepts such as NEC, to achieve the overall UK Defense Aim: "to deliver security for the people of the United Kingdom and the Overseas Territories by defending them, including against terrorism; and to act as a force for good by strengthening international peace and stability." NEC enables a flexible acquisition strategy to establish coherent acquisition programs, these programs adopting an incremental approach to realize rapid technology insertion to achieve a net-ready force which exploits a network infrastructure to enable shared awareness. Shared awareness underpins flexible working to deliver synchronized effects that address the dynamic mission which is undertaking the defined UK military tasks to achieve military and non-military effects. NEC requires an information infrastructure to provide secure and assured information access to support the network infrastructure and facilitate shared awareness. NEC also employs Effects Based Planning across Government which requires a dynamic planning system supported by distributed tools and models to manage Agile Mission Groups, thus enabling Flexible Working. NEC improves equipment integration of weapon systems, intelligence, surveillance, target acquisition and reconnaissance systems, and command and control nodes to facilitate the Agile Mission Groups. NEC also enables Networked Support across public and industry to sustain Agile Mission Groups used to enable Flexible Working.*

11.3.4 Step 4: Systemigram Design

A systemigram model is created as designed from the structured text to capture and represent the essence of the original conceptual thinking (Figure 11.6). Systemigrams

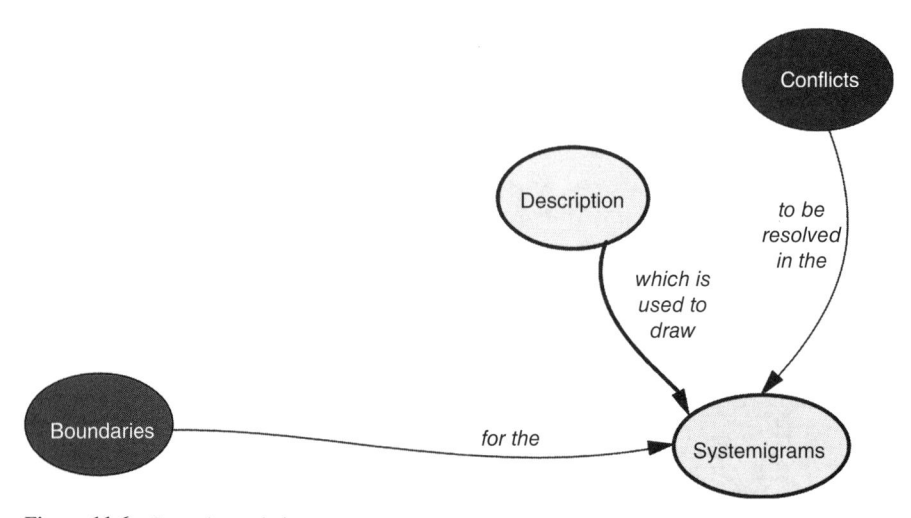

Figure 11.6 Systemigram design.

were developed by John Boardman in the late 1980s (Boardman, 1990a) and originated from a Pan-European IT project. In this instance, the systemigram was employed to communicate business strategy to a project team using written prose found in a project notebook whose sole author had left the company. The evolution of systemigrams may be considered in three phases: its development as a form of visual language, its adaptation as a methodology for business architecting, and its refinement as a learning system. Communicating strategic intent, especially when it is intelligently written, can neither be entrusted to writings alone nor to the presentations of the author. Additional support is needed, support that is faithful to the statements expressing the strategic intent but value adding in ways the author points to in the written prose. As a system, a systemigram exhibits parts, wholes, emergence, boundary, flows, inputs, outputs, transformations, process, and networks—those elements necessary to systems thinking and systems practice (Blair et al., 2007).

Systemigrams are an effective choice to provide the ability to understand motivations, viewpoints, and interactions and address qualitative dimensions of problem situations. A systemigram becomes a picture for understanding and identifying the significant elements within a system of interest. The systemigram characterizes the problem along with the complex interrelationships and diverse expressions of stakeholder concerns and needs. A systemigram structures the complex interrelationships into a graphical presentation and encapsulates the problem into a coherent problem description (Boardman and Sauser, 2013). Systemigrams, like any model, help "to reason about the problem, understand the complexities, and to communicate with others" (Cloutier, 2011a). Systemigrams allow one to bridge the problem representation to an actionable product. They assist in finding alternatives to the problem and further advance the efforts toward a solution. Initially, the systemigram is an informal capturing of the perspective from the end user's viewpoint. Later, these informal systemigram representations transition to a more formal approach and take on the viewpoint of the stakeholder with traceability to the systemigram. Systemigrams enable the exploration of diversity in perspectives while maintaining a single objective. The pragmatic merit of systemigrams is established by virtue of its successful employment in a variety of interventions, for example, Table 11.1.

A systemigram is a network, having nodes and links, flow, inputs and outputs, and beginning and end. Key concepts, noun phrases specifying people, organizations, groups, artifacts, and conditions are nodes. The relationships between these nodes are verb phrases (occasionally prepositional phrases) indicating transformation, belonging, and being. Some nodes can contain other nodes, for example, to indicate breakout of a document or an organizational, product, process, or structure. The network must be legible so this limits the number of nodes and links. There should be no crossover of links, improving clarity. The primary sentence (mainstay), which supports the purpose of the system, will read from top left to bottom right of the systemigram. The final node of the mainstay, in the bottom right, should be the system's goal or objective. The other segments of the systemigram flow out of and back into this mainstay, connecting as needed with its landmark noun phrase nodes. As the systemigram is developed, it should capture the systems transformations that have a structure and a process.

Creation of a systemigram model as designed from the structured text captures and represents the essence of the original conceptual thinking. To accomplish this, systemigram

Table 11.1 Sample Systemigram Usage

Author	Description
Boardman (1990a)	Project management controller
Boardman (1990b)	Analysis of large-scale project planning problem
Boardman (1994)	Design of a process that has the ability of couple systems engineering functionality with project management processes
Cole et al. (1995)	Development of project life cycle, project phase, project subphase, and project task
Sherman et al. (1996)	Reinforcement of a program of cultural change in a medium-sized systems engineering business
Ramsay et al. (1996)	Creation of an organization learning platform through organizational knowledge creation and effective distribution
Clegg and Boardman (1996)	Process integration and improvement
Bulbeck and Clegg (1996)	Analysis and simulation of manufacturing processes
Bulbeck et al. (1997)	Modeling business process simulation
Clegg and Boardman (1997a)	Material resource planning modeling
Clegg and Boardman (1997b)	Development of synergy between stakeholders in a process
Sagoo and Boardman (1998)	Analytical modeling in conjunction with Petri net
Blair et al. (2007)	Modeling of U.K. Ministry of Defence network-enabled capability concept for communication with stakeholders
Baldwin (2008)	Used in conjunction with other systems thinking tools for identification, elaboration, and validation of paradox existence
Mansouri et al. (2009)	Study critical of properties of the Maritime Transportation System of Systems through systemic interrelationships
Meentemeyer et al. (2009)	System of systems characterization analysis
Frittman and Edson (2009)	Holistic systems analysis of written ConOps document
Bayuk (2010)	Analysis of systems security
Mehler et al. (2010)	Creation of composite definitions of systems, for assessing system characteristics and for informing collective decisions
Squires et al. (2010)	Definition of body of knowledge and curriculum to advance systems engineering
Mussante et al. (2010)	Tool for system comparison in the disaster management domain
Randall et al. (2011)	Systems perspective to explaining the effectiveness of performance-based logistics
Sauser et al. (2011a)	Identification of organizational and communication bottlenecks, understanding of architectural structure
Sauser et al. (2011b)	Formulation of defining resilience in maritime homeland security
Bayuk et al. (2012)	Architects' current organizational cybersecurity policy issues on a global scale
Polacek et al. (2012)	Present a principle orientation as compared to rules orientation of a complex systems
Bayuk and Mostashari (2013)	Survey practice of security metrics from both a technical and historical perspective
Salado and Nilchiani (2013)	Propose a contextual- and behavioral-centric approach for stakeholder identification

Table 11.2 Systemigram Guiding Principles

Principle	Systemigram guidance
Correctness	Mainstay that supports the purpose of the system reads from top left to bottom right
	Ideally, there should be 15–25 nodes
	Nodes must contain noun phrases
	Links should contain verb phrases (to reduce trivial links)
	No repetition of nodes
	No crossover of links
Relevancy	Remember that the model is really "theirs"
	Remember that the model is not really "theirs"
	Remember that the model is not reality
Comparability	It should compare to reality and the original system description
Clarity	It should read well
	Beautification (e.g., shading and dashing of links and nodes) should help the reader read the sentences in the diagram
	Exploit topology to depict why, how, what (who, when, and where it is built into system description)
Systematic design	Is it a system in its own right?
	Does every node (except for the beginning and ending nodes) have an input and an output?
	Can you follow any node to the end node?

design has seven simple rules. Despite the simplicity of the rules, they often create challenges in systemigram design. The metapurpose of these rules is to make sure the systemigram maintains parsimony. Systemigrams do not remove the complexity from systems, but they can make complex systems understandable. The rest of this section will articulate the seven rules while continuing to use the systemigram of the BSSM to demonstrate and explain the rules:

Rule [1] The primary sentence (*mainstay*) that supports the purpose of the system will read from top left to bottom right (Figure 11.7):
  a. This is the anchor for the entire visualization.
  b. Until one understands the picture as a whole, there is always the chance that confusion, not clarity, will arise.
  c. The other segments of the systemigram flow out of and back into this mainstay, connecting as needed with its landmark noun phrase nodes.

Rule [2] Ideally, there should be 15–25 nodes (less can make for a trivial systems description, and more can create clutter and illegibility):
  a. For a node to be linked to another node, n/2 links are necessary for an even number of nodes and (n + 1)/2 links for an odd number.
  b. For all nodes to be linked, n − 1 links are necessary.

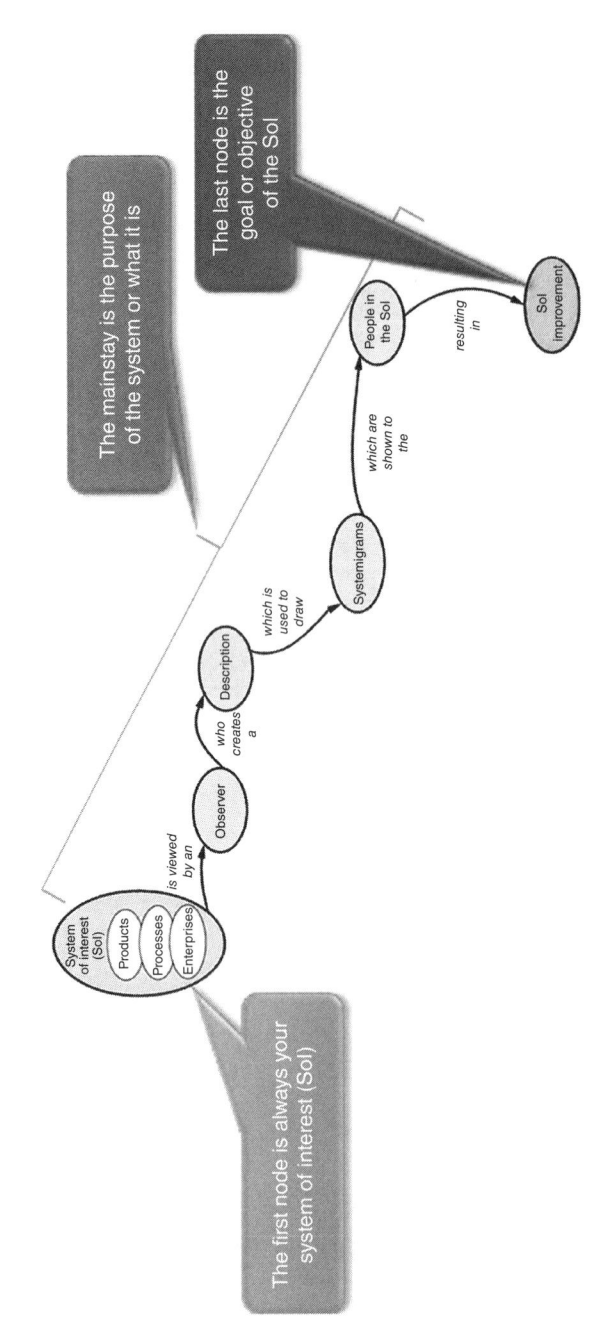

Figure 11.7 The mainstay.

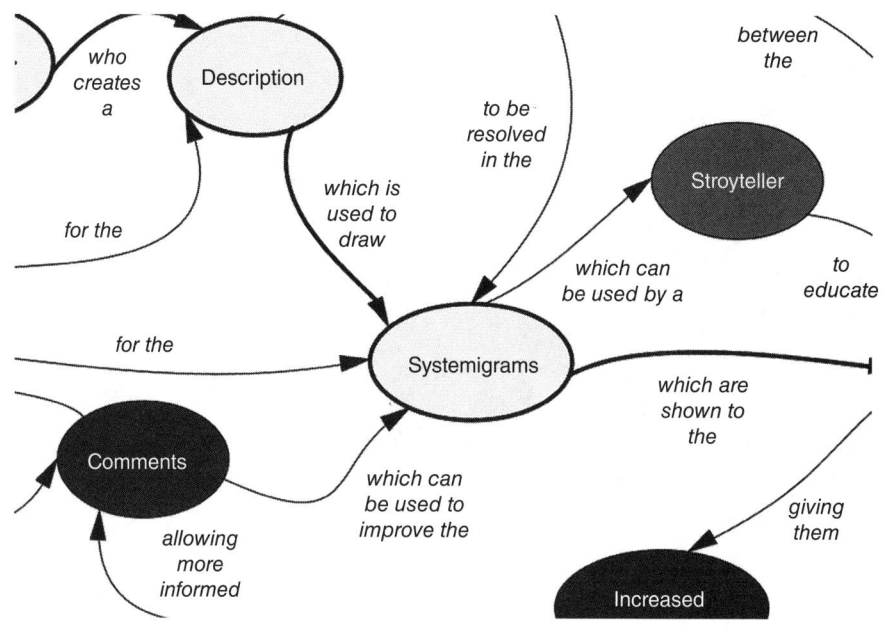

Figure 11.8 Node–link relationships.

  c. For all nodes to be linked to every other nodes, n(n − 1)/2 links are necessary for a nonplanar graph.
  d. With the number of nodes, it is possible to find the number of possible combinations: $2^{n(n-1)/2}$.

Rule [3] Nodes must contain noun phrases (people, organizations, groups, artifacts, and conditions), and links should contain a verb or verb phrase (transformation, belonging, and being) (Figure 11.8):
  a. A system has transformations that have a structure and a process.
  b. The structure is an arrangement of parts and relationships.
  c. Parts will tend to be nouns or noun phrases, while relationships are best expressed by verbs or actions.

Rule [4] No repetition of nodes. Redundant nodes lose the essence of relationships (Figure 11.9).

Rule [5] No crossover of links (Figure 11.10):
  a. A design rule that not only makes the systemigram cleaner and clearer to view but also leads to the observance of an important heuristic in systems design.
  b. For a systemigram of 20 nodes, the total number of possible links is 190, whereas the actual number will be about 30.
  c. This ratio is about 15%, which is held to be the optimal ratio of interfaces in a system relative to how many there could be (Rechtin, 1990).

Rule [6] Beautification (e.g., shading and dashing of links and nodes) should help the reader read the sentences in the diagram.

Rule [7] Exploit topology to depict why, how, and what.

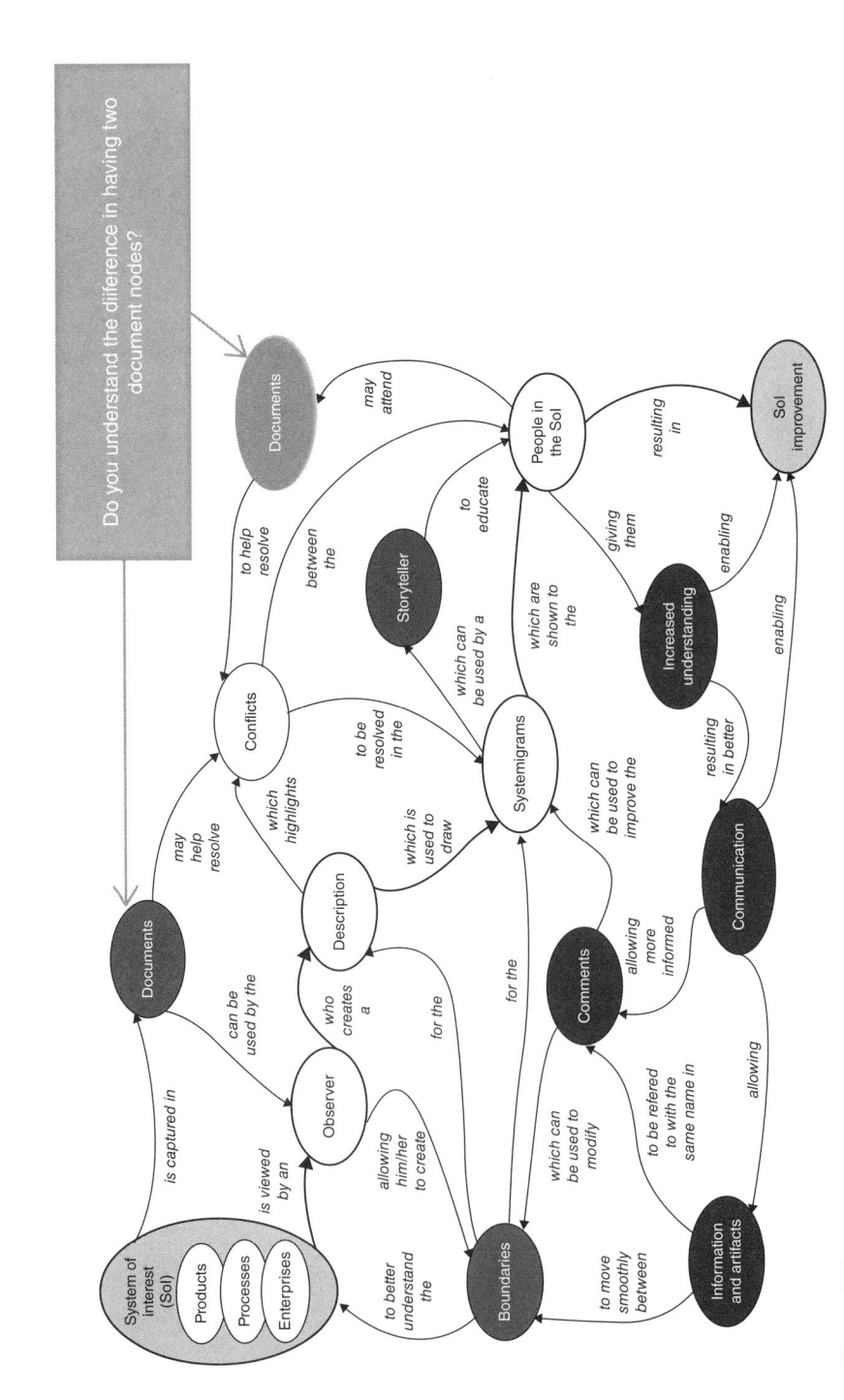

Figure 11.9 No repetition of nodes.

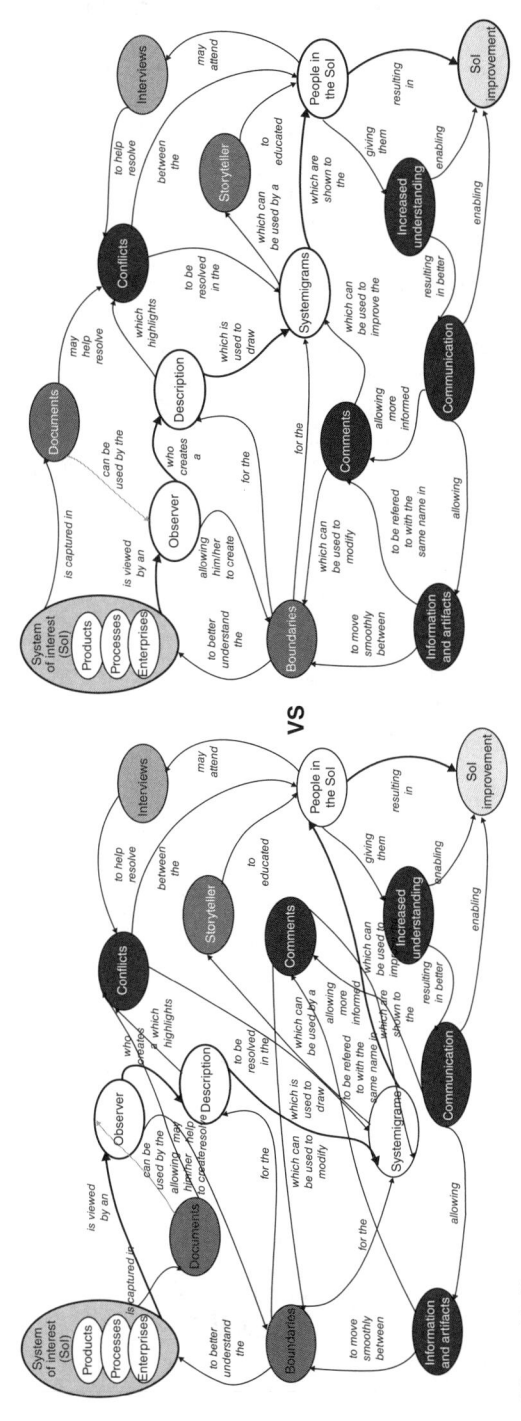

Figure 11.10 No crossing of links.

287

11.3.5 Steps 5: Dramatization and Dialogue

Key to systemigram modeling is the ability of the modeler to decompose the model into scenes that can articulate a story of what the systemigram represents (storyboarding) (Figure 11.11). A completed systemigram then is not the end of the story, but the basis for telling a story. The story can be told in a variety of ways, but all have the same generic format—to create a storyboard using carefully selected scenes, which are subnets of the systemigram. A subnet is a child systemigram that further elaborates some detail(s) of the parent systemigram. This is a very important step for verifying the systemigram with respects to its ability to capture the multiple views of the stakeholders. Each scene represents a key part of the message, but by the same token, it begins to tell a more detailed message, which can only be amplified by having the right people listen to the systemigram story. The systemigram scenes are then dramatized via storyboarding to key stakeholders. This is done so that the model and reality can be compared and contrasted. The differences become the basis for discussion: how things work and might work and what the implications are. It is important at this point to engage multiple stakeholders. Multiple stakeholder engagement can:

- Accumulate more knowledge and facts and thus generate more and better alternatives
- Display superior judgment when evaluating alternatives, especially for complex problems

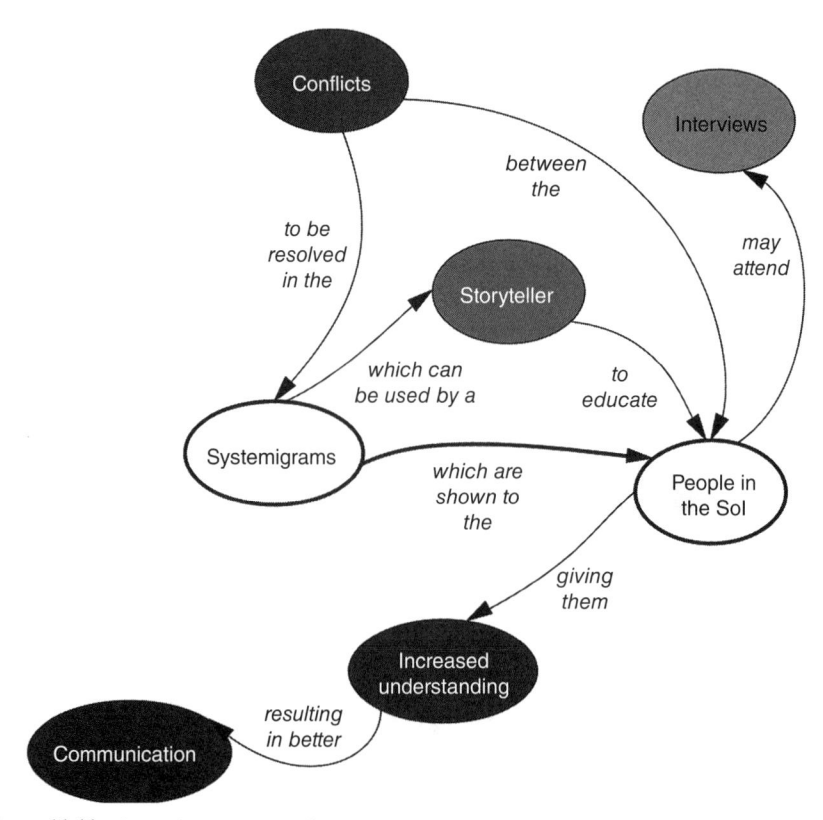

Figure 11.11 Systemigram storyboarding.

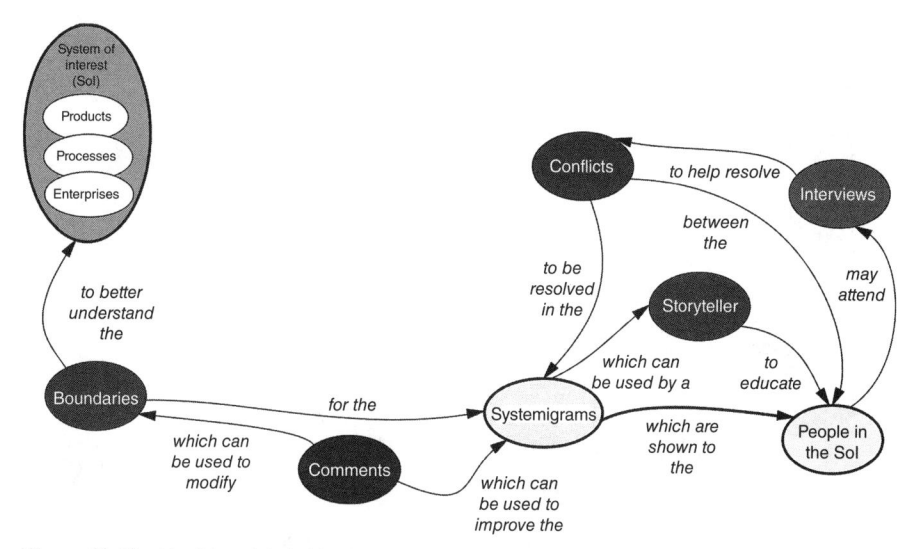

Figure 11.12 Feasible and desirable changes.

- Lead to a higher level of acceptance of the decisions and satisfaction
- Result in growth for members of the group (Hitt et al., 2005)

Rules for Storytelling

Rule [1] Ensure key sentences are recovered in specific scenes (*fidelity*).

Rule [2] Use background to link previous sentence structures where this helps current scenes (*emphasis*).

Rule [3] Attempt a storyboard effect so that each scene leads naturally to successors thereby conveying added meaning to the "drama" (*insight*).

Rule [4] Look for new aspects that the whole might suggest that would not be obvious from prose (*value added*).

11.3.6 Step 6: Feasible and Desirable Changes

At this step, the identification of feasible and desirable changes is deciphered from the previous step, understanding that they are likely to vary (Figure 11.12). There is a question for what is desirable, "Is the change (technically) an improvement to the problem (system)?" and a question for what is a feasible, "Does the change fit the culture?"

11.3.7 Step 7: Action to Improve the Problem Situation

Every individual or collective input that is deemed desirable or feasible is incorporated into the systemigram (Figure 11.13). Only changes that have an answer of "no" to one or both of the two questions presented in step 6 are dismissed.

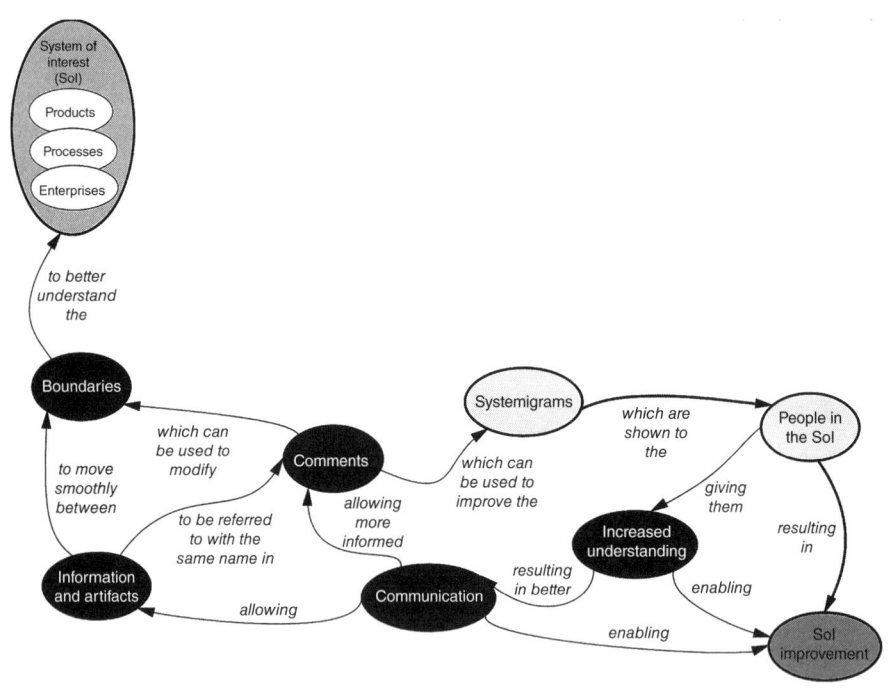

Figure 11.13 Action to improve the problem situation.

Steps 1–7 are then repeated until a successful outcome of a BSSM is achieved. Success is defined as:

1. The people concerned, that is, stakeholders, feel that the problem has been solved
2. The problem situation has been improved
3. Insights have been gained

11.4 CASE EXAMPLES

11.4.1 The Network-Enabled Capability

The following case was published in Blair et al. (2007). This excerpt is reprinted with permission from Wiley and Sons and the *Systems Engineering* journal.

Following the attacks on September 11, 2001, the United Kingdom revisited the standing Strategic Defence Review (SDR) of 1998 and published the SDR New Chapter (SDR-NC) in 2002 [UK MoD, 2003b] to address the changing security environment brought on by the global war on terror. The Defence White Paper, "Delivering Security in a Changing World" [UK MoD, 2003a], was built on the SDR-NC by laying out the policies for NEC, realizing that operations against international terrorism require increased precision and rapid delivery of military effect and capabilities achieved only through a networked force employing shared situational awareness. A number of policy and guidance initiatives have spun off from the SDR-NC, all feeding the future U.K. operational concepts for space, air, land, maritime, and logistics, setting the operational context for NEC. These various operational concepts must adopt NEC as an enabling capability crossing all Defence Lines of Development (training, equipment, people, infrastructure, doctrine, organization, information, and logistics) to remain coherent with the overall U.K. Defence Strategy. This is no small order indeed and represents an unstructured view of the problem situation. Figure 11.14 is the systemigram that was produced from the structured text presented in Section 11.3.3 (Step 3: Structured Text).

11.4.2 U.S. Army Small Combat Units

The following case is based upon work supported, in whole or in part, by the Systems Engineering Research Center (SERC). SERC is a federally funded University Affiliated Research Center (UARC) managed by Stevens Institute of Technology. Any opinions, findings, and conclusions or recommendations expressed about this material are those of the authors and do not necessarily reflect the view of the SERC, Stevens Institute of Technology, University of North Texas, NSRDEC, and/or any agency or entity of the U.S. government. A more detailed description of this work can be found in Sauser et al. (2012).

The Department of Defence (DoD) is vigorously pursuing greater efficiency and productivity in defence spending to provide the armed forces with superior capabilities in an environment of flat defence budgets. Toward that end, the Office of the Secretary of Defence (OSD) has issued new acquisition guidance that places increased emphasis on early life cycle Systems Engineering to balance operational performance with affordability (Public Law 111–123). However, both contingency basing (CB) and soldier load, when realized as an SoS or enterprise, have extremely complex and uncertain missions with deployments into extremely hostile environments from combat zones to human relief efforts. The technology mapping dynamic and the number of requirement constraints needed to capture prior uncertainties are unmanageably large. Modern design theory suggests systems design is better served by methodologies that focus on constructing objective functions with penalties that capture value and uncertainty, as opposed to attempting to capture the unmanageable large number of requirements—constraints. Consistent with the U.S. Army Research, Development, and Engineering Command's (RDECOM) vision and mission to be the Army's primary source for integrated research, development, and engineering capabilities to empower, unburden, and protect the warfighter, there is a need for the creation of an early collaborative and systems modeling methodology to express and characterize soldier load and CB at the patrol base, combat outpost, and SCU (company minus) level.

SCU (consisting of a company [300 soldiers] and below) currently establish nonstandardized base camps for contingency operations (CB), potentially limiting their ability to project and efficiently employ full-spectrum operations [the Army defines full-spectrum operations as the combination of offensive, defensive, and either stability operations overseas or civil support operations on U.S. soil (DoD, 2010)].

The SCU may not be able to support the modern full-spectrum battlefield demands unless CB capabilities specific to the SCU are combined as a single, integrated, agile, force projection platform. A contingency base should provide soldiers with an effective, logistically supportable, affordable, and rapidly deployable environment to project force across the full-spectrum of operations. The user, trainers, RDECOM, various Program Executive Offices (PEO), logistics, and other stakeholders working together must develop

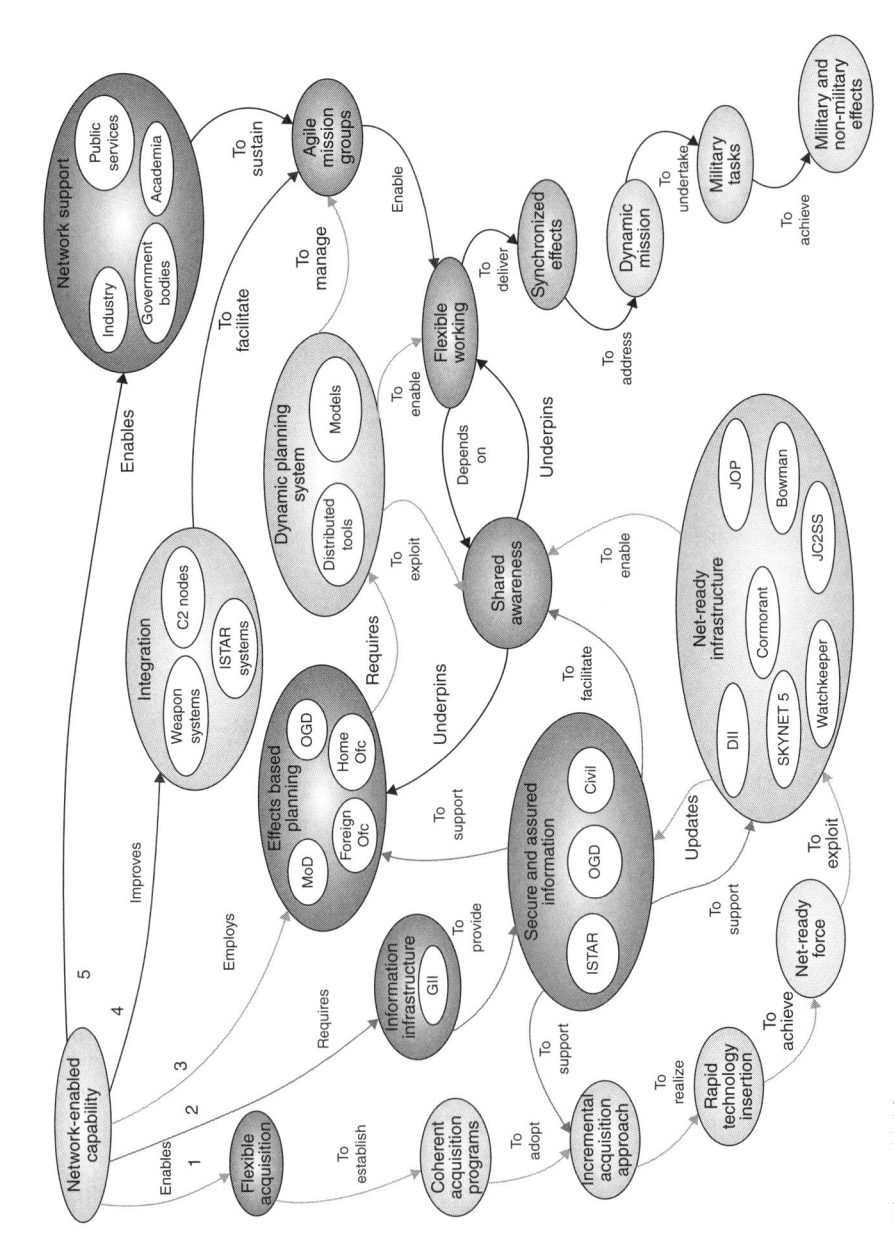

Figure 11.14 The network-enabled capability.

a CB capability that accomplish planning to disposal of the base and how soldier will employ. The development planning process should enable an Army enterprise approach to deliver to the tactical edge with total systems integration aligned to Army modernization strategies and Army Force Generation (ARFORGEN).

This case study focused on specific aspects of SCU-CB, as a force projection platform and a potential means to address the interrelated individual soldier and SCU load (cognitive and physical). The problem is being able to develop a set of interrelated processes, mechanisms, and tools to capture, explain, and manage the complex operational and systems interaction posed by the dynamic nature of the SCU operations, along with a means to measure progress. The SCU exhibits a complex, pluralistic set of requirements across a number of factors ill suited to standard Systems Engineering practices. Novel means to optimize SCU-CB need to be considered.

The systemigram created in Figure 11.15 was the result of a series of joint workshop between systemigram modelers and subject matter experts at the U.S. Army RDECOM—Natick Soldier Center RD&E Center (NSRDEC). From these workshops, a top-level systemigram was constructed and decomposed into several lower-level systemigrams to assist the U.S. Army RDECOM in defining the problem.

11.4.3 Systems Biology

The following case is an excerpt from *A Systems Biology Approach for Medical Understanding: An Exercise in Systems Thinking* [Christopher Oster (2011). Stevens Institute of Technology: Hoboken, NJ].

Within the medical research community, the concept of Systems Biology does not remove the need for detailed investigation of the genome, protein interactions, molecular biology, and the like (referred to as frequently in that community as "omics"—i.e., genomics (detailed study of genes), proteomics (detailed study of proteins), etc.). Instead, they are intended to help make sense of the data and add a focus on the interplay between components and subsystems within the organism that cannot possibly be understood at the genome or protein level. As is illustrated in Figure 11.16, this detailed analysis and experimentation forms a basis on which Systems Biology can stand. As such, the interplay between traditional medical researchers and those trained in reductionist approaches is a crucial part of the Systems Biology puzzle—without the physical experimentation being performed by these teams, Systems Biologists will not have sufficient data from which to construct valid models. Furthermore, without application of a systems-level approach to medical research, many diseases currently afflicting the human population (diabetes, cancer, etc.) will be difficult to prevent and treat as they typically are not isolated to a reducible component.

In some sense then, the boundaries of Systems Biology can include the data providers (i.e., "omics"), traditional researchers, drug researchers, as well as the Systems Biologists themselves. Systems Biology is a study of relationships, interconnections, and cause and effect. The medical industry is facing increasingly hard problems where reductionist approaches alone are not cost effective or efficient and a new holistic methodology can provide a mechanism for considering the whole of an organism rather than simply its list of parts. Just as engineers working on increasingly complex problems began applying Systems Engineering methods to deal with complexity, biologists and medical researchers too need to find new ways of handing complexity.

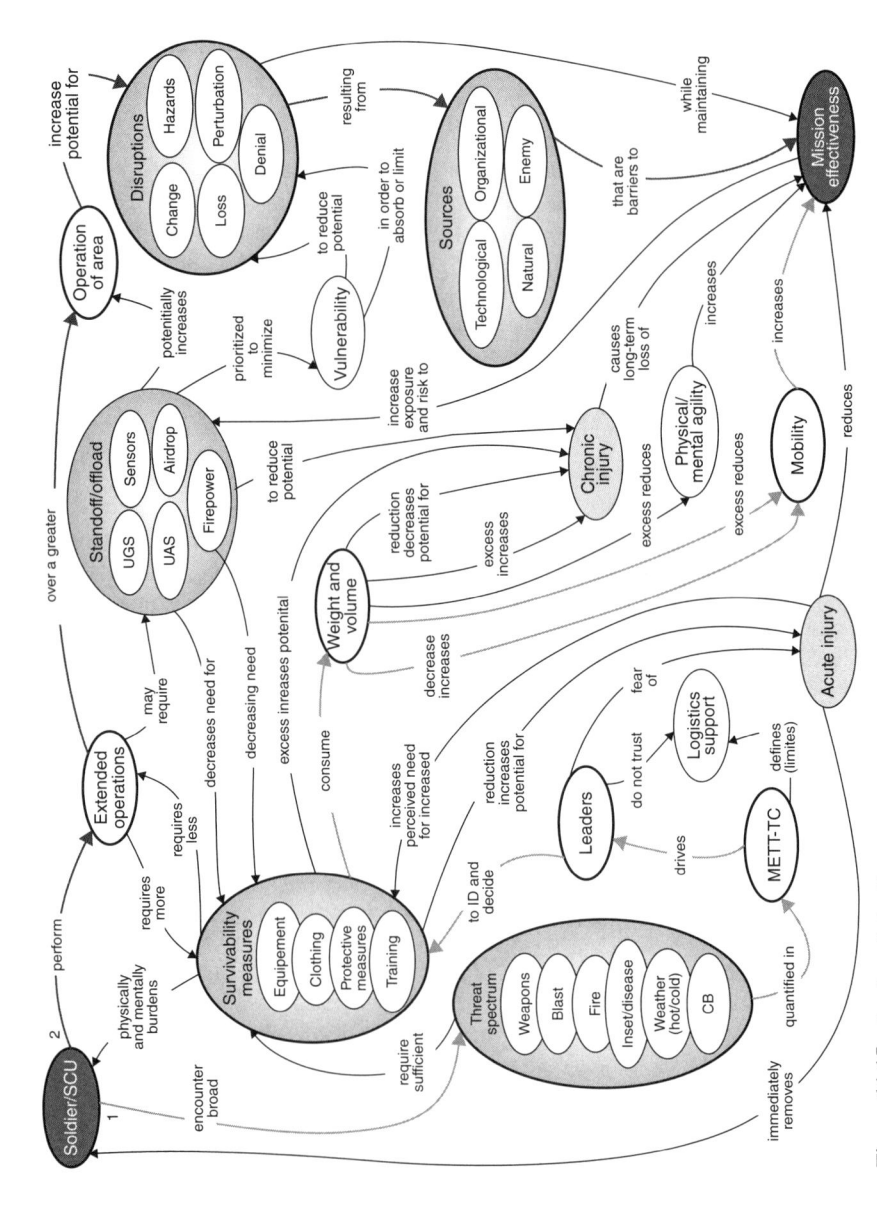

Figure 11.15 Small combat unit.

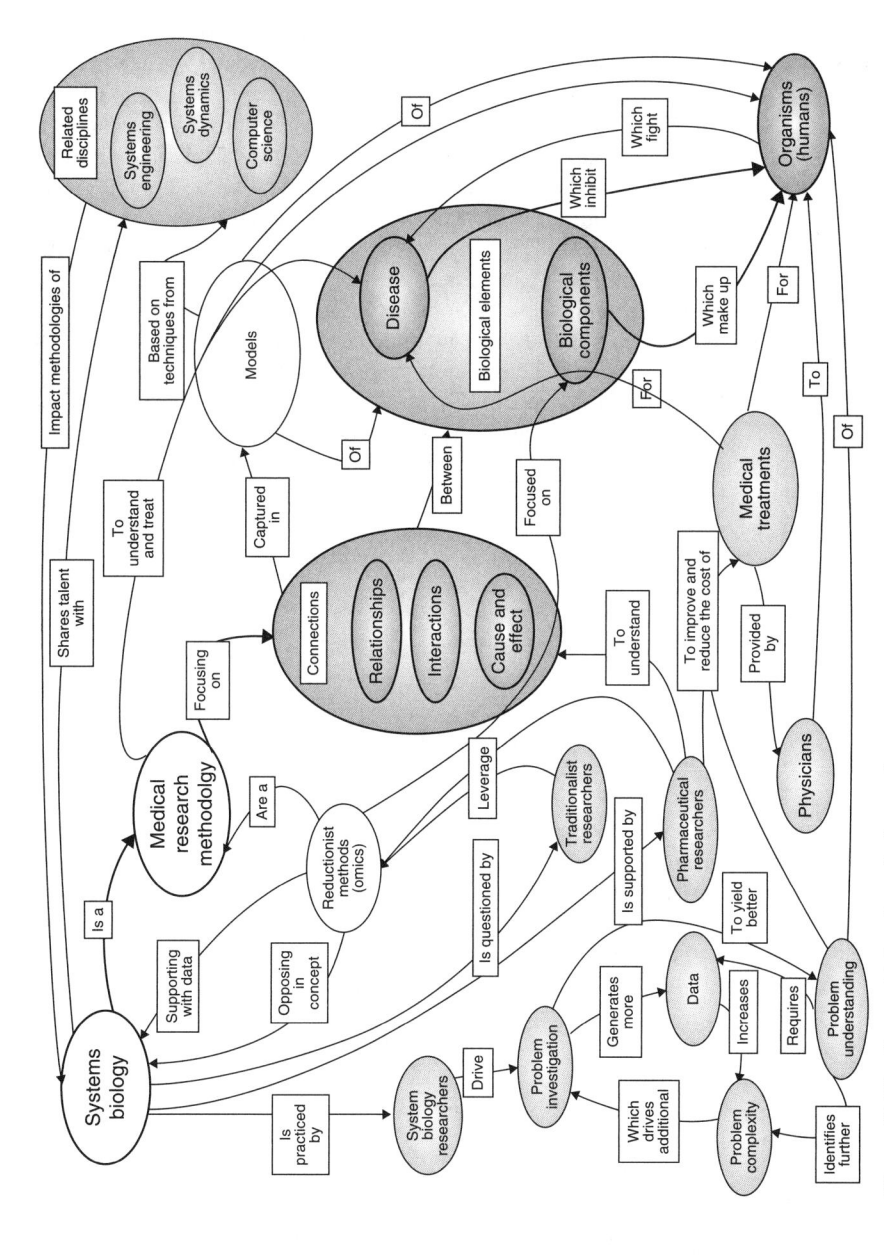

Figure 11.16 Systems biology in medical research.

Furthermore, this challenge with increasing complexity is not just limited to the biology and medical fields. Similar trends can be found in software development, engineering, and even organizational planning. With an increased ability to store and process data, we've discovered complexity underlying each of our hard societal problems and require systems-level methods for dealing with them. Systems Biology professionals must look beyond the bounds of biology into engineering and computer science to identify methods, tools, and techniques to apply systems-level concepts to the problem at hand, and similarly, engineers and computer scientists should look toward biology. Technology has mirrored biological understanding, and biological analogies have historically mirrored popular technology of the day. The interconnections between engineering and biology have existed for some time, but Systems Biology brings to the forefront the requirement for cross-pollination of ideas. Systems Biology cannot succeed without infusion of talent from other disciplines, and similarly, other disciplines such as Systems Engineering will only mature with a diversity of perspectives as well.

The field of Systems Biology itself is in some sense an application of systems thinking techniques to biology and medicine; however, this case demonstrates a systems thinking approach back onto the concept itself and its position in the established medical and scientific body. The system of interest is captured in a systemigram (Figure 11.16) resulting from an assessment of Systems Biology in medical research.

11.4.4 Alignment of Test and Evaluation with Program Management

The following case was published in Eigbe et al. (2010). This excerpt is reprinted with permission from Wiley and Sons and the *Systems Engineering* journal.

The FAA WJHTC is developing a strategy to achieve world-class systems integration and test capability that supports effective management of test and evaluation (T&E) programs to meet sponsors' expectations and achieve customer satisfaction. The strategy must address the unification of T&E technical and program management processes to facilitate effective T&E management required to achieve program integrity, accountability, and the deployment of quality air traffic control (ATC) systems that make up the National Airspace System (NAS). The NAS is used to provide safe and efficient air transportation services to the aviation industry and the flying public. Test program sponsors' expectations are characterized by performance metrics, such as on-time systems delivery, on budget, and meeting the operational requirements (quality). The T&E strategy must address budget overruns and late systems deployments by programs that do not meet customer satisfaction. It must minimize critical defects found in deployed systems that cause rework leading to cost and schedule variances. Critical defects are caused by inadequate test and evaluation of systems as a consequence of the overall program management constraints.

The strategy must address the need for a unified supportive infrastructure, which comprises a TSB and a Program Management Excellence Center (PMEC). To be credible, the supportive infrastructure must, at a minimum, be staffed with qualified test engineers and program management experts. The TSB will develop and institutionalize technical T&E processes, while the PMEC will coordinate, develop, and institutionalize program management processes (foundational) to facilitate the technical processes. The PMEC must address training, quality management, configuration management, and project management support services that are essential for consistent and predictable program performance metrics to achieve customer satisfaction. The FAA systems integration and test organization must ensure that the supportive infrastructure collaboratively provides the necessary services to accomplish the strategic objective of achieving customer satisfaction.

Taking a holistic view of both T&E and program management activities, using the tools of systems thinking can provide new insights to understand and create a more effective T&E execution strategy. Using the Boardman's Soft Systems Methodology, Figure 11.17 is a systemigram that was the results of evaluating the unification of a FAA's T&E program management into its T&E strategy.

11.5 CONCLUSIONS

Systems Engineering has predominately focused on technical solutions for bounding a system's scope. With increased complexity in systems, the inappropriate bounding of a systems problem becomes a notable source of failure (Mitroff, 1998). Bounding the problem space is the first step in any problem solution-seeking process. Therefore, our tools for bounding or scoping the problem for SoS should be based on core systems (thinking) principles. The BSSM and systemigrams are based on a complete respect for systems thinking and systems theory. In this chapter, we presented our context of SoS thinking and a tool to move thinking to problem definition and ultimately lead to problem resolution.

If requirements are the foundation that systems are built upon, then prose is the corollary foundation for systemigrams. Prose then deserves to find graphical expression and, in that graphical expression, inspire further detailed grammatical exhibition leading to more detailed graphical description. The existence of systemigrams as a value-adding proposition is to reveal the inner meanings of strategic intent and help build a greater shared understanding in a growing community of people. The two should go hand in hand—excellent prose and great graphics—together supporting the translation of strategy into tactics. The progress of systemigrams, over almost 30 years of development, has followed an evolutionary process, involving several PhD students, faculty, and industry champions. This can be summarized by three distinct phases in the evolutionary process:

1. Concentration on graphical portrayal of structured prose
2. Development of methodologies that use systemigrams for enterprise architecting purposes, for example, extended enterprises or business process architectures
3. Development of systemigram technique (i.e., BSSM) for drilling down from architectural vantage points into detailed consideration of solution implementation

The top-level requirements that guide the systemigram construction process are:

- To faithfully interpret the original structured text as a diagram in such a way that with little or no tuition, the original author, at the very least, would be able to perceive his/her writings and additionally their meanings.
- To create a diagram that is a system, or could at least be considered a system *in its own right*—thus, if the original structured text could be considered a system, then its faithful interpretation as a new object should also be systemic but with features not possible with prose alone, but quite amenable as a graphic (or picture).

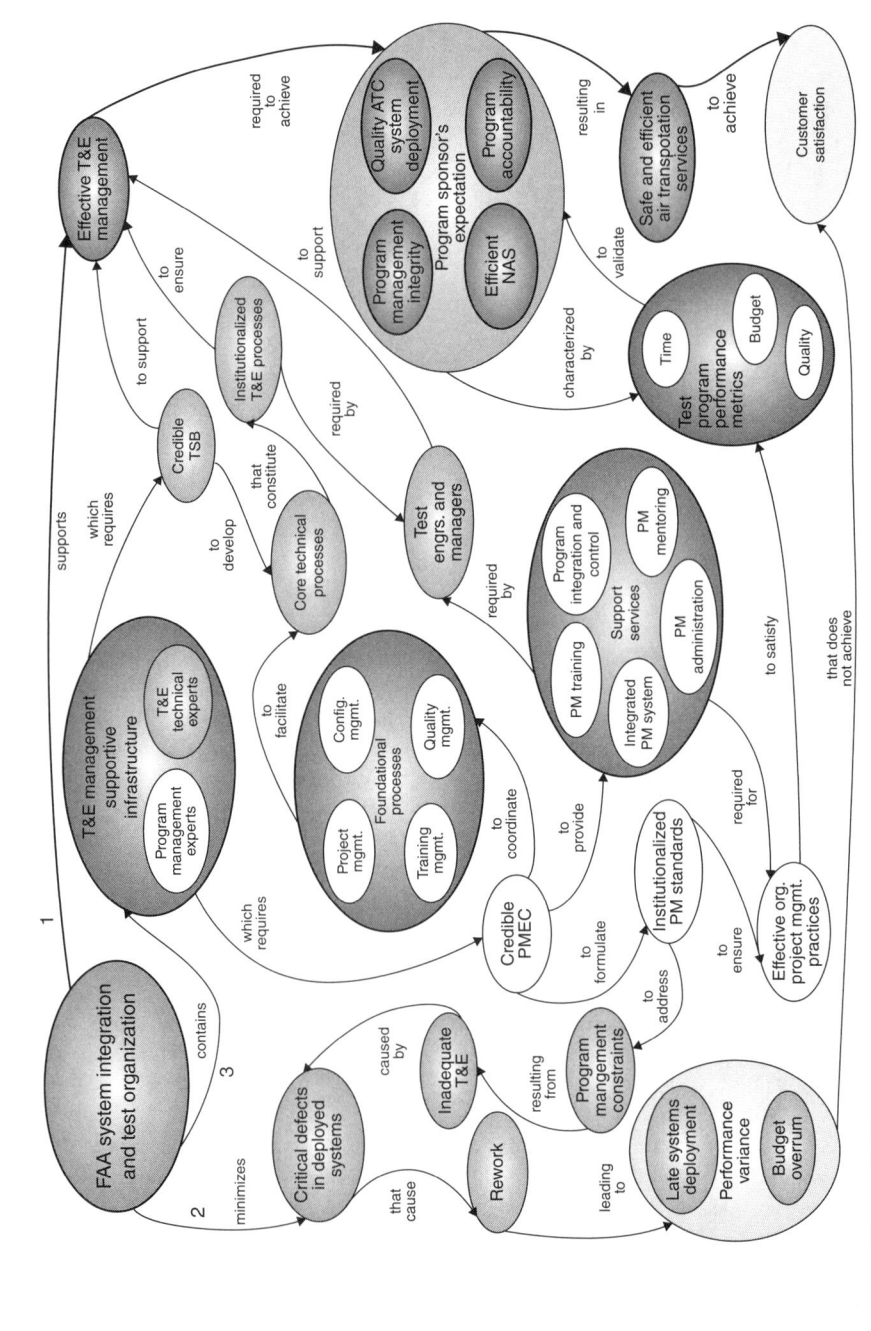

Figure 11.17 Test and evaluation program management model.

- To ensure not only compatibility between the graphic object and structured text but also synergy so both objects could evolve into more potent instances, capable of improved dissemination, and community building, development, and mobilization—thus, it is not a case of either/or, but both.

We have propositioned BSSM and systemigrams not as a problem solutioneering tool, but one that allows us to better understand the context, complexity, and perspectives that will realize SoS solutions. Systemigrams are most effectively created with the software tool SystemiTool. For more information on the Boardman's Soft Systems Methodology, how to build systemigrams, or how to acquire a copy of SystemiTool, see Boardman and Sauser (2013) or visit http://www.WorldsofSystems.com.

REFERENCES

Bahill, A. T., and B. Gissing. 1998. "Re-evaluating systems engineering concepts using systems thinking." *IEEE Transactions on Systems, Man, and Cybernetics, Part C: Applications and Reviews* 28(4):516–527.

Baldwin, W. C. 2008. Modeling paradox: Straddling a fine line between research and conjecture. Paper read at *IEEE International Conference on System of Systems Engineering*, June 2–4. Singapore.

Baldwin, W. C., T. Ben-Zvi, and B. Sauser. 2012. "Formation of collaborative system of systems through belonging choice mechanism." *IEEE Transactions on System, Man, and Cybernetics-Part A: Systems and Humans* 42(4):793–801.

Bayuk, J. L. 2010. The utility of security standards. *IEEE International Carnahan Conference Security Technology (ICCST)*, October 5–8, San Jose, CA.

Bayuk, J. L., J. Healey, P. Rohmeyer, M. H. Sachs, J. Schmidt, and J. Weiss. 2012. *Cyber Security Guidebook*. Hoboken, NJ: John Wiley & Sons, Inc.

Bayuk, J. L., and A. Mostashari. 2013. "Measuring systems security." *Systems Engineering* 16(1):1–14.

Blair, D., J. Boardman, and B. Sauser. 2007. "Communicating strategic intent with systemigrams: Addressing the network-enabled challenge." *Systems Engineering* 10(4):309–322.

Boardman, J. T. 1990a. "Control in project management." *IEE Colloquium on Generic Control Systems* 115:19–32.

Boardman, J. T. 1990b. "A methodology for strategic plan modelling." *IEE Colloquium on Goal-Driven Planning in Complex Environments* 131:1–14.

Boardman, J. T. 1994. "A process model for unifying systems engineering and project management." *Engineering Management Journal* 4(1):25–35.

Boardman, J. T., and B. J. Sauser. 2006. System of systems: The meaning of. Paper read at *IEEE International System of Systems Engineering Conference*, April 24–26, Los Angeles, CA.

Boardman, J. T., and B. J. Sauser. 2008. *Systems Thinking: Coping with 21st Century Problems*. Boca Raton, FL: CRC Press/Taylor & Francis Group.

Boardman, J. T., and B. J. Sauser. 2013. *Systemic Thinking: Building Maps for Worlds of Systems*. Hoboken, NJ: John Wiley & Sons, Inc.

Brocklesby, J. 1995. "Using soft systems methodology to identify competence requirements in HRM." *International Journal of Manpower* 16(5/6):70–84.

Bulbeck, J. M., J. T. Boardman, and J. S. Sagoo. 1997. "Business process simulation using soft systems modelling." *Fifth International Conference on Factory 2000-The Technology Exploitation Process*, April 2–4, Cambridge, 435:437–442.

Bulbeck, J. M., and B. T. Clegg. 1996. "A systems methodology for analysing and simulating manufacturing systems." *IEEE International Conference Systems, Man, and Cybernetics* 1:778–783.

Bustard, D. W., R. Sterritt, A. Taleb-Bendiab, and A. Laws. 2006. "Autonomic system design based on the integrated use of SSM and VSM." *The Artificial Intelligence Review* 25(4):313–327.

Checkland, P. 1999. *Systems Thinking, Systems Practice: Includes a 30-Year Retrospective*. Chichester, UK: John Wiley & Sons, Ltd.

Checkland, P. 2000a. *Systems Thinking, Systems Practice*. Hoboken, NJ: John Wiley & Sons, Inc.

Checkland, P. 2000b. "Soft systems methodology: A thirty year retrospective." *Systems Research and Behavioral Science* 17(1):11–58.

Checkland, P., and S. Holwell. 1998. "Action research: Its nature and validity." *Systemic Practice and Action Research* 11(1):9–21.

Checkland, P., and J. Scholes. 1999. *Soft Systems Methodology in Action*. Hoboken, NJ: John Wiley & Sons, Inc.

Clegg, B. T., and J. T. Boardman. 1996. "Process integration and improvement using systemic diagrams and a human-centred approach." *Concurrent Engineering Research and Applications* 4(2):119–135.

Clegg, B. T., and J. T. Boardman. 1997a. "Systemic analysis of concurrent engineering practice." *Fifth International Conference on Factory 2000-The Technology Exploitation Process*, April 2–4, Cambridge, 435:464–471.

Clegg, B. T., and J. T. Boardman. 1997b. "A systems approach to process improvement in design and manufacture." *IEE Colloquium on a Systems Approach to Manufacturing* 3:1–9.

Cloutier, R. 2011a. "An introduction to the jet special issue of journal of enterprise transformation: Enterprise modeling." *Journal of Enterprise Transformation* 1(3):4.

Cloutier, R. 2011b. *SYS750 Advanced System and SoS Architecture Modeling & Assessment*. Hoboken, NJ: Stevens Institute of Technology.

Cole, A. J., S. J. Wolak, and J. T. Boardman. 1995. Computer-based process handbook for a systems engineering business. Paper read at *Seventh International Workshop on Computer-Aided Software Engineering*, July 10–14, Toronto, ON.

Derro, M. E., and C. R. Williams. 2009. Behavioral competencies of highly regarded systems engineers at NASA. Paper read at *Aerospace Conference, 2009 IEEE*, March 7–14, Big Sky, MT.

DiMario, M., J. Boardman, and B. Sauser. 2009. "System of systems collaborative formation." *IEEE Systems Journal* 3(3):360–368.

DoD. 2010. The army operating concept 2016–2028. *TRADOC Pamphlet 525-3-1*, August 19. Fort Monroe, VA: US Army Training and Doctrine Command.

Duczynski, G. 2004. "Systems approaches to economic development for indigenous people: A case study of the Noongar Aboriginals of Australia." *Futures* 36(8):869.

Eigbe, A. P., B. J. Sauser, and J. T. Boardman. 2010. "Soft systems analysis of the unification of test and evaluation and program management: A study of a federal aviation administration's strategy." *Systems Engineering* 13(3):298–310.

Frittman, J., and R. Edson. 2009. Systems thinking based approach to writing effective concepts of operations (CONOPs). *Conference on Systems Engineering Research,* March 17–19, Hoboken, NJ.

Gorod, A., M. DiMario, B. Sauser, and J. Boardman. 2009. " 'Satisficing' system of systems (SoS) using dynamic and static doctrines." *International Journal of System of Systems Engineering* 1(3):347–366.

Gorod, A., B. Sauser, and J. Boardman. 2008. "System of systems engineering management: A review of modern history and a path forward." *IEEE Systems Journal* 2(4):484–499.

Haskins, C., K. Forsberg, and M. Krueger. 2008. *INCOSE Systems Engineering Handbook Version 3.1*. San Diego, CA: International Council on Systems Engineering.

Henderson, R. M., and K. Clark. 1990. "Architectural innovation: The reconfiguration of existing product technologies and the failure of established firms." *Administrative Science Quarterly* 35(1):9–30.

Hitt, M. A., C. C. Miller, and A. Colella. 2005. *Organizational Behavior: A Strategic Approach*. Hoboken, NJ: John Wiley & Sons, Inc.

Jackson, M. C. 2003. *Systems Thinking: Creative Holism for Managers*. San Francisco, CA: Jossey Bass.

Jacobs, B. 2004. "Using soft systems methodology for performance improvement and organisational change in the English National Health Service." *Journal of Contingencies and Crisis Management* 12(4):138.

Jansma, P. A., and M. E. Derro. 2007. If you want good systems engineers, sometimes you have to grow your own! Paper read at *Aerospace Conference, 2007 IEEE*, March 3–10, Big Sky, MT.

Jenkins, G. M. 1969. "The systems approach." In *Systems Behaviour*, edited by J. Beishon and G. Peters, p. 82. London: Harper & Row.

Keating, C., R. Rogers, R. Unal, D. Dryer, A. Sousa-Poza, R. Safford, W. Peterson, and G. Rabadi. 2003. "System of systems engineering." *Engineering Management Journal* 15(3):36–45.

Lehaney, B., and R. Paul. 1996. "The use of soft systems methodology in the development of a simulation of out-patient services at Watford General Hospital." *The Journal of the Operational Research Society* 47(7):864–870.

Maier, M. 1998. "Architecting principles for system-of-systems." *Systems Engineering* 1(4):267–284.

Mansouri, M., B. Sauser, and J. Boardman. 2009. "Applications of systems thinking for resilience study in maritime transportation system of systems." *IEEE Systems Conference*, March 23–26, Vancouver, BC, pp. 211–217.

Meentemeyer, S. M., B. J. Sauser, and J. T. Boardman. 2009. "Analysing a system of systems characterisation to define system of systems engineering practices." *International Journal of System of Systems Engineering* 1(3):329–346.

Mehler, J., S. McGee, and R. Edson. 2010. Leveraging. Systemigrams for conceptual analysis of complex systems: Application to the US National Security system. *Conference on Systems Engineering Research*, March 17–19, Hoboken, NJ.

Mingers, J., and J. Rosenhead. 2004. "Problem structuring methods in action." *European Journal of Operational Research* 152(3):530–554.

Mitroff, I. 1998. *Smart Thinking for Crazy Times*. San Francisco, CA: Berrett-Koehler.

Molineux, J., and T. Haslett. 2007. "The use of soft systems methodology to enhance group creativity." *Systemic Practice and Action Research* 20(6):477–478.

Munro, I., and J. Mingers. 2002. "The sue of multi-methodology in practice—results of a survey of practitioners." *Journal of Operations Research Society* 53(4):369–378.

Mussante, J., J. Frittman, and R. Edson. 2010. The use of systemigrams in system template development: An example in disaster management. *Conference on Systems Engineering Research*, Hoboken, NJ.

Neves, L. M. P., A. G. Martins, C. H. Antunes, and L. C. Dias. 2004. "Using SSM to rethink the analysis of energy efficiency initiatives." *The Journal of the Operational Research Society* 55(9):968.

Polacek, G. A., D. A. Gianetto, K. Khashanah, and D. Verma. 2012. "On principles and rules in complex adaptive systems: A financial system case study." *Systems Engineering* 15(4):433–447.

Ramsay, D. A., J. T. Boardman, and A. J. Cole. 1996. "Reinforcing learning, using soft systemic frameworks." *International Journal of Project Management* 14(1):31–36.

Randall, W., D. Nowicki, and T. Hawkins. 2011. "Explaining the effectiveness of performance-based logistics: A quantitative examination." *International Journal of Logistics Management* 22(3):324–348.

Rechtin, E. 1990. *Systems Architecting*. New York: Prentice Hall.

Reisman, A. 2005. "Soft systems methodology: A context within a 50-year retrospective of OR/MS." *Interfaces* 35(2):164–178.

Ryschkewitsch, M., D. Schaible, and W. Larson. 2009. "The art and science of systems engineering." *Systems Research Forum* 3(2):81–100.

Sage, A. P. 2003. Conflict and risk management in complex system of systems issues. Paper read at *IEEE International Conference on Systems, Man and Cybernetics*, October 5–8, 4:3296–3301.

Sagoo, J. S., and J. T. Boardman. 1998. "Towards the formalisation of soft systems models using Petri net theory." *IEE Proceedings: Control Theory and Applications* 145(5):463–471.

Salado, A., and R. Nilchiani. 2013. "Contextual- and behavioral-centric stakeholder identification." *Procedia Computer Science* 16:908–917.

Sauser, B. J. 2006. "Toward mission assurance: A framework for systems engineering management." *Systems Engineering* 9(3):213–227.

Sauser, B. J., R. Cloutier, L. Layman, F. Shull, K. Sullivan, and U. Becker-Kornstaedt. 2012. *Systems Engineering for Contingency Basing, TR-033*. Hoboken, NJ: Systems Engineering Research Center.

Sauser, B. J., Q. Li, and J. Ramirez-Marquez. 2011a. "Systemigram modeling of the small vessel security strategy for developing enterprise resilience." *Marine Technology Society Journal* 45(3):88–102.

Sauser, B. J., M. Mansouri, and M. Omer. 2011b. "Using systemigrams in problem definition: A case study in maritime resilience for homeland security." *Journal of Homeland Security and Emergency Management* 8(1)1–19.

Senge, P. 1990. *The Fifth Discipline: The Art & Practice of the Learning Organization*. New York: Currency Doubleday.

Shalhoub, Z. K., and J. Al. Qasimi. 2005. "A soft system analysis of nonprofit organizations and humanitarian services." *Systemic Practice and Action Research* 18(5):457.

Sherman, D. G., A. J. Cole, and J. T. Boardman. 1996. "Assisting cultural reform in a projects-based company using Systemigrams." *International Journal of Project Management* 14(1):23–30.

Squires, A., A. Pyster, B. Sauser, D. Olwell, D. Gelosh, S. Enck, and J. Anthony. 2010. Applying systems thinking via Systemigrams for defining the body of knowledge and curriculum to advance systems engineering (BKCASE) project. Paper read at *Conference on Systems Engineering Research*, March 17–19, Hoboken, NJ.

Taylor, M. J., M. Baskett, G. D. Hughes, and S. J. Wade. 2007. "Using soft systems methodology for computer game design." *Systems Research and Behavioral Science* 24(3):359.

Warfield, J. N., and G. H. Perino. 1999. "The problematique: evolution of an idea." *Systems Research and Behavioral Science* 16(3):221–226.

Yin, R. K. 2008. *Case Study Research: Design and Methods*. 4th ed. Thousand Oaks, CA: Sage Publications.

Chapter 12

Using Modeling and Simulation for System of Systems Engineering Applications in the European Space Agency

Joachim Fuchs and Niklas Lindman
European Space Agency, ESTEC, Noordwijk, the Netherlands

12.1 INTRODUCTION

12.1.1 Agency Context

The European Space Agency (ESA) is Europe's gateway to space. Its mission is to shape the development of Europe's space capability and ensure that investment in space continues to deliver benefits to the citizens of Europe and the world. It is an international organization with currently 20 member states. By coordinating the financial and intellectual resources of its members, it can undertake programs and activities far beyond the scope of any single European country. In order to achieve this, the Agency draws up the European space program and executes it.

ESA's programs are very diverse and cover all application and science areas linked to space. They are designed to find out more about Earth, its immediate space environment,

---

Modeling and Simulation Support for System of Systems Engineering Applications, First Edition.
Edited by Larry B. Rainey and Andreas Tolk.
© 2015 John Wiley & Sons, Inc. Published 2015 by John Wiley & Sons, Inc.

our solar system, and the universe, but they also develop satellite-based technologies and services, as well as the required launchers to ensure the access to space and infrastructure to operate the space systems once in orbit.

For its work, ESA cooperates closely with other organizations within and outside Europe. In its daily working, the Agency has to take into account a great variety of stakeholders. At a prominent place are the member states (represented at highest level by the corresponding ministers), providing the political and financial context for its operations. Academic partners are influencing the scientific agenda for future systems and represent the end users for scientific missions. However, operators and service providers are also users of ESA developments. Industries/companies are usually contracted to design and build the required systems, and other national and international space agencies become partner for the most ambitious programs, as is the European Commission.

12.1.2 Complexity Problems

12.1.2.1 System of Systems

ESA has recently been challenged by the developments of systems that can no longer be considered to be space systems only. New advanced user needs require systems that provide services consisting of capabilities delivered by a combination of independently developed and operated space and nonspace systems. Examples of such large-scale applications or system of systems (SoS) are advanced navigation services, air traffic control systems, search and rescue, and disaster management to mention some. The development of these applications imposes great challenges in terms of program management and Systems Engineering. Assessing these features, they generally expose some or all of the characteristics accepted as defining an SoS, as also elaborated in Chapter 2 of this book (Maier, 2014):

- Operational independence
- Managerial independence
- Geographic distribution
- Emergent behavior
- Evolutionary development

12.1.2.2 Program Types

Space programs are long-term activities typically running from 2–3 years up to 20–25 years. A decade passes easily between the decision to develop a world-class scientific satellite and the moment that it returns long-awaited data from space to universities and research centers. In the last years, programs have changed in character. From being partially driven by technology, they have evolved to much more product- and service-oriented missions and programs today. The complexity of systems to be built increases, and external partners are more and more required to be able to achieve the ambitious scientific or technological goals.

An existing program "par excellence" to illustrate this complexity is the International Space Station. ESA is working together with the National Aeronautics and Space Administration (NASA), the Russian Federal Space Agency (Roscosmos), and the

Japanese Space Agency (JAXA). In this program, Europe has a substantial share and contributes as main hardware elements Columbus (a multipurpose science and technology laboratory) and the Automated Transfer Vehicle (ATV, a service vehicle essential for the ISS). In addition, ESA is also providing two of the three nodes of the ISS, the Cupola, the European Robotic Arm (ERA), the data management system for the Russian Service Module, and various facilities for scientific research. Although the ISS has certainly some aspects of an SoS as described in Section 12.1.2.1, it was conceived as one complete (new) system, under one design and operational authority. In addition, it is in its operational phase, and therefore, only limited benefit is expected from dedicated new modeling activities.

Apart from the increased complexity, more and more programs rely on already existing elements and try to enhance them in order to achieve ambitious goals in an efficient manner. Legacy aspects have to be taken into account, heritage in design and developments have to be considered, and heterogeneous concepts and interfaces are in the way of an easy integration. Contrary to the ISS described earlier, existing systems are combined as elements of an SoS, usually augmented by new developments. These elements are not under a single design authority, nor subject to centralized operations. Products and services need coordination across diverse authorities. Despite differing life cycles of individual elements, services need to be maintained and evolve to meet new requirements.

Representatives for this type of system are, for example, the system for Global Monitoring for Environment and Security (GMES) (European Commission, undated) and the Space Situational Awareness Program (SSA) (ESA, undated). Both combine existing space and ground assets to a larger system and augment them with new ground- and space-based elements to address their overall objectives. These assets are not under a unique designing or operational authority. As will be further detailed and illustrated in the following, these two programs have all major characteristics of SoS, and modeling has been applied for specific aspects of the architecting and engineering activities.

Other programs requiring substantial external cooperation are the Integrated Applications Program (IAP) and the navigation programs (Galileo and EGNOS and their respective evolution programs). The IAP program deals with applications of space systems that combine different types of satellites and services such as telecommunications and Earth observation and navigation, also taking into account ground-based assets. The envisaged services are implemented in combination with semipublic organizations (e.g., air traffic), inherently exposing complex external interfaces and needing tight integration and coordination at this interface level. The navigation programs have extensive external interfaces, and the complexity of the ground system and the number of elements create an important number of internal interfaces. Similar challenges exist for a potential Solar System Exploration Program, which relies on major international collaborations. However, in these programs, formal architecture modeling has not yet been applied at the SoS level.

12.1.2.3 Problem Types

It is obvious that these programs expose major challenges on technical level. In addition, major issues exist in the area of programmatics (financing streams, funding, exploitation), ownership and agreements, governance, and data policy, security, risk, and

Table 12.1 Mapping of Problem Types to Agency Programs

	GMES	SSA	IAP	Navigation	Exploration
Funding	Technology development and initial space segments by ESA. Operations and maintenance through the European Union (EU)	Space and ground segments existing, additional elements and services through ESA programs	Space segment elements existing. Usage for demonstration by ESA, in combination with application community (e.g., air traffic control, emergency services). Operations through application community	Technology development and in-orbit validation through ESA. Full system capability and operations by the EU. ESA and the EU have different funding schemes and procurement rules In addition, a potential commercial service is possible	Collaboration of national and international space agencies due to the ambition and complexity of any program
Governance	Development authority and initial operations ESA. Further operations after deployment not yet fully defined but potentially in a separate entity. Different possibilities for civil and defense applications	Assets owned by different entities (space agencies and institutes). Service component not yet defined	Space assets owned by Agency and/or operators (telecom). Application services by third parties/companies, governed by contracts and agreements	Development and commissioning ESA. Operations under the EU, with a separate organization. Ownership by the EU	Multinational cooperation, usually based on barter agreements and memoranda of understanding. Ownership usually on element level, with operational responsibility allocated to one partner only. Exploitation of data distributed

Security	Relevant on operations and on data level. Depending on the data, different levels of security might be involved (civil and strategic) and might be different between participants (nations)	Relevant at data level. Users will have different levels of authority	Very much dependent on the individual application/service	By the provision of a civil, a commercial and a strategic service different levels of security are implied. For monitoring and controlling major driver for system	In general not security critical
Risk	As a system to support environmental monitoring and disaster support requiring low-risk approaches for development	No particular risk at element level. Main risk in inter-dependency between different governance schemes.	Depending on the application	Integrity must be assured for safety of life applications (descoped for initial deployment)	For development increased risk due to high innovation content of related technology
Dependability	As a potential source for governmental decisions and actions requiring a high degree of dependability	If introduced in government decisions will require a high degree of dependability	Some of the envisaged applications have major implications and require high degree of dependability	Driven by some services, extremely high level of dependability	System problems usually do not lead to immediate-risk situations if not crewed, but recovery must be possible in difficult environmental conditions

dependability. These problem types have not been an integral part of traditional Systems Engineering processes, which focused much more on the purely technical aspects. The fact that issues go beyond the technical dimension is closely linked to the fact that complex systems are initially often a patchwork of existing elements, owned and controlled by different stakeholders. The interests of these stakeholders need to be taken into account appropriately in order to achieve an efficient synergy of individual systems for new services. The link between these programmatic and the technical dimension of any development or integration needs to be made explicit and ideally formalized in order to be manageable.

These aspects will be more detailed and illustrated in Section 12.2.

The scope of ongoing and future projects of ESA and the related problems exceed the traditional complexity and domains of Systems Engineering approaches used in the past. Communication and interaction with more stakeholders needs to be ensured on all levels in an unambiguous way. The traditional methodology of modeling and simulation (M&S) can play an important role to support this process. The formal description (modeling) of elements and their relations is the basis for a more structured information exchange and can be used to support the use of this information in different contexts without loss of semantic meaning.

Table 12.1 illustrates the mapping of problem types to Agency programs.

12.2 APPLICATION CASES

To illustrate the use of the architecture framework in the context of Agency projects, two application cases have been identified for illustration. The two programs, GMES and SSA, will be introduced with some more details.

12.2.1 GMES (or Copernicus)

The system for GMES (recently renamed Copernicus) is a complex set of systems with the objective to monitor the Earth. It is based on data coming from different sources, such as Earth observation satellites, *in situ* measurements, and airborne and seaborne sensors. It is processing the available data in order to provide accurate, timely, and easily accessible information to improve the management of the environment in six thematic areas, being land, marine, atmosphere, climate change, emergency management, and security. The services covering these themes support numerous applications, from environment protection over disaster management to civil security.

In this program, ESA coordinates the delivery of data from about 30 satellites, and the European Environmental Agency (EEA) is responsible for data from airborne and ground sensors, while the EC, acting on behalf of the European Union (EU), is responsible for the overall initiative, setting requirements and managing the services.

Managed by ESA, the Space Component is in its preoperational stage, serving users with satellite data currently available through the so-called GMES Contributing Missions at national, European, and international levels. A number of dedicated operational missions (Sentinels) are presently under development, and GMES will start the transit to a fully operational phase after the launch of the first of this Sentinel missions.

As the ground segments contain the necessary interfaces for requesting and accessing data from national and commercial assets of different types, a comprehensive Earth observation approach for EU civil security could be obtained by networking them, so ideally enhancing efficiency and sustainability through the coordination of existing or planned systems. This is the fundamental principle of the civil program GMES.

12.2.1.1 Architecting Support

In this context, ESA has studied architectures to integrate this SoS as large-scale integrated systems that are heterogeneous and independently operable on their own, controlled and owned by different actors, but that are networked together for a specific goal. Interoperability of the different ground segments needed to be studied not only at technical but also at programmatic and political levels. Different ownership and independent control of those assets by different user communities pose a particular challenge for the designing and interfacing of responsive ground segments in order to generate value adding and near-real-time data and products. The missions considered in this study are European and non-European, both optical and radar. In support of these architecture definitions and analyses, dedicated SoS models have been developed, using the available ESA Architecture Framework (ESA-AF). The focus was lying on the ground segment elements, and an overall ground segment architecture definition was the objective.

12.2.1.2 Considered Scenarios

In order to define the use cases, scenarios relevant for civil security have been defined. These scenarios can be classified into three groups with different requirements on the services to be provided by the overall systems. These groups cover different stages of support as follows:

1. **Reference mapping**: A more or less continuous mapping exercise has to be done to establish reference material for assessments of any changes introduced (e.g., by a disaster). This is usually data to be provided for archiving and also the basis for strategic decisions concerning particular geographical areas.

2. **Area of interest monitoring**: The main purpose of monitoring of specific areas is the detection of changes and evolution over time. The regularity (periodicity) is most important in this case, but the delivery time (latency) of the data is less critical. Operational aspects (deployment of operational means) can be based on information derived from monitoring activities.

3. **Rapid mapping**: In the case of a crisis, fastest response time (high responsiveness) is needed. The objective is to get new information about an area in the shortest possible time, and therefore, availability and reliability are of highest importance. Overall, delivery time (tasking and delivery) needs to be considered together and optimized. The relevant information has an impact on tactical level concerning the use of resources.

For the architectures supporting these scenarios—starting from the existing infrastructure and aimed at an evolution—different governance schemes have been identified as options, and the corresponding trade-off has been supported by the analyses based on the architecture models.

12.2.1.3 Modeling Activities

Before engaging into a detailed modeling exercise, it was crucial to define the type of questions/investigations to be supported by the modeling. The following list is taken directly out of the architectural summary justification and description document (also called AV-1 as explained in Section 12.3.3) of the study:

- **Federation**: What EO GS elements could be modified to better support a federated cooperative approach to civil security missions?
- **Parameters**: What are the programmatic and technical impacts of modeled systems' captured parameters (especially regarding VHR and high responsiveness assets) on overarching SoS performance?
- **Gaps**: What are options for filling capability and interoperability gaps by technical improvement to elements of the EO GS SoS?
- **Interoperability**: What are the nontechnical options for improving interoperability, within the identified operational constraints on the governance and access of EO assets and data?
- **Procurement**: How can approaches to planned EO GS element and service procurement be modified and aligned to improve SoS integration?
- **Joint Capability Management System (JCMS)**: Which options (including centralized and distributed solutions) for the JCMS should be studied in detail?

As a result of the analysis of the scenarios and the underlying questions, the following performance parameters are considered of major interest. They are partly technical and partly programmatic, although the big advantage of the exercise is the possibility to model and demonstrate the interdependencies between these areas, constituting the basis for analyses made at architectural level:

- **Technical**: Responsiveness, tasking time, data latency, availability, quality of products, revisit time, and coverage
- **Programmatic**: Data policy, interoperability, and funding/business cases

On one hand, analyses have been performed addressing the more technical parameters and in particular the more dynamic aspects of the scenarios. The underlying question was concerned with the time needed for providing particular space imagery products to support emergency response. For this case, an earthquake in Haiti example scenario was used to establish a concrete context for comparison of the different JCMS options.

On the other hand, the concern was on the structural and programmatic aspects, addressing the complexity of the process for delivering space imagery products for emergency response. In this context, a particular (existing) rapid mapping process was used to assess the structural differences when the process is performed according to the different JCMS options.

For each example scenario, a mapping to the different architecture options was achieved by modeling the programmatic aspects (agreements and participating organizations), the systems (generic models and concrete participating missions), and the services (service orchestrations).

The end-to-end responsiveness analysis was supported by computing the total duration over all service orchestration paths in an end-to-end process.

To illustrate the approach, a few diagrams and exploitations from the architecture analysis are provided in the following (Figures 12.1, 12.2, 12.3, and 12.4).

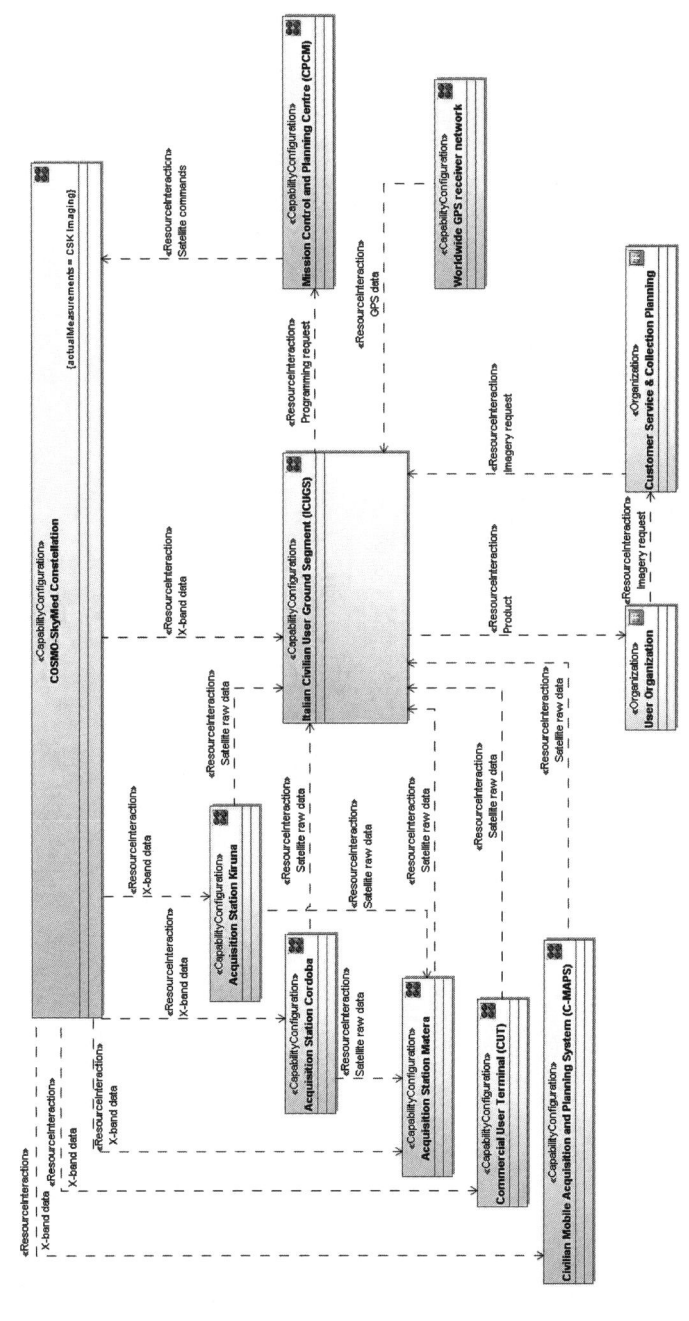

Figure 12.1 Resource interaction specification for one of the considered existing systems.

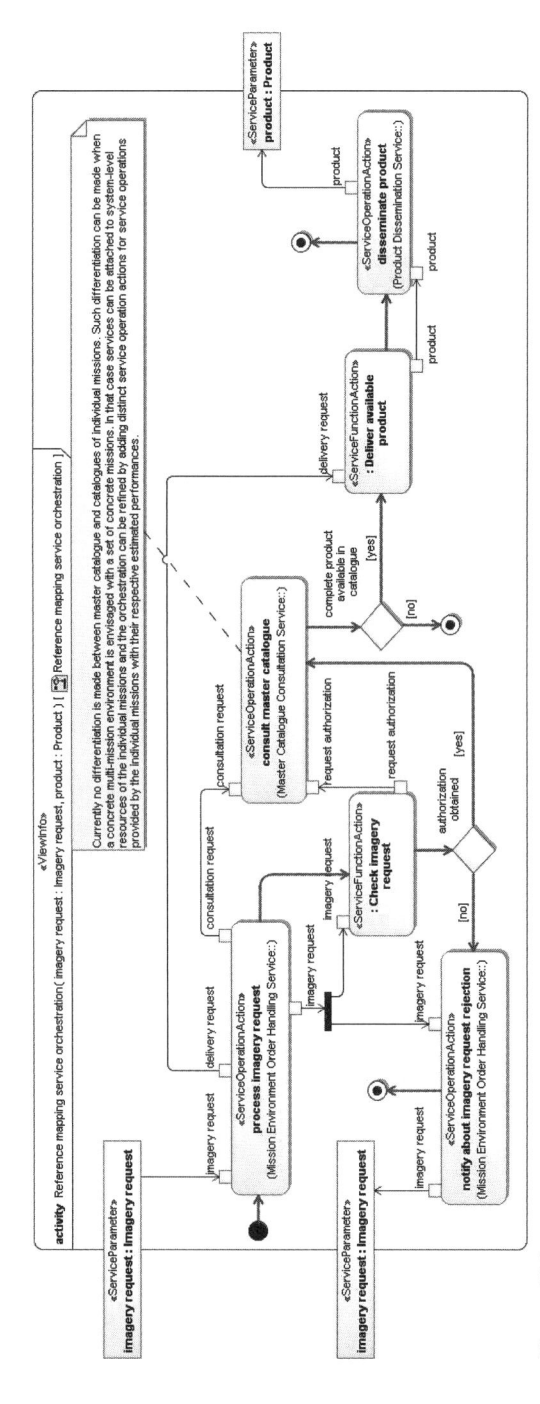

Figure 12.2 Service functionality flow for the reference mapping service to capture the dynamic service.

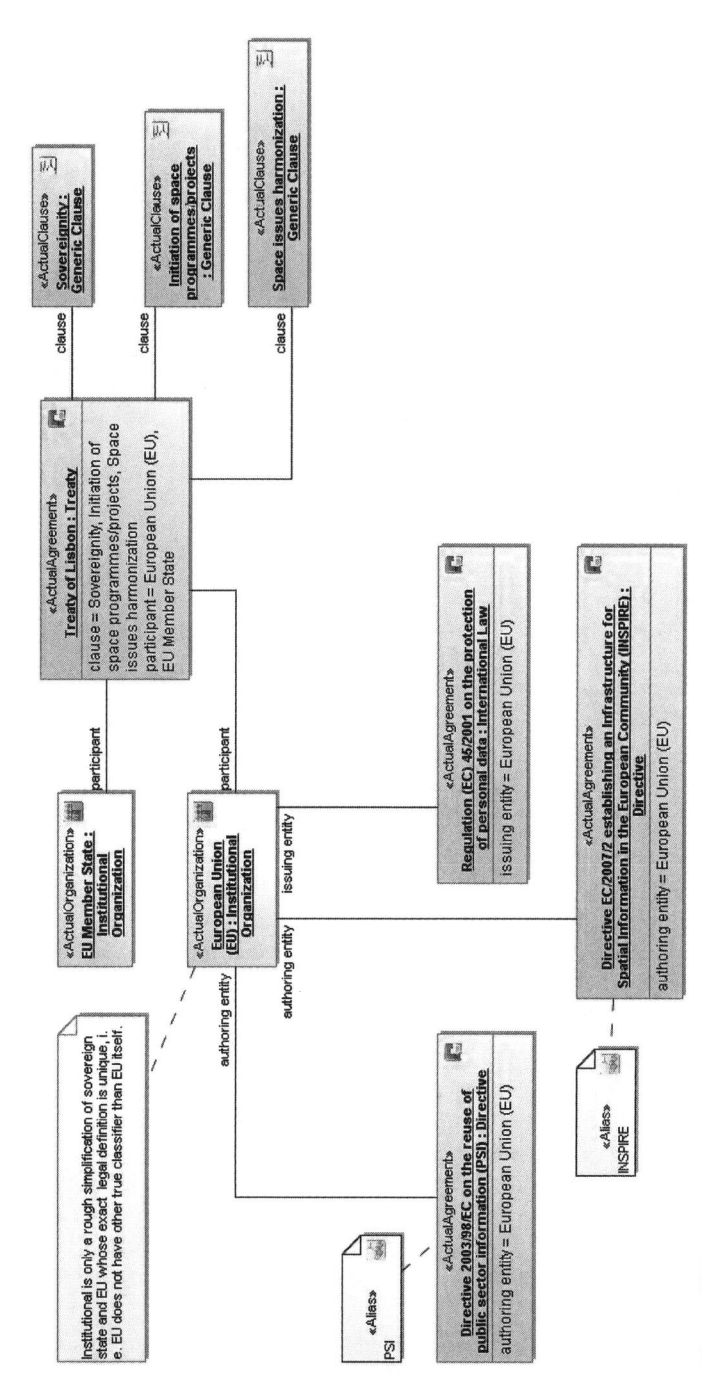

Figure 12.3 EU-related agreement structure and their dependencies.

Path	Activity	Called operation	Called activity	Time (h)
1				
	Haiti emergency response service orchestration	Process imagery request		0
	Haiti emergency response service orchestration		GeoEye-1 imagery delivery: Reference mapping service orchestration	6
	Haiti emergency response service orchestration	G-MOSAIC reprocess image data: reprocess image data		0
	Haiti emergency response service orchestration	Develop reference maps and basic cartography		8
			Total duration: 14	
2				
	Haiti emergency response service orchestration	Process imagery request		0
	Haiti emergency response service orchestration		GeoEye-1 imagery delivery: Reference mapping service orchestration	6
	Haiti emergency response service orchestration	G-MOSAIC reprocess image data: reprocess image data		0
	Haiti emergency response service orchestration	Perform trafficability analysis		12
			Total duration: 18	

Figure 12.4 Example result for responsiveness analysis.

12.2.2 SSA

The ESA-AF has been and is currently successfully used in the SSA program in support of the Systems Engineering activities at ESA and by European industry. In particular, the following main activities have been supported by a formal modeling approach using the ESA-AF (detailed in Section 12.3):

- **Data Policy and Security Methodology**: SSA has very stringent data policy and security constraints, in particular in the Space Surveillance and Tracking (SST) segment. These constraints originate from the fact that when surveying space, sensitive military objects in space can be detected. The characteristics of these objects can under no circumstances be disseminated as it is considered a threat to national and European security. The ESA-AF has been used to develop a method to verify that the systems design satisfies the data policy and security requirements. This has been done as an internal activity at ESA.
- **Functional Analysis and Functional Systems Requirement Consolidation**: As part of the systems requirements definition and specification, the ESA-AF has been used to provide a functional decomposition to properly structure the

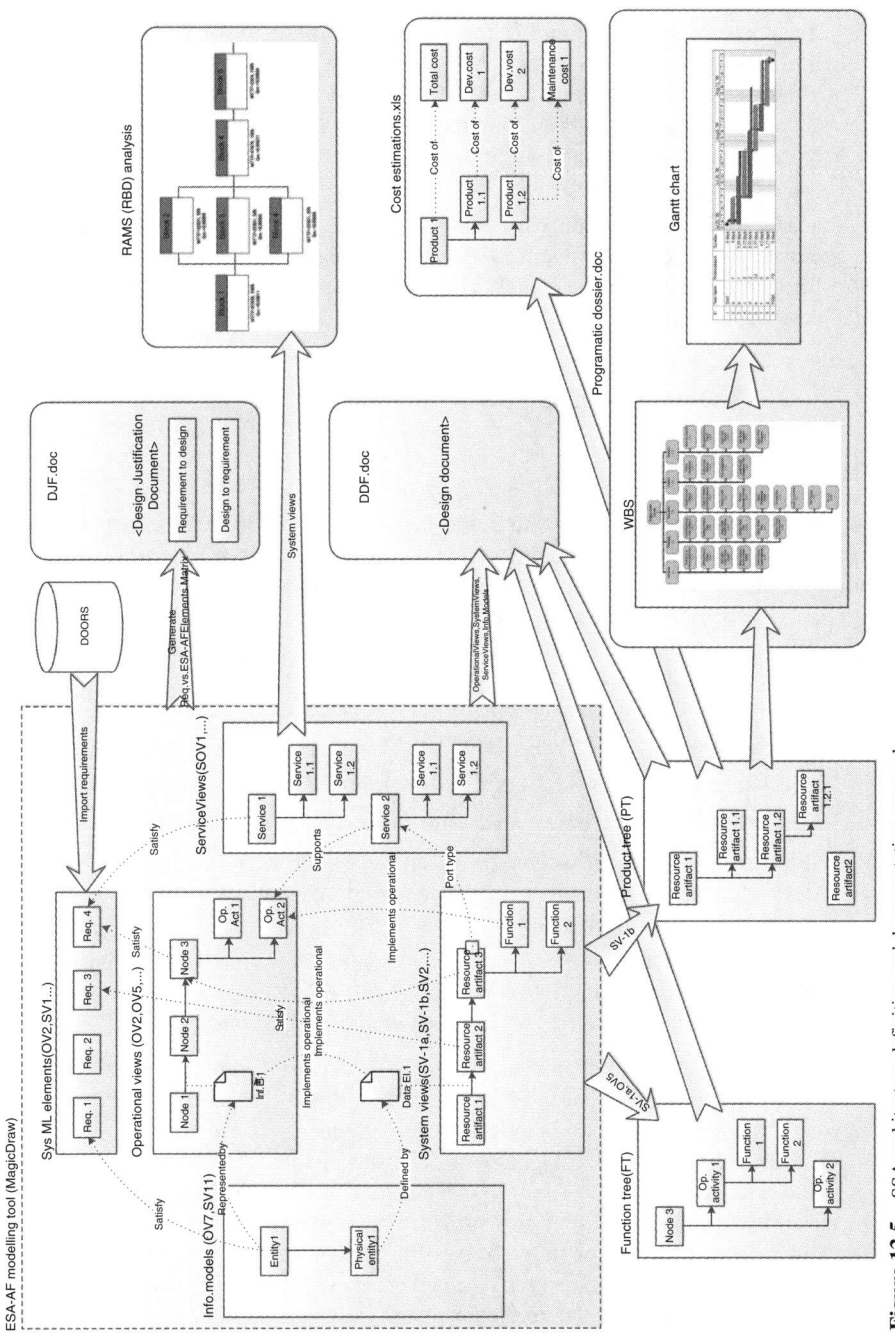

Figure 12.5 SSA architecture definition and documentation approach.

functional requirements as well as ensure the completeness. This has also been done as an internal activity at ESA.

- **Architecture Trade-Off and Definition:** As part of the SSA architectural design studies, the ESA-AF is being used by industry for the architecture trade-off analysis as well as the design specification. The complete architecture including different architecture options has been captured in the ESA-AF, and then exploitation and reporting tools have been used to optimize the architecture based on predefined parameters such as performance, risk, cost, etc.
- **Model to Documentation**: The ESA-AF has also been used to capture the architecture in a model and to derive the required design artifacts and documentation. Figure 12.5 illustrates the model-driven process that has been used to capture the design and to include the model artifacts into the design documents.

12.3 METHODS AND TOOLS: THE ESA-AF

12.3.1 Overview

The ESA-AF and its metamodel are based on the Ministry of Defense Architecture Framework (MODAF) (U.K. Government, 2012) with numerous ESA-specific extensions. Additional supporting functionality, implemented in a custom exploitation tool, provides diagramming and reporting capabilities beyond those of the predefined architecture views. The scope and flexibility of the ESA-AF exploitation framework is also being enhanced to take advantage of the increased ESA-AF capabilities. In the light of the foregoing consideration of the requirements on a suitable architecture framework for ESA, four major aspects of the ESA-AF need to be pointed out:

1. It is based upon established industry standards (UPDM, SoaML, and SysML).
2. ESA-AF's design and exploitability are under exclusive ESA control.
3. ESA-AF's metamodel extensions specifically address the domains of special interest to ESA: SoS governance (ownership, agreements), finance (cost and funding), risk, and availability.
4. The ESA-AF has its own associated exploitation framework, with dedicated diagramming and reporting functionality that particularly supports the compilation of summary and derived data by interrogation of architecture models.

Figure 12.6 shows a high-level overview of the ESA-AF and its different parts—framework governance, modeling, and exploitation. These parts are introduced briefly in the following text and in more detail in the following sections.

The ESA-AF governance comprises the definition and management of the ESA-AF metamodel and process model. The ESA-AF metamodel specifies the conceptual structure for all ESA-AF models. It further provides the baseline for the generation of an exploitation metamodel and of different modeling components. The ESA-AF process model specifies the glossary to be used during the elaboration of enterprise architectures as well as the procedures for the development and management of architecture models including the involved user roles and their corresponding tasks. The ESA-AF architecture process is aligned with The Open Group Architecture Framework (TOGAF) Architecture

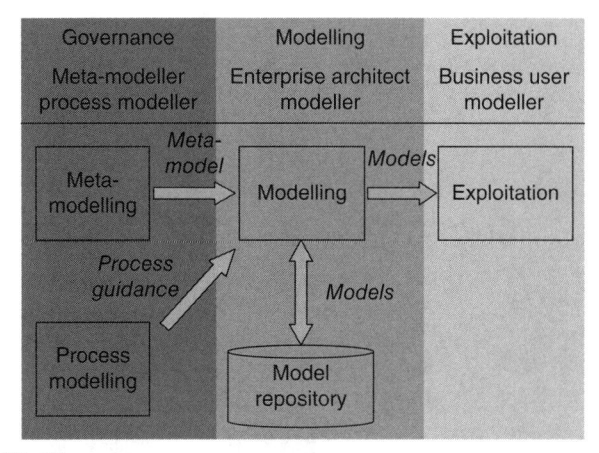

Figure 12.6 ESA-AF overview.

Development Method (ADM) (The Technology Open Group Architecture Framework (TOGAF), undated). The process model is implemented through the Eclipse Process Framework (EPF) (Eclipse, 2013c). Two conceptual roles are envisioned at the governance level: the metamodeler, responsible for upgrading and maintaining the ESA-AF metamodel, and the process modeler, responsible for upgrading and maintaining the ESA-AF glossary and process definitions.

ESA-AF modeling refers to the elaboration of enterprise architecture models based on the ESA-AF metamodel and supporting tools. Models can be stored to and retrieved from a model repository to enable their exchange and reuse. The conceptual roles envisioned at the modeling level are modeler and enterprise architect. The modeler is a person with deeper technical expertise, who is able to perform detailed modeling of technical components, interfaces, and functionalities. The enterprise architect on the other hand is focused on higher-level concerns on strategic and operational level.

The ESA-AF exploitation is delivering sophisticated reporting and diagramming capabilities that go beyond the basic views of standard architecture frameworks. The exploitation is intended to support business users (e.g., project and program managers, section heads) and in general decision makers, for example, in impact and trade-off analysis and evaluation of different architecture options. The exploitation goes beyond the strict boundaries of the modeling environment and enables the creation of custom overviews that visually highlight certain aspects of interest related to a given architecture. As a result, stakeholders in the defined architecture do not need to struggle to interpret standard architecture views and are enabled to see the "big picture" in a rather user-friendly, visual manner. In addition to the conceptual role of the business user, the role of a modeler is envisioned at this level to support the definition of exploitation tools and components.

12.3.2 Process

This section describes the high-level approach for enterprise architecting, summarizing the process defined for the ESA-AF in its architecture methodology, itself based on TOGAF best practice.

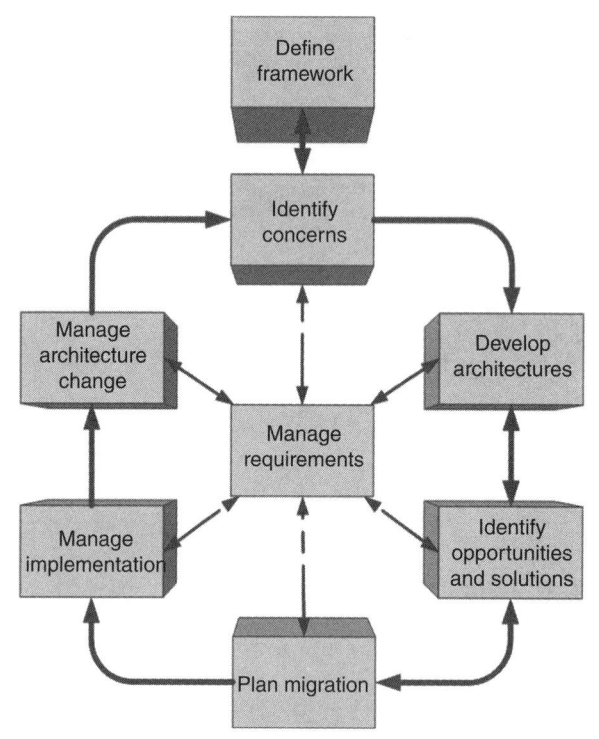

318 Theoretical and Methodological Considerations with Applications and Lessons Learned

Figure 12.7 The cyclic enterprise architecture life cycle.

Figure 12.7 shows the version of the TOGAF ADM that is used for the ESA-AF enterprise architecture life cycle, which is a guide for SoS model development, exploitation, and maintenance. The steps are very briefly summarized below:

1. **Define framework**: This step has been completed by providing the ESA-AF. However, it is a step that has to be repeated once concerns from previous iterations of model development and exploitation are identified or a new program has some specific requirements not currently covered by the ESA-AF. As ESA has control of the ESA-AF evolution, the membership and responsibilities of the architecture framework board will take on more importance (if the framework is to remain useful and responsive to changing demands).
2. **Identify concerns**: The concerns to be addressed by any ESA-AF model determine the information that needs to be captured and analyzed. They therefore determine, for example:
    - Whether existing models can be reused (as core reference information or as a baseline)
    - Which views are to be populated and what exploitations should be used
    - Whether "as is" and "to be" (possibly for variant solutions and different future phases on a roadmap) versions of diagrams are needed
    - What measurements need to be attached to which model elements

- Whether operational and technical services need to be captured
- How deep the decomposition of operational activities or solution assembly should be

The concerns are normally determined by the sponsors of the architecture tasking. For example, project managers may wish to evaluate the interoperability of their planned delivery. However, other stakeholders may help identify concerns to address. For example, an accreditor may require specific evidence before allowing the solution to interoperate over a shared network, or the manager of the architecture program may require the model to be specified in way that supports reusability. Note that these concerns may be collected before the information that populates the model.

3. **Develop architectures**: The population of the ESA-AF models depends on information gathered from documents and stakeholders. Documentation may include user or systems requirements, technical architecture or specifications, project summary, and management information or use cases and presentations that demonstrated envisioned operations. Stakeholders may include the project or program's sponsor, its managers and staff, technical specialists, representatives of end users, accreditors, or other solution review authorities, suppliers, and stakeholders from related projects.

   The architecture model itself is developed by creating elements in the repository, placing them on diagrams and relating them to each other. The elements must have appropriate information, typically including descriptions, values for measures, and other attributes, and the provenance of the information. Diagrams may be constructed across the perspectives of the ESA-AF, proceeding top-down from Strategic Views (StV) to the technical solutions that meet them or bottom-up from the available solutions to the models of the collaborative operational activities and capabilities they could support. As noted, there may be different versions of the diagrams for "as-is" and "to-be" pictures of the project or SoS. There may be several "to-be" collections of diagrams for variant alternative solutions being evaluated or a succession of future phases on a program roadmap.

4. **Identify opportunities and solutions**: In TOGAF, which primarily applies enterprise architecture to business change programs, it is at this stage that the architecture modeling effort generates recommendations. In the context where TOGAF is typically used, these may be opportunities for rationalization following a merger, new market growth from an identified product line, or selection of one of the evaluated technical solutions to specific requirements.

   In the ESA-AF, which is intended to address the broader concerns of ESA SoS program delivery, this step should be interpreted as the exploitation of the architectural models. Custom reports and exploitation views may be given to key stakeholders and decision makers, or summary diagrams may be taken directly from the model for presentations. Preferred options (supported by summarizing measurement values), issues to be addressed, or other evidences for recommendations may be identified. These may specifically feed into the project management information, such as the risks, assumptions, issues, dependencies, and opportunities.

5. **Plan migration**: In TOGAF, this step represents the recommendations made at the previous step being taken forward. This typically coincides with the handover of responsibility from the architecture team to the project team. For example, once they have evaluated possible solutions, the architecture team only provides evidence for a preferred option. An executive decision needs to be taken to proceed with that option, but it will not then be the architecture teams' responsibility to deliver that option. The term "plan migration" is used because it is typically where the hypothetical roadmap captured in the architecture turns into a concrete plan for business change. The document focuses on the aspects of the architecting approach and omits aspects that are covered by common project management practices.

   For ESA and teams using the ESA-AF, this step may be taken as a point at which feedback is given from a program team on the model. It may indicate which versions of variant options have been approved, making one candidate the new baseline truth for other models to reference. It may also be that this step is just one of the gates in a development program, immediately feeding back to further architecture development to support the next phase.

6. **Manage implementation**: This is the stage in which the planned migration, itself based on the solution roadmap identified in the architecture, is executed. The original architecture model should be used as a reference throughout implementation, and divergence from the conceptual model (made for expediency or to work around unforeseen issues) should be noted.

7. **Manage architecture change**: This stage is shown following implementation and represents the reconciliation of the actual solution with the model. In fact, the model may be continuously updated to reflect reality throughout the development, transition, and operation of a solution. Just as for Systems Engineering or service management best practice, there must be a good change control procedure in place. This should avoid lost updates or conflicts and help maintain confidence in the model as an accurate representation of the actual architecture.

   Though the architecture model may guide SoS change, as well as represent the evolution of the SoS, the ESA-AF does not provide a mechanism for architecture change governance through the model itself. It does not implement an issues management system, define element types for change requests, or enable the capture and reporting on change records as their status is updated (e.g., from submitted via assessed and approved to implemented, reviewed, and validated). These aspects should be handled by regular project management tools (or service management tools, if the SoS architecture being modeled is already operational).

   Note that it may be that the SoS change management process has already started during architecture development. ESA projects and programs have long lives, and the architecture task may be initiated to address specific concerns other than solution option selection. In this case, the governance mechanisms must ensure that changes that arise while the architecture is being developed are also well managed.

8. **Manage requirements**: TOGAF places requirements management at the heart of the enterprise architecture life cycle and indicates the bidirectional flow of

information from requirements to architecture development and all other stages. The requirement set contains the user and systems or technical requirements for a given implementation project. It also includes the requirements for the architecture task itself; for project management; for governance and accreditation; for testing, validation, and verification; and so on.

12.3.3 Modeling

The perspectives of the ESA-AF, as shown in Figure 12.8, address the points of view of different classes of SoS stakeholder. Each perspective defines a group of views that are designed to address these stakeholders' concerns. Each perspective is described subsequently and includes aspects of their views' architecture development to illustrate their purpose.

The following subsections therefore function as a brief introduction to the recommended steps within the architecture development stage of the architecture life cycle. It can be read as a suggestion of which views to start with, what to develop next, and how to link their elements together.

12.3.3.1 *Strategic* and Related Views

The Strategic Views (StV) of the ESA-AF are a suitable starting point for an architecture model with strong programmatic aspects. The overarching purpose of the enterprise, SoS assembly, or program of interest can be captured and presented in a few diagrams. The essential dates and deliveries of those programs are captured in Acquisition Views (AcV).

Capabilities are the key metamodel elements shown on the StV. They also link the StV together and link other perspectives to the StV. The capabilities of an enterprise are abstract expressions of what the business achieves. They are typically enduring and expressed in a general way, so that different solutions (projects' deliveries) may meet the same capability (or variations of it) over time.

Like most modeled elements (e.g., except relations), when capabilities are created, they should be given short descriptive names (up to six words, as a guideline), copying

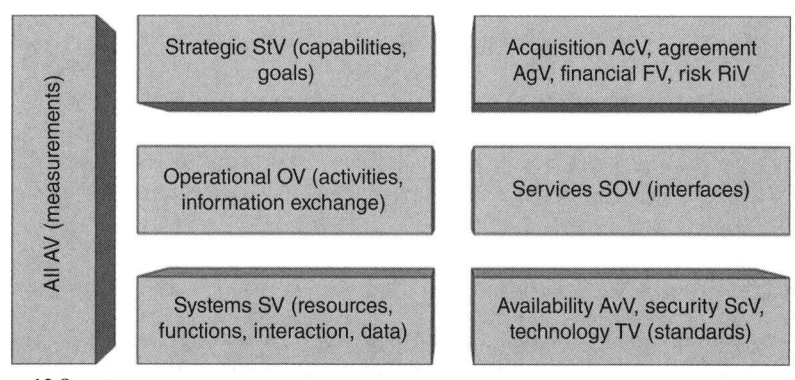

Figure 12.8 ESA-AF views are grouped into perspectives.

established phraseology if possible. They should be recorded with a more detailed description (between two sentences and two paragraphs) and that information's origin (a source document or website or a subject matter expert's name with the date when their opinion was recorded).

As modeling progresses, if what had been captured as a capability is associated to a specific role or becomes part of a sequence (and a specific information flow), it may need to be remodeled as an operational activity in the Operational Views (OV). Alternatively, if it is associated with quite specific qualities of service or constrained to well-defined providers and consumers, it should be remodeled as a service in the Service-Oriented Views (SOV).

It is also possible that a project's deliverable is represented as a capability. For example, the Envisat capability may be used as shorthand for the set of Earth observation capabilities that are supported by that satellite's instruments and supporting infrastructure. The capability configuration model element type, used especially with the Systems Views (SVs), should be used in this case. Operational activities, services, and capability configurations can all be linked to capabilities.

Some specific uses of the StV clarify the value of capability modeling and serve as examples of architecture development:

> **Identifying and Relating Required Capabilities**: The StV-2 Capability Taxonomy is primarily used to capture capabilities and their relationships. The specialization relationship is a useful way to group capabilities into classes. A hierarchic taxonomy can help to drive out the essential purpose of an enterprise, to identify underdeveloped areas, or to expose aspirations that are poorly aligned to the core business (orphan capabilities that lack a common parent with other capabilities).
>
> The StV-2 also supports capability composition and dependency relations. Composition is different to specialization, as each child in a specialization hierarchy may achieve the parent capability (albeit in a refined way), while the parts of a composition breakdown alone cannot support the aggregate capability.
>
> Metrics may be associated with capabilities on StV-2 diagram using measurement elements. It is useful to record capability metrics so their support by the operational performance measures, service attributes, and resource performance parameters used in the other views can be demonstrated. It may be that different metric values distinguish capabilities that share a common parent via the specialization relationship.
>
> **Goals and Capabilities for Enterprise Phases**: The capabilities captured in StV-2 diagrams may be linked to enterprise goals (which are broader than the capabilities) on StV-1 Enterprise Vision diagrams via enterprise phases. The phases typically represent quite long intervals in which several programs will be delivered. They may represent the epochs of a roadmap toward successively higher CMMI levels of organizational performance, for example. The StV-1 captures the vision statement for each phase and allows goals and capabilities to be associated to that vision.

12.3.3.2 Acquisition and Related Views

The Acquisition Views (AcV) of the ESA-AF may be populated alongside the StV, as the projects and programs they capture should align to the enterprise phases of StV-1. The key model element type of the AcV is the project. The ESA-AF does not distinguish programs from projects, but it can capture subproject relations, so when a program has oversight of a number of related delivery projects, this can be modeled. Organizations' ownership of projects and the resources they deliver can also be captured on the AcV.

Related to the AcV are four views in three perspectives that also largely lie within the program and project management focus of concern. They are the Agreements Views (AgV-1), Financial Views (FiV) FiV-1 Cost and FiV-2 Funding, and RiV-1 Risks. Agreements are between organizations and are typically fulfilled by projects' deliverables. These deliverables are also typically what is modeled as incurring costs, and it is organizations (possibly representing national governments) that provide the funds to support those costs. It may also be the deliverables that are subject to identified risks, and again, it is modeled organizations that own the risks.

Programs to Deliver Capability StV-3 Capability Phasing diagrams are used to capture capabilities' links to capability configurations. AcV-2 Program Timeline diagrams are used to capture the dates that projects deliver their capability configurations, which represent the solutions that realize capabilities. The AcV-2 also allows for a project to deliver different capability increments. Project milestones may also represent the out of service date of a capability configuration. Additionally, projects may be decomposed to subprojects, allowing major programs' work packages to be represented, for example.

The capabilities that are realized by programs' delivered capability configurations may also be shown on the AcV-2. However, for model clarity, it is recommended that AcV-2 diagrams are used for just program timelines and StV-3 is used for the capability link.

In a similar way to the AcV-2 and StV-3 linkage of programs' delivered capability configurations to capabilities, the AcV-1 Procurement Clusters shows how programs contribute to the higher-level enterprise goals. This view therefore allows clusters of programs that support a shared goal of an enterprise phase (shown in StV-1) to be shown together. Programs may also be grouped together if the same institution is responsible for them, as represented by links to instances of the actual organization element type.

12.3.3.3 Operational

The OV are an essential aspect of EA modeling that sets it apart from Systems Engineering models like the UML-based SysML. The OV diagrams present, in an abstract way, what an enterprise or program needs to do, capturing stakeholder expectations and requirements. This presentation can be analyzed independently from the solutions, which are captured in SV diagrams that show how the operational needs are delivered. By decoupling user requirements from solution specification, different solutions can be evaluated against logical needs and common solutions can be found for diverse operational scenarios.

The split between OV and SV perspectives is also highlighted by the OV using information, while technical solutions deal with data. Information has meaning to people, is needed for operational activities, and makes its providers valuable, while the underlying data exchange and storage are neutral to these human concerns.

With or without the higher-level strategic modeling, the operational perspective can be a useful entry point to an architecture model. The authors and interpreters of the model should be able to put themselves in the described picture and understand the needs and merits of the operational parts. The separation of needs from solutions may also help stakeholders appreciate the scope of their responsibilities.

Notably, the documentation published for the initial version of the MODAF was explicit about the views used by capability stakeholders for user requirements versus those used by procurement teams and suppliers for systems requirements. The suite of SV diagrams produced by the latter would be evaluated for compliance with the OV suite of the sponsors at early program development points.

To enable an abstract view of operations, OV diagrams use the operational node model element type. Operational nodes may represent human organizations or roles, physical systems or assets, or the capability configurations that programs deliver that combine systems and skilled operators. Links to these resources can be shown with the known resource model element type. However, the operational nodes themselves should be thought of as abstract components, which could be realized by anything; they are somewhat similar to UML actors or the external entities of data flow diagrams (DFD). Though they are abstract logical entities, they are still the locus of operational activities and the subjects of the information flows shown on OV diagrams.

The following specific uses of the OV clarify their use and further demonstrate architecture development:

Scenarios Illustrating Essential Operation Operational scenarios are a useful way to think about how an enterprise operates or what a program should achieve. These can be a short story of the enterprise's main business, or the employment of a capability, supported by a graphical picture of major operational parts and their relations. This graphic may capture the interaction of major components within the enterprise and their connection to external entities.

The OV-1 High-Level Operational Concept may be used for these scenario diagrams. OV-1 diagrams are less formal than other parts of the model—a diverse mixture of model elements may be shown on it, and the links drawn between them have arbitrary meaning.

The OV-1 can also be used to show operational performance attributes, using the measurement element type (which can be attached to any ESA-AF element). These measurements can capture values for the required measures of effectiveness (MoEs), separated from solutions' measures of performance (MoPs) and supporting the StV-2 capability metrics. It is also possible to use different OV-1 diagrams to illustrate the difference in operational effectiveness between the as-is enterprise and the envisioned to-be phases.

Operational Information Need Lines The operational nodes that may be implied on OV-1 overviews are more rigorously defined and connected on OV-2 Operational Node Relationship diagrams. The OV-2 is important for understanding of the end-to-end operation of the enterprise or the scenarios in which a capability is employed. Even without connections, the inventory of the operational nodes implied by the high-level operational concept or other statements of stakeholders' needs is useful. Additionally, operational nodes may also be associated with a location, which may be a generic environment (e.g., space).

The OV-2 connections are defined as need lines, which summarize the information flows between pairs of nodes. These relational model elements highlight the reason why operational nodes interact. This relation further reinforces the difference of the OV and SV levels of abstraction—the needs that are the reason for information exchanges disappear when the technical solutions are linked by resource interactions that enable data exchange.

Note that though semantically information consumers need their providers, need lines are conventionally shown going from providers to consumers (in the direction of the information flow). The information exchanges that support a need line are themselves captured on the OV-3 Operational Information Exchange view.

Modeling Operational Activities The activities that are performed by operational nodes are captured on the OV-5 Operational Activity Model. This follows the standard UML activity model convention, representing the possible flow of control from each activity to the next (with start and end points, as well as conditional and concurrent paths, if required). Swim lanes are used to indicate which operational node performs the activities, and the flow of control may pass between the activities of one operational node, or across swim lanes to the activities of other nodes. The OV-5 can also capture the information flowing between activities; any activity that takes as input information entities that are output by other activities must necessarily be downstream of them.

Other views of operational nodes' behavior are available in the ESA-AF. The OV-6b Operational State Transition Description also follows the UML convention. Like UML, therefore, OV-5 diagrams can be formally linked to OV-6b diagrams, as operational nodes' activities are equivalent to state transitions. Additionally, the OV-6c Operational Event Trace Description shows behavior as a message sequence chart, with the lifelines representing operational nodes. ESA-AF modelers may choose either OV-5 or OV-6c for an architecture model, depending on which is clearer to the model's audience. Alternatively, they may use OV-5 diagrams for summary views of possible activity flows (with conditional branches, including both normal operation and exceptions) and OV-6c diagrams for specific interaction scenarios (without branches).

Behavioral modeling is useful in highlighting constraints on activities and the sequence of events. For example, there may be preconditions that must be true before an activity begins (such as information being available and an operational node being certain state), postconditions that must be met before the activity completes, invariants that must be upheld throughout the activity, and constraints on coupled flows or sequences (e.g., specifying minimum and maximum times between events). These may be brought together on the OV-6a Operational Rules Model along with any other constraints from the OV (e.g., concerning the OV-7 Information Model, discussed in the following).

Modeling Operational Information Flows and Entity Relations Information exchange between operational nodes is implied by the need lines of OV-2 diagrams, the activity flows that cross swim lanes in OV-5 diagrams, and the messages on OV-6c diagrams. These information exchanges are shown explicitly on the OV-3 Operational Information Exchanges that indicate which operational nodes the information exchanges go from and to. For a given pair of nodes, several information exchanges may support each need line that links them, and each information exchanges can carry several information elements. The OV-3 can show the information exchanges' relation to both these model element types. Additional properties can be allocated to information exchanges using measurable properties, which may be different from the attributes of the information elements themselves. For example, frequency (say, once per day), priority, and language may be indicated. Other technical qualities, such as radio-spectrum frequency, data rates, encryption, and digital versus analogue encoding, are better suited to the equivalent SV-6 Systems Data Exchange.

While the OV-3 shows the information exchanges and therefore packaged message content, the OV-7 Information Model is used to show the structure of operational information at rest. Specifically, it can show the relations between information entities that could define a logical information model for a database. OV-7 diagrams can therefore be used in the same way as conventional ER diagrams, though by using standard UML class diagrams additional relations (such as specialization hierarchies) can be shown.

12.3.3.4 Services

The service-oriented approach, as used in Service-Oriented Architecture (SOA), is itself a powerful way of abstracting away from solution details. It allows the components of an SoS to be treated as black boxes, described just by the interfaces they present. Those interfaces invoke functionality or trigger the production of information, but they also hold the quality of service (QoS) that the component provides. By presenting a description of operations with an advertised QoS, consuming components can be fitted to the most suitable providers in an end-to-end delivery chain. There may even be automated mechanisms for dynamic binding and contractual guarantees of service level, including payment. Such mechanisms can be implemented for software through the Web service (WS) set of standards, but the service-oriented approach can be extended beyond this class of technology to apply to logical business operations. It allows organizations to recognize the essential components of their enterprise (and those they interact with), enabling escape from inflexible legacy arrangements toward agile interoperation of decoupled capabilities.

The ESA-AF SOV are independent of the OV and SV levels but link to both of them, enabling a service-oriented approach to be applied to both business operations and technical solutions. Services may be provided by and consumed by either operational nodes or resources. Note that it is convenient to keep operational services separated from technical services. There are no formally defined service classes in the ESA-AF though, as additional numbers service layers or tiers, or other classifications (e.g., of business, information, management, and enabling services), may be required in certain architecture models.

Service Taxonomy and QoS Constraints Just as StV-2 diagrams are useful for the capture and arrangement of capability descriptions, SOV-1 Service Taxonomy diagrams can be used to present a classification hierarchy of services. In the ESA-AF, the concepts of capabilities and services are closely related, and services can be shown as aiming to achieve capabilities on SOV-3 Capability to Service Mapping diagrams. The line between capabilities and services may be drawn differently for different architecture models, but as a rule of thumb, if details of the information products delivered can be identified, then modeling is at the level of operational services.

As services are refined further, they may be identified as supporting interfaces defined by data standards. In this case, technical services are identified. These are not necessarily specializations of the operational services though. Instead, there may be orthogonal service taxonomies, with the operational hierarchy broken down into classes of enterprise activity while the technical is structured by the types of solution. For example, groups of operational space mission services may include classes for scientific data, operations mission plans, and telemetry, while technical communications services may be grouped into fixed network versus radio solutions. However, once operational and technical service taxonomies are defined, SOV-1 diagrams may be used to show which technical services are used by the operational services using service composition relations, which are available in addition to the taxonomic specialization relations.

The SOV-1 also allows constraints on service attributes to be captured using measurable properties for service levels. For communications services, attributes may include data rate and availability, for example, and a satellite's radio frequency and orbit may constrain these. Just as the specialization relationship means children have at least all the attributes of their parents, if service-level constraints are indicated for the parents, children must also meet at least those levels.

12.3.3.5 Systems and Other Solution Resources

Though titled System View (SV), SV diagrams are generated to describe any type of technical solution. Operational needs may be implemented by people (represented by actual organizations, organization types, posts, and roles), physical assets, or software as well as regular engineered systems. These things are all represented by model element types that are specializations of the generic resource type.

Capability configuration elements, mentioned earlier as the deliverable products of programs, are also resources. As noted, capability configurations can contain both systems and the people that operate those systems, as well as other physical assets and hosted software resources. This makes it explicit that any capabilities associated with the delivery will not be achieved if only equipment is delivered. Competent operators are also needed, as well as responsible organizations, physical assets to host the system, and software to run on it, if appropriate. The capability configuration resource type may also be used where no details about that enterprise component need to be captured, except that it needs to be linked to other components to support end-to-end operations. This still implies that there is an organization (with programs and funds) to support delivery of that capability configuration.

As the implementation details of technical solutions are captured in SV diagrams, the EA model can overlap with systems design (and software or technical architecture). In general, the EA model should not reproduce the actual solution architecture, as it normally serves a different purpose from the Systems Engineering high-level design in a development life cycle. If the technical architecture exists, those aspects of it that support the concerns of the EA analysis should be summarized in the SV diagrams. Therefore, if, for example, interoperability of different programs' delivered components in an SoS is being assessed, just the external interfaces' characteristics of the components should be captured (without their internal assemblies, functional breakdowns, or data models).

Resource Composition and Interaction SV-1 Resource Interaction Specification diagrams are typically the starting point for modeling the resources that support the identified operational picture; they may even be the only SV used in an EA model. Often, a useful reference model is created by just using SV-1 diagrams to bring together a reference inventory of the resources in an SoS, holding descriptions and provenance from different sources.

Reasonable compositions of resources may be shown on SV-1 diagrams—for example, equipment may host software, but software cannot host a physical asset. These relations allow subsystem decomposition and physical asset part assembly to be shown. Note that a model of a solution's physical construction may be quite different from its subsystem breakdown, which identifies functional blocks. For example, the physical assets of a satellite may include parts on a core framework, with separate antennae and solar arrays, but the communications and power systems cut across these.

The resource composition relations can also capture organizational hierarchy, with the posts' hosted in an organization and the roles taken by people in those posts. While this hierarchy may be shown on the SV-1, the OV-4 Organizational Relationship Chart is also available for this; the OV-4 doesn't hold other resource types. Versions of the OV-4 for actual and typical organizational structures are available. Program owners and organizations on AcV-1 diagrams are normally drawn from the actual OV-4, while actors in operational scenarios (known resources realizing OV-2 nodes and the resources in the SV picture of solutions) are drawn from the typical view. The OV-4 is also used to show which roles have competency to use a given resource.

As well as resource composition, SV-1 diagrams show interactions between model elements in the resource family. This makes them similar to OV-2 diagrams. Therefore, those entering and examining the model can verify that the required operational node need line links are supported by resource interactions. Note that while resources are associated to the operational nodes they realize, their resource interactions are actually associated with the information exchanges bundled by need lines. The resource interactions carry data elements, just as information exchanges carry information elements.

Systems Interface Technical Detail For systems resources specifically, more detailed pictures of how resource interactions are achieved can be shown on SV-2 Systems Communications Description diagrams. Systems port model elements indicate interfaces, and systems port connectors link the ports for pairs of systems to realize their resource interactions. Each systems port can be modeled as implementing standards, possibly representing layers of a protocol stack. Therefore, independently designed and modeled systems can be brought together in the architecture to check for compatibility at all applicable layers of Open Systems Interconnection (OSI) reference model, for example.

The standards themselves, including those modeled as protocols, can be presented together on TV-123 Technical View diagrams. These can also show the institutions responsible for the identified standards, with the dates that standards are ratified or withdrawn, if appropriate.

Solution Behavior and Data Views There are equivalent technical solution views for each of the behavioral and information views described for operations. Though the definition of these views are almost the same as their operational equivalents, the points made earlier about the separation of operational user needs and technical solutions' implementation details should be kept in mind when populating them. The pictures of systems' (and other resources') functions and data should capture how the low-level enterprise components collaborate. This description of physical composition and interaction shouldn't overlap with the logical view of what is operationally achieved (arising from the human stakeholder needs that determine operational value).

The technical behavioral and data diagrams may necessarily be more detailed than the logical operational model. The exact sequence of data exchanges, with exceptions, may need to be defined for a protocol. However, a complete EA model needn't necessarily be bottom heavy, as one technical solution may meet several operational scenarios.

12.3.3.6 Technology Standards, Security, and Availability

Typically, associated with the concrete SV perspective onto the architecture are some narrowly focused perspectives and views. They are:

- AvV-1 Availability—This dedicated ESA-AF view picks out one key performance metric, availability, to highlight values for alternative modeled solutions, possibly including equipment clusters designed for redundancy in the case of failure.
- ScV-1 Information Assets and Security Requirements and ScV-2 Security Solutions—These new ESA-AF views are not yet fully defined. Security may also be seen as a key nonfunctional requirement in the SoS to be pulled out on dedicated views. However, even before these are fully defined, security requirements and solutions can be captured in the ESA-AF. Security analysis addresses information value, threats, vulnerabilities, controls, and risk management. Value itself depends on the qualities confidentiality, integrity, and availability, which can be indicated by measures on information entities. Threats, vulnerabilities, and controls can be modeled in connection diagrams at both the OV and SV levels. For risk management, the residual risk (after the controls against identified threats and vulnerabilities are implemented) can be tracked from the RiV perspective.
- TV-123 Technical View—As introduced in "Systems Interface Technical Detail" earlier, TV-123 diagrams primarily collect information standards and protocols.

12.3.3.7 All

The All Views (AV) perspective, typically shown as orthogonally cutting across the other perspectives, is not intended as a perspective on the architecture model itself, but rather a container for information about the architecture. Three AV diagram types are specified by the ESA-AF:

1. AV-1 Overview and Summary Information—The AV-1 contains information about the architecture modeling task. Initially, it should indicate the scope of the work and the concerns to be addressed, the approach to be taken with regard to views and contextual information to be used, and other planning information. Once the architecture is completed, the findings of the architecture task may also be recorded here. Throughout the architecture task, the status of the model (e.g., baselined, draft, in review, or approved) and other management information may be tracked. In the ESA-AF, this information may be collected in an architectural description model element, which may be linked to an enterprise phase and additional metadata.
2. AV-2 Definitions—The AV-2 is the integrated dictionary for the model, and in MagicDraw, it is represented by the explorer panel for the repository (with a navigable hierarchy of packages containing the diagrams and the model elements and relations that appear on them, as well as the metamodel).
3. AV-3 Measurements—The ESA-AF identifies the AV-3 for the measurements that can be attached to any element in the architecture model. However, the modeler will typically only enter and show measurements on the views that hold the

associated element (e.g., StV-2 Capability Taxonomy for capability metrics). The ESA-AF uses measurement sets to group measurements together and separates the measurement set types from actual measurement sets. Thus, a measurement set for a launcher MoPs could be defined to include mass to orbit, for example. The actual measurement set for an Ariane 5 ES would be an instance of this measurement set types, with an actual measurement value set to 21 tons to low Earth orbit (LEO).

Though not recommended, the AV-1 can be used to indicate deviations from the ESA-AF metamodel in a given model. The associated additional metadata can represent metamodel extensions, which should be described in the architectural description element. Such deviations and extensions impair the ability to merge and reuse architectures, as models with frameworks that diverge from each other and cannot easily be reconciled in a federated repository for the SoS model. This therefore undermines the value of using the ESA-AF as a methodology for SoS engineering.

12.3.4 Modeling Tool

MagicDraw is the current modeling tool in the ESA-AF. Apart from its enhanced customization and extensibility capabilities, the tool also supports collaborative development of architecture models and model exchange over a model repository provided through the MagicDraw Teamwork Server. The ESA-AF modeling is supported particularly through the profiles and customizations generated from the ESA-AF metamodel. Several profiles are generated from the Ecore metamodel—for ESA-AF, UPDM, and the accompanying standards, SysML and SoaML, packaged in a single ESA-AF module. The profiles are used to apply ESA-AF-specific stereotypes to standard UML entities in MagicDraw. The tool customizations additionally provide ESA-AF-specific menu items, diagram types, etc. These customizations facilitate the modeling, for example, by displaying only relevant elements to be used in the various ESA-AF-specific diagrams, restricting the ends of relationships or new attributes, behavior, operations, ports, etc. created for a specific element only to allowed types just to name few.

 A further feature addressing better model readability especially for instance-level elements is provided through processing of element containments on a diagram. For example, an Actual Project can have a parent project (denoted through a "whole" attribute) and subprojects (denoted through "part" attribute).

 The ESA-AF plug-in delivers enhanced extensibility for the containment feature to be able to respond to metamodel updates or specific user preferences. Concretely, it allows definition of containments in an XML-based configuration file, where the container and contained types are defined through their qualified names, the corresponding properties to be updated (e.g., whole/part for the earlier example), as well as whether the containment should be applied also for subtypes.

 The ESA-AF plug-ins further enable enhanced formatting of diagrams and elements, for example, for setting different colors for elements coming from specific viewpoints (profiles). This formatting can also be adjusted through an XML-based customization file.

Overall, the modeling plug-ins provide significant improvements to the MagicDraw modeling environment addressing the complexity of the metamodel (based on UPDM) and delivering improved user experience.

12.3.5 Exploitation

In addition to the modeling, the ESA-AF provides enhanced exploitation functionalities with dedicated diagramming and reporting that particularly support the compilation of summary and derived data by interrogation of models published from the modeling tool. The exploitation is aimed at users with general business knowledge who are more focused on the information contained within the models rather than on the formal model representations in architecture views and modeling constraints.

The general approach taken in the ESA-AF is to build the exploitation framework and tools based on the Eclipse framework to make use of its inherent extensibility. The diagramming functionality is based on the Eclipse Graphical Modeling Framework (GMF), which utilizes the Graphical Editing Framework (GEF) to build a graphical editor automatically based on a given Ecore metamodel (Eclipse, 2013b). GMF provides an industry-proven, model-driven approach for generating diagram-based graphical editors. Thereby, the development of the diagram tools is limited to the definition of XML-based GMF models and does not involve any actual programming. Modifications of the generic GMF diagram tool functionalities needed for the ESA-AF exploitation tools are made automatically during diagramming tool generation by using tailored GMF generator templates. Additional ESA-AF-specific functionalities for the diagramming tools are provided through a common exploitation framework diagramming plug-in. References to the latter are inserted automatically in the generated diagramming exploitation tools by adding appropriate extension points in the plug-in descriptor during diagram tool generation through the custom GMF generator templates. The additional ESA-AF-specific functionalities enable, for example, showing selected references for a given model element on an exploitation diagram so that the user can establish complex relationships and explore the context of a model element in a step-by-step manner. A further functionality enables producing a report input file by extracting only those elements from an explored published SoS model that are visible on a given exploitation diagram. This functionality allows the user to report only on a subset of the SoS model that is of particular interest.

Figure 12.9 shows an ESA-AF exploitation diagram exploring GMES detection and monitoring responsiveness and latency. The view presented is more pictorial than the "boxes and lines" diagrams of typical architecture framework views. The view brings out specific performance characteristics (here responsiveness and latency) through simple, standardized, and informative icons, so that the nature of the various elements and their complex interrelationships can be explored "by eye" by nonspecialists.

The reporting part of the exploitation framework is built on top of the Eclipse Business Intelligence and Reporting Tools (BIRT) (Eclipse, 2013a). Models published from the ESA-AF modeling environment can be imported (through simple file copy/paste) in a project in the BIRT environment and used as data sources for a report. To enable using EMF models as BIRT data sources, the Ecore Open Data Access Driver has been used and extended with additional exploitation framework-specific features, for example, for handling appropriately the models according to the exploitation metamodel.

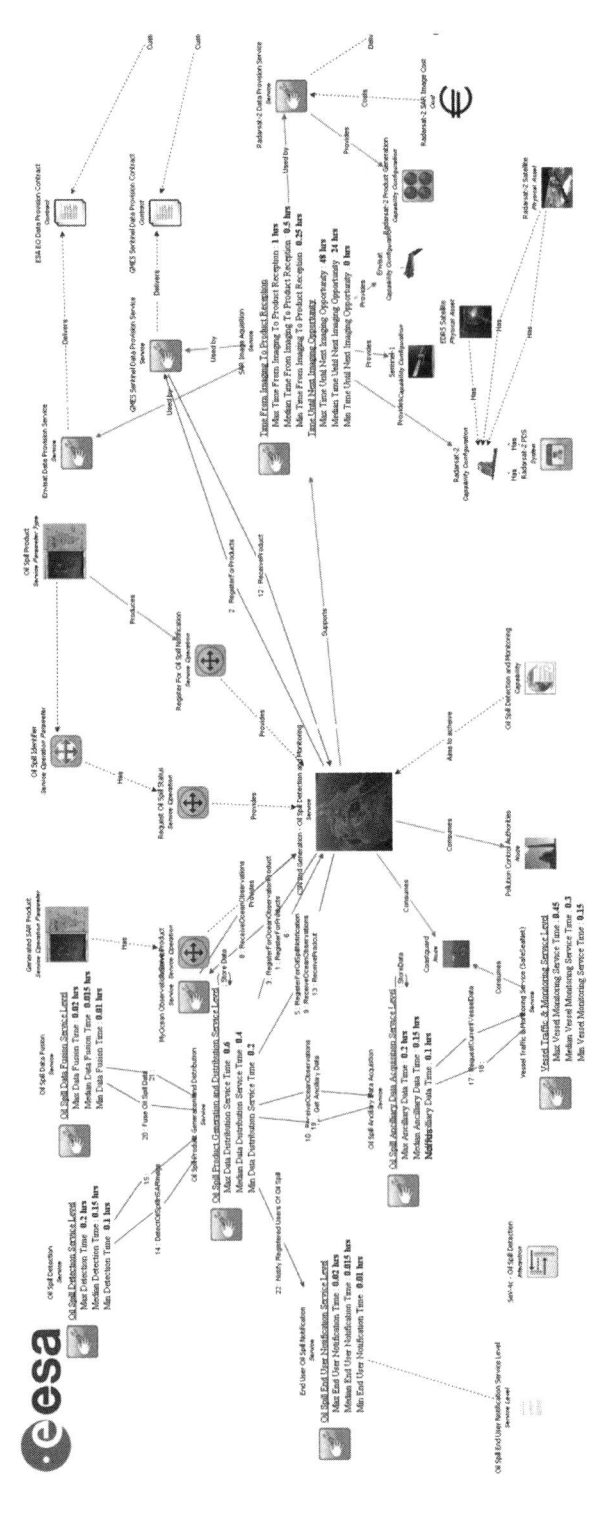

Figure 12.9 ESA-AF exploitation diagram for GMES detection and monitoring responsiveness and latency.

Multiple models originating from different programs or different architecture options can be published from the modeling tool and used as data sources in a single report. This allows direct comparison of architecture options and projects, which is highly relevant in ESA context. Multiple data sets can be defined for a given data source in BIRT, defining what data needs to be extracted from the data source. The relevant elements and their attributes can be browsed visually in a hierarchical structure generated according to the type hierarchies in the exploitation metamodel. Here, it is important to stress that the definition of the data sets will not be performed by a business user, for example, a program manager or analyst, but rather by modelers or technical users with knowledge of the metamodel and the conceptual relationships. End users of the reports are enabled to use report templates and to customize reports in a lightweight manner, for example, through setting parameter values. In general, no software development is required for report definition.

Further, joint data sets can be defined by linking different data sets (including other joined data sets). For example, a data set containing all ActualProject elements in a model can be joined with a data set containing all "CostOf" dependencies. The resulting joined data set can then be joined with a data set containing all ActualCost elements thus delivering a data set containing all actual costs for projects. The results of this data set can be arranged in a cross tab to provide cost breakdown for projects. Figure 12.10 shows an example of a sophisticated report of project costs. The report enables also parameter input for specifying a project to report the costs for, filters for coloring values in a specific manner if certain conditions are met, charting, etc.

Figure 12.11 shows another example based on the scenario in the diagram in Figure 12.9.

Cost Overview:

- The total project cost should be no grater than 50 000 (otherwise colored red)
- A total operational cost above 10 000 is questionable (colored orange)

Cost Type	Project 2		Grand Total
	Sub Project 2-1	Sub Project 2-2	
Development Cost	30000	20000	50000
Maintenance Cost		3000	3000
Operations Cost	4000	8000	12000
Grand Total	34000	31000	65000

Project 2 Cost Distribution

Sub Project 2-1 Development Cost:30,000
Sub Project 2-1 Operations Cost:4,000
Sub Project 2-2 Operations Cost:8,000
Sub Project 2-2 Maintenance Cost:3,000
Sub Project 2-2 Development Cost:20,000

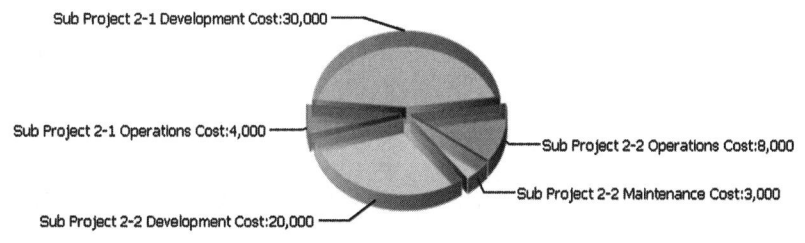

Figure 12.10 Example of a project cost report.

End-to-End Responsiveness and Latency Report
System of Systems Design and Analysis Environment

Interaction	Services in End-to-End Interaction	Min Time (hours)	Median Time (hours)	Max Time (hours)	Information Sources
SeV-4c - Oil Spill Detection	Oil Spill Product Generation and Distribution	0.2	0.4	0.6	Speculative data, for illustration purposes only
	CSN 2nd Generation - Oil Spill Detection and Monitoring	Unspecified			
	GMES Sentinel Data Provision Service	Unspecified			
	MyOcean Observation Service	Unspecified			
	Oil Spill Ancillary Data Acquistion	0.1	0.15	0.2	
	Oil Spill Detection	0.1	0.15	0.2	http://cleanseanet.emsa.europa.eu/Remote_Sensing/detection.html
	Vessel Traffic and Monitoring Service (SafeSeaNet)	0.15	0.3	0.45	
	Oil Spill Data Fusion	0.01	0.015	0.02	
	End User Oil Spill Notification	0.01	0.015	0.02	http://www.ksat.no/Downloads/Oil-spill-example.pdf

Figure 12.11 ESA-AF exploitation responsiveness and latency report.

BIRT is a powerful framework and enables visual report definition through dragging data sets on a report sheet and using different visualization elements such as tables, charts, cross tabs, custom images, etc. It further provides advanced scripting capabilities (including JavaScript), which enable an increased level of report customization and even programmatic report generation. The reports can be exported to a number of formats such as Microsoft Word, Excel, and Powerpoint, HTML, and PDF, thus producing documentation straight from a model. An important advantage is that by being part of the Eclipse framework, BIRT provides enhanced extensibility so that advanced exploitation components and tools can be developed on top of it.

12.4 LESSONS LEARNED

Complex systems modeling was able to support discussions and decisions on programmatic level by providing a well-structured representation of the concerns and the links between the different systems elements at all levels. However, some aspects were clearly identified as crucial in this process and need to be addressed very early when deploying such an approach for a new SoS:

- It is primordial to identify the architectural concerns early in the process. If this is not done carefully, much modeling effort is actually wasted by not providing the right information and not modeling the appropriate artifacts and relations and therefore not modeling them efficiently.
- Also at an early stage, it is necessary to identify if all concepts of interest are represented in the underlying metamodel. In both presented cases earlier, the "standard" UPDM profile did not offer adequate representation of programmatic and policy-related information. This needed to be included into the metamodel, leading to the extended ESA-AF. In the case of data policy, quite some effort had to be made to capture the relevant concepts, leading to a better common understanding of these concepts within the project team independent of the actual modeling exercise.

- At the present stage of tool support, it is still absolutely necessary to include an expert modeler in the team. The systems architect in general is not able to use the richness of possibilities and exploit the strength of the metamodel in an appropriate way. In this context, it has also been proven very useful to separate the actual model and the usage/exploitation of the model. With the modeling environment being based on the graphical representation of UML/UPDM, it was necessary to find a representation relevant for the architect and program management. This has been achieved by the exploitation framework of the ESA-AF, allowing also to define new representations/reports as required without changing the model definition or even the metamodel.
- Since the modeling effort is at first sight an additional effort on a project, it is primordial to have a "champion" in the team, that is, a believer in the benefit of the modeling exercise at sufficiently high level to provide the resources to do this. Resources in that case do not only entail modeler expertise but also the unhindered access to the expert knowledge to be represented in the corresponding model.

12.5 CONCLUSIONS

ESA Architecture Framework has been used in the context of two major Agency programs. They have contributed positively to the understanding of the systems architectures and the communication between all stakeholders and supported the trade-off analyses necessary to define a baseline architecture for the projects in question. These tools were an addition to the tools already used traditionally and never meant as a replacement. They need to be integrated in existing workflows and interface with other domain-specific tools. For the ESA-AF, this link still needs to be established on a manual modeling basis, but further works are ongoing to ensure a more systematic link between systems information coming from external sources.

REFERENCES

Eclipse (2013a). BIRT project. Available at http://www.eclipse.org/birt/phoenix/(accessed September 24, 2014).

Eclipse (2013b). Graphic modelling project (GMP). Available at http://www.eclipse.org/modeling/gmp/ (accessed September 24, 2014).

Eclipse (2013c). The eclipse process framework (EPF). Available at http://www.eclipse.org/epf/(accessed September 24, 2014).

European Commission (undated). Copernicus: The European earth observation programme. Available at http://www.copernicus.eu (accessed September 24, 2014).

European Space Agency (undated). SSA programme overview. Available at http://www.esa.int/Our_Activities/ Operations/Space_Situational_Awareness/SSA_Programme_overview (accessed September 24, 2014).

Federal Enterprise Architecture Program Management Office, USA (2007). FEA practice guidance. Available at http://www.whitehouse.gov/sites/default/files/omb/assets/fea_docs/FEA_Practice_ Guidance_Nov_2007.pdf (accessed September 24, 2014).

Jamshidi, M. (2009). *Systems of Systems Engineering: Innovation for the 21th Century*, John Wiley & Sons, Inc., Hoboken, NJ.

Maier, M. (2014). *The Role of Modeling and Simulation in Systems of Systems Development*, John Wiley & Sons, Inc., Hoboken, NJ.

Stoitsev, T. et al. (2011). *System of Systems Engineering with the ESA Architecture Framework*, 62nd International Astronautical Congress, Cape Town, SA.

The Technology Open Group Architecture Framework (TOGAF) (undated). TOGAF introduction. Available at http://pubs.opengroup.org/architecture/togaf9-doc/arch/(accessed September 24, 2014).

UK Government (2012). MOD architecture framework. Available at https://www.gov.uk/mod-architecture-framework (accessed September 24, 2014).

Chapter 13

System of Systems Modeling and Simulation for Microgrids Using DDDAMS

Aristotelis E. Thanos, DeLante E. Moore, Xiaoran Shi, and Nurcin Celik
Department of Industrial Engineering, University of Miami, Coral Gables, FL, USA

13.1 INTRODUCTION

This chapter takes the principles and concepts of system of systems (SoSs) and applies them to microgrids. This chapter will define a microgrid as it pertains to the five principles of SoSs, use an SoS architecture to develop a novel way to model such a distributed electricity system, and provide a case study in the modeling and simulation of a microgrid. The first section defines a microgrid focusing on the characteristics that make it an SoS as well as how it relates to the levels of control concept. Section 13.2 introduces the dynamic data-driven application system (DDDAS) framework of a particular microgrid design. Section 13.3 provides detailed experiments and results of the DDDAS framework as it applies to microgrids. Section 13.4 forms conclusions about simulation techniques for SoSs such as microgrids and gives a foundation for further research in the areas of DDDAS and its relationship to agent-based simulation.

13.1.1 Five Key Lessons Learned from this Chapter

1. A microgrid is an SoS and uses the five factors that define an SoS. Distinguishing itself from typical distributed electricity systems, it enables the concept of different levels of control.
2. The behavior of a microgrid requires data and inputs from multiple independent components that are located in separate geographies to function properly. An adaptive simulation scheme such as a DDDAS is the best modeling choice for this because of its ability to take ever-changing data and make real-time changes to the overall system based on those inputs.
3. Using optimal computing budget allocation (OCBA) and fidelity selection techniques for a simulation reduces costs and dramatically increases its ability to meet the demand requirements of all customers in an electricity distribution network.
4. Developing an agent-based simulation to model complex systems illustrates how agents are related to the SoS construct.
5. Fault detection provides a fast and reliable tool to restore or isolate the point in an electrical system where an abnormality occurs. Therefore, developing a robust fault detection system in a simulation model is critical to the overall security and reliability of the network.

13.2 MICROGRIDS AND DDDAS

A microgrid is an energy distribution system consisting of both renewable and conventional power generation sources along with some form of energy storage. A microgrid system is capable of operating both alongside of a municipal power grid and as an "island" separate from the local utility grid. The purpose of a microgrid is to provide energy security as well as an uninterrupted source of energy for its prescribed customer. An added benefit of microgrids is that the use of renewable energy sources allows for significant cost savings in terms of utility bills. To put these savings in perspective, the military was able to realize an annual cost savings of approximately one million dollars with the installation of a solar array on Nellis Air Force Base. In addition to the cost savings, the Nellis solar project will reduce the base's carbon dioxide emissions by 24,000 tons annually. This is a reduction that equates to removing 185,000 cars from the road (Air Force Civil Engineering Center, 2012).

To review, in order to meet the requirements of an SoS, a collection of systems must possess operational independence, managerial independence, geographic distribution, emergent behavior, and evolutionary development. A microgrid combines software, hardware, policy, and technology to form a more reliable, secure, robust, and convenient alternative to its current traditional counterpart grids. Microgrids satisfy all of the properties of an SoS as they include several operational and managerial independent component systems such as distributed and renewable energy sources and energy storages that are geographically separate from each other. Additionally, the component systems present emergent behavior due to the containment of renewable distributed energy resources that are driven by weather and are subject to random outages to which a microgrid must adjust

(Johnson, 2013). They are also highly complex in that the control system must effectively manage all of the component systems, ensuring that each component works to the benefit of the overall system. Finally, a microgrid presents evolutionary development as it can accept new power generation as storage capabilities as they become available to either meet the increased demand needs of the customer or to further reduce their costs. As such, a microgrid fits perfectly into an SoS construct.

13.2.1 Microgrid Design Challenges and Levels of Control

The ever-increasing demands placed on the nation's power grid and local electrical utilities present significant challenges in the effective management and control of a microgrid (Kassakian et al., 2011). First, microgrids require sensored meters in order to measure the customers' consumption and patterns of demand. Without this, the microgrid would not be able to effectively detect faults in the system or provide secure power to critical assets in the event of an emergency. Secondly, microgrids require a much more robust means of communication than presently in use in the local utility transmission and distribution networks. The control system in a microgrid uses a great deal of telecommunications resources in order to send price and status signals and receive information from consumers in real time. The next issue is that very few local utility grids can properly handle renewable resources. This is due to renewable energy resources that are by nature intermittent and nondispatchable. Furthermore, the systems operators for existing transmission networks do not possess ability to manage power generation equipment not owned by their specific utility company. For example, an orange farm that normally receives service from a local utility decides to install a diesel generator to supply power to their farm. The local utility company cannot control how much electricity the generator produces at any given time or when the owner of the farm decides to turn the generator on or off. This causes problems for the utility company in that the generator on the orange farm reduces their ability to forecast how much electricity they need to generate to meet the needs of their customers. Also, with the generator running, customers may not be able to utilize all of the generated electricity leading to waste for the utility. Finally, existing local utility grids cannot support new technical developments such as vehicle-to-grid technology. Such technologies allow electrical vehicles to store energy when the grid's demand is low and give back to the grid when the demand is high.

Viewing a microgrid in terms of levels of control helps to address the implementation challenges associated with installing a microgrid in parallel to a local utility. There are three main levels of control when referring to an electrical network. These are strategic, tactical, and operational. The strategic level provides a high-level overview of the entire system and provides a long-term vision for the future of the overall system (Ela et al., 2011). This is where we address the problems associated with the robustness of communications systems, sensored meters, and operators' inability to control components outside of their network. At this level, the microgrid controller ties into an existing electricity substation and monitors how much electricity the utility company is providing as well as the demand of the customers serviced by the microgrid. The most important aspect occurring at this level of control is the decision to isolate the microgrid from the local utility in the event of a power outage or national emergency. Component systems inside a microgrid that operate at the strategic level of control would be the computer mainframes and servers that oversee the overall electricity distribution system. Below the strategic level is the tactical level. This level is where we address problems stemming from the introduction of renewable energy resources into an electrical utility network (U.S. Department of Energy, Office of Electric Transmission and Distribution, 2003). At this level, the microgrid determines how much electricity to purchase from the local utility based upon whether or not the renewable energy resources meet the current consumer demand. It also determines when it is most cost-effective to store excess energy or sell it back to the utility company using a form of net metering. The lowest level of control is the operational level. This is where we address the issues of integrating new technologies into an electrical distribution system. Component systems operating at the operational level would include the photovoltaic (PV) array, a wind farm, energy storage, and electric vehicles, all of which require integration with a microgrid controller as well as some form of tie into the local utility. Breaking down microgrid component systems into its different levels of control allows designers to effectively solve the major challenges associated with installing an electricity network operating both in parallel and separated from a local utility network.

13.2.2 DDDAS and Microgrid Modeling

Traditionally, modeling techniques for the strategic, tactical, and operational control of microgrids involve a multipronged approach where the development of a theory, a model, and a set of experiments happens in succession. In this work, we develop a novel DDDAS framework for the operational control of a microgrid SoS. DDDAS is a new modeling and control paradigm that adaptively adjusts the fidelity of a simulation model. This paradigm steers the measurement process for selective data update and incorporates the real-time dynamic data into the executing model. The film Terminator 2 portrays a good visualization of the DDDAS paradigm in its application of dynamically driven data flows into SoS, machine learning, and simulations. In the film, T100 describes a world in which computers and other systems no longer require direct human input of data to achieve their desired outcomes. The strategic level component systems dynamically receive new input data from their instrumentation while the overall system is running. The tactical-level component systems then use that data to make changes to the overall system, thus enabling the operational-level component systems to run themselves. T100 refers to when the military replaced the computer systems in stealth bombers that enabled them to fly fully unmanned. After the upgrades, the bombers fly with a perfect operational record at which point the operational-level component systems begin to learn at a geometric rate, ultimately becoming self-aware. For all intents, Skynet from the Terminator movies provide some great foresight into the future of DDDAS and its application to an SoS. With the recent advances in technology such as those relating to the timely management/processing of large volumes of data and info-symbiotics, faster and more reliable data connection services, simulation and modeling mechanisms, and the like, DDDAS will allow multiple independent systems to interconnect. Such interconnections enable independent systems to operate efficiently with little human input or interaction. To this end, this project contributes to the advancement of DDDAS in that the proposed framework achieves a significant reduction in the computational resources required to conduct real-time simulation.

We achieve this by only looking at the detailed specifics of the system whenever the microgrid detects an abnormality, otherwise having the simulation in a more overseer mode. Combining the concepts of DDDAS and microgrids produces the embodiment of an SoS operating at the operational level of control as each component system within a microgrid can all operate independently for their own individual purposes. Furthermore, the DDDAS running within the microgrids uses sensors, computational resources, and various algorithms to link all of these separate systems together with varying levels of control to effectively manage a complex power distribution system.

13.2.3 Relevant Applications of DDDAS to Microgrids

Recently, many researchers (Son et al., 2002; Darema, 2004; Douglas et al., 2004; Mandel et al., 2004; Carnahan and Reynolds, 2006; Frew et al., 2013; Han and DeLaurentis, 2013; Wang et al., 2013) began to study new applications for adaptive simulations such as DDDAS. Electric power distribution networks, more specifically microgrids, are one of the application areas to make use of the decidedly effective measurement and control processes available by utilizing DDDAS modeling techniques.

One challenge in modeling a microgrid using a DDDAS is automatically adapting simulations when experimental data indicates that a simulation must change. Carnahan and Reynolds (2006) draw attention to this challenge and determined that the goal of the first run of every simulation is to gain insight about a particular phenomenon. We then use this insight to determine what new observations to collect and then adapt the simulation to reflect these observations. Here, attempting to generalize software so that it is able to anticipate each possible way to change will significantly impact its performance and make its underlying code unmanageably intricate (Parnas, 1979). Carnahan and Reynolds, therefore, propose a semiautomated adaptation approach that exploits the flexibility and constraints of model abstraction opportunities to automate simulation adaptation. While their study does not involve manual modification of the code or application of optimization methods, which can make the software extremely complex to control, it is still in need of human intervention to determine the most likely places of the code that should be changed.

Modeling an electrical distribution system, especially one that utilizes renewable energy resources such as a microgrid, requires the collection and processing of a substantial amount of sensor-based data between a variety of different devices and interfaces. In Celik et al. (2013) and Thanos et al. (2013), changes in the level of detail of data acquisition and the choice of certain parameters over others allow the automatic multifidelity adaptation in the simulation model. In addition to the development of interfaces to physical devices, they also address the creation of an infrastructure to support the communication and data requirements. Through a networked sensor-driven control system and integrated grid architecture for distributed computing, they proposed a methodology for the timely transfer of up-to-date data along the layers of the simulation model in real time.

Integrating all of the component systems of a microgrid into a managed electrical network capable of tying into a local utility requires a great deal of computing power. One of the significant benefits of a DDDAS is its ability to reduce the computational power required to accomplish its tasks. Wang et al. (2013) proposed a wide-ranging planning and control structure for unmanned aerial and ground-based vehicles based on adaptive simulations. In their research, Wang's team proposed processes to update both intelligence and surveillance data dynamically. Additionally, their research allows operators of such vehicles to merge and interpret time and geographical data to facilitate improved use of unmanned systems for surveillance and crowd control. Their use of integrating real-time ever-changing data into a simulation model seamlessly ties into these methodologies for simulation in distributed electricity systems. This allowed them to reduce the computational power required for simulation without sacrificing accuracy. Moreover, Frew et al. (2013) applied a DDDAS to continuous detection in intricate atmospheric conditions. Their research used novel onboard and remote wind monitoring techniques and capabilities. Furthermore, they developed a control framework that autonomously adapts their system to maximize efficiency of environmental, sensory, and computational resources. Their discoveries proved quite useful in efforts to combine multiple renewable energy sources into a dynamic model to reduce electricity costs to the customer.

The primary purpose of a microgrid is to combine multiple energy generation and storage capabilities and make them seamlessly work together to provide electricity to the customer. For a microgrid to be completely effective, it is critical to know a service interruption in one energy source impacts the overall microgrid performance (and potentially other energy sources). Han and DeLaurentis' (2013) research into managing the complicated interdependencies between component systems also proved extremely useful in mitigating such issues, which may result in exceeding the cost and time constraints of the customer. In relation to research with distributed energy systems, Han and DeLaurentis (2013) highlight the need to understand the capabilities and limitations of each component system that composes the overall system. Additionally, they emphasize the need to develop protocols to mitigate the risks of disruption to the overall system propagating from deficiencies in one or more component. In other words, when designing and building a microgrid, it is critical to understand not only how component systems act when combined as a whole but how each component system interacts with every other. The aforementioned research provides a great deal of insight into how to effectively model an SoS as well as methods to reduce the computational resources needed to model real-time systems data. Our approach differs from the previous research in that in addition to collecting sensory data at given intervals, it collects more detailed data when it detects an abnormality in the system (not meeting demand, not enough load on a bus, etc.). Furthermore, this model is the first to look at different portions of the grid in different levels of detail. There is no need to devote significant computational resources to portions of the grid that the previous computations have already determined are operating efficiently and effectively.

13.3 AGENT-BASED SIMULATION

A great way to simulate the behavior of an SoS such as a microgrid is by using an agent-based simulation model. Agent-based simulation is a cutting-edge technique for modeling systems composed of several autonomous "agents" that interact with one another. An agent-based simulation must contain three components: a collection of agents, the relationships between the agents, and the environment in which the agents interact. The manner in which the individual agents interact with one another influences the overall behavior of the system. Modeling agent interactions with each other allows patterns and behaviors to emerge that were not necessarily present or preprogrammed into the base model. The emergent and evolutionary behavior of these multiple independent agent interactions distinguishes agent-based simulations from other simulation techniques such as discrete-event and systems dynamics. Agent-based modeling offers a novel approach to the simulation of electricity distribution systems. This is because multiple independent agents (power generation resources) that interact with and feed into a larger network create the structure of an electricity distribution system. This larger network learns from their capabilities and adapts the overall systems behavior to better suit the needs of its customers.

13.3.1 Agents

Within this collection of agents, the most important characteristic of an agent is its ability to act without any external input. In general, agents act to achieve their own individually specified ends without regard for the actions of the agents in their environment. In terms of building a simulation model based on agents, each agent must possess certain characteristics. First, an agent must be individually distinguishable. This means that a person can determine whether a component in the simulation model is part of a specific agent. All agents in the environment have features and characteristics that distinguish them from but make them recognizable to all other agents. Next, agents are autonomous. Each agent is capable of performing its prescribed function independently in their given environment. Individual agents detect information and translate it into outputs (behaviors) that impact the entire system. Agents also possess states that vary over time. Similar to how the overall system has a state that is the collection of its state variables, an agent also has associated variables that determine its present behavior over time (Macal and North, 2010). Lastly, each agent has rules defining its interactions with other agents. These rules can include but are not necessarily limited to how different agents communicate with each other and their ability to respond to changes in the environment or other agents. In addition to these requirements for all agent-based simulation models, particular models, such as those for a microgrid, require that agents be heterogeneous. For an agent-based microgrid simulation to function properly, the model must consider the full spectrum of diverse agents operating within the environment. The simulation model gives each agent a different amount or resources and will in turn produce varying amounts of energy as a result of the different agent interactions.

13.3.2 Agent Relationships

When designing agent-based simulations, it is important to model how each agent interacts both with other agents and the overall system. There are two concerns when dealing with agent interactions: (1) whom each agent interacts with and (2) how the agents interact. In dealing with these concerns, it is critical to note the decentralized nature of agent-based simulation models. Furthermore, each agent is only privy to information specific to that agent. There is no centralized information available to all agents to optimize the performance of the overall system. When required, agents will interact with each other to receive information. However, just as in real-world systems, not all agents interact directly with all other agents all the time. Agents typically only interact with a subset of other agents from its environment.

The model's topology determines how agents interconnect and thereby how each agent interacts with others in the environment. The term topology refers to the mapping of information transfer among agents. In a microgrid, the energy-generating resources (agents) do not share information with each other, but all share information with the microgrid controller agent, which also receives information from the demand agents. Also, when referring to the topology of the model, it is important to note which relationships are static and which ones are dynamic. With static relationships, the model predefines the agent interactions, and they do not change, whereas externally occurring situations programmed into the model determine dynamic relationships. The microgrid model developed in this work includes agents interacting both statically and dynamically.

13.3.3 Agent Environment

Agents interact both with their physical surroundings and the other agents within the simulation model. The environment may simply be used to provide the location of an agent relative to other agents on a map or, in the case of a microgrid, more detailed information such as wind speed, solar radiation, and wave height.

13.4 CASE STUDY

A real-world application of a DDDAS framework to a microgrid is shown in Figure 13.1. In microgrids, the ultimate goal is to simultaneously minimize the operating cost and emissions while ensuring that the power demand is met at all times. In order to achieve this goal, our proposed DDDAS framework identifies the optimal mix of energy generation/storage resources to meet the demand in a distributed energy system. The proposed framework consists of a real sensor system, a database for storing the information collected by the sensors, a state estimation algorithm that uses fault detection and isolation restoration procedures, a fidelity selection algorithm, and an optimization algorithm.

13.4.1 DDDAS Framework

The data collected for this study comes from sensors located at different sources. There are sensors that measure the electrical parameters, that is, voltage magnitudes, bus injections, line flows, and sensors for the environmental factors, that is, solar radiation, wind speed, and temperature. To measure solar irradiance, we use the Kipp and Zonen CMP-11 pyranometer. To measure wind speed, the sensor used by the National Renewable Energy Laboratory (NREL) is the model WS-201 that records both the wind speed and the wind direction; and for the temperature, the sensor used by NREL is the model T-200A. Figure 13.1 depicts a method to apply a DDDAS framework to a real-world microgrid SoS. As we receive the sensory data, we filter and store it in a structured database. This database contains the topology information of the microgrid as well as the current cost of generating power from various sources. It then sends this filtered sensory data to the state estimation algorithm to trace any potential faults and outages. This method offers a faster

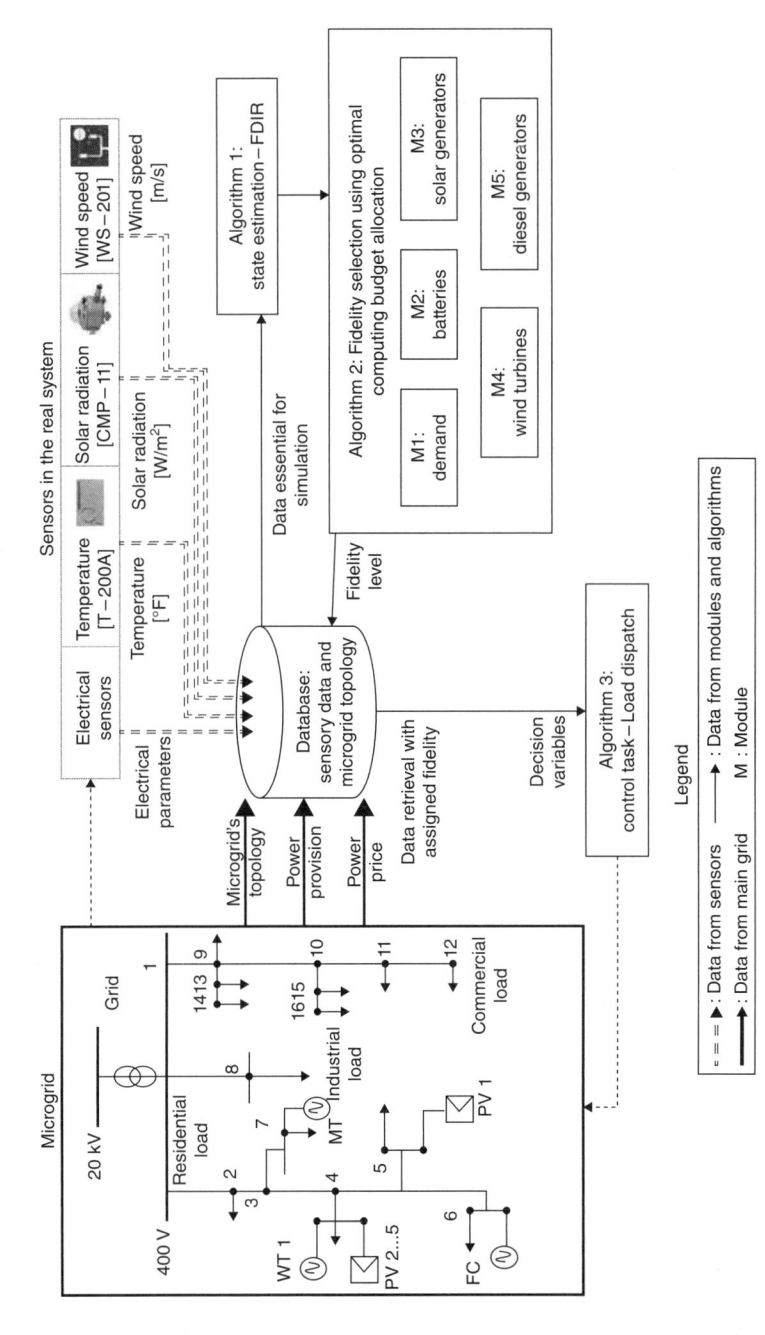

Figure 13.1 DDDAS framework applied to a real-world microgrid.

"come back" to the microgrid's original formation after the completion of maintenance. If the state estimation algorithm detects a potential fault or state abnormality, the fidelity selection algorithm initiates in order to determine the most suitable level of detail for the simulation of the system. An interesting characteristic of the fidelity concept is that they are not selected globally for the entire simulation model. Conversely, the grid structure splits the microgrid into regions based upon fidelity level. In a concept first introduced by Chen (1995), the fidelity selection algorithm uses OCBA algorithm to optimize the use of computational resources. Given the availability of computational resources and the time span before the next crucial decision set, the microgrid system simulates under different scenarios. These scenarios are the different levels of fidelity within the predefined regions. The OCBA algorithm assigns each region a fidelity level based on which scenario has the most significance. The optimization algorithm then uses the fidelity information to compute the optimal usage of the system's sources in terms of cost and emissions in addition to performing the load dispatch. To this end, the optimization algorithm sends a signal to the database and request data for each region with respect to the fidelity level. Finally, the optimization algorithm solves this multiobjective nonlinear problem and sends the decisions to the real microgrid. However, if the simulation does not reach the computational threshold and the available time span gives the opportunity for further study, the optimization algorithm updates the fidelity selection algorithm with the new decisions. We note here that the fidelity selection algorithm requires the dispatch decisions of the optimization algorithm in order to perform the simulation of the system under the combination of the different fidelities within the regions. This methodology is unique for a variety of factors. Primarily, the proposed framework achieves a significant reduction in the computational resources by utilizing an OCBA algorithm to track the available computational resources. This enables the algorithm to manage the number of replications of the simulation under different scenarios. To this end, the system optimizes based upon currently available resources. Furthermore, this framework looks at the system in an overseer mode when the system is running effectively and only looks at the system in great detail whenever the microgrid detects a potential issue with the system.

13.4.2 Military Base Microgrid

Military bases in the United States provide an excellent case study in the potential benefits of microgrids in terms of costs and security. Given that, we created a model to simulate a microgrid at a generic military base along the coast of the United States. The simulation includes renewable energy sources (solar, wind, and ocean energy), a large-scale or grid energy storage system, and a connection to the municipal grid. These energy sources all feed into the installation electricity network via feeders (buses) that distribute electricity to the different customers on the base. The sources all have switches that connect and disconnect them from all or a portion of the base electrical system based upon control signals sent from the microgrid.

For the purpose of the simulation, we broke down the electrical needs of the base into three categories: critical, priority, and noncritical. Critical demand areas include areas on a base that directly impact national security as well as first responders and healthcare facilities (Van Broekhoven et al., 2012). Priority demand includes the headquarters' buildings for the major supporting elements on the base. Noncritical demand areas on the base include the recreation, housing, and shopping facilities. The model also includes a bank of backup diesel generators to provide emergency electricity to the critical and priority areas in the event of a total power failure.

In the simulation, the microgrid selects the most efficient combination of energy sources to supply power to the base when all systems are operating normally. In the event of a national emergency or electrical disruption from the municipal utility, the microgrid can disconnect the base from the municipal grid and rely only on its renewable and storage resources. When an extended disruption occurs, the microgrid can switch off the noncritical demand from the base preserving resources from the critical and priority demand. In the event that a combination of renewable energy and energy storage resources is unable to meet the demand for both critical and priority loads, the microgrid disconnects the priority loads. When this occurs, affected loads rely on their backup diesel generators. Ultimately, this microgrid design ensures that the most critical assets of our national defense infrastructure have sufficient sustainable power capable of lasting the duration of any emergency that would cause them to isolate from a municipal grid while simultaneously saving the government millions of dollars annually. Figure 13.2 shows the microgrid topology for this case study.

13.4.3 State Estimation

Efficient state estimation is crucial in microgrids due to its significant impact on the control of the power flow and security of the system. While researchers have studied state estimation techniques extensively for transmission systems, such techniques are not suitable for distribution networks because the measurement data and models lack reliability. Evaluating the electricity consumption at each node allows for the effective estimation of states in distribution-level, medium-voltage networks. This is not a trivial task, collecting data on past demand and using that information to develop a model to forecast anticipated future demand requires a significant amount of computational resources. Furthermore, electricity can potentially vary a great deal between time periods due to weather or other factors beyond human control. In addressing these challenges, the use of a smart two-stage particle-filtering (PF) approach appears to be most suitable (Thanos and Celik, 2013). In that work, by using previously estimated states (voltage magnitudes and phase angles in the buses of the system) and incoming dynamic measurements obtained from historical data and electrical sensors, their algorithm updates estimations for states of each node in the network at each decision cycle.

The proposed state estimation algorithm incorporates a PF algorithm embedding different density selection rules. The core idea behind the PF is the procedure of the importance sampling, in which we estimate the properties of a desired distribution using an alternative distribution, called importance density, rather than using the distribution of interest. Acknowledging the fact that selection of different importance densities may present variances in the performances of PF algorithms, we consider two density selection rules that are structurally dissimilar. The first one is a density rule most commonly used in filtering literature (Doucet and Johansen, 2008). This rule collects samples from the previous rather than the subsequent iteration and assumes that the impact of the recent evidence y_n is trivial. Embedding this rule, a PF algorithm becomes easier to compute and implement, but may encounter degeneracy problems, in which all the mass may be

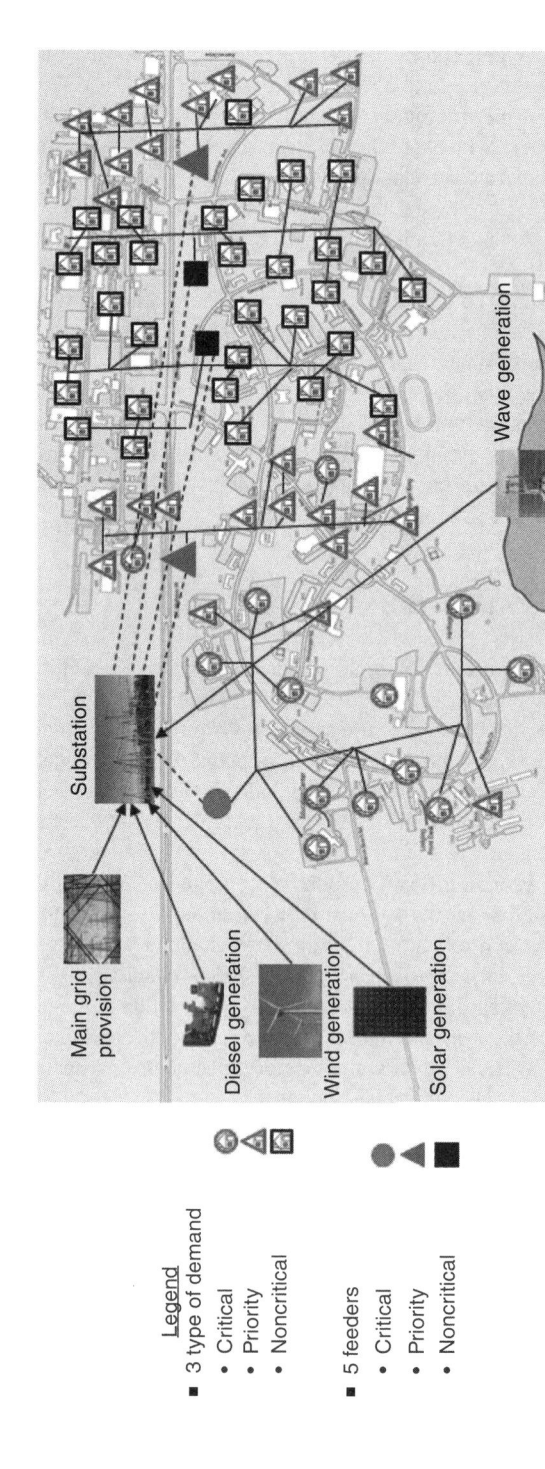

Figure 13.2 System topology of notional military base microgrid.

concentrated on a few random samples with most of particles having negligible weights, as n (iteration number) increases (Shi and Celik, 2013). In addition, only drawing samples from a previous transition prior ignores the measurements. This may lead to the poor performance on the estimation, especially when observations have significant effects on the posterior states. In order to exceed these barriers in performances of PFs, an improved importance-density selection rule, originally proposed by Shi and Celik (2011), is utilized as a second rule in this work. This rule focuses on preventing the degeneracy problems while taking both the previous state (the prior) and the probability that the next state is the same as the previous state (the likelihood) into the consideration simultaneously. Figure 13.3 shows how PF impacts the operation of the state estimation algorithm mentioned previously. When new measurements affected the posterior states or, as in our case study, when the likelihood is too peaked compared to its prior, the second rule is preferred.

In order to provide high-quality predictions, we develop a state-space model for daily electricity demand forecasting, based on a time-varying process (i.e., historical average demand and peak demand). This case study forms the electricity consumption using two parts: average demand and peak demand. The average hourly demand for each

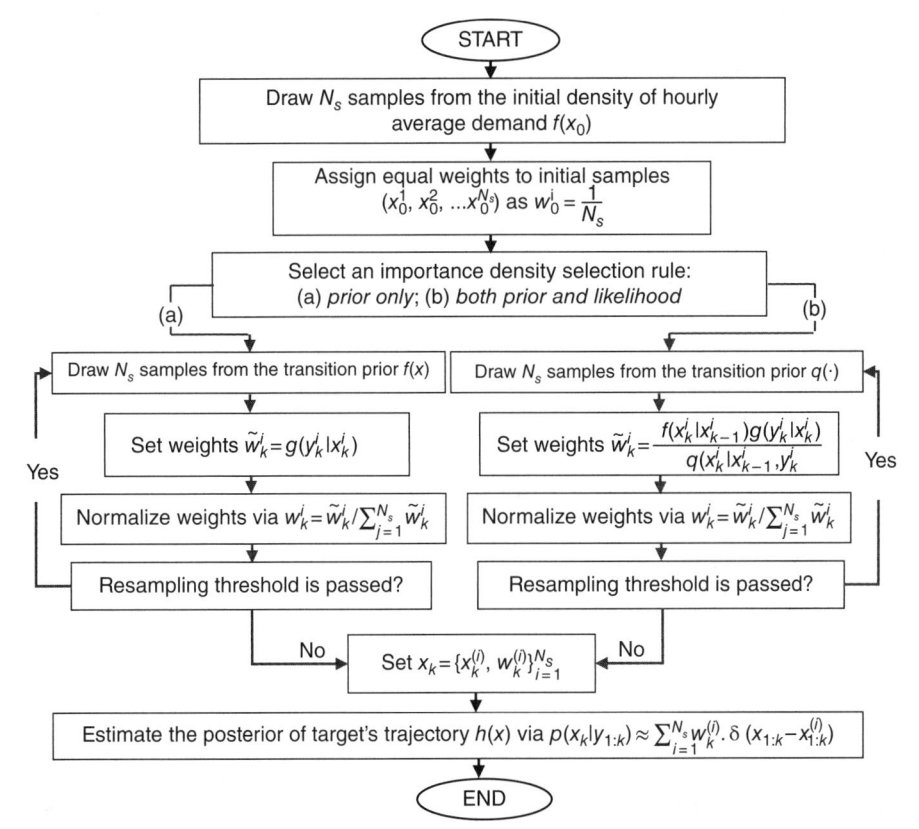

Figure 13.3 Operation of the proposed state estimation algorithm incorporating with PF embedded with different density selection rules.

day is a function of the peak demand on that day and the time (hour) that the process makes a prediction. To this end, the state-space model for electricity demand forecasting in our case study is detailed in the following equations:

$$D_{h,d+1} = \alpha_{h,d} D_{h,d} + U_{h,d} \text{ and } P_d = \beta_{h,d} D_{h,d} + V_{h,d}$$

where $D_{h,d}$ is the electricity demand at hour h of day d measured in kilowatts, $P_{h,d}$ represents the peak demand measured in day d, $U_{h,d}$ and $V_{h,d}$ denote the process and observation noises, and $\alpha_{h,d}$ and $\beta_{h,d}$ are parameters related to the state evolution and observation functions, respectively. Of note is that these functions update periodically in a time series manner as the parameters change hourly. While it is possible to obtain forecasting results for any time spanning the 24 h period, this analysis focuses on four particular times (6 A.M., 11 A.M., 4 P.M., and 9 P.M.) for ease in representation.

13.4.4 Fidelity Selection Algorithm

The purpose of this algorithm is to set the level of detail in which the simulation checks for abnormalities for each predefined region (critical, priority, and noncritical). In this way, it will manage the trade-off between the computational resources and accuracy of the decisions. Specifically, it will decrease the computational burden in steady states, while it will increase the effectiveness of the segregation mechanism and enhance the fault detection in abnormal instances. In general, higher fidelity means more frequent collection of data, more frequent update to decision variables, shorter cycle time of dispatch update, increased user control of the microgrid, improved segregation mechanism, enhanced fault detection, and reduced time to respond to the crisis situations that may arise.

We define five unique fidelity levels. We set isolation points as the points where we are measuring the demand and we allow the microgrid to control the switches in order to isolate specific areas of the microgrid from the municipal utility grid.

In fidelity 1, everything in the microgrid runs smoothly. There are no faults detected. The main grid is working properly and can provide the area controlled by the microgrid with energy. Thus, the control is limited to only one isolation point, and the demand is considered only as the total demand measured in the substation level.

On the contrary, in fidelity 5, we have detected a fault using the fault detection algorithm, and the system is observed and controlled at the maximum detail (at the buildings level). Here, the control is extended to every single node in the system (in our case with 186 isolation points—41 critical, 61 priority, and 84 noncritical nodes).

Fidelities 2 through 4 are selected using the OCBA algorithm in a nondeterministic way. The level of observation and control varies from the feeder level for the noncritical feeders to the node level for the critical and priority feeder. We consider fidelities 2–4 when the main grid does not function properly and cannot supply energy to the microgrid. Such conditions of the microgrid when an incident occurs are sent to the OCBA for a more sophisticated decision. These conditions include the demand within each individual feeder, generation of each renewable source, and diesel generators' real-time capability. Figure 13.4 depicts the different fidelity levels considered in our case study model.

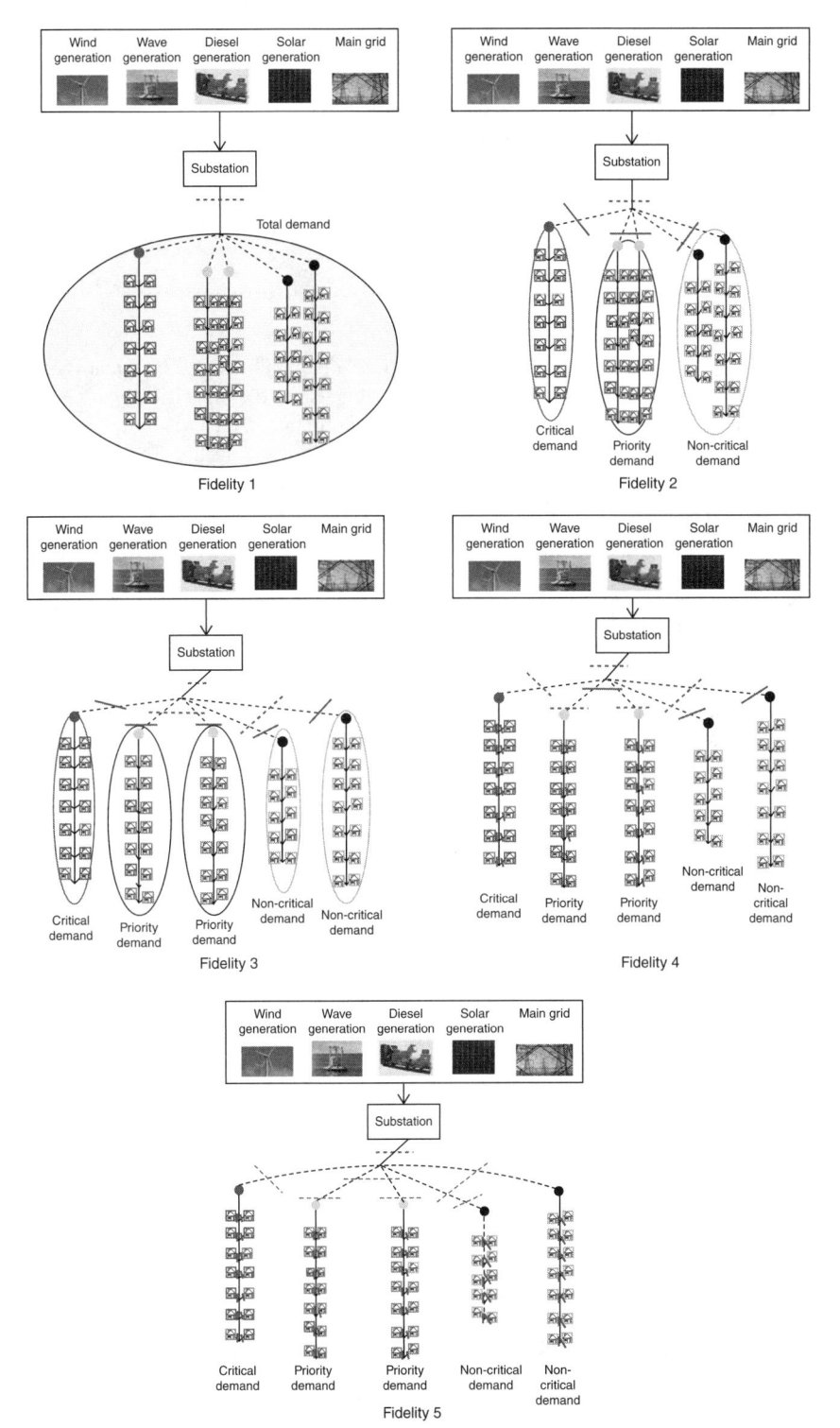

Figure 13.4 Fidelity levels for the proposed simulation model.

13.4.5 Agent-Based Simulation Model for the Considered Microgrid

In our case of a microgrid simulation model, agents are designed for the loads, transformers, wind turbines, PV arrays, diesel generators, wave generators, and batteries. Details of the major agents are as follows.

13.4.5.1 Demand Agent

For each building on the military base, we define a demand agent. The agent reads its parameters from the corresponding demand tab of an excel file. Specifically, it reads the coordinates of the buildings on the map (*x*- and *y*-coordinates), the type of demand (critical, priority, noncritical), the peak demand of this building, and the transformer that feeds this building. By using a power factor plot within a typical day for an industrial load, we made an hourly schedule for the power factor.

Following the values of the power factor in the hourly schedule, we compute the current demand (power) that follows a normal distribution with mean the peak demand times the current power factor and variance 1/12 of the mean. The current demand is updated each minute. This parameter is changed when there is a new fidelity in this area. Using the current demand, we compute the electricity (Wh) by dividing by 60 since the current demand corresponds to an interval of 1 min. In this agent, we have two events. The first one is triggered each minute and calculates the new demand power electricity and total power. The second one is triggered only once during the initialization of the simulation to set the initial value of the power factor (this is because the action in schedule that sets the power factor variable is triggered only when there is a transition to the schedule).

13.4.5.2 Solar Agent

A PV cell is a device that converts the solar irradiance into electricity. This work designs the solar generator agent to simulate each solar farm in the system consisting of several PV cells. To this end, the agent compiles the area and efficiency of each cell along with the total number of cells. It then uses the embedded functions to calculate the power output. Each minute, the simulation triggers an event and calculates the power output for the solar farm using the solar radiation measurements.

This case study is limited to 44 acres of usable land for a solar farm. The simulation will make use of SunPower E19/425 Solar Panels with a rated efficiency of 19.7 (Sun Power, 2010). Due to the high costs of installation and the renewable energy credits available to commercial companies willing to invest in solar power, this particular agent will utilize a power purchase agreement between the military base and the commercial utility company. In this way, the military base will purchase the electricity generated from the solar farm at a cost of 2.23 cents/kWh.

The following equation allows for the calculation of the power output from PV system with an area $A\,(m^2)$ when total solar radiation of $Ir\,(kWh/m^2)$ is incident on PV surface: $P = Ir \cdot \eta \cdot A$, where the systems efficiency η and module efficiency η_m are $\eta = \eta_m \cdot \eta_{pc} \cdot P_f$ and $\eta_m = \eta_r \cdot (1 - \beta(T_c - T_r))$, where η_r is the manufacturer's reported efficiency, η_{pc} is the power conditioning efficiency, P_f is the packing factor, β is the array temperature coefficient, T_r is the reference temperature for the cell efficiency, and T_c is the current temperature. Figure 13.5 is the state chart for solar agent used in this simulation.

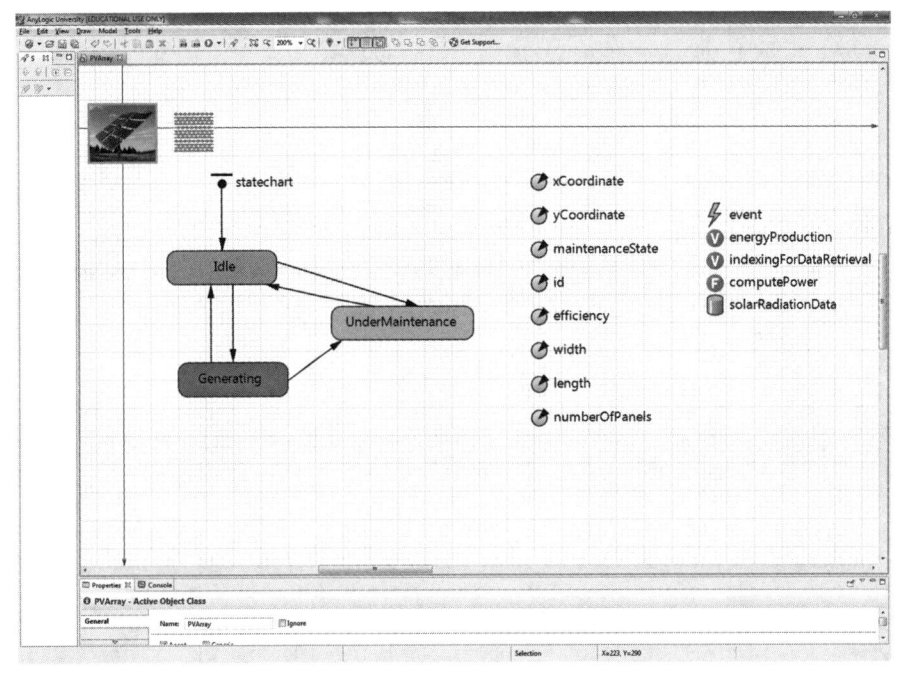

Figure 13.5 State chart and considered parameters in the solar generator agent.

13.4.5.3 Wave Generator Agent

One of the latest technologies in renewable power generation is wave energy. This technology extracts energy from either surface waves or the pressure created below the surface of the water due to waves. Unlike other renewable energy resources, which only generate electricity in the kilo or gigawatt range, wave energy is capable of generating electricity in the terawatt range. To put this in perspective, 1 TWh is capable of powering close to 94,000 homes for an entire year (United States Department of the Interior, 2006).

Waves can generate electricity by either offshore or onshore generation systems. Offshore systems typically require a depth of more than 40 m of water to function properly. There are two major types of offshore generation systems. The first utilizes the up and down motion of waves to power pumps that create the electricity. The other uses hoses connected to floats that ride along waves. The rise and fall of the waves creates pressure inside the hose that in turn rotates a turbine.

Wave energy is not available in all locations due to different existing uses of the ocean such as shipping and commercial fishing. Environmental considerations also significantly reduce the available wave energy sites. This work employs a wave generator agent to simulate the power output of each wave generation construction by using the real-time measurements of the wavelength and wave period. This case study will simulate Ocean Power Technologies' Power Buoy™ system (Ocean Power Technologies, 2011). This system makes use of the up and down movement of waves at a fixed point. A floating buoy inside the main compartment is free to move vertically depending on the movement of the waves at that location. Each buoy is 150 ft tall and 40 ft wide with a maximum power output of 1.5 MW at a cost of $4,000,000 each. Environmental impact studies show that the region we are studying can accommodate no more than 6 Power Buoys without significant environmental impact. The simulation will use the following equation to determine the energy potential from waves in watts per meter of wave height:

$$P = \frac{\rho g^2 T H^2}{32\pi}$$

where ρ is the density of seawater (1025 kg/m³), g is the gravitational acceleration (9.81 m/s²), T is the period of the wave in seconds, and H is the wave height in meters.

13.4.5.4 Wind Generator Agent

Wind power is growing in popularity chiefly due to it producing no greenhouse gas emissions and using little space. Wind turbines are responsible for converting the wind energy into electricity. This work uses a wind generator agent to calculate the power output of the wind farm over short periods of time (on the order of seconds). The agent takes into consideration the number of wind turbines, their efficiency and rated power output, and the real-time measurements of the wind speed to make the calculations. This equation gives the actual power output of a wind turbine, $P = P_W A_W \eta$, where A_W is the total swept area, η is the efficiency of the wind turbine generator and the corresponding converters, and P_W is the power output. A wind turbine generates electricity only when the wind velocity V is within a standard range: $V_{min} \leq V \leq V_{max}$. The following equations give the power output P_W of a turbine with rated output P_r:

$$P_W = \begin{cases} 0 & V < V_{min} \\ aV^3 - bP_r & V_{min} \leq V < V_r \\ P_r & V_r \leq V \leq V_{max} \\ 0 & V > V_{max} \end{cases}$$

where V_r is the rated wind speed, $a = \dfrac{P_r}{V_r^3 - V_{min}^3}$, and $b = \dfrac{V_{min}^3}{V_r^3 - V_{min}^3}$. In the simulation model, each minute, the simulation triggers an event and calculates the power output of the wind farm using the wind speed measurements and the given parameters of the wind turbines. This case study will simulate GE Class I 1.5-77 wind turbines (General Electric, 2011). Each turbine is 80 m tall with 37 m-long blades. A two-turbine system costs 9.6 million dollars with an estimated $25,000 in operating costs.

13.4.5.5 Energy Storage Agent

The energy storage agent refers to a method to store excess electricity within an electrical grid. Utilizing storage eliminates the need to scale up or down electricity production during periods of increased or decreased demand. This allows planners to create a constant supply of electricity for its customers while reducing on costs of ramping up or down production. Grid energy storage is particularly useful in systems that rely on renewable energy sources such as solar and wind as grid storage systems solve the problems of intermittency (wind and solar energy production vary significantly with time of day, seasons, and the randomness of the weather). In hybrid energy grids, those that use both renewable and fossil fuels, lack of a grid energy storage system creates a need to ramp up production of fossil fuel energy during periods of intermittency.

13.4.5.6 Diesel Generator Agent

Diesel backup generators supply an uninterruptible source of electricity in the event that all other sources of electricity fail. In this work, the diesel generator agent acts as a standby system that activates automatically whenever the system detects a loss of power. When a power loss occurs, the microgrid controller sends a signal to start and transfer loads to the generator. When the utility restores power or the renewable resources are once again able to meet demand, the controller automatically transfers the electrical load back to the utility. The microgrid then signals for the generators to power down and return to standby mode to await the next outage.

13.4.6 Net Metering and Storage

Net metering is a process of accurately measuring the amount of electricity that a utility company buys back from its customers that own renewable energy facilities. These customers, instead of only using the energy they generate at the time of generation, can bank the surplus of energy generation for future credit. Thus, under net metering, a customer with renewable generation obtains retail credit for a percentage of the produced electricity. However, the regulations for customers that own renewables differ significantly by country and state, meaning it may not be available in some areas, the credits for the excess of energy banked may expire, and the price of the credits varies.

Net metering usually applies to residential customers that are absent during the day while their PV arrays in the roof of their house generate electricity. At the time that the renewables generate the highest, they (the customer) consume the lowest. The residential customer sells this excess to the utility company to offset their production during peak periods of the day. However, for the commercial and industrial customers, this may not be feasible. When the renewables produce the highest during the day, they consume the highest, thereby providing little opportunity to bank credit for excess energy.

In many instances, intermittent sources of generation come equipped with storage capabilities. This allows storage of the excess energy for future use. Combining both storage and net metering becomes inefficient since the excess of energy in most cases is not enough for both storing and giving back to the main grid (Office of Electricity Delivery and Energy Reliability, Smart Grid R&D Program, 2011). Commercial and industrial customers amplify the phenomenon since excess of energy on average is less. Having this in mind, the model takes into consideration two different scenarios trying not to combine net metering and storage capabilities. In the first one, we are using only storage, and in the second one, we are using only net metering.

13.5 RESULTS

13.5.1 Performance of the State Estimation Algorithm

In the experiments conducted in this section, the process and observation noises are assumed to be generated with Gaussian (0, 10) and Gaussian (1, 2), respectively. Experiments are run with 400 initial particles and each run contains 50 replications. Table 13.1 shows the comparison between actual electricity demands and estimated results in terms of the posterior mean four times of a day where demand is relatively high (9 A.M., 11 A.M., 4 P.M., and 9 P.M.). It is noted that the PF embedded with the minimum relative entropy-based importance-density selection rule (MREIS-PF) outperform the PF embedded with its previous state (referred to as "prior") as importance density (PRIOR-PF) in all four experiments conducted at different time quarters.

13.6 SIMULATION RESULTS

Using actual atmospheric data from May 18, 2013, the simulation runs under the highest fidelity (fidelity 3) during an induced 3 h blackout. To account for the intermittency of renewable energy resources at different times of the day, the simulation obtained results for three different times of the day, late morning, late afternoon, and night. For the purposes of simulation, solar and wind penetration is set to 10% of the average daily demand for the base. Additionally, the critical loads make up 25% of the total daily demand, and priority loads comprise 35% with the remaining 40% of the demand allocated to noncritical loads. The simulation uses solar radiation and wind speed data obtained from a weather station approximated 60 miles from the military base. From the simulation results (see Table 13.2), it is clear that the total demand varies significantly at different hours throughout the day. Also, at 2:30 P.M., the wind speed is lower than the cut-in speed of the wind turbines, and therefore, there is no wind energy generation for that time period. Furthermore, around noon, the solar radiation dropped significantly signifying heavy cloud cover or a thunderstorm. This period experienced a significant drop-off in solar energy production, forcing the diesel generators to pick up the difference (approximately 5 MW).

13.7 CONCLUSIONS AND FUTURE VENUES

Based on the five principles of an SoS, a microgrid presents itself as a dead-on example in that it possesses operational independence, managerial independence, geographic distribution, emergent behavior, and evolutionary development. The study of the military base reveals that DDDAS and agent-based simulation are promising methods to model such systems as they provide a means to find the most efficient method for multiple independent systems to interoperate. The results obtained from the simulation have shown that in each of the blackout scenarios, if the DDDAS framework were not in place, the base would not have been able to meet a significant portion of its electricity demand. More specifically, when the individual systems (renewable generation resources and local utility) are unable to collectively satisfy the demand requirements of the base, the

Table 13.1 Estimation Results for Daily Power Demands Obtained versus the Actual Demands at Time Points: 6 A.M., 11 A.M., 4 P.M., and 9 P.M.

Scenario	PRIOR-PF RMSE mean	RMSE var	MREIS-PF RMSE mean	RMSE var	Screenshot
6 A.M.	12.6593	5.9735	6.7761	0.5695	
11 A.M.	24.7763	19.6049	11.0643	1.0543	
4 P.M.	24.2722	5.6755	10.4195	0.5925	
9 P.M.	15.0971	6.9076	7.7362	0.6015	

Table 13.2 Simulation Results

	Scenario 1	Scenario 2	Scenario 3
Blackout time	11:30 A.M.–2:30 P.M.	4:30 P.M.–7:30 P.M.	9:30 P.M.–12:30 A.M.
Total demand (kWh)	69,241.84	49,876.00	27,881.45
Demand satisfied (%)	25.53	31.69	53.86
Total solar energy generated (kWh)	2,141.91	691.65	0.10
Total wind energy generated (kWh)	147.99	0.00	0.00
Total critical demand (kWh)	16,006.91	11,551.17	6,452.48
Critical demand satisfied (%)	93.63	94.70	99.27
Total priority demand (kWh)	24,751.16	17,818.90	9,968.45
Priority demand satisfied (%)	10.49	28.03	85.68
Total noncritical demand (kWh)	28,483.77	20,505.94	11,460.52
Noncritical demand satisfied (%)	1.10	1.10	1.57
Average number of buildings isolated	139.78	129.28	90.72

framework eliminates the time associated with the human decisions of where and how to employ the backup generators. Furthermore, proposed work is quite promising in determining the optimal level of detail to analyze the system while minimizing the simulation time and required computational resources. In the future, this work could expand to apply these concepts to multiple bases simultaneously or a series of microgrids within a city. Also, as new renewable energy sources become available and more efficient, the simulation can add a robust storage element to store excess electricity for future use.

ACKNOWLEDGMENT

This project is supported in part by the AFOSR via the 2013 Young Investigator Research Award (Award No: FA9550-13-1-0105).

REFERENCES

Air Force Civil Engineering Center, (2012), Renewable energy case study, Available from http://www.afcec.af.mil/shared/media/document/AFD-121207-056.pdf (accessed March 4, 2014)

Carnahan, J.C. and Reynolds, P.F., (2006), Requirements for DDDAS flexible point support, in *Proceedings of the 2006 Winter Simulation Conference*, Monterrey, CA, pp. 2101–2108.

Celik, N., Thanos, A.E., Saenz, J.P., (2013), DDDAMS-based dispatch control in power networks, in *Proceedings of the 2013 International Conference on Computational Science*, Barcelona, Spain, pp. 1899–1908.

Chen, C.H., (1995), An effective approach to smartly allocate computing budget for discrete event simulation, in *Proceedings of the 34th IEEE Conference on Decision and Control*, New Orleans, LA, pp. 2598–2605.

Darema, F., (2004), Dynamic data driven applications system: A new paradigm for application simulations and measurements, in *Proceedings of the 2004 International Conference on Computational Science*, Kraków, Poland, pp. 662–669.

Doucet, A. and Johansen, A.M., (2008), A tutorial on particle filtering and smoothing: Fifteen years later, in *Handbook of Nonlinear Filtering*. Cambridge, MA: Cambridge University Press.

Douglas, C.C., Shannon, C.E., Efendiev, Y., Ewing, R., Ginting, V., Lazarov, R., Cole, M.J., Jones, G., Johnson, C.R., and Simpson, J., (2004), A note on data-driven contaminant simulation, *Computer Science - ICCS*, Part 3: 3038, pp. 701–708.

Ela, E., Milligan, M., and Kirby, B., (2011), Operating reserves and variable generation: A comprehensive review of current strategies, studies, and fundamental research on the impact that increased penetration of variable renewable generation has on power system operating reserves. NREL/TP-5500-51978. Available from http://www.nrel.gov/docs/fy11osti/51978.pdf (accessed September 11, 2014).

Frew, E. et al., (2013), An energy-aware airborne dynamic data-driven application system for persistent sampling and surveillance, in *Proceedings of the 2013 International Conference on Computational Science*, Barcelona, Spain, pp. 2008–2017.

General Electric, (2011), 1.7–77 wind turbine, Available from http://www.ge-energy.com/products_and_services/products/wind_turbines/ (accessed September 11, 2014).

Han, S.Y. and DeLaurentis, D.A., (2013), Development interdependency modeling for system-of-systems (SoS) using Bayesian networks: SoS management strategy planning, in *Proceedings of the 2013 Conference on Systems Engineering Research*, Atlantic, GA, pp. 698–707.

Johnson, M., (2013), Military microgrids and SPIDERS implementation, Available from http://www.apec-conf.org/wp-content/uploads/2013/09/is2.5.4.pdf (accessed March 18, 2014).

Kassakian, J. et al., (2011), The future of the electric grid, Massachusetts Institute of Technology Study, Available from http://mitei.mit.edu/system/files/Electric_Grid_Full_Report.pdf (accessed September 11, 2014).

Macal, C. and North, M., (2010), Tutorial on agent-based modelling and simulation, *Journal of Simulation*, 4, pp. 151–162.

Mandel, J., Chen, M., Coen, J.L., Douglas, C.C., Franca, L.P., Johns, C., Kremens, R., Puhalskii, A., Vodacek, A., and Zhao, W., (2004), Dynamic data driven wildfire modeling, in *Dynamic Data Driven Application Systems*, edited by Darema, F. Amsterdam:Kluwer Academic Publishers, pp. 725–731.

Ocean Power Technologies, (2011), PowerBuoys, Available from http://www.oceanpowertechnologies.com/mark3.html (accessed September 11, 2014).

Office of Electricity Delivery and Energy Reliability, Smart Grid R&D Program, (2011), Department of Energy Microgrid Workshop Report, Available from http://energy.gov/sites/prod/files/Microgrid%20Workshop%20Report%20August%202011.pdf (accessed March 12, 2014)

Parnas, D.L., (1979), Designing software for ease of extension and contraction, *IEEE Transactions on Software Engineering*, 5, pp. 128–138.

Shi, X. and Celik, N., (2011), Importance density selection for particle filtering using improved relative entropy, in *Proceedings of Annual Industrial and Systems Engineering Research Conference*, Orlando, FL, pp. 769–778.

Shi, X. and Celik, N., (2013), Sequential Monte Carlo-based radar tracking in the presence of sea-surface multipath, in *Proceedings of Annual Industrial and Systems Research Conference*, San Juan, Puerto Rico, pp. 2100–2109.

Son, Y.J., Joshi, S., Wysk, R.A., and Smith, J.S., (2002), Simulation based shop floor control, *Journal of Manufacturing Systems*, 21(5), pp. 380–394.

Sun Power, (2010), E19/425 solar panel, Available from http://us.sunpowercorp.com (accessed March 18, 2014).

Thanos, A.E. and Celik N., (2013), Online state estimation of a microgrid using particle filtering, in *Proceedings of Annual Industrial and Systems Research Conference*, San Juan, Puerto Rico, pp. 316–325.

Thanos, A.E., Shi, X., Sáenz, J.P., and Celik, N., (2013), A DDDAMS framework for real-time load dispatching in power networks, in *Proceedings of the 2013 Winter Simulation Conference*, Berlin, Germany, pp. 1893–1904.

U.S. Department of Energy, Office of Electric Transmission and Distribution, (2003), Grid 2030: A national visions for electricity's second 100 years, Available from http://www.ferc.gov/eventcalendar/files/20050608125055-grid-2030.pdf (accessed March 5, 2014).

United States Department of the Interior, (2006), Technology white paper on wave energy potential on the U.S. outer continental shelf, Available from http://www.camelottech.com/CMFiles/Docs/OCS_EIS_WhitePaper_Wave.pdf (accessed September 11, 2014).

Van Broekhoven, S.B. Judson, N., Nguyen, S.V.T., and Ross, W.D. (2012), Microgrid study, energy security for DoD installations (Technical Report 1164), MIT Lincoln Laboratory, Lexington, MA.

Wang, Z., Li, M., Khaleghi, A.M., Xu, D., Lobos, A., Vo, C., Lein, J., and Son, Y.J., (2013), DDDAMS-based crowd control via UAVs and UGVs, in *Proceedings of the 2013 International Conference on Computational Science*, Barcelona, Spain, pp. 2028–2035.

Chapter 14

Composition of Behavior Models for Systems Architecture

Clifford A. Whitcomb, Mikhail Auguston, and Kristin Giammarco
Naval Postgraduate School, Monterey, CA, USA

14.1 INTRODUCTION

The specification of a system's architecture has emerged in the last two decades as one of the fundamental concepts in systems and software engineering. ISO (2011) defines architecture as the "fundamental concepts or properties of a system in its environment embodied in its elements, relationships, and in the principles of its design and evolution." The current interest in understanding architecture and applying the methods across new disciplines as a basis for systems design and evaluation can be tied to recent systems failures and the fact that many can be traced to problems in their early stage definition (Maier and Rechtin, 2000). Architecture development methods have been used for many complex situations, even when the designers were not aware that this was the case. "Architectural methods, similar to those formulated centuries before in civil works, were being used, albeit unknowingly, to create and build complex aerospace, electronic, software, command, control, and manufacturing systems" (Maier and Rechtin, 2000). Indeed, the concept of developing architecture is very old, predating engineering, and continues to this day. "Architecting, the planning and building of structures, is as old as human societies—and as modern as the exploration of the solar system" (Rechtin, 1991). Architecture provides structure for stakeholders of a system of interest to express their

---

Modeling and Simulation Support for System of Systems Engineering Applications, First Edition.
Edited by Larry B. Rainey and Andreas Tolk.
© 2015 John Wiley & Sons, Inc. Published 2015 by John Wiley & Sons, Inc.

respective needs and wants and plays a role as the bridge between those needs, requirements, and implementation of a system. Decisions made at the architecture level during the earliest conceptual stages propagate through to detail design and then beyond into the implementation and operation. Errors exposed during conceptual architecture development and design can be corrected less expensively using models than those discovered during later life cycle phases of testing and implementation. Earlier discovery of design problems, especially those related to stakeholder needs and engineering feasibility, is the motivation behind a new approach to formal systems and software architecture specification presented in this chapter.

Consider this definition of a system of systems (SoS): "a set or arrangement of systems that results when **independent** and task-oriented systems are **integrated into a larger systems construct**, that delivers unique capabilities and functions in support of missions that cannot be achieved by individual systems alone" (Vaneman and Jaskot, 2013, boldface added by author). For our SoS modeling approaches to support **independent** systems behavior models and their subsequent **integration**, they must address each system and the systems interactions as separate concerns. Separation of concerns is a design principle adopted by the software engineering and computer science communities to write highly cohesive software modules such that each module is associated with exactly one main function and to reduce unnecessary coupling among modules within a software program, such that a given module needs to access only a minimum of other modules to perform its functions. Just as programmers use this principle to keep their code organized and maintainable, systems and SoS engineers may use this concept in structuring their systems behavior models.

The modeling community uses many terms to describe the physical manifestation of natural and technological objects. Among these terms are system, SoS, object, component, performer, actor, asset, participant, and so on. These natural language terms may take on different meanings in different communities who adapt a term for their use. For simplicity, and to amplify the hierarchical nature of physical entities, the term **component** is primarily used throughout the remainder of this chapter, with a few exceptions where it is necessary to make a particular point about a component being a part of a larger system. When the term **component** is used, this is a reference to any type of physical entity at any level of abstraction, from an SoS to a configuration item, from the universe to a subatomic particle, from a human person to human-created technology, from a concept to a creation, and from hardware to software to organization. In other words, a **component** can be anything that exhibits behavior. If another term that fits this description is preferred, the reader is encouraged to make the substitution, since the hierarchical nature of these behavior-exhibiting building blocks is more important than the label given to the building blocks.

The new behavior-modeling approach, called Monterey Phoenix (MP), is predicated on the following foundational premises:

- A component's **behavior** is central to stakeholder satisfaction. Behavior is the way in which a component acts on its own and responds to stimuli. For human-designed components, predicting functional and dysfunctional behavior during the design stage reduces the risk of stakeholder dissatisfaction and lack of engineering feasibility during the component's operation. Modeling the behavior of components of complex natural and technological systems increases human understanding of overall systems behavior.

- Modeling a component in the context of its environment is necessary, but not sufficient, for predicting the full range of component behaviors before physical implementation. Modeling the behavior of **each** interacting component rather than just one component's interactions with external components has the potential to expose many more design flaws and tacit assumptions pertaining to the component operation in a larger construct.
- Describing component interactions at a high level of abstraction, orthogonal to descriptions of component behavior, enables automatic solving for distinct instances of behaviors (scenarios, use cases) from an exhaustive superset of possible behaviors and early testing of systems behavior against stakeholder expectations/requirements using scenario inspection and assertion checking. The assumption related to an "exhaustive superset" is supported up to the scope limit. This concept is addressed in more detail later in the chapter.

14.2 COMMON CHARACTERISTICS FOR ARCHITECTURE DESCRIPTIONS

Architecture is concerned with the selection of architectural elements, their interactions, and the constraints on those elements and their interactions necessary to satisfy the requirements and serve as a basis for the design (Perry and Wolf, 1992). An architecture description has converged on the concept of architectural elements, such as component, connector, and relationships among them. "When designers discuss or present a software architecture for a specific system, they typically treat the system as a collection of interacting components. Components define the primary computations of the application. The interactions or connections between components define the ways in which the components communicate or otherwise interact with each other" (Abowd et al., 1995). A conclusion in Rozanski and Woods (2012) states: "Every system has an architecture, whether or not it is documented and understood."

The following aspects have emerged as characteristic for architecture descriptions (Perry and Wolf, 1992; Bass et al., 2003):

- An architecture description belongs to a high level of abstraction, ignoring many of the implementation details, such as algorithms and data structures.
- An architecture specification should be supportive for the refinement process and needs to be checked carefully at each refinement step (preferably with tools).
- There should be flexible and expressive composition operations for the refinement process.
- The architecture specification should support the reuse of well-known architectural styles and patterns. Practice has provided several well-established, reusable architectural solutions.
- An architecture of a system should be considered in the context of the environment in which it operates, as suggested in the international standard ISO/IEC IEEE 42010 "Systems and Software Engineering Architecture Description" (ISO, 2011).

- The software architect needs a number of different views of the software architecture for the various uses and users (Kruchten, 1995) (including visual representations, like diagrams).

MP utilizes these characteristics and complements existing languages and notations by extending them to include an abstract interaction specification capability.

14.3 RELATED WORK

The following ideas of behavior modeling and formalization have provided inspiration and insights for the MP approach.

Literate programming introduced in Knuth (1984) set the directions for hierarchical refinement of structure mapped into behavior, with the concept of pseudocode and tools to support the refinement process.

Campbell and Habermann (1974) and Bruegge and Hibbard (1983) have demonstrated the application of path expressions for program monitoring and debugging. Path expressions in Perry and Wolf (1992) have been used (semiformally) as a part of software architecture description.

Hoare's Communicating Sequential Processes (CSP) (Hoare, 1985; Roscoe, 1997) is a framework for process modeling and formal reasoning about those models. This behavior-modeling approach has been applied to software architecture descriptions to specify a connector's protocol (Allen, 1997; Allen and Garlan, 1997; Pelliccione et al., 2009).

Rapide (Luckham and Vera, 1995; Luckham et al., 1995) uses events and partially ordered sets of events (posets) to characterize component interaction.

Statecharts (Harel, 1987) became one of the most common behavior-modeling frameworks, integrated in broader modeling and specification systems (Unified Modeling Language (UML) (Booch et al., 2000) and AADL (Feiler et al., 2009)). UML has four behavior diagrams: activity, sequence, state machine, and use case.

Wang and Parnas (1994) have proposed to use trace assertions to formally specify the externally observable behavior of a software module and presented a trace simulator to symbolically interpret the trace assertions and simulate the externally observable behavior. The approach is based on algebraic specifications and term rewriting.

The Alloy modeling framework (Jackson, 2007) has strongly influenced this work through ideas of integration of sets and first-order predicate logic within the relational logic framework; inheritance structure; emphasis on lightweight formal methods as opposed to the full-scale theorem proving, with the fundamental concept of small scope hypothesis; and the principles of immediate feedback and visualization during model design.

The concept of software behavior models based on events and event traces was introduced in Auguston (1991, 1995) and Auguston et al. (2002, 2006) as an approach to software debugging and testing automation. The early draft of MP has appeared in Auguston (2009a, b).

14.4 THE MP APPROACH TO BEHAVIOR MODELING

The behavior of the system is usually the main concern for the developer, and the presence of unintended behaviors manifests errors in the design and ultimately the implementation and operation of the system. Many detectable errors made early in the systems design go undetected until later in the development life cycle, when they are more expensive to fix. For example, systems architects may be interested in detecting errors in a system's interaction with the operational environment, for example, by querying a systems model to find scenarios that contain potential hazard states. SoS architects are concerned with detecting emergent behaviors resulting from the interactions of subsystems, some of which may lead to undesirable behavior.

The considerations in Section 14.2 suggest the importance of architecture models and the practical need to test and verify the systems architecture early in the design phase. Behavior modeling is at the core of the MP systems and software architecture modeling framework, which has the following main principles:

- A view of the architecture as a high-level description of possible systems behaviors, emphasizing the behavior of subsystems and interactions between subsystems.
- The concurrency of actions is a default, unless ordering is imposed (thus representing a design decision introducing a dependency between activities).
- Specifying the interaction between the system and its environment is important. A model of the system and its environment behaviors and interactions can be a contribution to the system's requirements specification.
- The event grammar provides a view of the behavior as a set of actions (event trace) with two basic relations, where the PRECEDES relation captures the dependency abstraction, and the IN relation represents the hierarchical relationship. Since the event trace is a set of events, additional constraints can be specified using set-theoretical operations and predicate logic.
- The behavior composition operations support architecture reuse and refinement toward design and implementation models.
- The MP architecture description is amenable to deriving different views, including a structural view (traditional architecture box-and-arrow diagrams) or those desired by the Department of Defense Architecture Framework (DoDAF) (DoD, 2009).
- The executable architecture models provide the possibility to automatically generate examples of behaviors (use cases) for early systems architecture testing and verification with tools.

The main objective of the MP approach is to provide a formal framework for specifying behaviors of the system, its parts and environment, and the interaction between them. From a Systems Engineering point of view, the following two main principles of MP are the key for complex system and SoS behavioral analysis:

- In addition to modeling the behavior of the system along with its interfaces to external systems, also model the behavior of the environment in which the system operates.
- Model component interactions abstractly and separately, rather than instantiated in specific use cases.

The MP approach provides extensions to current modeling notations to significantly expand the coverage of the design space explored. MP does this by applying a *separation of concerns* that has previously not been done in behavior modeling. Specifically, the MP approach leverages the power of abstraction to model internal and external *interactions* among components as a separate concern from the *behavior* of each component to extract the overall possible behaviors of all components acting together. Separation of concerns is a design principle adopted by the software engineering and computer science communities that aids in the writing of highly cohesive software modules such that each module is associated with exactly one main function and to reduce unnecessary coupling among modules within a software program such that a given module needs to access only a minimum of other modules to perform its functions. This same concept is accessible to systems architects through MP to structure and organize systems behavior models, just as programmers have used it to keep their code organized and maintainable. The partitioning of component behavior models and the component interaction specification into separate concerns enables the component behaviors and interactions to be woven together during model execution, automatically generating use cases from the separate behavior and interaction specifications.

MP models may be used for early design testing and verification, for early performance and safety assessment estimates, and for generating examples of scenarios (use cases), which in turn can be used to support test case construction and monitor for systems implementation testing. MP architecture models can be integrated into standard frameworks, like UML, Systems Modeling Language (SysML), and DoDAF, providing the level of abstraction convenient for architecture models with the emphasis on behavior and interaction aspects (see Example 14.7 in Section 14.5 for more details).

14.5 MODELING COMPONENT BEHAVIOR

In a certain sense, an executable architecture model is a compact description for a set of required behaviors. The architecture model—a finite object by itself—may specify a potentially infinite number of execution paths. Computers are used to solve problems usually by finding an algorithm that describes these possible execution paths and mapping it on the appropriate computational platform, that is, by applying a step-by-step procedure to design a behavior to solve the problem at hand. A component operates in a certain environment, which has its own behavior that interacts with the system and causes systems responses. In the MP approach, behavior of the environment in which a component operates is described in addition to the system itself to increase the likelihood of predicting these responses.

The behavior of a system of components is defined in MP as a set of events (event trace) with two basic relations: precedence and inclusion. The structure of an event trace is specified using event grammars and other constraints organized into schemas. Behaviors for both system and its environment are specified within the same framework. Suggested composition operations on schemas are based on event pattern matching and provide for behavior merging and abstract interface specification. The schema framework is amenable to stepwise refinement, reuse, visualization of multiple architecture views, and application of automated tools for consistency checks and systems behavior verification early in the design process.

14.5.1 Event Concept

The MP behavior model is based on the concept of an event as an abstraction of activity. The event has a beginning and an end and may have duration (a time interval during which the action is accomplished). The behavior of a system is modeled as a set of events with two binary relations defined for them: precedence (PRECEDES) and inclusion

(IN)—the event trace. One action is required to precede another if there is a dependency between them, for example, the send event should precede the receive event. Events may be nested, when a complex activity contains a set of other activities. Imposing one of these basic relations on a pair of activities represents an important design decision. Usually, systems behavior does not require a total ordering of events. Both PRECEDES and IN are partial ordering relations. If two events are not ordered, they may occur concurrently. Appendix 1 in Auguston and Whitcomb (2012) provides more details specifying the properties of the basic relations.

14.5.2 Event Grammar

MP uses an event grammar that allows for the compact specification of behavior for each component. Events are abstractions of activities that may be experienced from the perspective of system or its environment. Data inputs and outputs are not modeled in a separate class as in an Enhanced Functional Flow Block Diagram (EFFBD) and other data flow-oriented notations, but are represented by actions (events) that may be performed on that data, following the concept of abstract data types (ADT) introduced in Liskov and Zilles (1974). Behavior is modeled in MP as an algorithm for each component, describing the step-by-step procedure by which it achieves a well-defined goal.

Events have two main binary relations used to construct *event traces*, or particular instances of behavior. Sequencing of events is denoted using the PRECEDES relation, and decomposition of events is denoted using the IN relation. An event grammar rule specifies structure for a particular event type in terms of these two relations and has the form

```
A: right-hand-part;
```

where **A** is an event type name. Event types that do not appear in the left-hand part of rules are considered atomic and may be refined later by adding corresponding rules. More details about event grammar notation can be found in Auguston (2009a). For brevity, this chapter only describes the composition operations that appear in the example models.

Events are composed to describe possible event traces using composition operations in the right-hand part of the event grammar rule. The composition operations comprise an algorithm for each root event. Behavior is described using composition operations such as ordered sequence of events **A B C**; alternative (**A** | **B** | **C**); ordered iteration (* **A B C** *) (A B C repeated zero or more times); (+**A B C** +) (one or more times); optional event [**A**]; {**A, B, C**}, set of unordered (potentially concurrent) events; {* **A** *}, set of zero or more of unordered events A; and {+**A** +}, set of one or more of unordered events. An event grammar, as in Example 14.1, is essentially a graph grammar, which specifies directed acyclic graphs of events with the arcs representing relations IN and PRECEDES.

Similar to context-free grammars, event grammars can be used as production grammars to generate instances of event traces, as in Example 1.

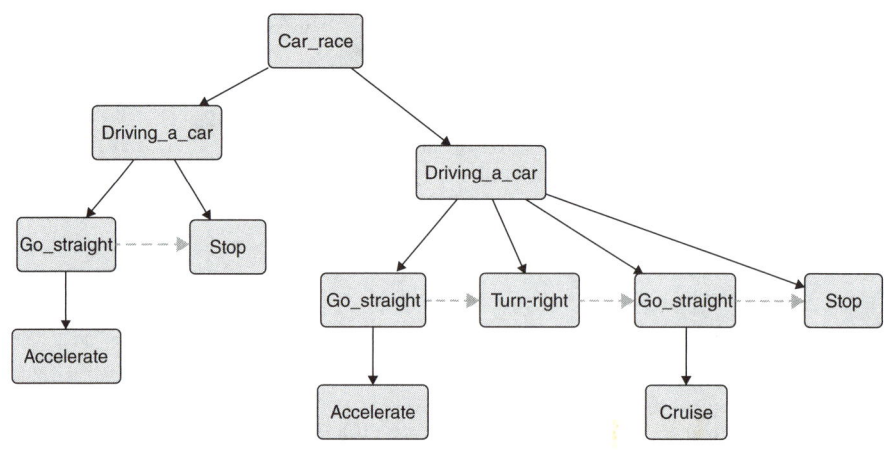

Figure 14.1 An event trace derived from the event grammar in Example 14.1.

Example 14.1 An event grammar for car race scenarios.

```
car_race:       {+ driving_a_car +};
driving_a_car:  go_straight (* ( go_straight |
turn_left | turn_right ) *) stop;
go_straight:    ( accelerate | decelerate | cruise );
```

An instance of an event trace satisfying the grammar can be visualized as a directed graph with two types of edges (one for each of the basic relations) (Figure 14.1).

14.6 MODELING COMPONENT INTERACTION AND ARCHITECTURE VIEWS

The behavior of a particular system is specified as a set of all possible event traces using a *schema*. The concept of the MP schema has been inspired by the Z schema (Spivey, 1989). The purpose is to define the structure of all possible event traces (in terms of IN and PRECEDES relations) using event grammar rules and other constraints. A schema usually contains a collection of events called *roots* representing the behaviors of parts of the system (e.g., components and connectors in common architecture descriptions), *composition operations* specifying interactions between these behaviors, and additional constraints on root behaviors.

There is precisely one instance of each root event in any trace. The schema also may contain auxiliary grammar rules defining composite event types used in other rules. Roots in turn may be defined as schemas, thus providing for architecture reuse and composition. A schema may define both finite and infinite traces, but most analysis tools for reasoning about a system's behavior assume that a trace is finite.

The schema represents instances of behavior (event traces), in the same sense as Java source code represents instances of program execution. Just as a particular program execution path can be extracted from a Java program's source code by running it on a Java virtual machine (JVM), a particular event trace specified by an MP schema can be generated from the event grammar rules by applying behavior composition operations and constraints.

In addition to describing specific systems behavior, MP can also be used to describe behavior patterns. In order to establish coordination between sending and receiving messages, we use the behavior composition operation **COORDINATE**. In Example 14.2, the composition operation takes two traces and defines a modified event trace (merges behaviors of Task_A and Task_B) by adding the **PRECEDES** relation between the selected **send** and **receive**.

Example 14.2 A simple pipe/filter architecture pattern.

```
SCHEMA simple_message_flow
ROOT Task_A:   (* send *);
ROOT Task_B:   (* receive *);
COORDINATE     (* $x: send *)   FROM Task_A,
   (* $y: receive *)   FROM Task_B ADD $x PRECEDES $y;
```

The first part of composition operation (the source) uses event patterns to specify segments of root traces that should be selected. The **(* $x: send *)** pattern identifies the sequence of totally ordered **send** events (with respect to the transitive closure of **PRECEDES** relation—**PRECEDES***). Use of the **(* P *)** pattern for selection means that all events P in the source root should be ordered, both iterations should have the same number of selected elements (**send** events from the first trace and **receive** events from the second), and the pair selection follows this ordering (**synchronous coordination**). Labels **$x** and **$y** provide access to the events selected within each iteration. The **ADD** composition completes the behavior adjustment, specifying that an ordering relation will be imposed on each pair of selected events. Behavior specified by this schema is a set of matching event traces for Task_A and Task_B with the modifications imposed by the composition.

The composition operation may be considered as an abstract interaction description for root behaviors. In the case when **asynchronous coordination** is needed, an iterative set pattern can be used. For example,

```
COORDINATE  {* $x: E1 *} FROM A, {* $y: E2 *} FROM B ADD
$x PRECEDES $y;
```

In this case, matching root traces for A and B still should contain an equal number of selected events of types E1 and E2, correspondingly. But now the resulting merged traces will include all permutations of events E2 from B matching events E1 from A, with the **PRECEDES** relation imposed on each selected pair. This assumes that other constraints, like the partial ordering axioms from Appendix 1 in Augston and Whitcomb (2012), are satisfied. Each permutation yields one potential instance of a resulting trace for the schema deploying this composition. In order to reduce the exponential explosion, optimizations similar to symmetry reduction in model checking tools should be considered. Changing (* ... *) for {* ... *} in Example 14.2 may increase the number of composed traces in the schema.

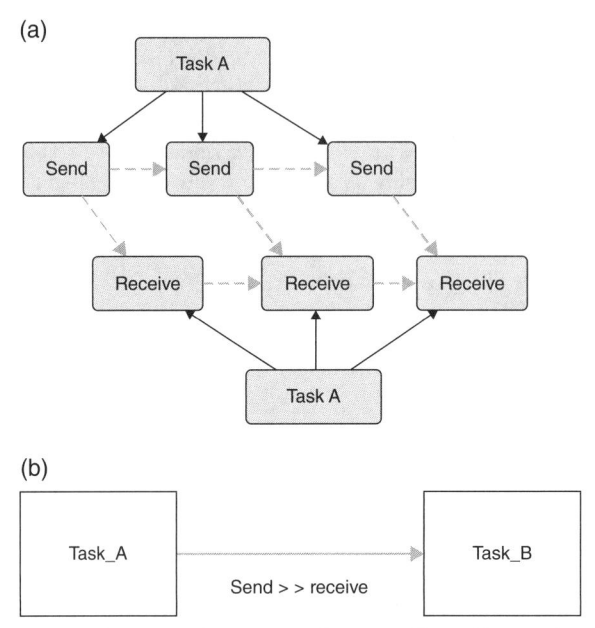

Figure 14.2 An example of a composed event trace and corresponding architecture view for the simple_message_flow schema. (a) The composed event trace for the simple_message_flow schema is labeled. (b) The architecture view for the simple_message_flow schema is labeled.

Different views for different stakeholders can be extracted from MP schemas. For example, each root may be visualized as a box (Figure 14.2), and if there is a composition operation specifying an interaction between root behaviors, the boxes are connected by an arrow marked by the interaction type. The root behavior may be visualized with UML activity diagrams (Booch et al., 2000) (see Figure 14.6). The MP developer's environment may have a library of predefined views providing different visualizations for schemas.

Data items in MP are represented by actions (events) that may be performed on that data. This principle follows the ADT concept introduced in Liskov and Zilles (1974), as in Example 14.3.

Example 14.3 Data flow.

```
SCHEMA Data_flow
ROOT Process_1:  (* work write *);
ROOT Process_2:  (* ( read | work ) *);
ROOT File:       (* write *) (* read *);
Process_1, File SHARE ALL write;
Process_2, File SHARE ALL read;
```

Behavior of the file requires that all **write** operations should be completed before any **read** operations. The view of this schema in Figure 14.3b renders root interaction with a line where the shared event name is attached as a label.

(a)

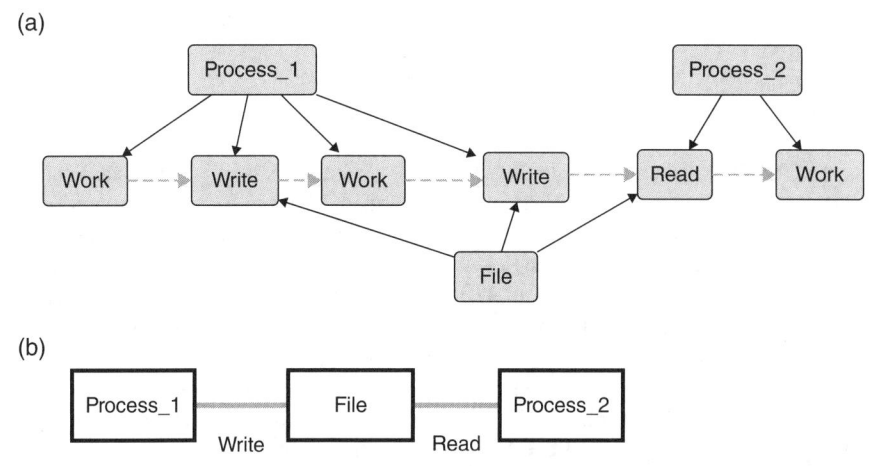

(b)

Figure 14.3 An example of a composed event trace and corresponding architecture view for the Data_flow schema. (a) The composed event trace for the Data_flow schema is labeled. (b) The architecture view for the Data_flow schema is labeled.

The schema in Example 14.4 specifies the behavior of a stack in terms of stack primitive operations.

Example 14.4 Stack behavior.

```
SCHEMA Stack
ROOT Stack_operation:   (* ( push | pop ) *);
SATISFIES FOREACH $x:   pop FROM Stack_operation
   (Number_of (pop) before ($x) < Number_of (push)
before ($x));
```

Let **IN*** denote the transitive closure of the **IN** relation (similarly as **PRECEDES*** is a transitive closure for **PRECEDES**). The domain of the universal quantifier is the set of all **pop** events **e** such that (**e IN* Stack_operation**). The function **Number_of (pop) before ($x)** yields the number of **pop** events **e** such that (**e PRECEDES* $x**). The set of event traces specified by this schema contains only traces that satisfy the constraint. This example presents a filtering operation as yet another kind of behavior composition.

The reuse of a schema is demonstrated through Example 14.5.

Example 14.5 Reuse of a schema.

```
SCHEMA Two_stacks_in_use
INCLUDE Stack;
ROOT Main:   {* (do_something | use_S1 | use_S2) *};
   use_S1:   (push | pop);
   use_S2:   (push | pop);
ROOT S1:     Stack;
ROOT S2:     Stack;
```

```
S1, Main SHARE ALL $x:   (pop | push) SUCH THAT
Has_enclosing (use_S1)($x) WITHIN Main;
S2, Main SHARE ALL $x:   (pop | push) SUCH THAT
Has_enclosing (use_S2)($x) WITHIN Main;
```

The **INCLUDE** statement brings the schema Stack into the scope. This means that all constraints specified in the Stack also will be included. The rule for Main is intentionally left lax without imposing any specific ordering on embedded activities. Roots **S1** and **S2** represent the presence of two independent stacks as data items. The ordering of **pop** and **push** events inside **use_S1** and **use_S2** in each stack behavior is ensured and will be brought into the resulting trace by the included Stack behaviors as a result of sharing these events with the Stack behavior. The **SHARE ALL** composition operation uses event patterns and context conditions to accomplish the necessary event trace construction. The predicate **Has_enclosing(T)(e1)** is true iff there exists an event e2 of the type T in the trace specified by the **WITHIN** clause such that **e1 IN* e2**.

Predicates and functions like **Has_enclosing(T)(e)** and **Number_of (T) before (e)** are used for convenient navigation in the event graphs.

Connectors and components, which are core elements in an architecture description, can be uniformly modeled in MP as behaviors. The idea that connectors should be elevated to the first-class-citizen status on a par with components is often discussed in the literature, for example, in Taylor et al. (2010), as in Example 14.6.

Example 14.6 Connectors and components.
Suppose that the communication between the components is implemented via a buffer of size **max_buffer_size** and not necessarily all sent messages are consumed, that is, some of them could stay in the buffer indefinitely. Each message may be consumed no more than once, and the ordering of receiving does not necessarily correspond to the ordering of sending. The root **Buffered_channel** simulates the behavior of a connector between **Task_A** and **Task_B**. This behavior model does not provide details about what happens after a buffer overflow event:

```
CHEMA Buffered_transaction
ROOT Task_A::  (* Send *);
ROOT Task_B::  (* Receive *);
ROOT Buffered_channel: {* (Send [Receive]) *}
(Overflow | Normal);

Task_A, Buffered_channel SHARE ALL Send;
Task_B, Buffered_channel SHARE ALL Receive;
SATISFIES FOREACH $x: Receive FROM Buffered_channel
        ( Number_of (Send) before ($x) - Number_of
(Receive) before ($x) ) <= max_buffer_size;
SATISFIES FOREACH $x:  Overflow FROM Buffered_channel
        ( Number_of (Send) before ($x) - Number_of
(Receive) before ($x) ) > max_buffer_size;
SATISFIES FOREACH $x:  Normal FROM Buffered_channel
        ( Number_of (Send) before ($x) - Number_of
(Receive) before ($x) ) <= max_buffer_size;
```

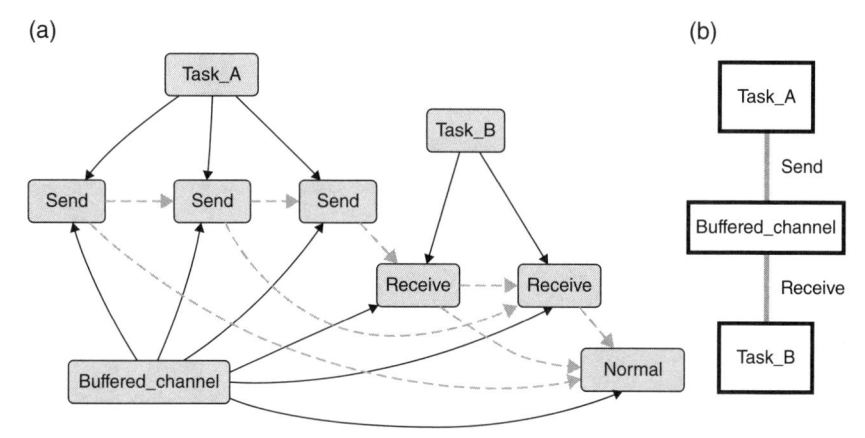

Figure 14.4 An example of an event trace and corresponding architecture view for the Buffered_transaction schema. (a) The event trace (without overflow) for the Buffered_transaction schema with max_buffer_size = 3 is labeled. (b) The architecture view for the Buffered_transaction schema is labeled.

If the schema should satisfy only behaviors without buffer overflow, the three **SATISFIES** conditions above can be replaced by the following constraint (and the **Overflow** event can be removed from the schema):

```
SATISFIES FOREACH $x:  Send FROM Buffered_channel
   Number_of ($y: Send) before ($x) SUCH THAT ( ¬
Has_next(Receive)($y)) < max_buffer_size;
```

Note that **PRECEDES** relation is defined explicitly either in the grammar rule or by **ADD** composition operation and is a proper subset of its transitive closure **PRECEDES***. The predicate **Has_next(T)(e1)** is true iff there exists an event **e2** of the type **T** in the trace such that **e1 PRECEDES e2** (Figure 14.4).

Example 14.7 demonstrates how to integrate the behavior of an environment with the behavior of a system (Figures 14.5 and 14.6). The ATM_withdrawal schema specifies a set of possible interactions between the Customer, ATM_system, and Data_Base. An event trace generated from this schema can be considered as a use case example.

Example 14.7 Withdraw money from ATM.

```
SCHEMA ATM_withdrawal
ROOT Customer:    (* insert_card
   ( ( identification_succeeds
         request_withdrawal
         ( get_money | not_sufficient_funds ) )  |
       identification_fails )    *);
ROOT ATM_system: (* read_card  validate_id
    ( id_successful check_balance
       (  (sufficient_balance dispense_money) |
          unsufficient_balance )   |
     id_failed )   *);
ROOT Data_Base: (* (validate_id | check_balance) *);
```

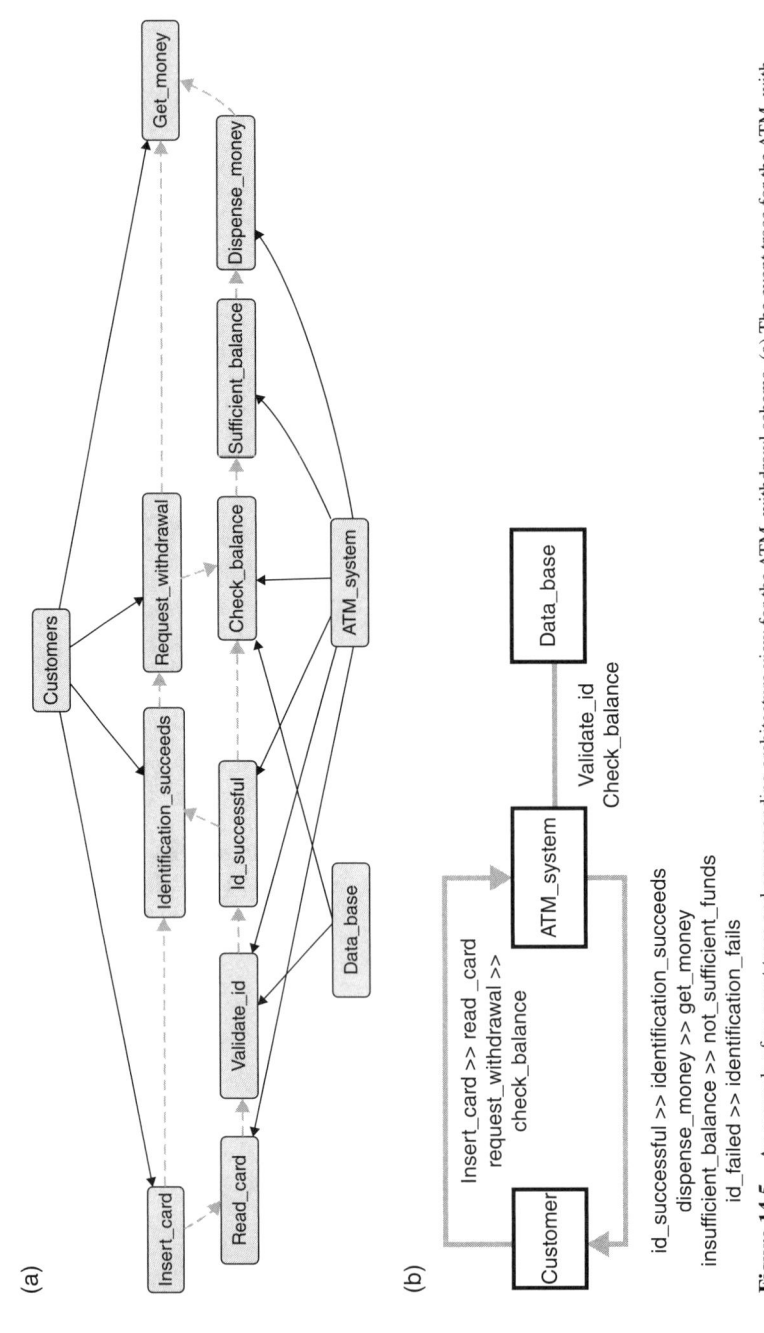

Figure 14.5 An example of an event trace and corresponding architecture view for the ATM_withdrawal schema. (a) The event trace for the ATM_withdrawal schema is labeled. (b) The architecture view for the ATM_withdrawal schema is labeled.

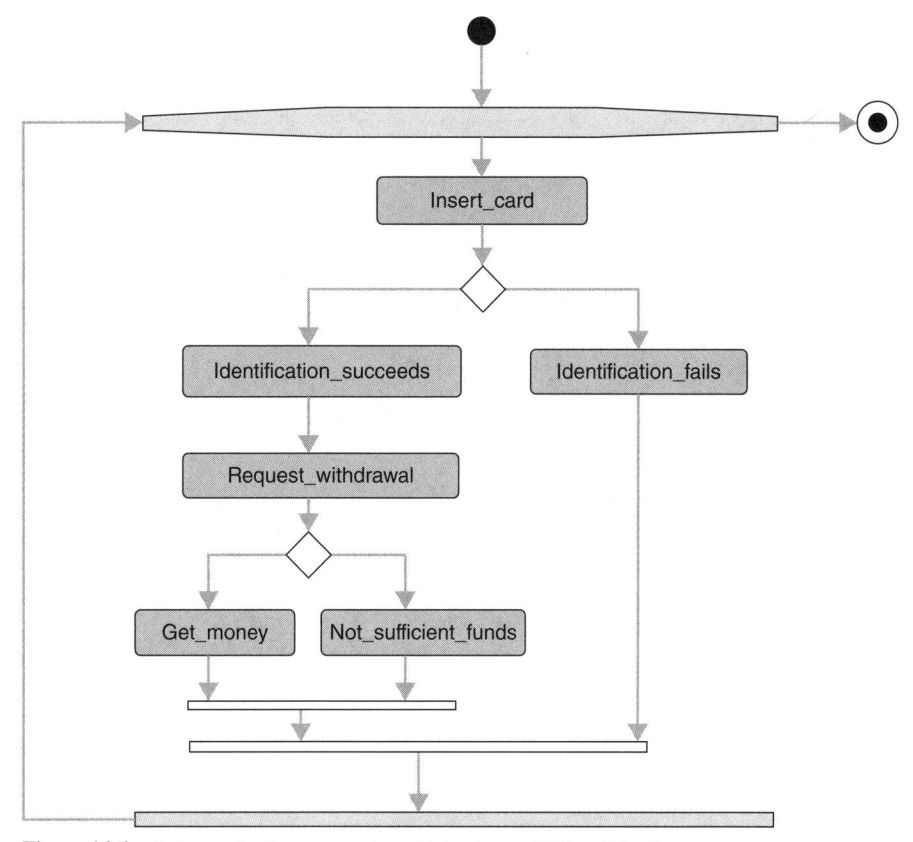

Composition of Behavior Models for Systems Architecture   **375**

Figure 14.6 A view on the Customer root event behavior as a UML activity diagram.

```
Data_Base, ATM_system SHARE ALL validate_id, check_balance;

COORDINATE (* $x: insert_card *)  FROM Customer,
   (* $y: read_card *) FROM ATM_system ADD $x PRECEDES $y;
COORDINATE (* $x: request_withdrawal *) FROM Customer,
          (* $y: check_balance *)       FROM ATM_system
ADD $x PRECEDES $y;
COORDINATE (* $x: identification_succeeds *) FROM Customer,
   (* $y: id_successful *) FROM ATM_system ADD $y
PRECEDES $x;
COORDINATE (* $x: get_money *) FROM Customer,
   (* $y: dispense_money *) FROM ATM_system ADD $y
PRECEDES $x;
COORDINATE (* $x: not_sufficient_funds *) FROM Customer,
   (* $y: unsufficient_balance *) FROM ATM_system ADD $y
PRECEDES $x;
COORDINATE (* $x: identification_fails *) FROM Customer,
   (* $y: id_failed *) FROM ATM_system ADD $y PRECEDES $x;
```

If the view of the whole system's behavior emphasizing the interaction between the parts (components) can be visualized as in Figure 14.5b, the view of root's stand-alone behavior can be visualized as a UML activity diagram (Figure 14.6 provides an example for the Customer root behavior). Since event aggregates (iterations, alternatives, sets) in MP are well structured, it is possible to use Nassi–Shneiderman diagrams (Nassi and Shneiderman, 1973) as yet another kind of view. The event trace on Figure 14.5a can be viewed as an analog of UML sequence diagram's "swim lanes" for the Customer and ATM_system interaction. This example demonstrates that MP models can be integrated into standard frameworks, like UML, SysML, and DoDAF, providing the level of abstraction convenient for architecture models, where, in particular, MP focuses on the interaction aspects.

14.7 MERGING SCHEMAS

So far, we have seen examples of assembling schemas using previously defined schemas (Example 14.5). Each schema in the assembly holds its own roots and composition operations (**SATISFIES** filter and interaction constraints, like **COORDINATE** and **SHARE ALL**) within its scope.

The join operation for schemas looks like:

```
SCHEMA A EXTENDS B
Roots for A
Constraints and composition operations involving roots
from both A and B
```

The resulting schema A joins roots defined in A and roots defined in B, merges within its scope all constraints and composition operations defined in B, and may have additional constraints and composition operations involving all roots. A typical use of such schema composition may be for assembling the architecture of an SoS from the architectures of its constituent systems.

14.8 COMPARISON OF MP WITH COMMON SYSTEMS ENGINEERING NOTATIONS

MP complements and extends Systems Engineering behavior-modeling notations. The Functional Flow Block Diagram (FFBD) notation was developed in the 1950s to show systems functions and their chronological order of execution (NASA, 2007). The EFFBD was developed in the 1990s to show information flow on the diagrams as inputs/triggers and outputs (Long, 2000). The notion of an xFFBD has been proposed to extend EFFBD with additional formalisms to make it more expressive (Aizier et al., 2012). The SysML (OMG, 2012) was developed to extend UML for application on the systems scale and directly reuses all behavior diagrams except the activity diagram, which has been modified from UML for consistency with the EFFBD and to support a continuous flow of matter or energy (Friedenthal et al., 2006). Although these notations have been successfully used in modeling slices of systems behavior and interaction, none are presently used to model the behavior of each component and the interaction of each component with other components in its environment, as separate concerns, nor do existing frameworks such as the DoDAF (DoD, 2009) address this separation of concerns when describing event-based interactions.

Many component models describe only a subset of possible behaviors with assumptions about possible component interactions in specific scenarios or use cases, since behavior of external components may be outside the scope of the component under design. This practice prevents the opportunity to observe behaviors that result from combinations of interactions that fall outside the scope of the assumptions made about external component behavior.

Example 14.8 considers a simple user authentication scenario done internal to a system.

Example 14.8 User authentication.
User provides a general identification.

1. System requests unique identification.
2. User provides a unique identification.
3. If the credentials are valid, the System authorizes the User to access the services; otherwise, the System notifies the User that credentials are invalid and the user may reattempt access up to two more times.
4. The User or the System ends the session.

This narrative gives rise to at least two possible use cases: user authentication succeeds and user authentication fails. The EFFBD activity model in Figure 14.7 is a first attempt at graphically depicting behavior for the scenario shown earlier, minus the access reattempts. The EFFBD uses functional **activities** transforming inputs into outputs, **exit conditions** documenting possible outcomes of an activity, and **inputs/ triggers and outputs** consisting of matter or energy consumed or produced by an activity. The approach taken in the diagram illustrates some generally accepted conventions when modeling with EFFBDs, such as allocating the activities of each main component taking part in the thread (in this case the User (an "external" component) and the System (component under design)) onto its own primary branch, similar to the use of swim lanes on a UML or SysML activity diagram (Long, 2000; Long and Scott, 2011; Armstrong, 2013).

In this example, conditions leading to different possible behaviors based on the outcome of the credential verification are specified on the System branch. The User functions in this example, however, do not exhibit any structured logic for User behavior. All User functions instead simply serve as source or sink for information interactions with the System. The main limitation of this approach is that only a limited set of use cases can be generated from it since the User behavior is "hard coded" to respond the same way each time the model is executed.

A revision to this model takes a slightly different approach, compressing the User functions onto one main branch and placing C.3 Access Services and C.4 Process Access Failure as alternative exit conditions, since only one of these functions would be selected depending on whether the supplied credentials are valid or invalid (Figure 14.8).

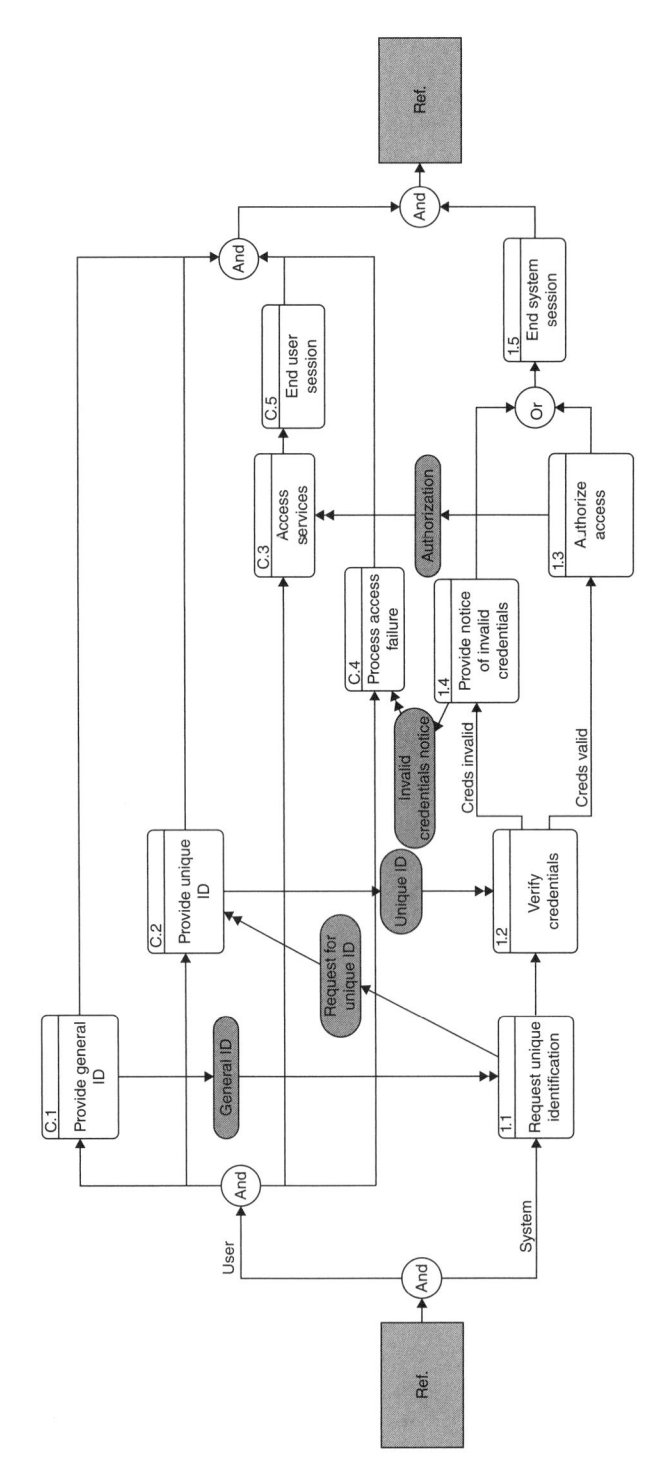

Figure 14.7 An example of EFFBD depicting an authentication behavior model that excludes dependencies on the behavior of an external system (the User).

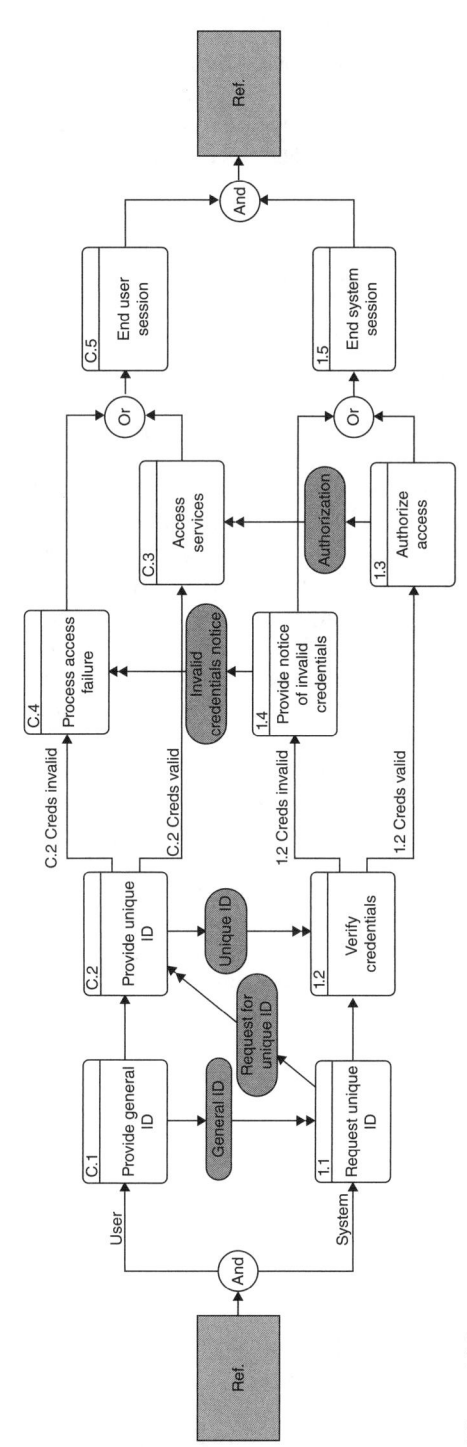

Figure 14.8 An example of EFFBD that includes a description of behavior for both systems in the authentication scenario.

In the revised model, two of the User activities are related to exit condition selections, consistent with corresponding activities on the System branch. If, after providing a unique ID, the System determines the credentials to be valid, the User receives authorization to Access Services. If, on the other hand, the System determines the credentials to be invalid, the User receives notice of this to Process Access Failure. To implement this approach correctly in simulation, a specification must be added to coordinate the branch selections, such that when exit condition "1.2 creds valid" is selected, "C.2 creds valid" should always occur and likewise for the case of invalid credentials. Without such a specification, a simulator has no way to know that this is a requirement and selects exit conditions on different branches at random.

To incorporate additional possible use cases, the EFFBD model could continue to be expanded to encompass more behaviors for both the User and the System. For example, consider the possibility that the User does not respond to the request for unique ID, as tacitly assumed in Figure 14.8. How shall the System behave then? An important consideration is that an EFFBD showing *all* potential behaviors for *all* components interacting in a use case will likely become unwieldy and prone to human error, the larger it grows. This is one likely rationale for the scoping mechanism employed in Figure 14.7.

MP is an approach that resolves this conundrum by employing a divide-and-conquer strategy involving the creation of a *separate* behavior model for each component and specifying interactions between the components as a *separate* concern from that of the behavior of each component. This concept would be akin to creating a separate EFFBD or SysML activity model for each component, allowing elaboration on each component's behavior without concern about adding clutter to a diagram already busy with multiple components and their interactions. This approach allows an architect to focus on describing behavior for one component at a time and then separately specify the general rules (interaction patterns appearing in *all* use case instances) for interaction among components. These component interactions (e.g., the general ID trigger the request for unique ID) may be captured abstractly in a specification of general interaction rules that apply in many similar use cases. Such a specification is what is missing from contemporary Systems Engineering approaches, notations, and frameworks. By separating these concerns, the component behaviors and interactions can be woven together during model execution, automatically generating use cases from the separate behavior and interaction specifications, thereby achieving increased coverage of predictable component interaction.

In MP event grammar, the authentication scenario is described as follows. Each component's behavior is specified separately as a root event in the left-hand part. For example, root events (lines 01 and 08) specify the behaviors of the User and the System, correspondingly. The User's behavior is described in lines 02–07:

```
01  ROOT User:
02  (* request_access
03  (* creds_invalid request_access *)
04  (creds_valid (run_services | abandon_access_request)|
05  creds_invalid (attempt_exhausted |
    abandon_access_request))
06  end_User_session *);
07  request_access: provide_general_ID provide_unique_ID;
```

First, the user requests access (line 02). If the credentials are invalid, the user repeats the request for access (line 03). Line 04 specifies what the user does when credentials are valid: the user may run services having been granted authorization or may abandon the access request for some reason (e.g., experiences an interruption). Line 05 specifies more events that can occur when credentials are invalid: the number of allowable attempts may be exhausted (the number of access attempts is constrained in the systems model), or perhaps the user may abandon the access request. The User session ends (line 06) at the conclusion of event traces for both valid and invalid credentials. In line 07, request_access is decomposed into provide_general_ID followed by provide_unique_ID, to demonstrate the ability to create a hierarchy of events similar to a hierarchy of functions.

The System's behavior is specified in lines 09–17:

```
08  ROOT System:
09  (* request_unique_ID
10  [ creds_invalid request_unique_ID
11  [ creds_invalid request_unique_ID
12  [ creds_invalid attempt_exhausted
13  invalid_creds_notice cancel_access_request]]]
14  [(creds_valid ( authorize_access run_services |
15  long_wait_for_User cancel_access_request ) |
16  creds_invalid long_wait_for_User cancel_access_request)]
17  end_System_session *);
```

The first event in the System for this authentication scenario is request_unique_ID (line 09). If invalid credentials are supplied, the System requests the unique ID up to two more times (lines 10–11). If invalid credentials are supplied for a third time, the number of attempts is exhausted (line 12), and the System provides an invalid credentials notice and cancels the access request (line 13). If valid credentials are supplied, then the System may authorize access and run services (line 14) *or* cancel the access request after a long time elapses while the System is waiting for input (line 15). Yet another alternative is that if invalid credentials are supplied, then there is a long wait for User input; in that case also, the System will cancel the access request (line 16). Regardless of the presence or absence of valid or invalid credentials, the system will always end the session (line 17).

Note that each of these models describes events independent of interactions between the User and the System. The separation of concerns about component behavior and component interaction allows the development of detailed algorithms for every component in the environment and furthermore allows the clean specification of access attempt repetition.

The concept of **abstract** interaction specification is a crucial missing link in current notations and frameworks. As seen in Figures 14.7 and 14.8, systems interactions are often manually embedded in specific use cases or instances of behavior by hard-coding sequenced interactions through multiple components on the same diagram. Many use cases are slight variations of another (such as an authentication scenario resulting in success or failure), so changes to the decomposition or sequence of activities in one use case thread may trigger changes in all affected threads. For example, one may wish to specify that in **any** authorization scenario, the general ID from the User always precedes a request for a unique ID from the System. In MP, this is accomplished using the **COORDINATE** composition operation:

```
18  COORDINATE (* $x: provide_general_ID *) FROM User,
19  (* $y: request_unique_ID *) FROM System
20  ADD $x PRECEDES $y;
```

This composition operation adds the **PRECEDES** relation between selected **provide_general_ID** and **request_unique_ID** events. The first part of composition operation uses event patterns to specify segments of root traces that should be selected. The (* **$x: provide_general_ID** *) pattern in line 18 identifies the sequence of totally ordered **provide_general_ID** events (with respect to the transitive closure of the **PRECEDES** relation). Use of the (* **P** *) pattern for selection means that all events P should be ordered, both iterations should have the same number of selected elements (**provide_general_ID** events from the first trace and **request_unique_ID** events (line 19) from the second), and the pair selection follows this ordering (synchronous coordination). Labels **$x** and **$y** provide access to the events selected within each iteration. The **ADD** composition in line 20 completes the behavior adjustment, specifying that an ordering relation will be imposed on each pair of selected events.

Likewise, one can state that the request for a unique ID from the System always precedes the providing of the unique ID from the User:

```
21  COORDINATE (* $x: request_unique_ID *) FROM System,
22  (* $y: provide_unique_ID *) FROM User
23  ADD $x PRECEDES $y;
```

Note that both the User and System behavior algorithms have event names in common. A constraint must be written to explicitly state that the User and the System share all instances of those events when they occur. For example, there should be no event traces in which credentials are valid from the User perspective but not from the System perspective—such a trace would be invalid. The **SHARE ALL** composition ensures that the schema admits only event traces where corresponding event sharing is implemented:

```
24  User, System SHARE ALL creds_valid, creds_invalid,
25  attempt_exhausted, run_services;
```

Event sharing is in fact yet another way of behavior coordination. Shared events may appear in the root event at any level of nesting.

MP is an executable architecture modeling framework. Event traces (use cases or examples of behavior) can be generated by automated tools from the MP models. Events may be visualized as boxes, and dependencies between pairs of events as arrows marked by the relation type (Figures 14.9, 14.10, and 14.11). Each **PRECEDES** relation may correspond to a control flow or trigger commonly used in flow-oriented notations (e.g., Figure 14.8). Architecture views can also be extracted from MP schemas for different stakeholders to answer typical questions. The root behavior may be visualized with UML activity diagrams (see Figure 14.6). An MP developer's environment may have a library of predefined views providing different visualizations for schemas.

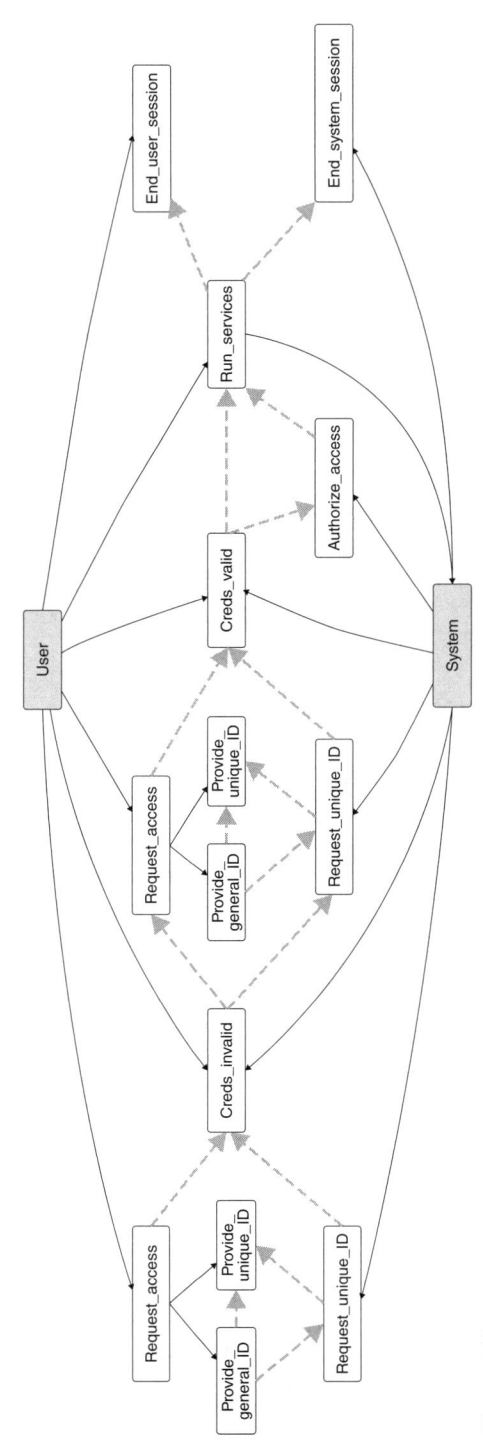

Figure 14.9 An example of event trace (use case) where the User gets access to the System after one unsuccessful attempt. Solid arrows denote IN relations, and dashed arrows depict PRECEDES relations.

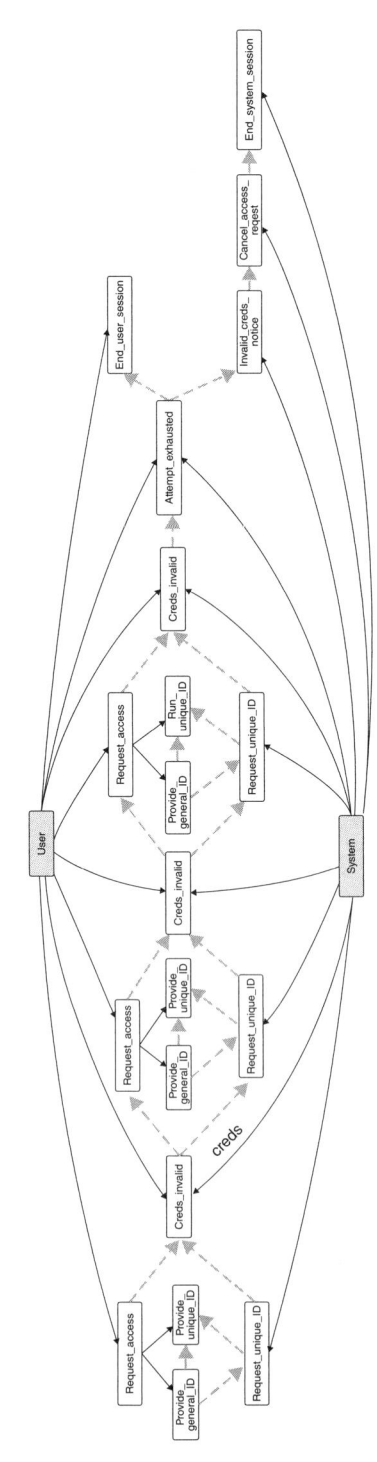

Figure 14.10 An example of event trace (use case) where the User is denied access after three unsuccessful attempts.

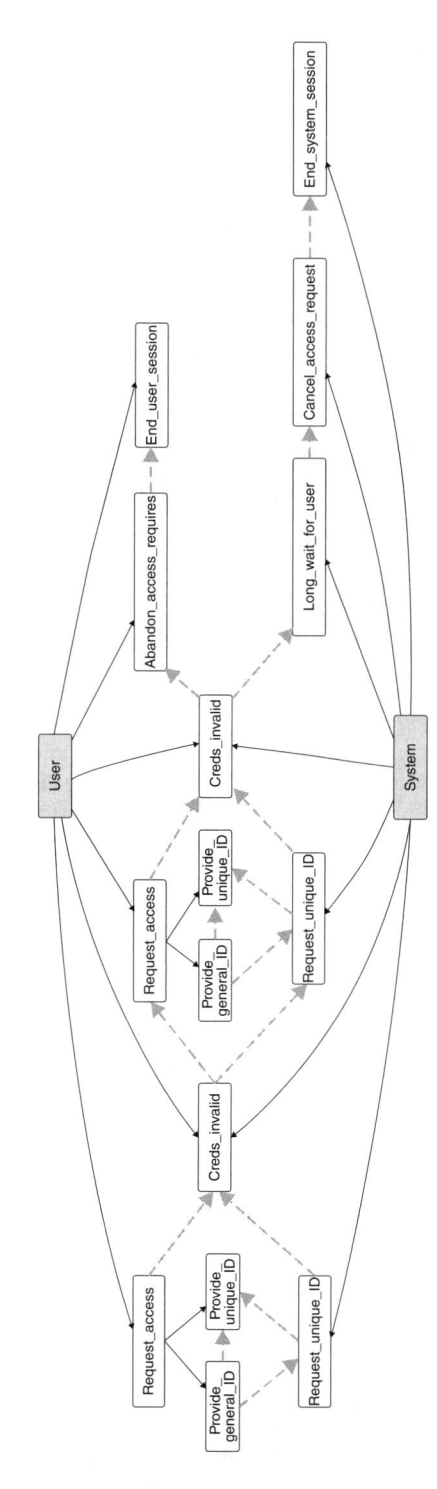

Figure 14.11 An example of event trace (use case) where the User abandons the access request after two unsuccessful attempts.

14.9 ASSERTIONS AND QUERIES

An event trace represents an example of particular execution of the system (or use case, especially if the behavior of the environment is included) that can evolve from the architecture specified by a schema. Event traces can be effectively generated from the event grammar rules and then adjusted and filtered according to the composition operations in the schema. This justifies the term **executable architecture model** for MP. It is possible to obtain all valid event traces within a certain limit. Usually, such a limit (*scope*) may be set by the maximum total number of events within the trace or by the upper limit on the number of iterations in grammar rules (recursion in the grammar rules can be limited in similar ways). For many purposes, a modest limit of a maximum three iterations will be sufficient. This process of generating and inspecting event traces for the schema is similar to the traditional software testing process.

In the case of MP models, it is possible to automatically generate all event traces within the given scope (exhaustive testing). Careful inspection of generated traces (scenarios/use cases) may help developers identify undesired behaviors. Usually, it is easier to evaluate an example of behavior (particular event trace) than the generic description of all behaviors (the schema). The small scope hypothesis (Jackson, 2006) states that most errors can be demonstrated on relatively small counterexamples.

Certain properties of behavior can be formalized as assertions about traces (similar to the **SATISFIES** constraint in Example 14.4 and Example 14.6) and verified exhaustively for all event traces within the scope, yielding the counterexamples when the assertion is violated. For example, hazard states can be specified as a result of certain interactions between the system and its environment, and then the traces within scope can be searched for a trace that matches the hazard scenario. An example of such assertion checking performed on an MP prototype is given in Auguston and Whitcomb (2010). Since assertion checking is performed on a complete event trace, it becomes possible to refer to events following a given event, for example, to specify fairness conditions. This brings the expressiveness of MP assertions closer to temporal logic (Pnueli, 1981).

In a similar fashion, queries can be performed on the traces, providing different kinds of statistics. For example, events may have **attributes**, such as estimated duration, and system's performance estimates can be obtained from collecting a representative amount of event traces and calculating durations for event sets of interest.

Another example of an event attribute may be the probability of an event in alternatives, like **(A [0.3] | B [0.7])** establishing that **A** happens with the probability of 0.3 and **B** with probability of 0.7. Now, it becomes possible to estimate probabilities of certain event traces, for example, probability for the system to get into a hazard state. This opens a whole direction for systems simulation and statistical experiments based on executable systems architecture models and their environment models.

Using MP to automatically generate use cases from component behavior models and abstract interaction specifications, a much larger set of systems behaviors can be predicted. Inspection can be used to expose design errors early in the life cycle by examining each generated use case for logic flaws or undesirable sequences of events. The maximum benefits are gained with assertion checking. The small scope hypothesis provides that the scope of use case generation may be limited by simulating only a specified number of loop iterations for every event trace. MP leverages the small scope hypothesis to provide a solution to expose far more design errors than do current approaches alone, without requiring specialized skills. If an assertion results in a counterexample (an event trace that contradicts the assertion), it can be used to observe precisely why the assertion is false and, if needed, help the architect write a constraint to prevent the sequence of events that makes the assertion false. MP consequently provides a means for observing and correcting design errors in a modeled architecture, so that an architect can weed undesired behavior from the specification through the addition of abstract SoS interaction constraints.

14.10 IMPLEMENTATION PROTOTYPES

The first MP prototype (Auguston and Whitcomb, 2010) has been implemented as a compiler generating an Alloy model (Jackson, 2006) from the MP schema and then running the Alloy Analyzer to obtain event traces and to perform assertion checks. It has benefited from Alloy's relational logic formalism and visualization tools. Performance depends on the performance of SAT solver used by the Alloy Analyzer.

Direct trace generation from the event grammar can be accomplished quite efficiently, and the process of generating all traces for the given schema and within a given scope can be roughly described by the following procedure:

1. Generate all possible traces within the given scope for each root in the schema.
2. Select one trace from each root's collection. Apply all the schema's composition operations and filters. If the resulting composed trace is consistent with the schema's filters and composition operations, it is included into the schema trace collection. Otherwise, proceed with the next selection.

This process may lead to an exponential explosion, but it has potential for optimization by applying early pruning whenever possible. The main optimization ideas stem from the considerations that composition operations (**COORDINATE** and **SHARE ALL**) usually require an equal number of selected events in the matching traces. Root traces can be sorted according to the number of required events to avoid selection of inconsistent root traces in step 2. Careful rearrangement of composition operations and filters may also provide a significant speed up in the trace assembly.

Other examples using this technique and an online demo of MP automated tools are found in Rivera (2010) and Rivera Consulting Group (2013), respectively. A prototype trace generator has been built to convert MP schemas into a C++ code and then compile and run it. This architecture solution is similar to the one that has been implemented, for instance, in the SPIN/PROMELA model checker (using C as a target language) (Holzmann, 2004).

Several optimizations similar to the one mentioned earlier have been implemented. A sample run on an iMac with 2.8 GHz/4 GB yields the following performance for a schema example with approximately 60 lines of MP source text, 31 event types including 9 roots, 10 composite event types, 12 atomic event types, and 12 **SHARE ALL** compositions and for a maximum scope of 3 for iterations (actually it is an architecture model for the MP → C++ prototype itself):

- Total of 1,328 traces generated, with total of 79,836 events, average of 60.1175 events/trace, and max trace length of 69
- Initial search space (number of all root trace selections before filtering) of 35,100

- Selection ratio of 3.78348% and generation speed of 18021.8 events/s
- Elapsed time (including compilation of the generated C++ code) of 4.42997 s

14.11 CHAPTER SUMMARY

The MP executable architecture models provide a high level of abstraction for testing, verifying, and documenting systems architecture early in the conceptualization and design phases. The main advantages may be summarized as follows:

- The use of MP focuses the attention of developers early on the behavior of the system and provides tools to verify the assumptions.
- The schema framework is amenable to stepwise architecture refinement, reuse, composition, visualization, and application of automated tools for consistency checks.
- The executable architecture models integrated with the environment behavior models can be helpful for identifying emerging behaviors.
- The ability to generate use cases for requirements specification and for testing the system's implementation.
- The ability to create abstract views on the interfaces, composition, and coordination within the system.
- The ability to develop performance estimates based on statistics obtained from the event traces.
- The possibility to extract different architecture views, for example, based on stakeholder viewpoints, from the architecture model.

MP provides a uniform way to extract use cases from a single architecture model composed of component behavior algorithms and an abstract interaction specification—the latter being a capability that is absent from current Systems Engineering approaches. Use cases are based on generic descriptions of systems behavior, rather than on a limited number of use cases. This approach allows architects to expand their definitions of a "representative" set of use cases to increase the design space explored early in the life cycle and to correct undesired behaviors prior to the implementation. It also transfers the burden of maintaining consistency among similar use cases to automated tools.

Architecture modeling touches on the very fundamental issues in Systems Engineering and software design processes and has substantial consequences for the next phases in software systems design in particular. There are many threads of future research based on the ideas described earlier:

- Monitoring whether the behavior of an implemented system matches the MP architecture model (testing automation). If the source code of implementation can be marked up to indicate which segments of code start and end corresponding MP events, it becomes possible to log actual execution traces and to check them for consistency with expected behaviors.
- Developing methods and techniques for early performance, throughput, and latency estimates based on duration and frequency estimates for events within components and connectors.

- Developing methods and techniques for an architecture model's static analysis, for example, by verifying MP models with a model checking tool (Zhang et al., 2012).
- Introducing architecture metrics for MP models for systems cost estimates.
- Developing a library of reusable architecture patterns.
- Developing a library of reusable architecture views.
- Developing a collection of reusable environment behavior models, including business process models in MP.
- Extending the MP approach to the meta-architecture level to support software product lines and domain-specific architectures by representing the variation points as macroconditions in schemas. The same mechanism may be used for architecture configuration management.

Because of its high abstraction level, application of the MP approach should not be considered limited to the improvement of human-designed software intensive and complex adaptive systems. Design flaws manifesting themselves at inopportune times in these classes of systems were merely the original motivation for developing this approach to behavior modeling. Future research may explore its application to the improvement of human understanding of emergent behavior in economic, biological, and ecological systems and to study the causality of events from patterns in cellular behavior to sustainable food and energy production.

Existing software engineering tools have codified the concepts described herein; the next step is to integrate them into notations and modeling environments used by systems engineers and other professionals concerned with complex technological and/or natural systems. The MP approach is a force multiplier for systems architects that is open for implementation in any academic, government, or commercial modeling tool or environment whose objective involves architecting complex systems.

This chapter described a novel approach for modeling and predicting systems behavior resulting from the interactions among subsystems and among the system and its environment. The approach emphasizes specification of component behavior and component interaction as separate concerns at the architectural level. MP provides a new capability for automatically verifying systems behaviors early in the life cycle, when design flaws are most easily and inexpensively corrected. MP extends existing frameworks and allows multiple visualizations for different stakeholders and has potential for application in multiple domains.

REFERENCES

Abowd, G., Allen, R., Garlan, D., 1995, Formalizing style to understand descriptions of software architecture, *ACM Transactions on Software Engineering and Methodology*, 4(4):319–364.

Aizier, B., Lizy-Destrez, S., Seidner, C., Chapurlat, V., Prun, D., Wippler, J.l., 2012, xFFBD: Towards a formal yet functional modeling language for system designers, In *Proceedings of the 22nd INCOSE International Symposium*. Rome, Italy, July 9–12.

Allen, R., 1997, A formal approach to software architecture, Ph.D. Thesis, Carnegie Mellon University, Pittsburgh, PA. CMU Technical Report CMU-CS-97-144, May 1997.

Allen, R., Garlan, D., 1997, A formal basis for architectural connection, *ACM Transactions on Software Engineering and Methodology*, 6(3):213–249.

Armstrong, J.R., 2013, Functional architecture's mental roadblocks and other things your mother didn't tell you, In *Proceedings of the 23rd Annual INCOSE International Symposium*, Philadelphia, PA, June 24–27.

Auguston, M., 1991, FORMAN—Program formal annotation language, In *Proceedings of 5th Israel Conference on Computer Systems and Software Engineering, Herclia*, IEEE Computer Society Press, Herclia, Israel, May 27–28, pp.149–154.

Auguston, M., 1995, Program behavior model based on event grammar and its application for debugging automation, In *Proceedings of the 2nd International Workshop on Automated and Algorithmic Debugging*, Saint-Malo, France, May 1995.

Auguston, M., 2009a, Software architecture built from behavior models, *ACM SIGSOFT Software Engineering Notes*, 34:5.

Auguston, M., 2009b, Monterey phoenix, or how to make software architecture executable, *OOPSLA'09/Onward Conference, OOPSLA Companion*, Orlando, FL, October 2009, pp. 1031–1038.

Auguston, M., Jeffery, C., Underwood, S., 2002, A framework for automatic debugging, In *Proceedings of the 17th IEEE International Conference on Automated Software Engineering*, Edinburgh, UK, IEEE Computer Society Press, September 23–27, pp. 217–222.

Auguston, M., Michael, B., Shing, M., 2006, Environment behavior models for automation of testing and assessment of system safety, *Information and Software Technology*, 48(10):971–980.

Auguston, M., Whitcomb, C., 2010, System architecture specification based on behavior models, In *Proceedings of the 15th ICCRTS Conference (International Command and Control Research and Technology Symposium)*, Santa Monica, CA, June 22–24.

Auguston, M., Whitcomb, C., 2012, *Proceedings of the 24th ICSSEA Conference (International Conference on Software and Systems Engineering and their Applications)*, Paris, France, October 23–25.

Bass, L., Clements, P., Kazman, R., 2003, *Software Architecture in Practice*, 2nd Edition, Boston, MA: Addison-Wesley.

Booch, G., Jacobson, I., Rumbaugh, J., 2000, OMG unified modeling language specification, http://www.omg.org/spec/UML/ (accessed October 15, 2014).

Bruegge, B., Hibbard, P., 1983, Generalized path expressions: A high-level debugging mechanism, *The Journal of Systems and Software*, 3:265–276.

Campbell, R.H., Habermann, A.N., 1974, The specification of process synchronization by path expressions, *Lecture Notes in Computer Science*, 16:89–102.

Department of Defense, 2009, *DoD Architecture Framework*, version 2.0, Washington, DC: ASD(NII)/DoD CIO.

Feiler, P., Gluch, D., Hudak, J., 2009, The architecture analysis & design language (AADL): An introduction, Technical Note CMU/SEI-2006-TN-011, http://www.sei.cmu.edu/publications/documents/06.reports/06tn011.html (Accessed June 2009).

Friedenthal, S., Moore, A., Steiner, R., 2006, OMG systems modeling language (OMG SysML™) tutorial, Presented at the 2006 *INCOSE (International Council on Systems Engineering) International Symposium*, Orlando, FL, July 11.

Harel, D., 1987, A visual formalism for complex systems. *Science of Computer Programming*, 8(3):231–274.

Hoare, C.A.R., 1985, *Communicating Sequential Processes*, Englewood Cliffs, NJ: Prentice-Hall.

Holzmann, G., 2004, *The SPIN Model Checker*, Boston, MA: Addison-Wesley.

ISO, 2011, International Organization for Standardization. ISO Standard ISO/IEC 42010:2011, *Systems and Software Engineering—Recommended Practice for Architectural Description of Software-Intensive Systems*.

Jackson, D., 2006, *Software Abstractions: Logic, Language, and Analysis*, Cambridge, MA: The MIT Press.

Jackson, M., 2007, Consultancy and research in software development. Past research topics. http://mcs.open.ac.uk/mj665/topics.html (Accessed October 15, 2014).

Knuth, D., 1984, Literate programming, *The Computer Journal*, 27(2):97–111.

Kruchten, P., 1995, Architectural blueprints—The 4 + 1 view model of software architecture, *IEEE Software*, 12(6):42–45.

Liskov, B., Zilles, S., 1974, Programming with abstract data types, *ACM SIGPLAN Notices*, 9(4):50–59.

Long, D., Scott, Z., 2011, *A Primer for Model-Based Systems Engineering*, 2nd Edition, lulu.com.

Long, J.E., 2000, Relationships between common graphical representations used in system engineering, In *Proceedings of the SETE2000 Conference (Systems Engineering and Test and Evaluation)*, Brisbane, Queensland, November 15–17.

Luckham, D., Augustin, L., Kenney, J., Vera, J., Bryan, D., Mann, W., 1995, Specification and analysis of system architecture using Rapide. *IEEE Transactions on Software Engineering, Special Issue on Software Architecture*, 21(4):336–355.

Luckham, D., Vera, J., 1995, An event-based architecture definition language, *IEEE Transactions on Software Engineering*, 21(9):717–734.

Maier, M., Rechtin, E., 2000, *The Art of Systems Architecting*, Boca Raton, FL: CRC Press.

NASA, 2007, *Systems Engineering Handbook*, Washington, DC: NASA/SP-2007-6105 Rev1.

Nassi, I., Shneiderman, B., 1973, Flowchart techniques for structured programming, *ACM SIGPLAN Notices XII*, pp. 12–26.

Object Management Group, 2012, *Systems Modeling Language Specification*, version 1.3, http://www.omg.org/spec/SysML/ (Accessed July 22, 2014).

Pelliccione, P., Inverardi, P., Muccini, H., 2009, CHARMY: A framework for designing and verifying architectural specifications, *IEEE Transactions on Software Engineering*, 35(3):325–346.

Perry, D., Wolf, A., 1992, Foundations for the study of software architecture, *ACM SIGSOFT Software Engineering Notes*, 17(4):40–52.

Pnueli, A., 1981, A temporal logic of programs, *Theoretical Computer Science*, 13:45–60.

Rechtin, E., 1991, *Systems Architecting: Creating and Building Complex Systems*, Englewood Cliffs, NJ: Prentice Hall.

Rivera, J., 2010, Software system architecture modeling methodology for Naval Gun Weapon Systems, Doctoral Thesis, Naval Postgraduate School, Monterey, CA, December 2010.

Rivera Consulting Group, 2013, *Eagle 6 Modeling*, http://eagle6modeling.riverainc.com/ (Accessed July 22, 2014).

Roscoe, B., 1997, *The Theory and Practice of Concurrency*, London, UK: Prentice Hall International Series in Computer Science, pp. 580.

Rozanski, N., Woods, E., 2012, *Software Systems Architecture*, 2nd Edition, Upper Saddle River, NJ: Addison-Wesley.

Spivey, J.M., 1989, *The Z Notation: A Reference Manual*, 2nd Edition, Englewood Cliffs, NJ: Prentice Hall International Series in Computer Science, pp. 1992.

Taylor, R., Medvidovic, N., Dashofy, E., 2010, *Software Architecture, Foundations, Theory, and Practice*, Hoboken, NJ: John Wiley & Sons, Inc.

Vaneman, W., Jaskot, R., 2013, A criteria-based framework for establishing system of systems governance, In *Proceedings of the 7th Annual International IEEE Systems Conference*, Orlando, FL, April 15–18, pp. 491–496.

Wang, Y., Parnas, D., 1994, Simulating the behavior of software modules by trace rewriting, *IEEE Transactions on Software Engineering*, 20(10):750–759.

Zhang, J.Y., Liu, M., Auguston, J.S., Dong, J.S., 2012, Using monterey phoenix to formalize and verify system architectures, In *19th Asia-Pacific Software Engineering Conference APSEC 2012*, Hong Kong, December 4–7.

Chapter 15

Joint Training

James Harrington[1], Laura Hinton[1], and Michael Wright[2]
[1] *The MITRE Corporation, McLean, VA, USA*
[2] *US Army Project Executive Office for Simulation Training and Instrumentation, Orlando, FL, USA*

15.1 JOINT TRAINING INTRODUCTION

The U.S. Department of Defense (DoD) performs specific and measured training at all levels from individual airman, sailor, soldier, and marine to the training of small units such as Army platoons or companies and up to the training of general officers commanding large complex organizations and operations. For example, though it's relatively straightforward to construct an environment to train rifle marksmanship, it can be daunting to create a training environment that replicates conditions, situations, and stimuli faced by the commander of a large military organization executing a complex Joint task force mission.

In the marksmanship example above, trainers can create an actual rifle range and target array or, increasingly, computer-generated virtual environments like gaming or virtual ranges augmented by computer-generated imagery. However, moving up the scale of unit size from individual soldier to Army platoon, company, battalion, etc., it becomes increasingly impractical to build and manage physical training environments. So, virtual environments are becoming more prevalent.

In the days before widespread use of computerized modeling and simulation (M&S) for training, several rudimentary modeling approaches approximated the environments faced by large-unit commanders—board games and sand tables where military units maneuvered within a crudely modeled battlespace. With such simple models of the battlespace, it was difficult to represent the large number of variables that impact the

---

Modeling and Simulation Support for System of Systems Engineering Applications, First Edition.
Edited by Larry B. Rainey and Andreas Tolk.
© 2015 John Wiley & Sons, Inc. Published 2015 by John Wiley & Sons, Inc.

conditions and performance of large military units in wartime. Computer-based M&S tools have made it easier to create such training environments.

At first glance, it seems a straightforward challenge to build a computer program that models the environment important to commanders of large military units and to employ this program to provide a context to support training. And, in fact, M&S software programs to accomplish this task were developed throughout 1980s and 1990s by the DoD individual armed services and Joint training providers. An example of an early training model initiated in the mid-1980s was the Joint Exercise Support System (JESS). These programs designed to model the broad environment and complex mix of variables and conditions encountered by military commanders are referred to as constructive simulations.

The challenge that the DoD now faces is that individual constructive simulations are bound by practical limitations. Modern complexities include specialized equipment, missions, and functions of both combat and noncombat units that comprise fully staffed and supported military units (in addition to enemy or opposing forces), physical environment (terrain and weather), civilian population, and infrastructure. Thus, even very low levels of modeling fidelity can exhaust the breadth of modeling that can be implemented in a single monolithic software system.

Another way to address this limitation would be to model all of the systems and environments of a large military organization in a highly abstract manner. However, the U.S. military is a demanding customer and has often not accepted this "depth for breadth" trade-off. They argue that sometimes mundane, low-level details can affect battlefield outcomes that matter greatly to commanders and staffs, details like fuse type on individual artillery rounds. One way to deal with the depth-versus-breadth modeling challenge is to define a software and communications infrastructure that allows a group of specialized models to collectively model the operational environment within a synchronized framework.

In its simplest form, each Service could contribute a constructive simulation that provides detailed modeling of its individual Service activities to support Joint commander and staff training. To reduce the overall complexity, each simulation could operate as an independent system, interfacing or interoperating with other Services' simulations in a loosely coupled fashion, exchanging data only when absolutely necessary. This form of system of systems (SoS) employment of simulation capabilities enables high-fidelity modeling of large, highly complex organizations, environments, and interactions for training DoD commanders and their staffs at the individual Service and Joint levels.

The objectives for this chapter are to present an example of M&S capabilities applied in an SoS context to enable training of DoD commanders and their staffs. This example will be used to describe the SoS engineering practices that manage development, integration, and support of this complex M&S SoS. The focus of this case study will be the U.S. Army's Joint Land Component Constructive Training Capability (JLCCTC) and its Systems Engineering (SE) team's practices.

15.2 SoS CHARACTERIZATION FOR ARMY AND JOINT TRAINING M&S

The DoD has long acknowledged that an SoS federation of high-fidelity models is needed to represent the complex environment of large military operations for Joint training. However, there have been significant shifts in type and manner in which M&S SoS capabilities have been developed and employed over the past 20 years.

To trace the progression of M&S SoS for Joint training, we should first review some fundamental definitions. The DoD identifies four distinct categories of SoS (OUSD(AT&L) SSE, 2008):

- **Virtual**: The SoS lacks central management and a centrally agreed-upon purpose.
- **Collaborative**: Component systems within the SoS interact more or less voluntarily to fulfill agreed-upon central purposes.
- **Acknowledged:** The SoS has recognized objectives, a designated manager, and resources, while the component systems retain their independent ownership, objectives, funding, development, and sustainment approaches.
- **Directed**: The SoS is built and managed to fulfill specific purposes. Component systems operate independently, but their operational mode is subordinate to central management purposes.

In the early 1990s, the first M&S SoS for Joint command and staff training was the Aggregate Level Simulation Protocol (ALSP) federation of models (Wilson and Weatherly, 1994). This federation developed and employed its namesake data exchange protocol and operated in a highly collaborative, partly acknowledged manner.

The DoD decided that collaborative and acknowledged forms of SoS lack unity of command and control when compared with the directed form of SoS management. So, in the late 1990s, the DoD set out to create a directed M&S SoS for Joint command and staff training. The result was the Joint Simulation System (JSIMS) project. It defined a common infrastructure, a common set of federation services, and a common set of individual M&S components to be used within the individual Services' federated models. JSIMS was never fielded—the program was cancelled in 2002, prior to completion.

Based on the experience of JSIMS, the Joint training community agreed that there is a limit to how much centrality an M&S federation of independently funded Service models can achieve. This is due in part to the autonomous nature of the Armed Services' responsibilities under Title 10 of the U.S. Code to organize, equip, and train forces for the conduct of their missions (US Code Title 10, 2012). This code, originally enacted in 1956, mandates that each Service develop its own training models to satisfy Service Title 10 training requirements. Limited resources dictate that these Service models must also double as the Services' contribution to a Joint training M&S SoS. Therefore, within the Joint training SoS context, contention and friction between individual Service and Joint SoS priorities will stymie any attempt to execute a directed form of SoS for training models. Many alumni of the JSIMS program attribute the program's cancellation to this root cause.

An example of the directed nature of JSIMS was its mandate that individual systems use common internal M&S components, including a single JSIMS simulation engine, a synthetic natural environment, and a common simulation role-player workstation. While use of common software components in an SoS can be desirable, lack of central authority of the JSIMS project over the individual Service development efforts led to unfulfilled dependencies and resulting limitations. In other words, the directed, tightly coupled JSIMS SoS architecture created technical and programmatic interdependencies that could only be resolved if all parties were completely subordinate to central SoS management, which they were not.

One insight that emerged in the aftermath of the JSIMS program was that in previously successful large, complex Joint training SoS employments, Service contributors generally formed "Coalitions of the Willing." Service-developed M&S capabilities were integrated into a Joint SoS in a loosely coupled manner to minimize interdependency and cost. This approach does not produce the optimal capability, but due to statutory, funding, and political constraints, it has proved to be the most pragmatic to date.

After JSIMS cancellation, the Joint and Service's M&S command and staff training communities each chose a more loosely coupled, less centralized approach for command and staff training capabilities. As mentioned earlier, this chapter presents a case study of the Army's M&S training SoS capability established in the wake of JSIMS. But note that a number of the SoS engineering practices and principles described in the Army context were also adopted by or have been borrowed from the Joint and other Services M&S training communities.

15.3 COMPLEXITY OF LAYERED M&S SoS

M&S SoS are employed at multiple levels. Instead of contributing a single simulation to a Joint training SoS federation, individual Services often contribute their own complex M&S SoS, one that they also use to separately train their commanders and staffs. For example, the Army has a collection of interoperable M&S components that support training of Army commanders in an Army-only training environment. When Army organizations come together with other Service forces for Joint training, the Army M&S SoS is federated or connected to other Service M&S SoS to form an "Uber-SoS." As multiple Service SoS are integrated to form a higher-level Joint SoS, the resultant complexity can rise dramatically.

This explosion of complexity in Joint and Service M&S training toolkits led the Joint M&S community to rethink and retool their Joint M&S SoS toolkit, data exchange, and communications infrastructure. They envision a cloud-enabled, Web service model that reduces complexity, operating costs, and sustainment costs. They refer to this future capability as the Joint Live, Virtual, and Constructive (JLVC) 2020 system. This next generation SoS approach addresses the limiting factors of previous Joint training SoS, such as complexity and cost as well as governance limitations when the Joint community and Services share capabilities.

The remainder of this chapter focuses on the practice and procedures of SE applied in an SoS environment to plan, develop, integrate, execute, and maintain the Army's current M&S SoS toolkit for command and staff training—JLCCTC.

15.4 JLCCTC OVERVIEW

The complexity of the operational environment facing Army commanders increases at each level of command. Notwithstanding the current counterinsurgency environment, a traditional Army company commander leads about 100 soldiers whose mission falls within limited focus areas, such as armored operations, chemical operations, or military intelligence. By the time an officer progresses to the rank of lieutenant general and takes command of an Army corps, he commands all of the Army unit types previously mentioned and many more. So the training environment for a corps commander and his staff requires modeling all of the Army combat and support organizations and all the operations that a corps commander might oversee. No single Army simulation provides the breadth and depth of modeling required to train a corps commander and staff.

Therefore, one configuration of JLCCTC focuses on training battalion and brigade commanders and comprises:

- A primary combat model
- A logistics model
- An intelligence model
- A civilian model
- An artillery model
- An aerial drone model
- An interface to real-world Mission Command (MC) computer systems
- An after-action review (AAR) system
- Technical control tools

This variation of the JLCCTC SoS is based on a high-fidelity representation of battlefield entities—individual vehicles and people are modeled and represented as individual objects—and is referred to as the Entity Resolution Federation (ERF).

An alternative configuration of JLCCTC is used for division, corps, and echelons above corps training. It comprises a similar set of components but substitutes a larger-scale, more abstract ground combat model and adds a fixed-site infrastructure model. This configuration is referred to as the Multi-Resolution Federation (MRF). It employs a hybrid of aggregate modeling (modeling units rather than individual vehicles and people) and entity-based modeling. Thus, JLCCTC can be configured as an ERF or MRF depending on the scale required, which corresponds to the highest level of command within the training audience.

An example of the composable systems architecture to meet the needs of the warfighter is shown in Figure 15.1. This MRF architecture is used in Army and Air Force combined training and known as the Joint Training Transformation Initiative (JTTI). In this SoS, most of the core federates operate at a centralized location, with distributed sites for technical control, logistics, and Air Force operations.

An SE team manages not only newly emerging developmental baselines but also the fielded versions of ERF and MRF. When the Army JLCCTC is augmented with non-Army Service models for Joint and international coalition training, five unique configurations of JLCCTC are possible. So, rigorous SE principles and practices must be employed to keep these SoS training federations on track to meet Army and Joint training requirements.

For the JLCCTC configurations, the SE team manages federation requirements and allocates specific tasks and requirements to individual component or federate developers.[1] Federate developers are development agents for the JLCCTC model components, that is,

---

[1] **Federate developer**—This term refers to any of the government or contractor organizations that are responsible for the design, implementation, and integration of individual components or systems that comprise the JLCCTC SoS.

398 Theoretical and Methodological Considerations with Applications and Lessons Learned

ground model, logistics model, AAR tool, etc., and include contractor and government technical and programmatic personnel. Note that not all JLCCTC federates are managed by the same development agent or military project manager (PM). In fact, JLCCTC is supported by a cross section of development agencies within and external to the Army and DoD. For example, the Department of Energy's Lawrence Livermore National Labs contributes a combat model to the Army JLCCTC, while the Air Force provides a fixed-site target model. So, JLCCTC is far from a homogeneous federation of models built from the ground up for low-level, detailed interoperability. Rather, it is a loosely coupled compilation of independently developed capabilities that provide significant training utility when linked together.

The Army Training and Doctrine Command (TRADOC) chartered a single user representative[2] organization, the National Simulation Center (NSC), to solicit, consolidate,

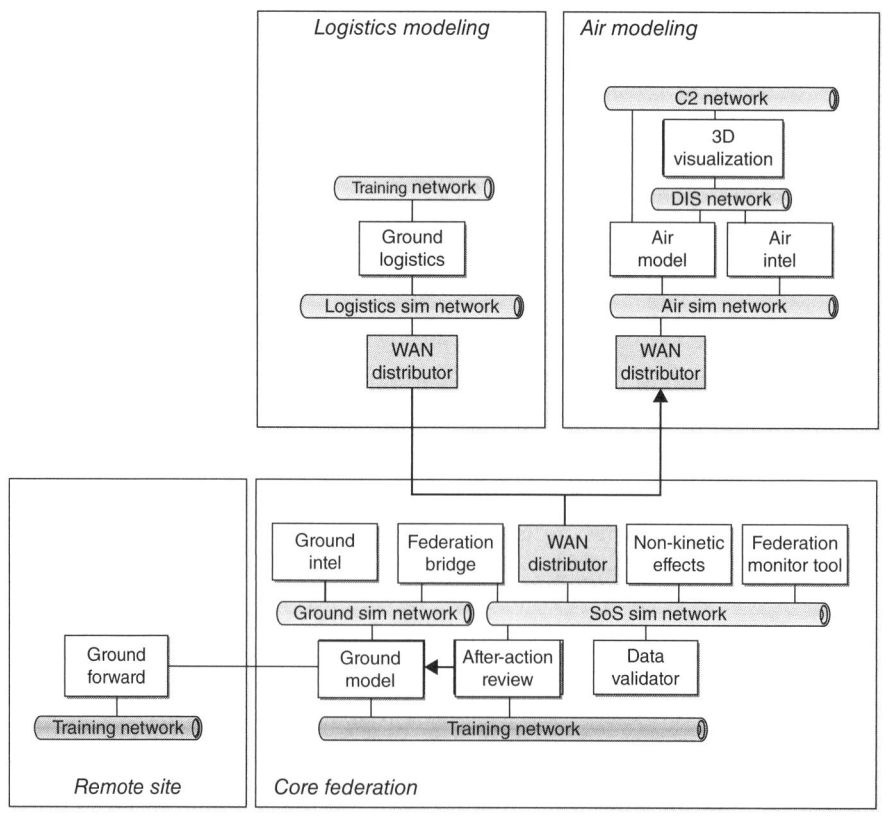

Figure 15.1 JTTI M&S system of systems architecture.

---

[2] **User representative**—The user representative is responsible for soliciting and evaluating capability needs from its constituent end users and translating these needs into SoS requirements to be addressed by the PM. Further, the user representative is responsible for validating the SoS to ensure that newly developed capabilities satisfy the requirements. In the context of JLCCTC, the user representative organization is the NSC.

prioritize, and issue requirements to the JLCCTC PM[3] or development agent—the Program Executive Office for Simulation, Training, and Instrumentation (PEO STRI). Within PEO STRI, the Project Manager for Constructive Simulation (PM ConSim) is responsible for developing (certain components), integrating, fielding, and supporting the fielded JLCCTC. PM ConSim employs government engineering staff and systems engineers from MITRE (a federally funded research and development center (FFRDC)) to lead the SoS SE effort for JLCCTC. The JLCCTC SE team[4] works directly for PEO STRI and acts as the SoS SE team for the entire system that includes other U.S. government labs and contractors working in support of other U.S. government and DoD organizations.

Because many of the JLCCTC development agents do not belong to a single DoD organization or report through a single command structure, the role of the SE team is more critical and challenging. Similar to the Joint training example described earlier where the directed form of SoS management was not practical, it is also not practical in JLCCTC because, again, all development agencies do not work for and are not funded by PEO STRI. However, some JLCCTC components like the ground model, intelligence model, and logistics model are developed by PM ConSim. So JLCCTC could be considered a hybrid SoS where some of its developers report to PM ConSim, some work for Army organizations outside PEO STRI, and some report to external organizations. JLCCTC, therefore, contains elements of a directed, acknowledged, and collaborative SoS—a "coalition of the willing"—which adds organizational and interpersonal complexity on top of the technical complexity of an SoS environment.

Regardless of whether JLCCTC is operating on its own or as part of a live, virtual, and constructive or Joint training SoS, it always interoperates with real-world Army and Joint MC systems used by the training audience. Because MC systems were not built with simulation interoperability in mind, time and budget constraints forced simulation and MC systems interoperability to be loosely coupled and limited to fixed format data exchanges. JLCCTC supports about 50 tactical message formats that form the basis of communication with MC systems. For example, simulated unit and entity positions can be shown on MC systems map displays, and call-for-fire missions initiated on real-world fire control computers can cause fire missions to be modeled in the simulation battlespace.

Army MC systems form an SoS unto themselves, so JLCCTC interoperating with Army MC systems forms another higher-level SoS that can be thought of as a collaborative SoS. For the past 20 years, the DoD M&S training and MC systems communities have focused and prioritized their architecture efforts toward meeting their respective individual requirements. Standard text and binary military message formats like the U.S. Message Text Format (USMTF) and Variable Message Format (VMF) permitted loosely coupled interoperability between these communities but with limited capability.

---

[3] **PM**—The PM is ultimately charged with fielding the SoS that satisfies the requirements that were originally identified by user representative, agreed upon by both the user representative and PM based upon resource constraints, and implemented by the SE team and federate developers. The PM is the principal governing authority for the task definition through integration and test phases of the SE process. In the context of JLCCTC, the PM organization is PEO STRI.

[4] SE **team**—The SE team is comprised of PM government engineers and engineering staff from the MITRE FFRDC. This combined team provides leadership and oversight for the end-to-end SE process described throughout this chapter.

With the emergence of Web service-based architectures in the MC systems community, new interfaces and capabilities are being exposed that were not previously available to simulation developers. Some in the M&S community expect that exposed Web services will enable a richer data exchange between M&S and MC systems without adding to the level of SoS management. Others within the Army believe that a more acknowledged or directed SoS for all Army software systems is needed. At this time, the latter thought process is prevailing with establishment of the Army Common Operating Environment initiative that aims to bring Army MC and training systems together in a mandated, overarching SoS. While the JLCCTC program continues to monitor impacts of this initiative, the balance of this chapter describes JLCCTC capabilities and practices to date.

15.5 SoS CHARACTERIZATION FOR JLCCTC

The emphasis and focus in this chapter is not on how M&S supports the JLCCTC SoS, but rather on providing an example of an SoS comprising interoperable M&S systems. Stated another way, much of the content of this book demonstrates how M&S tools and practices can help manage an SoS. In contrast, this chapter focuses on the Army's JLCCTC, which is a complex system of M&S systems where M&S is not a means to an end but is the purpose of the SoS under examination.

Constructive simulations in a large DoD SoS have been used for training and testing since the early 1990s. More recently, this was extended to encompass live and virtual simulations. Live simulations include maneuver and firing ranges; virtual simulations include cockpit or vehicle crew simulators that employ elaborate out-the-window computer-generated visual displays.

Like constructive simulations, Army live and virtual simulations are increasingly linked together in SoS to increase the breadth and depth of capability and to achieve higher scaling. For example, eight virtual Apache helicopter simulators can be linked to simulate company-level operations. Similarly, Army aviation simulators can be linked with virtual call-for-fire artillery trainers to train air observation for indirect fires. Thus, linkage of live, virtual, and constructive assets often implies live, virtual, and constructive SoS and the complexity of multilayered SoS. The point is that while JLCCTC is itself a complex SoS used for Army command and staff training, it is also the constructive component of the Army's overarching Live, Virtual, and Constructive Integrating Architecture (LVC-IA) SoS. Therefore, the JLCCTC SE team has the role of SoS integrator when JLCCTC plays the role of stand-alone constructive training tool. But when JLCCTC is employed as the constructive component of LVC-IA, the same group also acts as an individual systems representative to a larger SoS.

15.6 JLCCTC SoS ENGINEERING FUNCTIONS

As a preamble to the sections of this chapter that follow, this section provides a brief introduction to the main functions of the JLCCTC SE team.

Despite the myriad of relationships that exist within the overall JLCCTC team, the JLCCTC SE team provides the consolidated SoS technical and architecture perspective when communicating with users, user representatives, and other DoD stakeholders. This information is often consolidated through PM ConSim and PEO STRI staff, but the JLCCTC SE team gathers and aggregates federation technical and functional status and acts as the entry point for most external inquiries on existing or potential federation capability.

The primary function of the JLCCTC SE team is to ensure that the loosely coupled, interoperable capability among JLCCTC federation components or models:

- Meets user functional requirements at both the individual model and SoS level
- Conforms to Army and DoD information assurance policies and regulations
- Is planned, executed, and delivered within independent funding limitations and priorities of the JLCCTC development agents (both internal and external)

The JLCCTC SE team also tracks and reports development and integration progress to PM ConSim throughout the development and integration cycle and provides technical support to the NSC as they test and validate the SoS on behalf of JLCCTC users.

The basic development cycle is as follows: (i) NSC issues requirements to PEO STRI; (ii) the JLCCTC SE team analyzes and allocates requirements to development agents; (iii) developers enhance their products; (iv) integration, test, and validation events occur; and (v) the system is fielded to worldwide Army locations.

JLCCTC was envisioned as a phased evolution of an existing M&S SoS—the Joint Training Confederation (JTC). Therefore, an SoS SE approach was a natural fit. The goal was to evolve the JTC capability over a series of SE life cycle executions to gradually replace legacy systems with state-of-the-art M&S systems while periodically releasing the evolving system to the field. The JLCCTC SoS brings together existing Army simulation system and JSIMS, which themselves are employed as stand-alone training systems or as part of other SoS. Although some of the constituent systems are owned and managed by the JLCCTC PM, others are managed by other government organizations. Furthermore, some of the component systems are integrated into different SoS with competing schedules and requirements. So the JLCCTC PM cannot unilaterally direct that changes be made to all of its component systems. Because of this, the PM established an independent SE team and tasked it with negotiating among stakeholders to achieve requirements levied on the SoS.

15.7 JLCCTC SE KEY THEMES

Several aspects of the JLCCTC SE process have contributed to successful fielding and ongoing evolution of the JLCCTC ERF and MRF systems over the past decade. Through periodic evaluation, refinement, and documentation of JLCCTC SE processes, certain key themes have emerged: (i) an emphasis on coordination and communication among SoS stakeholders, (ii) a clear and methodical requirements management strategy, (iii) complexity management, and (iv) experimentation and innovation. The following is a brief description of these four key features, each of which is highlighted throughout the chapter.

15.7.1 Coordination and Communication

The success of the JLCCTC SE process is due, in large part, to its emphasis on developing strong working relationships among stakeholders. If a spirit of territorialism or "my way or the highway" were to exist, even the most seasoned SE process would struggle. Three means of coordination and communication are noteworthy.

1. First, the JLCCTC community—a gathering of all the principal stakeholders—meets weekly to review schedules, current development and integration statuses, and emerging requirements. This weekly meeting, planned and led by the SE team, ensures that all stakeholders are apprised of the latest information and that significant issues are identified before snowballing into larger system-wide problems.
2. Second, to tackle the large and varied number of systems requirements in an execution of the SoS life cycle, the SE team established **functional working groups** that meet regularly from concept and design through integration and test. This allows a small group of experts to focus attention on solving a set of related functional requirements.
3. Third, and perhaps most important, is the strong **user representative–PM relationship** that has been fostered over the life of the JLCCTC program. The user representative gathers and prioritizes warfighter training requirements and evaluates delivered systems; the PM builds and delivers the SoS that satisfies these requirements. It is natural that they would find themselves at odds from time to time. The opportunity for contention is inherent to this evaluator–evaluatee relationship. The JLCCTC SE process includes synchronization points at which the user representative can confirm that what the PM intends to deliver satisfies the intent of the requirements.

15.7.2 Requirements Management

The JLCCTC requirements management approach has been refined over multiple iterations of the SE life cycle, and its effectiveness has been demonstrated through consistent systems validation and fielding. Attention to requirements management runs throughout the entire SE process from requirements definition to fielding with several loopbacks. The requirements management approach is briefly described as follows:

- End-user organizations submit training capability needs to the user representative for consideration.
- The user representative synthesizes and consolidates the end-user capability needs and translates them into SoS requirements.
- The user representative, PM, and SE team work together to evaluate requirements in light of resource constraints and winnow the list of requirements.
- The SE team evaluates concepts and designs that are put forth by the federate developers or its own SE team members to ensure they satisfy user representative requirements.
- Federate developers implement enhancements to their own components and work with the SE team to integrate them into the SoS
- The user representative validates that new capabilities satisfy the requirements.

15.7.3 Complexity Management

A key responsibility of the JLCCTC SE team is management of complexity across the SoS. The user representative's primary responsibilities are to define requirements and to evaluate whether these requirements are satisfied by SoS delivered by the PM. Therefore, the user representative does not dictate how to address the requirement technically. Instead, this responsibility lies with PM in collaboration with the SE team. To guide the PM in its decision making, the SE team evaluates potential design solutions with an eye toward minimizing overall complexity. As a rule of thumb, the most cost-effective solution is one that uses existing components already in the SoS and minimizes the impact to system-wide interfaces.

15.7.4 Experimentation and Innovation

To ensure that the JLCCTC SoS continues to leverage evolving technologies and improve in stability and performance, the SE team dedicates a portion of its resources to experimentation and innovation. Often, such SE experimentation leads to development of tools to facilitate the SE life cycle itself or diagnostic tools used to verify conformance to SoS standards, to evaluate data exchanges, or to measure systems performance. Frequently, these diagnostic tools become fielded systems components that are used for troubleshooting at end-user[5] sites. Among the products that became fielded tools are system and performance monitoring tools, data alignment and cross-checking tools, a cross-domain security solution, and a test harness that acts as a surrogate for other SoS components when they are not present.

15.8 JLCCTC SE PROCESS

The JLCCTC SE team defined an SE process (see Figure 15.2) that is tailored to the specific demands and challenges of managing the end-to-end development of JLCCTC. This process also serves as the foundation for SE of related training SoS. The JLCCTC process incorporates best practices of the DoD *Systems Engineering Guide for System of Systems* (OUSD(AT&L)SSE, 2008) and elements of the *IEEE Recommended Practice for Distributed Simulation Engineering and Execution Process (DSEEP)* (IEEE Std 1730-2010, 2011). Those processes were adapted to satisfy the unique requirements associated with defining, implementing, and deploying the JLCCTC SoS. The complete process, from requirements definition through fielding, takes approximately 12–18 months.

Although the process steadily advances toward the fielding phase, there are several points along the way that loop back to earlier phases. For example, issues discovered during integration and test may require revising tasks resulting from the task definition phase and require further functional development and integration. The remainder of this section describes each phase in the SE process.

---

[5] **End user**—This term is used to describe two categories of users collectively. End user can refer to Army or Joint commanders and their staff who are actually being trained using the JLCCTC SoS. End user can also refer to the staff associated with the Mission Command Training Centers (MCTCs), TRADOC training facilities, and other customer locations that provide the training.

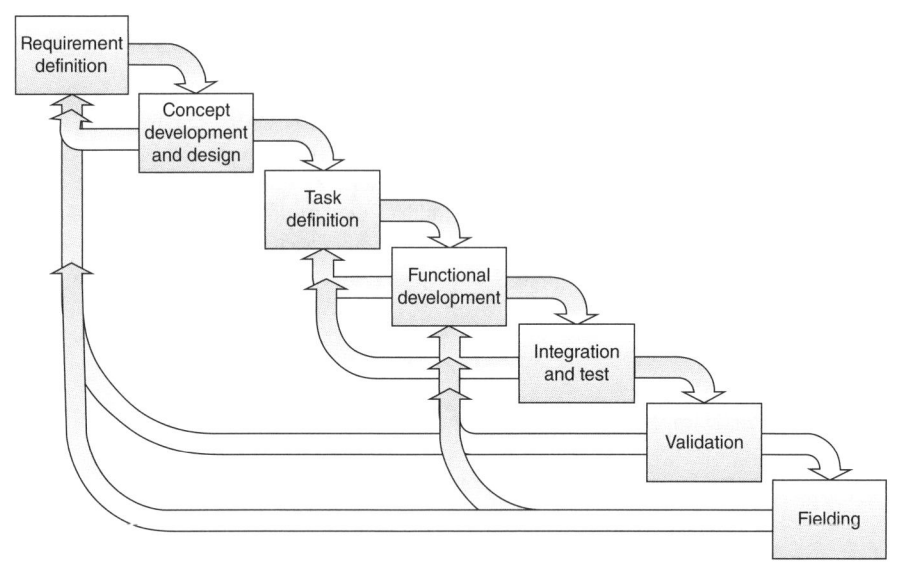

Figure 15.2 JLCCTC SoS engineering process.

15.8.1 Requirement Definition

The requirements definition phase starts the SoS SE cycle. Capability objectives are identified here. The SE team in this phase establishes the initial SoS baseline and supports defining the SoS engineering efforts needed for the next phase.

New requirements for the JLCCTC SoS originate from several sources, including (i) new capabilities requested by end-user organizations, (ii) changes to existing requirements from concept development and design, or (iii) problems identified during training or systems validation. Requirements for improvements and new functionality are documented in this phase.

The user representative prioritizes requirements. In addition, the SE team identifies potential technical enhancements (performance, tools, and infrastructure). Federate developers may also identify enhancements based on their knowledge of software features.

The SE team develops a framework for tasking the federate developers. This is the start of the integrated task list (ITL) to capture requirements from the user, SE team, and developers. The requirements definition process is shown in Figure 15.3.

To organize federation requirements, tasking, development, and progress, the JLCCTC program is decomposed into functional areas. Having a set of functional areas throughout the SoS development cycle provides several benefits. Most importantly, functional area decomposition provides the framework for the ITL. The ITL is a fundamental working document used by all participants at all stages of the cycle. It ensures traceability of development tasks to the original requirements. The ITL captures the user, SE, and developer requirements and, with the decomposition by functional area, provides transition into the next phase.

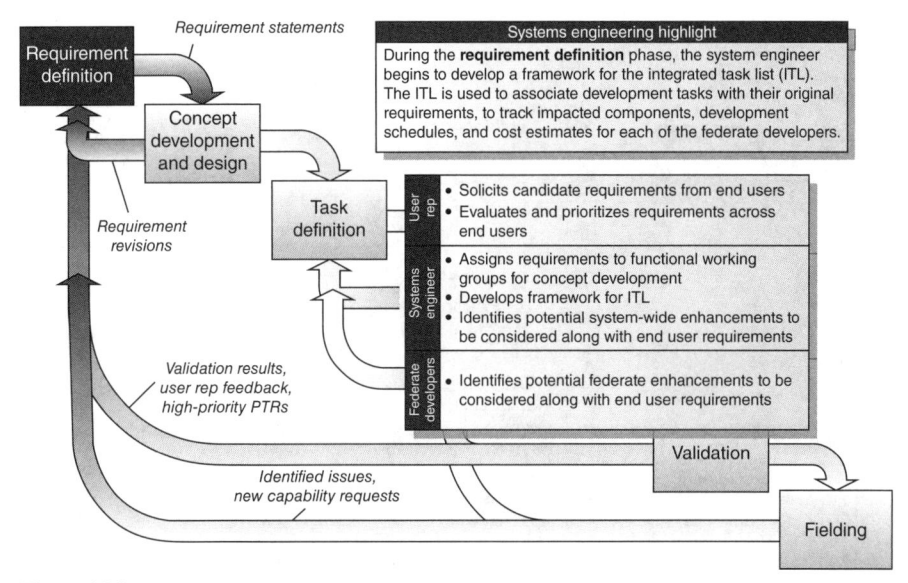

Figure 15.3 Requirements definition phase.

15.8.2 Concept Development and Design

After requirements have been defined, the M&S capabilities needed to achieve these requirements must be assessed and translated into SoS designs. Federate developers and the SE team propose initial concepts, and the entire JLCCTC community evaluates them to ensure they address the intent of the user representative's original requirements.

After these basic concepts are approved, federate developers identify the needed modifications to their software and prepare rough order of magnitude (ROM) estimates of the cost and time needed to implement them in their components. The user representative, PM, and SE team evaluate the ROM estimates and identify the simulation needs that can reasonably be accomplished within the development cycle time frame and the PM's resource constraints. Federate developers and the SE team then prepare detailed designs for the surviving requirements and present them to the community for concurrence.

The SE team, using the functional area organization, brokers solutions to federation requirements, promotes ideas for improving performance, and revises the ITL as requirements are clarified in preparation for task definition. Most importantly, the SE team assists with the assessment of the impact of the user requirements on the federation complexity and recommends solutions for implementation (Figure 15.4).

15.8.3 Task Definition

Based on the detailed designs from the concept development and design phases, the SE team consults with the federate developers and decomposes the chosen new requirements into actionable tasks that can be developed, integrated, and tested. The SE team augments the ITL by associating requirements with their corresponding development tasks and schedule.

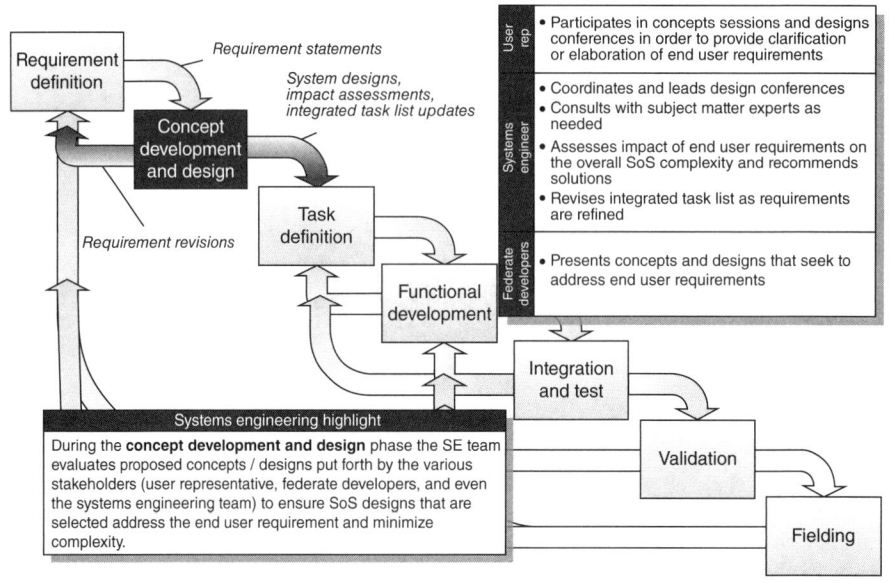

Figure 15.4 Concept development and design phase.

The refinement of the ITL during task definition is the most critical stage of its development. Tasks are defined by functional area, and the ITL captures and names actionable tasks for the developers. At this point, scheduling becomes important as developers provide input to an integration schedule, providing feedback to the SE team on when particular tasks can occur. The SE team uses the ITL and the integration schedule to begin planning integration events. Functional area leads review requirements with the user representative and designs with the developers to ensure that the result will produce the required capability. Figure 15.5 shows where task definition fits in the overall SE process.

The result of task definition is a set of actionable tasks, with incremental development subtasks scheduled to occur during the development of the software. The SE team enters all the scheduling and task data into the ITL and begins to prepare for integration events, while the developers begin software modification to the base M&S code.

15.8.4 Functional Development

As part of the task definition phase, the development tasks were grouped into functional areas. Functional working groups—composed of agents from the user representative, PM, SE team, and federate developers—guide and monitor software development tasks assigned to it. Federate developers implement enhancements to their components or systems according to the ITL schedule. Federate developers test new or updated functionality at their home facilities before bringing it to an SoS integration events.

Functional development is the software development period that occurs before and during integration and test. As progress occurs, reevaluation of tasking may change the development focus for the next integration and test event. Throughout a single JLCCTC SE cycle, task definition, functional development, and integration and

Joint Training **407**

test are continuously monitored and reevaluated, and changes are made to the ITL as needed. The ITL is a working document that acts as a common, synchronizing bond throughout all phases from the user requirements definition until the resulting functionality is delivered. As noted in Figure 15.6, functional development is midway through the JLCCTC phases.

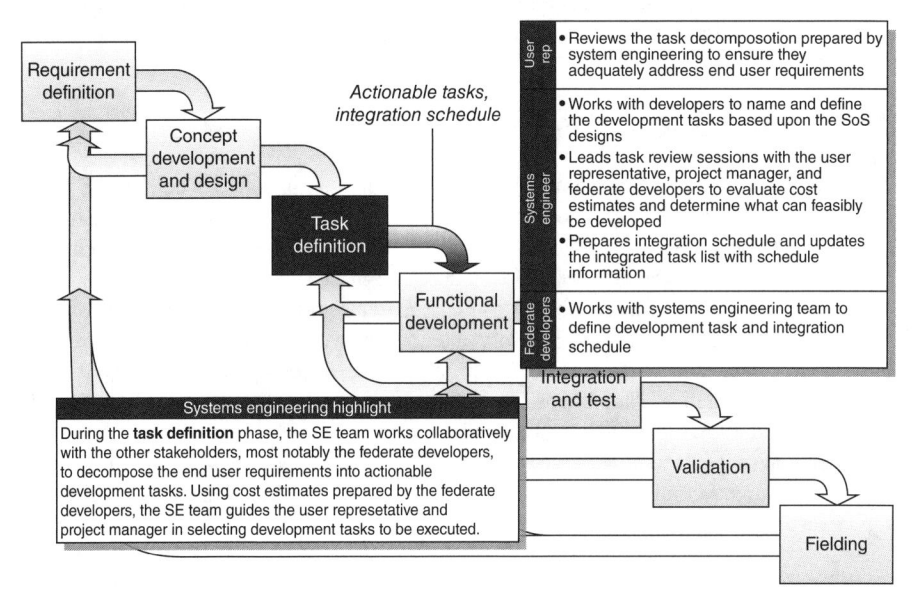

Figure 15.5 Task definition phase.

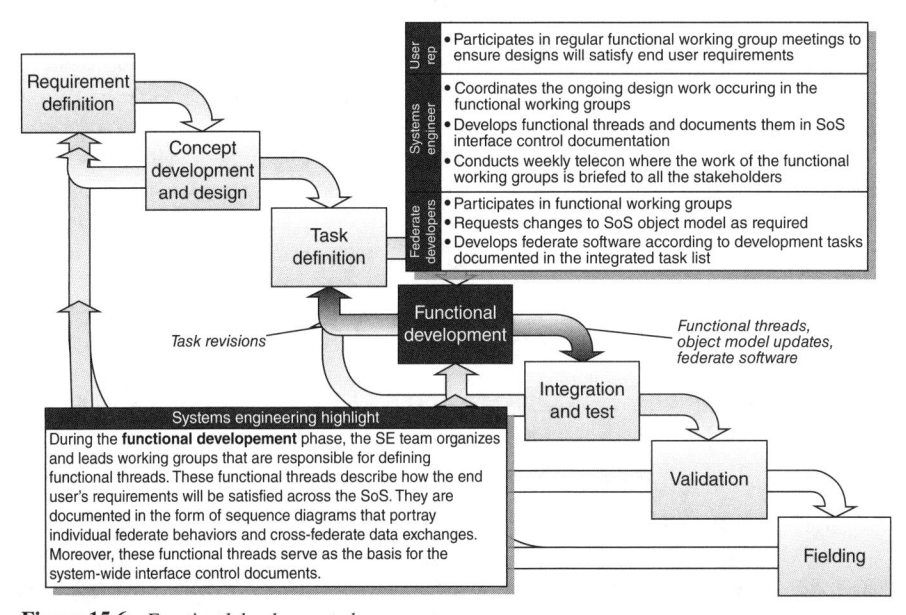

Figure 15.6 Functional development phase.

Functional working groups meet frequently during this phase, with all members (user, SE team, and developers) participating. The user representative attends to ensure that the design meets their original requirements. Developers describe their new functionality and request changes to data exchanges and federation agreements as needed. The SE team reviews these federation change requests, coordinating with other working groups as appropriate, and briefs the larger community when complete. The SE team prepares functional and technical test cases in preparation for integration and test.

15.8.5 Integration and Test

The integration and test phase of the JLCCTC SoS process is where users, the SE team, and developers work together to test new functionality. The SE team consults with the PM and federate developers to define a series of integration events during which new functionality is integrated into the SoS baseline. The SE team must confirm that this new functionality satisfies the target objectives and that it does not create negative side effects. Regression testing is conducted at every integration event. At the conclusion of each integration event, the SE team evaluates progress and produces a status report to the PM and user representative. This keeps the PM and user representative aware of schedule deviations and allows them to adjust accordingly.

Developers test their software within their own labs, but the integration and test events bring together multiple systems that comprise the SoS for interfederate and cross-model testing that can't be done in development labs. For JLCCTC, the test laboratory is a centrally managed suite of common hardware platforms with a defined processing capability. Before integration events begin, the SE team holds Integration Readiness Reviews with the developers to assure that development is complete and that test objectives are clear. Lab equipment is prepared with preinstalls of user databases, federate software, and scenarios. During integration and test, the SE team coordinates testing among federates, tracks progress against scheduled ITL objectives, and troubleshoots violations of federation-wide protocols and agreements. The integration and test phase is depicted in Figure 15.7.

As indicated by the complexity of data flow into and out of the integration and test phase, the cyclical nature and relationship to task definition and functional development is important to the success of the integration events. JLCCTC conducts approximately five integration and test events per SE cycle, with the later events preparing for functionality testing at validation.

At the conclusion of the last integration and test event, the federation is expected to be fully functional, with all Army training requirements satisfied. Software is delivered to the user representative for testing at the NSC's constructive simulation laboratory. If any Army training functionality is incomplete, the SE team prepares a report for the testers, identifying limitations and defining work-arounds for partial or incomplete requirements.

15.8.6 Validation

Validation is a one-time, end-of-cycle event where the user representative assesses federation functionality. At this step in the SE process, the PM delivers the SoS

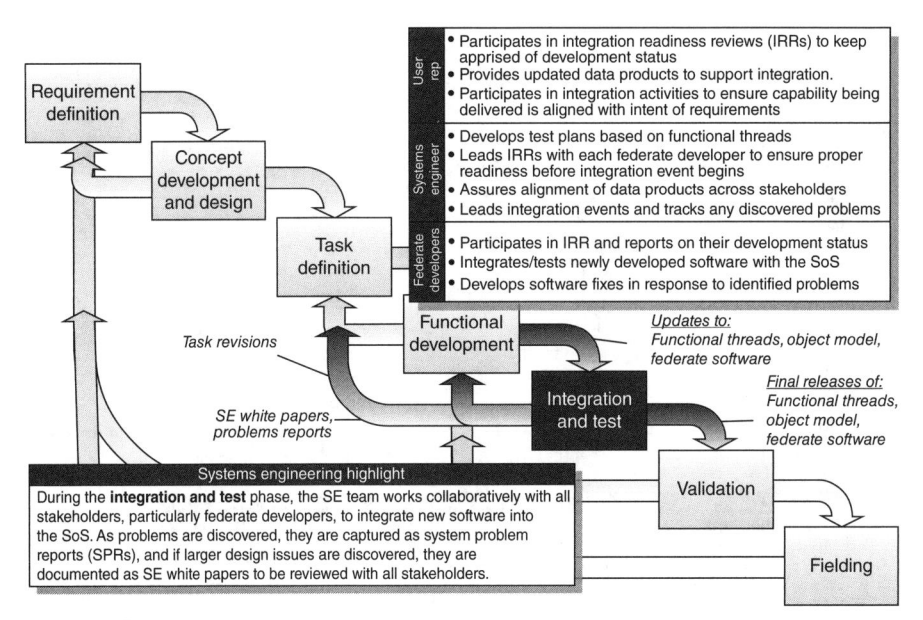

Figure 15.7 Integration and test phase.

software to the user representative who in turn installs the software in its own facility. Using its own technical and functional test cases, the user representative evaluates the SoS capabilities to confirm that the system satisfies requirements. Issues identified during validation are documented as problem tracking reports (PTRs). These PTRs represent one of the feedback loops to the requirements definition phase; they are considered at the beginning of the next JLCCTC SE life cycle iteration. If the SoS satisfies the original requirements, the user representative accredits the system, authorizing it to be fielded to end-user sites and employed for Army and Joint training.

The validation event is conducted in the user's laboratory where the NSC staff build and operate a **production federation** with complete Army training functionality. Military operators execute basic training scenarios, report problems with test cases, and approve any changes to baseline, configuration-managed software. The validation event involves tests that have been constructed to evaluate the federation functionality and operations against the original requirements. Tests are structured in specific threads by the user representative and in an operational setting reflective of its intended use. All tests are measured against a set of standardized criteria to determine success or failure.

Software modifications are tightly controlled during validation, and not applied to the production federation unless absolutely necessary, and then only by the validation team (not the developers or SE team).

To mitigate risk of injecting problems into the production federation with a software change, the SE team maintains and operates a **test federation**. All software patches are thoroughly reviewed in the test federation and recommended for production only after successful testing in the test federation. Federate developers are usually on-site during a

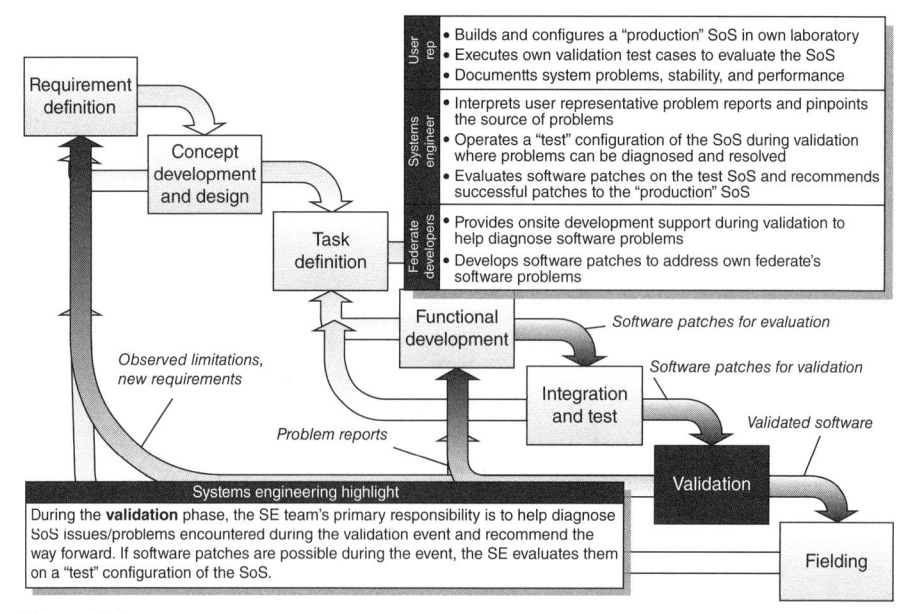

Figure 15.8 Validation phase.

validation event and typically support software patch testing on the test federation. Figure 15.8 shows the validation phase of the JLCCTC cycle.

As a result of successful validation, the user representative produces a letter of authorization that tasks the PM with collecting official software versions and documentation.

15.8.7 Fielding

Based on the letter of authorization, the PM ensures that the federation is successfully installed at user sites. The PM directs that validated software be prepared for delivery to the field. Federate developers deliver installation and user operating manuals. The SE team prepares operation manuals, federation agreement documentation, and interface control documents that define interfederate functions and data flows.

As depicted in Figure 15.9, the PM provides the validated software to its designated fielding contractor who in turn delivers and installs it at the training sites. The PM, via its fielding contractor, tracks where specific versions of the SoS software have been installed. Generally, the SE team provides on-site support for the first use of each new systems version release.

SE tasks continue through this phase, providing technical support to the PM and reviewing end-user training event results. PTRs can be generated from fielded sites, as the end users exercise the capabilities of the SoS. The SE team assists in troubleshooting site-related problems or training scenario technical issues.

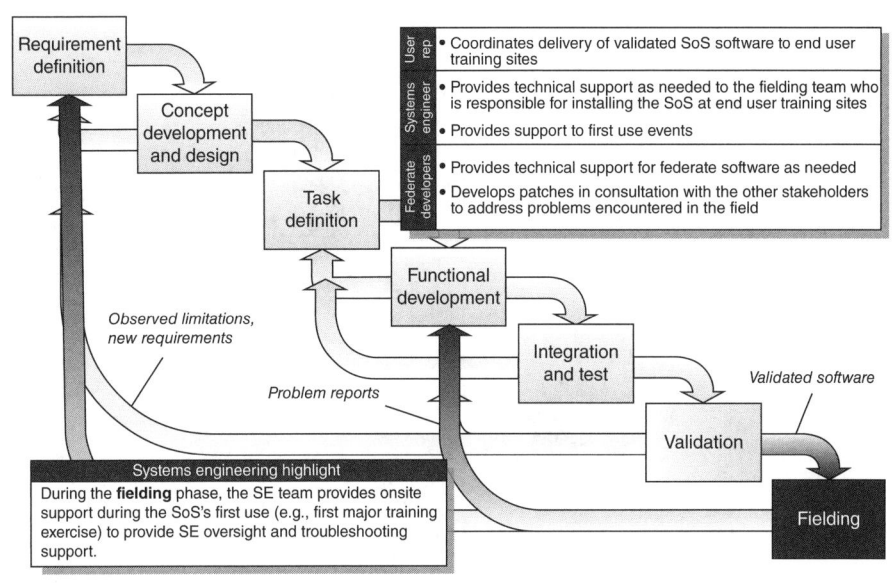

Figure 15.9 Fielding phase.

The PM for JLCCTC (PM ConSim) uses the validated SoS to acquire, field, and sustain the Army. "Constructive simulations and integrated multi-domain environments are the most effective and efficient means to train commanders and staffs from company to theatre level" (PEO STRI, 2013). Figure 15.10 summarizes JLCCTC-fielded training stations worldwide, denoted by gray shading. As noted in this figure, warfighter training units use integrated simulation environments to support Army mission training objectives all over the globe.

This final stage of the SoS SE process is not the end. Systems evolve, better hardware is introduced, more requirements are identified, and new systems are built—and the SE team's responsibility is to be proactive and forward thinking in preparation of repeating the cyclical process from the beginning again.

15.9 CONCLUSION

This chapter described some of the challenges involved with developing and using M&S systems for training of commanders and staff of large DoD organizations or units. Over the past 20+ years, in order to maintain the requisite level of modeling breadth and fidelity, the DoD has combined various M&S capabilities into federated SoS using a combination of collaborative, acknowledged, and directed approaches. The Army's JLCCTC is one example of a successful training SoS whose SE practices and procedures have been detailed here. As the Army and DoD evolve its future

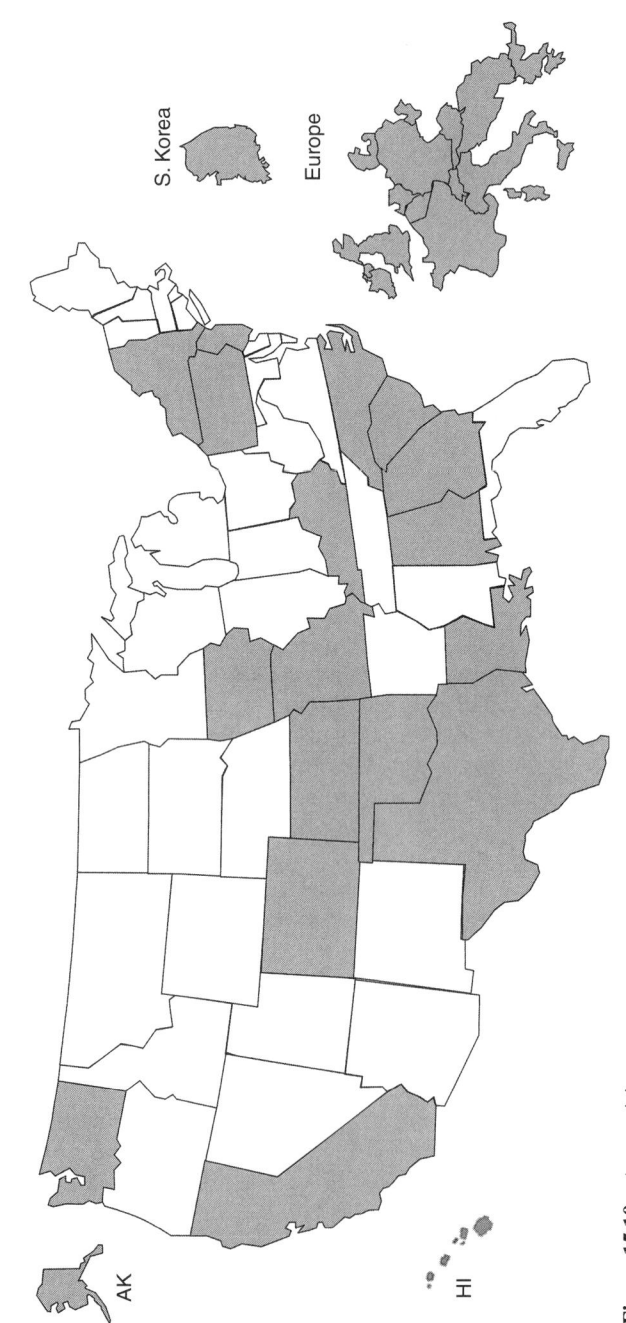

Figure 15.10 Army training.

training SoS to more service-oriented, cloud-accessible architectures, the authors have documented JLCCTC SE practices for evaluation and potential use with future M&S SoS efforts.

REFERENCES

IEEE Std 1730–2010. (2011). *IEEE Recommended Practice for Distributed Simulation Engineering and Execution Process (DSEEP)*. IEEE Computer Society, Institute of Electrical and Electronics Engineers, Simulation Interoperability Standards Organization. New York, NY.

OUSD(AT&L)SSE. (2008). *Systems Engineering Guide for Systems of Systems, Version 1.0*. Washington, DC: Office of the Under Secretary of Defense (Acquisition, Technology and Logistics).

PEO STRI. (2013). *Desk-side Reference Guide*. Orlando, FL: US Army PEO STRI.

US Code Title 10. (2012). *Armed Forces*. Washington DC. US House of Representatives, Sections 3001, 5001, 8001.

Wilson, A.L., and Weatherly, R.M. (1994). The aggregate level simulation protocol: An evolving system, *WSC '94 Proceedings of the 26th Conference on Winter Simulation*. Orlando, FL, pp. 781–787.

Chapter 16

Human in the Loop in System of Systems (SoS) Modeling and Simulation: Applications to Live, Virtual, and Constructive (LVC) Distributed Mission Operations (DMO) Training

Saurabh Mittal, Margery J. Doyle, and Antoinette M. Portrey
L-3 Communications, Link Simulation and Training, Air Force Research Lab, Wright-Patterson Air Force Base, OH, USA

16.1 INTRODUCTION

In the 1990s, a prototype Training Research Testbed was developed at Williams Air Force Base by scientists and engineers from the U.S. Air Force Research Lab (AFRL), 711 Human Effectiveness Directorate, Warfighter Training Research Division (WTRD), in Mesa, Arizona. Now known as the Warfighter Readiness Research Division (WRRD) at the 711 Human Performance Wing at Wright Patterson Air Force Base, Dayton, OH, their mission is to research, develop, demonstrate, evaluate, and transition leading edge technologies and methods through use of open collaborations between government, academia, and industry partners to train warfighters as they pertain to training and research

---

Modeling and Simulation Support for System of Systems Engineering Applications, First Edition.
Edited by Larry B. Rainey and Andreas Tolk.
© 2015 John Wiley & Sons, Inc. Published 2015 by John Wiley & Sons, Inc.

on training. Since then, WRRD has led the development of live, virtual, and constructive (LVC) distributed mission operations (DMO) training and rehearsal environments.

The continual advancement of modeling and simulation (M&S) technology has facilitated the ability for warfighters to participate in a continuous training cycle and maintain combat readiness by using cost-effective simulation alternatives in conjunction with live-fly operations and training missions. Current development of LVC systems for research, training, and mission rehearsal and the rapid advancement of networking technologies including stateless solutions, new multilevel security restrictions, and protocol standards/architectures such as Distributed Interactive Simulation (DIS) and High-Level Architecture (HLA) have all contributed to an environment where highly distributed training, mission rehearsal, operations support, and multiforce LVC DMO joint/coalition exercises have become a reality (Portrey et al., 2007). DMO is the Department of Defense's (DoD) answer to provide complete integration of LVC systems for training, mission rehearsal, and operations support in a theater of war environment and to address the U.S. DoD Strategic Plan for Training Transformation and Joint National Training Capability (JNTC) objectives as directed by the Office of the Secretary of Defense (Ales and Buhrow, 2006; OSD 2006). "The objective is to 'train the way we intend to fight,' enabling Air Force warfighters to maintain combat readiness and conduct mission rehearsal in an environment as operationally realistic as possible" (Chapman and Colegrove, 2013).

To advance the training methodologies and technologies within LVC DMO to address economic, resource, and training challenges for the future, synthetic immersive environments need to create a battlefield space where virtual and constructive entities present realism and provide rapidly adaptive training scenarios to warfighters. However, many behavior-modeling architectures in the synthetic environments today are predefined, rule-based, inflexible instantiations that are confined, in large part, by the type of modeling approach it depends upon. Traditional scripts (i.e., rule-based methods) are generally static and predictable, leading to an inability to scale, adapt, or react to changes in the environment (e.g., fly around mountains or die when shot down) or deal with novel situations. In addition, many rule-based scripts contain weaknesses, which are often "gamed" by operators who learn very quickly how to exploit and defeat rule-based models, creating a negative learning situation. These problems, which are common even in state-of-the-art game AI technologies, hamper the quality of training human operators can receive (Doyle and Portrey, 2011; Doyle et al., 2014).

AFRL WRRD initiated the Not-So-Grand-Challenge (NSGC) project (Doyle and Portrey, 2011) to evaluate the current generation of modeling architectures to determine their viability to execute LVC DMO training with accurate, believable, rapidly developed models that measure or mimic human behavior. Through collaboration with industry participants, Aptima, Charles River Analytics, CHI Systems, SoarTech, Alion and Stottler Henke, and AFRL's Cognitive Models and Agents Branch, the following activities were pursued that defined the scope of the larger problem set and addressed the integration and interoperability issues required for the use of rapid adaptive believable agents in LVC DMO training environments. The problem set is consolidated into the following issues:

- Integrate various modeling approaches with AFRL WRRD simulation systems.
- Identify methodologies and technologies brought to bear for rapid human behavior modeling.
- Bring the aforementioned issues together as an SoS engineering challenge.

The subject of our domain is complex combat systems (CCS), specifically the area that provides adversaries [computer-generated forces (CGFs)] for pilot training. The red force agent (RFA) pilots are the adversaries, and the blue force pilots are the trainees. One of the critical elements of such a system is how it models adaptive RFA behavior and how a simulation scenario unfolds because of such adaptive behavior. An ideal RFA is a human adversary pilot, for they can learn and adapt to a blue force behavior, thereby becoming a more formidable adversary. However, such a case is impractical and expensive. Modeling intelligence so that an intelligent RFA is believable enough to be a successful adversary is common but resource and time intensive. So the main question is, how do we model intelligent behavior in a simulated combat environment that has limited resources and hard real-time requirements?

Sternberg's definition of human intelligence is "(a) mental activity directed toward purposive adaptation to, selection and shaping of, real-world environments relevant to one's life" (Stemberg, 1985), which means that intelligence is how well an individual deals with environmental changes throughout their lifespan. In some types of complex systems, intelligence can be defined as an emergent property (Brooks, 1990). According to Brooks (1991a,b), intelligent behavior can be generated without explicit representations or explicit reasoning, of the kind that a symbolic artificial intelligence (AI) agent possesses. The prime driver of Brook's research has been that intelligent behavior arises from an agent's interaction with its environment. We propose that Sternberg and Brook's perspectives are complementary and an intelligent goal-directed behavior is "in the eye of the beholder," an observer phenomenon from a vantage point. To realize both of these definitions in artificial systems, the "observed" emergent phenomenon has to be semantically tagged intelligent and be available for exploitation by the interacting agents (Mittal et al., 2013). Emergent phenomenon is a system's behavior, defined by outcomes that cannot be reduced to the behavior of constituent agent–environment interactions (Deguet et al., 2006). Put simply, emergent behavior is detectable only at a level above local interacting agents (Bass, 1994; Banabeau and Dessalles, 1997; Doyle and Kalish, 2004; Mittal 2012) situated in their environment. This is the case because objects and environments contain perceptual information, information that indicates the potential actions an object or situation aligned with an active agent's goals and capability affords (Dohn, 2006, 2009; Bloomfield et al., 2010), that is, the capability to capitalize on the emergent affordances before them.

This chapter describes the pursuit of designing intelligent combat system of systems (SoS) for LVC DMO training research using modeling architectures that are based on behavioral/cognitive psychology and systems theory. NSGC phase I brings various behavior-modeling approaches, where the agents may subscribe to a particular cognitive architecture, and Brooks and Sternberg's perspectives, where believable behavior emerges from a group of interacting agents in a situated environment (live or synthetic) together. We will elaborate on situation awareness (SA), the emergent aspects of situation affordance facilitated by the concept of environment abstraction (EA) for intelligent observer agent behavior (Mittal et al., 2013) leading to advanced SA. Here, we define the concept of agent platform system within the **next-generation intelligent combat (NGIC) system**. We present a methodology to formally detect emergent intelligent behavior and capitalize upon the emergent properties/affordances of a dynamic complex system, such as Combat Air Force (CAF) DMO, through discrete-event systems (DEVS) specification formalism and system entity structure (SES) theory (Zeigler, 1984; Zeigler and Zhang,

1989; Zeigler et al., 2000) in a net-centric enterprise environment (Mittal and Martin, 2013b). After describing NGIC architecture and requirements, we propose to associate it with MECs and the AFRL LVC DMO Testbed.

The chapter is organized as follows. Section 16.2 provides background and an overview on systems theory, LVC DMO environment, linguistic levels of interoperability, and how the theory of M&S aids the development of interoperable systems. Section 16.3 describes the Model-Based Systems Engineering (MBSE) process as applied to human-in-the-loop systems for training in LVC DMO and the overall NSGC project. Section 16.4 presents an overview of the NSGC phase I that concluded in 2013. Section 16.5 presents the ongoing NSGC phase II and describes the agent-platform-systems architecture, introducing the EA concept to achieve complete semantic interoperability through a shared ontology for CCS. Section 16.6 elaborates on the engineering aspects of EA and discusses its goals, research issues, technical and operational requirements, and the methodology using DEVS and SES to build an EA SoS. The EA technical architecture is described as a net-centric SoS. It also provides an overview on existing combat systems, enumerates various requirements for NGIC system, and eventually proposes to integrate EA. Section 16.7 concludes with ideas for future work.

16.2 BACKGROUND AND SCOPE

This section begins with an overview of various definitions to eliminate any confusion with respect to similar concepts.

16.2.1 Working Definitions

- *System*: A system is defined as an area of interest with a defined boundary. Logically, this boundary is knowledge based, and technically, it takes the form of an interface, an interface through which a given system communicates with another system or environment either directly or indirectly. In component-based modeling paradigm, a component is a representation of a system, and we shall use the notion of a component and a system interchangeably.
- *Environment*: Any factor that influences a system outside its boundary is considered an environment. It may constitute other systems.
- *Agent*: An agent is any type of system (as defined earlier) that brings about a change in the environment. An agent may be reactive or proactive. A reactive agent only responds to environmental changes sensed at the agent's boundaries, and a proactive agent is a reactive goal-pursuing agent.
- *Dynamical system*: A system that can be in many states (i.e., a system has a state space) and that describes how it moves from its current state to the next state in a given time interval. Such systems are usually modeled with difference equations, differential equations, or discrete-event formalisms (Zeigler et al., 2000).
- *Complex system*: A system that displays emergent behavior from a set of interacting systems/subcomponents. The emergent behavior may be either weak or strong depending upon the role of constituent systems/subcomponents. In such systems,

the components are hierarchically organized to manage complexity. The structure and behavioral function of a complex system is *closed under composition* (i.e., a black box) that can display the complete behavior of its subsystems taken together.

- *Closed system*: A system defined within a boundary that is impervious to the intake of new knowledge or exchange energy. A complex system can be classified as a closed system. A closed system is *closed under composition*.
- *Open system*: A system that has no boundary and its behavior cannot be *closed under composition*. It can exchange energy and new knowledge can be generated, or synthesized within the system or externally injected such that the system evolves over time. Such a system may have a self-similar nature at different levels of hierarchy.
- *Complex adaptive systems (CAS)*: A CAS is a complex system constituting a persistent environment inclusive of persistent agents that adapt/learn/evolve. CAS is an *open* system. Any complex system with a human in the loop is a CAS.
- *Complex combat system (CCS)*: A CCS is a complex dynamical system that displays emergent intelligent (meaningful/purposeful) behavior in a combat exercise between combat agents in a dynamic environment. Taken together, it is generalized and referred to in this chapter as an **agent platform system**. It continues to be a closed system if there is no human in the loop. An LVC with a human-in-the-loop system becomes an *open* system: a CAS.
- *Threat system*: A CCS that allows the development of CGFs as threats for military training and education.
- *Intelligent threat system*: A threat system that displays believable adversary behavior by modeling or mimicking human behavior or intelligence.
- *SoS*: An SoS is an open system fulfilling a desired requirement or a common goal. It comprises systems that are operationally independent, are under multiple managerial controls, have evolutionary nature, and are geographically dispersed.
- *Human-in-the-loop system*: An open system involving human where the human has variable degrees of autonomy and may steer systems dynamics in undefined areas, resulting in strong emergent systems behavior.

16.2.2 SoS Characteristics in LVC DMO Environment

The CAF DMO program includes Air Force's operational fighter, bomber, and Command and Control, Intelligence, Surveillance, and Reconnaissance (C2ISR) weapon systems within a common simulated environment (Chapman and Colegrove, 2013). The major components include:

1. *Mission Training Centers (MTCs)*: Various physical elements such as operational fighter, bomber, and C2ISR bases. They house locally networked high-fidelity mission simulators, instructor cockpits, debriefing stations, CGF stations, and manned threat stations.
2. *DMO Network (DMON)*: Standards-based secured 24/7 wide area network (WAN) for CAF DMO mission briefing, execution, and debriefing supporting multiple training events.

3. DMO *Center*: Supports various national and international joint/coalition exercises and training events.
4. *Distributed Training Operations Center (DTOC)*: Supports DMO training for daily, multiple, and simultaneous distributed training events on DMON.
5. *Distributed Training Center (DTC)*: Supports operational fighter, bomber, and C2ISR units in planning and executing small- to large-scale distributed training events. It coordinates unit simulator schedules with DMON operations, develops scenarios, and arranges the resources needed for complex training events.

Clearly, LVC DMO environment is a heterogeneous environment and classifies as an SoS, per our definition. Its various characteristics are in accordance with Maier (1998) in the fact that they are operationally and managerially independent, have evolutionary nature within their own systems development life cycle, and are geographically distributed. What emerges in an LVC DMO SoS is a highly *immersive* experience that achieves the training goal at an individual, team, and interteam training in an operational (live or synthetic) environment. "Immersive is an attribute of realism that emerges from combining high-fidelity simulators, simulations, and models based on physical properties, and interactions of their real-world counterparts" (Chapman and Colegrove, 2013).

16.2.3 Simulation Technologies and Interoperability Considerations

From the success of the SIMNET program (Miller and Thorpe, 1995) to its evolution as a DIS standard (Hofer and Loper, 1995) and the development of HLA (Numrich, 2006) by the Defense Modeling and Simulation Office (DMSO), distributed simulation offers a technical solution to conduct large-scale simulation exercises for training and simulation research studies. However, DIS was too limited and HLA performance was unacceptable for live testing as it suffered from reliability issues (Henninger et al., 2008). These limitations led to Test and Training Enabling Architecture (TENA) that provided low-latency, high-performance services in the hard real-time application of integrating live assets in the test-range setting (Powell, 2005) and Common Training and Instrumentation Architecture (CTIA) (Army n.d.). DIS, HLA, TENA, and CTIA largely form the technical infrastructure of LVC training and rehearsal. While each of the aforementioned distributed simulation architectures is successful, these architectures are not inherently interoperable with each other. When a mixed-architecture approach is necessary, in a larger simulation exercise, the integration is largely addressed by the development of gateways at the syntactic level and middleware at the message exchange levels, introducing increased risk, latency, complexity cost, and level of effort. For additional information, review the LVC Architecture Roadmap (LVCAR): an exhaustive study that summarized various aspects of the current LVC architectures (Henninger et al., 2008). Further, the DoD has not mandated the use of any of the aforementioned architectures for Joint training and simulation exercises, creating a bottleneck at multiple levels of interoperability at higher levels of abstraction. The Levels of Conceptual Interoperability Model (LCIM) (Tolk and Muguira, 2003) is an example of the extensive interoperability research over the last decade (Tolk et al., 2013).

16.2.4 Linguistic Levels of Interoperability

To address the systems interoperability challenge in M&S, three levels of interoperability have been identified and can serve as guidelines to discuss information exchange. Two systems can be configured to work together, that is, they can be integrated. In order to interoperate, two systems must have a shared understanding of data at higher levels of abstraction. Systems integration facilitates interoperability and is the prime driver to achieve interoperability at various levels (Figure 16.1).

Interoperability is considered at three primary levels: syntactic, semantic, and pragmatic:

- *Syntactic*: At this level, the data structures are exchanged between two systems without the understanding of the meaning of the data. At this level, systems integration is realized and various data exchange standards are implemented (e.g., DIS/HLA, TCP/IP, XML).
- *Semantic*: At this level, the two integrated systems share a common meaning of the data. The data is transformed into knowledge (i.e., data with context), and new abstractions can be introduced. Technically, data is specified using metamodels and integration occurs at the metamodeling level, leading to the shared meaning required for semantic interoperability. For example, in a coalition exercise, an order from a commander is understood in its true meaning across national boundaries and gets translated to its appropriate syntactic version.
- *Pragmatic*: At this level, the integrated systems share a common context and purpose. For example, in a coalition exercise, a pilot acts on the "intent" of the order by a commander/instructor, received in a timely manner, preserving its context.

In order to have interoperability at multiple levels, the M&S conceptually also consists of the following layers (Mittal et al., 2008) as described in Table 16.1.

As illustrated in Figure 16.2, at the syntactic level, we associate network and execution layers. The semantic level corresponds with the modeling layer—where we have included ontology frameworks as well as dynamic systems formalisms as models. Finally, the pragmatic level includes use of the information such as experiment scenarios, objectives, and practical use. This use occurs, for example, in design and search, decision making, and collaborating to achieve common goals. Indeed, such computational/cognitive type activities, along with the associated actions in the world, provide the basis for enumerating pragmatic frames that are potentially of interest, as matched/applied to the context of use.

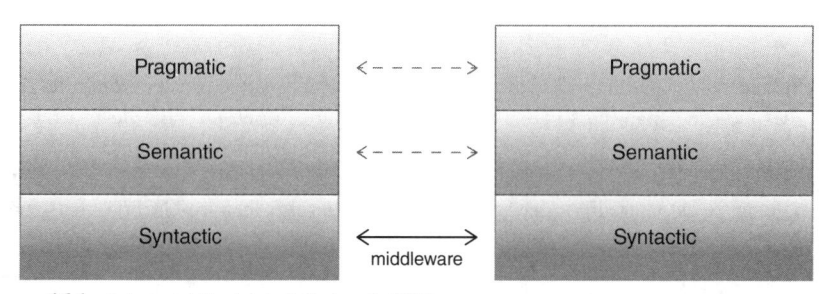

Figure 16.1 Interoperability levels (Mittal et al., 2008).

Table 16.1 M&S Framework Supporting Linguistic Levels of Interoperability

Interoperability levels	Role
Collaboration	Enables humans or intelligent agents to bring in partial information about the system for a common goal
Decision	Capability to make decisions based on the outcome of search layer
Design and search	Investigation of a large family of alternative architectures, models using frameworks like the Department of Defense Architecture Framework (DoDAF) or any other systems design methodology such as the system entity structure (SES) theoretical framework. Artificial intelligence (AI) and simulated natural intelligence are employed to search and simulate large sets for the purpose of determining course of action and optimization criteria
Modeling	Development of model formalisms, behavior representation, and knowledge abstraction by use of ontology description and system structure
Execution	Software systems (including data stores, analytics, and simulators) that execute the model in logical or real time to generate dynamical behavior in a local or distributed environment
Network	Actual hardware and support the execution aspects of M&S life cycle

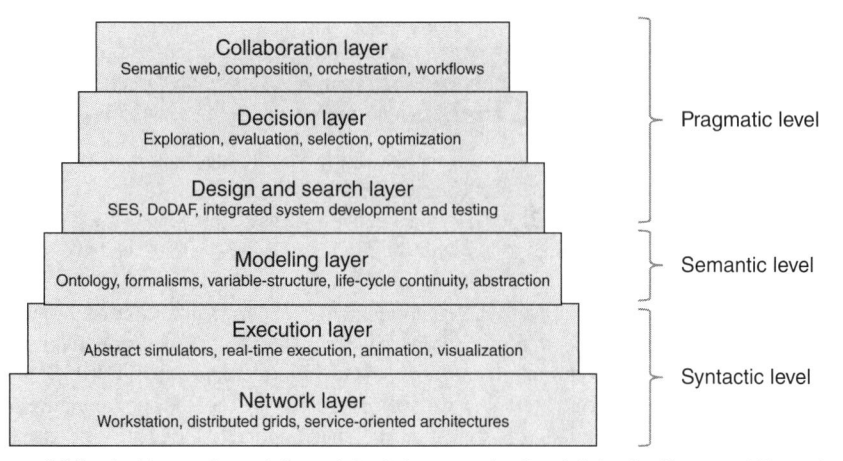

Figure 16.2 Architecture for modeling and simulation mapped to linguistic levels of interoperability (Mittal et al., 2008).

With some caveats, the LVC community has achieved syntactic interoperability; however, despite the capabilities of these LVC architectures, none provide the full range of capabilities required to support emerging warfighter requirements. For instance, integrating LVC with the C4ISR community in a manner consistent with the Service-Oriented Architecture (SOA) paradigm that underlies the Global Information Grid (GIG) is fundamentally absent (Henninger et al., 2008). Efforts from the LVC community are underway to bridge the gap (Allen et al., 2010). On another front, the M&S community has provided basic ideas in developing interoperability solutions (Sarjoughian and Zeigler 2000; Tolk and Muguira, 2004; Mittal et al., 2008a; Mak et al., 2010). These efforts lay the foundation for achieving interoperability at multiple levels; for example, while Tolk conceptualized six levels in LCIM (conceptual, dynamic, pragmatic, semantic, syntactic, and technical), Zeigler and Hammonds (2007) and Mittal et al. (2008b) consolidate systems interoperability within three levels (pragmatic, semantic, and syntactic) that are also incorporated in a much larger LCIM scope. However, these efforts have not provided an example case to illustrate how to achieve interoperability beyond the syntactic level, that is, semantic interoperability, within any simulation architecture. Further, to the best of our knowledge, semantic interoperability in LVC has not been documented and addressed formally or successfully, although considerable effort is spent by the Simulation Interoperability Standards Organization (SISO) and LVC communities toward achieving technical interoperability (SISO, 2010). This chapter will demonstrate the integration of an LVC environment with GIG/SOA toward the pursuit of semantic interoperability between multiple architectures and independent systems, which may or may not be LVC compliant.

16.3 MBSE PROCESS AS APPLICABLE TO LVC DMO TRAINING

Today's combat environment is the standard by which aircrew performance must be measured against and to which it must be trained to presently and ultimately expanded upon in the future. The logistical and economic challenges facing the DoD to maintain warfighter proficiency in an ever-changing environment have stimulated a significant need for realistic and immersive simulation-based training. Training systems must be dynamic to respond quickly and adapt effectively to present the warfighter with accurate real-world information to reinforce competency-based training criteria (Portrey et al., 2007).

This section describes the migration from a requirements-based/event-based DoD training program to a more competency-based/performance-based program. Section 3.1 describes the Ready Aircrew Program (RAP) and Mission Essential Competencies (MECs). Section 3.2 describes MBSE for competency-based training and the hardware/software tools associated with AFRL WRRD and DMO.

16.3.1 Migrating to Competency-Based Training and Assessment through LVC DMO

The current training specification being transformed is known as the RAP and was primarily developed by the operational community through trial and error rather than through periodic, thorough investigations and analyses of training effectiveness (Chapman and Colegrove, 2013). It suffers from the following limitations:

1. It is event and sortie based: When an aircrew member completes a required number of training events or sorties, it is declared "combat ready," and there exists a possibility that the member is still nonproficient in a combat task.
2. Oriented primarily to weapons systems proficiency and does not substantially address mission integration and information-based warfare.
3. Does not alleviate the training demands of an expeditionary, hi-tempo Air Force.

In today's environment, RAP is being phased out and replaced with a competency-based LVC DMO environment as new training methodologies and technologies are validated in such synthetic environments. The USAF Combat Air Forces are migrating their readiness training to a new capability that uses theory-driven, competency-based, and operationally defined specifications for the knowledge, skills, and experiences that are needed for mission readiness performance in complex combat environments, known as MECs (Portrey et al., 2007). MECs are "higher-order individual, team, and inter-team competencies that a fully prepared pilot, crew, or flight requires for successful mission completion under adverse conditions and in a non-permissive environment" (Colegrove and Alliger, 2003). Competency-based training in a DMO environment has proven to contribute to increased expertise in MEC skills and other critical air combat knowledge, which is shown to correlate directly to the success in end-of-training performance, in subsequent unit and exercise performances, and is also indicative of the likelihood of success during a combat mission (Bennett Jr. et al., 2002; Schreiber et al., 2003). Each mission area, such as air or ground superiority, has certain mission essential competences that, when followed, enable mission success. Figure 16.3 illustrates how DMO bridges the gap between proficiency developed in traditional flying training programs and the level of proficiency described by the MEC process in a combat environment. Simulation, when combined with a competency-based training program and live-flying training, can reduce the distance between continuation training and combat mission readiness (Colegrove and Alliger, 2003; Portrey et al., 2007).

From a design perspective, two approaches have been used in developing competency-based scenarios for DMO. The first approach involves using the MEC knowledge and skills as design parameters for scenarios. A second approach, and one that is uniquely tied to the MEC process, is to use the identified developmental experiences as the frame for designing scenarios. Not only do the experiences help to frame the

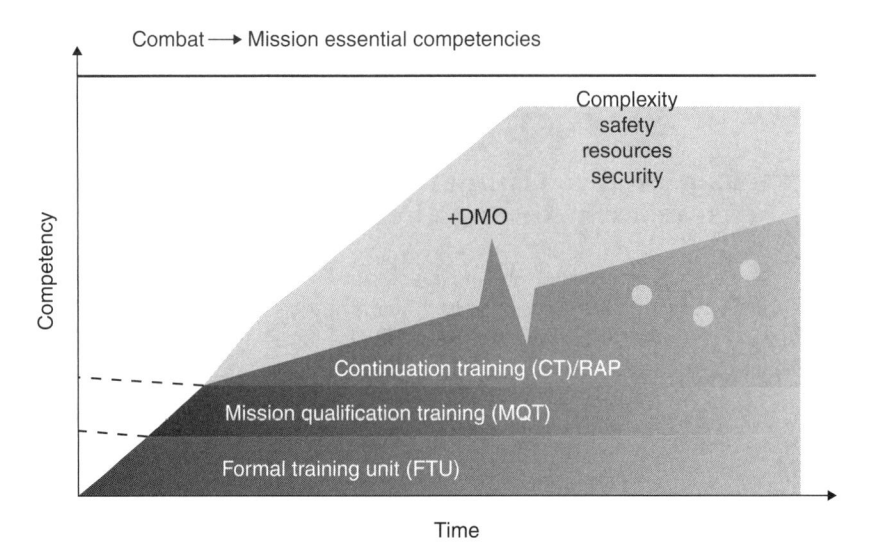

Figure 16.3 Competency-based training and DMO contribution (Portrey et al., 2007).

scenarios, but they also dictate the specific characteristics of each scenario that need to be included for the experience to be realized. AFRL WRRD has implemented both approaches with very good success and has also developed a common measurement approach for the training outcomes from either design approach. This measurement approach is driven by the MEC process and uses two methods for systematically gathering performance information. The first method is a subjective approach that uses behaviorally anchored rating scales to evaluate aspects of performance that are not easily derived from more objective data. The second method is one that uses the MEC specifications to gather more objective data from the DMON, instrumented range data, and also data from the actual aircraft.

16.3.2 MBSE for Competency-Based DMO Training

LVC DMO training is a human-in-the-loop system in the fact that the presence of an instructor and especially the trainee, even though bounded by the semantics of a training event, is apparently open system because it is impossible to define DMO training as a closed system. A training event tries to limit trainee's responses within an acceptable range to help them gather the required experience designed for a particular training event. As indicated in Figure 16.3, training in DMO is by far the most advanced of the available methods that prepares the airmen based on competencies required for a particular mission or scenario in a complex immersive environment.

Systems Engineering is a discipline, the prime objective of which is to ensure that systems satisfy their requirements across their entire development life cycle. Various processes such as requirement specifications (functional and technical), systems development, test, and evaluation are incorporated in the development life cycle. MBSE is the application of models across the entire Systems Engineering life cycle (Zeigler, 1984; Wymore 1993). While a system can be both hardware and software based, it is a general consensus that the model (a virtual and/or constructive entity in our case) is largely a software artifact. Consequently, MBSE as applicable to competency-based training in LVC DMO is a hardware–software systems codesign process, and its various elements are enumerated in Table 16.2.

Training in DMO is a collaborative effort as execution of a scenario (often called a training event) in LVC DMO incorporates all the players in a DMO SoS (Section 2.2). AFRL WRRD is making headway in all the training steps identified in Table 16.2 by development of various software and hardware products (Portrey et al., 2007) such as:

- SPOTLITE: Scenario-Based Performance Observation Tool for Learning in Team Environments that satisfies steps 2 and 7.
- PETS: Performance Evaluation and Tracking System is a modular, multithreaded application that has the ability to measure performance at team, interteam, and team-of-team levels. It satisfies step 5.
- NICE: Network Integrated Combat Environment is a virtual environment that interfaces with various virtual and constructive entities on a DIS/HLA backbone. It partially satisfies steps 3 and 4.

Table 16.2 Applying MBSE to Competency-Based Training in LVC DMO

S. no.	LVC training exercise phases	Hardware	Software	Comments	MBSE phase
1.	MEC-based scenario development	—	X	Subject matter experts (SMEs) help design scenarios and associate them with MEC and other subjective measures	Requirements elicitation
2.	Scenario engineering	X	X	SME transfer designed scenarios to computational environment	Requirement design and transformation
3.	Entity generation and asset planning	X	X	Virtual entities in LVC	Model instantiation from model library
4.	Constructive behavior	—	X	Constructive red air agent for threat modeling	Behavior composability and model-driven engineering
5.	Performance tracking in near real time	X	X	All the virtual entities and constructive models are simulated on real hardware	Model execution (simulation) and instrumentation
6.	Briefing/debriefing	X	X	After active review (AAR) for an LVC DMO exercise	Model validation
7.	Competency assessment	—	X	Scenario impact and usability evaluation	Scenario library

- PFPS: Portable Flight Planning System is a deployable brief/debrief station comprising of networked smart boards, Extron recorders, HD displays, and computer hardware. This satisfies step 6.

To strengthen the constructive entities element, AFRL WRRD initiated the NSGC project (Doyle and Portrey, 2011) to bring in various modeling methodologies based on disparate scientific theories and architectures. The NSGC project is divided into three phases (Table 16.3). The MBSE process, phases, and the architecture as applied to LVC DMO training and rehearsal objectives can be effectively summarized in Figure 16.4.

Human in the Loop in System of Systems (SoS) Modeling and Simulation   **427**

Table 16.3 NSGC Phases and Current Status

NSGC phases	Objective	Status
Phase I	Evaluate multiple behavior modeling approaches for constructive entity modeling in complex air combat systems in LVC domain at AFRL	Successfully completed
Phase II	Develop environment abstraction for advanced situation awareness and semantic interoperability	In process
Phase III	Invite various participants to display advanced constructive entity behavior for maximum combat realism	Planned

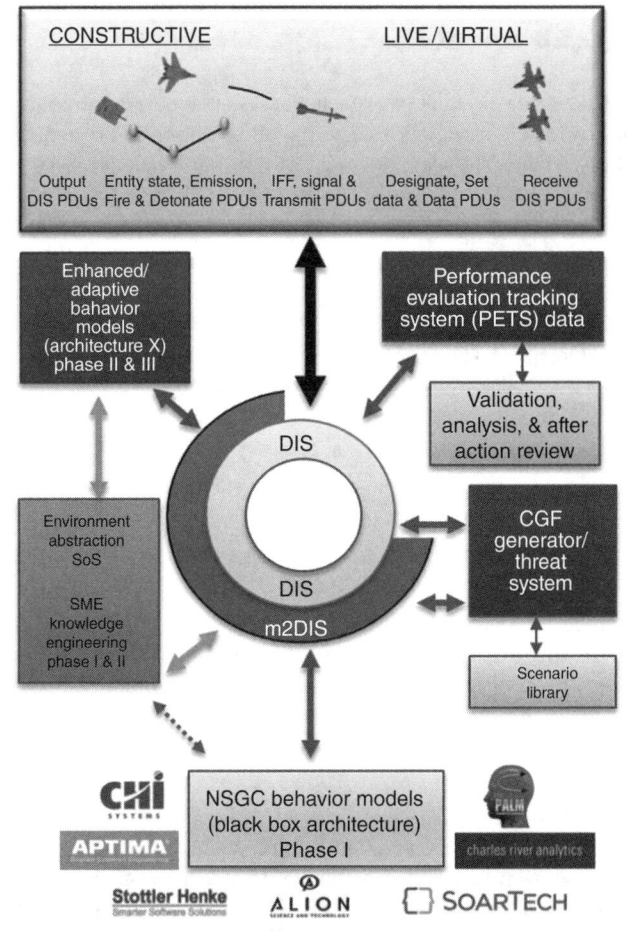

Figure 16.4 MBSE as applicable to LVC DMO training and rehearsal environment.

16.4 NSGC PHASE I: INTEGRATING VARIOUS MODELING ARCHITECTURES WITH AFRL SYSTEMS

This section describes NSGC phase I, reported in Doyle et al. (2014). We begin this section with an illustration of two scenarios that each team participating in the challenge was asked to demonstrate with their chosen modeling architecture. Section 4.1 describes two scenarios, Visual Identification 6 (VID-6) and Sweep-2. It also describes some other desired constructive entity behaviors that induce believability in a dynamic combat environment. Section 4.2 provides a brief overview on various modeling tools/approaches from various participants for constructive entity modeling. Section 4.3 describes the technical architecture that allows multiple agent modeling architectures to interoperate with the AFRL system. Section 4.3 provides an overview of each of the modeling architectures employed by the participants. Section 4.4 briefly describes model-based testing (MBT) used in the project. Section 4.5 discusses the systems verification and validation process. Section 4.6 summarizes the outcomes of this effort and an overall impression.

16.4.1 Requirements

The requirements are specified at two levels, that is, at a more general level where some of the behavior by a CGF are applicable across all scenarios and at a more specific level through carefully engineered scenarios with SMEs. For a detailed list of requirements, refer to Doyle et al. (2014).

16.4.1.1 Desired Believable Behavior

In addition to the aforementioned functional requirements, a constructive entity could better approximate believable behavior doing the following:

Prosecuting local numerical advantage: The red group should attack the nearest blue player and exhibit radar warning receiver (RWR) awareness.

1. *Building blue formation picture awareness:* If RWR no longer indicates a blue radar lock and picture SA indicates no blue player in position to immediately fire, a red player would turn to recommit/attack.

2. *Testing blue radar search*: We simulate this behavior with formations preemptively opening in azimuth.

3. *Influencing turn direction*: When pursuing a cold/retreating blue player, red forces should attempt to influence a blue turn into a cooperating wingman or element. If two or more red entities are chasing a cold blue player, they should open in azimuth so the blue player must assess turn direction, cannot turn hot and threaten both, and may not be able to preserve minimum desired range from both.

4. *Splitting defenses*: Red strikers should sometimes attack from different axes, attempting to divide the attention of blue defenders. Undetected striker groups should lean away from blue defenders if aspect indicates that blue forces are pursuing red counterair forces; such a maneuver avoids blue radar search and, typically, opens range against the defenders.

Human in the Loop in System of Systems (SoS) Modeling and Simulation   **429**

These are some of the functional mechanisms that introduced dynamism and unpredictability into the constructive entity behavior; and as we shall see ahead, such objectives are actively pursued in phase I efforts.

16.4.1.2 Test Scenarios

The functional requirements are laid out as two scenarios. The constructive entities developed by the various teams were expected to be able to perform in these scenarios as follows:

Visual Identification 6 (VID-6) The purpose of the Visual Identification scenario is to allow models to interact with nonreactive blue force players who are configured to drive straight ahead in a formation. A description of the VID scenario is shown in Figure 16.5. The objective of this scenario is to have red air perform a specific maneuver called a stern intercept. A stern (or "baseline") intercept is designed to build lateral offset, exit the target aircraft radar search volume, and close distance on the target while maintaining lateral and vertical turning room. The objective is to arrive within VID or Eyeball (ID via targeting pod) distance and fire on the target from the beam or stern Weapon Engagement Zone (WEZ). The specific red air objectives for this scenario are to have the northern red formation drive toward the blue air formation and perform a stern intercept maneuver. A typical stern intercept will result in the attacking (trailing) of aircraft arriving at 6 o'clock to the target at 6000–9000 foot range. The southern red air group will initially perform a flank maneuver for a period of 2 min and then perform a stern intercept maneuver. It will be necessary for the model to adjust airspeed accordingly to maintain the proper trail distance. Following completion of the stern intercept maneuver, the red air group will fire upon the blue air players.

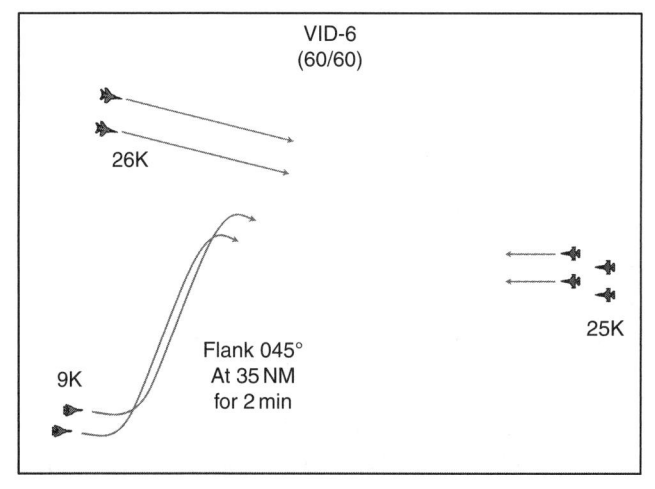

Figure 16.5   VID-6 scenario.

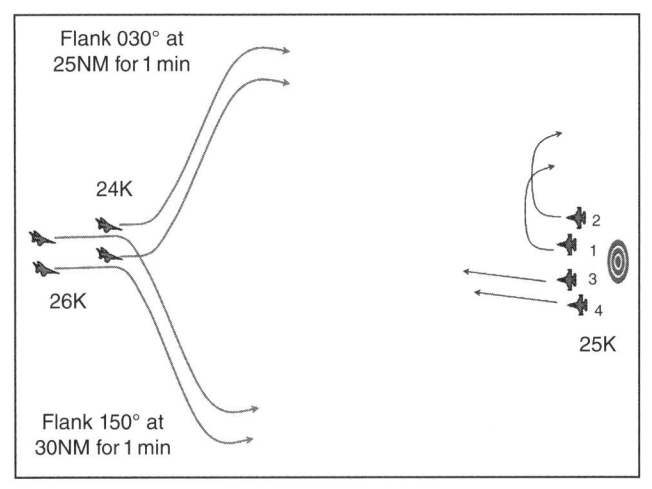

430 Theoretical and Methodological Considerations with Applications and Lessons Learned

Figure 16.6 Sweep-2 scenario.

Sweep-2 Figure 16.6 is an illustration of the "Sweep-2" scenario. The purpose of the Sweep-2 scenario is to allow red forces to initially present a range problem for blue forces; blue would typically attack this presentation by executing a grind tactic (one element continuing hot to target the leading edge of the red formation with the other blue element turning cold to gain a position in trail.) By executing the mirror flank maneuver, the red forces reduce closure and decrease the blue force WEZ and, at minimum, delay blue shots. If blue forces fire missiles far enough away, the flank maneuver may even defeat blue shots. By flanking, the red forces alter the tactical picture from a range problem to an azimuth problem. Red forces would reduce range to the target and reduce maneuver room for the Vipers (i.e., blue forces). As defensive range decreases, the blue forces have fewer tactical options and less room to run cold to remain outside the red aircraft WEZ. The blue force will react in this scenario.

16.4.2 Modeling Tool/Architectures

This section provides a description of the various participating architectures and modeling approaches for constructive entity behavior modeling.

16.4.2.1 Alion

Alion used the Human Behavior Architecture (HBA) (Warwick et al., 2008), an integrated development environment that combines the Micro Saint task network modeling tool with portions of the Adaptive Control of Thought—Rational (ACT-R) (Anderson, 2007) cognitive architecture. Micro Saint (and its government-owned variants IMPRINT and C3TRACE) provides a framework for representing human performance as a "network" of tasks; more precisely, Micro Saint allows the modeler to represent human performance by way of hierarchical task decomposition.

16.4.2.2 Aptima

The Aptima team leveraged their team's recent insights from the Mixed-Initiative Machine for Instructed Computing (MIMIC) approach to intent modeling to address rapid modeling for training (Riordan et al., 2011). They approached the problem of intent inference using the framework of Bayesian inverse reinforcement learning (Ng and Russell, 2000; Ramachandran and Amir, 2007), where the goal is to estimate the reward functions of agents through observations of their behavior and knowledge of the dynamics of the environment.

16.4.2.3 Charles River Analytics

Charles River Analytics proposed the inclusion of the AgentWorks platform as a constructive modeling architecture. AgentWorks has been used to develop models of human decision-making behavior that have been embedded as CGFs in a range of military and civilian simulation platforms.

16.4.2.4 CHI Systems

CHI Systems used the ASIST-AT architecture that consists of a Speech Recognition Client, a Speech Synthesis Client, a Personality-Enabled Architecture for Cognition (PAC) Client, a Simulation Client, and an OpenAnzo server hub used to pass scenario status updates throughout the system.

16.4.2.5 SoarTech

Soar has primarily been used as an efficient execution engine that supports rapid, associative, and adaptive knowledge-rich decision making, which emulates the tactical reasoning of human warfighters. The most sophisticated intelligent agent system developed in Soar is TacAir-Soar, a model of tactical air combat pilot decision making. TacAir-Soar runs in a variety of simulation systems and generates realistic humanlike decision making.

16.4.2.6 Stottler Henke

Stottler Henke used their SimBionic behavior-modeling tool, an authoring environment and runtime system that allows end users to specify the logic needed to implement intelligent agent behaviors without programming.

16.4.2.7 Performance and Learning Models (PALM): AFRL-RHAC

PALM used a cognitive modeling methodology coupled with formal systems engineering practices to produce RFAs that are designed to behave according to defined training objectives that solicit specific training behaviors from manned blue air pilots. The RFA agent is designed using the systems theoretical principles as implemented in the DEVS formalism (ACIMS, 2009).

16.4.3 SoS Integration with Agent-Platform-Systems Architecture

The first area of inquiry in this collaborative effort, that is, to create believable behavior in CGFs, was typically being handled by each collaborator through use of a premier, often proprietary, architecture or human behavior-modeling/mimicking approach. Most of the solutions can be denoted as fairly mature architectures, models, and methods. However, in order to collaborate effectively and have the capacity to evaluate any architecture and the specified agent model in an adaptive constructive training environment (such as LVC), each collaborator and the evaluator must account for integration and interoperability at three levels (i.e., the syntactic, semantic, and pragmatic). Since the critical path to success and the problem set in NSGC phase I is first primarily defined as a technical integration task, a task to integrate with DIS and the proprietary AFRL WRRD infrastructure, it was assessed that, currently, almost all collaborators comply or are working to comply with syntactic interoperability by using the DIS Protocol Data Units (PDUs). Additionally, almost all of the collaborators possess, or were capable of developing, proprietary DIS adapters and have a proprietary, uniquely engineered semantic layer as well.

Therefore, in order to facilitate semantic interoperability of each participant's architecture with the AFRL infrastructure, a model-to-DIS (m2DIS) application programming interface (API) (Watz, 2012) was developed (Figure 16.7, a simplified version of Figure 16.4). The m2DIS API provided a set of common terminology/nomenclature. The objective of the m2DIS API was twofold. First, it served as a starting point to introduce semantics and additional abstractions over the DIS PDUs. Second, each collaborator was free to publish and subscribe to DIS entities through their proprietary middleware. However, publishing to the prescribed NICE Testbed, inherently ensuring semantic interoperability at a partial (publish) level.

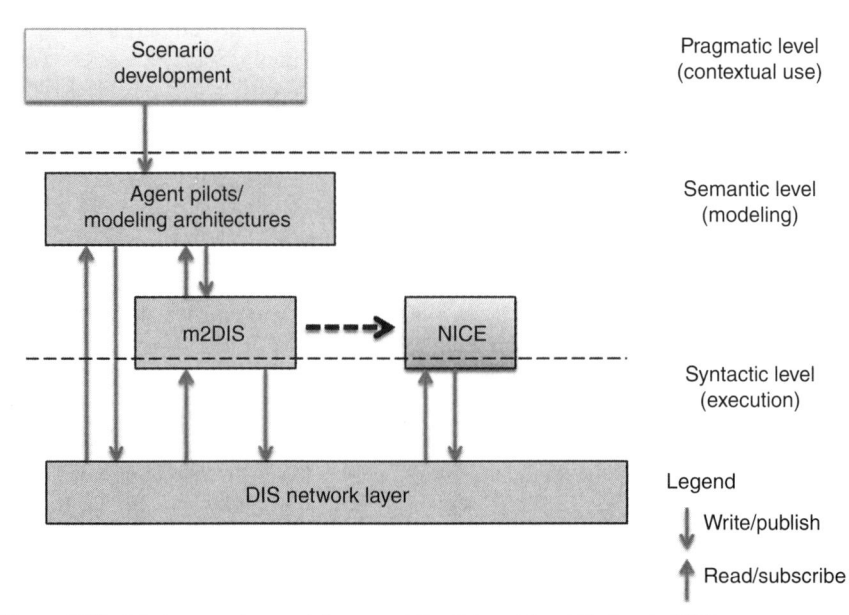

Figure 16.7 Moving toward semantic interoperability (Adapted from Mittal et al., 2013).

16.4.4 Model-Based Testing and Evaluation

The integration of various modeling architectures was facilitated with the m2DIS API that provided access to the F-16 threat platform to ensure realistic physics. While the m2DIS API was unit tested, the DEVS-based RFA agent model at both the individual level and at the team level was also used to test the m2DIS API and make it more robust for seamless integration with other modeling architectures for final integration exercises.

16.4.5 Verification and Validation

> Validation answers the question: "Did you build the right thing?"
> Verification answers the question: "Did you build it right?"

The validation and verification (V&V) studies were considered at two levels. The first level is at the level of physics, that is, at the airframe level, where it is very important that the agent and its vehicles do not violate the natural laws of physics. The second level is at the psychological and conceptual modeling of the constructive entity. While the first level has been addressed by the current combat environment infrastructure, NICE, and is accounted for in a separate initiative, NSGC phase I V&V efforts focused on the behavior modeling of constructive entities.

The validation studies amounted to evaluating RFA behaviors by SMEs, as elaborated in Section 4.6.1. The agent models demonstrated these behaviors as required by the two scenarios, VID-6 and Sweep-2. Both the scenarios were described in Section 4.1.2 earlier.

The verification studies amounted to testing the implementation of conceptual models to executable code. As there are varied architectures and no participant was mandated to use any specific software environment for his or her model execution, this effort did not pursue verification of each architecture. The responsibility for such verification, for the moment, rests with the participants, although a part of the verification is done through use of the m2DIS API that already abstracts a lot of low-level functionality that must be implemented in a constructive entity behavior. All the participants were mandated to use the m2DIS API to enable their constructive entities to execute behaviors with the current combat environment infrastructure, but they must ensure at the end that the action they initiated through m2DIS API was actually realized in the LVC environment. Currently, the m2DIS API is not designed to provide feedback to the modeling architectures in phase I.

16.4.6 NSGC Phase I: Summary

NSGC phase I is summarized through two perspectives. The first perspective allows for a capability assessment of the constructive entity as evaluated by an SME. The second perspective allows MBSE evaluation on how various model-based engineering practices are utilized for developing a constructive entity.

16.4.6.1 Pilot SME Capability Assessment

Four F-16 SMEs completed observation forms to rate (via a 4-point Likert scale) the extent that each model met the predefined SME expectations (i.e.,

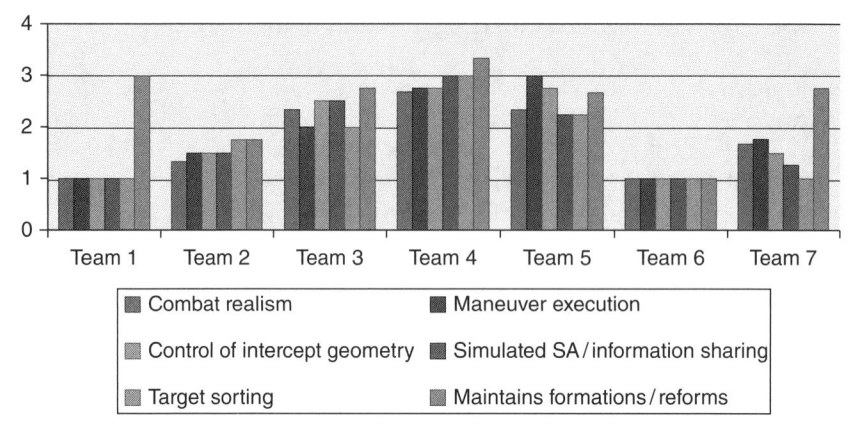

Figure 16.8 Constructive entity overall behavior (Adapted from Doyle et al., 2014).

high-level requirements) on variables such as combat realism displayed in Figure 16.8 (participating team names have been generalized and randomized to maintain confidentiality of results). Overall, the models as a group received an average anonymity rating. Meaning, the models, at least partially, met the objectives SMEs were looking to achieve. However, this also means that the models have room for improvement. As a whole, the models rate the highest on the ability to maintain formations and/or reform and the lowest on combat realism.

16.4.6.2 MBSE Evaluation

This perspective allows us to evaluate each of the modeling approaches on how it fits within the MBSE paradigm (Table 16.4).

16.4.6.3 NSGC Phase I Overall Impression

Considering that phase I was carried out primarily to determine what was needed for the models to be utilized by AFRL, creating a process/architecture for various models to hook up and interact in the world, and then testing the models under a basic "hello world" paradigm, having met any of the SMEs expectations means that the effort, overall, was a success. In addition, anecdotally, those who attended the final NSGC phase I meeting seemed impressed with the way industry-modeling teams opened up and explained in detail the theory, science, and engineering methods involved in their particular approach, divulging information that had been mostly kept under lock and key. Also, many who attended were pleasantly surprised as to the culture of collaboration and corporation that had been cultivated though this process, ultimately leading to a more harmonious state of affairs, never if rarely witnessed before with this particular group of industry teams, considered direct competitors. On this day though, each team freely revealed their process and technologies. Details are further available in Doyle et al. (2014).

Table 16.4 Summary of Modeling Approaches Satisfying MBSE Paradigm

S. no.	Criteria	Team 1	Team 2	Team 3	Team 4	Team 5	Team 6	Team 7
System structure								
1	Architecture/technology	Info. withheld	Info. withheld	Info. withheld	Info. withheld	Info. withheld	Info. withheld	Info. withheld
2	Modularity	—	—	Yes	Yes	Yes	Yes	Yes
MBSE process								
3	Requirements handling	Yes	—	Yes	Yes	Yes	Partial	Yes
4	Scenario engineering	Info. withheld	Info. withheld	Info. withheld	Info. withheld	Info. withheld	Info. withheld	Info. withheld
5	CGF behavior model	Simple	Simple	Near complete	Near complete	Near complete	Moderately complex	Simple
6	Instrumentation	Yes	Yes	Yes	Yes	Yes	Yes	Yes
7	Testing and evaluation methodologies	Yes	Partial	Yes	Yes	Yes	Yes	Yes
8	Integration with AFRL NICE (model execution)	Success	Success	Success	Success	Success	Success	Success
9	V&V and competency assessment	Present	Present	Present	Present	Present	Present	Present

16.5 NSGC PHASE II: ENVIRONMENT ABSTRACTION FOR ADVANCED SA AND SEMANTIC INTEROPERABILITY

Syntactic interoperability or technical integration was achieved in phase I efforts. The m2DIS API contained just a few semantic constructs such as *chase*, *fight*, *doFlank*, and *doDrag* that abstracted low-level actions and maneuvers, which can be implemented in the current combat environment's infrastructure. These initial steps toward enumerating semantic tokens that all the modeling architectures subscribed to paved the way to a more formal semantic interoperability that is being pursued in phase II of the NSGC project. In this section, we will elaborate on the ontology for CCSs that display weak emergent behavior. In Section 16.5.1, we will look at strong and weak emergence and discuss how they are paramount to model intelligent behavior in a complex dynamic environment with multiple interacting entities. We will then leverage these concepts toward specification of the improved agent-platform-systems architecture in Section 16.5.2.

16.5.1 Strong and Weak Emergence for Intelligent Behavior

Complex systems are characterized by the presence of an emergent behavior. CAS are characterized by usage of properties resulting from emergent systems behaviors back into the complex systems. An emergent phenomenon or behavior is defined as a novel phenomenon that was not explicitly encoded in the system but emerges as a result of concurrent interactions between multiple agents within the system. Put simply, an emergent behavior in a multiagent system (MAS) is an observed behavior at a level above the level of interacting agents (Mittal, 2012; Mittal et al., 2013). To understand emergent behavior, it has been classified as *strong* and *weak*, based on the behavior of the agent in the complex system (Chalmers, 2006). In *weak* emergence, an agent situated in an environment is simply a reactive agent and is not proactive at all. That is, such an agent has no apparatus to perceive the emergent behavior and its potential utility, of its own and its peer constituents create at a higher abstraction level. Because doing so would require knowledge of both the emergent behavior and the accompanying new affordances only available to the higher level observer. Affordances can be seen as opportunities for action as they align agent's goals with observed emergent opportunities. Affordances refer to the perceived and actual properties of objects and surrounding environments by animals or humans (Gibson, 1986). While in *weak* emergence affordances brought to bear through emergent phenomena are not available to the agent, in *strong* emergence, a situated agent displays a proactive role by taking advantage of emergent affordances. Affordances that match to an agent's capabilities and limitations, given the goals, will often guide an agent's actions and future goals. Consequently, an agent acts as an observer of the emergent phenomena while also being imbued with the capability to capitalize on an awareness of what the environment might afford an agent, affordances that may be aligned with an agent's goals. Additionally, in *strong* emergence, the notion of an information boundary becomes critical as the proactive agent, through weak links with other agents or objects in the environment or its own inherent capability or limitation, continues to evolve, learn, and adapt its behavior. Even more critical is the affordance of new information through a global observer that is at a higher level of abstraction than the interacting agent. This observer can either be a part of the agent's sensory apparatus or an external one that makes the agent aware of new affordances (Mittal, 2012). This new knowledge makes the proactive agent more competitive in the same environment, as the agent is more aware of its environment and can process information in a hierarchical way. Bringing together the ideas from Stemberg (1985) and Brooks (1991a, b) with Mittal's (2012) perspective, an emergent intelligent behavior has three elements, the agent, an environment, and an observer that situates the agent in that environment and calls the behavior as intelligent.

Having described the fundamental concepts related to emergent behavior, let us now look at the agent platform system that is capable of displaying the intelligent behavior as a strong emergence phenomenon. A CCS in LVC domain is a CAS, and emergent behavior engineering is implemented by means of an EA system realized as an observer agent (to display *weak* emergence). We will begin with a passive omniobserver and then later enhance this agent with causal powers (to display *strong* emergence). Categorically, an EA system is:

1. An observer at a higher level of abstraction
2. An information fusion/processing system that recognizes emergent behavior
3. An aid that can influence agent's behavior such that the system transitions toward displaying *strong* emergence
4. An *open* system that has a malleable knowledge boundary as the knowledge for identifying emergent behavior can be dynamically injected or updated by an SME
5. Oriented toward the recipient agent so that it alters its perception with new semantic knowledge and situates the agent in its contextual environment

16.5.2 NSGC AFRL WRRD Phase II Agent-Platform-Systems Architecture with EA

Rendering semantic interoperability formally between the modeling architectures was not possible in phase I, primarily because semantic tokens (shared meanings, terms), required for agent's consumption and use, or the domain these models are being applied to did not exist in entirety. At the functional level, the m2DIS layer did not restrict the DIS PDUs available to any model, allowing access to all world/situation knowledge (Omni-data), data that can be used for an agent's decision making, in every situation, at every instant. However, access to Omni-data could, and likely would, result in unrealistic behavior by the agent. Therefore, to gain/retain some semblance of realistic behavior, each model had and will have to determine what DIS information is relevant and what information it should act on, acting "as if" it resides within real-world physical and tactical limitations, and capabilities that are real. That is to say, if a human pilot cannot normally see or act as if they can see, over the horizon, then an agent expected to emulate a human-type response or a response a human operator can believe should not act as if it can see over the horizon when properties of the environment and situation do not afford such an action. An agent can appear superhuman if it utilizes Omni-data (world data) rather than a realistically limited perspective view of information for its decision making

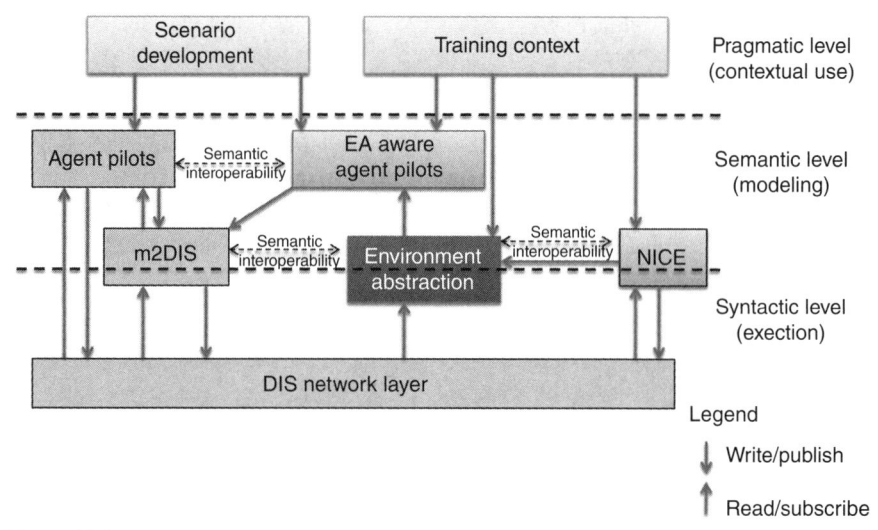

Figure 16.9 Agent-platform-systems architecture and interoperability levels (Mittal et al., 2013).

and actions. Although more information and knowledge can sometimes be a good thing, having access to, or acting to Omni-data most certainly, would degrade an agent's realism.

To alleviate this problem, the EA system was conceptualized as a modular agent system that limits the amount of semantic information available to any agent, based on the airframe (vehicle platform), LVC environmental context, and the nomenclature/taxonomy/ontology of the domain.

The EA system is geared to achieve complete semantic interoperability at the agent-platform-systems level. Figure 16.9, enhanced from Figure 16.7, shows the interoperability at the modeling *architecture* level. A knowledge-based agent is architected with pragmatic, semantic, and syntactic levels in terms of knowledge processing for a particular domain. Figure 16.10 shows various levels of interoperability in an agent platform system. An agent's perception and action capacity is bounded by the capabilities and limitations of the platform it inhabits (i.e., its body in the air combat system), a body that typically is rule based and obeys the laws of real-world physics. The platform (an agents' acting body) in this case is provided by the current combat environment infrastructure that allows various types of air vehicles (platforms) to act in this LVC world. Consequently, an agent's perception is bounded by the information that is visible to the agent within this platform's cockpit or through external radio communication with other team members. Similarly, the agent's actions are bounded within domain knowledge and by the actions that can currently be taken through the proprietary AFRL infrastructure.

16.5.3 Engineering EA

In an LVC DMO setting, the EA system sits between the platform and the LVC DMO Network. Based on the structural–functional capabilities and limitations of a platform and its geolocation in a combat exercise, the EA system, while aware of the

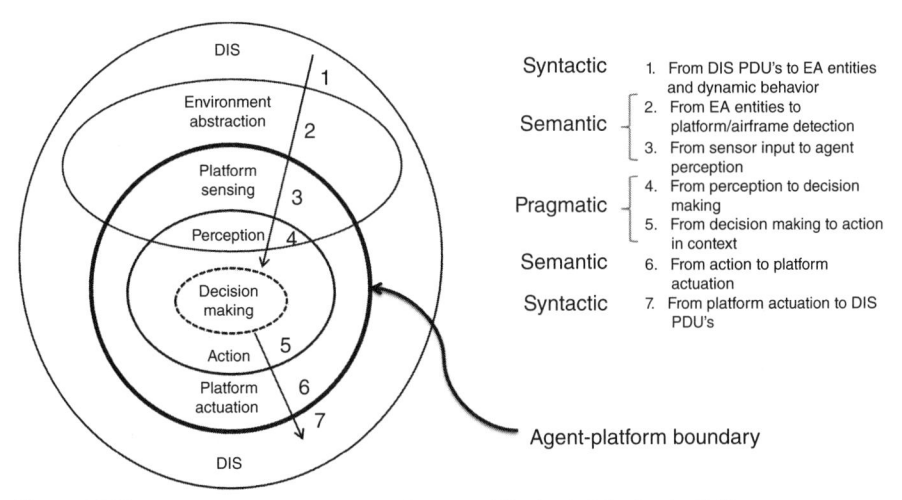

Figure 16.10 Agent-platform architecture and interoperability levels (Mittal et al., 2013).

entire combat exercise, makes available only the relevant information to the requesting (subscribing) agent at its platform level. The agent's actions (realized in the AFRL infrastructure) are published by an EA-aware agent through the m2DIS API as shown in Figure 16.9.

The EA system deals with sensing/perception and does not contribute to any actuation capabilities for the specific platform in this phase of the project. In order to understand the "sensing" capability of emergent affordances made available through the use of the EA system, consider the following example. Assume that a 4 X 4 scenario is in execution. Four blue pilots are flying in formation, and in response, agent-based CGF red forces start executing a maneuver in a coordinated fashion, responding to blue force tactics. In the current architecture, the entire worldview of information about any actions taken by any entity is available to all other entities via the DIS network. However, this is not realistic, because human agents and/or technology-based sensors do not typically have the capacity to function in such a manner, meaning that it would be erroneous to assume that real-world agents (i.e., humans) can see all world data. In minimalist terms, red agents show up on a particular blue agent's radar (visual or sensor based) in a sequential time-based manner so a red force can be "perceived" through the unfolding scenario or through communication between blue force agents based on their platform's sensing capabilities or their own visuals (i.e., a pilot's line-of-site capacity).

In the proposed architecture, the EA system interfaces between the DIS traffic and the air vehicle platform to limit the information available to any pilot agent. It is worth noting that information is defined as "semantically rich" data grounded in a domain-specific formal ontology and is provided by the EA system, that is, domain-based information about tactics, techniques, and procedures previously elicited from subject matter experts through a task analysis. Done so, an agent will appear to behave in a rational manner in response to human/agent actions supporting any collaborating team to freely access and leverage the EA-provided semantic information to develop pragmatic context for their agents. Once the semantic interoperability is ensured, the training aspect of this research effort will look to achieve the pragmatic interoperability between the training system's scenario and an agent-based system's pragmatic context.

16.5.3.1 EA Goals

The EA is designed to achieve the following objectives:

1. To detect emergent behavior supporting the rise and recognition of affordances (i.e., intelligent behavior, behavior of value) in a MAS constituting RFA. This portrays *weak* emergence
2. To incorporate the results of the EA system by dynamically taking into consideration a human in the loop (i.e., live agent/trainee) as emergent affordances and to display intelligence and behaviors based on *strong* emergence
3. To formalize interoperability at syntactic and semantic levels in an event-driven system performing as a CAS
4. To develop a domain ontology for air combat system using SES theory and computationally realize in a formal M&S framework such as DEVS

For detailed EA requirements, refer to Mittal et al. (2013). Let us look at the methodology to develop an EA system.

16.5.3.2 Methodology

The EA system is developed using two established formal engineering methods:

1. Formal DEVS theory
2. Formal SES theory

Both of these formalisms (Zeigler, 1976, 1984; Kim et al., 1990) are based on mathematical set theory and general systems theory (Wymore, 1976) and have been in use for over 30 years in the specification of complex dynamical systems, knowledge-based systems (Rozenblit et al., 1990; Rozenblit and Zeigler, 1993), advanced ontology fusion applications (Lee and Zeigler, 2010), and combat knowledge-based systems with DEVS dynamical behavior (Momen and Rozenblit, 2006). These two formalisms pioneered the concept of model-based design and engineering (Wymore, 1993; Zeigler and Chi, 1993), out of the many flavors of model-based and model-driven practices (Mittal and Martin, 2013b).

EA system is a hierarchical component comprised of many components. The design and hierarchical organization of these components is performed as follows:

1. Identify entities, messages, and events that constitute inputs and outputs to various entities. An entity is defined as a data structure that can be physically realized in a real-world component. A message is a type of entity used for data exchange. An event is realized as a type of message that may have a complex data type. In addition, an event is defined as a quantized change in the entity state and is marked by either the generation of message or update of entity state. The identification of these tokens/primitives is guided by the semantic domain knowledge and taxonomy developed with the help of SMEs. These tokens also constitute the emergent behaviors that are detectable.

2. Develop minimal input/output (I/O) pairs that characterize "state" of a particular entity. Such pairs define various other component "primitives" at a higher level of abstraction that can be assembled together using DEVS levels of systems specifications (Zeigler et al., 2000; Zeigler and Hammonds, 2007; Mak et al., 2010). This also paves way for a T&E framework.
3. Develop dynamic behavior of DEVS observer agent utilizing these primitives at multiple levels of hierarchy.
4. Develop the domain metamodel of CCS utilizing the constructed primitives with SES theory (Zeigler and Hammonds, 2007; Mittal and Martin, 2013a).

   Instantiate CCS metamodel as a CCS instance with respect to the subscribing pilot agent with respect to its geolocation and air vehicle platform specifications. In SES parlance, an SES instance is also called a pruned entity structure (PES) (Zeigler et al., 2000; Zeigler and Hammonds, 2007; Mittal and Martin, 2013a).
5. Push CCS event notifications to the pilot agent model, making it an EA-aware agent.

Figure 16.11 summarizes the above process.

16.5.3.3 EA SoS

The EA system is implemented as an enterprise SoS that is executed on a net-centric DEVS virtual machine (VM) mirroring the functionality of a DEVS-based event-driven architecture (EDA) (Mittal and Martin, 2013a, b). A DEVS-based EDA is an architecture that integrates DEVS complex dynamical models and the defined events over an SOA network or a DMON. The acceptance of event-based applications in

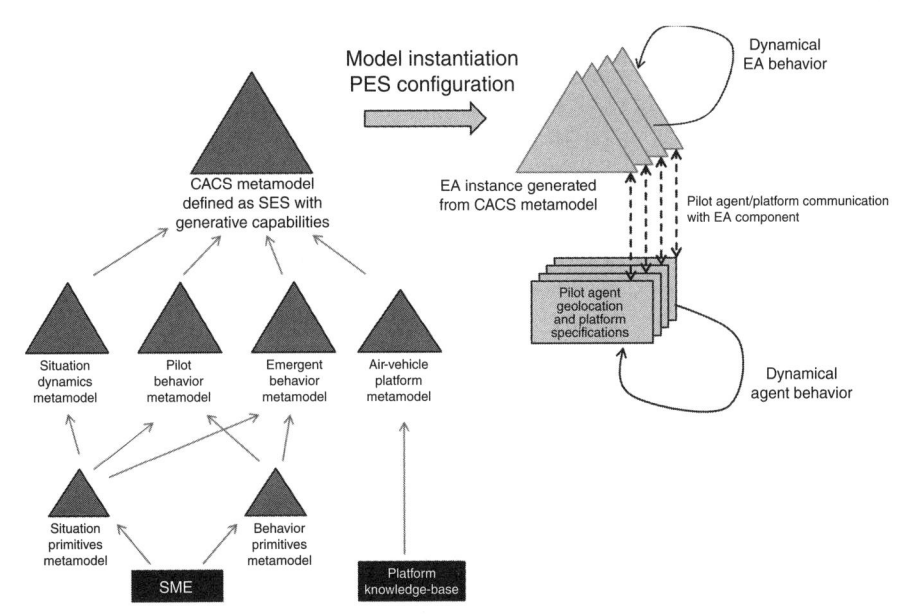

Figure 16.11 EA development and usage methodology (Mittal et al., 2013).

multiple venues of our society (Hinze et al., 2009). Consequently, EA is a net-centric component that interfaces with other systems. These systems can be radar systems, ionosphere propagation systems, sensor systems, etc. that are independent systems implemented in different languages, integrated as component systems (Mittal et al., 2006) that provide various services to the EA system. Figure 16.12 shows the internal design of an EA SoS.

Functionally, from the NGIC System's perspective, EA SoS has three inputs. The first input comes from SMEs that help define the agent-platform-systems ontology for EA knowledge base. The second input comes from the LVC DMO environment that subscribes to any LVC communications mechanism (such as DIS, HLA, TENA, etc.). This input is likely abstracted through the m2DIS API as discussed in Section 3. The third input comes from various platform systems that model airframe physics and other real-world constraints. As stated earlier, in an agent platform system, an EA system sits between the platform and the agent providing rich SA in a contextual setting. Again, from the NGIC perspective, EA SoS has only one output, proprietary DIS PDUs or complex data types in XML that are communicated to the pilot or agent models or any other information sink (e.g., visualization component) over the DMON.

Internally, an EA SoS has the following components. The EA Analytics toolbox is a set of various algorithms and estimation components that analyze the situation dynamics based on the recipient (Stanners and French, 2005; Doyle, 2008; Raj et al., 2011). The situation models/ontology/recommender system is a complex dynamical model that makes the CCS ontology executable on a net-centric DEVS platform. The methodology to automate SES to a DEVS executable simulation system is described at length by Mittal and Martin (2013a). Recently, DEVS/SES methodology was used to develop a decision support system (DSS) for Advanced Tactical Architecture for Combat Knowledge System (ATACKS) where the DSS module actuates a combat knowledge base with DEVS dynamical models (Momen and Rozenblit, 2006) toward providing a better understanding of the combat scenario. However, ATACKS is not recipient oriented, unlike EA, which provides contextual environment information per recipient. The enterprise service bus (ESB), middleware, data stores, and dashboard have their usual meanings in a net-centric system. An agent-proxy refers to a proxy for an agent that is implemented in an external modeling architecture and is not part of an EA system. An agent-proxy is an agent that acts as a conduit between the EA SoS and the actual agent (model/human). The agent-proxy receives the result of EA Analytics component and communicates it back to the actual agent. Further, this agent-proxy also acts as an interfacing agent that actuates the platform for the actual agent. Thirdly, it also maintains abstract state of the actual agent based on the CCS ontology and ensures that there is true mapping between the metacognitive user states in the agent-proxy and the actual agent.

16.6 APPLYING EA TO NGIC SYSTEMS

Sections 5 described the concept of EA. In this section, we look back at the adaptive training requirements described in Sections 2 and 3 and delineate the feature set of our proposed EA-enabled NGIC system.

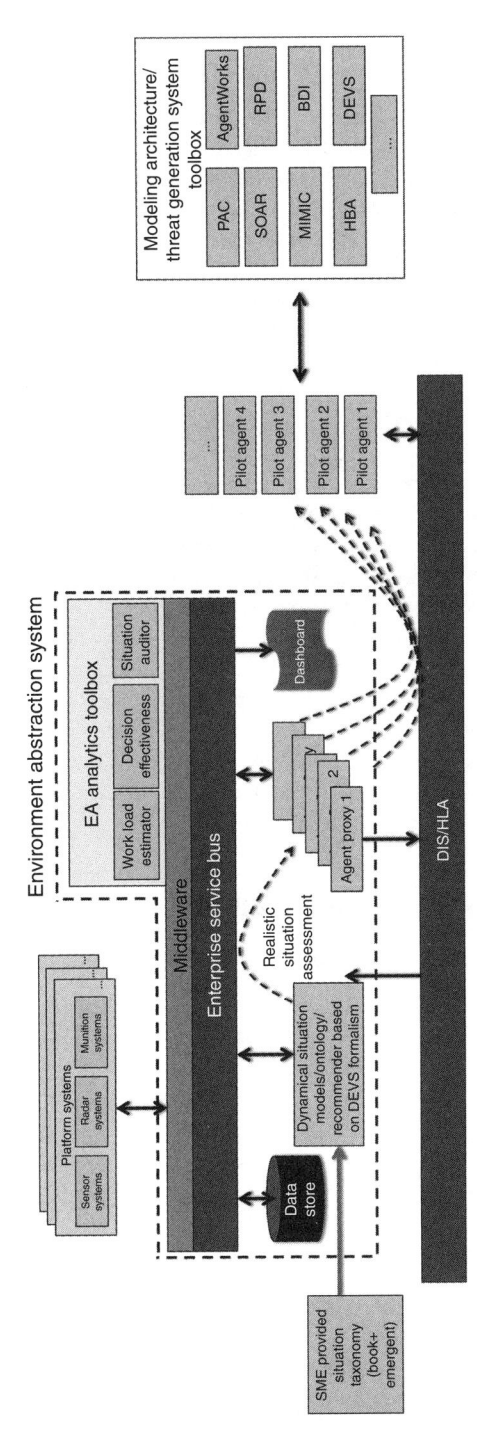

Figure 16.12 EA system as a system of systems for next-generation intelligent combat system.

16.6.1 Functional Requirements for NGIC Systems

EA-enabled NGIC is positioned to run in an LVC DMO training and rehearsal environment; however, LVC DMO DIS communications backbone is not a limitation for EA-NGIC system. If needed, the DIS backbone can be readily replaced with any enterprise service bus or GIG/SOA. The EA is independent of any communications medium as it uses platform-independent XML for all its communications needs. The technological leap being made by the use of an EA system is attributed to the phenomena of strong emergent behavior displayed by humans in a dynamic environment. A strong emergent behavior is labeled as an intelligent behavior. This kind of behavior only exists at an observer level, a level above the interacting agents creating what the observer agent perceives as something of value (Mittal, 2012). Consequently, any combat system that fails to account for this facet of a complex system will end up with a *weak* emergent system that only provides evidence of "programmed" behavior. In our pursuit of next-generation intelligent threat system, modeling strong emergence is at the heart of modeling an intelligent adversary. Associated with the *strong* emergence capability is the capability to change the knowledge boundary of the system at multiple levels. A *strong* emergent behavior always leads to new nomenclature, new patterns/sequences, and new semantics validated only by an SME. An NGIC system must be based on an *open* systems concept wherein the persistent system is able to extend its body of knowledge and use the new information in a running scenario at multiple levels of the system. Table 16.5 summarizes the requirements for intelligent behavior at three levels.

Apart from the strong emergent behavior aspect of intelligent combat systems, an NGIC system is fundamentally an M&S system and, in addition, must have the following features as listed in Table 16.6. The following features were developed based on the results of the NSGC effort, SME observations, initial customer requests, and team research.

16.6.2 Quick Review of Existing Combat Systems

The features identified in Section 6.1 were used to do a quick review of some of the current combat systems developed by government and industry via literature review and observations. These systems included Big Tac™ by the Boeing Co; the Automated Threat Engagement System (ATES), the NICE, and eXperimental Common Immersive Theater Environment (XCITE) developed by AFRL WRRD; and the One Semi-Automated Forces (OneSAF) developed by Science Applications International Corporation (SAIC).

What we noticed is that none of the existing combat systems addressed the emergent behavior at any of the individual, team, and team-of-team levels, making them less useful for competency-based training at the three levels described in Section 16.2. At best, they only modeled agent behavior at an individual level lacking the apparatus for responding to emergent affordances. Consequently, the existing combat systems only model weak emergence, resulting in predictable behavior that is easily gamed by the human pilots. In order to move to a truly adaptive training environment, these combat systems must capitalize on complex emergent behavior to bring novelty and intelligence in adversary behavior.

Human in the Loop in System of Systems (SoS) Modeling and Simulation 445

Table 16.5 Knowledge Boundary and the Strong Emergent Behavior at Multiple Levels of CCS

Levels/complex adaptive behavior	Knowledge boundary effects	Resultant strong emergent behavior
Individual/agent level	Agent is a modular component and is aware of new novel observations and possesses an "adaptive" interface that is programmable at runtime. This allows the input of new knowledge into agent's behavior	A persistent agent learns or provides for new behaviors and new environment structure at runtime. This allows for introducing novelty in the agent behavior that results from the actions based on emergent affordances in the dynamic environment
Strategy/team level	Agent and the interacting team members learn about new strategies of their adversaries with the help of SME to expand their understanding of the dynamic environment, not contained in playbook	An agent in a team setup responds by developing new strategies as a result of novel team behavior observed in a combat simulation environment. Agent leverages adversary strategies or their own team strategies
Tactical/team-of-team level	Team leaders synthesize new tactics at the red force level (topmost level) and then collaborate at the team strategy level as new information about their tactical capacities is enriched in a dynamic environment	At the force readiness level, a red force responds by responding with new tactics that are synthesized from observing emergent behavior of either their adversaries or of their own teams

16.7 CONCLUSIONS AND FUTURE WORK

Since 1997 when the first LVC MTC was established, LVC DMO has come a long way. AFRL WRRD continues to pioneer advanced technologies for better warfighter training and readiness. Networking multiple, real-time simulators together with CGFs in a competency driven and instructionally valid manner provides the warfighter the unique opportunity to operate in a team-oriented, multiplayer environment. MECs prove to be more tractable in an LVC DMO SoS where the system is configured on demand for a particular training event, thus increasing the effectiveness of the training and training program. The aircrew member requires mission-oriented training in addition to its formal, mission-qualification, and RAP training. Such mission-based training in a dynamic LVC DMO environment changes the nature of the entire training system from a man-in-the-loop system to an agent-in-the-loop system with varying degrees of autonomy. This autonomy and the resulting dynamic behavior are naturally embedded in a system when the agent is a human and must be engineered when the agent is a computational model.

Syntactic interoperability has been achieved by the LVC community today but it introduces performance overhead. Despite the capabilities of these LVC architectures, none provide the full range of capabilities required to support emerging warfighter requirements. LVC architectures in their current state have not been able to interoperate with GIG/SOA in a seamless manner. Interoperability research is advancing from syntactic interoperability to semantic interoperability using knowledge sharing, but no concrete example exists in the LVC domain. This chapter documents AFRL WRRD NSGC's efforts toward achieving semantic interoperability in LVC training and rehearsal environments in the area of NGIC systems.

Intelligent behavior, according to Sternberg, is about adapting to one's local environment. According to Brooks, intelligent behavior is an emergent phenomenon and arises as a result of agent's interaction with an environment. Mittal's (2012) perspective adds that the intelligent behavior is absent without an observer at a different vantage points, labeling the behavior as intelligent. In all the cases, there is an implicit assumption that an agent interacts and is bound with its environment (which includes other agents) based on its goal structure and available affordances. An intelligent act is a product of an environment, the agent's behavior, and an observer that labels it as intelligent. The fundamental question is, does the environment provide structural–functional affordance for the agent to act? We posed this question as an interoperability problem and that the agent system and its body (platform) should first semantically acknowledge what affordance is made available to it, based on its near-, mid-, or long-term purpose/goals, by the environment. The agent platform system must semantically interoperate with the environment. The semantic nature of environment is captured in the proposed EA component designed to achieve the said interoperability.

We applied the aforementioned concepts to the domain of CCSs and developing primitives toward a CCS metamodel incorporating the semantics of the agent behavior, the platform structure, the emergent phenomena, and the situation dynamics. The dynamical interplay of these elements filtered by the agent's geolocation and platform specifications yields an agent with EA-aware capability. An agent with such capability is an EA-aware agent capable of displaying strong emergence. As agent-in-the-loop systems require this capability to display strong emergence, having such a framework allows development of agents that can produce and recognize intelligent and believable behavior at multiple levels of CCS such as at the individual, team, and team-of-team level and contribute meaningfully to the MECs in LVC DMO training. Any human-in-the-loop system must address the issue of strong emergence.

Through the AFRL WRRD NSGC project plan, we provided an overview on the recently completed phase I where multiple behavior-modeling architectures seamlessly interoperated with the AFRL LVC DMO environment through MBSE practices. We employed the MBSE process that involved scenario specification, model engineering, model-based testing, behavior validation, and competency assessment of various constructive red force agent behaviors by SMEs. We also provided an overview of the ongoing phase II where EA SoS is being developed to capture emergent behavior for displaying strong emergent trainee behavior in the LVC DMO environment. We described the EA concept at length and presented a systems theory-based methodology to develop an EA SoS that provides semantic knowledge about the environment from a recipient's vantage point.

Our approach to an NGIC system is that the proposed EA SoS can be plugged in with existing modeling architectures provided that the CCS ontology that EA implements is also shared with the interoperating modeling architecture. Having this

Table 16.6 Summary of NGIC Features (Not a Complete List)

S. no.	NGIC feature	Description
1	Adversary behavior modeling	Whether a cognitive model or a CGF, the system must allow modeling of believable behavior. An NGIC system should have a modular way of describing behavior of an agent
2	Adaptive adversary behavior	An NGIC system should have the capacity for adaptation in the dynamic combat environment at an individual, strategic, and team-of-team level (Table 16.5)
3	Threat modeling	An NGIC system should be able to model the threat at an individual, strategic, and team-of-team levels (Table 16.5)
4	Scenario authoring and design	A basic requirement. Any combat system certainly has a scenario designer at individual and strategy levels. An NGIC system should also be able to design scenarios at tactical levels where different strategies are played in a combat exercise
5	Environment customization	An NGIC system should be able to change the agent's environment at runtime, which includes platform configuration, EA scope, etc. that impact the agent's decision-making ability
6	Platform realism	An NGIC system must accurately represent the vehicle platform that agents inhabit for a combat exercise. The platform models should have high fidelity so that EA can accurately evaluate the capacity of agent's behavior in an agent platform system
7	Training usability	An NGIC system should be user-friendly for trainee purposes and scenario designers such as SMEs
8	Agility for rapid prototyping	An NGIC system must be modular for rapid prototyping
9	Simulation platform (LVC, GIG/SOA, etc.)	An NGIC system must be interoperable across LVC architecture and the GIG/SOA environment
10	Repeatability/reproducibility	An NGIC system must be reliable in a training environment
11	Verification and validation	An NGIC system must have verification and validation studies at subsystems and at SoS levels
12	Systems Engineering approach	An NGIC system is an SoS and must adhere to Systems Engineering practices in a net-centric environment
13	Serve to improve LVC DMO training outcomes	An NGIC system ultimately has to address the LVC training at large

underlying framework integrated with DMON successfully provides a capability wherein multiple agent architectures seamlessly interoperated at both the syntactic and semantic levels. Recognizing that, this demands the modeling architecture/combat system have a fundamental capability to dynamically alter the behavior at individual, team, and team-of-team levels at runtime to account for the emergent affordances.

We reviewed the state of the art in combat generation systems and found that currently none of them caters to the subject of "knowledge boundary" for utilizing the emergent phenomenon in combat systems. We believe that the integration of EA with existing combat generation systems could position them to utilize emergent behavior for DMO training.

With these advances in technology and synthetic immersive environments, creating adaptive agents within a battlespace using the proposed NGIC would benefit the warfighter greatly. It would better prepare our warfighters by challenging them with more realistic RFAs. It's more economic and resource efficient in the long run because the RFAs would require less manpower to build and maintain. Due to the nature of the architecture, it would limit the integrations issues with various types of systems, often built with proprietary information. This system could ensure that we are training our warfighters to the way they intend to fight by maintaining combat readiness through more realistic and efficient mission rehearsal.

ACKNOWLEDGMENTS

The opinions expressed within are those of the authors and do not necessarily reflect the views of the sponsoring/employing organizations. We would like to express our thanks to various government, contractor, and industry partners and to Dr. Winston Bennett Jr. of the Air Force Research Laboratory, Warfighter Readiness Research Division, that enabled this research effort on WRSTP Contract # FA8650-05-D-6502. Our sincere thanks to the men and women in uniform whose courage and dedication continue to push the human performance envelope, thereby inspiring and leading us to new research frontiers.

REFERENCES

ACIMS. 2009. *DEVSJAVA Software*. http://acims.asu.edu/software/devsjava. Accessed July 19, 2014.

Ales, R., and S. Buhrow. 2006. The Theater Air Ground System Synthetic Battlespace. *Proceedings of the I/ITSEC Conference*. May, Orlando.

Allen, G.W., R. Lutz, and R. Richbourg. 2010. Live, Virtual, Constructive, Architecture Implementation and Net-centric Environment implications. *ITEA Journal* 31: 355–364.

Anderson, J.R. 2007. *How Can the Mind Exist in the Physical Universe?* Oxford: Oxford University Press.

Army, US. n.d. http://www.peostri.army.mil/PRODUCTS/CTIA/. Accessed June 6, 2013.

Banabeau, E., and J. Dessalles. 1997. Detection and Emergence. *Intellica* 2 (25): 85–94.

Bass, N. 1994. Emergence, Hiearchies and Hyperstructures. *Artificial Life* 3: 515–537.

Bennett Jr., W., B.T. Schreiber, and D. Andrews. 2002. Developing Competency-Based Methods for Near-Real-Time Air Combat Problem Solving Assessment. *Computers in Human Behavior* 18: 773–782.

Bloomfield, B.P., Y. Vurdubakis, and T. Latham. 2010. Bodies, Technologies, and Action Possibilities: When Is an Affordance? *Sociology* 44 (3): 415–433.

Brooks, R. 1990. Elephant's don't play chess. In *Designing Autonomous Agents*, by P. Maes ed., 3–15. Cambridge: The MIT Press.

Brooks, R. 1991a. Intelligence without reason. *Proceedings of the Twelfth International Joint Conference on Artificial Intelligence*. August 24, Sydney, Australia. 569–595.

Brooks, R. 1991b. Intelligence without Representation. *Artificial Intelligence* 47: 139–159.

Chalmers, D.J. 2006. *Strong and Weak Emergence: In The Re-emergence of Emergence*. Oxford: Oxford University Press.

Chapman, R., and C. Colegrove. 2013. Transforming Operational Training in the Combat Forces. *Military Psychology* 25 (3): 177–190. doi:10.1037/h0095980.

Colegrove, C.M., and G.M. Alliger. 2003. Mission essential competencies: defining combat mission readiness in a novel way. *NATO RTO Studies, Analysis and Simulation (SAS) Panel*. April, Brussels, Belgium.

Deguet, J., Y. Demazeau, and L. Magnin. 2006. Elements about the Emergence Issue: A Survey of Emergency Definitions. *Complexus* 3: 24–31.

Dohn, N.B. 2006. Affordances—a Merleau-Pontian account. *Proceedings of the Fifth International Conference on Networked Learning*. April, Lancaster.

Dohn, N.B. 2009. Affordances Revisited: Articulating a Merleau-Pontian View. *International Journal of Computer-Supported Collaborative Learning* 4 (2): 151–170.

Doyle, M.J. 2008. Modeling the interaction between workload and situation awareness: an overview and future course. *Symposium on Human Interaction with Complex Systems and 2nd Sensemaking of Complex Information Annual Conference*. Norfolk.

Doyle, M.J., and M. Kalish. 2004. Stigmergy: Indirect Communication in Multiple Mobile Autonomous Agents. In *Knowledge Systems for Coalition Operations*, by M. Pechoucek and A. Tate (Eds.). Prague, Czech Republic: Czech Technical University Press, October.

Doyle, M.J., and A.M. Portrey. 2011. Are current modeling architectures viable for rapid human behavior modeling? *Interservice/Industry Training Simulation, and Education Conference (I/ITSEC)*. April, Orlando.

Doyle, M.J., A.M. Portrey, S. Mittal, E. Watz, and W. Bennett. Jr. 2014. *Not So Grand Challenge: Are Current Modeling Architectures Viable for Rapid Behavior Modeling? (AFRl-RH-OH-2014-xxxx)*. Dayton: Air Force Research Laboratory, Human Effectiveness Directorate, Warfighter Readiness Research Division.

Gibson, J. 1986. *The Ecological Approach to Visual Perception*. Hillsdale: Lawrence Erlbaum Associates.

Henninger, A.E., D. Cutts, M. Loper, R. Lutz, R. Richbourg, R. Saunders, and S. Swenson. 2008. *Live Virtual Constructive Architecture Roadmap (LVCAR): Final Report*. Institute for Defense Analyses.

Hinze, A., K. Sachs, and A. Buchmann. 2009. Event-based applications and enabling technologies. *Third ACM Conference on Distributed Event-Based Systems*. July.

Hofer, R.C., and M.L. Loper. 1995. DIS Today (Distributed Interactive Simulation). *Proceedings of the IEEE* 83 (8): 1124–1137.

Kim, T.G., C. Lee, E.R. Christensen, and B.P. Zeigler. 1990. System Entity Structuring and Model Base Management. *IEEE Transactions on Systems, Man, and Cybernetics* 20 (5): 1013–1024.

Lee, H., and B.P. Zeigler. 2010. SES-based ontology process for high level information fusion. *Proceedings of the 2010 Spring Simulation Multiconference*. April.

Maier, M.W. 1998. Architecting Principles for Systems-of-Systems. *Systems Engineering* 1 (4): 267–284.

Mak, E., S. Mittal, M.H. Ho, and J.J. Nutaro. 2010. Automated Link 16 Testing Using Discrete Event System Specification and Extensive Markup Language. *Journal of Defense Modeling and Simulation* 7 (1): 39–62.

Miller, D.C., and J.A. Thorpe. 1995. SIMNET: The Advent of Simulator Networking. *Proceedings of the IEEE* 83 (8): 1114–1123.

Mittal, S. 2012. Emergence in Stigmergic and Complex Adaptive Systems: A Formal Discrete Event Systems Perspective. *Journal of Cognitive Systems Research, Special Issue on Stigmergy*. 21: 22–39 Available online at http://dx.doi.org/10.1016/j.cogsys.2012.06.003. Accessed on July 22, 2014.

Mittal, S., M.J. Doyle, and E. Watz. 2013. Detecting intelligent behavior with environment abstraction in complex air combat systems. *IEEE Systems Conference*. April, Orlando.

Mittal, S., E. Mak, and J.J. Nutaro. 2006. DEVS-Based Dynamic Modeling & Simulation Reconfiguration Using Enhanced DoDAF Design Process. *Special Issue on DoDAF, Journal of Defense Modeling and Simulation* 3 (4): 239–267.

Mittal, S., B.P. Zeigler, and J.L.R. Martin. 2008a. Implementation of formal standard for interoperability in M&S/System of Systems integration with DEVS/SOA. *International Command and Control C2 Journal, Special Issue: Modeling and Simulation in Support of Network-Centric Approaches and Capabilities.* 3 (1): 1–61. Available online at http://www.dodccrp.org/files/IC2J_v3n1_01_Mittal.pdf. Accessed on July 22, 2014

Mittal, S., B.P. Zeigler, J.L.R Martin, F. Sahin, and M. Jamshidi. 2008b. Modeling and Simulation for System of Systems Engineering. In *System of Systems Engineering for 21st Century*, by M. Jamshidi (Ed.). Hoboken: Wiley.

Mittal, S. and J.L.R. Martin. 2013a. *Netcentric System of Systems Engineering with DEVS Unified Process*. Boca Raton: CRC Press.

Mittal, S., and J.L.R. Martin. 2013b. Model-driven systems engineering in a netcentric environment with DEVS unified process. *Winter Simulation Conference*. December, Washington, DC.

Momen, F., and J. Rozenblit. 2006. Decision Support System in the Advanced Tactical Architecture for Combat Knowledge System. *Journal of Defense Modeling and Simulation* 3 (1): 11–26.

Ng, A., and S. Russell. 2000. Algorithms for inverse reinforcement learning. *17th International Conference on Machine Learning.* July.

Numrich, S.K. 2006. Models and methods: catching up to 21st century. *Summer Computer Simulation Conference.* July, Calgary.

OSD. 2006. *Strategic Plan for Transforming DoD Training*. Washington, DC: Office of the Under Secretary of Defense for Personnel and Readiness.

Portrey, A., W. Bennett, C. Colegrove, and B. Schreiber. 2007. DMO today and tomorrow: science and technology support for competency based training and Ops. *Spring Simulation Interoperability Workshop*. Norfolk: SISO. 07 W-SIW-072.

Powell, E. 2005. Range system interoperability. *Proceedings of Interservice/Industry Training, Simulation and Education Conference*. December, Orlando.

Raj, A.K., M.J. Doyle, and J. Cameron. 2011. Psychophysiology and performance: Considerations for human-centered design. In *The Handbook of Human-Machine Interaction: A Human-Centered Design Approach*, by B. Guy (Ed.). Burlington: Ashgate Publishers.

Ramachandran, D., and E. Amir. 2007. Bayesian inverse reinforcement learning. *20th International Joint Conference on Artificial Intelligence.* January.

Riordan, B., S. Bruni, N. Schurr, J. Freeman, G. Ganberg, N. Cook, and N. Rima. 2011. Inferring user intent with Bayesian Inverse Planning: Making sense of Multi-UAS Mission Management. *20th Behavior Representation in Modeling and Simulation*. March, Sundance Resort.

Rozenblit, J.W., J. Hu, T.G. Kim, and B.P. Zeigler. 1990. Knowledge-Based Design and Simulation Environment (KBDSE): Foundational Concepts and Implementation. *Journal of the Operations Research Society* 41 (6): 475–489.

Rozenblit, J.W., and B.P. Zeigler. 1993. Representing and constructing system specifications using the system entity structure concepts. *Proceedings of the 1993 Winter Simulation Conference.* December. 604–611.

Sarjoughian, H.S., and B.P. Zeigler. 2000. DEVS and HLA: Complimentary Paradigms for M&S? *Transactions of SCS* 17 (4): 187–197.

Schreiber, B.T., E. Watz, and W. Bennett Jr. 2003. Objective human performance measurement in a distributed environment: tomorrow's needs. *Interservice/Industry Training, Simulation and Education (I/ITSEC) Conference*. December, Orlando: NDIA.

SISO. 2010. *Live-Virtual-Constructive (LVC) Architecture Roadmap Implementation Workshop*. Presentation 10S-SIW-081. Spring Simulation Interoperability Workshop LVC Special Session,

Orlando: SISO. Available online at http://www.sisostds.org/workshops/pastworkshops/2010SpringSIW.aspx. Accessed on November 20, 2014.

Stanners, M., and H.T. French. 2005. *An Empirical Study of the Relationship between Situation Awareness and Decision-Making*. Commonwealth of Australia: DSTO Systems Sciences.

Stemberg, R.J. 1985. *Beyond IQ: A Triarchic Theory of Intelligence*. Cambridge: Cambridge University Press.

Tolk, A., L.J. Bair, and S.Y. Diallo. 2013. Supporting Network Enabled Capability by Extending the Levels of Conceptual Interoperability Model to an Interoperability Maturity Model. *Journal of Defense Modeling and Simulation* 10 (2): 145–160. doi:10.1177/1548512911428457.

Tolk, A., and J.A. Muguira. 2003. The levels of conceptual interoperability model (LCIM). *Proceedings of the Fall Simulation Interoperability Workshop*. September.

Tolk, A., and J.A. Muguira. 2004. M&S with Model-Driven Architecture. *I/ITSEC*. December.

Warwick, W., R. Archer, A. Hamilton, M. Matessa, A. Santamaria, R. Chong, L. Allender, and T. Kelly. 2008. Integrating architectures: dovetailing task network and cognitive models. *17th Conference on Behavior Representation and Simulation*. April, Providence: SISO.

Watz, E. 2012. *Interface Design Document for the Not So Grand Challenge (NSGC) Project, Revision C V4 Sep. 28*. Dayton: Air Force Research Laboratory, Human Effectiveness Directorate, Warfighter Readiness Research Division.

Wymore, W. 1976. *Systems Engineering Methodology for Interdisciplinary Teams*. New York: John Wiley & Sons.

Wymore, W. 1993. *Model-Based Systems Engineering*. Boca Raton: CRC Press.

Zeigler, B.P. 1976. *Theory of Modeling and Simulation*. Wiley Interscience.

Zeigler, B.P. 1984. *Multifaceted Modelling and Discrete Event Simulation*. London: Academic Press.

Zeigler, B.P., and S.D. Chi. 1993. Model-based architecture concepts for autonomous systems design and simulation. In *An Introduction to Intelligent and Autonomous Control*, by P.J. Antsaklis and K.M. Passino (Eds.), 57–78. Boston: Kluwer Academic Publishers.

Zeigler, B.P., and P.E. Hammonds. 2007. *Modeling & Simulation-Based Data Engineering: Introducing Pragmatics into Ontologies for Net-centric Information Exchange*. New York: Academic Press.

Zeigler, B.P., H. Praehofer, and T.G. Kim. 2000. *Theory of Modeling and Simulation: Integrating Discrete Event and Continuous Complex Dynamic Systems*. New York: Academic Press.

Zeigler, B.P., and G. Zhang. 1989. The system entity structure: knowledge representation for simulation modeling and design. In *Artificial Intelligence, Simulation and Modeling*, by L.A. Widman, K.A. Loparo and N. Nielseen (Eds.), 47–73. Hoboken: John Wiley & Sons.

Chapter 17

On Analysis of Ballistic Missile Defense Architecture through Surrogate Modeling and Simulation

Tommer R. Ender, Philip D. West, William Dale Blair, and Paul A. Miceli
Georgia Tech Research Institute, Atlanta, GA, USA

17.1 INTRODUCTION

Military and political leaders have desired to develop an effective Ballistic Missile Defense System (BMDS) since the earliest threats were launched into London during World War II. Contemporary examples on the world geopolitical stage—such as North Korea's initial testing of long-range ballistic missiles and Iran's suspected ongoing nuclear weapons development programs—provide us with justification to field an effective BMDS. Designers are faced with the magnitude of one of the most complex system of systems (SoS) and the challenge of analyzing numerous constituent subsystems. The full-scale flight testing of realistic scenarios is not feasible due to this complexity, and large-scale testing would be cost prohibitive. Flight tests alone would not provide parametric trends that may give decision makers insights on all possible scenarios, even those not actually tested. Historical information is therefore rare, limiting the insights that the analyst or designer might derive about integrated BMDS performance characteristics.

Systems designers must rely on modeling and simulation (M&S) tools to provide the data necessary to perform trade studies and sensitivity analyses, which otherwise would

---

Modeling and Simulation Support for System of Systems Engineering Applications, First Edition.
Edited by Larry B. Rainey and Andreas Tolk.
© 2015 John Wiley & Sons, Inc. Published 2015 by John Wiley & Sons, Inc.

be provided by integrated test and evaluation (Office of the Deputy Under Secretary of Defense for Acquisition and Technology, Systems and Software Engineering, 2008). M&S is typically performed at various elements of the SoS hierarchy and is used to achieve different development objectives. For example, detail digital models at the engineering level are often used to support the design and analysis of BMDS subsystem components (e.g., a particular sensor). Various types of live, virtual, or constructive (LVC) models are used to study the interoperability of interconnected systems, sometimes using extant communications networks and sensor feeds. These example M&S tasks generally support systems analysis at different ends of the technology readiness level (TRL) continuum, with the LVC testing often supporting tests of in-place components at higher (7 and above) TRLs. The engineering level of detail modeling often supports TRL levels of 3, 4, and 5.

M&S, however, is not a simple solution as the digital simulation of a complete BMDS scenario requires execution of many high-fidelity physics-based tools. A thorough simulation should include everything in a standard kill chain, such as target detection, tracking, development of a fire control solution, interception, and kill assessment (Ball, 2003). These tools usually require lengthy execution times, making it difficult to assess the dependence of overall SoS performance on lower-level systems parameters. This limits our ability to evaluate decisions made about the architecture of the integrated system. We have been working to develop engineering tools integrating various disciplines that enable high-level Systems Engineering trade-offs based on analysis tools representing elements of the BMDS kill chain simulation, as we documented in Ender et al. (2010) and Blair and Miceli (2008). Emulating this command and control/battle management kill chain within an analytical simulation framework will very likely expose nonlinear, emergent, and heuristic behaviors.

17.1.1 The Iterative Nature of Systems Engineering

The architecting, implementation, and integration processes associated with classic Systems Engineering prescribe a methodology to design a system that meets requirements that are ultimately verified and validated (Haskins et al., 2010). Information is scarce during the early phases of design but has a great impact on the life cycle development of that system. Program managers and systems engineers therefore place a premium on methods that inform those design-related processes such that iteration with verification and validation is minimized. For example, high-fidelity computer-based simulations may be directly integrated within that iterative design process early on. This is possible only when combined with appropriate subject matter expertise. In this chapter, we summarize a project where we were able to leverage our resident expertise in BMDS target tracking and multisensor track fusion to create a unique systems-level M&S tool that supported interactive collaborative systems trade-off analysis.

This study was developed to create an interactive tool that decision makers can use to evaluate the potential effectiveness of BMDS architectures through rapid interactive M&S, including target detection through intercept kill assessment. We specifically included key battle management issues such as asset location and sensitivity to changing BMDS architecture, which serves as the primary proof of concept for this study. The primary enabler for this rapid manipulation of M&S is through the use of response

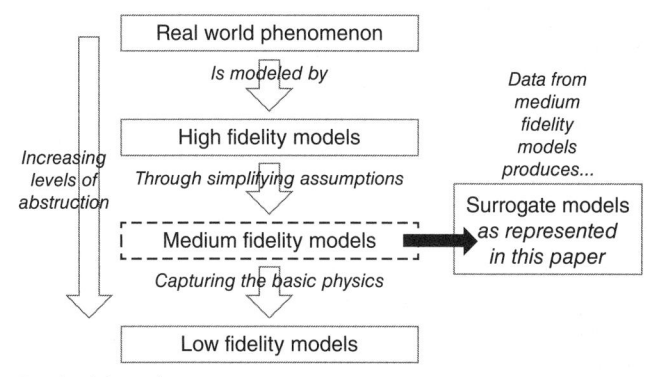

Figure 17.1 Levels of abstraction.

surface surrogate models. These surrogate models are multidimensional equation regression representations that approximate the results of the more complex M&S tools (Myers and Montgomery, 1995). The surrogate models we developed for this study show negligible loss of accuracy when compared to the original M&S tools, as we discuss later in this chapter. Surrogate modeling is an empirical technique that can be used at any level of abstraction of a real-world phenomenon, as given in Figure 17.1. For the purpose of the BMDS effort we present in this chapter, this technique is applied at the medium fidelity level. These surrogate models therefore represent the accuracy of detailed M&S, but by their very nature as closed-form equations, they can be rapidly executed with much less runtime than the original models on which they are based.

17.1.2 SoS Architecture Analysis

The term "system of systems" is generally applied to any problem that may be represented by a complex hierarchy of interacting elements. Maier (1999) characterizes an SoS as one that is composed of elements that are operationally and managerially independent of one another. The Department of Defense (DoD) defines an SOS as a set or arrangement of systems that results when independent and useful systems are integrated into a larger system that delivers unique capabilities (Office of the Deputy Under Secretary of Defense for Acquisition and Technology, Systems and Software Engineering, 2008). The individually useful elements of an SoS operate independently, collaborating only through information exchange. For the case of the BMDS problem, the targets, interceptors, radars, and launch platforms all evolve according to these descriptions, and their independent operation leads to emergent behaviors.

SoS architectures are concerned with architectures of systems created from other autonomous systems (Cole, 2008). Proper SoS analysis requires engineering at the architecture level, which deals with allocation of functionality to components and intercomponent interaction, rather than the internal workings of individual components. The SoS architecture therefore acts as a framework that directs the interaction of components with their environment, data management, communication, and resource allocation (Clements et al., 1996). The functions or activities that each system has to perform must be considered within this architecture for a successful analysis process. A model of an element within an SoS may be used to represent that element but provides no guidance for its structure within an SoS architecture. The SoS architecture defines the interfaces and composition that guides its implementation and evolution (Chen et al., 2004).

17.1.3 Current Approaches of BMDS Architecture Analysis

Current DoD guidance states that M&S should be adopted by Systems Engineering teams because it is difficult or infeasible to completely test or evaluate fully integrated SoS capabilities (Office of the Deputy Under Secretary of Defense for Acquisition and Technology, Systems and Software Engineering, 2008). This guidance, however, recognizes that implementing M&S could be particularly challenging given the complexity of most SoS problems.

Because of the complexity of high-fidelity BMDS effectiveness analysis and limited computational resources, current approaches consist of an ad hoc analysis with few degrees of freedom available. We review several documented efforts to create related Systems Engineering decision-making tools; however, each one lacks the ability to incorporate the benefits of M&S in rapid decision making. For example, Wilkening (1999) introduces a simple model for quickly determining required BMDS effectiveness by modeling the integrated system as series of Bernoulli trials. This requires that trusted values of key parameters are given a priori, including interceptor single-shot probability of kill, as well as various sensor-related issues represented as probabilities of successful detection and tracking of a target. Wilkening shows that he is able to quickly determine required numbers of interceptors needed to defeat all incoming threats as a function of the given key parameters, target properties, and rules of engagement (e.g., shoot–look–shoot, shoot–shoot–look). However, the introduction of this simple model implies a significant sacrifice in accuracy.

Ben-Asher (2004) claims that applying "design principles of system of systems" when defining the systems architecture will "enhance the probability of creating successful [BMD] systems." He stresses, however, that an "SoS treatment" of the BMDS problem is limited to defining responsibility sharing rules and interfaces between systems-level components. Parnell et al. (2001) states that methodologies defining or evaluating BMDS architecture should enable interaction between decision makers, operators, and developers through real-time "what-if" scenarios. However, the "robust methodology" that he introduces sacrifices modeling accuracy for rapid results. It is critical to note that the analyses of both Ben-Asher (2004) and Parnell et al. (2001) are not rooted in M&S. Additionally, BMDS is meant to integrate elements through organized command and control; therefore, modeling realistic interaction must come from an emulated command and control process.

We know that such methods are common in the BMDS analysis community despite the lack of academic references for analysis methods rooted in M&S. Many of these methods typically assume idealized systems track performance, where any target in a sensor's field of view is presumed to be under track with sufficient accuracy. However convenient this may be, this assumption conveniently ignores reality! The impact of the tracking system has far-reaching consequences on SoS performance since battle management decisions are heavily dependent on the tracking system to correctly detect and track targets. Other methods model the local tracking systems based on performance prediction techniques and sensitivity parameters, but only at the local sensor level. While a good step forward over idealized models, these methods usually make no attempt to model performance at the systems track level where battle management decisions are often made.

17.1.4 Novelty of Approach and Chapter Overview

We note a running theme across many of the current approaches for architecture analysis inclusive of the entire BMD problem: accuracy represented by an M&S tool's fidelity must be sacrificed in order to produce rapid trade study results. Our primary motivator to conduct this research was to enable timely simulation at the SoS level while leveraging physics-based M&S. This process requires two primary steps. First, surrogate models are generated based on physics-based M&S. Here, a Design of Experiments (DoE) is used to minimize the number of required samples given that runtimes are not unbounded. Second, those surrogate models are used for rapid analysis of the BMDS decision trade space. The models we used for this work account for performance at the systems track level, which is unique as compared to most related M&S methods.

With a particular motivation established, we will first describe the approach taken to infuse M&S within rapid BMDS analysis. The BMDS problem will be functionally decomposed into its fundamental elements. We will then introduce a notional scenario in which an actual application of a BMDS architecture and battle manager would be in place to defend against threats. This scenario is used to scope the M&S required to represent BMDS battle management elements used in this study. We then describe the various M&S simulation codes developed and used for this effort and then describe the method for representing those codes as surrogates within a larger decision-making framework. The technical core of this chapter ends with a discussion of results representative of the type of analysis enabled by this process. We close this chapter with recommendations for future work extending the M&S environment developed for this work and thoughts on applying these concepts in practice.

17.2 PROPOSED APPROACH

The goal of a BMDS simulation at its highest level is to model the protection of some area from incoming threat missiles, often referred to as targets, or threats, by engaging them with interceptors. This cannot be accomplished unless actions are taken against the incoming targets. The success of these actions is generally measured in terms of the likelihood of neutralizing all incoming threats. This likelihood is often referred to as the probability of engagement success, P_{ES}; the probability of raid annihilation, P_{RA}; or the probability of zero leakers (failed intercepts), P_{OL}. All of these terms essentially have the same meaning, with one wording favored over another depending on the type of scenario being studied. We use the probability of zero leakers, P_{OL}, as the top-level metric for BMDS architecture effectiveness.

We scoped this effort such that several key elements of the BMDS kill chain together determine the set of all possible intercepts (battlespace). A given battlespace contains many possible combinations of interceptor/target pairings that lead to a wide range of systems performance. The battle manager coordinates the means through which this battlespace is condensed to actions. The battle manager is further responsible for making decisions about which interceptor(s) will be launched against which target(s) based on the information provided by the sensor and tracking systems.

With these concepts in mind, we assembled an M&S environment to integrate those various analysis elements of the BMDS kill chain. We chose to include a sensor model for target detection and tracking, a multisensor track fusion model for systems tracking, and a fire control model for engagement planning. This integrated M&S environment uses a time-stepped simulation to calculate the expected number of leakers and the probability of zero leakers. We then create surrogate models of this M&S environment that reproduced top-level metrics without the lengthy computation of the sublevel metrics and other data. We show how this process allows a user to vary the placement of BMDS assets and rapidly determine architecture effectiveness.

17.2.1 Functionally Decomposing the Problem

The BMDS problem can be functionally decomposed into a series of battle management elements that execute in a serial process, as shown in Figure 17.2. This is comparable to the standard combat kill chain as suggested by Ball (2003), where the execution of any element is conditional on the success of the previous element. You should take special note that each of these elements is a field of study in and of itself, requiring various complex M&S codes to be accurately represented. This section will identify the elements necessary to represent the BMDS battle management problem; the very next section will describe those elements in a context relevant to a specific notional scenario. Only certain elements of the BMDS process are represented through M&S to aid in the proof of concept introduced in this chapter. Therefore, we do not claim that the complete BMD problem is represented in this study.

The BMDS battle management process starts with a sensor searching an area for potential targets. After tracks are established at each individual sensor, they are reported to the systems track processor that forms a multisensor systems track. There are a host of methods for forming systems tracks, including selecting the best reported track for each target (track selection), fusing all available track information for each target to further refine the targets' positions (track fusion), or fusing measurements from the sensors (measurement fusion). The systems tracks are prioritized based on various criteria such as criticality of the predicted impact point and the type of warhead being carried, and then available candidate interceptor platforms may be launched against those targets. The final fire control solution is formed when it is determined that the track quality on a given target is sufficient to maximize the potential for a successful intercept.

With a successful fire control solution formed against a target, an interceptor is launched to execute the engagement phase of the process. As the interceptor proceeds along its flight path, it may receive in-flight target updates (IFTU) as the radar tracking information is improved. In the terminal phase, the interceptor uses an onboard seeker to guide itself through the endgame intercept phase and make any guidance and control corrections necessary to overcome errors provided by the radar. An assessment must be made as to whether the hit resulted in a successful target kill once the interceptor is presumed to hit the target.

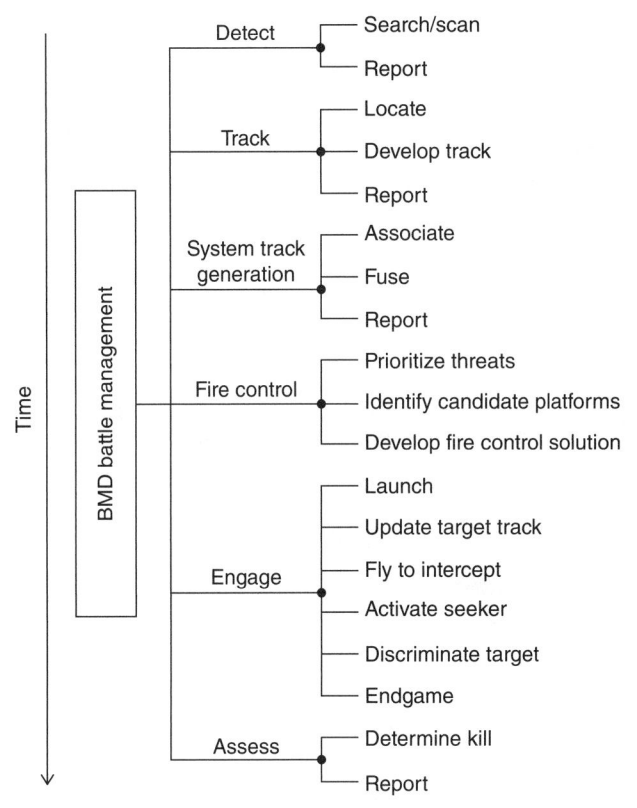

Figure 17.2 Functional decomposition of BMDS battle management.

17.2.2 Proof-of-Concept Architecture

We must define a scenario in order to analyze a BMDS battle management architecture using an integrated M&S environment. The notional scenario we define for this study, as shown in Figure 17.3, includes four targets that are near-simultaneously launched from a hostile country with the intent of attacking a friendly country across a narrow sea. Two BMDS capable warships are stationed in the sea between the two countries to defend the friendly nation and may operate within a diamond-shaped patrol area. A ground-based interceptor is also available and located at an arbitrary fixed location to supplement the sea-based defense. Each of the three platforms carries an interceptor battery as well as a radar sensor that can detect and track targets. A single-point analysis is defined by target trajectories, interceptor locations and inventories, sensor performance, and the various battle management issues described earlier in this chapter.

The various objects interacting in this complex BMDS SoS present many potential degrees of freedom available for architecture-related decision making. One of the primary drivers in architecture definition is the passing of information among each of the actors, which supports the formation of a single integrated track picture. We consider two procedures for the BMDS example: track selection and track fusion. We assume that

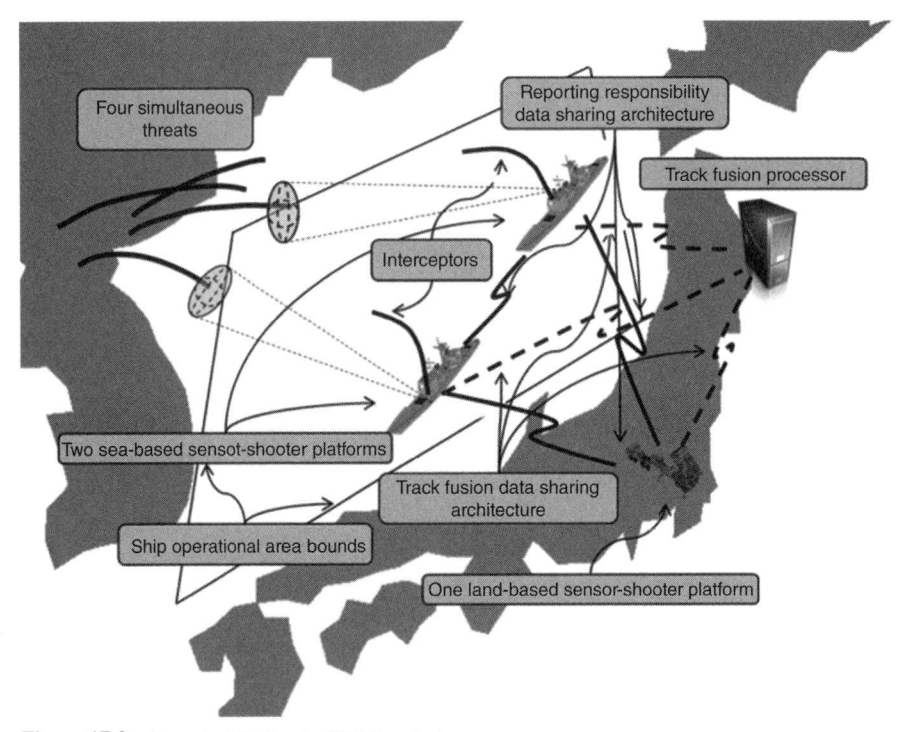

Figure 17.3 Scenario definition for BMDS analysis.

interceptor and target properties are fixed; therefore, we only examine the radar track-related elements of the problem decomposition.

17.2.3 M&S of the BMDS Elements

M&S of multisensor and multitarget tracking systems typically involves complex computer algorithms. Due to the nondeterministic nature of the algorithms, Monte Carlo simulations are typically required to accurately assess expected systems performance (Bar-Shalom and Blair, 2000; Bar-Shalom et al., 2001). Monte Carlo trials of such simulations are devastatingly time consuming, making them unsuitable for studying higher-level SoS performance for BMDS.

Rather, many analysis methods common in the BMDS community employ either truth-based or performance prediction models. Truth-based methods typically assume any target in a sensor's field of view is presumed to be under track with sufficient accuracy. These methods are insufficient since the performance of the tracking system is a critical driver of battle management decisions. Other methods common in the community model the local tracking systems based on performance prediction techniques and radar sensitivity parameters. Although these models make many assumptions, they have fast runtimes when compared to complex Monte Carlo simulations and are more appropriate for SoS studies. Therefore, we base the target detection and tracking M&S components in this study on performance prediction techniques.

The measure of performance we chose to drive battle management decisions is systems track accuracy for each target under track. Performance prediction techniques are available to us, so actual track states are not required. We created lower-fidelity performance prediction models based on a set of disclosed assumptions, which have fast runtimes, do not require Monte Carlo simulations, and can be benchmarked against higher-fidelity validated models. The BMDS analysis community widely accepts these comparatively lower-fidelity analysis methods—but only at the local sensor level. While a good step forward over idealized models, these methods typically make no attempt to model performance at the systems track level where battle management decisions are often made. We expanded performance prediction techniques for this study, which were used to handle multisensor tracking systems where local tracks are fused or selected at the systems level.

The remainder of this section will describe the M&S elements of the BMDS analysis tool developed for this study and the assumptions behind them.

17.2.3.1 Sensor Model

We consider two phased array radars with three-dimensional monopulse measurements: a notional ground-based X-band radar with a single array face and a notional ship-based S-band radar with four array faces. We developed a generic sensor model such that all sensors of the three-dimensional phased array varieties can be represented with the same code, where a unique parameter file defines the sensor type.

Given the predicted target and sensor trajectories and sensor parameters, the sensor performance can be expressed as a measurement covariance, which can in turn be used to drive predicted track performance. We found that we must use predicted trajectories since it is only reasonable to make decisions about future events with data up to the present time. Fortunately, the laws of ballistic flight allow us to make very accurate extrapolation given accurate track state estimates. Therefore, the predicted track accuracy based on the extrapolated trajectories is very useful for engagement planning. This method also implies that we did not consider engagements during boost phase, where the missile acceleration, number of stages, and burn time make extrapolating the missile trajectory much more difficult.

We made several major assumptions to generate a predicted measurement covariance. First, we assumed that the sensor has perfect resolution, which implies merged measurements are not an issue and sensor errors can be characterized as Gaussian. Second, we assumed no sensor biases, which imply the mean of the Gaussian distributed errors are centered about the predicted position of the target. The errors in each coordinate are presumed independent, which means the measurement covariance can be fully represented by a variance estimate in each of three coordinates: range, cross-range horizontal, and cross-range vertical. Third, there are no false alarms present, and the probability of detection is always one given sufficient signal to noise ratio.

Pending the development of more advanced performance prediction techniques, some of these assumptions will need to be relaxed. For example, the first assumption ignores the reality of closely spaced objects (CSOs). When two or more target objects fall within the same radar resolution cell, then only one measurement is returned by the sensor, confounding the track picture. As these CSOs separate and become resolved, the measurements are often so close together that an accurate track picture cannot be created. Common to the problem of tracking ballistic missiles, CSOs result anytime one object splits into two or many more objects (e.g., tank and reentry vehicle separate). Besides the reentry vehicle, these separations often involve debris, countermeasures, spent tanks, etc. The ability to correctly detect and classify (e.g., discriminate) the reentry vehicle from a slew of harmless debris and other objects is absolutely critical to the effectiveness of the BMDS.

Given sensor parameters and the target and sensor geometry, we can compute a signal to noise ratio based on the radar range equation. Assuming monopulse processing, a simple set of equations presented in Blair and Brandt-Pearce (1988) can be used to express the variance in each sensor frame coordinate as a function of signal to noise ratio, range to target, monopulse parameters, and off-boresight steer angle. The measurement variance in each coordinate and the measurement rate are the only sensor model output used to drive predicted local tracker performance.

17.2.3.2 Local Track Performance Prediction

Techniques can be used to estimate tracker performance in each coordinate of the sensor during the coasting stage of flight prediction. The performance can be expressed as a noise-only covariance and a maneuver lag (Blair, 1993). These terms can then be additively combined to produce a prediction of the track filter covariance.

A number of assumptions are implicit using this method of track performance prediction. We first assume there is perfect measurement-to-track association; measurements are always assigned to the track on the truth object from which the measurement originated. Second, the tracker is presumed to be in steady state, and the update rate is presumed to be constant. In reality, CSOs will eventually stray far enough apart so they become resolved by the sensor. Once resolved, new tracks with poor state estimates are spawned and usually result in some degree of incorrect measurement-to-track association. The missed or incorrect associations imply that update rates are not always periodic and measurements from the wrong object may contribute to track updates. In either case, performance degrades. Further research on the performance of tracking systems in the presence of data association errors will need to be conducted before the assumption of perfect measurement-to-track association can be relaxed.

17.2.3.3 Systems Track Performance Prediction

We modeled two network-level track fusion schemes for this study. The first is a centralized multisensor track fusion that combines tracks from multiple sensors. The second is a reporting responsibility network that selects the systems track as the "best" track from the set of all local tracks.

Given the target and sensor trajectories, sensor parameters, and filter process noise, we used performance prediction techniques in Blair and Miceli (2008) to predict the quality of the fused tracks from multiple sensors. The combined sensor-noise-only covariance fuses with a smaller covariance and the maneuver lags fuse with a gain proportional to the inverse to the covariances for the individual sensor tracks. As with the local sensors, this method employs some important assumptions. In particular, this algorithm assumes there are no sensor biases. Ideally, the fusion of multiple covariance ellipses results in an error equal to the union of the ellipses where the union is smaller than the smallest individual covariance. However, this only holds true if all covariances are centered about the same mean. In reality, relative biases between sensors create a misalignment in the means, which results in a union of the covariances that is larger than one, many, or all of the covariances of the individual tracks.

The second network track fusion technique we implemented is a reporting responsibility protocol similar to that used on the Link 16 networking system. In this paradigm, the local tracker with the best track is responsible for reporting the track to all other participants on the network.

17.2.3.4 Engagement Planner Analysis

The engagement planner computes the set of all possible launch solutions (battlespace). The battlespace is computed for each interceptor/target pair for all times at which the target is presumed to be in coast phase. For a candidate interceptor launch time, we can determine the potential intercept solution by fly-out fans, which are nothing more than a set of precomputed missile trajectories for a number of launch angles. Given the launch point of the interceptor and the predicted target trajectory produced by ballistic extrapolation, we can interpolate the fly-out fans to compute the time to intercept, flight path angle, and intercept point. While these fans may be used to find the fly-out trajectory and predicted intercept time, they do not guarantee that the systems track is accurate enough for launch.

We developed an additional algorithm to compare the maneuverability of the interceptor to a critical threshold, which is a function of the predicted systems track accuracy. This check is performed at three key decision points: interceptor launch time, interceptor burnout time, and interceptor seeker activation time. Each decision point determines if the interceptor has enough maneuverability to cover the error region defined by the predicted systems track accuracy at the decision time. If all three decision points are positive, the battle manager determines that a launch solution exists. Finally, the battle manager computes a probability of kill (P_k) for the launch solution based on the predicted track accuracy at seeker activation, interceptor/target geometry, and seeker parameters.

17.2.3.5 Fire Control

To facilitate this study, we created the battle manager to enforce a "shoot early" strategy. In this doctrine, a single battle manager supervised the interceptor/target assignments from all platforms. Targets were engaged as soon as the track accuracy of a target leads to a predicted P_k that exceeded a threshold value. To simulate the defense system's willingness to accept lower P_k as the target approaches, the threshold value decays with time from a default value of 0.85 at the time that the target becomes exoatmospheric to a minimum acceptable P_k at the time that the warhead enters the atmosphere. In order to mimic the physical limitation on a BMDS simulation, we constrained platforms to supporting a maximum of two interceptors in the air at a given time. This ensures that the sensors will be able to track the interceptors during fly-out and provide them with the necessary IFTU.

For a given candidate launch time reported in the battlespace, the battle manager checks to determine if another platform has already assigned interceptor to the current target. If the target has not been engaged, the battle manager then checks to see if the platform associated with the given candidate launch time can support the interceptor.

Lastly, if the platform is able to support the launch, the battle manager checks the P_k associated with the launch time to determine if it exceeds the threshold value. If a candidate launch time is successful in all these checks, it is chosen by the battle manager as an engagement and is reported in the output. At the end of a simulation, the battle manager has completed this task for all candidate launch times and has reported the launches taken against the corresponding targets.

17.2.4 Using DoE and Surrogate Models

Surrogate models enable rapid manipulation of complex M&S within a higher-level analysis tool. You might assume that building surrogate models of the highest-fidelity models available would be the ideal case. However, due to computer processing limitations, it is not feasible to run several thousand cases at several hours per run that would be required given a DoE to generate those surrogate models. Therefore, while surrogate modeling techniques can be used to build response functions that produce almost instantaneous outputs, we accept that the runtime of the simulations from which the surrogate models are created is not unbounded.

Surrogate models, based on response surface methodology (Myers and Montgomery, 1995), are equation regression representations of more complex M&S tools while maintaining a fairly high level of accuracy when compared to those original tools. A surrogate model is made by statistical regression of a set of data. In theory, a surrogate model may be built around data collected from a large number of M&S runs with random selection of values within the bounds of input variables. However, for a very complex systems model requiring time-consuming computer codes to run, a structured method for data sampling with the minimum number of simulation runs (or "experiments") is needed.

A statistical approach to experimental design can be useful in drawing meaningful conclusions from data. A statistical DoE is such an approach, which plans simulation execution cases such that meaningful conclusions can be drawn. A DoE therefore governs how the M&S tools are executed, given in Figure 17.4, which illustrates the process of constructing a surrogate model of some physics-based analysis that maps the n-vector x of inputs to the m-vector of outputs.

These equation-based surrogate models can take most any form based on assumptions of the way a given response varies as a function of given variables. A common form of surrogate models is the polynomial format and has proven successful for the design and optimization of a single system (Myers and Montgomery, 1995; Forrester et al., 2008). However, more complex design spaces cannot be approximated with polynomial equations, such as in a complex SoS integration problem where one may not necessarily be interested in the optimization of a particular systems component but on being able to quantify the interactions between the individual systems.

Neural networks can be used to generate surrogate models of multimodal, discontinuous, or otherwise highly nonlinear design spaces, which are commonly encountered when modeling complex systems. The authors used a set of single-layer feedforward neural networks with back propagation to create the surrogate models used in this study. Weaver (2008) provides further detail on the creation of these nested neural networks. Neural networks have been found useful in complex integrated systems analysis because they

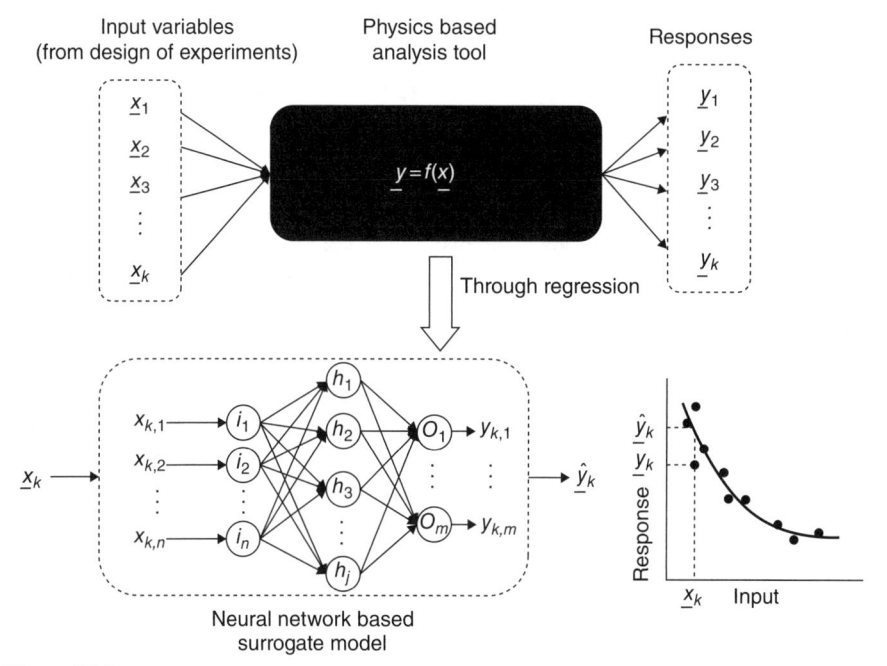

Figure 17.4 Surrogate model generation process.

enable the rapid population of a design space with thousands of cases. The investigator can then levy constraints anywhere a posteriori to dynamically filter out infeasible regions of a design space (Ender, 2006).

The assumptions regarding how a given response varies as a function of select variables, for example, linear or multimodal, govern the appropriate DoE to use in an analysis. We created a customized DoE that combines a Central Composite Design (CCD) used to capture the design space extrema, a Latin Hypercube Sample (LHS) to capture multimodal effects within the design space, and a random set used for validation but not used for regression. We then created neural network-based surrogate models by regression of the DoE data.

The neural network surrogates are then validated by means of coefficient of determination (R^2), also known as the correlation coefficient, and measurements of the model fit error and model representation error (Walpole et al., 1998). Values of R^2 range from 0 to 1, with 1 being a perfect regression. Forrester et al. (2008) recommend that R^2 values of 0.8 and above indicate sufficient predictive capabilities; however, some applications may require more stringent criteria. In order to construct these predictive representations, we assume that the error of the model is normally distributed with a mean about zero and has a standard deviation dependent on the purpose of the model.

In summary, because surrogate models are equations, albeit complex ones, they can be rapidly executed many times and provide a user the ability to access the analysis capabilities of M&S without the computational delay. Once these surrogate models are created, a design space can be explored by rapidly generating thousands of cases, each with small (but measurable) loss in fidelity from the original M&S environment.

17.3 RESULTS

We created neural network surrogate models to predict the probability of kill for each threat missile, the detail of which is beyond the scope of this chapter; for an in-depth treatment of the nested network process we used, see Weaver (2008). The models used had R^2 values ranging from 0.87 to 0.91 and an average error of approximately 2%.

We can now study questions that, by their nature, require hundreds if not thousands of data points to be run for trends to emerge. This would have been impossible to study by using the computationally expensive M&S environment directly but now is possible using surrogate models of the M&S tools. Recall from the introduction that one of our primary goals was to create an environment that could be used to make real-time decisions about both the BMDS architecture to be used in a given scenario and how to best use that architecture once it is in place. We therefore used surrogate models of the M&S tool for the two-ship, one ground-based interceptor scenario (described earlier) to perform several probabilistic architectural trade studies.

17.3.1 Asset Placement Trade Study

One of the more fundamental questions you might ask about a given BMDS architecture is about the most effective location of the individual assets. To begin answering this question, we created a graphical user interface (GUI) in MATLAB to facilitate a more intuitive use and manipulation of the created surrogate models. This GUI allows the user to place the first of two ships anywhere within the allowed operational area by simply selecting it and dragging it to the desired location (within the operation bounds previously given in Figure 17.3). Then, using a user-defined number of cases, the integrated tool repeatedly evaluates the surrogate models to generate a graphical performance map, which is then overlaid on the operational area. This performance map displays what the top-level performance (one of several measures of performance may be chosen) of the system would be if the second ship were to be placed at any given location within the allowed operational area.

Figure 17.5 shows the performance map resulting from 10,000 evaluations of the surrogate models, which are evaluated instantaneously on a desktop computer. An equivalent image can be created directly from the original high(er)-fidelity M&S environment; however, computational time would be on the order of days (on the same desktop computer). The low computational time allows for a real-time interaction with high-fidelity data by systems operators and high-level decision makers.

The location of the first ship (S1) in Figure 17.5 is denoted by the large white circle in the northeast corner of the operating area. The forward deployed ground unit is marked by a black triangle (S3) in the southeast corner of the defending nation. The shading of the smaller points represents the overall systems performance, in terms of probability of zero leakers (P_{OL}), if the second ship (S2) were to be placed at that point. Dark areas denote space where it is not possible to intercept all incoming targets (i.e., $P_{OL}=0$). Increasing systems performance is denoted by an increase in the brightness of the smaller points, with perfect systems performance ($P_{OL}=1$) represented by white.

Note that each point shown in Figure 17.5 is the result of an evaluation of the surrogate models, comparable to a complete BMDS simulation within the bounds of the M&S environment introduced in this study. These points are randomly chosen from

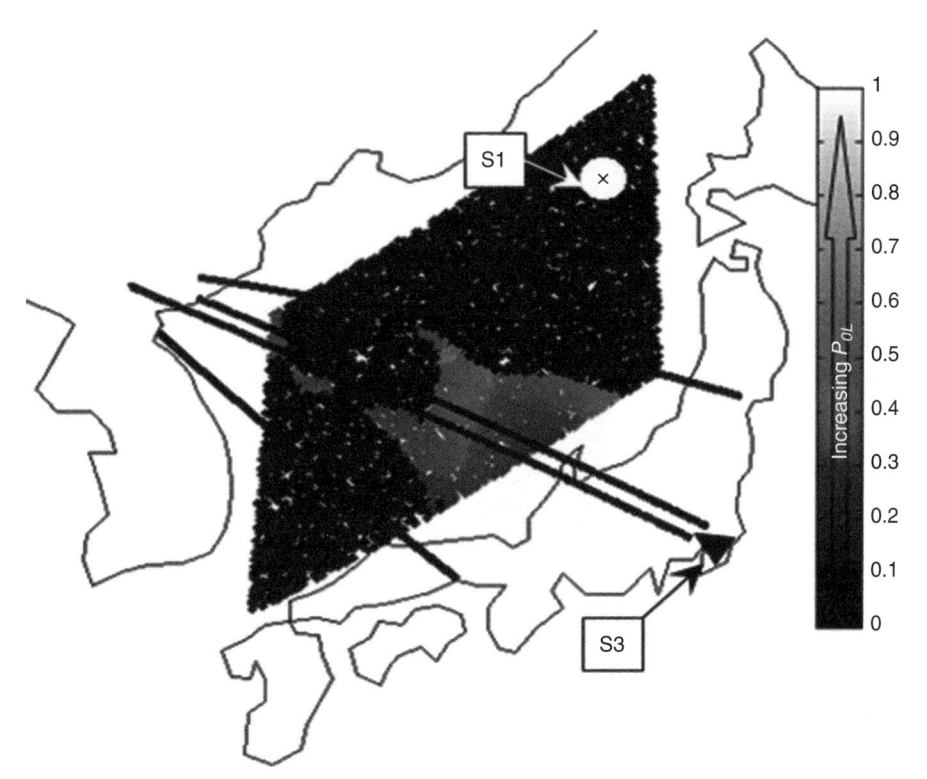

Figure 17.5 Operational area performance map.

the allowed operating area in order to show performance contours. Small white spaces can be seen between evaluations, which are not simulation results, but rather gaps between evaluated points.

Figure 17.5 shows us that an area of nonzero systems performance exists beneath the center trajectories. However, the performance associated with this area is far from ideal with an average performance around $P_{OL}=0.35$. The relatively poor systems performance can be explained by the location of the assets. Notice from Figure 17.5 that platform S1 is situated away from the threat trajectories in the northeastern corner of the operational area. This separation degrades the track quality of the sensor while also increasing the flight time of interceptors from S1 (if they are even in range). In general, this platform is not significantly contributing to the overall systems effectiveness while in this location, so in order to achieve a high P_{OL}, the remaining two platforms must be capable of intercepting all threats with a high probability of kill; due to the tracking and firing constraints, this may not be feasible.

Moving S1 to a new location, as shown in Figure 17.6, we find a larger band with performance superior to the previous case. Platform S1 is able to contribute to the track picture as it is closer to the threat trajectories, while the interceptor is able to reach multiple trajectories in less time than from the previous location. The combination of the increase in track picture accuracy and shorter flight times allows for a larger window of candidate intercept points, resulting in a larger operational area with higher confidence.

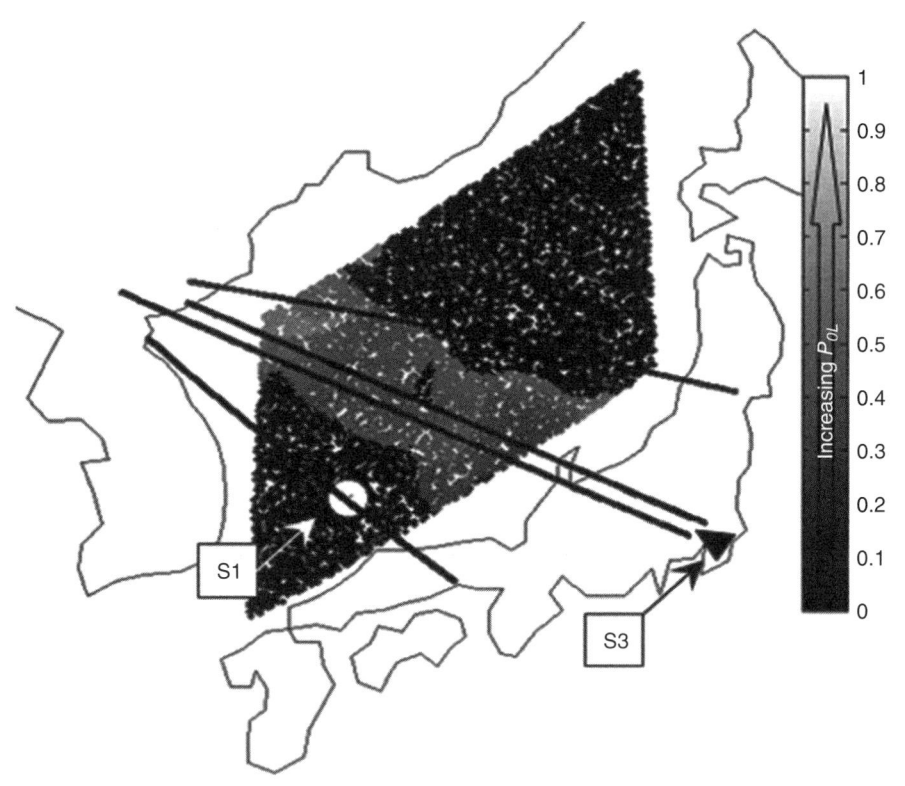

Figure 17.6 Performance map with new S1 location.

The band of nonzero systems performance through the middle of the design space shows overall systems performance between $P_{OL}=0.25$ and $P_{OL}=0.6$, with an average systems performance of approximately $P_{OL}=0.36$ ($P_{OL}=0.11$ for the entire design space). We can also see that the performance of the overall system increases as S2 is moved closer to the launch points of the targets.

17.3.2 Tracking Method Trade Study

Compare Figures 17.5 and 17.6, and you will note a sizable portion of the overall operational area is made feasible by changing the location of S1. However, neither figure has a high confidence of success, indicating room for improvement in architecture-level decisions regarding the sensor, interceptor, and fire control systems. The relatively tight band of performance in Figure 17.6 shows that the systems effectiveness drops off as the platforms move further from the threat trajectories; however, the reason for the decrease in performance is not explicitly known at this stage and requires further investigation.

The interceptor and fire control systems were held constant and surrogate models used to explore the systems performance under different sensor architectures in order to determine the cause of this poor performance. Figures 17.5 and 17.6 show results using a best track selection method, similar to real-world Link 16 reporting responsibility.

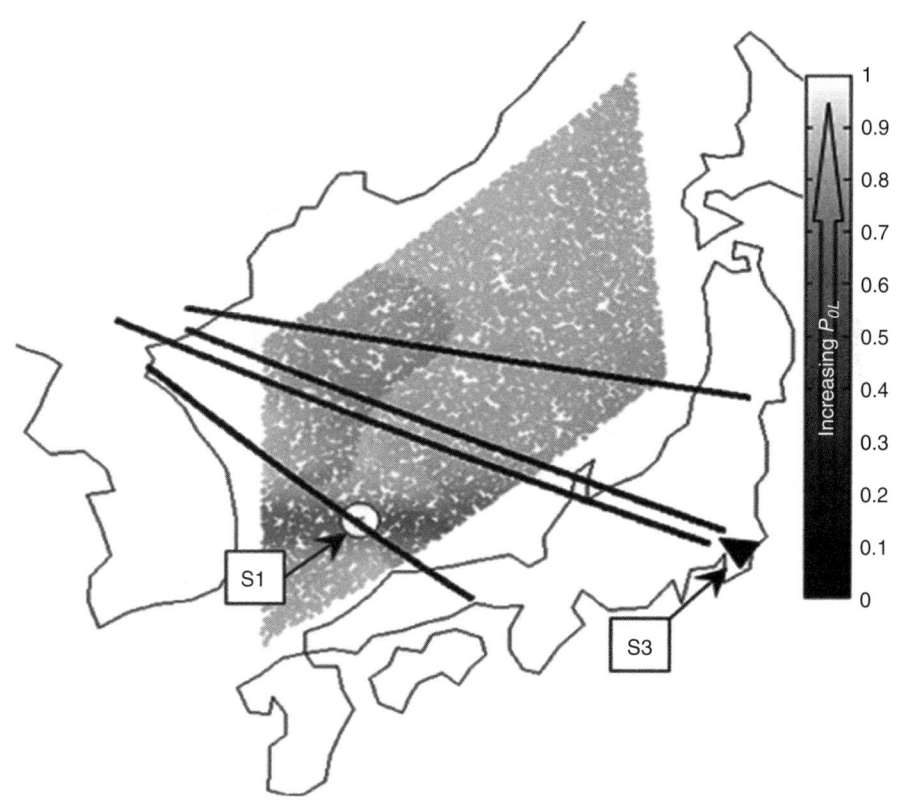

Figure 17.7 Performance map showing track fusion with three sensors.

Figure 17.7 shows results for the same southwest location of S1, now utilizing multisensor data fusion where tracks on a given object from all sensors are fused for a single track. Here, both the best possible performance of the system and the desirable operational area for ship two increase dramatically with systems performance ranging from $P_{OL}=0.53$ (in a small area immediately adjacent to the first ship) to $P_{OL}=0.87$ and a mean systems performance of approximately $P_{OL}=0.79$. The increase in P_{OL} and operating area results only from implementing data fusion, showing that the restricted operating area shown in Figure 17.6 was due to limitations in the track accuracy rather than fire controller or interceptor limitations.

17.3.3 Number of Sensors Trade Study

The results presented in Figure 17.7 show that the overall systems performance is not highly sensitive to the location of S2 within the designated area (for the given S1 location) when multisensor data fusion is used. This high level of robustness in the operational area of S2 may result from an unnecessary participant of the systems architecture. We desire to achieve the objective confidence level while using the minimum necessary resources due to the extremely high cost related to operating a BMDS capable platform. We used the

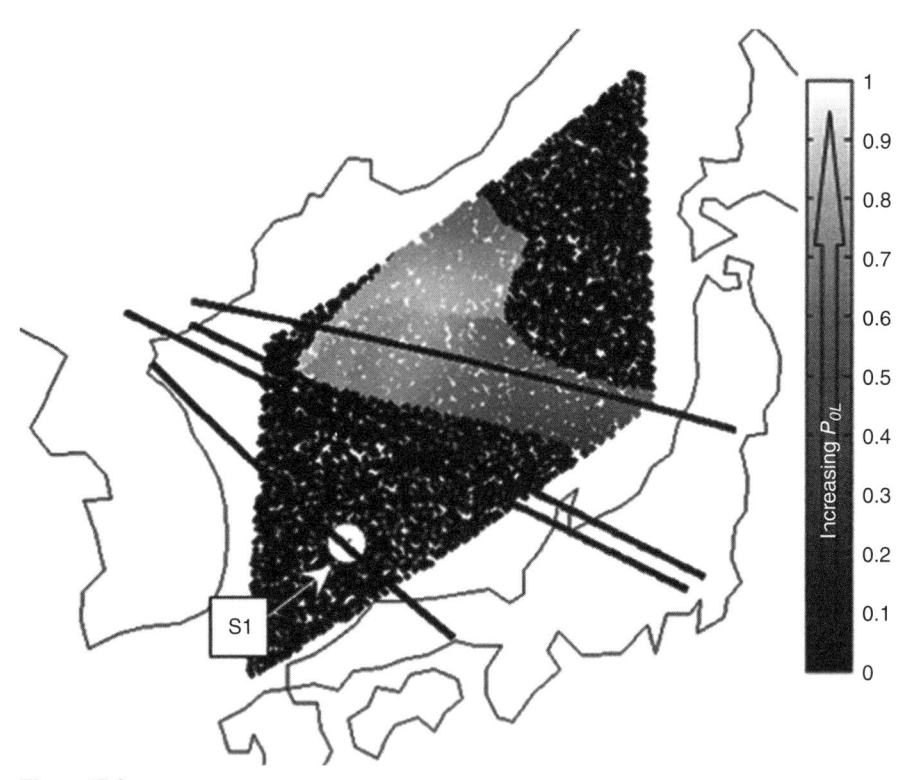

Figure 17.8 Performance map with track fusion with two sensors.

surrogate models to explore the architecture size and provide insight into the minimum required resources. Therefore, we generated surrogate models for the systems performance with the forward deployed ground-based interceptor/sensor platform removed.

Figure 17.8 shows the results using track fusion with S1 in the same southwest location and S3 removed from the scenario. The operational area for the two-sensor architecture has decreased greatly from the three-sensor data fusion results presented earlier to a size similar to that in Figure 17.6. Although comparable in size, the performance of the new architecture given in Figure 17.8 has improved significantly compared to that of the original situation described in Figure 17.6.

Table 17.1 summarizes the results of these trade studies, and for the two-platform data fusion architecture, shows a 50% increase in average systems performance over that in the three-platform track selection architecture with a maximum comparable to that in the three-platform data fusion architecture. The minimum value for the systems performance varies significantly between the two- and three-sensor architectures. Based solely on the metrics used in this study, the three-sensor track fusion architecture produces superior results over a larger operational area, which may or may not be worth the added cost of operating the third platform. The two-sensor track fusion architecture produces a smaller area of similar performance to that of the three-sensor track fusion architecture; however, it is up to the high-level decision makers to determine if the feasible region is satisfactory for the remainder of the goals not directly stated in this chapter.

Table 17.1 Architecture Trade Summary

Architecture	Min P_{OL}	Max P_{OL}	Average P_{OL}	Average P_{OL} (nonzero)
3 platforms using track selection	0.00	0.60	0.11	0.36
3 platforms using track fusion	0.53	0.87	0.79	0.79
2 platforms using track fusion	0.00	0.83	0.20	0.64

The introduction of track fusion shows to have a significant impact on probability of zero leakers as compared to individual track selection.

17.4 RECOMMENDATIONS FOR FUTURE WORK AND PRACTICAL APPLICATION

The research we present in this chapter is that of a work in progress, leaving us opportunities to extend these foundations and to improve the state of the art. We also find ourselves wanting to apply these concepts to improve the way those in the BMDS community do their jobs.

17.4.1 Future Work

There are many opportunities to extend the work we initiated, and we certainly recommend you not limit yourself to those recommendations in this section. We however can recommend a few key, critical "low hanging fruit" possibilities, some of which we've already started investigating. These include improving the BMDS-specific M&S environment, extending the trade space capabilities to enable mission planning utilizing the various SoS assets, and leveraging distributed computing to increase the breadth of the analytical power of the methods we introduced. An extension of distributing computing may leverage major architectural patterns in a distributed systems design.

17.4.1.1 M&S Environment

For future efforts, the breadth and depth of the M&S tools should be extended to include:

- Detection and tracking of multiple CSOs in a threat complex
- Misdetections and measurement-to-track association errors
- Sensor biases
- Dynamic sensor resource allocation
- Multisensor track correlation
- Multibody vehicle dynamics to capture hit-to-kill guidance laws governing the interceptor and target
- Assessment of endgame performance

17.4.1.2 Integrated Trade Study Capabilities

Future studies should invest the use of the tools developed in an operational or mission planning context. Wilkening (1999), as previously discussed, introduces a simple model for quickly determining required BMDS effectiveness by modeling the integrated system as series of Bernoulli trials. Instead of assuming probabilities of successful detection and tracking of target(s), the work presented in this chapter could be extended to compute these probabilities, which may be used as input to the algorithms in Wilkening (1999). In this sequence, the effectiveness of a BMDS system could better be tied to specified scenarios and serve as guidance for mission planning.

The rapid evaluation capability we presented also opens the door to the use of more computationally expensive applications such as systems optimization. Weaver (2008) initiated studies to investigate linking the developed surrogate models with a genetic algorithm optimizer to assess the effects of required operational maneuvers of mobile assets on the performance of the BMDS. This would enable us to find ways of compensating for any ensuing performance degradation by simultaneously moving other mobile assets within the theater. Preliminary results show promise, with no obvious barriers to successful application. This was, however, a basic proof of concept not yet leveraging the breadth of M&S tools available to the BMDS community.

17.4.1.3 Distributed Computing

The proof-of-concept M&S environment we presented in this chapter was designed to run on a single computer. However, we recognize that the runtime of the simulations from which the surrogate models are created is not unbounded. Even a cursory review of the open literature regarding distributed computing shows tremendous promise for increased efficiency in our approach. There is of course increased complexity in setting up a distributed computing environment, vice a running all our codes on a single computer. But think back to the chapter introduction and the section on modeling the BMDS elements (Section 17.2.3). Many assumptions were made to simplify the constituent systems characterization in order to develop an integrated M&S within trade study environment, executable on a single computer (surrogate models notwithstanding). The suggested improvements to the M&S environment, given earlier in this section, would benefit from a distributed computing environment; you could argue that it would not be feasible to implement all those improvements and run on a single computer.

17.4.1.4 Cyber Considerations

A BMDS architecture will include the cyber components of command and control, which is highly dynamic, allowing us to "tune" it on demand to address the mission (or missions) at hand. If the BMDS has to be used in a "real" missile attack, it will almost certainly be subjected to a cyberattack simultaneously and possibly during prelaunch. We can't ignore this given today's threat environment, so we recommend that future work include a thorough treatment of cyber considerations.

17.4.2 Application to the Relevant Workforce

You may be asking yourself, "this all sounds great, but what do I practically take away and how do I apply this?" We review some relevant heuristics and discuss your role as an SoS engineer.

17.4.2.1 Heuristics

Experience gives us the test first and the lesson afterward (Maier, 2009). We can leverage heuristics as collective experience of situational insight, stated in as simple and concise a manner as possible. Here, we examine how several high-level heuristics apply to BMDS.

"No complex system can be optimum to all parties concerned, nor all functions optimized" (Maier, 2009). BMDS is by all measures a "complex system," and there are certainly many stakeholders with individual interests. Your technical understanding includes the functions, interrelationships, and dependencies of the constituent systems within the SoS (e.g., sensors, shooters, battle management). However, you must not overlook the objectives, motivations, and plans of those constituent systems! Decisions about functionality are based on schedule and funding constraining those constituent systems—not technical optimization of the SoS as a whole. Therefore, keep in mind that "The greatest leverage in architecting is at the interfaces" (Maier, 2009). You can't optimize the individual systems, but as we show in Section 17.3, you sure can optimize how they all work together. With a dynamic and adaptable architecture, we can optimize how all the elements work together, ideally mix and match "on demand."

"Don't assume that the original statement of the problem is necessarily the best, or even the right, one ... Extreme requirements, expectations, and predictions should remain under challenge throughout systems design, implementation, and operation" (Maier, 2009). BMDS requirements suffer from the need for ultraquality, which is a level of quality so demanding that it is impractical to measure defects. Examining the actual requirements for BMDS is well beyond our scope, but we can examine a notional example. The BMDS system may be targeting a nuclear threat. Political leaders may impose a requirement to shoot down 99.9% of all incoming threats, given the grave impacts of even one reaching its destination. But how do you practically verify this? Would you agree that it may not be feasible to launch a 1000 ballistic missile salvo and prove that only one slips through? Perhaps, you can ask yourself what a realistic salvo would be (probably far <1000), but it is your job to show the impacts of those requirements on the BMDS architecture. This need not be a detailed assessment and leverage qualitative means as we discuss in the following.

"Modeling is a craft and at times an art....one insight is worth a thousand analyses" (Maier, 2009). As engineers, we love running the highest-fidelity model we can get our hands on, but this may not always be feasible for BMDS. The SoS should be modeled at as a high a level as possible; then the level of abstraction should be reduced progressively as needed. There are qualitatively different problem-solving techniques required by high and low complexity levels. Those lower levels would certainly benefit from purely analytical techniques. But those same techniques would overwhelm the higher levels, which would benefit from heuristics derived from experience.

17.4.2.2 Role of the SoS Engineer

We must address the organizational in addition to the technical issues when we make trades. We presume that you are reading this through the perspective of the SoS engineer. Acknowledge the different roles of the systems engineer and the SoS engineer (Rebovich, 2008). SoS engineering deals with planning, analyzing, organizing, and integrating the capabilities of a mix of existing and new systems into an SoS capability greater than the sum of the capabilities of the constituent parts. So this means that you need to leave systems-level issues to their respective systems engineers! SoS engineers focus on interfaces and cross-system risk, synchronizing the development of individual systems tied to an overarching mission (Office of the Deputy Under Secretary of Defense for Acquisition and Technology, Systems and Software Engineering, 2008). This responsibility would naturally extend to taming the behaviors emerging from SoS integration.

The SoS engineer typically does not have control over the requirements or funding of those constituent systems, which may or may not align with the SoS. Additionally, participation in the SoS may not be seen as a net gain from the individual systems perspective (due to additional obligations and constraints). You therefore may find a premium on influencing rather than directing.

So how do we all work together? Striking a balance in terms of technical management is key. It is not feasible to have systems engineers from each constituent system be involved in all aspects of the SoS. As an SoS engineer, you therefore balance of sustainable participation only after trust is established—and you do this through your own balance of experienced leadership and domain expertise.

Technically, we want the SoS architecture to be based on loose coupling, meaning that individual systems maintain independent operation, and open systems, meaning that they comply with standardized interfaces. We want to maintain flexibility to address changing needs that drive applying technology suited to meet those needs without impacting the SoS. Avoiding impact implies extensibility, meaning you strive to avoid constraints to extending its capabilities as new applications are discovered and developed.

17.5 CONCLUSION

The ability for decision makers to accurately and efficiently evaluate the effectiveness of various BMDS architectures has in the past been limited. The computational expense of executing all the models necessary to emulate the performance of a realistic BMDS has limited architecture-related analysis to simplified representations lacking the accuracy afforded by detailed M&S of each system within the larger SoS context. This has significantly limited the ability to study the high-level effects of changing features of a given architecture.

Surrogate modeling has enabled us to use an integrated M&S environment for real-time, high-level BMDS architecture trade studies. These methods allowed us to rapidly generate simulation-based data, enabling high-level SoS architectural trades to be performed without lengthy delays associated with typical simulations. We presented several example trade studies demonstrating this notion throughout the chapter. This included the characteristics of the BMDS architecture such as asset locations, tracking methods, and finally number of sensor/interceptor platforms that could be quickly related to the overall SoS-level probability of zero leakers, P_{OL}.

Our hope is that you take away several key points. First, Systems Engineering is by its very nature iterative; we formulate the problem as we formulate the solution. Any insights on the solution—even one that may be an SoS—can benefit the problem definition. Second, model fidelity can be both an enabler and a constrainer on the quality of decisions we make. Aiming for the highest fidelity possible across the SoS would severely limit your ability to conduct trade studies; however, aiming too low would limit the trust that decision makers have in your models. Third, we trust you appreciate the value of DoE and surrogate modeling in helping us navigate the interfaces of the BMDS SoS. Finally, you should understand your role as an SoS engineer working on BMDS (or similarly complex SoS problems), balancing the needs of the constituent systems with the overall capability desired from the integrated solution.

ACRONYMS

BMDS   Ballistic Missile Defense System—the architecture that integrates the sensors, shooters, and command and control elements to defeat ballistic missile threats (a good overview is available at URL: http://www.dote.osd.mil/pub/reports/FY2011/).

CCD   Central Composite Design—a form of DoE used to create polynomial surrogate models.

CSOs   closely spaced objects—refers to a situation when two or more target objects fall within the same radar resolution cell, resulting in only one measurement returned by the sensor.

DoE   Design of Experiments—a statistical approach for sampling to extract the most amount of information with minimal experimentation.

GUI   graphical user interface—that allows users to interact with software using images rather than text commands.

IFTU   in-flight target updates—that are sent to an interceptor as it proceeds along its flight path and radar tracking information improved.

LHS   Latin Hypercube Sample—a form of DoE used to capture multimodal effects within the design space.

LVC   live, virtual, or constructive, a taxonomy for classifying M&S—live M&S involves real people operating real systems; virtual M&S involves real people operating simulated systems; constructive M&S involves simulated people operating simulated systems.

M&S   modeling and simulation—a model is a static representation of a system or process, and a simulation is the execution of that model through time or a series of events.

P_{ES} probability of engagement success—the probability that the interceptor is able to engage a target missile (but not necessarily kill).

P_k probability of kill—the probability that an interceptor hits and destroys a target.

P_{RA} probability of raid annihilation—in a situation when multiple targets are launched, this is the probability that all are destroyed.

P_{OL} probability of zero leakers—identical to P_{RA}.

R^2 coefficient of determination—a measure of regression quality on a scale from 0 to 1.

SoS   system of systems—as the topic of this book.

TRL   technology readiness level—a 1 through 9 technology maturity scale developed by NASA (a great easy-to-follow reference may be found at URL: http://www.nasa.gov/topics/aeronautics/features/trl_demystified.html).

ACKNOWLEDGMENTS

The Georgia Tech Research Institute sponsored this work through internal research and development funding. We wish to specifically acknowledge Ryan Leurck, Jason Kramer, and Brian Weaver for technical contributions to the original research supporting this publication. Finally, we wish to thank John F. Sarkesain of the Aerospace Corporation for taking the time to review this chapter and provide valuable feedback.

REFERENCES

Ball, RE. (2003). *The Fundamentals of Aircraft Combat Survivability Analysis and Design*. Reston, VA: AIAA.

Bar-Shalom, Y, and Blair, WD, editors. (2000). *Multitarget-Multisensor Tracking: Applications and Advances III*. Norwood, MA: Artech House.

Bar-Shalom, Y, Li, XR, and Kirubarajan, T. (2001) *Estimation with Applications to Tracking and Navigation*. New York: John Wiley & Sons, Inc.

Ben-Asher, JZ. (2004). Systems Engineering Aspects in Theatre Missile Defense-Design Principles and a Case Study. *Systems Engineering* 7:186–194.

Blair, WD. (1993). Fixed-gain, Two-stage Estimators for Tracking Maneuvering Targets. *IEEE Transactions on Aerospace and Electronic Systems* 29:1004–1014.

Blair, WD, and Brandt-Pearce, M. (1988). Statistical Description of Monopulse Parameters for Tracking Rayleigh Targets. *IEEE Transactions of Aerospace and Electronic Systems* 34:597–611.

Blair, WD, and Miceli, PA, (2008). Performance Prediction of Multisensor Tracking Systems for Maneuvering Targets. *Journal for Advances in Information Fusion* 7:28–45.

Chen, P, Gori, R, and Pozgay, A. (2004). Systems and Capability Relation Management in Defense Systems-of-System Context. *Proceedings of the 9th International Command and Control Research and Technology Symposium*, September 14–16, 2004, Copenhagen, Denmark.

Clements, P, Krut, R, Morris, E, and Wallnau, K. (1996). The Gadfly: An Approach to Architectural-Level System Comprehension. *Proceedings of the 4th IEEE Workshop on Program Comprehension*, March 29–31, 1996, Berlin, Germany.

Cole, R. (2008). SoS architecture. In Jamshidi M, editor. *Systems of Systems Engineering: Principles and Applications*. Chicago, IL: CRC Press, p 37–70.

Ender, TR. (2006). *A top-down, hierarchical, system-of-systems approach to the design of an air defense weapon [dissertation]*. Atlanta, GA: Georgia Institute of Technology.

Ender, TR, Leurck, R, Weaver, B, Miceli, P, Blair, W, West, P, Mavris D. (2010). Systems-of-Systems Analysis of Ballistic Missile Defense Architecture Effectiveness through Surrogate Modeling and Simulation. *IEEE Systems* 4:156–166.

Forrester, A, Sóbester, A, and Keane, A. (2008). *Engineering Design via Surrogate Modelling: A Practical Guide*. West Sussex, UK: John Wiley & Sons, Inc.

Haskins, C, Forsberg, K, Krueger, M, Walden, D, Hamelin, RD. (2010). *Systems Engineering Handbook: A Guide for System Life Cycle Processes and Activities v3.2*. San Diego, CA: INCOSE.

Maier, MW (1999). Architecting Principles of Systems-of-Systems. *Systems Engineering* 2:267–284.

Maier, MW. (2009). *The Art of Systems Architecting*, Third Edition. Chicago, IL: CRC Press.

Myers, RH, and Montgomery, DC. (1995). *Response Surface Methodology: Process and Product Optimization Using Designed Experiments*. New York: John Wiley & Sons, Inc.

Office of the Deputy Under Secretary of Defense for Acquisition and Technology, Systems and Software Engineering. (2008). *Systems Engineering Guide for Systems of Systems, Version 1.0*. Washington, DC: ODUSD(A&T)SSE.

Parnell, GS, Metzger, RE, Merrick, J, and Eilers, R. (2001). Multiobjective Decision Analysis of Theater Missile Defense Architectures. *Systems Engineering* 4:24–34.

Rebovich, G. (2008). Enterprise systems of systems. In Jamshidi M, editor. *Systems of Systems Engineering: Principles and Applications*. Chicago, IL: CRC Press, p 165–190.

Walpole, R, Myers, R, and Myers, S. (1998). *Probability and Statistics for Engineers and Scientists*. Upper Saddle River, NJ: Prentice House.

Weaver, B. (2008). *A methodology for ballistic missile defense systems using nested neural networks [dissertation]*. Atlanta, GA: Georgia Institute of Technology.

Wilkening, DA. (1999). A Simple Model for Calculating Ballistic Missile Defense Effectiveness. *Science & Global Security* 8:183–215.

Chapter 18

Medical Enhancements to Sustain Life during Extreme Trauma Care

L. Drew Pihera[1], Nathan L. Adams[1], Tommer R. Ender[1], and Matthew L. Paden[2]

[1] Georgia Tech Research Institute, Atlanta, GA, USA
[2] Children's Healthcare of Atlanta, Emory University, Atlanta, GA, USA

18.1 INTRODUCTION

Extracorporeal membrane oxygenation (ECMO) is a method of providing temporary heart or lung support to severely ill patients. While lifesaving for the patient, it is a complex and demanding therapy for the caregivers, involving blood traveling outside of the body in tubes and then passing through an artificial lung before returning to the patient. Complication rates remain high and require a dedicated trained person for continuous monitoring in addition to the patient's nurse. ECMO is a therapy, not a device, and there are multiple ways to achieve this goal of adequate heart and lung support. There is no Food and Drug Administration (FDA) device approved to provide ECMO in the United States. Individual centers develop their own ECMO system from parts that are FDA approved for uses in other clinical settings. This leads to wide variation in systems between institutions. Additionally, these part choices are often made with very little engineering input and with little regard to how these individual components will work when assembled together.

In this chapter, we investigate several ECMO circuits, characterize the system of systems (SoS) architecture using Model-Based Systems Engineering (MBSE), and offer

---

Modeling and Simulation Support for System of Systems Engineering Applications, First Edition.
Edited by Larry B. Rainey and Andreas Tolk.
© 2015 John Wiley & Sons, Inc. Published 2015 by John Wiley & Sons, Inc.

recommendations for the medical community in implementing Systems Engineering (SE) best practices. We think that in an SoS that consists of not only the circuit circulating blood through the patient (which itself is an SoS of operationally and managerially independent systems) but also of the family, hospital staff, legal teams, and patient themselves these best practices will lead to the implementation of a safer ECMO circuit and better patient outcomes in the long term. We conducted this study through the collaboration of physicians from across the United States who specialize in ECMO, Georgia Tech researchers, and students of the Professional Master's in Applied Systems Engineering program.

18.1.1 What Is ECMO

Death due to organ failure is incredibly common in the intensive care unit. An overriding goal of bioengineering and medicine is to replace the function of failing organs and improve the mortality rate of organ failure. The history of organ replacement has been well reviewed elsewhere, including the development of hemodialysis for replacement of kidney function (Kurusz and McKellar, 2005). While a substantial breakthrough, patients continued to die of both heart and lung failures. A series of innovations in the cardiac arena led to the development of extracorporeal (outside the body) circuits that could provide temporary heart and lung support.

In general, these therapies involve the surgical placement of large cannulas, which are tubes inserted into the blood vessels, to drain blood out of the body. The blood is pushed through an extracorporeal circuit made up of tubing and an artificial lung via a pump. The artificial lung provides gas exchange (oxygenation and carbon dioxide removal) to the blood, after which it is returned to the patient through an additional cannula. There are two modes of ECMO, venovenous and venoarterial. In venovenous mode, the processed blood is returned into the venous system prior to the pulmonary arteries. It provides solely lung support and no direct cardiac support. In contrast, during venoarterial ECMO, the processed blood is returned into a large artery that provides both respiratory and cardiac supports. Doctors provide anticoagulation, usually via heparin, to ensure adequate function of the extracorporeal circuit and to reduce chance of blood clots.

Initially, these devices were limited to short-term (a few hours) use in the operating room. However, in 1972, JD Hill et al. published a report of a 24-year-old trauma patient with multiple organ failure who was supported during his course for 75 h using a bedside extracorporeal oxygenation circuit and survived (Hill et al., 1972). This success, ushered in the era of ECMO, and use expanded to other patient populations.

While the first ECMO patient was an adult, ECMO has been most used in the neonatal and pediatric populations. Refractory respiratory failure due to persistent pulmonary hypertension of the newborn or meconium aspiration syndrome had mortality rates approaching 80% in the most severe cohorts. ECMO was first used in this population by Dr. Robert Bartlett. Having previously supported a 2-year-old after a congenital heart surgery in 1972, his first respiratory failure patient was a neonate with meconium aspiration syndrome and resultant persistent pulmonary hypertension. The ECMO machine was brought straight from the laboratory and, after consent was obtained, was hooked up to the child for 72 h. She survived and has gone on to live a normal life. The story of this first

Table 18.1 ECLS Cases Reported to the ELSO Registry, July 2012

Survival to discharge or transfer			
Type	Total cases	Number	%
Neonatal respiratory	25,746	19,232	75
Cardiac	4,797	1,912	40
ECPR	784	304	39
Pediatric respiratory	5,457	3,061	56
Cardiac	5,976	2,913	49
ECPR	1,562	630	40
Adult respiratory	3,280	1,808	55
Cardiac	2,312	891	39
ECPR	753	207	27
Total	50,667	30,958	61

neonate, named Esperanza (Spanish for "hope"), parallels themes seen in the development of ECMO technology (Bartlett, 1985).

Over the next decade, multiple single-center publications demonstrated approximately 75% survival in neonates with predicted survival of approximately 20%. After multiple clinical trials, ECMO has gone on to be considered the standard of care for neonatal respiratory failure refractory to traditional therapies (O'Rourke et al., 1989; UK collaborative ECMO trial group, 1996). Similarly, the publication of the Conventional Ventilation or ECMO for Severe Adult Respiratory Failure (CESAR) trial in adults demonstrating a significant improvement of survival without severe disability in patients randomized to an ECMO center as compared to those sent for conventional respiratory management has led to an increase in ECMO use in the adult population (Peek et al., 2009).

In 1989, the Extracorporeal Life Support Organization (ELSO) was founded by nine centers with a goal of following and optimizing outcomes of this new therapy. One of the key functions of ELSO was to maintain a registry of ECMO cases. This registry records patient demographics, disease and ECMO course-related variables, clinical and mechanical complications, and limited outcome data. Growth in ECMO use has been rapid, with 170 centers and nearly 51,000 patients with overall 61% survival reported to the ELSO registry as of July 2012 (Paden et al., 2013). Overall outcomes, stratified by age and indication for Extracorporeal Life Support (ECLS), are shown in Table 18.1.

From a bioengineering perspective, ECLS therapies around the world share an unusual developmental history as devices. For example, there is no ECMO device approved by the FDA in the United States. While described earlier as a relatively standardized therapy, there is great center-to-center variation in how those principles are carried out at the bedside. This is in large part due to the fact that these ECLS therapies were not developed by any singular company, but rather developed by researchers and clinicians interested in the field sharing principles of therapy and utilizing both off-the-shelf technologies that were designed for other uses and specific devices for ECLS from the laboratory. While this is laudable from an innovation standpoint, it makes intercenter comparisons of outcomes and complications more difficult.

18.1.2 An SoS Characterization of ECMO

SE methodologies and considerations can vary considerably if you are engineering a system as opposed to an SoS. Treatment of the interfaces is one such consideration. In a typical system, the interfaces are often far better defined than in an SoS, and complexity of these interfaces must be managed differently. We must also take redundancy into consideration. Often in a system, redundancy is easier to incorporate. In an SoS, it may be difficult to even determine where redundancy can be inserted. Management of the life cycle of a subsystem is again often easier than for a system in an SoS, where the life cycle may belong to another entity with their own agenda for the constituent system. Finally, in the case of an SoS, unintended and unexpected interactions that result from integrating the constituent systems force us to be wary of emergent behavior. We must develop a plan to recognize these behaviors and mitigate them if necessary. With this in mind, we take great care to correctly characterize the system of interest properly.

Characterization by Maier's Criteria. Maier (Maier, 1999) describes an SoS as "an assemblage of components which individually may be regarded as systems" and which possess two additional properties. The first is that the component systems possess "operational independence": if we disassemble the SoS into its component systems, the component systems individually provide useful stand-alone operation. That is, the components fulfill their customer–operator purposes on their own. Secondly, the component systems exhibit "managerial independence": they can and do operate independently. An SoS builder separately acquires and integrates the component systems, though the components maintain a continuing operational existence independent of the SoS. Maier further describes three types of SoS based on managerial control: directed, collaborative, and virtual SoS. SoS of the directed variety are built and managed to fulfill a particular purpose. A collaborative SoS is one in which the constituent systems collaborate voluntarily without being directed to do so. Virtual SoS do not have central management with authority over the SoS, nor is there an agreed-upon purpose for the SoS as a whole. Maier also argues that geographic distribution and complexity of subsystems are not good discriminators for the SoS taxonomy, though they are often associated with the classic SoS concept. Using Maier's criteria, we classify ECMO as an SoS because:

- ECMO is an assemblage of components (systems).
- The components appear operationally independent.
- The component systems are managerially independent.
- The SoS is of the directed variety.

The ECMO SoS is an assemblage of components. Some of the systems included in this assemblage are a pump (Children's Healthcare of Atlanta (CHOA) currently uses a roller-head pump, though the move to a centrifugal pump is currently underway), a heater for keeping the blood at body temperature, an oxygenator for removing carbon dioxide from and infusing oxygen into the blood, pressure sensors for monitoring the pressures in the circuit (both pre- and postoxygenator), and an arterial filter for trapping air and blood clots before the blood is returned to the patient. Many of these can be described as systems unto themselves.

These components appear as well to meet the requirement of operational independence. Although the FDA does not certify ECMO circuits, the FDA does approve each subsystem for another use outside of ECMO and therefore must fulfill a useful purpose outside the confines of the ECMO SoS. For example, the heater used in ECMO therapy will heat any liquid to a predefined temperature as long as the liquid (be it blood, water, or anything else) is delivered to the heater through the required interface. It may be used in cardiopulmonary bypass (CPB) or any other medical treatment calling for heating of a liquid. The FDA approved the oxygenator for 6 h use in the CPB setting, and physicians use it in an off-label manner for ECMO. A similar argument can be made for the rest of the components in a typical ECMO setup.

The constituent systems in the ECMO SoS are also managerially independent. Each system does in fact operate independently of each of the others in the SoS. Each component has its own acquisition plan and may be switched out with other components fulfilling the same purpose as new models emerge or new methods are found to be better, as with the switch in pump types. Again, as there is no FDA-approved circuit design, each center integrates the circuits as they best see fit and as their lawyers will agree to. Practically speaking, each center purchases a collection of tubing and connectors and pieces together their circuit. The centers integrate the components in the same manner for all circuits (unless a new technology is being trialed), but the integration from facility to facility will generally differ (in both the general design of the tubing circuit and the components used). The surgeons dictate the cannula choices and manage what gets ordered, what is put in the patient, and how they are cared for during the case. Respiratory therapy deals with both the purchase and maintenance of the gas flow meters, such as keeping them up to specification. The ECMO department itself has tubing, oxygenators, and related systems. Each component has an operational existence outside of the ECMO therapy given that no approved components are made specifically for ECMO. This can be seen in the case of the membrane oxygenator and pump in their use in CPB for much shorter durations (hours) than a typical ECMO treatment (weeks).

ECMO is of the directed variety as defined by Maier's SoS management types. As opposed to a collaborative or virtual SoS, a group builds and manages a directed SoS for a specific purpose by his definition, and ECMO certainly falls into this category. Medical staff integrate the systems specifically to perform the ECMO therapy, and the systems are cataloged and maintained in the same manner. The components meant for ECMO are typically stored in a separate unit dedicated to the therapy, though they can be used for other purposes in healthcare.

Bjelkemyr describes another definition that warrants comparison (Bjelkemyr et al., 2009). These authors classify an SoS as belonging to one of two types based on level of redundancy, which will be described in the following. In addition, the authors describe properties of an SoS, which include evolutionary behavior, self-organization, heterogeneity, emergent behavior, and network topology. We argue that ECMO currently possesses all of these except for network topology.

Evolutionary behavior is described by Bjelkemyr as a "trial-and-error process of variation and natural selection of systems" (Bjelkemyr et al., 2009). The ECMO SoS evolved due to the observations of the medical staff administering the therapy over time, rather than automatically evolving on its own. Circuits evolved and continue to evolve as new models of components become available and as new research suggests better ways of configuring and integrating circuits. As previously stated, different facilities typically have different circuit configurations. These "genomes" of circuits lead to new understanding of how certain configurations perform. Facilities share this tribal knowledge among themselves leading to further advancements in circuit design. A best-of-breed analysis on these circuit designs could eventually lead to an FDA-approved ECMO circuit. The authors also argue that there is no top-level decision maker for an SoS. In the case of ECMO, there is a top-level decision maker for circuits at a particular facility, but this is certainly not the case for the therapy on a state, country, or global scale. This leads to experimental designs that can eventually replace the current best design in use by a particular facility. In addition, the top-level decision maker does not make the decisions for the constituent systems of the SoS but rather for the local SoS as a whole. The constituent systems follow their own design–realization–operation–recycling phases.

Bjelkemyr et al. describe self-organization as "primarily decomposed into operational and managerial independence." As we previously discussed, the components in the ECMO therapy appear to meet the definitions of these terms. Therefore, the ECMO SoS possesses the property of self-organization.

For the property of heterogeneity, Bjelkemyr et al. define this to mean SoS "consist of a multitude of dissimilar subsystems, structures and agents." This is absolutely true of the ECMO SoS. The SoS consists of several dissimilar subsystems described previously. In addition, there are dissimilar structures including the rack containing the heater, pump and oxygenator, the electronic medical records station, and perhaps a continuous veno-venous hemofiltration (similar to dialysis) unit. Some of the dissimilar agents acting in the SoS are the medical staff, the patient, the legal staff, the family, and a failsafe designed to shut the operation of the therapy down in cases of emergency (such as an air bubble in the circuit). Other dissimilar elements of the SoS are the lights, sounds, and displays of the constituent systems, which may seem similar in display or tone but which may mean very different things. The heterogeneity of an ECMO circuit is further illustrated by the fact that the systems were not developed with integration in mind, as they were not designed to communicate with one another and displays are located in a multitude of locations at the bedside. Medical staff typically enter data (periodic circuit checks, some patient vitals, bladder pressures, etc.) manually, and in the case of an emergency, data that cannot be captured manually is lost.

Finally, ECMO displays emergent behavior. Bjelkemyr et al. define emergent behavior as "the added behaviors that arise due to the interactions between its subsystems and parts, which cannot be directly attributed to an individual subsystem or part." They further describe the existence of strong and weak emergence. Strong emergence, defined as "high-level behaviors are autonomous from the systems and elements on lower levels," does not appear to apply to ECMO in its current state as much as weak emergence, which is defined as that "which can be predicted by experience or extensive modeling and simulation." An example of weak emergence in ECMO would be the behavior of blood clotting when traveling through the circuit. Depending on the length of the tubing raceway and the materials used for various components of the SoS, the blood will clot in different amounts of time. Because of this, heparin is used to prevent clotting as part of the ECMO therapy. You could likely imagine that with sufficient modeling and simulation (M&S), a model for blood clotting might be obtained and that this M&S could lead to improvements in the materials used in the components.

We suspect that strong emergent behavior could occur in the future. If ECMO achieves a state where it is considered much safer than it is currently, physicians may come to rely on ECMO as a platform for treating underlying illnesses for which it is not currently considered. You might argue this increase in demand would constitute strong emergence.

18.1.2.1 Characterization by Redundancy

Bjelkemyr also introduces a concept by which an SoS can be characterized by redundancy in its constituent systems (Bjelkemyr et al., 2009). Redundancy in an SoS is the case in which constituent system cannot complete a task (a failure); the other systems can reconfigure to fulfill the function satisfied by that task. This requires increased coordination between constituent systems. This may not result in the same exact task execution, but should result in function being satisfied. Redundancy is certainly a welcome attribute in any SoS, especially in one concerning healthcare and human lives. However, integrating multiple heterogeneous constituent systems greatly increases the complexity of that SoS, where complexity is understood as the measure and type of relationship between those systems. Redundancy becomes a trade-off between cost of failure and cost of implementing the redundancy itself. Bjelkemyr et al. propose the following SoS classifications:

- Type 1—SoS with multiple constituent systems delivering the same or similar functionality. Contain redundant constituent systems (i.e., high-level subsystems).
- Type 2—SoS that do not have redundant constituent systems delivering the same or similar functionality. A system of different systems interlinked traditional "systems," which, through increased demands, have been forced to "interlink," increasing size and complexity.

A CHOA ECMO circuit falls into the type 2 characterization for SoS (Pihera et al., 2013). There are no automatically redundant components in the circuit itself. If a pump fails, an ECMO specialist must manually crank the pump until a working replacement is found and integrated. Spare parts for circuits may be available as long as all circuits are not in use and there have not already been failures in other in-use circuits. Doctors may fully remove a patient in the case of an oxygenator or pump failure until a replacement can be obtained and integrated. The current circuit design generally has no failure mode other than catastrophic. However, given that the specialist is part of the SoS, one could argue that there indeed does exist redundancy with degraded performance.

18.1.3 Chapter Overview

In this chapter, we introduce modeling efforts undertaken for the improvement of ECMO, a fielded and complex medical system that had been constructed in an ad hoc and unstructured manner. Additionally, we describe why the various models were used and how they fostered communication from the engineers to the medical staff and vice versa, from the medical staff to the engineers. We will continue this chapter with a brief overview of MBSE and an introduction to the modeling techniques we used to characterize this stage of ECMO. Next, we will describe the first phase modeling process in detail, followed by a description of the refinement and extension of the model in the second phase. Finally, we provide recommendations on where future M&S can be applied.

18.2 TAMING THE PROBLEM THROUGH MBSE

The traditional SE discipline focused heavily on document-based methods in the past. Some examples of this approach you would likely encounter include engineering drawings and other diagrams made on paper, requirements typed into a spreadsheet for tracking, and expected performance calculations done by hand or with the aid of specialized software. Recently however, work in the wider SE community, specifically the International Council on Systems Engineering (INCOSE), shows a trend toward MBSE methods (INCOSE-TP-2004-004-02, 2007). Partly as a result of this trend, we adopted this approach for our ECMO SE characterization effort.

MBSE is defined by INCOSE (INCOSE-TP-2004-004-02, 2007) as "the formalized application of modeling to support system requirements, design, analysis, verification and validation activities beginning in the conceptual design phase and continuing throughout development and later life cycle phases." One major area where MBSE differs from the document-based approaches of the past is in the potential of integrated models, which allow for traceability of properties throughout the entire systems model. A system may be comprised of hundreds or more models, but MBSE tools often allow for one to follow the allocation of requirements (from the requirements model, for instance) to components of the system (possibly in their own models) and onto performance models showing the ability to meet requirements. This also allows for a more rigorous verification of the system as a whole as well as of components of the system.

Many methods and frameworks exist for the application of MBSE. Some methods allow for greater interaction between models, and some are more appropriate as stand-alone models. In the case of this ECMO effort, we chose a mixture of existing frameworks and languages including elements of the Department of Defense Architecture Framework (DoDAF), the Systems Modeling Language (SysML), and related modeling methods we describe later in this section.

The DoDAF is a framework used by the U.S. Department of Defense (DoD) to model enterprise architecture descriptions in a common manner across the DoD. The DoDAF breaks the architecture of a system down into a series of eight viewpoints (as of version 2.x) and further subdivides the viewpoints into various representations including timelines, tables, and other graphical formats. In total, there are 52 individual representations divided among the various viewpoints (DoD Architecture Framework, 2010). In general, a DoDAF architectural description will not require the use of all possible representations. In the case of ECMO, we chose only to create the highest-level graphic model of the operational viewpoint, known as the Operational Viewpoint 1 (OV-1). We selected this model to give an overall operational context for the desired long-term outcome of their engineering effort and to describe this end state to the medical staff.

A visual language for modeling systems that has gained prominence in the last decade is SysML. SysML is a seven-diagram subset of the Unified Modeling Language (UML) that also adds two new diagram types and is implemented as a UML profile (Friedenthal et al., 2009). Rather than being geared toward more software-centric applications like UML, SysML is more appropriate for the modeling of SE applications. We performed the bulk of the ECMO modeling in SysML, using of many of the available diagram types and making great use of the ability to represent "crosscutting relationships" (Friedenthal et al., 2009) showing the traceability, satisfaction, and verification of requirements to and by elements of the modeled system. We chose SysML in part because it allows creation and extension of the model over time. We therefore focused the initial phase on the characterization task, and in future phases, we could extend and refine the model while keeping track of the relationships between requirements and the constituent systems of the ECMO SoS.

In addition to the two methods for modeling the architecture and operational viewpoint of ECMO, we included two additional modeling methods. One is the N^2 chart, used for showing components of a system and the feedforward and feedback loops that exist between the interfaces of those components. For this effort, we selected the N^2 to show an example of where bottlenecks could be identified in current practices and how these could be eliminated with future improvements. Finally, we also created a software prototype. The prototype models a possible realization of a single data-visualization display. The prototype shows how the information currently shown in a multitude of bedside displays could be reduced to a single display and how the information could be tailored to the needs of the medical practitioner viewing the data.

18.3 MBSE AND ECMO PHASE 1: UNDERSTANDING THE PROBLEM

With a long-term goal of ongoing collaboration, CHOA and Georgia Tech first partnered to bring a phased approach of improvement to ECMO facilitated by SE in 2011 (Pihera et al., 2012). Students from the Professional Master's in Applied Systems Engineering would spend their capstone project for the program performing SE to further the needs of the staff at CHOA in relation to the improvement of ECMO. The phased approach fits this goal well, as the capstone projects are only 12 weeks' long.

Applying SE to an already fielded system, especially one not originally engineered in a traditional sense, can be an arduous task. To complicate matters, the initial partnership defined only one requirement: to "improve" the ECMO therapy. Given the vagueness of the requirement, we chose to spend the first phase characterizing the ECMO domain and the architecture of the existing CHOA circuit in various models. We also proposed a roadmap of potential future projects that could all lead to various forms of improvement. Through iteration with the CHOA stakeholders, colleagues in future phases can decide on a direction of improvement to undertake at the start of a phase and then perform SE related to that improvement for the remainder of the phase.

We think this approach is useful in many ways. First, you cannot hope to improve something as complex as ECMO without first having a good understanding of what it is and where it currently stands. The first phase documentation provides this. Second, activities such as requirements elicitation cannot be performed without having a clear idea of what is to be achieved. Having a set of possible future projects and the inputs and outputs of each provides this. Third, by leaving the improvement choice up to the SE practitioners and sponsors at the start of each phase, they may choose the appropriate life cycle (waterfall, iterative, evolutionary, etc.) for the task at hand instead of having one, possibly inappropriate, dictated from the beginning.

In regard to the greater SE process, specifically requirements elicitation, the first phase barely scratched the surface. It is, however, an important and necessary first step toward such an end.

18.3.1 Modeling Methodology

In order to simply understand an SoS of this complexity, let alone begin to model it and approach derivation of requirements for improvement, you must first talk to those that administer the therapy and live ECMO on a daily basis as well as study the existing documentation. Therefore, the first and second phases both included activities of this type.

One of the first phase knowledge capture activities consisted of interviews of CHOA ECMO staff. We also attended ECMO training events normally reserved for medical staff. We interviewed seven CHOA personnel including a pediatric cardiologist, a pediatric intensivist, four advanced technology specialists (highly trained ECMO practitioners that can also construct and prime an ECMO circuit with blood), and one ECMO specialist (highly trained ECMO practitioner at the bedside 24/7) as well as a representative from the pump apparatus manufacturer used by CHOA. One missing class of stakeholder that would have been very valuable was that of the biomedical engineer who is responsible for maintenance of the system. However, meeting with the biomedical engineer was not possible in the time available.

Other knowledge capture events for the first phase involved having us attend training events provided for both ECMO specialists and doctors that would have to interact with ECMO in some capacity in their daily work. All ECMO specialists are required to attend annual training involving the performance of ECMO at CHOA. The physician training is a hands-on training session that is broken into multiple sessions. One trains physicians to recognize the failure conditions that arise during the ECMO therapy and requires them to correct the emergency. Another session covers the roles of the ECMO specialists and describes proper circuit procedures. We also recorded these sessions for later study and for sharing with future phases.

18.3.2 High-Level Understanding: Getting on the Same Page

When attempting to merge the understandings of teams from disparate backgrounds (in this case medical staff and engineers), high-level models of the system often help in framing the discussion. In this case, we required a high-level graphic model to describe the need of the medical staff in the improvement of ECMO to a potential future state. For this model, we chose the OV-1 from the DoDAF.

One would use this model to present a high-level view of the system of interest along with its operational elements and major data flow. The OV-1 model used in this study provides a snapshot of ECMO in its future state, from the viewpoint of the CHOA staff. This diagram, shown in Figure 18.1, represents the ECMO SoS from this viewpoint and includes the key components, operators, and environment after future data synthesis and visualization improvements. Notice the inclusion of specific CHOA staff, activities performed by the staff, current data of interest to these stakeholders, and historical data all being displayed. Let's explore these in more detail.

First, there is a timeline embedded in this diagram. Time zero occurs at the location of the ECMO circuit. To the left of this point, we have historical ECMO run data, shown from 24h in the past up to 5min in the past. Future time data exists to the right of the ECMO circuit. As you can see, the historical side of the data depicts graphs of notional

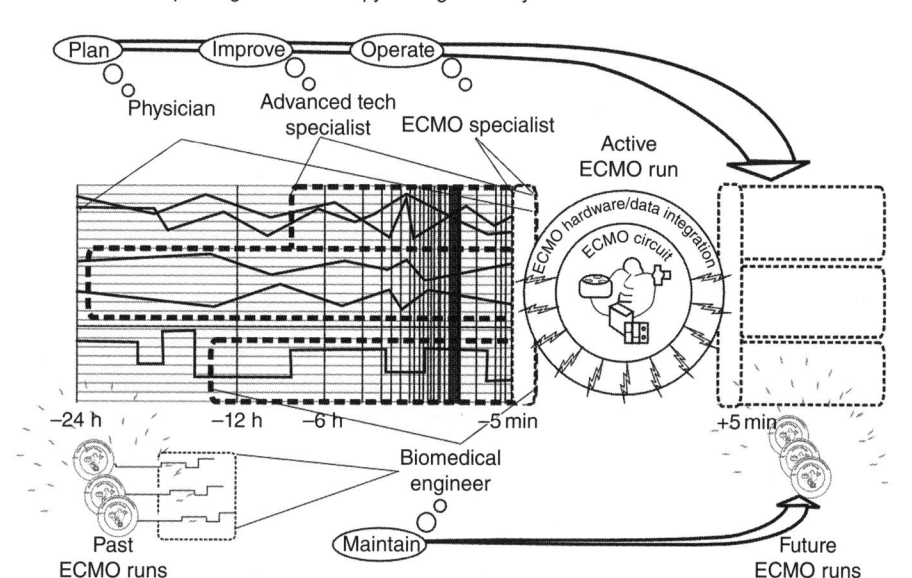

Improving ECMO therapy through data synthesis and visualization

Figure 18.1 Operational view of the proposed ECMO future state.

time series data, while the future side contains no such information. These graphs represent data made available to specific staff members in a potential future system, filtered to show only what is relevant to a particular user's current activity. The lightning bolts emanating from the ECMO circuit and the hardware/data integration that surrounds it represent the exchange of data from the circuit to the proposed displays.

The location of the CHOA staff on the diagram (above and below the timelines) is significant due to their placement with respect to the timeline. Staff members concern themselves primarily with data from the time below their box's left edge up to the current time. To the far left, we place the physician, who generally cares about data from the last full day. Next is the advanced technology specialist who cares about a much smaller time frame, such as the last 8 h or so. Then we have the ECMO specialist, who cares most about what happened in the last hour, up until the current time, so they are placed to the right of the advanced technology specialist. The biomedical engineer, located below the timeline, needs data about the circuit going back slightly further in time than what is required by the ECMO specialist.

We also include activities performed by the staff in the model to describe why the time slice is important to that class of user. A physician needs to plan the therapy for the patient both in the near term and for the next 12–24 h. They require data from the last day in order to make diagnoses for the patient and determine the orders that others will follow for the next time interval until their return. The advanced technology specialist is more concerned with the improvement of circuits and their operational integrity, so much less historical data is required. The ECMO specialist considers the patient's current status and the up-to-the-minute operation of the circuit and therefore requires an even smaller time slice. The biomedical engineer performs the maintenance for ECMO circuits, and his interest is about data that could show failure conditions in the circuit itself, which is depicted as a 4 h window.

To the left of the biomedical engineer is a new data store of historical ECMO runs. This would not contain patient-specific data, but would instead focus on the data related to circuit operation, including which components were integrated together, any equipment failures, and how long equipment had been in operation. The biomedical engineer could use this data during maintenance phases to plan the next integration of a circuit, as well for informing a predictive maintenance schedule for components of the system. We depict this by the arrow leaving the maintenance activity and ending in future ECMO runs.

After describing the intent of this model to the sponsors, we reached consensus that this is a desired potential state to mature ECMO. The CHOA staff agreed that filtering data views to what a particular staff member requires would be a welcome feature. Though developed specifically for and with CHOA, this model is generic enough to apply to any ECMO circuit. If no single circuit is approved for use in the United States, the difference in implementation would be at the circuit and hardware/data interface levels, rather than the entire future state.

18.3.3 Examples of Improvement at the Interfaces

Once the stakeholders agreed at least at a high level to the benefit of the potential future state of ECMO as described in the OV-1, we set out to create graphical models describing where long-term improvements could be attained in the ECMO therapy. We envisioned these models leading to future projects in the ECMO space, and we wanted to give the medical staff a flavor for what these might look like. One model such used is the N^2 diagram, developed in this case to show the current and future states of ECMO in relation to data integration with the patient records system (called EPIC).

The N^2 diagram is a tool that identifies interactions of major functional and physical interfaces for a system of interest. Though typically used for describing software interfaces, it is also a valuable way to show the flow of information throughout ECMO. The diagonal of the diagram consists of the components of a system or SoS. Arrows between components in the diagram represent feedforward (arrows to the right and down) and feedback (arrows to the left and up) loops of data transfer or required physical interfaces. Minimizing the number of feedforward and feedback loops reduces the complexity of the system. These interrelationships may also change as the long-term effort evolves in the case of ECMO. Figures 18.2 and 18.3 present the N^2 diagrams for the current state and proposed future state, respectively.

The small dash outlines (top left and bottom right) show organizational boundaries between CHOA and ELSO, while the larger dash outline (center) shows the area targeted for improvement. As you can see, there exist many feedforward and feedback loops in both figures. Some of these will always be necessary and are performed by interacting people, such as the data transfer from physicians to other staff members and then back to physicians. We intended to show where some of this could be automated, allowing for machines to handle some of the load, and thus freeing medical staff to focus on patients.

Looking at the current state N^2, within the area targeted for improvement, you can see feedforward loops from the medical staff into EPIC, a stand-alone Microsoft Access

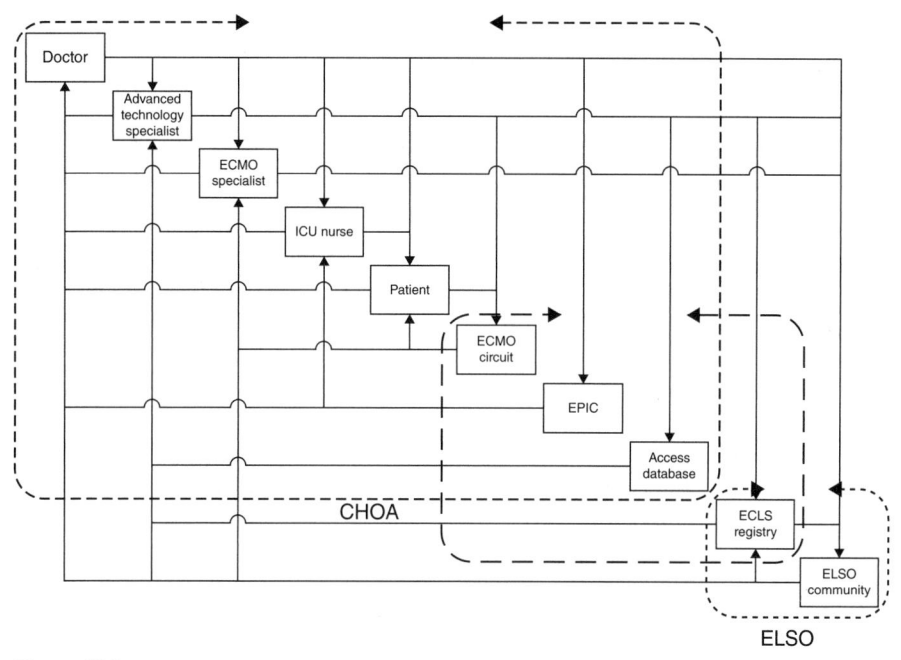

Medical Enhancements to Sustain Life during Extreme Trauma Care   **491**

Figure 18.2 Current ECMO information flow N^2.

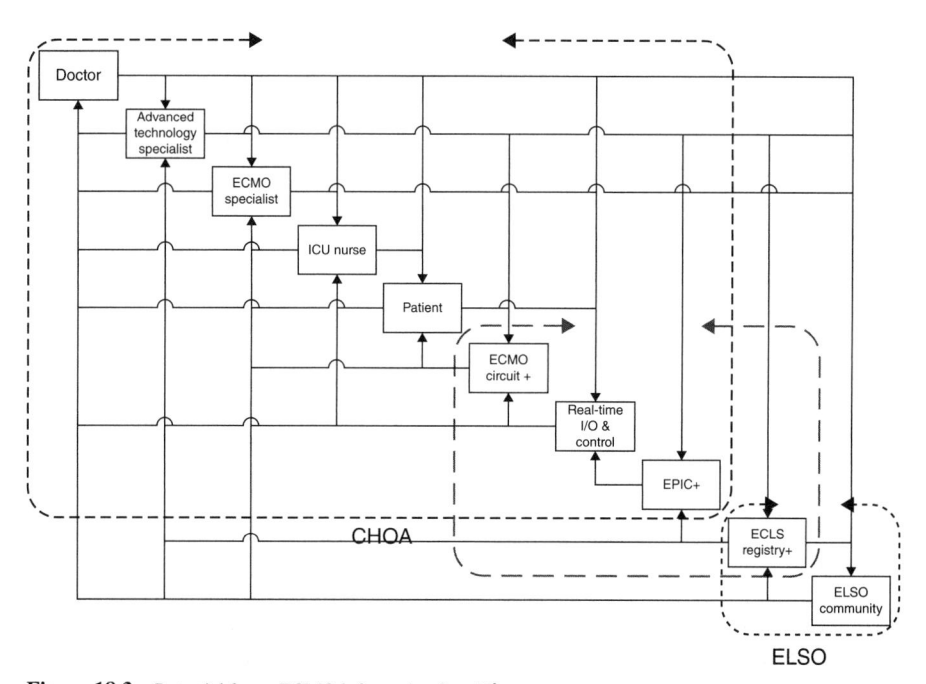

Figure 18.3 Potential future ECMO information flow N^2.

database, and the ECLS registry. Each of these also has feedback loops to members of the CHOA medical staff. Because we wish to minimize the number of these loops, we sought ways to minimize the manual entry of data into these systems. We capture this result in the future state N^2.

When comparing the current state and future state N^2 diagrams side by side, the area of improvement of the future state exhibits more automation in the system than the current design does, specifically with respect to data entry. We represent this by the new feedforward and feedback loops from ECMO circuit+ to real-time I/O and control, EPIC+, and ECLS registry+. We envision these + systems as being augmented with functionality such that they can communicate data between themselves rather than relying on a person to do so manually. As a result, more time could be spent with the patient rather than at the computer.

In this particular case, we used the N^2 model as a tool to quickly identify an area of improvement that is easily understood by the decision makers at CHOA. The N^2 diagram allowed for a realization that this information transfer could be much more timely and accurate using a real-time I/O and control component for data entry in both CHOA data stores and the ECLS registry.

18.3.4 Stakeholder Identification with Use Cases

In any engineering effort, proper identification of the stakeholders of the system is paramount. Often, the initial stakeholder identification is incomplete and requires iteration with known stakeholders to complete the picture. One model that aids in this process is the SysML use-case diagram. A major argument for using this format in the case of ECMO is the ease in which others can understand the model even with no formal training in SysML. The simplicity of the diagram clearly conveys the stakeholders as well as why they are included.

In a use-case diagram, stick figures or blocks with labels identifying them represent actors in the system, and ovals with the title of an action embedded within them represent the use cases actors perform. Rectangular areas group use cases that belong to the same domain. Lines connect actors to their use cases and signify the actor performing the use case. Lines also connect use cases that include or extend other use cases. This type of diagram also captures classification of actors and use cases. Figure 18.4 is the final use-case diagram from the first phase and shows examples of all of these concepts.

As stated, initial version of use-case diagrams is often incomplete and ours was no exception. The initial version was too simplistic and contained little more than the project team, the lead physician, the ECMO specialist, and the patient. Through iteration with the sponsoring CHOA representatives and other members of the ECMO staff, we greatly improved the model and our domain knowledge and incorporated this into the SE process. Through this iteration, we learned of other stakeholders that were not initially part of the model, including surgeons, equipment manufacturers, biomedical engineers, and ELSO.

It is clear from the diagram that there are three concurrent activities: the continued patient treatment performed by the CHOA staff, the system improvement we performed, and the ELSO registry maintenance. The diagram also shows some other interesting information.

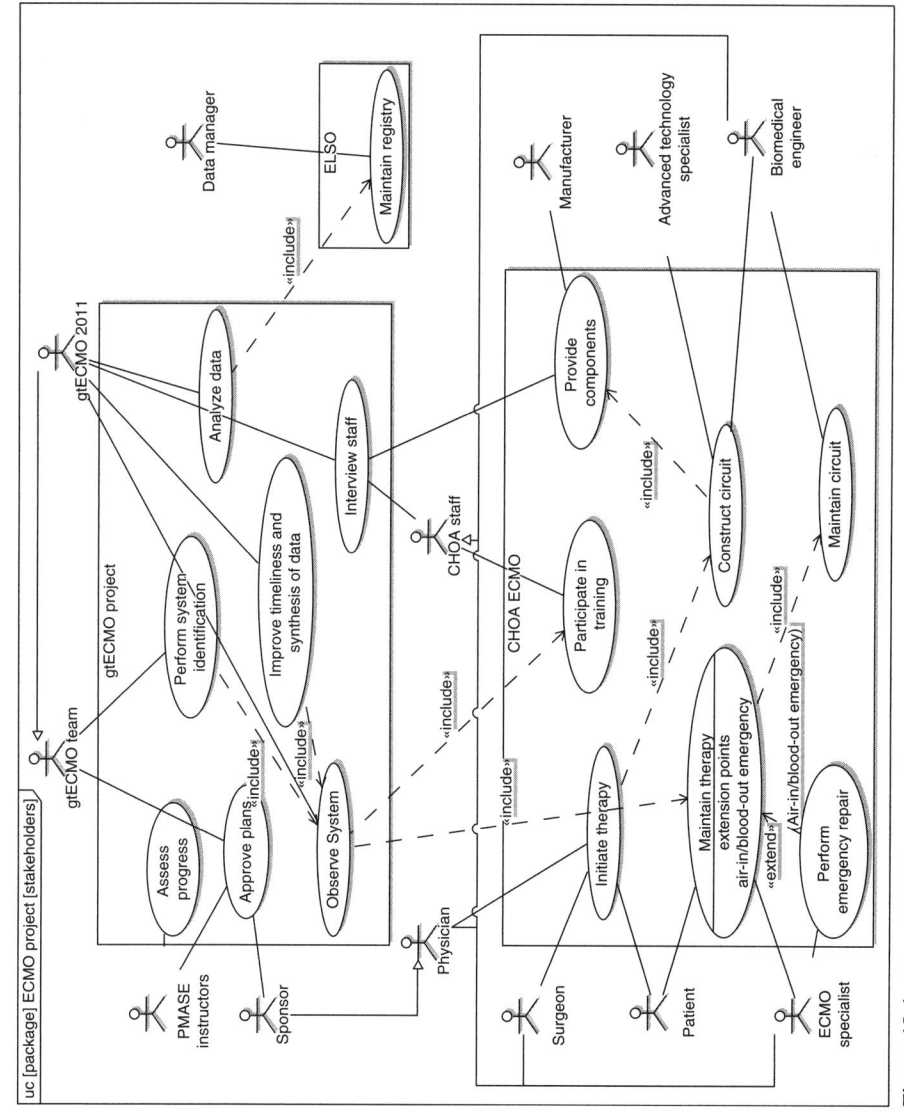

Figure 18.4 Phase 1 ECMO use case diagram.

First is an inheritance hierarchy of the actors, also known as a classification hierarchy (Friedenthal et al., 2009). The gtECMO2011 actor is a gtECMO Team, and the hollow arrow line drawn from the former to the latter (called generalization) depicts this. This means that whatever use cases a gtECMO Team actor participates in, the gtECMO2011 actor will also participate in. The gtECMO2011 actor may participate in some use cases that the more general team will not. For example, the gtECMO Team participates in the "Perform System Identification" use case, so the gtECMO2011 actor would as well. The gtECMO2011 actor performs the "Interview Staff" use case, but not all gtECMO Teams would necessarily participate. A similar generalization can be seen for physicians, surgeons, ECMO specialists, advanced technology specialists, and biomedical engineers, which are all specializations of the CHOA staff actor.

You can see in the model that some use cases include or extend other use cases. For example, "Perform System Identification" includes "Observe System." This means that in order to do the former, you must also do the latter. "Perform Emergency Repair" shows a different relationship in that it extends "Maintain Therapy." This means that during the normal case of maintaining therapy, some exceptional behavior (in this case "Air-In/Blood-Out Emergency") can trigger the performance of the extension use case. Though not shown in the diagram, use cases can also specialize other use cases.

Though relatively simple conceptually, the use-case diagram conveys a wealth of information. The use cases themselves may contain other use cases, allowing you to generate very complicated interactions in fairly simple linked diagrams. This model very effectively aided the communication between the medical staff and the engineers.

18.3.5 Modeling Structure, Architecture, and Interfaces

Given the need to characterize the architecture of ECMO as it exists at CHOA, we chose to capture the structure in a SysML Block Definition Diagram (BDD). A BDD captures structural elements of a system using blocks (drawn most simply as labeled rectangles) as well as the composition and classification of the elements (Friedenthal et al., 2009). Part and value properties of each block capture the composition of elements in the systems. Similar to the use-case diagram, this diagram captures the classification hierarchy using generalization. Figure 18.5 shows the first iteration of the ECMO structure.

The blocks of the BDD are the easiest part to understand. Each block represents a structural component of the ECMO SoS. The model contains a block for the ECMO circuit as you would expect. We define the circuit in this case as any part of ECMO that blood actually passes through or that is permanently attached to anything that blood passes through such as the pressure transducers. In addition, blocks exist for parts of the circuit such as the tubing, the oxygenator, and the arterial filter. You can tell that they are parts of the circuit by the black diamond connector on the circuit block end of the line drawn to them, and they are referred to as part properties in SysML.

Other blocks connect to the circuit block using white diamond connectors; SysML calls these reference properties. These elements interact with the circuit, but are not actually a part of it. Take, for example, the drug pump. This component attaches to the circuit for infusion of heparin or other required drugs. While it is attached to and used in conjunction with the circuit, it is not physically part of the circuit and can be moved from bedside to bedside as needed.

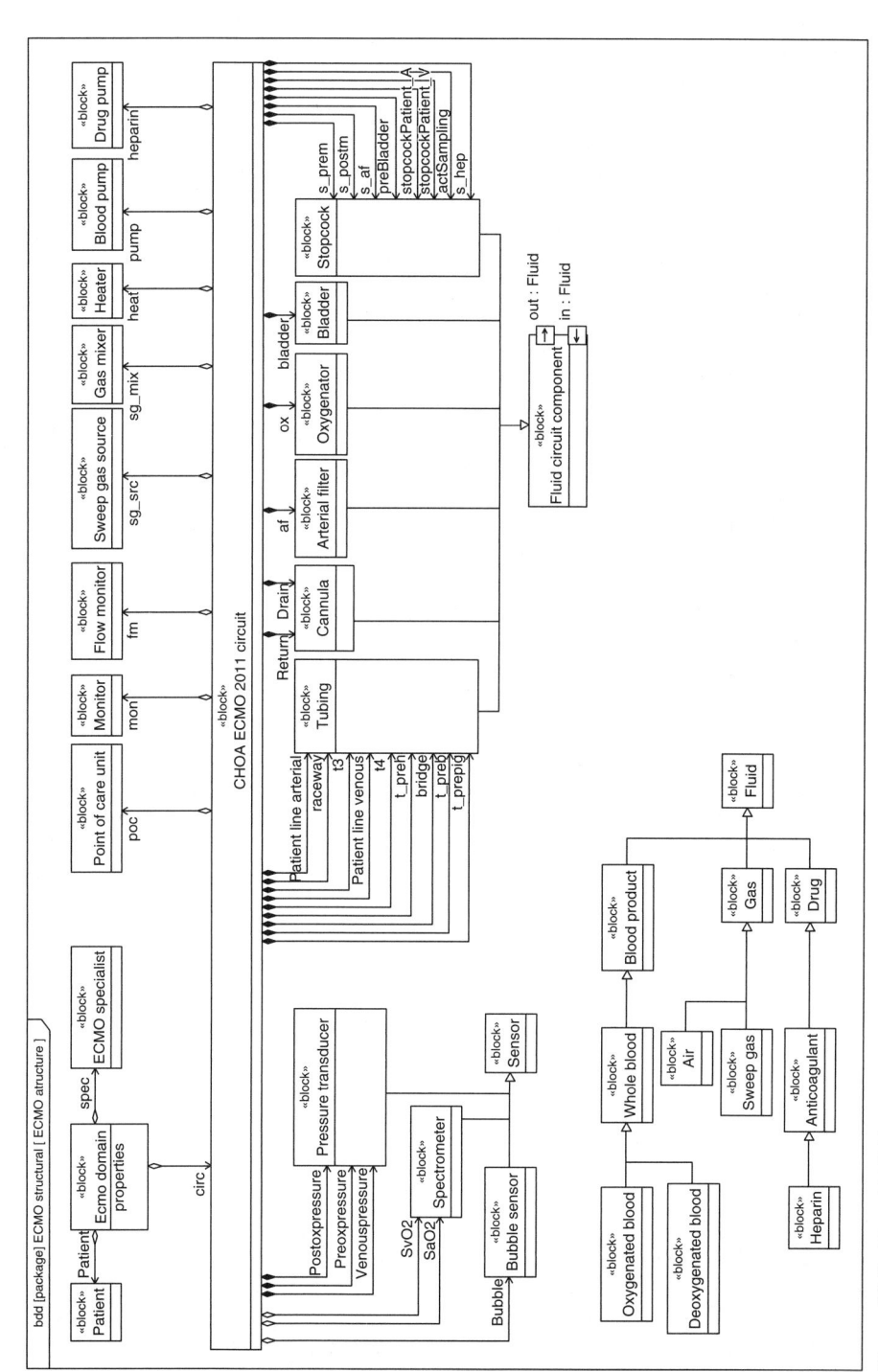

Figure 18.5 Phase 1 ECMO structure BDD.

Again, we have examples of generalization classification hierarchies shown in the part and reference properties. You can see this in the case of the pressure transducer, spectrometer, and bubble sensors, which all specialize the sensor block. We include another classification hierarchy in the lower left of the model. This shows the hierarchy of the fluid types in the ECMO SoS. You can see, for example, that oxygenated blood specializes whole blood, which specializes blood product, which specializes fluid.

We further decomposed the ECMO domain of the architecture BDD into a SysML Internal Block Diagram (IBD). You would use an IBD to capture the contents of a block at the lowest level, showing how parts interconnect and what the interfaces exist between the parts of the block (Friedenthal et al., 2009). Figure 18.6 shows the first iteration of the ECMO domain IBD.

The IBD in this case shows the interconnections and interfaces of the three reference properties of the ECMO domain: the patient, the ECMO circuit, and the ECMO specialist. The patient connects (physically in this case) to the circuit at the drain and return cannula interfaces. The interfaces for the ECMO specialist to the circuit exist both at the points where they exert control on the circuit and the points where they consume sensory information from the circuit.

The circuit itself encompasses the majority of the content of the diagram. Looking at the first two-thirds of the circuit from left to right, you will notice the start of the flow of blood from the drain cannula (top left) through the circuit denoted by the labels of "Deoxygenated Blood." The deoxygenated blood continues through the circuit components until it reaches the oxygenator, located approximately halfway down the vertical at about the two-thirds mark in the horizontal of the circuit. At this point, it becomes oxygenated blood and continues through the rest of the circuit components until it exits the return cannula.

The model also captures the data, sensory information, control, and other flows in the circuit. These flows are important to capture for complete documentation of the system and for informing future efforts. Data, for instance, flows from measuring components (usually sensors) to monitors. At that point, monitors convert the data into sensory information for display to a user. This information is useful in determining what kind of information would be important for improved future displays. We show the flow of control from the specialist to components of the system in order to understand where human/machine interfaces exist, as these could lead to areas of study in human/system integration.

Much like the use-case model, these formal SysML structural models represented our initial and imprecise understanding of the ECMO SoS. Through iteration with the stakeholders, we refined these models into a state where the stakeholders agreed on the content and fidelity of the model. While initially foreign to the medical staff, they easily understood the final SysML characterization of the domain and systems models once we explained some basic concepts of SysML. The CHOA representatives received the end result well since a formal visual representation of the ECMO systems architecture was not something they previously had.

18.3.6 Modeling User Interfaces with Prototypes

Another form of model that aids in expressing technical concepts to others is a prototype. In this case, we wanted to convey the idea of a single, fused, dashboard-like display for all of the various measurements being taken in the ECMO SoS. A multitude of displays

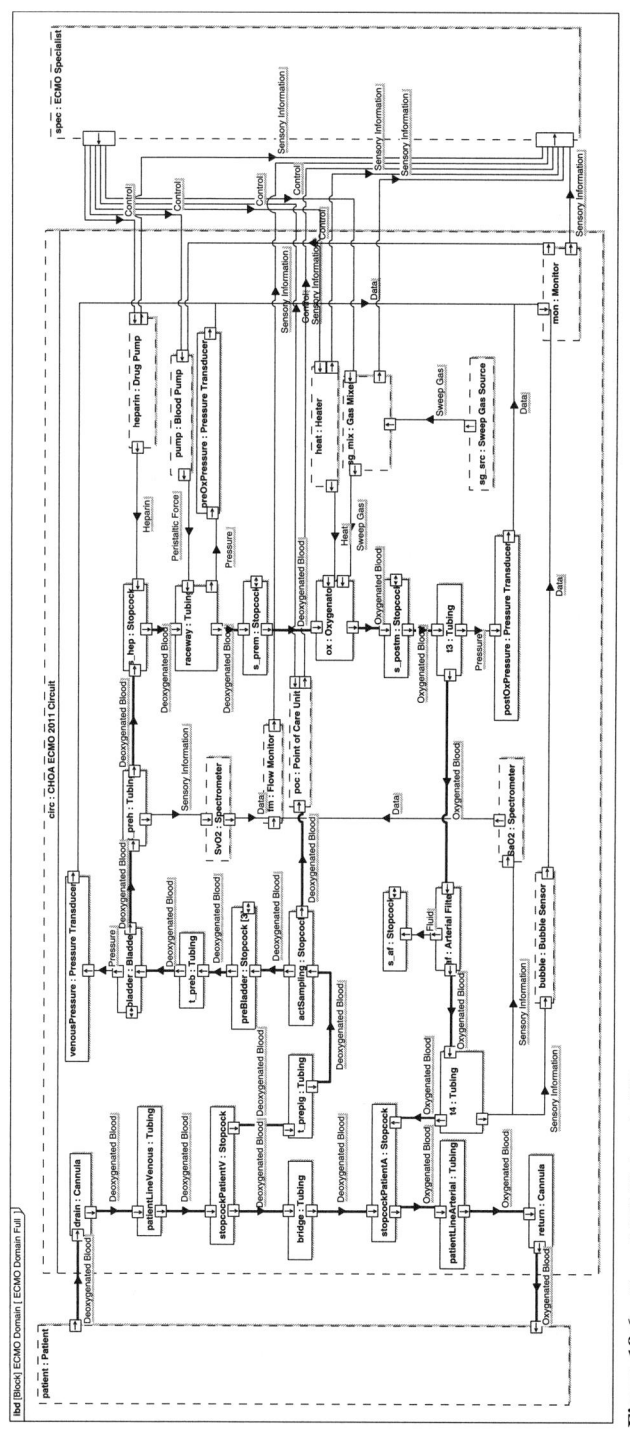

Figure 18.6 Phase 1 ECMO domain IBD.

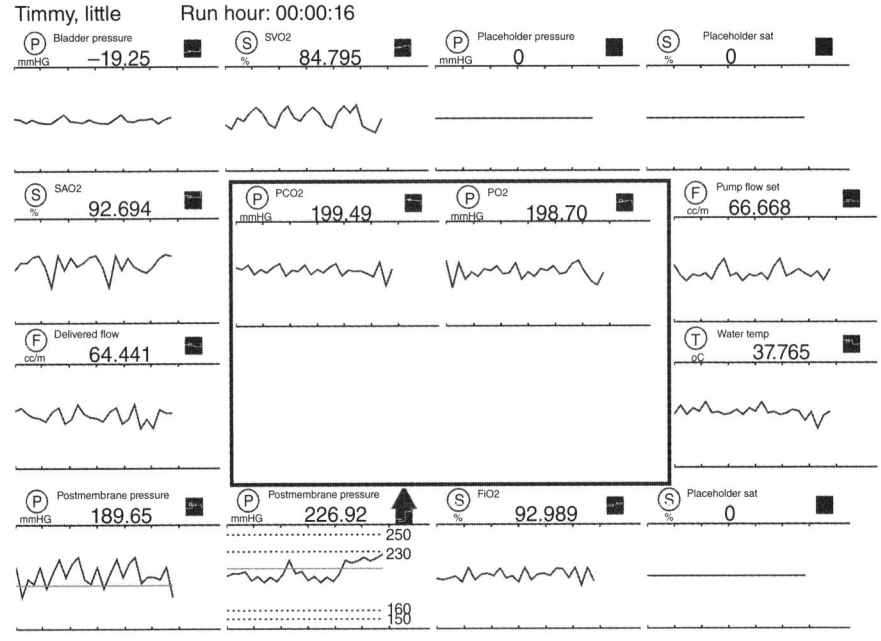

498 Theoretical and Methodological Considerations with Applications and Lessons Learned

Figure 18.7 Proposed ECMO dashboard data-visualization prototype.

currently show these values in various locations at the bedside and show the same information regardless of the viewer. Recall in the OV-1 we wanted to show the information of interest to a particular class of medical staff depending on who was viewing. A configurable display like this would facilitate exactly that.

The short timeline for developing the prototype did not allow for an iterative approach with the stakeholders like we followed for the other models. We did iterate on the prototype internally, but the sponsor's schedule limited the time they had available to review the prototype. They received the resultant model well, but the prototype requires further cooperation and iteration to get it to a state worthy of considering for implementation. The current state of the prototype is shown in Figure 18.7.

In this prototype, we portray each individual measurand as a "widget" in the dashboard. The data displayed by default represents one hour of runtime. We organize the widgets both inside and outside a central box that represents the patient. The widgets outside of the box represent measurands of the circuit itself, and widgets inside the box are specific to the patient vitals. The layout of the circuit widgets (clockwise starting from the top right) follows the order of events that occur in the ECMO circuit.

Each widget displays the name of the measurand and has a symbol representing the type of data being measured, such as P for pressure. The widget display shows the maximum, minimum, and average values for the last time increment. In addition, it shows the minimum and maximum set points (min and max values to keep a measurand between) and a graph of the measurand over the last time increment. Finally, a small trend symbol shows the general trend of the measurand over a much smaller time increment (5 min by default). The trend symbols consist of a red arrow (upward or downward trend currently outside desired set points), blue arrow (upward or downward trend currently within desired set points), or a black square (no discernable change). This allows the minute-to-minute operators to see the immediate trends, while a physician can have a more generalized overview of the last time increment.

As mentioned previously, in its current state, the prototype serves primarily as a point of departure for further discussion with the target users and will help determine the requirements for a future design iteration. This is a Web-based application with a customizable scenario generator. The graphs, numeric values, and trend symbols all update during a run and demonstrate the ability to have a single display for all relevant measurands. We envision the dashboard as being customizable, so ideally a specialist could remove widgets deemed unnecessary for their viewpoint, while a physician could select a completely different set for their needs.

18.4 MBSE AND ECMO PHASE 2: REFACTORING THE MODELS TO BETTER SUPPORT TRADE STUDIES

With the start of a second research phase in May 2012, we identified a need to provide a framework for performing trade-off analysis of various technology insertion projects at CHOA (Adams and Pihera, 2012). These included the introduction of a new centrifugal blood pump to create a second CHOA ECMO circuit design and the need to replace the heating component that was scheduled for end of life within the next calendar year. We also recognized the need to capture engineering artifacts depicting the full life cycle of ECMO therapy operations including acquisition costs, maintenance, training, and knowledge sharing with the wider ELSO community. We were particularly concerned with best practices that could lead to improved outcomes for ECMO patients. Because the 3-person Georgia Tech-based SE team was geographically spread between Atlanta, GA; Tucson, AZ; and Long Beach, CA, we also identified an opportunity to include additional ECLS medical centers in Arizona and California in the project. The Kapi'olani Medical Center for Women and Children in Honolulu, Hawaii, also participated in large part due to the opportunity provided by the business travel schedule of one of the team members.

We leveraged the Defense Acquisition University (DAU) SE process (Department of Defense, 2001) to update and extend the ECMO MBSE framework. We:

1. Used the previous cohort's concept, need statement, use cases, and ECMO domain capture as the input to the process and updated and refactored existing SysML diagrams as needed.
2. Developed a minimal set of requirements at the SoS level and decomposed those into system (ECMO circuit) and then component-level requirements.
3. Performed a functional analysis of ECMO circuit systems behaviors. We set a primary goal to establish at least two levels of abstraction in the model: one level of abstraction captured the functions and behaviors of an ideal ECMO circuit, and another level of abstraction captured the functions and behaviors of existing ECMO implementations.
4. Synthesized structural models of the ECMO circuit and its encapsulating domain. Again, we used a layered approach to capture both an idealized or abstract ECMO circuit and more concrete models of existing ECMO circuit designs.

5. Leveraged the capability of SysML to capture crosscutting relationships such as requirements traceability, requirements allocation to functions, and functional allocations to structural components.
6. Generated a comprehensive set of SysML models as the primary engineering artifact and output of the process.

We iteratively and incrementally developed requirements, performed functional analysis, and synthesized the structural model, often performing these phases in parallel. To accomplish the system analysis and control portion of the DAU process, we elicited regular feedback from stakeholders in the ECLS medical community, observed actual ECMO therapies and systems whenever possible, performed internal team reviews of work products, and consulted with the primary stakeholder and other Georgia Tech researchers.

Stakeholders in the medical community gave us generally good feedback on the process and its iterative nature. As we systems engineers incrementally expanded the breadth of our knowledge of the medical processes, the medical staff incrementally expanded their knowledge of the engineering artifacts and practices.

18.4.1 Requirements Development

Prior to our work, nobody had ever attempted to capture formal requirements for an ECMO circuit. We felt that capturing a set of requirements and getting buy-in from the larger ELSO community was key to eventually designing and deploying a best-of-breed ECMO circuit. While there is widespread agreement that such a circuit design is needed, there is disagreement as to what that circuit might look like. We contend that a well-vetted set of requirements (and the associated verification and validation tests) acts as an objective measure of what constitutes an ECMO circuit.

We set as our goal to capture a minimum set of requirements that would specify an ideal ECMO circuit. We first developed requirements at an SoS level using use cases, the ECMO domain model, and two reference texts (Van Meurs et al., 2005; ECMO Policy and Procedures Manual, 2011) as the primary inputs. We scoped SoS requirements to include the extension of patient life, systems certification, standardization of patient care, standardization of ECMO circuit use, safety, and costs. We then further decomposed these high-level requirements. For example, we decomposed the requirement to extend patient life into a requirement to lengthen the physician's available time to perform a diagnosis and a requirement to lengthen the available time for a patient's heart and lungs to heal.

We decomposed the certification requirement into separate certification requirements for ECMO specialists, physicians, and hospitals. We also identified an annual recertification requirement. The requirement for standardized patient care decomposed into requirements for standard ECMO patient selection criteria, standard methods for quantifying an ECMO patient's current status, standard methods for discontinuing ECMO therapy, standardized emergency procedures, and standards for patient treatment. The requirements for standardization of ECMO circuit use, safety, and costs decomposed in a similar manner.

We developed systems-level requirements for the ECMO circuit and grouped them into two basic categories: normal operation and abnormal operation. Normal operations included requirements to treat blood, circulate blood, and provide feedback to the attending physician and the ECMO specialist. We then decomposed these requirements to an appropriate level of detail. For example, we decomposed the requirement to treat blood into requirements to administer injections, control blood temperature, provide an interface for blood removal, and exchange carbon dioxide for oxygen in the blood. Requirements for abnormal operations included the requirement to provide alerts for abnormal conditions, for the detection and removal of blood clot formations (thrombosis), for detecting and purging gas bubbles during an "air-in occurrence," for detecting and resolving "blood-out occurrences," and for providing backup sources for power, required gases, and work lights.

Before attempting to develop component-level requirements, the team acknowledged that requirements at the component level typically assume that an architectural description has been developed for the system of interest. In the case of ECMO therapy, no such architectural description currently exists. In fact, a primary goal of the ECMO MBSE framework was to facilitate the necessary trade-off studies and analysis for creating a robust ECMO architecture. However, a secondary goal of the project was to provide the project's stakeholders in the medical community with working examples of the engineering processes. Therefore, we decided to generate an example set of component-level requirements against the abstract structure models described in the following. For instance, these example component requirements included a requirement for a circuit to include components such as a gas exchanger, tubing, gas bubble sensor, and blood-warming unit. We hope that future trade studies, systems M&S, industry standards, and rigorous analysis of best practices and lessons learned will be used to capture an exemplar ECMO architectural description. We will then use such an architectural description to drive the development of a robust set of component-level requirements.

We used the LaPlue method (LaPlue et al., 1995) throughout the requirements development process to ensure that behavioral requirements were developed in a rigorous manner. Under this methodology, engineers organize behavioral requirements by external nodes (e.g., systems-level components) and then by output. We loosely followed this organization in the development of the ECMO requirements. The methodology further prescribes a template for how each requirement is written. We found the requirements template to be the most useful part of the LaPlue methodology, and we used the template extensively to ensure that behavioral requirements were unambiguous, verifiable, consistent, necessary, and attainable. For example, we wrote the "Lengthen Physician Diagnosis Time" requirement as follows:

> "The ECMO system SHALL produce a patient with stable heart and lung functions
>
> for use by the diagnosing physician,
>
> if the patient has been identified as a suitable candidate for ECMO treatment,
>
> and if the patient suffers from rapid onset of cardiopulmonary failure before a definitive diagnosis is made,
>
> using the patient's blood and the specified blood pressure value and the specified oxygen saturation level and the specified maximum CO2 saturation level,
>
> where the resulting blood pressure is within 2% of the specified value,
>
> and where the oxygen saturation is within 2% of the specified value,
>
> and where the CO2 saturation is less than the specified maximum value,

and where the blood temperature is 37 +/−1 degC,

and where the patient's overall condition remains stable for a maximum of 30 days."

Three different ECMO experts from two of the partner medical organizations reviewed and validated the requirements. These reviews focused on ensuring that the requirements properly satisfied the stakeholders' needs and completely described the requirements of the entire ECMO domain. A separate team review of the requirements focused on whether the requirements provided enough information to generate an acceptable design solution and whether they were free of implementation decisions. In the end, the medical staff directly involved in the project expressed satisfaction that we had captured a proper set of requirements for the ECMO SoS. We will see if the larger ELSO community agrees with them!

18.4.2 Modeling Behavior

The team decided to employ a layered approach to capturing systems behaviors. One layer captures behaviors with the minimal amount of assumptions for the physical implementation of an ECMO circuit. This allows the framework to be used as a tool for future trade-off analysis without unduly biasing the results toward a particular type of circuit. For example, two major types of blood pumps exist: one type uses a roller head to squeeze the plastic tubing of the circuit thereby inducing a pressure, while another type uses a spinning fan blade to create a centrifugal force on the blood. Both pumps satisfy an abstract function (inducing a pressure and flow on the blood), each with its own set of trade-offs. The first layer of behavioral models does not assume a particular type of blood pump; instead, it describes the abstract functionality that a blood plump provides. The second layer of behavioral models captures specific actions and procedures that are employed at various partnering medical centers. These behavioral models apply to specific physical implementations of ECMO circuits.

We started the process of capturing behavioral models of the ECMO domain by creating two state machine diagrams. As the name implies, a state machine depicts the various states of being that an entity can achieve and also depicts events that can trigger a transition from one state to another. The entity can also perform behaviors while in each state. The first state machine we created modeled the various states of a patient from the time that they are a candidate for ECMO therapy to the time that they have completed treatment and are removed from heart and lung bypass. The second state machine modeled the state of a generic or abstract ECMO circuit from the time that it is constructed to the time that the patient is removed, leaving the circuit in a contaminated state. Both state machines captured major off-normal states such as a patient being rejected for ECMO therapy and a circuit that has catastrophically failed.

The ECMO circuit state machine, shown in Figure 18.8, became the basis for further behavior analysis. Once a patient is connected to an ECMO circuit, the circuit reaches the Connected and Circulating state and performs the Circulate Patient Blood behavior.

An integrated SysML model provides systems architects and designers with the opportunity to create rich descriptions of the systems architecture by allowing individual elements from one model diagram to be further described in depth in another model diagram. We leveraged this capability in SysML to further define the Circulate Patient Blood behavior. The Circulate Patient Blood activity diagram shown in Figure 18.9 depicts all

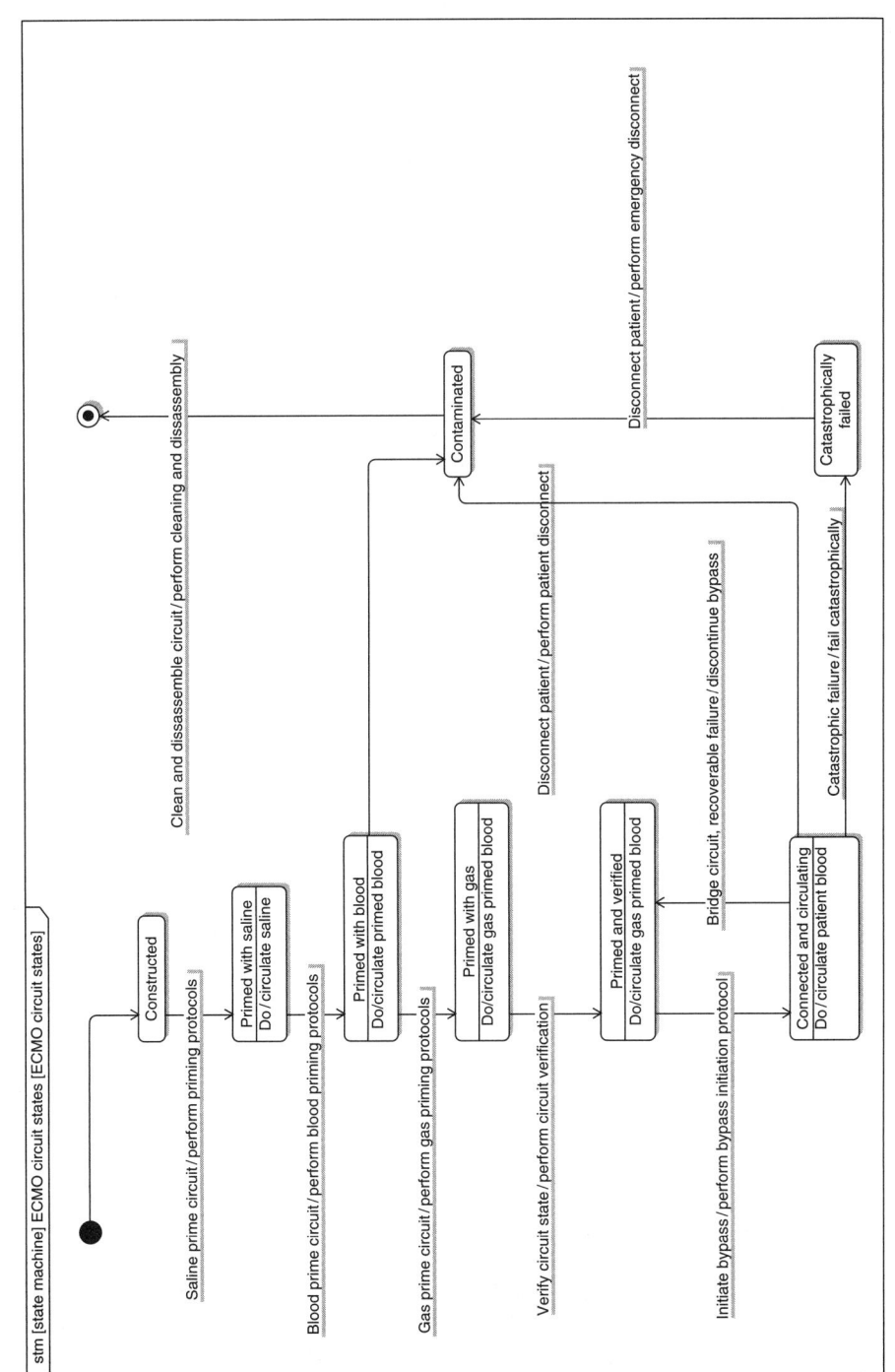

Figure 18.8 ECMO circuit state machine diagram.

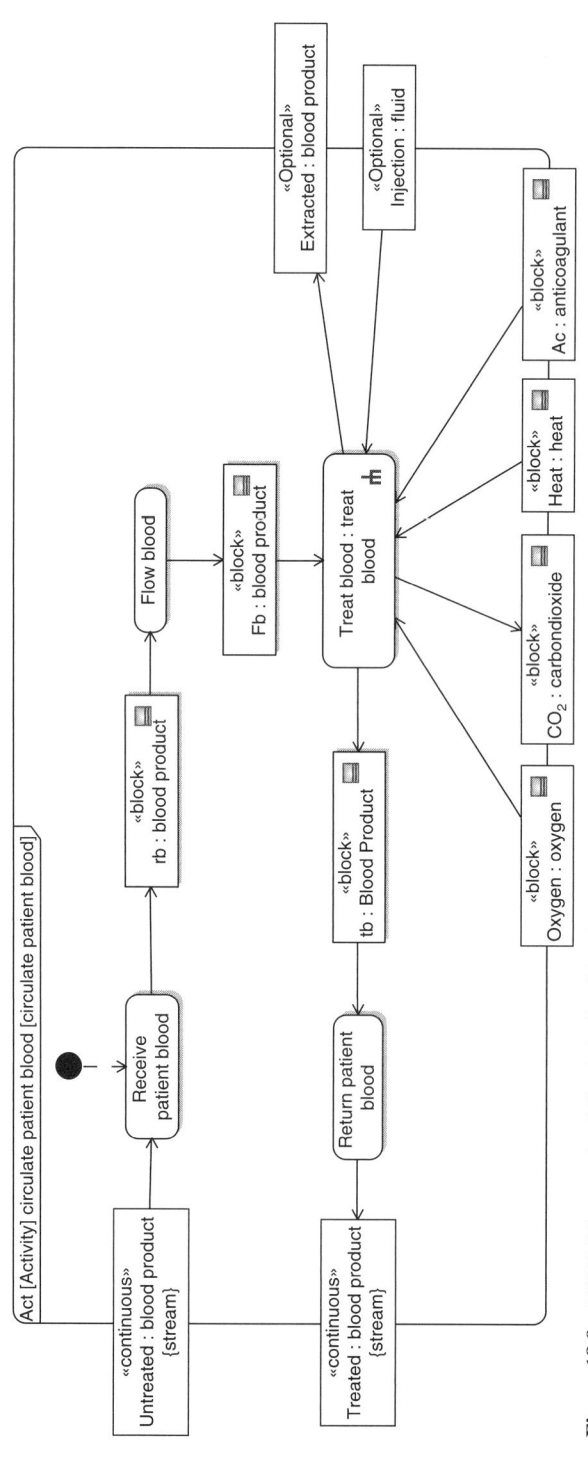

Figure 18.9 ECMO "Connected and Circulating" activity diagram.

of the behaviors that a generic ECMO circuit performs while in the Connected and Circulating state.

Activity diagrams depict behaviors that are flow based. In the Circulate Patient Blood activity diagram, a continuous stream of blood enters the activity, and the activity performs the Receive Patient Blood action. The Blood Product then enters the Flow Blood action (the heart function) and is in turn passed to the Treat Blood action (the lung function). At this point, five activity parameters come into play. Medical staff may extract the patient's Blood Product from the circuit for testing and analysis, or they may inject Fluid (drugs, nutrients, or other fluids) into the circuit. Anticoagulant is a critical part of any ECMO therapy and slows the blood's natural tendency to clot when it comes into contact with foreign surfaces such as the inside of an ECMO circuit. The Heat activity parameter ensures that blood is returned to the patient at the appropriate temperature. To complete the Treat Blood action, the Carbon Dioxide parameter is removed and Oxygen added. The activity then performs the Return Patient Blood action on treated Blood Product and treated Blood Product streams continuously out of the activity.

The Treat Blood action was further decomposed into the activity diagram shown in Figure 18.10. The activity details actions performed on Blood Product such as Prevent Coagulation, Inject Fluids, Exchange CO^2 for Oxygen, Warm Blood, and Extract Blood. This functional decomposition process was iterated several times, and all of the captured behavioral requirements were linked to a specific action in the model.

We generated sequence diagrams to capture detailed procedures and protocols at the various medical centers. For example, we captured the circuit priming protocols used at Rady Children's Hospital and the University of Arizona Medical Center as separate sequence diagrams. Other sequence diagrams were generated for circuit inspections, pump start-up, and patient disconnect procedures. We hope that by capturing protocols and procedures from various medical centers in a common language, SysML, we can perform future studies to compare these behaviors and discover best practices for safety and efficiency.

The behavioral models we generated impressed the medical staff that reviewed them. ECMO centers capture policies, procedures, and instructions almost exclusively in written form, and the SysML model diagrams we created were some of the first attempts to show ECMO therapy behaviors and interactions in graphical form. Medical staff were excited by the ability to show the complexity of certain interactions in a graphical form and to be able to compare procedures across multiple ELSO centers in standard graphical way. We believe that this sort of behavioral modeling could be expanded and revised to support analysis of various ECMO policies and procedures resulting in better operational efficiencies and, perhaps, safety.

18.4.3 A Return to Modeling Structure

Systems architects make a fundamental decision when architecting a system and performing a functional allocation: they either allocate a physical element to satisfy a desired function or abstract the element and delay the physical allocation until more information is known. For the ECMO framework, we wanted to provide a completely abstract model of the system and to capture the physical allocations that already existed in the numerous circuit designs currently in use. To accomplish this, we implemented two layers of abstraction for the structural models. The first layer consisted of generic, abstract blocks,

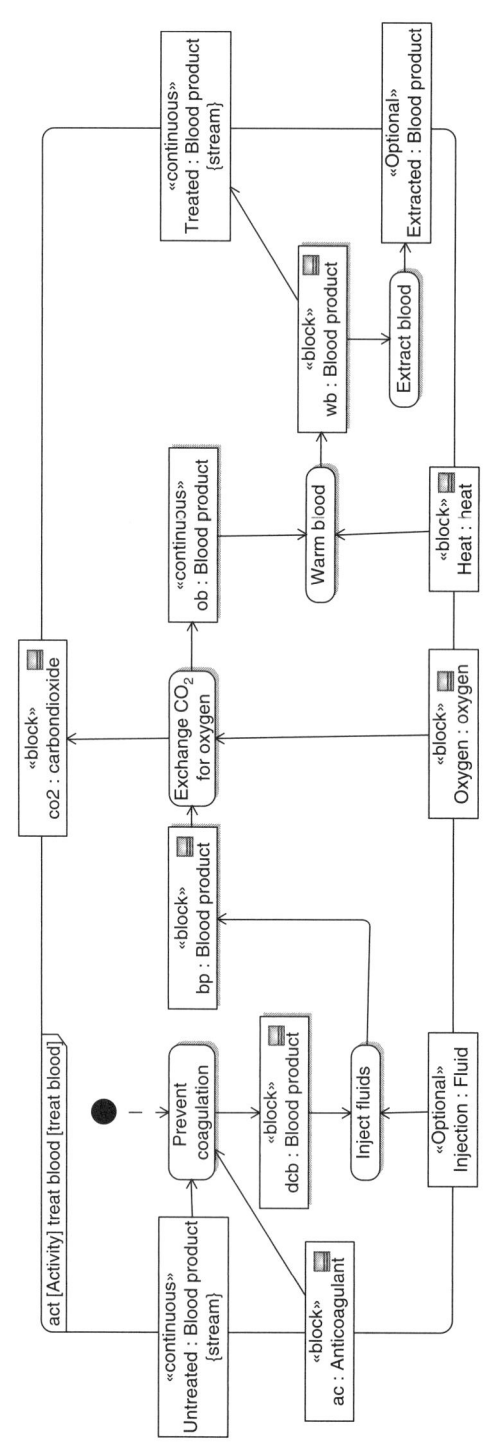

Figure 18.10 ECMO "Treat Blood" activity diagram.

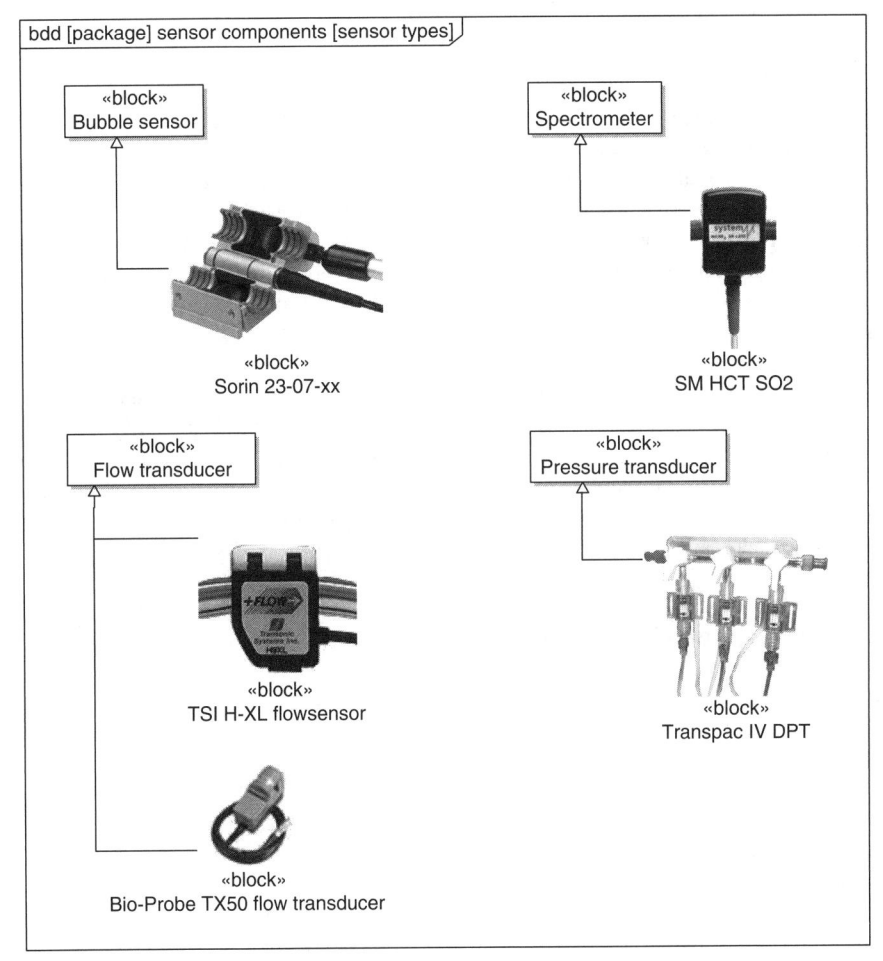

Figure 18.11 Phase 2 ECMO sensor classification BDD.

while the second layer consisted of more concrete elements. Each layer of abstraction has utility in the SE process. As described in the crosscutting section, we use the abstract structures for functional allocation during architectural design iterations. The second layer of abstraction models the many different circuit designs, both in use today and proposed, and maps those designs back to the abstract circuit model.

To accomplish this, we first refined the ECMO domain diagrams initially produced at the outset of this study. This task included expanding the BDD to include blocks such as Family and Laboratory, moving the generalization/specialization relationships to a separate diagram for clarity, and removing CHOA-specific implementation details from both the BDD and the IBD.

Next, we defined the blocks for specific hardware components as specialization of the abstract hardware component blocks. For example, the block representing the CSZ ECMO Heater, a blood-warming device common to many circuits in use today, was defined as a specialization of the abstract Heater block. Figure 18.11 shows examples of the generalization/specialization relationships defined for sensor components commonly found in ECMO circuits.

Finally, several existing circuit designs were captured and modeled as blocks that specialized the abstract ECMO Circuit block. The part properties for these blocks were typed using the hardware component specialization blocks. This allowed us to capture the design of circuits currently in use at partnering medical centers in as high detail as possible. The revised IBD for the roller-head circuit design used at CHOA is shown in Figure 18.12. This circuit design illustrates one of the major challenges of current ECMO circuit designs: monitor and control complexity. With this basic circuit setup, an ECMO specialist monitors six different displays and operates the circuit using four different control panels.

Prior to our work, few, if any, ECMO centers captured the physical layout of their circuits in any kind of graphical format. The closest artifacts we found were order sheets for the tubing used to connect the subsystems of an ECMO circuit. With a little coaching, the medical staff easily understood the SysML structural diagrams we created and were able to review them for correctness and completeness. Capturing the circuit layouts from the participating hospitals in a standard way allowed systems engineers and medical staff alike to compare and contrast the various ECMO circuits under analysis.

18.4.4 Traceability Using Crosscutting Relationships

One of SysML's strengths as a modeling language for systems and SoS is its inherent ability to create crosscutting relationships, or links, between various entities in the model. For ECMO, we wanted to systematically identify and capture these relationships and to systematically perform gap analyses on ECMO circuit implementations. SysML provides those capabilities.

In an effort to minimize our impact on the medical staff's already busy schedules, we generated the initial set of requirements using Van Meurs' "ECMO Extracorporeal Cardiopulmonary Support in Critical Care, 3rd edition" (Van Meurs et al., 2005) and the "ECMO Policy and Procedure Manual" from CHOA (ECMO Policy and Procedure Manual, 2011) as primary sources. We then spent time with stakeholders validating the content, scope, and organization of the draft requirements. To facilitate requirements traceability, we modeled each reference text as a SysML block with each chapter modeled as a part property of the appropriate reference text. We then created a "trace" relationship from each requirement to the part property representing the reference text that the requirement was developed from.

Once various systems behaviors had been defined, we performed a requirements allocation by creating a "satisfies" relationship between each modeled behavior and the appropriate requirement. We then generated a requirements satisfaction matrix to show which requirements had been satisfied by a defined behavior and which had not.

After revising and greatly expanding the structural models, we allocated each modeled behavior to an abstract structural component. For example, we allocated the Return Patient Blood action, which appears on the Circulate Patient Blood activity diagram, to the Cannula part property of the abstract ECMO Circuit block. With these "allocate" relationships in place, we generated a behavior allocation matrix and used it to verify that the SysML model was correct and complete; using this matrix, we

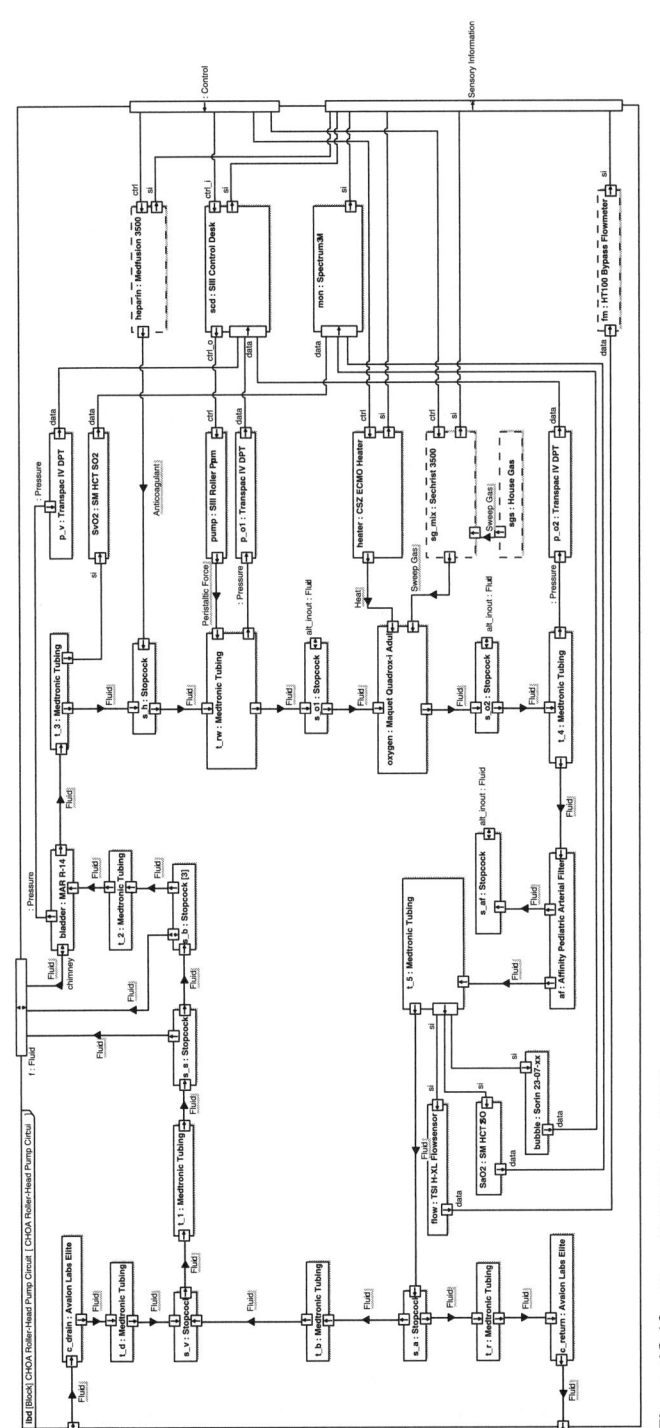

Figure 18.12 Phase 2 ECMO circuit IBD.

analyzed structural elements that did not have a behavior allocated to it to determine if the structural element was actually needed or if some behavior had not yet been identified and modeled.

Finally, we created a "redefinition" relationship between the part properties of the abstract ECMO Circuit block and the part properties of each specific circuit design. By doing so, we mapped each part property in each circuit design to the abstract structural component that it was meant to serve as. Take Figure 18.13 for example.

In Figure 18.13, the SCP Revolution 5 is a particular kind of blood pump that is used in the CHOA centrifugal pump circuit. In the SysML model, the part property for the SCP Revolution 5 pump redefines the abstract Blood Pump part property of the ECMO Circuit block. The Circulate Patient Blood activity has been allocated to the Blood Pump block and in turn satisfies the Circulate Blood requirement. That requirement traces back to the overarching ECMO Circuit Operation requirement. Finally, the ECMO Circuit Operation requirement traces back to the ECMO Circuitry and Equipment chapter of the CHOA ECMO Policy and Procedure Manual.

With these crosscutting relationships in place, we can trace any element of an ECMO circuit design from source materials to requirements, from requirements to behaviors, from behaviors to abstract structural element, and finally from abstract structural element to a specific make and model of a component. This end-to-end traceability provides an objective mechanism for determining whether a particular circuit design meets its requirements and a mechanism for performing impact analysis should requirements, best practices, or technology change. Future iterations of the model could incorporate more of the full life cycle aspects of the ECMO domain such as training and maintenance and could expand any impact analysis accordingly.

18.5 FUTURE WORK AND CONCLUSIONS

With an SoS such as ECMO, which did not benefit from up-front application of SE, we seem to always have more SE work to do with the SoS. The first two phases proposed areas of work to be tackled in the future. Let's highlight a two of those.

First, we propose using simulation methods to generate new models to further the understanding of some of the emergent behavior in the ECMO SoS. For example, we could develop a blood clot model and use it to determine the increased or decreased probability of blood clots given a particular circuit design or set of materials used. We could then use the model to reduce the failure rate of components like the oxygenator that can clog when blood clots naturally form inside the ECMO circuit.

Second, we propose using M&S to drive the development of a redundant ECMO circuit. As we mentioned before, there is no redundancy in ECMO. If an oxygenator or pump fails, the staff employ manual methods to keep the patient as stable as possible until a replacement can be deployed. We imagine adding a backup for each of these that could automatically switch over in the case of a failure. However, as simple as this sounds, a great many questions require study in order to see this through. One such question is "how could a second pump stay primed with blood (or potentially saline) and be switched to without introducing air or a clot into the circuit?" Does introduction of redundancy into ECMO increase cost to a level that it isn't worth it? Do failures happen often enough to

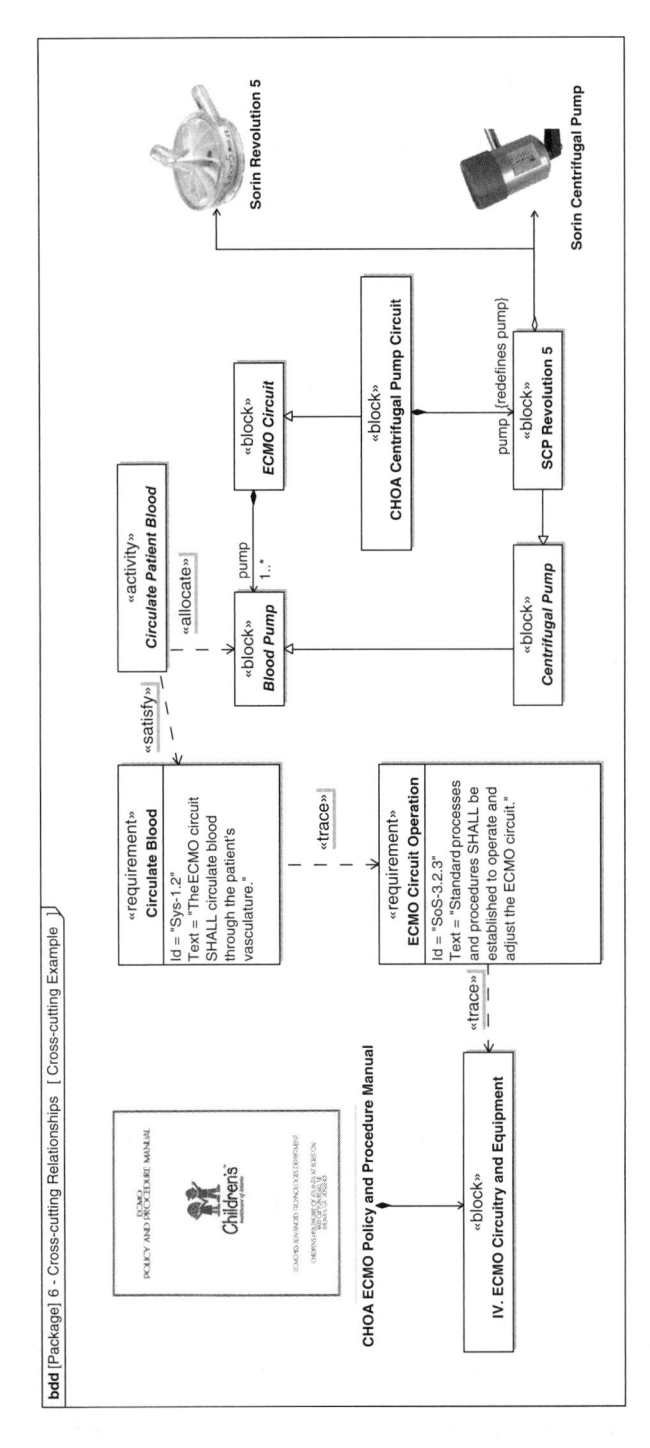

Figure 18.13 Traceability of requirement to activity to component.

warrant changing the configuration? We believe the answer to that question is yes, but work will be required to prove the benefit.

We believe that MBSE is an extremely valuable tool for not only characterization of a system or SoS but also for bridging the communications gaps that exist between engineers and nonengineers. We also hope that the examples shown in this chapter give you the confidence in the methodology to apply MBSE to your own SE efforts. Through formal modeling and constant iteration with stakeholders, you can achieve a much greater understanding of the problem at hand and can tame the complexity of systems and SoS much as we did with ECMO.

ACRONYMS

BDD	Block Definition Diagram
CHOA	Children's Healthcare of Atlanta
CPB	Cardiopulmonary bypass
DoDAF	Department of Defense Architecture Framework
ECLS	Extracorporeal Life Support
ECMO	Extracorporeal membrane oxygenation
ELSO	Extracorporeal Life Support Organization
FDA	Food and Drug Administration
IBD	Internal Block Diagram
MBSE	Model-Based Systems Engineering
M&S	Modeling and simulation
N2	N-Squared diagram or chart
OV-1	Operational View 1, from DoDAF
SE	Systems Engineering
SoS	System of systems
SoSE	System of Systems Engineering
SysML	Systems Modeling Language
UML	Unified Modeling Language

REFERENCES

Adams N, Pihera LD. (2012). A Systems Engineering Approach for Informing Extracorporeal Membrane Oxygenation (ECMO) Therapy Improvements. In Proceedings of CSER, March 20–22, Atlanta.

Bartlett RH. (1985). Esperanza. Transactions—American Society for Artificial Internal Organs 31: 723–735.

Bjelkemyr M, Semere DT, Lindberg B. (2009). Definition, classification, and methodological issues of systems of systems. In Jamshidi M, editor. Systems of Systems Engineering Principles and Applications. Boca Raton: CRC Press. p 191–206.

Department of Defense. (2001). Systems Engineering Fundamentals. Fort Belvoir: Defense Acquisition University Press.

Department of Defense, Architecture Framework (version 2.02). (Aug 2010). Available at http://dodcio.defense.gov/Portals/0/Documents/DODAF/DoDAF_v2-02_web.pdf. Accessed July 20, 2014.

ECMO and Advanced Technology Department. (2011). ECMO Policy and Procedure Manual (3rd Edition). Atlanta: ECMO and Advanced Technology Department, Children's Healthcare of Atlanta at Egleston.

Friedenthal S, Moore A, Steiner R. (2009). A Practical Guide to SysML. Burlington: Morgan Kaufmann.

Hill JD, Obrien TG, Murray JJ, Dontigny L, Bramson ML, Osborn JJ, Gerbode F. (1972) Prolonged extracorporeal oxygenation for acute post-traumatic respiratory failure (Shock Lung Syndrome). New England Journal of Medicine 286: 629–634.

INCOSE-TP-2004-004-02: INCOSE SE Vision 2020 (version 2.03). (2007). Technical Operations, International Council on Systems Engineering. Available at http://www.incose.org/ProductsPubs/products/sevision2020.aspx. Accessed July 20, 2014.

Kurusz M, McKellar S. (2005). An introduction to Project Bionics' "guide to collections relating to the history of artificial organs." American Society for Artificial Internal Organs 51(1): 128–132.

LaPlue L, Garcia RA, Rhodes R. (1995). A Rigorous Method for Formal Requirements Definition. In Proceedings of INCOSE, July 22–26, St. Louis.

Maier MW. (1999). Architecting Principles for Systems-of-Systems. Systems Engineering 2(1): 1–18.

O'Rourke PP, Crone RK, Vacanti JP, Ware JH, Lillehei CW, Parad RB, Epstein MF.. (1989). Extracorporeal membrane oxygenation and conventional medical therapy in neonates with persistent pulmonary hypertension of the newborn: a prospective randomized study. Pediatrics 84: 957–963.

Paden ML, Conrad SA, Rycus PT, Thiagarajan RR, ESLO Registry. (2013). Extracorporeal life support organization registry report 2012. American Society for Artificial Internal Organs Journal 59(3): 202–210.

Peek GJ, Mugford M, Tiruvoipati R, Wilson A, Allen E, Thalanany M, Hibbert C, Truesdale A, Clemens F, Cooper N, Firmin R, Elbourne D. (2009). Efficacy and economic assessment of conventional ventilatory support versus extracorporeal membrane oxygenation for severe adult respiratory failure (CESAR): a multicentre randomised controlled trial. Lancet 374: 1351–1363.

Pihera LD, Paden ML, Ender TR, Taylor B, Bollweg NR, Lopez A, King S. (2012). Application of Systems Engineering to Improve Extracorporeal Membrane Oxygenation (ECMO) Therapy. In Proceedings of INCOSE, July 9–12, Rome, Italy.

Pihera LD, Ender TR, Paden ML. (2013). Extracorporeal Membrane Oxygenation (ECMO)—A Systems of Systems Engineering Characterization. In Proceedings of IEEE SoSE–, June 2–6, Maui, Hawaii.

UK collaborative ECMO trial group. (1996). UK collaborative randomized trial of neonatal extracorporeal membrane oxygenation. Lancet 348: 75–82.

Van Meurs K, Lally KP, Peek G, Zwischenberger JB, editors. (2005). ECMO Extracorporeal Cardiopulmonary Support in Critical Care (3rd Edition). Ann Arbor: Extracorporeal Life Support Organization.

Chapter 19

Utility: Problem-Focused, Effects-Based Analysis (aka Information Value Chain Analysis)

Thomas W. O'Brien[1] and John F. Sarkesain[2]
[1]US Air Force, Colonel (Retired)
[2]SIOC Group, L.L.C. and The Aerospace Corporation, Ashburn, VA, USA

19.1 INTRODUCTION

In this chapter, we show how to architect distributed system of systems (SoS) based on optimizing end-to-end, enterprise information chain value and how to quantitatively determine that optimized value in terms of customer mission outcome-based effects and, along with our chapter on Cyber Command and Control (C^2), provide the framework necessary to analytically inform risk-managed acquisitions and investments and operational decisions on cybersecurity based on customer mission value and cost and risks. It also lays the foundation for the following chapter that goes into the details of the supporting frameworks.

19.2 THE NEED FOR A CYBERSECURITY FRAMEWORK

Powered by Moore's law (1965) and driven by the demands of an insatiable global marketplace, the ongoing information revolution is irrevocably changing the world as we know it. The virtual explosion of social media and smartphones, mobile devices, and the "Internet of things"—coupled with the tailored exploitation of *big data* and *cloud-based*

Modeling and Simulation Support for System of Systems Engineering Applications, First Edition.
Edited by Larry B. Rainey and Andreas Tolk.
© 2015 John Wiley & Sons, Inc. Published 2015 by John Wiley & Sons, Inc.

services—collectively is reshaping personal behavior and relationships in profound ways none of us yet fully understand. Experiencing millions of cybersecurity intrusions per day and unrelenting efforts to steal our personal identities, intellectual property, and national security secrets, coupled with vulnerabilities of critical national infrastructure to cyberattacks, the world has been irrevocably changed by cyberspace. Numerous real-world examples abound, from the heavy-handed takedown of Estonia's telecommunications infrastructure to the precision targeting of the Iranian nuclear fuel enrichment facility at Natanz; these inextricably intertwined technological and social–political developments are disrupting organizational behaviors and security relationships at every level—locally, regionally, and globally. An analytical framework is needed to quantitatively address how to deal with these issues to determine how to make some sense of it all. In short, we need a way to architect distributed SoS solutions that can simultaneously leverage the extraordinary advantages the IT revolution is offering us while simultaneously protecting and resiliently defending the very "cyberspace" that we operate in to provide us those advantages. This and our accompanying Cyber C^2 chapter together offer an analytically based approach that can enable us to successfully meet that challenge. This chapter sets the stage to doing so by showing how to quantitatively measure IT advantages in terms of mission effects-based value; and the related chapter addresses the cybersecurity architecture required to protect, secure, and dynamically defend the cyberspace that provides us those information advantages.

19.3 THE PROBLEM

To paraphrase President Bill Clinton's 1992 campaign quote, "It's the economy, stupid"— "It's the information, stupid." Motivating this work is a sense of urgency on the need to rapidly secure the nation from the likely devastation resulting from cyberattacks against our vulnerable critical national information-based infrastructure, such as financial, telecommunications, and power systems. The problem posed here is how and where we should make the appropriate investments necessary to protect against and actively mitigate the effects of such attacks while simultaneously confronting significant government downsizing and reductions in our military force structure. We must be prepared to defend our national IT superiority advantages against advanced, persistent disruption, exploitation and attacks posed in this new domain of "cyberspace." Chapters 19 and 20 address the highly interrelated foundational building blocks of *IT information value* and *active cyber defense*—which together provide the framework for providing the metrics necessary to inform *cyber resiliency* cost and risk trades.

The related Chapter 20 describes the *C^2 architecture design* crucial to *dynamically* managing the kinetic network and supporting cybersecurity networks to *resiliently* endure those unrelenting, advanced persistent cyber threats in contested environments. The objective of the current chapter is to set the stage for this framework and explain assumptions and constraints necessary to appreciate all its aspects from the operational as well as from the technical perspective.

Although there are no current quantitative measures associated with *cyber resiliency* as of this writing, the need is to protect and defend our networks against increasingly sophisticated threats and changing vulnerabilities, to be responsive to man-made attacks as well as natural network failures and disruptions, and to provide continuous *situational awareness (SA)* on the state of mission operations and its supporting *cybersecurity*. To that end, any metrics for *resiliency* we develop should be based on our relative ability to mitigate impacts of threats against critical mission functions. This demands that we provide an ability to measure the mission outcome consequences of cyber incidents and to accurately relate those consequences to *cybersecurity protection controls and dynamic defensive* actions. Mission critical functions need to be supported by passive and active *dynamic cyber defense* that contain the attackers, limit their exfiltration of our critical data, and enable us to operate through, as well as recover from attacks, that is, maintain critical operations in contested environments.

19.4 THE ANALYTICAL PROCESS

The application of sound Systems Engineering (SE) principles is required to develop and access alternative integrated SoS enterprises based on their net-centric, *information value* and the *dynamic cyber defense* capabilities necessary to "resiliently" mitigate the effects of cyber threats in such contested environments. The latter demands that we properly frame the problem and apply smart *distributed* Cyber/Kinetic C^2 architecture designs to optimally protect, operate in, and dynamically secure this unusual domain of "cyberspace."

Presented here is how to quantitatively measure *information value* in terms of mission outcomes and architecture design trades for *globally distributed dynamic cyber defense* that can ensure mission success in contested environments. Both capabilities are informed by decades of military "kill chain" analysis (O'Brien, 2004) and Cyber/Kinetic C^2 prototyping (Howes et al., 2004). Jointly, they provide an understanding on how to conduct *dynamic network allocation and cyber defense*. Their interrelationship constitutes cyber "resiliency," a topic of great community interest but of little understanding at the time of this publication—a principal finding of the Defense Science Board's Study on "Resilient Military Systems and Advanced Cyber Threat" (Kaminski, 2013).

19.5 THE APPLICATION OF THE SE AND OTHER DISCIPLINES

The amplifying power of IT comes from integrating and operationally synchronizing systems, people, and processes together to optimize enterprise-level, "net-centric" effects. The basic questions needed to properly frame this problem are then: what do we mean by integration, at what level, and to what effect? The simple example below underscores that problem framing (asking the "right question") is highly dependent upon perspective and the context.

Figure 19.1 shows the example of the three blind men and the elephant, presented in a recent popular book on SE (Meadows, 2008), which suggests that we can "see" and therefore fully appreciate the large gray animal in its entirety standing before us. In this case, the reader may be mildly amused by the parochial, tactical perspective each blind man has on what an elephant is vice the "true" perspective afforded we the seeing. However, before we hastily jump to framing the issue as we see it, an easy trap to fall into, we need to think more deeply about what insights we are trying to uncover. As we attempt to draw the dotted lines around what's in and what's out of the "elephant system," we

What's the elephant system?

Three blind men and the elephant
- Each has different perspective of what elephant is

- Unlike them, we "know" because we can see entire elephant

Figure 19.1 The three blind men and the elephant.

need to be reminded about the general properties of a "system." The first property is that we cannot know the behavior of a system just by knowing the elements that make up the system—thus today's limitations of static, point solution IA controls for cybersecurity, but more on that in our related chapter (see Chapter 20, this volume). In this regard, Ludwig von Bertalanffy (1976), a recognized pioneer in the development of SE, said that "A system is a dynamic, complex, and interdependent whole, interacting as a structured functional unit; is holistic, exhibiting emergent properties not possible to detect by analysis of individual parts; and is a community situated within an environment." In addition to SE discipline, as will be seen later, the application of other disciplines is necessary to support these network-based analyses—namely, control theory, information theory, and complexity theory. This includes Dr. Norbert Wiener's pioneering work on control, interestingly enough called "Cybernetics" (Wiener, 1961); his MIT cohort Dr. Claude Shannon's development of information theory, the mathematics for determining information value (Shannon, 1948); and the Santa Fe Institute's pioneering work on nonlinear "feedback" in systems (Waldrop, 1983) (aka chaos theory, complex adaptive systems, or "complexity theory").

The basic approach for proper problem framing requires us to carefully examine and uncover what parameters drive a system's behavior. In this example, the elephant is likely part of a herd; elephants rarely live alone. That is, the "elephant system" behavior involves a herd of elephants. Upon further investigation, we see that the herd itself makes up part of a more complex SoS that includes the "nonlinear," complex adaptive relationships it has with other parts of a larger ecosystem (Shannon, 1948). This is in keeping with Dr. Mark Maier's definition of SoS architectural principles: operational independence of the elements, managerial independence of the elements, evolutionary development, emergent behavior, and geographic distribution" (Waldrop, 1993). Depending upon the insights one is trying to uncover in the problem framing stage, one may want to further address each of these design principles in turn.

Figure 19.2 shows that the elephant SoS includes the herd's relationship to its local environment. Depending upon the context of the question, the analyst may want to extend

Or do we? Let's look further.
— Part of a larger group of elephants (herd — behavior)

- Input, output — what elephant breathe, eat, and deposit
- Local environment — key to their health and survival
- Imprints they leave on the jungle floor
- Puddles of water that are left in the imprints when it rains
- Bacteria that live in the rain puddles
- Insects that eat bacteria
- Birds that eat insects

Figure 19.2 System of systems: a group of elephants.

the dotted lines out to include the local environment, which in this case includes an intricate, interactive food chain affected by the herd's presence and behaviors. This ecosystem is made up of an elaborate food chain network of microorganisms, bacteria, plants, insects, birds, and so on. Although our interest here is in gaining insights into the elephant ecosystem, an ornithologist may be more interested in the "bird" ecosystem that also includes elephants—an issue of proper problem framing.

Problem framing is thus structured according to relevant insights that the analyst desires, the context of the question, and the analyst's perspective. When done well, problem framing process will iteratively identify and focus on "root causes" of behaviors, the driving parameters. It is not unusual to be surprised during such a "what's in, what's out" problem framing exercise—in fact, problems sometimes seemingly "solve themselves" when properly framed. As is often the case, others may judge the final definition of the problem as "intuitively obvious," but that may rarely be the case at the outset.

Although problem framing is pivotal to getting at real solutions to problems, as in our "information value, dynamic cybersecurity" problem, the process is often not done well for numerous reasons:

- Analysts are often in a hurry—feeling pressured to "get on with it"—to provide answers to an impatient customer.
- It can be hard to do; there are no simple checklists; it takes an open mind, broad experience, and extensive patience and can be highly iterative.
- The emerging behaviors of SoS involve multiple moving parts that include high interaction with reinforcing feedback and can thus be complex and heuristic—requiring agent-based vice classical decomposition, reductionist approaches.

520 Theoretical and Methodological Considerations with Applications and Lessons Learned

- Each stakeholder is appreciably biased to different degrees by his/her own background, education, experiences, perspective, and motivations:
    ○ Many often confuse the symptoms with the problem itself—vice searching for "root causal factors."
    ○ Customers often state problems in terms of notional solutions instead of obstacle(s) to obtaining their desired mission-based objectives.
    ○ Engineers, uncomfortable with working in the broad parametric "problem space," want to immediately focus on specific engineering designs in the "solution space," especially for designs that involve their own tools, experience, and expertise, that is, "looking for keys under the lamppost" syndrome.
    ○ Systems program managers are under day-to-day pressures to build "something" and focus on systems design specifications, cost, and schedule milestones vice delivering systems capabilities—much less enterprise-level, SoS "capabilities."

    Note: INCOSE defines "verification" as ensuring that you are building the system right (meeting systems specifications) and "validation" as assuring that you are building the "right system." (Maier, 2009)

    - However, "right system" might be better stated as building the "right capabilities" or—better still—the "right manifested behaviors."
        ○ Manifested behaviors are more appropriately addressed through "agent-based modeling," a topic discussed in depth throughout this book.

Figure 19.3 somewhat humorously shows that proper problem framing is the "sine qua non" for successfully conducting good analysis. "If you don't know where you are going any path will get you there…," the Cheshire told Alice (INCOSE). To meet the objective,

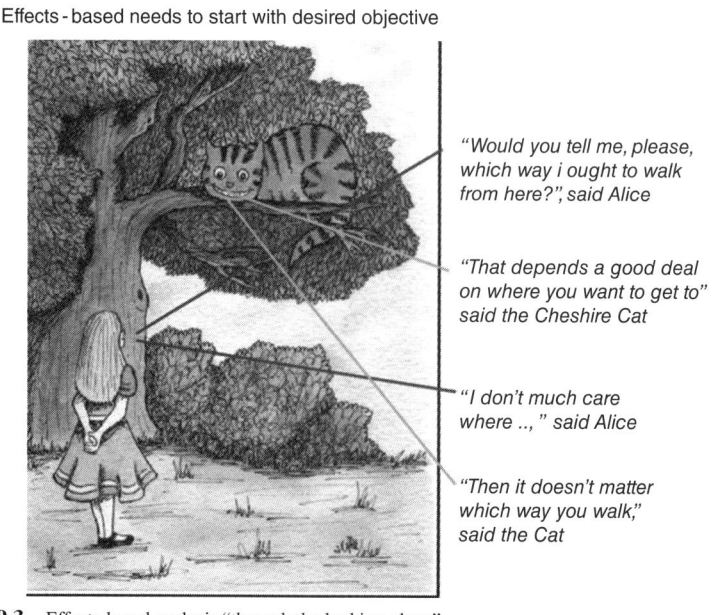

Figure 19.3 Effects-based analysis "through the looking glass."

the analytical team's development of a well-articulated *study terms of reference* can be indispensable to good problem framing—that is, agreeing to and communicating the overall study objectives of the analytical team, customers, and stakeholders; assigning and allocating resources to the respective leads and substantive area experts; establishing testable ground rules and assumptions; providing a common dictionary of terms to mitigate likely misunderstandings and confusion; providing top-level ConOps on how the overall SoS enterprise fits together and is expected to behave in its environment; and establishing study success criteria—so the team knows when they are done. This highly iterative framing process, however difficult, is fundamental to setting the stage for uncovering the "root causes" of the problem(s). It is the authors' experience that even large, well-financed studies have failed in the end due to neglect in properly developing and maintaining their study's "statement of work."

Thus, the proper framing of the "cybersecurity problem" requires an analytic approach that enables development and assessments of alternative solutions that best on optimizing mission effectiveness. The most illustrative example is DoD's initiative to increase combat effectiveness though "net-centric warfare" (NCW). Initiated in the mid-1990s by John Garstka and Admiral Zerbrowski on the Joint staff (Carroll, 1993), NCW is now reflected in major military systems acquisitions and operations, for example, Missile Defense Agency's global missile defense (Alberts et al., 1999). Based on information value, the military is reshaping its force structure around Joint Vision 2020's "Information Superiority" to transformation the operational capabilities of the joint force and the evolve Joint command and control. Taking it a step further after 9/11, the secretary of defense directed the DoD to move from its "Cold War," threat-based, systems-centric approach to a more flexible and responsive, capability-based approach. Secretary Rumsfeld stated, "…we do not know the true face of our next adversary or the exact method of engagement…this uncertainty requires us to move away from past threat-based view of the world and force development. We MUST change" (Joint Chiefs of Staff, 2003). The military is transforming to achieve "information superiority," and the uncertainty and complexity presented by the need to protect and secure cyberspace to make that happen underscore the criticality for a disciplined approach—now more than ever.

The basic premise of this transformation ("Revolution in Military Affairs") is that net-centricity translates information into a competitive advantage through the robust networking of well-informed, geographically dispersed forces. The DoD's Transformation Planning Guidance (2003) makes clear that "… we must achieve: fundamentally joint, network-centric, distributed forces capable of rapid decision superiority and massed effects across the battlespace. Realizing these capabilities will require transforming our people, processes, and military forces." In accordance with the DoD's guidance, net-centricity's basic tenets are that we can dramatically increase mission effectiveness through:

- *A robustly networked force enabled by improved information sharing*
- *Information sharing that enhances the quality of information and shared*
- *Shared SA that enables collaboration and self-synchronization and enhances sustainability and speed of command*

And, these, in turn dramatically increase mission effectiveness.

The application of the SE and other aforementioned analytic disciplines to the cybersecurity problem fundamentally involves developing and accessing alternative SoS architectures based on their mission outcome value and providing the C² architecture that can best synchronously coordinate kinetic and cyber operations in a contested environments. Information "value" can be determined in terms of customer utility or enterprise-level performance. The former is related to degree of mission accomplishment; the latter is measured in terms of enterprise efficiencies and economies of scale. In many cases, these two objectives complement each another; however, in some cases, trade-offs must be made between them; however, both need to be addressed.

Thus, for customer utility, SoS architecting involves measuring net-centric value, as Figure 19.4 shows in terms of mission outcomes.

Understanding that value is essential to establishing kinetic network resource allocation as well dynamic cybersecurity priorities for operating distributed, real-time C². The real-world example here highlights *information value* in terms of military mission outcomes. The data were taken from the second Schweinfurt Raid, and air mission flown during WWII. Launched in October 1943 against the precision ball bearing factories crucial to running the Axis warfighting machine, "Black Thursday," as it became known, suffered the highest Americans losses of any bombing mission of the war. Of 391 B-17ss flown, 77 were shot down and 121 damaged, and over 655 airmen were killed in action or taken as prisoner. Compare that to achieving similar measure of effectiveness (MoEs) during desert storm, where carpet bombing by large formations of bombers was replaced by single bombers surgically delivering bombs to a specific network's "center of gravity." The "information leverage" can be measured here in terms such as bombs dropped; the size, costs, and delays of the logistics tail; the degree of collateral damage; and other

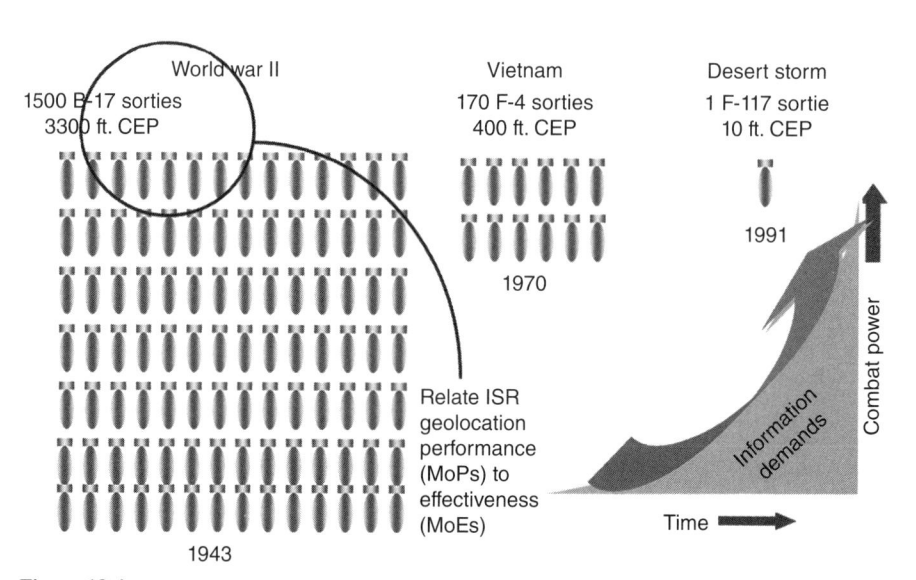

Figure 19.4 Measuring information value as mission outcomes.

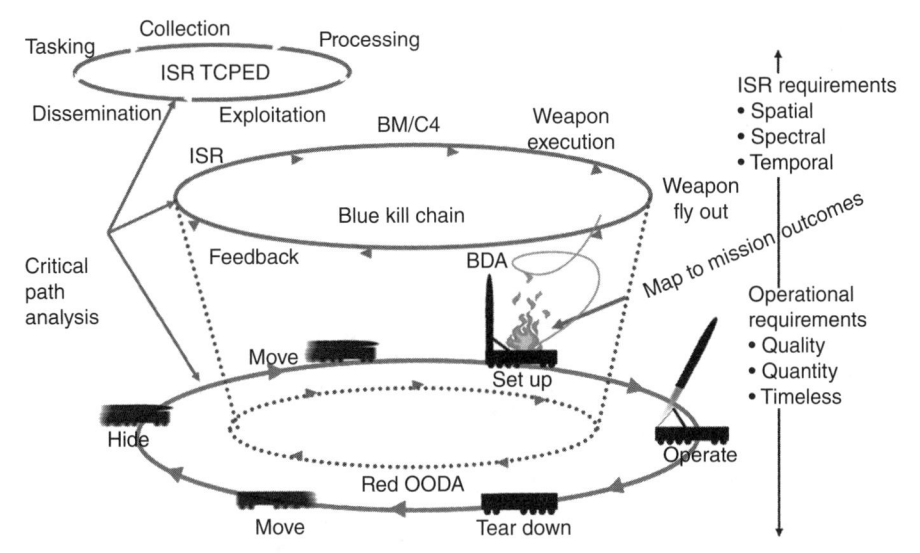

Figure 19.5 "Kill chain" effects-based analysis example.

metrics. In this case, the measures of SoS performance (MoPs for SoS) are based on bomb delivery circular error probable (CEP) and target geolocation precision. This example illustrates what a "pound of information" is worth in terms of mission outcomes.

Figure 19.5 provides a simple information value chain (aka "kill chain" in the tactical example) analytical construct for the missile defense mission, discussed again in our related chapter. In this specific example, the "kill chain tuning" process is applied to defeating mobile missiles. As shown, the red mobile missile threat goes through a process where the operator hides, moves, sets up to launch, fires a missile, tears down, moves, hides, and repeats the cycle—using mobility to enhance survivability. Blue however must rapidly execute his end-to-end processes that includes Intelligence, Surveillance, and Reconnaissance (ISR); Command, Control, Communications, and Computers (C4) and Battle Management (BM); weapons systems execution and fly out to the target; battle damage assessment (BDA); and SA feedback to the commander to modify the process to the degree and timing necessary to achieve his/her mission objectives. In military parlance, blue is attempting to get inside and defeat red's **o**bserve, **o**rient, **d**ecide, **a**ct (OODA) cycle (Boyd, 1995). Note that there are feedback loops within each processes and that there are processes within each process. For example, ISR is itself a process that involves sensor **t**asking, data **c**ollection, data **p**rocessing and **e**xploitation, and information **d**issemination, that is, the "TCPED" process. In fact, every step involves processes within itself. Also note that the red missile battery behavior is modeled—in this case—in terms of states and state transitions, reflected by metrics of quality, quantity, and timeliness, and the ISR processes are reflected in terms of spatial, spectral, and temporal metrics.

Thus, the disciplined modeling and simulation (M&S) process supports development of specific requirements that must be met and enables the analyst to explore large numbers of variations and permutations, to identify the parametric drivers as well as their sensitivities to degrees of variation, and to statistically bind a probable range of threat behaviors. What's more, the analyst can use actual data collected from test range experimentation, as

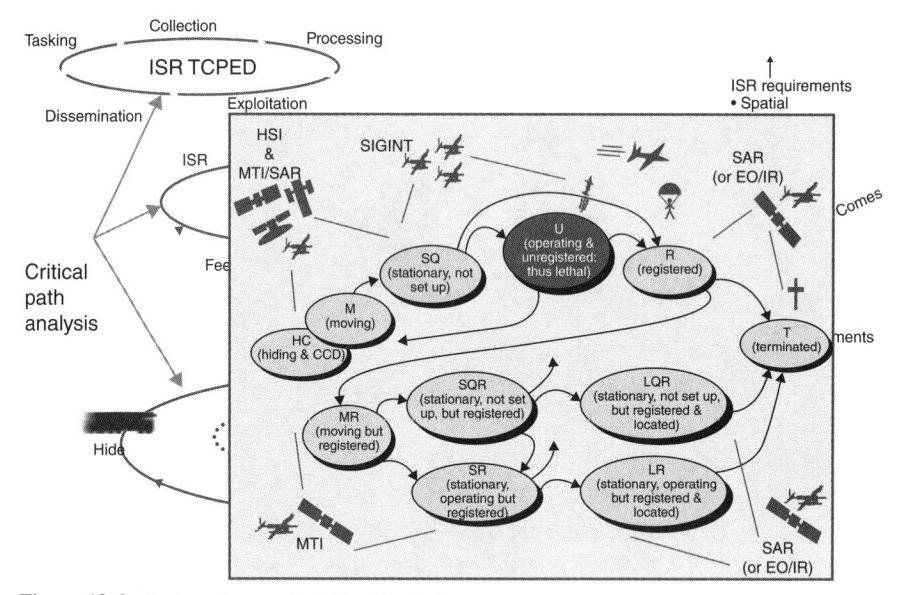

Figure 19.6 Systems view on effects-based analysis.

well as real-world experiences, as sources for calibrating the M&S. Figure 19.6 shows modeling of this process in terms of a DoDAF Systems View 10b state transition diagram. A more detained example is provided below for the mobile SAM problem.

One prime developer of this information value construct studied under Claude Shannon (the father of information theory) and Norm Weiner (the father of control theory and cybernetics) at MIT. He has been applying his rigorous analytical foundation and more than four decades of experience toward these type "information value-based" problems; and the approach has provided repeatable, variable resolution analysis that has successfully supported informed decision making for senior DoD and IC customers.

Different metrics are used to measure other types of enterprise performance, typically stated in terms of higher efficiencies and lower costs offered through enterprise-wide economies of scale. Illustrative is the current effort to achieve lower cost and provide superior information sharing and higher efficiencies through cloud computing. Cloud's basic premise is to provide hardware and software computing resources as a service over a network to achieve economies of scale (Figure 19.7). However, cloud introduces its own cybersecurity issues even as it offers economic advantages. For example, not all cloud deployment modes are appropriate for each service, customer, or all users; and it is unlikely to be cost-effective for all cloud providers to offer the same very high levels of expensive security to meet less stringent needs. Common security cloud issues include network availability, data control and privacy, provider viability, multilevel security, and different systems vulnerabilities.

The objective for coupling *information value chain* analysis with *distributed Cyber C² Architecting* is to determine how best to achieve, assure, and sustain *information*

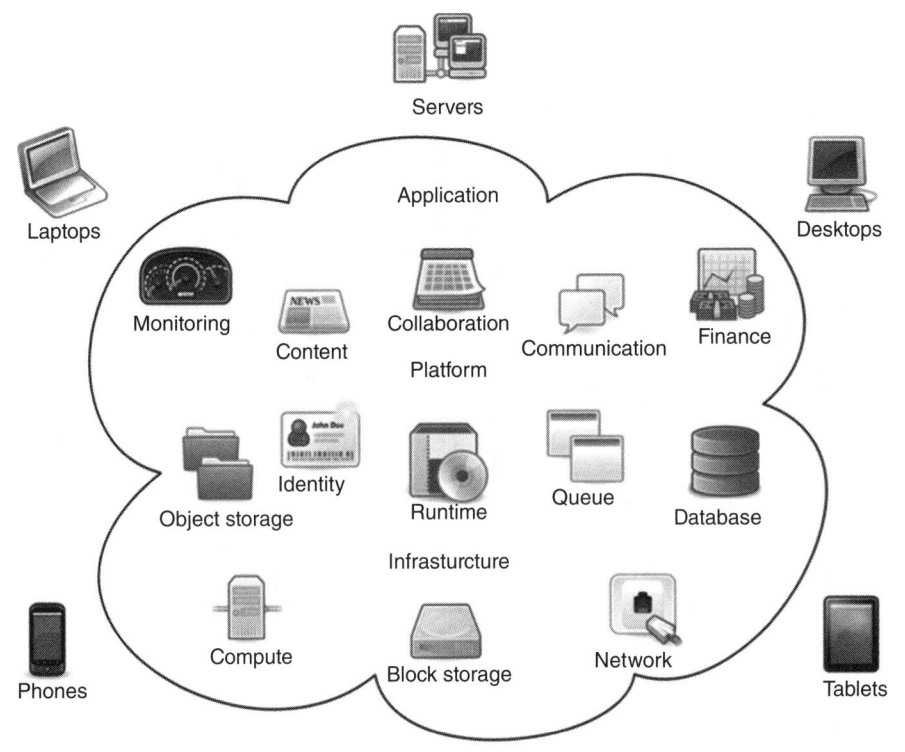

Utility: Problem-Focused, Effects-Based Analysis   525

Figure 19.7 Cloud computing.

superiority/optimum mission effectiveness during natural disruptions or man-made attacks, that is, resiliently sustain successful mission accomplishment in challenged or contested environments. In the military case, the effectiveness of defensive cyber is in direct proportion to the ability to mitigate an adversary's effort to deny, disrupt, or degrade the amplifying combat power that network centricity provides. Alluded to earlier, the meaning of NCW is further clarified by the following definition: NCW "…consists of networking the warfighting enterprise – shooters, decision makers, and sensors – to translate information superiority into combat power by effectively linking knowledgeable entities in the battlespace." In this case, the cyberattack is an effort by an adversary to deny, degrade, disrupt, or destroy our *information superiority*. In fact, NCW's downside is that the very domain of cyberspace it operates in represents its own "Achilles' heel"; that is, the supporting infrastructure was never designed to handle and is now proving to be inherently vulnerable to cyberattacks. In this case, changes to the cyber architecture performance can be measured by decreases in combat power, that is, kinetic mission outcomes.

Addressing information value as the basis for development and assessment of SoS alternatives (AoAs) is further amplified in Figure 19.8, which shows maximizing mission outcomes by optimizing end-to-end information flow; matching information needs to data, to tasking, to sensors, to targets, to C^2, to weapons delivery platforms, to weapons, and to targets—that is, "tuning the kill chain" by applying an analytic-based, outcome-driven, automated optimization control loop.

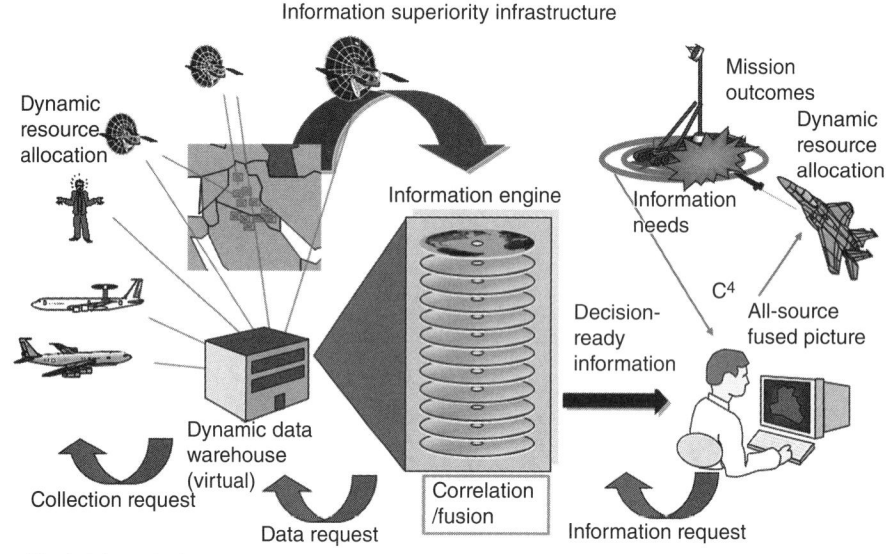

526 Theoretical and Methodological Considerations with Applications and Lessons Learned

Maximizing mission outcomes by optimizing end-to-end information flow matching info needs... data... tasking... sensors... targets... C4... platforms... weapons- -tuned by an analytic-based, outcome-driven, automated optimization control-loop.

Figure 19.8 Information superiority infrastructure.

In this example, the mission is to defeat mobile surface-to-air missiles (SAMs). Note that instead of handling each problem ad hoc as it arises, this approach developed by one of the authors in the mid-1990s would instead employ a "cybernetic" system that constantly "works problem solutions" by continually collecting and processing the multi-source data needed to solve it, able to be rapidly tasked for refinements as an actual event demand. The reader can appreciate the value of recent advancements with *big data* and *cloud computing* to crunching problems in this way. Note however that "cloud architectures" themselves need to be designed to accommodate real-time operations, not just less timely client–server approaches as they are today (more on the major Cyber C^2 architecture middleware implications in our related chapter).

One analytical example of *information value/effects-based* approach is the Markov chain modeling shown here for the mobile SAM problem. Figure 19.9 shows how blue's application of multi-ISR severely limits red's SAM threat capabilities by enabling blue to attack any SAM at every phase throughout its operational cycle. The state and state transitions are modeled, and the resulting curves quantitatively show how multi-intelligence (multi-INT) for tracking and targeting mobile SAMs reduces threat effectiveness under different conditions.

Let's now move on to describing the crucial metrics for examining both SoS architecture performance (MoPs) and customer utility (MoEs) and their important interrelationship to answer the "so what" question. First, according to the DoD Architecture Framework (DoDAF) process, architecture MoPs are metrics associated with the different viewpoints (e.g., capability, data and information, operational, project, services, standards, and systems). For simplicity, we will address only the operational and systems views here and, for this specific example, against time-critical targets (TCTs). As described earlier, systems

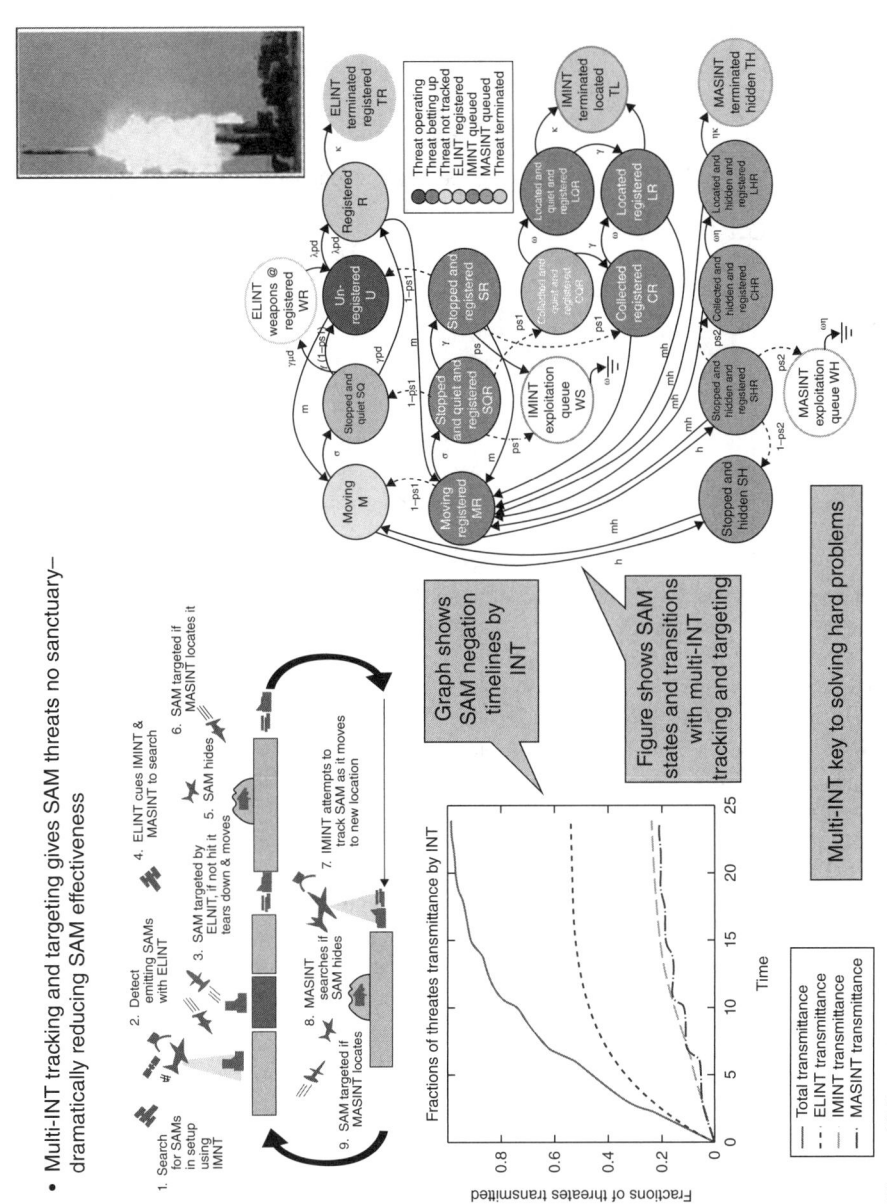

Figure 19.9 Multi-INT JSEAD.

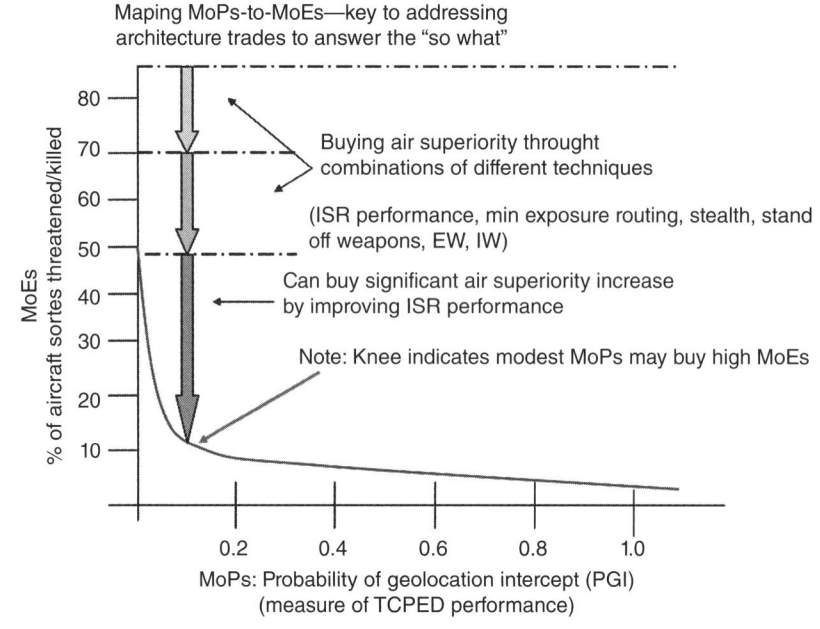

Figure 19.10 Measure of performance and measure of effectiveness.

architecture performance can be measured and assessed based on quality, quantity, and timeliness of the red's process and blue's ISR spatial, spectral, and temporal performance.

Next are MoEs. These are metrics based on customer's mission outcome-based criteria that are established mission by mission. For the air superiority mission example, the metric used here is friendly aircraft threatened, damaged, or destroyed. The architecture here can then be assessed in terms of customer mission outcome criteria associated with reducing threat effects. Figure 19.10 shows the "knees of the curve" that relate measures of performance (MoPs) to customer mission MoEs.

The MoPs-to-MoEs relationship then can inform the decision maker about the value of each architecture alternative in terms of contribution to improving "information superiority" or mission value. The relationship tells us "return of investment" in terms of how much is enough. In the case of the air superiority example, one can see that even a modest level of information (ISR TCPED) performance improvement can have a significant impact to the customer's mission outcomes (MoEs). Chapter constraints limit a fuller explanation; but before we move on to our other chapters, a quick summary of what has been discussed thus far is in order.

19.6 SUMMARY

This chapter focused on *information value* and explained what it is and how to measure it. It explained SoS architecting in terms of problem framing, showed quantitative metrics for architecture performance and mission utility and how their important relationship can help answer the "so what" question in terms of how much is enough, and showed how to measure net-centricity quantitatively in terms of combat power. Our related chapter underscores the implications of the new cyber threat environment to our ability to provide that IT-enabled combat power, focuses on the why dynamic C² must now be addressed in all SoS architecture designs, and provides the analytic framework to enable us to leverage the extraordinary advantages the information revolution while resiliently securing and dynamically defending those advantages from the cyber threats against them.

Key thus far is that information value can be measured in terms of net-centric effects. Our related Chapter 20 discusses the distributed architecture design patterns informed by over a decade and a half of analysis and operational prototype experimentation on Cyber/Kinetic Operations C². These patterns include *distributed operating systems services middleware, shared data structure, peer processing*, a *virtual operations architecture, and Cyber C² applications*.

Note that this chapter describes an essential but missing piece of the framework and Chapter 20 in this book shows the why and how of the C² architecture design. Together, they show how to quantitatively define cyber resiliency metrics that can be used to assure information superiority is best designed, secured, and leveraged for national defense.

REFERENCES

Alberts, DS., Garstka, JJ., and Stein, FP. *Network Centric Warfare: Developing and Leveraging Information Superiority*. Command Control Research Program Press, Washington, DC, 1999.

Carroll, L. *Alice's Adventures in Wonderland* (Dover Thrift Editions). Dover Thrift Editions. Dover Publications, Mineola, NY, May 20, 1993.

DoD's Transformation Planning Guidance, 2003. Available at: http://www.defense.gov/brac/docs/transformationplanningapr03.pdf. Accessed July 20, 2014.

Howes, NR, Mezzino, M, and Sarkesain, J. *On Cyber Warfare Command and Control Systems*, Command and Control Technology Research Symposium Proceedings, San Diego, CA, 2004.

International Council on Systems Engineering (INCOSE) Verification and Validation Working Group Charter. Available at: http://www.incose.org/practice/techactivities/wg/vvwg/. Accessed July 20, 2014.

Joint Chiefs of Staff. *Joint Operations Concepts*. Department of Defense, Washington, DC, 2003.

Kaminski, P. *Resilient Military Systems and Advanced Cyber Threat*. Office of the Under Secretary of Defense for Acquisition, Technology and Logistics, Washington, DC, 2013.

Maier, MW. *The Art of Systems Architecting*, Third Edition (Systems Engineering). CRC Press, Boca Raton, FL, January 6, 2009.

Meadows, DH. *Thinking in Systems: A Primer*. Chelsea Green Publishing, White River Junction, VT, 2008.

Moore, GE. *Cramming More Components onto Integrated Circuits*; Electronics, Volume 38, Number 8, April 19, 1965.

O'Brien, TW. *Tuning the Kill Chain Against Time Critical Targets*, presentation made to USJFCOM Conference, Kirtland Air Force Base, New Mexico, 2004.

Shannon, CE. A mathematical theory of communication. *The Bell System Technical Journal*, Vol. 27, pp. 379–423, 623–656, July, October, 1948.

Von Bertalanffy, L. *General System Theory: Foundations, Development, Applications*. George Braziller Inc., New York, 1976 [1969].

Waldrop, MM. *Complexity: The Emerging Science at the Edge of Order and Chaos*. Simon & Schuster, New York, 1993.

Wiener, N. *Cybernetics; or, Control and Communication in the Animal and the Machine*, Second Edition. The MIT Press, New York, 1961.

Chapter 20

A Framework for Achieving Dynamic Cyber Effects through Distributed Cyber Command and Control/Battle Management (C²/BM)

John F. Sarkesain[1] and Thomas W. O'Brien[2]
[1] SIOC Group, L.L.C and The Aerospace Corporation, Ashburn, VA, USA.
[2] US Air Force, Colonel (Retired)

"Imagination is more important than knowledge."

—Albert Einstein, a scientist

20.1 INTRODUCTION

20.1.1 Cyber, Not a Revolution in *Military* Affairs, but a Revolution in *National/Global Security* Affairs

The cyber warfare revolution extends far beyond the military to the very foundation of our national security. Driven by a revolution in information technologies and smart devices, a virtual revolution in global communications connectivity, and by the movement to the

---

Modeling and Simulation Support for System of Systems Engineering Applications, First Edition.
Edited by Larry B. Rainey and Andreas Tolk.
© 2015 John Wiley & Sons, Inc. Published 2015 by John Wiley & Sons, Inc.

"cloud" and the massive collection and rapid, tailored exploitation of "big data," the world is experiencing a revolution in human and national security affairs, impacting us all in profound ways that are not yet fully understood. With routine reports about the unrelenting cyber intrusions and espionage on massive scales against our national security and commercial and banking industries, as well as periodic reports about cyberattacks against the critical infrastructures (CI) of countries across the globe, such as the takedown of Georgian and Estonian telecommunications grids to the surgical attacks against the Iranian hardened nuclear fuel enrichment plant at Natanz, the world is being forever and irrevocably changed by cyberspace. These emerging events are highly disruptive to traditional government and nongovernmental organizational structures and relationships around the world—at all levels. A framework is needed to guide how we should respond to what is happening, to make some sense of it, and to architect flexible and resilient solutions to enable us to continue to fully leverage the advantages the information revolution offers us—despite the attacks.

From a national security perspective, the failure of "defense-only" information assurance (IA) security controls to protect us against the volume and increasing sophistication of cyber threats has promulgated issuance of new national cybersecurity strategies and policies, as reflected in the following: the creation of U.S. Cyber Command (USCC); two recent Presidential Directives; Defense Science Board Task Force (DSB, 2012), and findings; and the development of new Department of Defense (DoD) and DHS policies (DOD/DHS, 2010). To assure operational freedom in cyberspace, we must provide new forms of deterrence, passive and active defenses, and even retaliatory cyber capabilities if we are to adequately defend ourselves against these malicious activities in cyberspace. From a military perspective, this portends to the development and deployment of new material and nonmaterial system of systems (SoSs) solutions, that is, doctrine, organization, training, materiel, leadership education, personnel, and facilities (DOTMLPF) to protect and sustain the information advantages we've enjoyed—*a framework*. This chapter offers one step in the new development of such a framework and documents Cyber Command and Control/Battle Management (C^2/BM) as central to that process. "C^2 systems are the only systems that matter. Defense is command and control; everything else is a detail," said Paul Strassmann, a former Assistant Secretary of Defense. As our cyber knowledge, experience, and understanding evolve, the framework will need to be updated to continually inform R&D, SoSs architecture developments, acquisitions, and operations. For the purpose of our discussion in this chapter, our architectural framework is defined in terms of enterprise architecture as follows: *the policies, laws, principles, practices, taxonomy, and major architectural patterns developed, assessed, and used to create enterprise [global] architectural design views of a systems or SoSs.*

A zeroth-order diagram framing an enterprise Cyber C^2/BM Framework consistent with our definition and the discussion that follows could be represented by the diagram below. We will not elaborate further on this diagram, other than to say it was derived as Cyber C^2/BM complement to the National Institute of Standards and Technology (NIST) Framework for Improving Critical Infrastructure Cybersecurity, Version 1.0 (NIST, 2014). Figure 20.1 shows the readers the conceptual relationships consistent with our framework definition above and also serves as a framing reference for the discussions that follow in this chapter. We will argue that cybersecurity is one of many capabilities required for effectively defending an information network. Further, cybersecurity is a supporting capability to cyber operations (CyberOps) and can realize enhanced effectiveness (i.e., net-centric cyber effects) when integrated and managed by a distributed near-real-time Cyber C^2/BM system.

A Framework for Achieving Dynamic Cyber Effects through Distributed Cyber Command 533

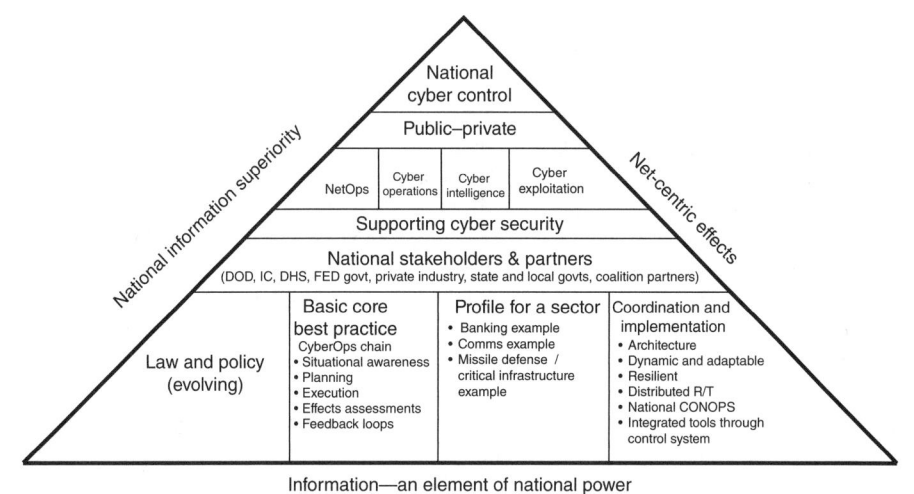

Figure 20.1 National Cyber C²/BM Framework.

C², a doctrinal requirement for the prosecution of all military operations, must be thoughtfully designed to accommodate this new and unusual cyberspace domain. It must dynamically control globally distributed SoSs within operational tempos and time-critical demands that are orders of magnitude faster than the most demanding kinetic operations (i.e., at network and computing speeds). An effective CyberOps C² architecture must be distributed and adaptable, perform in near real time, and dynamically respond to operational requirements "on demand" as they emerge. Cyber C²/BM systems must be designed to implement and manage security controls, other cyber capabilities, and standards across the globally distributed cyber battlespace to ensure interoperability with the supporting and supported commands, with the national CI stakeholders, and with our allies and coalition partners—it also must integrate with Kinetic C². The need to rapidly share information, dynamically reallocate resources, and redirect forces in near real time within a globally distributed battlespace, where executing engagements are at millisecond speeds, is paramount to mission success. *If defense is C²*, then getting C² *right* in the extreme and uniquely demanding environment of the cyber battlespace is all that matters; *everything else is detail.*

From a design perspective, pervasive evidence indicates continuing misunderstanding of what "distributed, real-time architectures" mean or, more to the point, what C²/BM design is best for rapidly controlling a globally distributed enterprise. In part, this is because many believe "distributed" refers to the physical network layer versus the logical, distributed operating service network layer; and that network "bandwidth" limitations are the cause for C²/BM performance shortfalls. However, network bandwidth seldom has proven to be the culprit; the majority of time, the significant performance problems are found with the design of the layers above the physical network—that is, with the design of the logical distributed system (DS) itself. For more on this, we recommend the reader to the IEEE ReadyNotes Series on Software Engineering of Distributed Systems, authored by Dr. Norm Howes (IEEE ReadyNotes Series, 2009–2011). The series provides an excellent discussion on problems and approaches for DS solutions: performance, DS project management, architecture and specification, security, and formal theory.

Exacerbating the situation are deep historical biases toward centralized, hierarchical C^2—that are negatively affecting distributed SoSs architecture designs, for example, centralized databases using request–reply messaging semantics. This is somewhat understandable since centralized C^2 has served the military well in the warfare for centuries. But, for time-critical kinetic warfare domains (e.g., global missile defense), centralized C^2/BM has demonstrated inferior DS performance, scalability, and flexibility for executing time-critical missions. "Speed of light" CyberOps across the new, unusual, and globally distributed cyber battlespace pose even greater demands on C^2/BM performance than the most demanding kinetic operations. Given these unique properties of the cyber battlespace, we argue that Cyber C^2/BM must be based on the SoSs networked computing architectures and highly congruent with the cyber battlespace itself.

In Scherrer and Grund (2009), the authors introduce three main models for C^2, namely, (1) the traditional model, currently employed by the DoD, (2) the cybernetics model, and (3) the cognitive–physiological model. The authors show that these models are linked to the advantaged and disadvantaged of the hierarchical, the heterarchical, and a hybrid organizational models that are potential candidates for C^2. We found their analysis of the models to be excellent, providing the strengths and weaknesses of each, and we strongly recommend their paper to our readers. In their conclusions, the authors recommend the "hybrid model" for the DoD's Cyber C^2/BM; we agree, but we caution that a Cyber C^2/BM implementation must be capable of all three forms "on demand" in near real time; that is, it must dynamically form C^2 structures that are "matched tuned" to cyber events as they emerge in near real time. We argued that no one form is right for all situations at all times or necessarily appropriate at all levels of warfare.

In this chapter, we present a Cyber C^2/BM architecture and prototype that can dynamically, "on demand," match tune its operational and systems architectures to form C^2 structures required at any given moment (e.g., create a Cyber C^2/BM structure that spans network and organizational boundaries—e.g., across the DoD and DHS). Further, with the use of a virtual organization (VO) as our operational architecture and a complementing high-performance DS architecture, we demonstrate the ability to maintain the hierarchical C^2 form with its control advantages while simultaneously operating as peers, thus providing the freedom of action and speed advantages of the other structural forms. This capability provides a powerful dynamic because cyberattacks, by their very nature, emerge instantly, can span network and organizational boundaries, and need tailored resources and forces on a moment's notice. Our Cyber C^2/BM architecture is therefore postured to engage unknown and emerging threats at all levels of cyber war (to be defined later), and it enables resiliency and is adaptive, distributed real time, and highly dynamic.

20.1.2 The Background and Foundation

Whereas, over 15 years of Cyber C^2/BM analysis, prototyping and experimentation have shown that systems architectures are best designed and implemented when they are operationally distributed, are agent based, employ publish-and-subscribe messaging and distributed shared data spaces, and employ peer-processing execution and deployment. This design has proven three to five orders of magnitude faster and simpler than client–server/request–reply architectures; has proven to be more scalable, efficient, adaptable, and robust; and has provided the more resilient "on-demand" properties and services required by CyberOps. We again recommend Howes (2009–2011) who expands on these architectural approaches and research findings in much detail.

Another critical architecture requirement for CyberOps is to enable kinetic commanders to remain continuously aware of cyber events impacting their kinetic warfighting systems performance. This suggests the need for the integration of Cyber C^2/BM and Kinetic C^2 to enable collaborative information sharing (e.g., situational awareness (SA) and coordination). The operational effectiveness, as well as ease of integrating cyber with kinetic systems, can be significantly enhanced by applying the same near-real-time, distributed Cyber C^2/BM systems architectural design patterns of the cyber systems to kinetic C^2 architectures. What's more, extensive prototype experimentation has demonstrated that we can further optimize Cyber C^2/BM effectiveness by applying the same distributed, real-time design to provide a complementary (or highly congruent) operational architecture. This enables a commander to dynamically adjust span of control "on demand" near real time, as well as redirect his/her forces and reallocate resources while delegating responsibilities to other supporting commanders as he/she deems appropriate to the unfolding battlefield conditions. Wholly consistent with the DoD Architectural Framework (DoDAF) architecting goal of harmonizing the operational and systems architectures, this approach can assure that implementations are "match tuned" to the operational processes. Integral to this framework is the customer mission outcome-based metrics methodology, described in Chapter 19, to continually tune the information value chain to optimize mission outcomes/effectiveness.

Our extensive Cyber C^2/BM prototype experimentation has also demonstrated that the design and implementation of a VO is a very effective operational architecture for Cyber C^2/BM problems. A VO enables operational execution that continuously harmonizes the Cyber C^2/BM processes with the SoSs execution. It enables operational federation of the SoSs elements and provides the ability to continually and selectively tune the integration of the distributed stakeholders and their functions across organizational and network boundaries.

20.1.3 The Beginnings, the Team, and Acknowledgments

The lead author of this chapter wrote the original concept papers utilizing a VO for distributed information [cyber] operations in the late 1990s. These early concept papers informed a funded DoD Advanced Concept Technology Demonstration (ACTD) project at the DARPA/DISA Joint Program Office (JPO) utilizing the VO as the operational architecture for distributed real-time intrusion detection (ID)—FY2001 start. The incentive for the ACTD was fueled by prescient concerns that traditional IA approaches could not adequately defend DoD networks against increasingly sophisticated cyber threats. ID was one of the first priorities at the time, and the ACTD focused on trying to resolve the "real-time, distributed" ID problem. Dr. Norman Howes, referenced earlier, joined the effort early, as the Institute of Defense Analysis' (IDA) lead—the IDA was the principal supporting contractor for the ACTD and the Cyber C^2/BM efforts described in this chapter. The author has known and worked with Dr. Howes on and off since the early years of the DoD Global Command and Control System (GCCS) initiative at the Defense Information Systems Agency (DISA).

In late 2000, the author transferred and brought members of his/her team and the ACTD concepts to the Missile Defense Agency (MDA) where he initiated, led, and was one of two principal architects (Dr. Howes was the other) and the overall government manager for a new project that extended the ACTD concepts to a full-spectrum distributed near-real-time Cyber C^2/BM architecture. The expanded project included Cyber C^2/BM ConOps, requirements generation, architecting, prototyping, and experimentation between 2000 and 2008. Dr. Michael Mezzino, a retired University of Houston Professor Emeritus, joined the team circa 2001 as the lead developer and an adjunct at the IDA. Mike got his/her daughter Meredith engaged; Meredith was primarily responsible for the design and development of the Cyber C^2/BM prototype GUI.

There were other members of the team from Computer Science Corporation, Chris Pettit and Scott Rasmussen, experts in CyberOps; Bill Fithen from Software Engineering Institute's Computer Emergency Response Team (CERT); members of the Naval Surface Warfare Center's IA Team; Mike Nassif, formally of the Air Force Research Laboratory and the MDA who was the chief of IA engineering branch in the author's IA group at the MDA; and Dr. Maarten Boasson, a former professor at University of Amsterdam and a former lab director of Thales, Netherlands, where he led the development of Splice, which is now owned by PrismTech. Maarten was a consultant to the IDA, an expert in C^2, and the principal DS infrastructure advisor to the team. Maarten also got his/her son Erik involved, who was also an expert with DS middleware and Splice applications.

The author felt honored and privileged to work with such an outstanding group of professionals and would describe those early years with the ACTD and Cyber C^2/BM R&D efforts as among the most enjoyable years of his/her federal career. The author is grateful to all those colleagues who supported the effort and the DoD who provided the environment and resources to pursue the work. There were numerous other team members who participated in the Cyber C^2/BM effort, and the author is very appreciative of their contributions as well.

The MDA team developed and addressed many Cyber C^2/BM systems questions whose answers informed requirements for congruent high-performance operational, systems, and software architectures for distributed real-time Cyber C^2/BM operations across the global Ballistic Missile Defense System (BMDS) enterprise. Some of the questions addressed included how to:

1. *Rapidly share information, collaborate, and coordinate activities,*
2. *Rapidly tailor data on demand as missions require,*
3. *Integrate cyber defense across organizational and network boundaries as a SoSs,*
4. *Rapidly integrate Cyber Engineering support and development with operations,*
5. *Implement agile acquisition to rapidly acquire new capabilities,*
6. *Rapidly manage, conduct threat and vulnerability assessments (VA),*
7. *Integrate cyber security controls managed by a Cyber C^2/BM in near real time,*
8. *Provide C^2 operational architecture that can maintain hierarchical C^2 structures while simultaneously operating as peers ("lateral"),*
9. *Implement a CONOPS that enables adaptable, dynamically reconfigurable, and decentralized operations, and*
10. *Continually integrate and synchronize Cyber C^2/BM with DoDIN C^2 and Kinetic C^2.*

The iterative architecture development and experimentation process with the Cyber C^2/BM prototype addressed these and other questions and derived a host of important insights and design requirements. Many new questions emerged as the team conducted its research and experimentation and are likely to continue to emerge as the mission of cyber warfare itself evolves.

At the Aerospace Corporation (2010–2012), the author was the principal investigator (PI) for IR&D experimentation utilizing the prototype developed at the MDA; the author has also collaborated for a number of years with Dr. Bret Michael at the Naval Postgraduate School, where Dr. Michael has used our Cyber C^2/BM architecture as a candidate topic for his/her graduate students theses. The formal Cyber C^2 construct defined later in this chapter extends and complements Ballistic Missile Defense (BMD) agent-based modeling work (Wijesekera et al., 2005) done by Dr. Michael and colleagues by adding a Cyber C^2/BM formal construct to their work.

The late Tom O'Brien, during his time with the Aerospace Corporation, joined the project in early 2012 and provided a methodology he and Dr. Martin Dixon developed and successfully applied to the DoD and the Intelligence Community (IC) problems since the early 1990s, called Problem-Focused, Effects-Based Analysis (PFEBA)—a.k.a. *"information value chain"* analysis, described in Chapter 19. He led the original government work and reconstituted his/her analytical process at the Aerospace to support multiple military and IC customers from 1996 to 2008, which included creation of the "kill chain" analysis. His/her work provides the foundation for integrated SoSs enterprise, net-centric value, and prioritization; establishes the basis for quantitative "resiliency" metrics; and fits hand in glove with "CyberOps Chain" construct discussed in this chapter.

20.1.4 Chapter Contents and Current Status of Research

As the Cyber C^2/BM architecture design and doctrinal needs to conduct CyberOps evolve, it is this authors' hope that the Cyber C^2/BM R&D and the complementary information value chain work will provide the community a foundational framework to build upon. The framework details are summarized in the following sections:

Section 1: Summary of the current state of IA: The Cyber C^2/BM problems, the DISA/ DARPA DARPA JPO, and the MDA team addressed; a brief history of the team's work and topics of the chapter

Section 2: A deeper dive into the failures of the IA and DinD strategy

Section 3: The operational architecture of the Cyber C^2/BM architecture

Section 4: The systems architecture of Cyber C^2/BM as well as a formal construct for dynamic Cyber C^2/BM to support agent-based modeling, simulation, and emulation

Section 5: A Cyber/Kinetic C^2/BM integration use-case scenario—BMD and Cyber

Section 6: Conclusion and some thoughts on the future

This framework is a work in progress and serves only as a snapshot in time and is hardly complete, as national policy; cyber strategy; tactics, techniques, and procedures (TTPs); rules of engagement (ROEs); new technologies; international laws governing cyber conflict; and operational art are all still evolving. Maturing these areas will continue to inform Cyber C^2/BM R&D, acquisition and operations, and other DOTMLPF requirements.

20.2 IA AND DEFENSE IN DEPTH: A FAILED STRATEGY

20.2.1 The DSB Stated, *"Defense Only Is a Failed Strategy"*

This cyber strategy of applying IA and defense in depth (DinD) was characterized by static point solution controls that implemented single security functions (e.g., authentication) that usually stood alone and independent of other functions in the security architecture. The resulting security implementation for a given system was then subjected to a certification and accreditation process that was executed by an independent certification team to ensure the appropriate security controls were implemented—compliance. Often, an informal risk judgment was then made based on the findings of the certifying team, which typically included a VA that identified vulnerabilities and inherent risks in the system. Finally, a decision to operate the system was made by a designated approval authority (DAA) based on the certification team's assessment and findings as well as inputs from others (e.g., a user community).

While this approach was intended to provide holistic security for DoD's computer *networks*, its effectiveness has proven marginal against the more sophisticated threats. This approach offered little in net-centric returns, since enhancing security through integration or federation of security capabilities is not an inherent property of "point solution" security control architectures. Figure 20.2 shows a contrast of security for the *sake of security* versus *cybersecurity implemented to enable net-centric cyber effects* (i.e., managed by Cyber C^2/BM processes).

20.2.2 Computer Network Defense

An operational activity involving computer network defense (CND) was implemented to complement the IA and the DinD strategy, but the ConOps was based on passive monitoring, reaction, and recovery forensic analysis; an effective Cyber C^2/BM capability had

Cyber security for the sake of security results in…	Cyber security as an integrated net-centric cyber capability results in…
• Lacks C^2 model • Point solutions • Functionally focused • System focused • Hard to create net-centric effects • Difficult to measure net-centric effects • Difficult to enable information sharing • Not designed for distributed R/T • Difficult to achieve and support advanced cyber properties / capabilities (e.g., dynamic and resiliency)	• By definition, enabled by C^2 model • Facilitates federated solutions • Synergistically focused • By definition, is SoSs focused • Naturally enables net-centric effects • Easier to measure net-centric effects • Easier to enable information sharing • Easy to design for distributed real-time • Net-centric naturally enables cyber security as supporting capability to cyberspace operations and advanced cyber properties / capabilities (e.g., dynamic, resiliency)… because C^2 is its integrating element

Figure 20.2 Cybersecurity/net-centric contrast.

not yet been developed. Integration of IA controls, as supporting capabilities to CND activities, was at the time crude at best, offering little synergistic, net-centric effects or architectural congruency between the operational CND and the IA systems architecture. One method for evaluating architectural congruency is to assess an integrated operational and systems architectures and relate architecture measures of performance (MOPs) trades to customer mission measures of effectiveness (MoEs) to determine customer utility—in general, better customer mission outcomes come from greater architectural congruence. This has been discussed further in Chapter 19.

Missing the proper analytic, net-centric, distributed SoSs and operational architectural construct, coupled with the lack of understanding on the need for architectural congruency, made cyber defense implementation awkward and ineffective. Furthermore, from an engineering and acquisition perspective, it appeared to be little understanding of DS or distributed SoSs architecture design, behaviors, or the implications of poor design to their system and operational performance. For example, using client–server architectures with request–reply messaging semantics and centralized databases for the so-called "real time," enterprise ID provided less than desired outcomes in terms of time-critical performance. Cyber engagements occur in milliseconds (or even faster), and client–server architectures employing request–reply messaging are at the low end of spectrum in terms of DS performance—often taking seconds to respond. Your browser serves as an example, where request–reply model is typical. Typically complementing these slow DS architectures are centralized operational C^2 structures, also having shown to have slow operational performance—coupling the low-performance *systems* architecture complete with a complementing low-performance *operational* architecture. The Cyber C^2/BM architect should strive for high congruency between the operational and systems architectures, establish metrics and conduct architectural trades, and perform other analysis based on assuring desired mission outcomes.

20.2.3 Information Insurance as Part of a More Holistic Cyber Strategy

Cybersecurity or cyberspace security is now emerging as the new term for IA in the DoD. However, at the time of this writing, both are still being used. We will avoid the debates over terminology, allowing the department to eventually settle on its terminology. As previously indicated, the "failure of defense-only" IA strategy to protect our networks against more sophisticated threats has compelled the cybersecurity strategy to move from one of static security controls complemented by CND to one that integrates passive and active security controls with full-spectrum CyberOps to create a more holistic cyber defense posture. To enable the new strategy, the President, Congress, and DoD have established new cyber policies and new military and civil organizations to defend the nation against cyberattacks. This includes the establishment of the USCC and its service components, two Presidential Directives (PPD-20 and PPD-21), the declaration of "cyberspace" as a battlespace or operational domain having equity with the other warfare domains, the recent DSB Task Force findings, and development and implementation of a new cybersecurity risk management framework (RMF).

Consistent with the new strategy, DISA's Strategic Goal 3 states a need to "… provide adaptive and innovative cyber/network C^2 to enable responsive operations and defense of a joint and coalition enterprise information environment in a contested or degraded cyber battlespace ensuring information superiority in defense of our Nation across the full spectrum of military operations" (DISA Strategic Plan 2013–2018). C^2 is the integrating element of net-centric warfare and an enabler of net-centric effects in the kinetic battlespaces. It's reasonable to assume that a similar strategy for cybersecurity will enable net-centric cyber effects in cyberspace. Another cited requirement of the new strategy is rapid information sharing, which can be enabled through appropriate C^2 architectures and processes.

A key objective of the new strategy is *cyber resilience.* The DoD's cyber strategy assumes attackers will gain access to DoD systems, will operate in those systems, and will cause disruption and degradation to their operations. For critical mission systems, this directly translates to loss of warfighting capabilities' operational effectiveness. Therefore, a fundamental goal of the new strategy is to operate through the cyberattacks to continually ensure mission success. The DSB recommended development of an escalation latter that would employ resiliency and cybersecurity, complemented with defensive cyber operations (DCO). Should cyberattacks progressively become more destructive, the DSB recommends an escalation to offensive cyber operations (OCO) and then the use of kinetic forces (DSB, 2012). This suggests that integration of cyber and kinetic capabilities, as we identified earlier, will be necessary to enable the recommendations of the DSB report (i.e., Cyber C^2/BM and Kinetic C^2 integration).

The challenge for the defense community, its partners, and other stakeholders is the implementation of security controls and DCO architectures to achieve high congruency and synergistic defensive effects in a SoSs enterprise or global environment; and meet the policy intent and objectives of net-centric operations (e.g., integrated single security architecture with near-real-time performance, and rapid information sharing). Implementers and users must recognize that cyber security being acquired is not only to protect and defend the system of interest to them, but may also have to share in the cyber defense of a larger SoSs enterprise, or even support the strategic cyber defense needs of allies and coalition partners. For example, the Cyber SA at one geographical location may have to support a commander at another geographical location. A use-case scenario in Section 5 illustrates such an example with SoSs integration of Cyber C^2/BM and Kinetic C^2.

It's important to note that the failure of the "defense-only strategy" doesn't obviate the need for security controls; it does however demand that those controls be implemented differently, that is, controls employing dynamic and adaptable properties as an enabler of resiliency, across the enterprise in a coordinated and federated way and managed by real-time distributed Cyber C^2/BM processes. Furthermore, many security controls are designated "operational," meaning operators conducting DCO or DoD information network (DoDIN) operations use them. When defending and securing an SoS (e.g., a BMDS as shown in Section 5), each system has to comply with its own set of security controls IAW NIST 800-53; but the program manager (PM) will likely have to consider how those controls can be implemented to cooperate with each other across desperate systems to enable net-centric cyber defense effects. Manifested SoSs behaviors will also have to be considered, as they are likely to be stochastic, nondeterministic; integration of systems into SoSs itself will likely create new cyber vulnerabilities and behaviors.

Like all military operations, DCO requires a Cyber C^2/BM structure (and system) to manage the new security controls (e.g., adaptable and dynamic), as well as other DCO cyber tools and capabilities. The two, DCO and the supporting dynamic and adaptable security control processes, must be continually coordinated, self-synchronizing, and reinforcing and in some cases automatically execute to best assure time-critical desired mission outcomes. Their combined effects can be notionally measured in terms of risk-managed, cyber-"*resilient*," net-centric effects.

20.2.4 A Metaphor for Adaptable, Dynamic, and Resilient Security Controls

While models and architectures for cyber resilience and dynamic security are evolving and there is not yet an agreed-upon approach, we suggest that a candidate cyber architecture should be at least characterized by "on-demand" adaptable and dynamic properties and be capable of near-real-time DS and SoSs performance (i.e., messaging, information sharing). Furthermore, it should be managed through distributed Cyber C²/BM processes. It's these dynamic properties, as well as cybersecurity controls implemented as we suggested earlier, that will help enable a more resilient, dynamic, and adaptive capability.

Figure 20.3 shows a familiar example "Trekies" will resonate with. It shows how the Starship Enterprise's defenses are dynamic and adaptable in near real time to enable resilience to an attack. Factiously, the Starship Enterprise had defensive capabilities that enabled it to be resilient against attacks, such as *deflector shields* and *cloaking capabilities*. Metaphorically, we are seeking similar properties and capabilities to make our distributed computer networks. The gray banner highlights *Captain Kirk* "on demand" orders to Scotty. The countermeasures were effective if they make the *Starship* resilient to attacks.

The DSB study on *Resilient Military Systems and the Advanced Cyber Threat* (DSB, 2012) could find no good quantitative metrics for cyber resiliency. However, extensive prototype experimentation indicates that inherent capabilities can be designed into a Cyber C²/BM architecture that enables SoSs resiliency. For example, the ability to maintain and

- "on demand" deflector shields
- "on demand" cloaking device
- "on demand" warp speed
- "on demand" escape shuttle pods

- "On demand" weapons
  - Pulse cannons
  - Photon torpedoes
  - DEW to kill or stun
  - Disrupters
  - Lasers
  - Ferengi energy whips
  - Thalaron radiation
  - Cobalt diselenide
  - Aceton assimilators
  - Other

"Scotty ... direct all power to starboard shields on my command."
"Scotty ... give me warp speed now!"

Figure 20.3 Adaptable, dynamic, and resilient—Starship enterprise metaphor.

ensure a protected, persistent state on a node. If that node is subjected to a fault or cyberattack, we can publish the last good state to a peer node running the same C^2/BM processes and continue operations—peer processing (Howes, 2009–2011). Resiliency is an important element of our Cyber C^2/BM Framework, and we look forward to future advances in the approaches and technologies. However, we also believe that robust resilience will be achieved through cyber forces executing DCO as well as the DSB's escalation ladder capabilities through Cyber C^2/BM; there is no silver bullet.

20.3 CYBER C^2/BM OPERATIONAL ARCHITECTURE

20.3.1 Levels of War

In kinetic warfare, geography and area typically govern the scope of military operations that will be conducted and sustained to achieve military objectives. Military organization, objectives, and operations are usually defined in terms of tactical, operational, or strategic. At the strategic level of war, activities deal with the determination of military strategic objectives and providing a framework for the use of military forces to accomplish those objectives. It involves the analysis and careful assessment of military capabilities and their deployment as one element of national power to achieve the strategic objectives (Vego, 1998).

At the operational level of war, operations are typically conducted in a given theater of operations, for example, a major regional conflict such as the Afghanistan campaign. The fundamental responsibilities here are to plan and execute a campaign as directed by a command echelon. The command provides guidance and direction to subordinates on the deployment and redeployment, conducts operational C^2 and intelligence, and provides operational fires, maneuver and movement, deception, and operational protection, as well as planning for posthostilities (Vego, 1998).

At the tactical level of war, operations are typically confined too much smaller areas, typically defined as within zones or sectors of an area of operations. The tactical level is almost exclusively focused on physical engagements or combat—that is, applying military force to achieve tactical objectives. The tactical commander is concerned with planning and executing battles, engagements, and so on to achieve tactical and sometimes [support] larger operational objectives. Figure 20.4 shows the U.S. doctrinal levels of war and their relationships (Vego, 1998).

A similar analogy to the kinetic levels of war can be applied to CyberOps, as Figure 20.5 shows. At the strategic level of cyber war, the global Internet and Global Information Grid (GIG) and the evolving Joint Information Environment (JIE) govern the scope of military operations that will be conducted and sustained to achieve military objectives in cyberspace. It involves activities that deal with the determination of strategic objectives and providing a framework for the use of CyberOps to accomplish those objectives. It requires the analysis and careful assessment of cyber capabilities and trade-offs and risks contrasted against the use of other elements of national power to achieve the strategic objectives. At this level of war, the use of information as an element of national power is fundamental to the objectives.

At the operational level of cyber war, operations are conducted in the networks of given theater of operations and usually are manifested within a major regional conflict. However, this can easily extend across theaters and areas of responsibility (AORs)

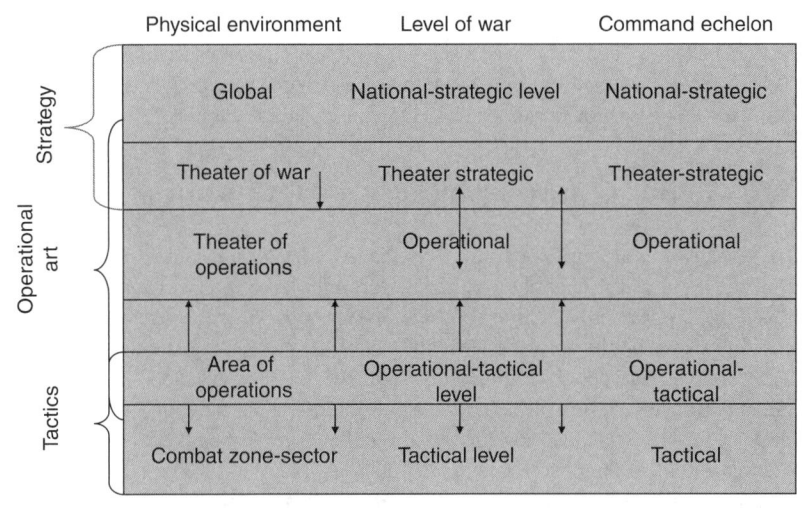

A Framework for Achieving Dynamic Cyber Effects through Distributed Cyber Command 543

Figure 20.4 Levels of war—U.S. Naval War College.

Physical environment	Level of war	Command echelon
Global	National-strategic level	National-strategic
Theater of war	Theater strategic	Theater-strategic
Theater of operations	Operational	Operational
Area of operations	Operational-tactical level	Operational-tactical
Combat zone-sector	Tactical level	Tactical

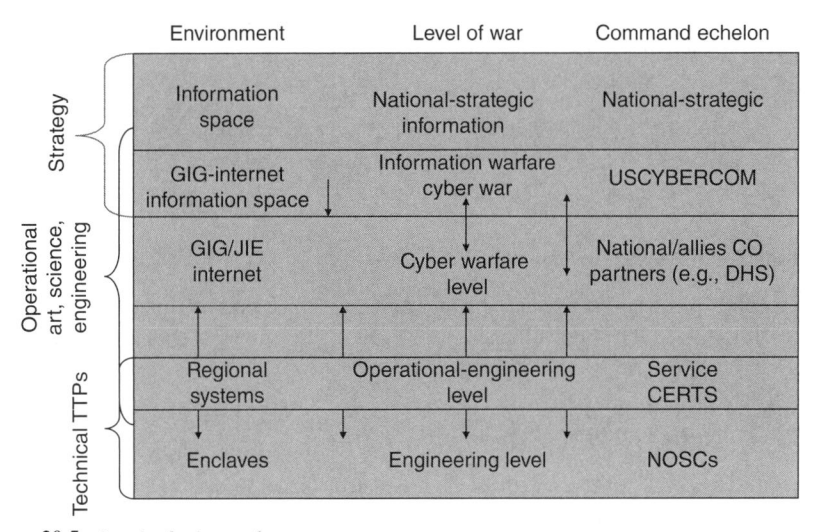

Strategy: Global, Theater of war
Operational art: Theater of war, Theater of operations, Area of operations
Tactics: Area of operations, Combat zone-sector

Figure 20.5 Levels of cyber warfare.

Environment	Level of war	Command echelon
Information space	National-strategic information	National-strategic
GIG-internet information space	Information warfare cyber war	USCYBERCOM
GIG/JIE internet	Cyber warfare level	National/allies CO partners (e.g., DHS)
Regional systems	Operational-engineering level	Service CERTS
Enclaves	Engineering level	NOSCs

Strategy: Information space, GIG-internet information space
Operational art, science, engineering: GIG-internet information space, GIG/JIE internet, Regional systems
Technical TTPs: Regional systems, Enclaves

because of the ubiquitous expanse of the cyber battlespace. The fundamental responsibilities at the operational level involve planning and executing a single DCO and OCO as directed from the command echelon (i.e., USCC and its service components). The command provides guidance and direction to subordinates, deployment and redeployment; establishes operational Cyber C^2 integration with Kinetic C^2 and intelligence; provides and conducts operational OCO and DCO; and directs maneuver and movement, deception, operational protection, and planning.

At the tactical level of cyber war, operations are conducted as technical engineering tasks. These areas are typically defined as within the local network environment; coordination can occur at higher levels of operations. The tactical level is almost exclusively focused on activities involving the use of computer applications and other tools designed to conduct CyberOps. The tactical commander is concerned with planning and executing DCO and OCO engagements to achieve tactical and sometimes larger operational objectives. Figure 20.5 shows an analogous level of war model, based on the earlier discussion, for CyberOps.

20.3.2 Cyber C²/BM Architecture Definition

DoDAF is the DoD's architecture framework that enables the definition of an architectural structure for a specific stakeholder concerns through viewpoints organized by multiple views. The views have expanded beyond the original operational, systems, and technical views and now include the software view and many other views. These views act as mechanisms for visualizing, understanding, and assimilating the broad scope and complexities of an architecture description through tabular, structural, behavioral, ontological, pictorial, temporal, or graphical means. In essence, the collection of stakeholder views helps the architect reason about the overall design and make architectural trades (DoDAF 2.0).

Our work utilized DoDAF view concepts. This section discusses aspects of the operational views of our Cyber C²/BM architecture. First, a few definitions:

20.3.2.1 CyberOps C²

The activities of a cyber commander in planning, directing, coordinating, and controlling forces in the accomplishment of the mission. The resources include personnel, equipment, communications, facilities, and procedures. The functions performed in the C² process are SA generation, planning, and COA development.

20.3.2.2 CyberOps BM

The activities of operators executing cyber defense, cyber exploitation, and cyberattack. Cyber BM involves the application of cyber TTPs, battle damage assessment (BDA), and feedback loops to create a new SA picture.

20.3.2.3 Cyber C²/BM Organizing Theme

In kinetic warfare, C²/BM is based on the concept of organizational *cells* that serve as the construct for the organization. The aggregated cells form the C² organization—its organizing theme. Cyber C²/BM organizations exist in all levels of war and are typically characterized by an *operations cell* (OPS cell), an *intelligence cell* (INT cell), a *logistics cell* (LOG cell), and so on. Kinetic warfare operators occupy these cells and perform the mission of each respective cell.

These are often *physical cells* that are located in different geographical locations whose time and distance encumber rapid coordination. Ideally, operators want immediate access from other cells to enable their rapid mission execution; but the ability to be in multiple physical cells simultaneously, or at least have detailed information and knowledge from multiple cells simultaneously, *does* not typically exist when employing physical cells. While there is interaction among physical cells in a traditional C² organization, typically through the use of technology, they each have detailed information pertinent to their local situation, not in the other cells. The other cells get summarizations of this information usually referred to as a "sitrep," or *situational picture* (e.g., the intelligence picture or the operational picture).

When cyberattacks occur in the cyber domain, vulnerabilities are discovered, or threats emerge, we have a need to disseminate their status rapidly to all stakeholders, operators, and supporting engineers alike. Further, there is a need to bring stakeholders together quickly and share information to assure accurate analysis. In the case of cyberattacks and threats, continuous updated characterization of the threat also needs to be disseminated in near real time to all stakeholders as soon as it becomes available.

20.3.3 CyberOps Chain

Cyber C^2/BM needs to provide rapid and continuously updated SA in near real time in a distributed or peer-to-peer manner across an enterprise or operational domain, for all stakeholders, simultaneously. This also requires a predefined set of C^2 command protocols, policies, and standard operating procedures (SOPs) for information sharing and related Cyber C^2/BM functions. Most importantly, the operational architecture must be able to accommodate these requirements on an "on-demand" basis enabled by a highly congruent DS architecture. The operational process or analytical construct executed by the Cyber C^2/BM system is the "CyberOps Chain," described in the following.

Figure 20.6 shows the key processes that are executed by the Cyber C^2/BM system and its operators. The CyberOps Chain is analogous to the familiar kinetic kill chain used for kinetic warfare C^2 operations:

- C^2: Cyber SA, planning, and COA development
- BM: Execution, BDA, and feedback to SA

20.3.4 The CyberOps Chain

The CyberOps Chain is a continuous, near-real-time cycle occurring across the cyber battlespace. Employing our Cyber C^2/BM prototype to implements the CyberOps Chain, discussed in further detail in the following, each of the operational cells can be mapped to the respective segments of the process, with each cell acting as executing elements of the CyberOps process, which are functionally assigned in the overall Cyber C^2/BM processes. Note some actions in the phases are autonomic or preplanned. Figure 20.6 shows the CyberOps Chain process and its integrated phases, to include preplanned autonomic responses.

20.3.4.1 SA

The first step in the CyberOps Chain involves perceiving information within the environment crucial to decision making. Local SA involves being aware of what is happening in one's local vicinity to understand how information, events, and one's own actions can impact goals and objectives, both immediately and in the near future. In the cyber battlespace, SA includes local, shared, and global SA pictures, generated simultaneously and in near real time. Having complete, accurate, up-to-the-date SA for each of these levels is essential, especially in light of the technological sophistication, complexity, and dynamically changing nature of the threats. Timely SA has been recognized as critical element of CyberOps, yet an often-elusive foundation for successful decision

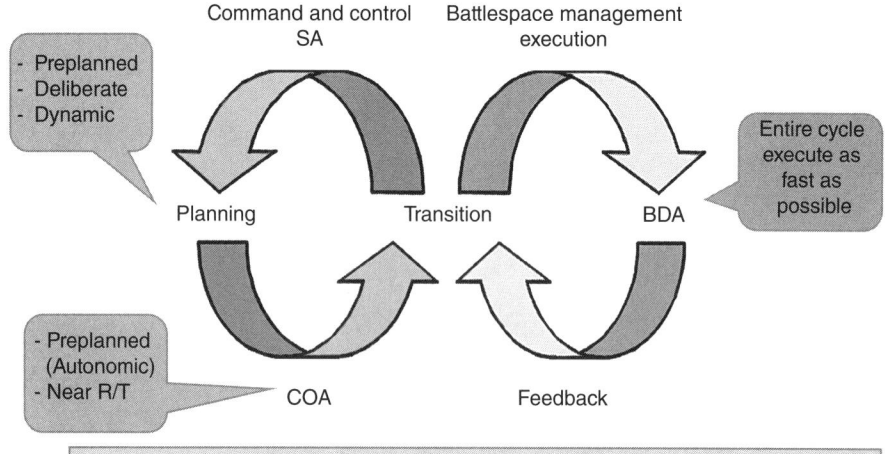

Figure 20.6 Cyber Operations chain with autonomic actions.

making considering the broad range of ever more complex and dynamically changing cyberattacks. Providing effective near-real-time distributed cyber SA has been a difficult challenge across the community. Fundamental to its successful implementation is a smart, near-real-time DS design.

20.3.4.2 Cyber C² Planning

C^2 planning can be deliberate or ad hoc (real time). Planning for cyberattack operations often has to be deliberate, whereas defense planning often has to be dynamic and near real time. Deliberate offensive planning requires exquisite detail on the target infrastructure in terms of hardware, connectivity, software, interfaces, and particular configurations, protocols, etc. The CyberOps operator needs to have a thorough understanding of the extent and consequences of the attack in order to optimize the effects while minimizing collateral damage—Stuxnet is a good example. However, because of the inherent nondeterministic nature of the battlespace, software precision, and Cyber C^2/BM, collateral damage may be difficult to predict or control. Research in this area is essential.

20.3.4.3 C² Course of Action

Course-of-action (COA) development requires the commander's staff develop and evaluate a list of important governing factors that form COAs, consider advantages and disadvantages of each, identify actions to overcome the disadvantages, make final tests for feasibility and acceptability, and weigh the relative merits of each. This step ends with the commander selecting a specific COA for further ConOps development or mission execution. Note that in some cases COAs are predetermined and execute automatically. CyberOps puts high demand on speed and performance of COA selection.

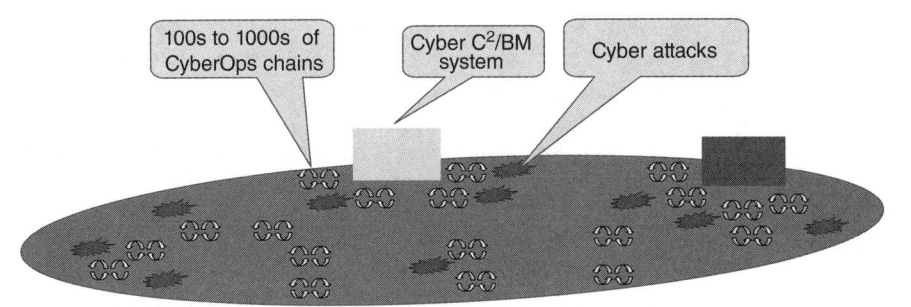

Figure 20.7 Deployed Cyber C²/BM with CyberOps Chains.

20.3.4.4 BM Execution

BM Execution begins once the commander has selected a COA and gives orders to the cyber forces to execute the COA. COAs apply to defensive, offensive, and exploitation actions or any combination of them, depending of the nature of the cyber engagement. COAs may include coordination and collaboration with other DoD and/or national stakeholders, depending on the scope of the cyber engagement.

20.3.4.5 BDA

BDA occurs after the COA has been executed, and the commander gets an assessment of its outcomes/effects—referred to as BDA. The details of the methods used to support the BDA assessment are covered in Chapter 19. However, from an information-gathering perspective, BDA is typically conducted with sensors and other techniques. In the conducting of CyberOps, it must be done in near real time. Results of the BDA are fed back into updating the SA picture via their respective feedback loops—this must happen in near real time as well to be effective.

20.3.4.6 Cyber C²/BM Deployment

Because of the systems and operational deficiencies of centralized C² discussed earlier, Cyber C²/BM systems have to be deployed as illustrated in Figure 20.7, which conveys the idea of hundreds, if not thousands, of CyberOps Chains distributed across an enterprise. Numerous CyberOps Chains could be managed simultaneously with a Cyber C²/BM system. The number of deployed Cyber C²/BM systems would be a function of the information network topology, information technology assets to be defended and other technical and operational requirements. Additionally, the Cyber C²/BM systems must employ an integrated and cooperative execution strategy; i.e., unity of effort.

20.3.5 A VO as the Cyber C²/BM Operational Architecture

20.3.5.1 The VO as the Operational Architecture for Cyber C²/BM

In the chapter introduction, we cite the impediments of centralized, hierarchical, reports-to chain-of- command structures and how they have proven inadequate for responding to real-time cyber-attacks. They are designed for centralized management of forces, assets, and recourses. Analysis and observation indicated that these traditional structures, the mainstay of the old CND C^2 operational approach, are unable to facilitate rapid dissemination of SA and information sharing within the necessary cyber planning and of COA generation within timelines to support effective Cyber Command decisions. Furthermore, they do not easily facilitate Cyber C^2/BM processes across a distributed peer environment.

An approach proven highly effective for overcoming the *operational architectural* deficiencies of hierarchical layers characteristic of traditional C^2 organizations and "reports to" C^2 is the "*VO*." A VO is inherently distributed and can be characterized by peer-to-peer operational interfaces able to span organizational and network boundaries—enabling more rapid C^2 and BM. The *CyberOps VO* approach facilitates rapid information sharing and cross-coordination and collaboration as the basis for conducting distributed information operations (Sarkesain et al., 2000) and its subset, Cyber C^2/BM operations. Furthermore, operationally effective and efficient "on-demand" capabilities are enabled by a cyber VO design when complemented with congruent systems architectural patterns. Through experimentation, analysis, and architectural trades, we discovered it can help solve many of the difficult operational problems encountered in CyberOps—such as the questions identified in this chapter's introduction. Furthermore, the VO can make operationally integrating SoSs across commands and organizational boundaries rather easily, as will be illustrated later in a use-case scenario in Section 5 of this chapter. Of course, polices and operational protocols must be established to deploy a VO, but in the cyber warfare domain, C^2 integration and cross network and organizational communication policies and protocols are critical to success.

20.3.5.2 The VO's Operational Cell Construct

The operational architecture of our Cyber C^2/BM prototype is based on *virtual cells* (also referred to as *logical cells*) as discussed and as follows from Howes et al. (2004). Virtual cells enable cyber operators and other critical personnel to exist in multiple cells simultaneously. A cyber warrior can be in multiple virtual cells simultaneously and thus have access to information from cells he *virtually* occupies. When properly implemented, they can shorten the cyber *observe, orient, decide, act* (OODA) loop timelines both locally and enterprise wide. However, it is unclear if Col Boyd's OODA loop can account for the nondeterministic behavior of distributed cyberattacks and the behaviors of the distributed battlespace; but having multiple views is highly valuable—an area for further CyberOps research.

The ability to be in multiple cells simultaneously avoids the need for hierarchical reporting of SA and facilitates rapid peer-to-peer information sharing—yet *allows for the traditional command hierarchies to be maintained, if so desired*. Cyber warfare commanders can be members of multiple lower-level virtual cells and multiple *peer cells* (virtual cells at the same level of command at other locations) and, if permitted by approved operational protocols, be members of higher-level virtual cells. In general, the VO operational architecture allows for any structure the commander wants, and new structures can be created "on demand" when complemented and enabled by highly congruent DS architecture.

Up to this point, we have defined our operational architecture as a VO, constructed with virtual cells. Let's now proceed to specifically define those cells. Each cell performs a specific cyber function (e.g., ID cell). Members of the cell perform the tasks associated with their particular, singular mission. For this reason, it's called a *mission* cell. Each cell also has a commander. In addition, each cell has a set of cyber tools tailored to the mission of that cell along with SA and other displays tailored to their mission. The cell is supported by a local virtual shared data space, which itself can be tailored to individual consoles in the cell. The original operational architecture had the following cells, with names assigned consistent with their mission functions. These cells are cyber commander cell, ID cell, VA cell, engineering cell, and the intrusion response (IR) cell. More cells have recently been added and include a legal cell (Michael et al., 2010) to inform commanders of legal issues related to CyberOps in near real time. Other new cells include OCO cells, cyber INT cells, and BDA cells. We have shown that the architecture is flexible enough to map to any organization's operational architecture; so the introduction of many new cells is anticipated as Cyber C^2/BM and military organizations evolve.

To demonstrate the flexibility of the architecture, in early the 2000s, we implemented the Cyber C^2/BM prototype at NASA's Johnson Space Center's campus under the auspices of a joint NASA and IDA small research grant. In this nonmilitary environment, the name of the commander's cell became the CIO cell. Other cell name changes and functions can easily be easily accommodated to align with an organization's cyber defense [or business] processes and policies. It took the development team about 2 months to characterize the environment and have NASA's initial capability up and running. In fact, the NASA implementation validated the ease of integrating our Cyber C^2/BM prototype into an unfamiliar environment.

We also developed and explored the concept of and capability to create dynamic cells. These come in two forms: a dynamic cell can be created by members of cells for the same mission function (i.e., mission cell), from their own or other AORs; or a *group cell* can be created from members of different mission areas (e.g., ID cell and IR cell). There can be members from n number of mission areas and levels in a group cell or from different organizations with each organization performing a critical role in the larger context of the problem (e.g., attacks across organizational and network boundaries). The dynamic cell is one way we can integrate across organizational and network boundaries both vertically and horizontally in near real time—either as a single mission cell or a group cell. This dynamic operational architecture property allows for the easy integration of traditional Kinetic C^2 and Cyber C^2/BM as well. It also enables relatively easy integration of the DoD with IC components or with allies or coalition partners if all are using interoperable Cyber C^2/BM systems, ideally employing a peer-processing model.

Figure 20.8 shows a dynamic and adaptable C^2 construct concept. As you can see, the BOLD OUTLINED dynamic cell is a group cell that includes the strategic commander and DCO ID cells from AOR 1 and AOR 2. This is our Cyber C^2/BM operational construct. You will note from the diagram that it is possible to maintain a hierarchical command structure while simultaneously operating as a peer-to-peer organization. In the systems architecture section of this chapter, we will develop a Cyber C^2/BM formal construct building on the work (Wijesekera et al., 2005) to support agent-based modeling, simulation, and emulation of this construct as Cyber C^2/BM integrated capability with BMD C^2/BM. Further, this approach can enable ad hoc command structure generation to meet the dynamic allocation demands of cross-organization and network boundaries of Cyber C^2/BM (e.g., DoD cooperating with DHS).

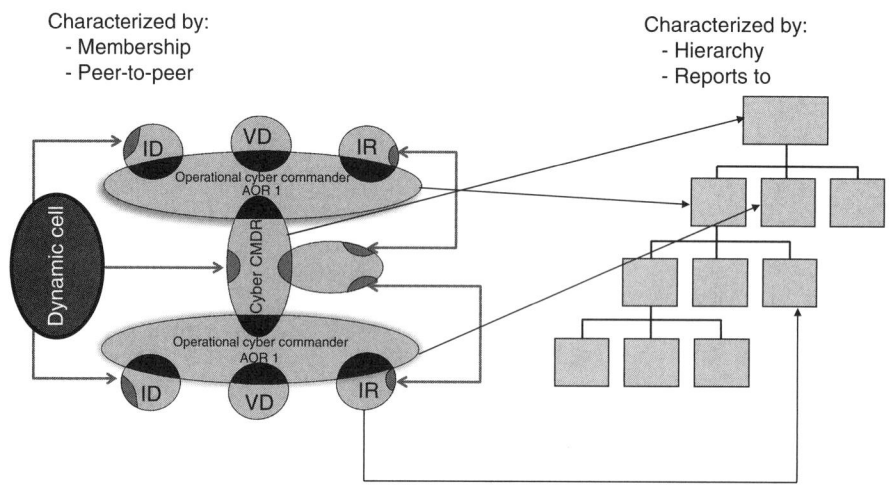

Figure 20.8 Hierarchical versus P2P C² structure.

20.3.5.3 The Virtual C²/BM Cells, Mission's Deployment, and Levels of War

From our discussion of military and cyber-analogous levels of war and Figure 20.6, we can map the ID, VA, and IR cells to the *engineering level* of cyber warfare. Similarly, the operational AORs can be mapped to the *cyber warfare* level of warfare and the strategic cyber commander (SCC) cell to the *strategic information* level of warfare. All of these mappings have equivalent mappings to the kinetic levels of war—further making cyber and kinetic military operational integration and organization familiar and conceptually straightforward.

At the tactical level of war, operators conduct tactical operations (e.g., ID) in their AOR, managed by their distributed Cyber C²/BM system in a manner depicted in Figure 20.5 and 20.8. Three tactical cells will be under the command of the regional or AOR cyber commander. We would characterize the cyber operational commander (COC) as having cyber responsibilities analogous to the responsibilities of the kinetic operational commander. Further, there must be close coordination between the two. It should be noted, depending on the number of networks and other factors in an AOR, that the number of tactical cells could vary. For example, there may be multiple ID cells and so on. What's important about this operational architecture is that it can be tailored and dynamically adapt to any organizational construct or network topology "on demand." What we've provided herein are a discussion and illustrations to explain the concepts. Furthermore, as further operational refinement occurs, new cells with new functions and C² processes will likely evolve. The operational architecture can easily accommodate extensions.

Employing a VO operational architecture as we've discussed will require new policy, doctrine, a Cyber C²/BM SoSs model, and organizational cultural changes enabling peer-to-peer, dynamic and adaptable cyber command and control operations. It should be noted however that the VO concept is not new to the DoD. A 2002–2005 DoD study entitled *The Joint Reserve Component Virtual Information Operations Organization (JRVIO)* (Duklis, 2002) describes just such an organization employing military reservist; in Sarkesain et al. (2000), the authors describe a virtual information operations agency

(VIOA) concept where our Cyber C²/BM VO instantiates virtual groups analogous to physical organization structure in a traditional Kinetic C²/BM organization.

Our VO group cells supported full-spectrum CyberOps by managing cybersecurity controls as a supporting capability to DCO. OCO can be integrated and managed in a similar manner to conduct full-spectrum CyberOps. OCO and DCO are managed as an integrated capability through Cyber C²/BM CyberOps Chain cycles, as discussed earlier, to enable net-centric cyber effects. We have already discussed cyber and kinetic integration, thus providing the commander the ability to leverage and continually tune his/her military forces and dynamically allocate his/her assets to any given mission.

20.3.6 Cyber C²/BM MoPs and MoEs

One of the new cells recently added to our Cyber C²/BM architecture is the BDA cell. Operators in this cell perform assessments of cyber operational effects. As described in Chapter 19 on PFEBA, the cyber adversary is trying to degrade our freedom of action in cyberspace (i.e., deny or distort information to commanders or critical kinetic systems). On the other hand, friendly forces are trying to maintain information superiority through Cyber C²/BM net-centric integration and operational management of cybersecurity, resiliency, DCO, OCO, and so on. The effects of cybersecurity, DCO, OCO, and cyber resiliency are measurable with architecture MoPs and customer mission MoEs—in terms of information superiority. Figure 20.9 shows the dynamic effects of CyberOps in the contested cyber

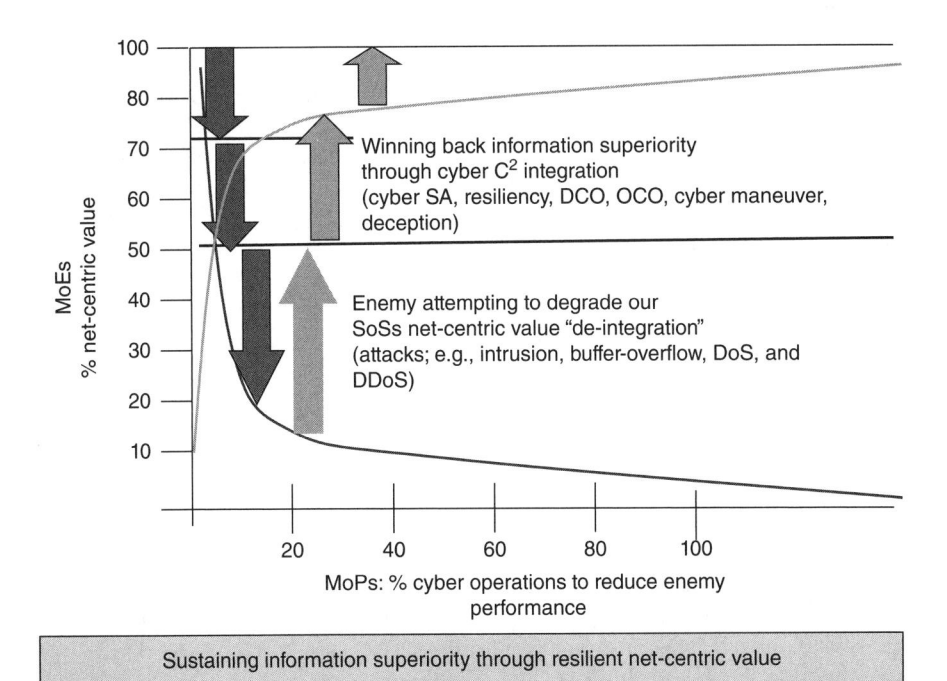

Figure 20.9 Cyber performance effect-based analysis.

battlespace, where a struggle for information superiority and dominance is continuous, requiring high-performance Cyber C^2/BM to counter and manage net speed attacks and obtain desired effects. In the next section, we'll discuss a Cyber C^2/BM systems architecture and its systems patterns that enable our high-performance, "on-demand" dynamic Cyber C^2/BM execution.

20.4 CYBER C^2/BM SYSTEMS ARCHITECTURE

20.4.1 The C^2/BM Systems Architecture

The operational architecture for our Cyber C^2/BM prototype is based on a VO with virtual cells and groups analogous to a typical military C^2 organization. Its dynamic and adaptable properties were discussed in the previous section. The supporting systems architecture will now be discussed. Figure 20.10 shows a high-level systems architectural view of our Cyber C^2/BM prototype. The application layer includes management of cybersecurity controls (e.g., ID) as integrated Cyber C^2/BM capability. The ID applications would be operated from the ID cell, discussed earlier. In the following, other applications supporting planning and COA, DCO, and OCO map to their appropriate cells.

Consistent with Howes' (2009–2011) DS design rules, our systems architecture employs an OMG Data Distribution Services (DDS) publish-and-subscribe messaging middleware standard, one of the Cyber C^2/BM systems services indicated as an infrastructure service. The initial version of our Cyber C^2/BM prototype, now referred to as the *"Cyber Command and Control and Information Operations Systems"* (C^3IOS), was written primarily in Java and ran over the DDS compliant infrastructure called *OpenSplice*, a product of PrismTech. Today, the prototype has also been deployed to Real-Time Innovations DDS product and TwinOaks Computing DDS technology. Java enabled the use of the *Java 2 Platform Security Model* (Gong, 1998), which provided security for individual distributed platforms. The Model enables the protected execution of computer code received from remote and potentially untrusted network locations (Howes et al., 2004). With that said, we recognize and have articulated the shortcomings of cybersecurity for the sake of cybersecurity.

From the systems perspective, the virtual cells are instantiated with virtual machines, much like what's now called a "cloud" architecture. However, our Cyber C^2/BM architecture is not a traditional cloud client–server architecture. The "cloud"-like properties and capabilities of the systems architecture enable the dynamic "on-demand" execution necessary to respond to distributed near-real-time cyber events (e.g., instantiate a dynamic cell or establish new organizational relationships) on the fly. This is an import capability when having to coordinate cyber activities across organizational and network boundaries or move cyber resources to support residency capabilities against emerging threats in near real time. Figure 20.10 shows the architectural organization of our Cyber C^2/BM prototype.

We have discussed cybersecurity as a supporting activity to CyberOps and managed by Cyber C^2/BM processes (i.e., CyberOps Chain). There are a number of NIST 800-53 cybersecurity controls that are particularly operational in their implementation and use and directly support DCO. An example is the IR family of controls (e.g., IR1–IR10) used in conjunction with ID capability.

Figure 20.11 shows a sample set of these IR cybersecurity controls (left side of the diagram) processing cyber SA, DCO C^2, and DCO Execution Management (EM) (right side of the diagram). The reader will note that this illustration shows a cybersecurity control

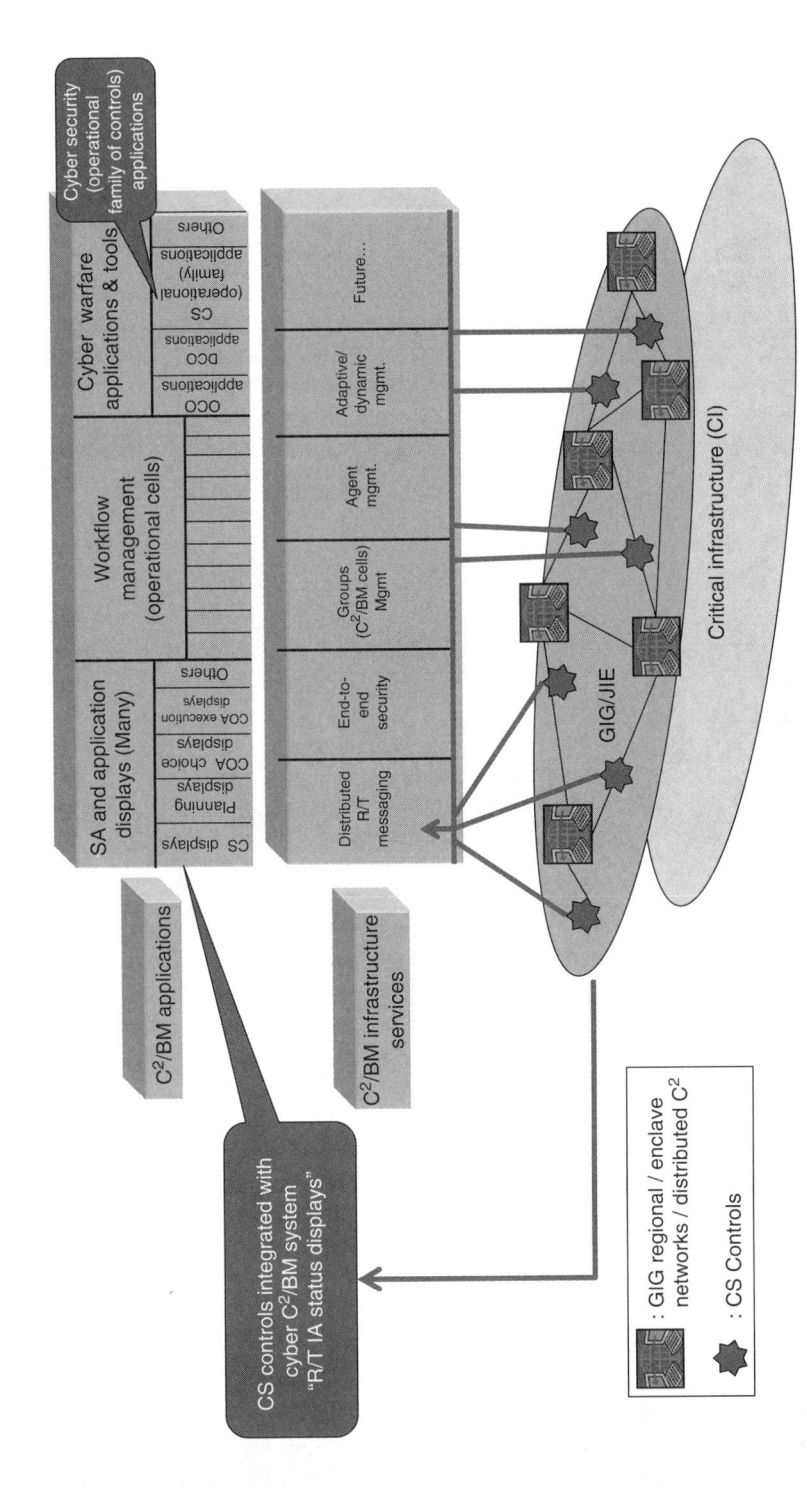

Figure 20.10 Cyber C^2/BM systems architecture (Ref: MDA Briefing, circa 2003 J.Sarkesain/N. Howes/M. Mezzino, revised).

554 Theoretical and Methodological Considerations with Applications and Lessons Learned

C² defense cyber operations (DCO)

- Cyber situational awareness
- Cyber C²/BM for DCO
  - Planning
  - COA development
  - COA selection
- DCO execution management
  - DCO COA execution
  - DCO execution assessment
- Feedback Loops

Update SA

Cyber security controls

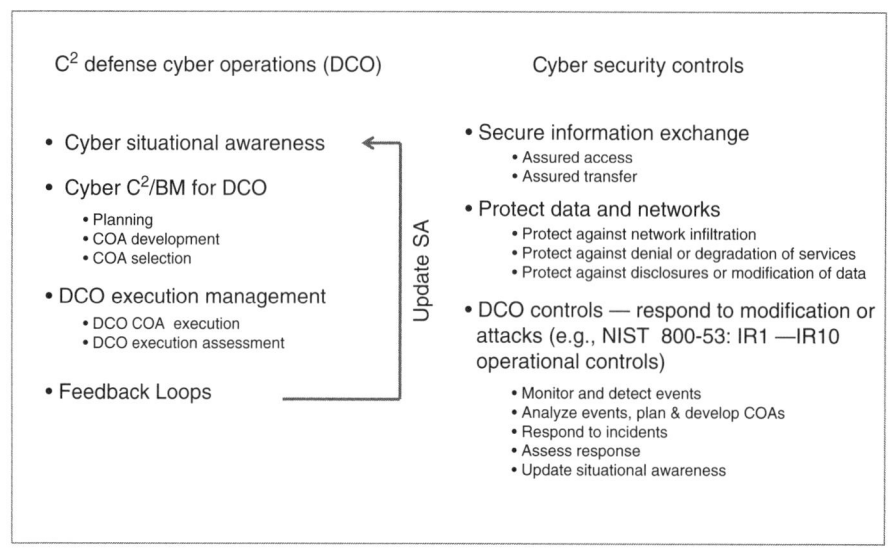

- Secure information exchange
  - Assured access
  - Assured transfer
- Protect data and networks
  - Protect against network infiltration
  - Protect against denial or degradation of services
  - Protect against disclosures or modification of data
- DCO controls — respond to modification or attacks (e.g., NIST 800-53: IR1 —IR10 operational controls)
  - Monitor and detect events
  - Analyze events, plan & develop COAs
  - Respond to incidents
  - Assess response
  - Update situational awareness

Figure 20.11 Intrusion response cybersecurity controls managed by Cyber C²/BM processes.

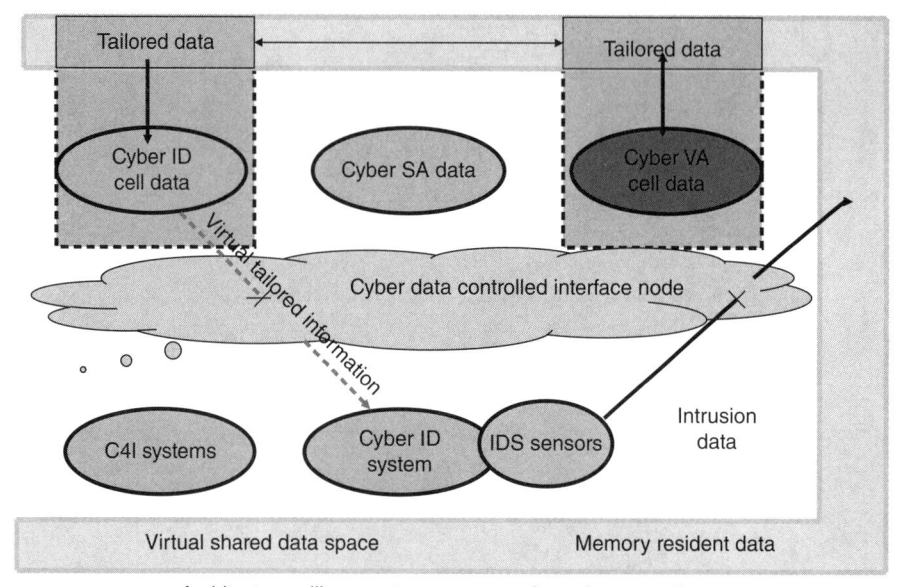

Architecture will support more aggressive cyber operations

Figure 20.12 Distributed shared data spaces (N. Howes/J. Sarkesain, revised, MDA R&D Plan, circa 2005).

being managed by the CyberOps Chain processes (i.e., Cyber C²/BM) we discussed earlier. This provides an example of cybersecurity controls as an integrated net-centric-capability (i.e., managed by distributed Cyber C²/BM) as illustrated in Figure 20.11.

Along with the publish-and-subscribe messaging infrastructure, the second major systems architectural pattern the prototype uses is a *distributed* (formally called *virtual*) *shared data space*. Figure 20.12 shows the distributed shared database concept employed with the prototype. The reader will note the distributed shared data space with locally tailored data that supports individual cells. All of the cells, including members in the Cyber C²/BM systems prototype, have their own slice (or tailored) data to perform their specific mission (e.g., ID cell). This distributed data management approach means that all cells have their own local data and do not have to query across the network to access data they need to perform their mission. As Figure 20.12 shows, the virtual shared data space makes it easy to integrate kinetic and cyber data (e.g., where data from both domains can be used to display a *combined* kinetic and cyber SA picture).

In the future, it is expected that sophisticated Cyber C²/BM will be integrated with Kinetic C² operations—providing a true SoSs architecture to enable net-centric effects across cyber and kinetic warfare operations. The third and final systems architectural pattern is a peer-processing architecture. This is a special case of P2P which theoretically employs *identical* Cyber C²/BM software suites across enterprise that operationally coordinate and cooperates in a federated manner. The concept was pictorially illustrated in Figures 7 and 10 and is discussed in Howes et al. (2004) and Howes (2009–2011).

20.4.2 Cyber C²/BM Modeling, Simulation, and Emulation

Our cells will be modeled and emulated as Cyber C²/BM systems agents, both individually and collectively. Currently, we are unaware of any official policy or operational processes governing Cyber C²/BM and cyber warfare activities, so our effort here will formally define our Cyber C²/BM command structure only. To support Cyber C²/BM modeling, simulation, emulation, and mathematical formulation, cells can be defined using set theory. Sets are defined by their membership relationships and operations, unions' intersections, and so on. Our cells are sets of CyberOps members. The cells operational behaviors can be characterized by set operations. For example, its cyber mission members define a cell's membership (e.g., ID cell), that is, a set consisting of members. Notation: "x 'is a member of' A..." means that x is a member of set A.

Figure 20.13 shows our Cyber C²/BM virtual cell structures as Venn *diagrams* in Howes et al. (2004). In general, we think of a virtual cell as a set that consists of the members in the cell. Notice that there are two BOLD OUTLINED cells in Figure 20.13. They are represented as dynamic cells in this case, which we defined earlier. Figure 20.13 has a mix of dynamic and persistent cells. These persistent cells, as the name suggests, are always up and operational. As you recall from our operational architecture discussion, persistent cells typically perform a single cyber mission (e.g., VA); its members will always have the same mission function. But that does not mean persistent cells cannot also be group cells working multiple missions. As is the case with missions cells, members may leave a group cell, which is not consistent with members leaving a set and still having the same set (i.e., a mission). Since a group cell may have members from different mission sets (cells), a member leaving could change the overall mission makeup of the group cell. For example, persistent mission virtual cells will operate 24 h a day. But there may be multiple working shifts through the course of a day, so the entire membership of a cell would change during shift change, but the mission of the cell would not change. But this is not the case because if a member leaves a group cell, the member leaving a group cell could change the skills sets in the group cell and, therefore, potentially its mission. Unlike sets, these dynamic cells and groups can include members performing a singular mission or members

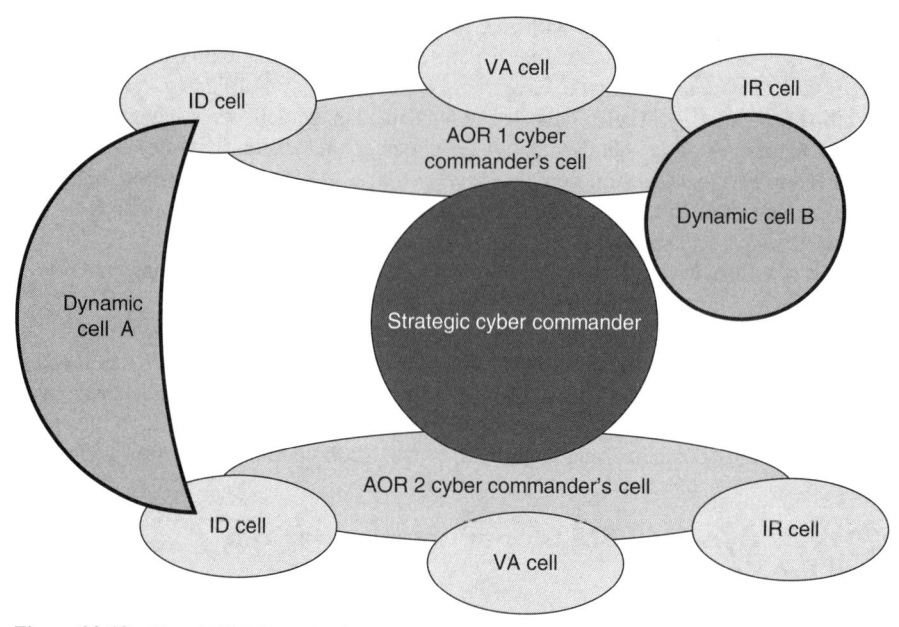

Figure 20.13 Virtual C²/BM organizations with integrated cells.

who perform different missions. But that's the purpose and power of the dynamic cell, its ability to change on demand. While it's possible for a dynamic cell or group to have a long life, it will eventually reach a decommissioned state (Howes et al., 2004).

For brevity's sake, the reader is directed to Howes et al. (2004) and Howes (2009–2011) for further information about DS and our Cyber C²/BM architectures and behaviors.

Now, we'll discuss a Cyber C²/BM formal command construct that can support Cyber C²/BM agent-based modeling, simulation, and emulation. Cyber C²/BM and Missile Defense C²/BM are similar distributed SoSs problems in that they are both globally distributed and require time-critical engagements to neutralize threats. Given the author's familiarity with missile defense and Dr. Michael's work, our formal C² construct is consistent with and augments work done at NPS (Wijesekera et al., 2005) by providing a cyber-analogous Cyber C² formal structure that can integrate with the formal BMD C² structures defined in Wijesekera et al. (2005). We now define a formal Cyber C² structure that can integrate as an SoS BMDS C²/BM with Cyber C²/BM. Lacking operational policy, processes, and ROEs for kinetic and cyber integration, we develop a Cyber C²/BM and BMD C²/BM integration use-case scenario to inform us and future work—this supports the framework's fundamental purpose as described in our introduction.

20.4.2.1 Cyber Cell Commander Agents

For the purpose of Cyber C²/BM modeling, simulation, and emulation, we will construct our cells and groups' commanders as agents. Within our Cyber C²/BM operational architecture, we define virtual cells that are analogous to cells in a typical military C² organization. Each cell has a commander. Cells are deployed at the strategic level of command by a SCC, at the operational level of command by a COC, and at the tactical level of command by a tactical cyber commander (TCC). As the reader recalls, our architecture also provides for the "on-demand" creation of dynamic cells and groups, and these cells and groups have a commander [or leader in civilian organizations] as well, i.e., a dynamic cell commander (DCC). Our formal command structure definitions are intended to be highly adaptable with "on-demand" dynamic structuring to meet the uncertainty of who may become an instant stakeholder because cyberattacks are crossing organizational and network boundaries. This requirement demands the operational capability to create "on-demand" Cyber C^2 structures with associated operational process, tools, data, protocols, and systems capabilities.

20.4.2.2 Composing Dynamic C^2 Structures

The section describes the agent-based Cyber C^2/BM formal structures that will be employed to emulate and model our Cyber C^2/BM system while executing the CyberOps Chain described earlier. While the steady state of the Cyber C^2/BM structure is usually hierarchical, with three tiers roughly equivalent to strategic, operational, and tactical level of command, we envision modeling and emulation, using which will provide our experimentation environment the ability to "on demand" dynamically tune the Cyber C^2/BM structures to maximize effects while engaging emerging or unknown cyber warfare threats (e.g., rapidly establish a C^2 structure to coordinate across organizational and network boundaries) in near real time.

20.4.2.3 Cyber C^2/BM Command Structures

In order to model, simulate, and emulate Cyber C^2/BM cells, we first define their formal structures. As we have discussed, the Cyber C^2/BM cells can be organized as a hierarchical structure familiar to kinetic warfare operators, as the right side of Figure 20.8 illustrates. However, as illustrated in Figure 20.8, there is an equivalent peer-to-peer virtual command structure that has a 1 : 1 mapping to the hierarchical C^2 structure. Because of the flexibility of the VO, the virtual C^2 organization can form a pure hierarchical structure with three tiers as we illustrated or manifest itself as a purely peer-to-peer structure or a hybrid with any desired levels of command.

Consistent with Wijesekera et al. (2005), our SCC has a set of OCCs and each OCC has a set of persistent TCCs. All TCCs read the cyber sensors net (*CyberSensorNet*) in their network AOR. They generate cyber SA and share cyber sensor net summaries and execution assessments (*ExeAssessSt*) summaries with their persistent OCCs. Execution assessments have three "states," namely, "successful," "partially successful," and "failed." And they apply at all levels of command too; summaries are tailored to the commander's needs at their respective levels. OCCs further summarize their cyber sensor net summaries and execution assessments from TCCs and forward to the persistent SCC where a strategic summary occurs from the OCC inputs. CyberOps Chain (*CyberOpsChnSt*) has a state value that maps to the each phase of the CyberOps Chain described earlier (e.g., in execution state) and again applies at all levels of command; again, the difference being the summaries is tailored to the commander's needs at their respective levels.

As the reader recalls, our C^2 structure can have dynamic cells and groups that can be created across all levels as operationally required, each with its own commander (e.g., enhance operational effects through tailored coordination). Dynamic cells or group DCCs can read all information needed to perform their dynamic mission. They provide summaries of their dynamic missions in the same manner and protocol as persistent cells, but their summaries are tailored to their dynamic mission. For the purpose of this iteration, our C² structure is bounded by the formal definition in the following. We recognize that as Cyber C²/BM processes and policies evolve, this formal definition will become much more complex. We purposely excluded the cyber weapons from the formal C² structure because this is not well defined or would likely be classified. However, it should be easy to add later.

20.4.3 Formal Structures

Definition: *A cyber-"persistent" C² structure is a 7-tuple (ID_p, scsID, occID, tccID, scc-Schema, occSchema, tccSchema) where ID_p is a finite set of identifiers that is unique to all entities in the Cyber C² model. ScsID, occID, and tccID are the identifiers of the SCC, OCC, and TCC and are satisfied by the following conditions:*

$ID_p = (tccID \cup occID \cup sccID) \cup \{CyberOpsChnSt, ExeAssessSt, CyberSensorNet,\}$ *where sets on the right side are disjoint.*

Definition: *A cyber "dynamic" C² structure is a 3-tuple (ID_d, dccID, dccSchema) where ID_d is a finite set of identifiers that is unique to the "dynamic" Cyber C² model. dccID identifies DCC: DCC is defined as n!/k!(n–k)!, where n is the number of cell commanders and k ≤ n are combinations created with "on-demand" dynamic cell instantiation across C² levels, and DCC is the highest ranking commander in the dynamic combination and is satisfied by the following conditions:*

$ID_{d=}$ [(tccID ∪ occID ∪ sccID) ∪ (tccID ∪ occID) ∪ (tccID ∪ sccID) ∪ (occID ∪ sccID)] ∪ {CyberOpsChnSt, ExeAssessSt, CyberSensorNet,} *where sets on the right side are disjoint.*

DCC combinations are defined as follows:

DCC = {DCC : (SCC ∈ DCC) ∪ (OCC ∈ DCC) ^ (TCC ∈ DCC)} ˅ {DCC : (SCC ∈ DCC) ^ (OCC ∈ DCC)} ˅ {DCC : (SCC ∈ DCC) ^ (TCC ∈ DCC)} ˅ {DCC : (OCC ∈ DCC) ^ (TCC ∈ DCC)}

Definition: *The CyberOpsChnSt identifier is satisfied as follows:*

CyberOpsChnSt(x, t) is a state variable x with values "SA" or "planning" or "COA selection" or "execution," "execution assessment," or "feedback" or "SA update." These states correspond to the phases (or states) of the CyberOps Chain at any point in time (t).

Definition: *The ExeAssesSt identifier is satisfied as follows:*

ExeAssess(y, t): y is a state variable with vales "failed" or "partial success" or "success" at any point in time (t).

Analogous to Wijesekera et al. (2005), sccSchema, occschema, and tccSchema are sets of well-typed instances of the following:

SCC (id, myPeers, CyberOpsChnSt, ExeAssess, CyberSensorNet)

OCC (id, mySuperiors, myTCCs, CyberOps, myChnSt, ExeAssess, CyberSensorNet)

TCC (id, mySuperiors, myOCC, CybetOpsChnSt, ExeAssess, CyberSensorNet)

DCC (id, myPeers(dccID), CyberOpschSt, ExeAssess, CyberSensorNet)

From D. Wijesekera and J. Michael et al., the well typedness of the schema instances is defined as follows:

1. *All instances of myPeers in sccSchema are subsets of sccID.*
2. *All subordinate instances in sccSchema are subsets of occID ∪ tccID and all subordinate instances of occID are subsets of tccID.*
3. *All superior instances in tccSchema are singleton subsets of occID ∪ tccID and all superior instances of occSchema are singleton subsets of sccID.*
4. *All superior instances in dccSchema are determined by singleton subsets such that sccID > occID > tccID.*

Again, derived from Wijesekera et al. (2005), Lemma 1 states some simple conditions satisfied by the Cyber C^2 structure.

Lemma 1: C^2 structures satisfy the following conditions:

1. *Every C^2 structure is a forest of trees and peer structures simultaneously (See Figure 20.8 for our related VO operational cell model).*
2. *Every tree in a C^2 structure can have n ≤ 3 levels, in which every path from a root to a leaf list the agents in the order [SCC > OCC > TCC].*
3. *When DCC agents exist in the C^2 structure, the order follows [SCC_d > OCC_d > TCC_d], starting with the highest-level commander in the dynamically created tree.*

In the next section, we walk through a Cyber C^2/BM and BMD C^2/BM integration scenario. An example of a formal integrated C^2 structure for Cyber/BMD would be defined as follows and again building on Wijesekera et al. (2005).

Definition: *A Cyber/BMC-"persistent" C^2 structure is a 13-tuple ($_{CB}ID_p$, scaID, rcaID, tcaID, sccID, occID, tccID, tcsSchema, rcaSchema, scaSchema, sccSchema, occSchema, tccSchema) where $_{CB}ID_p$ is a finite set of identifiers that is unique to all entities in the Cyber/BMD integrated C^2 model. sccID, occID, and tccID are the identifies of the SCC, OCC, and TCC in the cyber structure, and scaID, rcaID, and tcaID are the identifiers for the BMD structure. The integrated C^2 structure is satisfied by the following condition:*

$_{CB}ID_p$ = (tccID ∪ occID ∪ sccID) ∪ (scaID ∪ rcaID ∪ tcaID) ∪ {CyberOpsChnSt, ExeAssessSt, CyberSensorNet} ^ {sensorNet, weaponsNet} where sets on the right side are disjoint.

It follows that formal Cyber/BMD C^2 "dynamic" structures ($_{CB}ID_d$) can be defined in similar manner as they were in the previous text. We also suggest a dynamic C_2 structure that has utility in the BMD C^2 model (Wijesekera et al., 2005), for example, for rapidly establishing a C^2 structure across AORs to conduct crisis missile defense crisis planning.

20.5 CYBER C²/BM AND MISSILE DEFENSE SOSs USE-CASE SCENARIO

20.5.1 Scenario Description

The following use-case scenario can inform an SoSs cyber/kinetic agent-based integration for modeling, simulation, and emulation by employing the formal Cyber C^2 structural definitions defined herein and the BMD C^2 agent-based models defined in Wijesekera et al. (2005). As we stated earlier, further definitions are required of the Cyber C^2/BM formal models to be used for formal modeling effort, but we have provided a start.

Cyberattacks on military mission critical systems will require rapid coordination with cyber defense components. This includes cyberattacks on elements of the national CI, as it is often critical to military operations. CyberOps coordination and collaboration activities are also formally arranged between the DoD and DHS in accordance with the cooperation of the memorandum of agreement (DOD/DHS, 2010) for CI protection. In order to respond to cyber events across organizations, mission domains, AORs, and other boundaries effectively, C^2 integration capability is necessary to facilitate coordination, planning, COA development, and so on. Cyber defense of the BMD would be one such example, as it is an SoSs made up of desperate weapons platforms, sensors, and Cyber C^2/BM systems, owned and operated by different military organizations. The use-case scenario that follows illustrates such an example and is intended to provide the PM with a conceptual model to aid in the understanding of cyber support (e.g., cyber SA) to kinetic warfighting operators manning their system during a live engagement.

For this use-case scenario, it is envisioned a cyber warfare officer is an integrated member of a missile defense launch crew. This is easily accomplished with the virtual C^2 cell model employed by the Cyber C^2/BM system described (Howes et al., 2004).

In our scenario, we assume a Cyber C^2/BM system as described in this chapter executes the "CyberOps Chain." In this mock scenario, it's envisioned that the cyber warfare officers will coordinate and provide near-real-time cyber SA to their respective launch crew commanders and the overall missile defense cyber commander, who in turn would oversee the overall cyber state of the missile defense system. The launch crew cyber warfare officer would also serve as a cybercoordinating member with other cyber officers serving in other missile launch crews as well as coordinating with appropriate cyber commanders. By integrating cyber officers with missile defense launch crews, we create the distributed and operational interfaces for integrated Cyber C^2/BM and Missile Defense C^2/BM. Further, by defining missile launch crews as cells (in this case, they are *physical* cells), we create a natural integration and interface model for C^2/BM of the two operational domains (i.e., kinetic and cyber). The discussion in the following illustrates this concept with a hypothetical scenario describing a simultaneous cyberattack and missile attack against a global missile defense system.

Figure 20.14 shows graphically a hypothetical use-case scenario illustrating the integration of cyber defense and missile defense during a combined arms attack with cyberattacks and an incoming missile threat. It does not necessarily represent actual organizational alignments, command structures, or order of battle. However, it is intended to illustrate the operational tempo and speed of cyberattacks; the coordination, sharing of information, and collaboration; and most importantly, the national and joint integration needed to neutralize both cyber and missile threats that occur simultaneously against the global missile defense system. It's also intended to inform integrated agent-based modeling, simulation, and emulations in the Cyber/BMD domains. The scenario illustrates the integration of global missile defense operations and global CyberOps, both requiring simultaneous *real-time* C^2/BM and simultaneous threat engagements in their respective battlespace domains. The reader will note from Figure 20.14 the employment of both mission cells agents (e.g., dynamic cyber cell with all cyber operators) and group cell agents (e.g., launch crew with cyber and BMD operators).

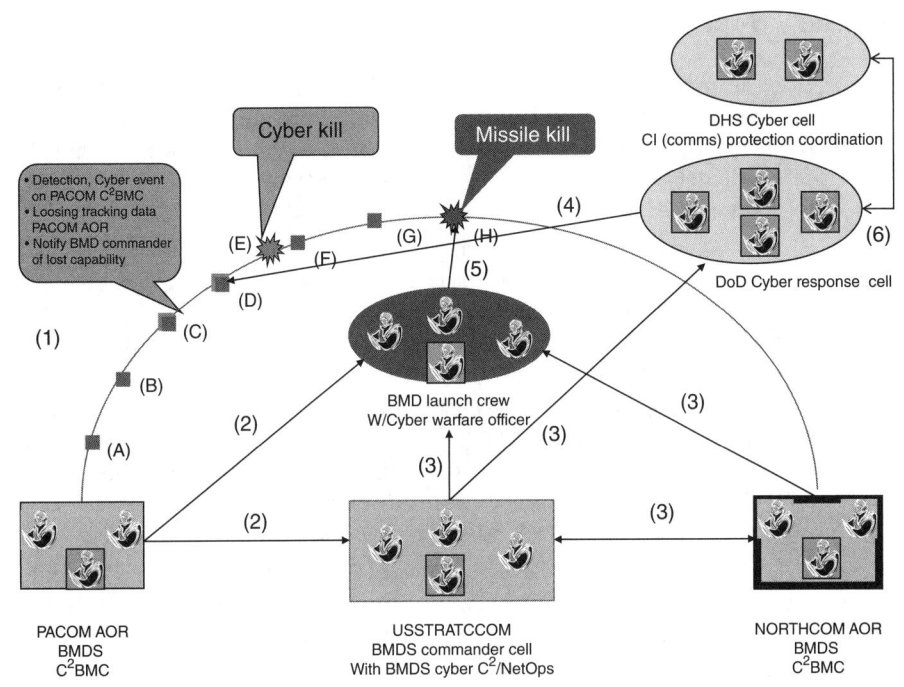

A Framework for Achieving Dynamic Cyber Effects through Distributed Cyber Command 561

Figure 20.14 SoSs Cyber C²/BMD and C²BMC integration (J. Sarkesain, General Dynamic/SIOC Group L.L.C., Staff Officers Cyber Planning Course, 2010–2011).

20.5.2 Scenario Interpretation

The symbols used in Figure 20.14 have to be interpreted as follows:
- Square boxes are major command centers.
- Ovals are operational cells associated with major command centers.
- Non-square-encased operators are missile defense-related actions and activities.
- Operators encased in a boxes are cyber operators and symbols, and callouts are cyber warfare-related actions and activities.
- ARROWS indicate notification, exchange of data, coordination, and other operational activities to ensure mission success.

It is assumed appropriate Cyber C²/BM systems are in place and will perform at real-time cyber "optional tempo" speeds to exchange data, perform information handoffs, coordination, collaboration and planning. The numbers 1–6 represent events of identification, sharing information, coordination, and collaboration, which results in actions to be taken. Letters A–H on the curve represent missile tracking in terms of elapsed time and both cyber and kinetic events (cyber and kinetic kills). The activities across both AORs, USCYBERCOM and DHS, occur in parallel, and they are not sequential.

20.5.3 Scenario Events

1. Cyber warriors detect a cyber event that is degrading missile defense at elapsed track time (C). PACOM AOR C²BMC operators (cyber and kinetic) begin analysis of missile defense and cyberattack implications on the immediate missile defense mission and their AOR. At elapsed time (D), the PACOM AOR BMD capabilities have been degraded.
2. PACOM AOR C²BMC begins sharing of cyber and missile defense information and coordination with BMD commander and BMD launch crew. Cyber SA and implications of cyberattack are shared and COA development begins. Additionally, cyber and network operators begin remediation of attack and begin to analyze the effects of the cyberattacks and begin recovery of degraded capabilities.
3. The BMDS cyber commander is coordinating missile defense COAs and cyber SA with launch crew. NORTHCOM is receiving cyber and missile defense SA updates and taking C²/BM control of the missile threat and launch. Simultaneously, USCYBERCOM is being engaged and COCOMS are sharing cyber SA and other relevant information to plan COAs and engage the cyber threat.
4. USCYBERCOM element engages the cyber threat and neutralizes the attack at elapsed time (E). At elapsed time (F), missile tracking has been restored. At elapsed time (G), tracking is confirmed and engagement of missile threat is resumed.
5. Missile launch crew engages with interceptor, and at elapsed time (H), the incoming missile threat is neutralized.
6. During engagement, DHS and DoD cyber cells sharing cyber data (e.g., CI cyber SA) with DoD cyber cell engaging cyberattackers that may be attacking CI communications vital to BMD.

As stated, this scenario is hypothetical, and it does not necessarily represent actual organizational alignments, command structure, or order of battle. However, it is intended to illustrate the operational tempo and speed of cyberattacks and the coordination, sharing of information, collaboration, and most importantly the national and joint integration that most occur to neutralize both cyber and missile threats that occur simultaneously. We must assume any missile attack targeting the United States, our allies and/or our coalition partners will be proceeded by a cyber-attack simultaneously targeting our missile defenses and CI. It is also important to note that the integrated missile and cyber coordination and actions do not occur sequentially, but rather occur in parallel. In our mock scenario, a Cyber C²/BM system must be tightly integrated with BMD C²/BM to achieve the successful outcome.

20.5.4 Scenario Summary

The earlier text is a hypothetical scenario and it does not necessarily represent actual organizational alignments, command structure, or cyber/kinetic order of battle. As we suggested above, we must assume any missile attack targeting the United States, our allies and/or coalition partners will be preceded by a cyber-attack simultaneously targeting our missile defenses and CI. By including a CI cell, we illustrate the *national* integration and coordination necessary for success during a live missile attack against the

United States. One could assume there are real scenarios where international coordination with allies and coalition partners would be required. It is also important to note that the BMD and cyber defense actions cannot occur sequentially, but rather must occur in parallel and as fast as the physical and logical Cyber/BMD environment will allow.

In a time-critical cyber/kinetic scenario as we have described, a Cyber C^2/BM system must be integrated with the Kinetic C^2 system and possess "on-demand" properties and capabilities (e.g., to instantiate a USCC dynamic cyber cell to engage the BMD cyber threat) we have described in this chapter (e.g., cybersecurity managed via near-real-time Cyber C^2/BM) if successful SoSs cyber/kinetic integration and desired combined arms operational outcomes are to be expected.

Figure 20.14 shows the SoSs integration of Cyber C^2/BM and BMD C^2/BM. From an SoSs architecting perspective, C^2 systems are the natural and most logical interface for integrating kinetic operations and CyberOps. This is consistent with the DoD's net-centric model of warfare. We believe that further research integrating CyberOps and kinetic operations will be critical to warfighting systems and SoSs in the future.

20.6 CONCLUSIONS

This framework is a work in progress and serves only as a snapshot in time. What we have provided in this first version is hardly complete, as national policy; cyber strategy; tactics, techniques, and procedures (TTPs); ROEs; new technologies; international laws governing cyber conflict; and operational art are all still evolving. Maturing these areas will continue to inform Cyber C^2/BM R&D, acquisition, operations, and other DOTMLPF requirements. We are in the midst of a revolution in *national security and global* affairs and the revolution extends far beyond the military to the very foundation of our national security. To address this threat, a Cyber C^2/BM Framework is needed to help define and make sense of this revolution, to better understand what is happening, and to architect flexible and resilient solutions that enable us to continue to fully leverage the advantages the information revolution offers us—despite the cyberattacks.

The defense-only strategy, characterized by static cybersecurity or IA controls and DinD, has failed, and we must now extend our strategy to full-spectrum operations in cyberspace. We argue that an effective cyber strategy must emphasize cyberspace operations with distributed, near-real-time Cyber C^2/BM, vice stand-alone point solution cybersecurity controls. When organizing the cyber strategy around CyberOps, cybersecurity naturally becomes a supporting activity to CyberOps. From an implementation perspective, cybersecurity becomes one of many integrated Cyber C^2/BM capabilities—analogous to Kinetic C^2 weapons, sensors, and platform integration (e.g., BMD). This leads to net-centric effects, much enhanced over current cyber approaches. Because of Cyber C^2/BM's overarching span of control, it is central to enabling holistic effective CyberOps, and for this reason, it is also central to the framework.

As we have discussed, Cyber C^2/BM must be congruent with and intrinsically linked to the cyberspace environment itself; a distributed SoSs congruent with cyberspace and capable, to the extent possible, of managing its manifested and emerging behaviors. Cyber C^2/BM and operations in cyberspace must be capable of adapting to, managing, and harmonizing with those behaviors. We have experimented and researched models for Cyber C^2/BM and have concluded that a decentralized and distributed approach is the most desirable form. Further, we have argued that Cyber C^2/BM must provide distributed near-real-time performance, be adaptable, enable resilience, and be capable of "on-demand" structural forms to dynamically tune its architecture to achieve maximum effects against unknown and emerging threats. Finally, Cyber C^2/BM provides the natural interface that links the cyber operators to cyberspace and the systems interface that links those Cyber C^2/BM processes to the logical and physical cyber battlespace itself. It is also serves as the interface for cyber and kinetic warfare integration.

REFERENCES

Department of Defense, *Department of Defense Architectural Framework*, http://en.wikipedia.org/wiki/Department_of_Defense_Architecture_Framework#Versio n_2.0_viewpoints, 2009. Accessed July 21, 2014.

Defense Information Systems Agency, *Strategic Plan 2013–2018*, Department of Defense, p. 12, 2013.

Defense Science Board, *Task Force Report: Resilient Military Systems and the Advanced Cyber Threat*, Department of Defense, p. 6, 2012.

P. Duklis, *The Joint Reserve Component Virtual Information Operations Organization (JRVIO); Cyber Warriors just a Click Away*, DLAMP, U.S. Army War College, 2002.

L. Gong, *Java 2 Platform Security Architecture, Version 1.0*, Sun Microsystems, Inc., 901 San Antonio Road, Palo Alto, CA, 1998.

N. Howes, *Distributed Systems Project Management*, Volume 1, Series in Distributed Systems Software Engineering, IEEE Computer Society, 2009.

N. Howes, *Distributed Systems Architecture and Specification*, Volume 2, Series in Distributed Systems Software Engineering, IEEE Computer Society, 2010.

N. Howes, *Distributed Systems Theory*, Volume 6, Series in Distributed Systems Software Engineering, IEEE Computer Society, 2011.

N. Howes, M. Mezzino, and J. Sarkesain, *On Cyber Warfare Command and Control*, 9th International IEEE Command and Control Research Technology Symposium, Copenhagen, 2004.

Department of Defense & Department Homeland Security MOA, Memorandum of Agreement between the Department of Homeland Security and the Department of Defense, Regarding Cyber Security, http://www.gwu.edu/~nsarchiv/NSAEBB/NSAEBB424/docs/Cyber-037.pdf, 2010. Accessed July 21, 2014.

J. Michael, J. Sarkesain, T. Wingfield, G. Dementis, G. Nuno Baptista de Sousa, *Integrating Legal and Policy Factors in Cyber Preparedness*, IEEE Security, April 2010, pp. 90–92.

NIST, *Framework for Improving Critical Infrastructure Cyber Security,* Version 1.0, 2014.

PrismTech, Data Distribution Services, *OpenSplice,* http://www.prismtech.com/opensplice. Accessed July 21, 2014.

J. Sarkesain, N. Wagoner, and D. Allain, *A Virtual Information Operations Organization: Concept Definition*, MILCOM (Classified Session), 2000.

J. Scherrer, W. Grund, *A Cyber Command and Control Model*, Air War College Maxwell Paper #47, Air University Press, Maxwell Air Force Base, Alabama, 2009.

RealTime Innovations, http://www.rti.com/. Accessed July 21, 2014.

TwinOaks Computing, Data Distribution Services, *CoreDX,* http://www.twinoakscomputing.com/. Accessed July 21, 2014.

M. Vego, United States Naval War College, *On Operational Art, 2nd Draft*, Joint Military Operations Department, Newport, RI, 1998.

D. Wijesekera, J. Michael, A. Nerode, BMD Agents: *An Agent-Based Framework to Model Ballistic Missile Defense Strategies*, 10th Annual International Command and Control Technology Symposium, Paper # 339, June 13–16, 2005, McLean, VA, 2005.

Chapter 21

System of Systems Security

Bharat B. Madan
Old Dominion University, Norfolk, VA, USA

21.1 INTRODUCTION

Geddes et al. (1998) informally define system of systems (SoS) to be "a collection of interacting systems embedded in a dynamic environment." Maier (1998) provides a more precise definition that requires constituent systems of an SoS to work collaboratively toward the common SoS objective while at the same time having operational and managerial independence. Such independence implies that in addition to performing SoS-related tasks, the constituent systems have the autonomy to perform other non-SoS-related tasks as well. The most beneficial attribute of SoS is one of "achieving a total force effect, which surpasses the sum of the individual systems" (Matthews and Collier, 2003) working collaboratively to achieve SoS functionalities. These benefits, however, come at the expense of unpredictable emergent behaviors arising from interactions and collaboration (Hall-May, 2007) between multiple systems, and it is important for an SoS to selectively rein in such destructive behaviors (Edwards, 1996). Such destructive behaviors can be the result of design deficiencies that may get manifested either due to nonmalicious actions resulting from mistakes and accidents or from malicious intrusions into the cyber systems (CSs) associated with individual constituent systems. Since our focus is on SoS security, the discussion in this chapter limits itself dealing with improper SoS behaviors caused by

---

Modeling and Simulation Support for System of Systems Engineering Applications, First Edition.
Edited by Larry B. Rainey and Andreas Tolk.
© 2015 John Wiley & Sons, Inc. Published 2015 by John Wiley & Sons, Inc.

malicious actions of attackers and how to prevent, detect, and recover from such actions. SoS security, however, presents special challenges as a result of operational and managerial independence of constituent systems (Maier, 1998), and such independence extends to the management of security as well.

SoS concept has been applied to a wide range of applications. Some of the well-known applications include Ballistic Missile Defense Systems (BMDS) (Caffall and Michael, 2005); ICSs used in energy, communication, transport, waste management, etc. (Jamshidi, 2007); and Aviation sector (Hosking and Sahin, 2009; Khosravi et al., 2009). Despite the diverse nature of these applications, there is a common architectural thread shared by all these applications. This involves collaborative integration of multiple independent and interacting systems to achieve desired system functionality. Boardman and Sauser (2006) provide a more formal definition based on the basic attributes of an SoS. These attributes are:

- Autonomy: A component system has the freedom to pursue its individual purpose while making its functionalities available to an SoS.
- Belonging: Component systems have the freedom to offer or decline their services to an SoS, depending on their needs and returns gained by accepting participation in an SoS.
- Connectivity: Mechanisms provided for communication and interactions between component systems necessary for achieving SoS functionalities.
- Diversity: SoS should exhibit diversity in terms of its functionalities as compared to the limited functionalities of its constituent systems (CS).
- Emergent behaviors: SoS should be capable of explorative operational modes that lead to the formation of previously nonexistent behaviors.
- Collaboration: SoS harnesses capabilities of multiple systems by orchestrating collaboration between constituent systems to achieve desired functional behavior at different epochs of time.

DeLaurentis and Callaway (2004) mention three additional traits to characterize SoS—physically distributed systems, dependency of overall system functionality on the linkages between the distributed systems, and heterogeneity of constituent systems. These characteristics require constituent systems to invariably incorporate embedded computer systems (henceforth referred to as CSs) that enable SoS to:

- Acquire data from sensors that sense its internal state and perceive operational environment
- Process and fuse data and information received from sensors
- Perform control algorithm computations using state information received from sensors and communicate control commands to controllers and actuators
- Communicate with peer constituent systems to affect collaboration between constituent systems
- Log activity data and important events

It is also relevant to mention that dependable communication and interaction between sensors and embedded computers as well as between CSs is a critical requirement for the proper functioning of an SoS. Communication may take place over mechanical, hydraulic, or electronic links. However, only the electronic communication is relevant in the context of SoS security. Each CS is assumed to consist of one or more embedded computers with communications links that connect a CS to its other peers. Figure 21.1 shows the logical architecture of a generic CS with emphases on inter-CS communication. There are three possible ways to architect the inter-CS communication:

- Tightly coupled dedicated point-to-point links between the CSs (Figure 21.1)[1]
- Local area network (LAN) communications architecture (Figure 21.2)
- Flexible internetworked wide area network (WAN) communications architecture (Figure 21.3)
- Loosely coupled CSs based on Service-Oriented Architecture (SOA)

All engineered systems, including SoSs, continue to become increasingly dependent on normal functioning of the embedded and visible CSs. Note that the CSs by themselves meet all the attributes of an SoS. However, we are specifically focused on SoSs that are primarily physical systems supported by CSs for communication, sensing, control, and management, that is, cyberphysical SoS. High-value functionalities, real-time requirements, and high emphasis on the safety aspects of the underlying physical processes have made the CSs used in such SoS an attractive target for cyberattackers. As a result, security of such CSs presents additional challenges, such as:

- Successful compromise of a single CS's functionality can translate into entire SoS's functionality being compromised.
- Increased attack surface area presented by multiple and interconnected cyber components.
- CSs provide critical computing, storage, and intersystem communications functionalities for SoS.
- Multiple interconnected systems with varying and autonomous security management make it easier for a cyberattacker to exploit inter-CS communication to move from a less secure CS to a more secure CS, thereby enhancing its privilege level.

The goal of a cyberattacker is to compromise SoS security in order to prevent it from delivering its functionalities. SoS designers and implementers, on the other hand, aim to

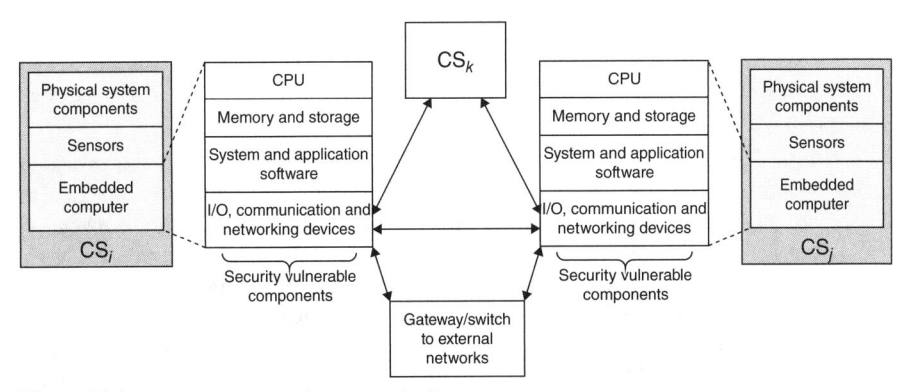

Figure 21.1 Point-to-point inter-CS communication.

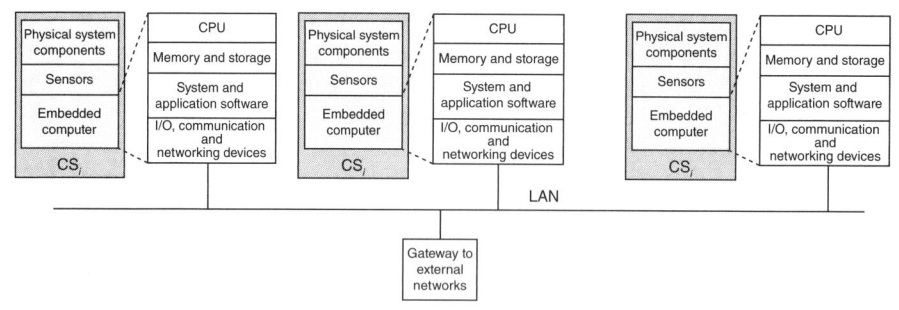

Figure 21.2 LAN-based inter-CS communication

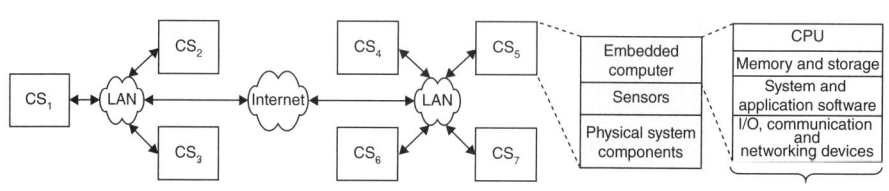

Figure 21.3 LAN/Internet CS communication

engineer robust CSs that minimize (ideally, zero) probability of a successful attack on an SoS. It is a fairly common practice to synthesize an SoS from available working systems. While it is a common practice to secure individual CSs and the associated networks, however, such mechanisms are designed to enforce the security policies of the organization or the domain that owns a particular CS. In general, an SoS may be structured as a distributed systems involving integration of multiple CSs owned and operated by different organizations with varying security architectures and priorities. Therefore, the available security mechanisms may not meet the security requirements of a particular SoS. These factors make the task of engineering SoS much more complex for the following reasons:

- SoS design philosophy emphasizes individual systems to have high degree of autonomy, which extends to security functions (e.g., authentication, authorization, encryption, etc.) as well (Bodeau, 1994). Consequently, if a particular CS_i has to interact securely with n other constituent CSs, its security architecture may have to deal with n different authentication mechanisms.
- In addition to protecting individual CSs, interface boundaries between interacting CSs are potential sources of malicious content being transferred from one CS to another (Maier, 1998), thereby substantially increasing the attack surface area (Hernan et al., 2006).
- Security requirements of a CS are primarily driven by the resources it is designed to protect. As an example, the security architecture of the history management system of an industrial control system (ICS) needs to give priority to prevention and detection of unauthorized attempts to modify the logged information over ensuring privacy and/or timely availability of such information. In contrast, the security architecture of the network connecting programmable logic controllers (PLCs) used for controlling an industrial process needs to prioritize on dealing with attacks capable of modifying, delaying, or preventing arrival of control commands sent to the PLCs, over maintaining privacy of such commands.

21.2 SoS SECURITY REQUIREMENTS

Despite best design efforts, CSs consisting of software-driven computer(s) and networking hardware used to store, process, and transport data between CSs are invariably left with known and unknown security vulnerabilities. These vulnerabilities are contributed by:

- Inadequate security policies (e.g., weak passwords, promiscuous file access policies)
- Noncompliance with security policies (e.g., connecting plug-and-play devices in defiance of a security policy)
- Need to secure interfaces used for communication and interaction independent systems with differing security requirements
- Software bugs (e.g., software code not setting and checking data buffer bounds)
- Security-deficient configuration of underlying networks (e.g., lack of network firewalls and other intrusion protection/prevention devices, which can block malicious incoming/outgoing network traffic, monitor/protect information leakage, and detect network anomalies)
- Accessing Web services and URLs embedded in e-mails and documents without validating their trustworthiness (e.g., invoking an online virus scanning service from unknown entities)
- Unintentional human error (e.g., poorly constructed SQL statements, sharing a password, etc.)
- Inadequate sanitization of insiders

Every cyberattack works by discovering and exploiting vulnerabilities, which allows an attacker to perform unauthorized actions. Typically, an attacker has to perform a multiple number of such unauthorized actions (henceforth referred to as atomic exploits), such that each atomic exploit takes the attacker closer to its final goal. Attack Graph (AG) (Phillips and Swiler, 1998; Sheyner and Wing, 2003), Privilege Graph (Ortalo et al., 1999), Attack Response Graphs (ARG) (Madan and Trivedi, 2004), Attack Trees (Schneier, 1999), and Attack Countermeasures Tree (ACT) (Roy et al., 2012) are some of the tools that can be used to model a full-fledged attack as a sequence of atomic exploits.

Security of a system and of an SoS is a multidimensional performance metric consisting of three primary attributes, *confidentiality*, *integrity*, and *availability*, and two secondary attributes, *authentication* and *nonrepudiation*. Fundamental aspect of every attack is to gain the ability to perform unauthorized actions in order to compromise one or more security attributes. Therefore, security performance is measured by the ability of the security architecture to assure high level of:

- *Confidentiality* (of data and information in-processing, in-storage, and in-transit)
- *Integrity* (of data and information in-processing, in-storage, and in-transit)
- *Availability* (of data and information to allow the SoS to perform authorized actions in time)

The *confidentiality* attribute of a system is concerned with protecting information privacy during the processing, storage, and transportation (communication) phases. The *integrity* attribute measures the robustness of a system in ensuring that attacks do not succeed in modifying information in unauthorized ways. In the context of SoS, *availability* is ultimately the most important attribute to be defended. The *availability* attribute of a system measures its ability to defend itself from attacks that seek to prevent authorized users and applications from getting access to its functionalities and services. It is relevant to point out here that vast majority of SoS applications (e.g., smart grid, ICS, air traffic control, military systems, etc.) involve integration of physical components with CSs. In all such systems, real-time response, safety, continuity of operations, etc., are of primary concern. Consequently, protection of system *availability* becomes the primary focus of the SoS security architecture. However, for the many examples of SoS that are made up of only cyber components, such as the Internet, Electronic Health Record systems, Electronic Banking, etc., making all three security attributes equally important is pivotal. Denial of Service (DoS) or Distributed DoS (DDoS) attacks, which force a CS to perform excessive, useless, and unauthorized work, are the main weapon for compromising system availability. While compromise of confidentiality, integrity may not be directly relevant to cyberphysical SoS, attackers may choose to first compromise these security attributes with eventual goal of compromising availability. For example, an attacker may carry out unauthorized modification to a critical configuration file on one of the CSs (i.e., compromising its *integrity*). If such a modification manages to successfully change the network (or the IP) address of the network interface, then this CS gets effectively disconnected from other CSs, thus compromising the *availability* of the targeted CS and that of the entire SoS. Similarly, if an attack succeeds in stealing the shadow password file (i.e., compromising the system *confidentiality*) and subsequently processes this file by a password guessing and dictionary attack tools to extract important passwords, access to these valid passwords will then enable the attacker to execute other unauthorized actions designed to make the targeted CS become unavailable to the other collaborating CSs, eventually compromising the SoS *availability*. This scenario also highlights the need for securing interfaces between the CSs of SoS. The worm malware is in fact designed specifically to migrate between CSs. Consequently, malware detection tools used for SoS security must also sanitize the communication that occurs over the interfaces between CSs.

Responsibility of security architecture is to prevent execution of such unauthorized actions by employing authentication, access control, authorization, and firewalls as preventive mechanisms. However, it is generally not possible to guarantee that preventive mechanism will not fail and, when these fail, the second line of defense is to employ intrusion detection tools, for example, malware scanners, Snort (Roesch, 1999), network traffic analyzers, etc. Detection mechanisms too have a finite probability of failing, and when these fail, the intrusion succeeds, resulting in loss of system confidentiality, integrity, and/or availability. It therefore follows that reactive security paradigms based on preventing, detecting, and recovery mechanisms are not well suited for SoS used for real-time and safety applications due to lack of time guarantees of such mechanisms. Instead, we need to device proactive intrusion-tolerant security architectures. Section 21.5 discusses the need for proactive attack tolerance in SoS and describes possible ways to provide attack tolerance capabilities to SoS in order to make them resilient by surviving attacks.

21.3 SoS SECURITY CHALLENGES

Despite rigorous system development practices, every CS is susceptible to security failures. This argument applies to CSs used in individual stand-alone systems and in SoS. Crucial SoS performance attributes that should be prevented from being compromised by cyber threats are safety, real-time response, and availability of an SoS and its constituent systems. It is worth reinforcing once again that the fundamental cause of these threats centers around the cyber components consisting of embedded computer systems within the constituent systems and the communication and networking systems used to sense, control, and manage an SoS. Irrespective of how and where a CS is used, the key asset that needs protection from threats is the data or the information derived from the data. We understand information to be derived by from data by processing, organizing, structuring, and presenting it to make it useful in the context of an application. From the security perspective, henceforth, we will use these two terms interchangeably. In this sense, SoS security is no exception and it too reduces to securing the associated CSs, which translates to the security of data—security of data-at-rest (storage), data-in-processing and data-in-transit (over communications networks). Securing data at rest requires protecting confidentiality, integrity, and availability of data storage and file systems. Data-in-processing requires protecting the data while it is being processed by a CPU of a computer by any type executable code, for example, control algorithm code, SQL code, system monitoring, and visualization code. Finally, security of data-in-transit implies protecting data as it travels over a network links.

As shown in Figures 21.1–21.3, SoS can be viewed as a synthesis of multiple systems residing in one or more autonomous network domains, which have to work collaboratively as a single system. Consequently, loss of functionality of any network domain or a CS can lead to the loss of SoS functionality. Even though on the surface, SoS may appear similar to a conventional network of CSs residing in interconnected autonomous network domains, the two can have quite different behaviors and security requirements. Unlike SoS, in conventional networked systems, an individual CS need not necessarily collaborate with other CSs to achieve its functionality. Therefore, failure of a particular CS compromises only its own functionality and possibly those of few other systems dependent on its services. Collaborative and interdependent functioning all participating constituent systems is what distinguishes SoS from conventional networked systems. As a result, SoS security requires integrated defense of individual and autonomous constituent systems, network communications links, and interfaces used for intersystem interactions.

At the individual system level, existing security technologies and measures (e.g., malware scanners, cryptography, authentication, authorization, and minimum privilege principles, VPNs and firewalls) are capable of providing acceptable security. However, security at the SoS level presents additional challenges that have not been dealt with adequately in the literature. Furthermore, since in general, individual CSs typically support multitasking, security compromises in a CS have to be managed so that compromises in one application do not propagate into other applications, particularly those supporting SoS. From the perspective of an attacker seeking to compromise an SoS, integrated nature of constituent systems becomes a valuable tool. Data flowing between two remote CSs A and E may have pass through intermediate systems or networks B, C, and D, which may or may not be part of an SoS with unknown security robustness (Zhou et al., 2008; Kennedy et al., 2010). Data flows between constituent systems allow an attacker to cause an atomic exploit in one constituent system, use this exploit to elevate its privilege level, and utilize the elevated privilege to cause a more serious atomic exploit into the next interconnected constituent system and so on. Byres et al. (2004) have modeled such phenomena with attack trees in SCADA type of SoS. It therefore follows that SoS security has to be managed at multiple levels—individual constituent systems, network communication, intersystem interface, and integrated SoS level.

SoS security also needs to cover situations in which a particular constituent system may require services of multiple other systems. For example, a military mission planner using C4ISR may require access to multiple sensor platforms that may be owned and operated by different military branches. Ensuring secure communication will require the mission planner to manage multiple identities and security associations (e.g., passwords, encryption keys, certificates, etc.), one for each sensor platforms. Besides increased overheads, requiring multiple identities also increases the possibility of identity/password theft. Distributed systems based on SOA, which allows very flexible coupling between distributed systems, have sought to address similar problems through the use of federated identity (Erl, 2007). Application of federated identity and the associated issue of single sign-on for SoS are discussed further in the next section.

21.4 SoS SECURITY SOLUTIONS

As compared to securing conventional networked information systems, managing SoS security is much more complicated because multiple constituent systems have to function properly, together as a single system. At the same time, it is important to realize that every new security technique or an algorithm needs to be rigorously and formally proven to be secure. Consequently, it is much more cost-effective to reuse existing security techniques and modify these incrementally to meet additional SoS security requirements. Additional effort required to secure SoS depends on the architecture of a particular SoS. Broadly, there are two ways to architect SoS, which may be differentiated from each other, based on the type of network coupling between constituent CSs—tight coupling and loose coupling. Tight coupling between interacting constituent systems implies that the SoS primarily is an integration of dedicated constituent systems with limited or no autonomy. An automated assembly line manufacturing system organized as a network of robotic systems according to Figure 21.1 or Figure 21.2 is an example of such an SoS. Since communication with the outside world may be minimal, security of such an SoS can be ensured by securing individual CSs using routine computer security techniques and communication with outside world by constructing a secure firewalled perimeter with only a single entry and exit point (Northcutt et al., 2005). In contrast, it is much more complex to manage the security of loosely coupled SoS architecture (Figure 21.3) used for designing military C4ISR (Dickerson, 2009), BMD (Caffall and Michael, 2005), SCADA-based ICS (Chandia, 2007), etc. Increased complexity stems from:

- Multiple constituent systems have to function properly, together as a single integrated system.
- Autonomy of constituent systems implies that the SoS will experience security, which will be least common denominator of the security of individual CSs. Therefore, if a particular CS does not meet the SoS security requirements, these may have to be met incorporating additional security mechanisms within the software code of the SoS application code running on this CS. As an example, consider the situation wherein a particular CS supports only the Digital Encryption Standard (DES), whereas SoS specifications mandate stronger Advanced Encryption Standard (AES). In this case, the SoS application software will need to incorporate additional AES. Therefore, in general, whenever certain SoS security requirements

cannot be adequately met by a particular autonomous security domain, the SoS application software to run on the corresponding CS must explicitly state such requirements and design the code to meet these requirements.
- Dynamic system boundaries due to the constantly changing network topology (Kennedy et al., 2010).
- Since individual systems may reside in autonomous network domains, it will require each network to have its own firewalled gateway to the outside world. This will require an SoS to deal with proliferation of autonomously operating gateways and security postures.
- A particular constituent system may require services of multiple other systems. Therefore, secure communication will require its security manager to manage multiple identities and security associations (e.g., passwords, encryption keys, certificates, etc.). Alternately, additional single federated identity (Gollmann, 2011) and single-sign-on systems based on Kerberos (Miller et al., 1987), public key infrastructure (PKI) (Adams and Lloyd, 2002), or Shibboleth (http://shibboleth.net/about/basic.html) may have to be integrated to meet SoS security needs.

For a loosely coupled SoS, security has to be dealt with at the individual constituent CS level, at the intersystem network communications level, and at the integrated SoS level. Security issues are further complicated by the fact that an SoS may be constructed by interconnecting existing working systems (Kennedy et al., 2011). A naïve way of meeting additional SoS security requirements would be to incrementally modify the security mechanisms (e.g., malware signatures, passwords, network security associations, intrusion detection system rules, firewall settings, etc.) provided in the release version of a system. However, such modification may not be acceptable due to the autonomous nature of individual systems. Bodeau (1994) instead argues for a more structured process, referred to as S^2 engineering, whose ultimate objective is to ensure mission-oriented secure functioning of an integrated SoS. The S^2 process outlined in Bodeau (1994) is driven by the following basic principles:

- Identification and mitigation of risks associated with end-to-end flow of information and control. If possible, do not focus risks internal to individual systems. It should be noted, however, that focus on intrasystem risks that have the potential to induce risks into other connected systems will need to be maintained, as it is highlighted in the literature on AT, AG, and PG.
- Focus first on boundaries between security domains.
- Focus second on interfaces between individual systems.
- Integrate security into target requirements and transition planning.

The S^2 engineering highlights that security SoS architecture needs to:

- Identify and mitigate risks resulting from connectivity between differing security domains
- Integrate security into the target SoS architecture by clear identification of long-term SoS security requirements
- Deal with security constraints imposed by legacy constituent systems and ensure that the integration and interactions between constituent systems do not make the SoS more vulnerable to threats identified on constituent systems.

The S² approach also lends itself to situations in which it is difficult or not feasible for a constituent system to meet the SoS security needs. Such situation can be met by letting the SoS application software code running of such a constituent system to take over such security needs. While the roots S² lie in military SoS applications, the prescribed security principles are applicable to any other SoS as well. It emphasizes on mitigating near-term risks by putting effective controls at policy levels between domain boundaries and at inter-system interfaces to ensure interoperability between differing existing security domains. It suggests that in the long term, risks can be mitigated more effectively by incorporating uniform enforcement mechanisms within and between different security policy domains.

Loose network coupling also makes the topological structure of an SoS highly dynamic since data flows between constituent CSs may not always follow a fixed network path. This causes the network perimeter to become ill defined and amorphous, making it difficult to defend the data flows over uncertain paths. Agarwal (2009) argues that traditional static security is not adequate to deal with security issues arising in dynamic uncertain environments. Scenarios dealt with in Agarwal (2009) fall into a class in which a user or a constituent system U_i wants access to an information object O_j, owned and managed by another autonomous constituent system U_j. Such scenarios are similar to the ones caused by dynamic and ad hoc data flows described earlier in this section and in Kennedy et al. (2010) and Zhou et al. (2008). The security problem arising in such scenarios is that once U_j grants access to O_j, it loses control over how U_i can use O_j. If U_i has been compromised, then it becomes possible for U_i to leak O_j to other users (i.e., loss of confidentiality) or modify it and then forward the modified version to the ultimate user U_k (i.e., loss of integrity). To deal with such problems, Agarwal (2009) proposes a new access schema that takes into account local security risk within constituent CSs and the aggregated global risks manifesting in an SoS due to dynamic integration of autonomous constituent CSs. The proposed schema suggests employing quantitative net cost–benefit analysis at individual system and integrated SoS level to arrive at suitable risk management policies when there is a finite possibility of one or more constituent systems being compromised. At this stage, while the schema is mostly theoretical, its application to practical SoS should facilitate rational decision making for arriving at SoS security engineering decisions.

Web applications running on the Internet also satisfy all the attribute of SoS. The basic request–response Web model has been enhanced into subscribe–publish model (http://www.w3.org/TR/2002/WD-ws-arch-20021114/). The enhanced Web model consists of service providers and service consumers. Providers publish their services in a service directory, while consumers search the service directory to locate services. When applied to information systems, the two entities then rendezvous to enable a provider to lease its services to a consumer through the use of remote procedure call (RPC) (Richard Steven, 1999) or remote method invocation (RMI) (http://docs.oracle.com/javase/7/docs/platform/rmi/spec/rmi-title.html). The approach to architecting highly flexible heterogeneous distributed systems is referred to as the SOA and is well suited for architecting next-generation SoS. Simanta et al. (2010) discuss synergy between SOA and SoS and suggest that dynamic collaborative nature of SoS makes SOA approach well suited for SoS. They also suggest that fail over mechanisms used in SOA can also be applied to SoS to increase their dependability. Tolk et al. (2007) discuss the suitability of integrating information technology assets of the U.S. Department of Homeland Security's spread across multiple departments, organizations, states, and nation information technology assets as SoS based on SOA principles.

In legacy SoS, security has often an afterthought, resulting in ad hoc solutions. SOA standards, in contrast, have kept the security requirements at the fore front, leading to well-structured SOA standards, for example, the SAML standard (http://docs.oasis-open.org/security/saml/v2.0/saml-core-2.0-os.pdf). SAML forms the basis of the Shibboleth (http://shibboleth.net/about/basic.html) single-sign-on technologies, which greatly reduces the overhead of managing multiple identities essential for secure access control to resources owned by different autonomous systems forming SoS. Another open standard, eXtensible Access Control Markup Language (XACML) (http://docs.oasis-open.org/xacml/3.0/xacml-3.0-core-spec-os-en.html), has been developed for securing flexible SoS based on SOA principles. Accessing data or more generally resources in a secure way is an important aspect of cybersecurity, which is equally important for SoS security. Role-Based Access Control (RBAC) and, more generally, Attribute-Based Access Control (ABAC) are the most common ways to enforce access control. One of the main goals of XACML is to support and express common access control terminology, which is human and computer understandable, in order to facilitate interoperability between different autonomous security domains. As a result, XACML has become an important component of SOA security (Kim and Jain, 2009) and for the security of SoS based on SOA.

Overlay network approach, termed as SABER, has been proposed by Keromytis et al. (2003) to manage the security of SOA in a holistic manner. Though not specifically meant for SoS, their approach can be directly applied to SoS based on SOA principles. SABER is a service-oriented network that is overlaid on a typical network protocol stack, such as the TCP/IP stack used in the Internet. It is built by creating a DoS-resistant overlay network consisting of distributed anomaly detection tools to detect intrusions and preintrusion malicious probing activities, for example, port scanning, system configuration probing, etc. The SABER design proposes to provide an integrated solution for managing security by automatically detecting, blocking, and evading attacks through collaboration between distributed and independent set of firewalls and intrusion detection systems. It is postulated that independent collaborating systems will facilitate multiple defensive strategies to be deployed in shorter time in detecting, evading, and resisting attacks. When a system is detected to be under attack, affected process is migrated to a new network location, thus hindering an attacker by effectively making the attacked target mobile. System is also designed to automatically apply software patches to address software vulnerabilities, for example, stack smashing buffer overflows, heap overflows, etc.

The ultimate goal of the SABER architecture is to provide inherent survivability, particularly against DoS attacks whose aim is to disrupt the *availability* of a CS, which makes it appealing for safety-oriented SoS. However, (Kennedy et al., 2010) points out that additional systems required by SABER to achieve its functionality add another potential vector.

21.5 INTRUSION-TOLERANT SoS

Fault tolerance is a widely accepted concept for developing safety critical systems to enable a system to keep providing its functionalities despite occasional failures. Typically, fault tolerance in such systems is implemented by employing multiple independent redundant components that operate in parallel. Redundancy may be provided passively or actively with the primary objective of preventing performance decline with minimal or no human intervention.

Passive redundancy uses excess capacity to reduce the impact of component failures. One common form of passive redundancy is the extra strength of cabling and struts used in bridges. This extra strength allows some structural components to fail without bridge collapse. The extra strength used in the design is called the margin of safety. Living creatures being endowed with two eyes (and ears) provide a good example of passive redundancy. Vision loss in one eye does not cause blindness but depth perception is impaired. Hearing loss in one ear does not cause deafness but directionality is impaired. Performance decline is commonly associated with passive redundancy when a limited number of failures occur (http://en.wikipedia.org/wiki/Redundancy_(engineering). In the context of an SoS, this may take the form of redundant sensors used to sense a state variable.

Active redundancy works by actively monitoring the performance of redundant components of a system continuously to identify any underperforming or a failed component using a switch that automatically swaps the identified component with a healthy component. Electrical power distribution is an example of such active redundancy. Multiple power lines connect each generation facility with customers. Each power line includes monitors that detect overload and circuit breakers, which disconnect a power line when a particular power line gets overloaded and power is redistributed across the remaining lines (http://en.wikipedia.org/wiki/Redundancy_(engineering). Multiple redundant components or subsystems can also be structured to work concurrently to produce multiple output streams that feed a voting system to vote out malfunctioning component(s), which are typically in minority (Srihari, 1982). Avionic and flight control systems used in contemporary fly-by-wire aircraft and space shuttle are examples of such active voting-based fault-tolerant systems.

Fault tolerance concept can be extended to deal with failures caused by cyberattacks for providing intrinsic intrusion tolerance (Pal et al., 2010). SITAR (Wang and Upppalli, 2003) is an example of a Web server that employs active redundancy and voting to provide intrusion tolerance. SABER system Keromytis et al. (2003), on the other hand, uses active monitoring to detect impending DoS attacks on a host providing services as part of an SoA system. If an attack is forecasted, then these services are moved to another redundant host. Resorting to such mobility maneuvers is a well-established tactics that has been used for thousands of years conducting warfare operations by militaries. Cyber maneuvering has recently been proposed as means to defend CSs (Applegate, 2012) and for providing network resiliency (Beraud et al., 2011). While we are not aware of the use of cyber maneuvering for tolerating intrusions in an SoS, it appears to be a promising technique for SoS.

Simple redundancy proposed in Pal et al. (2010), Wang and Upppalli (2003), and Keromytis et al. (2003) is limited to providing tolerance only against DoS type of attacks that are designed to cause *availability* failures. As argued previously, even though we may only be interested in protecting the *availability* attribute of an SoS, atomic attacks designed to compromise *confidentiality* and *integrity* are also often used to eventually compromise system *availability*. An attacker may achieve this goal by first stealing or breaking passwords (i.e., *confidentiality* attacks) and then altering a critical system configuration file (i.e., *integrity* attack) to compromise system *availability*. Consequently, we need a holistic approach capable of providing intrusion tolerance against attacks designed to compromise all three security attributes. The smart redundancy concept that combines fragmentation, coding, dispersion, and reassembly (FCDR) has been shown to successfully provide such intrusion capabilities for network routing (Madan et al., 2010) and network storage systems (Madan and Lu, 2013). These papers have shown that the FCDR technique is effective in providing intrusion tolerance against attacks designed to compromise *confidentiality*, *integrity*, and *availability* of a CS. We are currently exploring ways to incorporate the FCDR technique into the control system elements, such as the PLCs, to provide SoS used in safety critical cyberphysical applications intrinsic capabilities of tolerating wide range of intrusions.

21.6 MODELING, SIMULATION, AND EMULATION FOR SoS SECURITY

Modeling and simulation plays an important role as part of the system development life cycle of all complex systems, including SoS, associated CSs, and their security management systems. Security modeling effort for a CS starts with threat modeling and risk analysis to prioritize threat using cost–benefit risk analysis. Utility of threat modeling techniques (e.g., STRIDE (Hernan et al., 2006), AG (Sheyner and Wing, 2003), and AT (Schneier, 1999)) is limited to identifying cyber risks faced by a particular SoS. Stochastic modeling techniques (Ortalo et al., 1999; Madan and Trivedi, 2004; Madan et al., 2004), though capable of quantitative security analysis, are mostly of theoretical interest due to the difficulties involved in accurately estimating the modeling parameters (i.e., state transition probabilities and transition rates).

High value associated with information stored, processed, and transported in SoS requires deployment of dependable security mechanisms to meet safety constraints. Such mechanisms need to be capable of not only preventing intrusions but also of detecting and tolerating such intrusions. Fast pace of innovative exploits created by attackers requires these mechanisms to be constantly updated, mostly in the form of software updates, to keep pace with emerging threats. In the ensuing race, it is not uncommon for defensive mechanisms to fall behind in this race. Such situations lead to the so-called zero-day cyberattacks (O'Harrow, 2013), which installed security mechanisms may not have the algorithmic capabilities to deal with.

Providing dependable security requires extensive and periodic testing. However, once an SoS becomes operational, cybersecurity of the associated CSs testing is difficult to carry out due to the possibility of introducing unacceptable risks of causing disruptions to the critical systems managed and controlled by the CSs. Such situations can be mitigated by predicting unexpected attacks through effective training of SoS security professionals in conjunction with additional security-oriented collaboration between networked systems through sharing of malware and worm injection attempts. This raises the question of how to affect training and information sharing across multiple autonomous systems? Autonomy of individual collaborating systems implies that it may not be easy to get permissions to access these systems for training, experimentation, testing, and information sharing. The work being done by the Sandia National Lab (Urias et al., 2013; Van Leeuwen et al., 2009) suggests that simulation and emulation can play a useful role to address the issues of cybersecurity training, experimentation, and testing.

SSFNet is an open-source powerful discrete-event simulation (DES) framework for developing high-fidelity simulations of large networked systems based on standard networking protocols, that is, TCP/IP and associated routing protocols (e.g., RIP, OSPF, BGP, etc.). A networked system to be simulated is described using the Domain Modeling Language (DML) (http://www.ssfnet.org/SSFdocs/dmlReference.html). Components and their functional behaviors used in networked systems described are described as Java or C++ classes, which are instantiated as objects outlined in the DML description of the targeted system. Simulated implementation of DDoS and worm propagation attacks are also included in the standard SSFNet software package. In fact, any attack that can be described in terms of discrete event can also be easily simulated with SSFNet. Its open design makes it easy to add new application specific protocols (e.g., Modbus TCP, Distributed Network Protocol (DNP3), and IEC 60870 used in SCADA systems) to the SSFNet.

The work being done by the Sandia National Lab (Van Leeuwen et al., 2009; Urias et al., 2013) suggests that for building a cybersecurity Testbed for studying the security of networked systems, simulation alone is not sufficient. They instead highlight the need to allow integration of real and simulated components through emulation. For example, an SoS Testbed that integrates a large number of simulated constituent systems and small number of real network routers/firewalls would allow more realistic security training. The open-source Network Simulator-3 (NS3) (http://www.nsnam.org/wiki/Installation) and the OPNET (2009) software tools do in fact provide such capabilities. Combining emulation with virtualization hypervisors (e.g., VMWare Player (https://www.vmware.com/support/download-player.html), VirtualBox (https://www.virtualbox.org/), XEN (http://wiki.xen.org/wiki/Xen_Overview), etc.) can provide multiple and diverse CSs that can be integrated in an NS3 or an OPNET to create very realistic simulated SoS for cybersecurity training, testing, and experimentation.

21.7 CONCLUDING REMARKS

This chapter discusses some topics related to the security of CSs used to control and manage SoS. Based on the fundamental characteristics of SoS, we identified three security attributes—confidentiality, integrity, and availability, which need to be protected individually at each constituent CS and collectively at the SoS level. Since most SoSs are used for controlling and managing physical processes that have to meet safety constraints, the need to ensure high availability of SoS functionalities in the presence of cyber intrusions is highlighted. However, it is shown that attacker typically compromises availability by first compromising confidentiality and/or integrity of participating CSs. Next, the chapter deals with ways to leverage existing cybersecurity solutions to meet the security requirements of SoS. The challenges posed by the need to integrate the security architectures of individual CSs were identified and possible approaches for meeting challenges were discussed. The chapter also identified the need that SoS security solutions must shift from prevention, reactive detection, and mitigation paradigm to proactive intrusion tolerance to ensure that an SoS continues to provide services in the presence of successful intrusions. Finally, the chapter highlights the importance of modeling, simulation, and emulation as part of the SoS development life cycle for ensuring SoS security.

NOTE

1. CS_k denotes the *k*th CS.

REFERENCES

Adams, C., and S. Lloyd. *Understanding PKI: Concepts, Standards, and Deployment Considerations.* 2nd Edition. Boston: Pearson Education, Inc., 2002.

Agarwal, D. "A new schema for security in dynamic uncertain environments." *IEEE Sarnoff Symposium.* Princeton: IEEE, March 30 – April 1, 2009. 1–5.

Applegate, S.D. "The principle of maneuver in cyber operations." *4th IEEE International Conference on Cyber Conflicts*. Tallinn: IEEE, June 5–8, 2012. 1–13.

Beraud, P., A. Cruz, S. Hassell, and S. Meadows. "Using cyber maneuver to improve network resiliency." *IEEE Military Communication Conference*. Baltimore: IEEE, November 7–10, 2011. 1121–1126.

Boardman, J., and B. Sauser. "System of systems-the meaning of of." *IEEE/SMC International Conference on System of Systems Engineering*. Los Angeles: IEEE, April 24–26, 2006. 118–123.

Bodeau, D. "Systems-of-systems security engineering." *Proceeding of the 10th Computer Security Applications Conference*. Orlando: IEEE, December 5–9, 1994. 228–235.

Byres, E.J., M. Franz, and D.D. Miller. "The use of attack trees in assessing vulnerabilities in SCADA Systems." *IEEE International Infrastructure Survivability* Workshop. Lisbon, Purtugal: IEEE, December 5, 2004.

Caffall, D., and J. Michael. "Architectural framework for a system-of-systems." *IEEE International Conference on Systems, Man and Cybernetics*. Waikoloa, Hawaii: IEEE, October 10–12, 2005. 1876–1881.

Chandia, R., J. Jesus Gonzalez, T. Kilpatrick, M. Papa, and S. Shenoi. "Security strategies for SCADA networks." In *Critical Infrastructure Protection*, edited by E. Goetz and S. Shenoi. New York: Springer, 2007. 117–131.

DeLaurentis, D., and R.K. Callaway. "A system-of-systems perspective for public policy decisions." *Review of Public Policy Research* 21, no. 6 (2004): 829–837.

Dickerson, C.E. "Defense applications of SoS." In *Systems of Systems Engineering: Principles and Applications*, edited by M. Jamshidi. Boca Raton: CRC Press, 2009. 319–337.

Edwards, W.K. "Policies and roles in collaborative application." *Proceedings of the Conference on Computer-Supported Cooperative Work*. Cambridge, 1996. 11–20.

Erl, T. *SOA: Principles of Service Design*. Englewood Cliffs: Prentice Hall, 2007.

Geddes, N., D. Simth, and C. Lizza. "Fostering collaboration in system of systems." *IEEE International Conference on Systems, Man and Cybernetics (SMC '98)*, 1998. 950–954.

Gollmann, D. *Computer Security*. 3rd Edition. Hoboken: John Wiley & Sons, Inc., 2011.

Hall-May, M. "Ensuring safety of system of systems." PhD Thesis, University of York, York, 2007.

Hernan, S., S. Lambert, T. Ostwald, and A. Shostack. "Uncover security design flaws using the STRIDE approach." *MSDN Magazine*, November 2006.

Hosking, M., and F. Sahin. "An XML based system of systems agent-in-the-loop simulation framework using discrete event simulation." *IEEE International Conference on Systems, Man and Cybernetics*. Piscataway: IEEE, 2009. 3293–3298.

Jamshidi, M. (Ed.). *Systems of Systems Engineering: Principles and Applications*. Boca Raton: CRC Press, 2009.

Kennedy, M., D. Llewellyn-Jones, Q. Shi, and M. Merabtim. "System-of-systems security: a survey." *11th Annual Conf Convergence Telecommunication, Networking and Broadcasting (PGNet 2010)*. Liverpool, 2010.

Kennedy, M., D. Llewellyn-Jones, Q. Shi, and M Merabtim. "A framework for providing a secure systems of systems composition." *The 12th Annual Conference on the Convergence of Telecommunications, Networking & Broadcasting (PGNet 2011)*. Liverpool, 2011.

Keromytis, A.D., J. Parekh, P. Gross, G. Kaiser, V. Misra, A. Nieh, D. Rubenstein, and S. Stolfo. "A holistic approach to service survivability." *1st ACM Workshop on Survivable and Self-Regenerative Systems*. Washington, DC, October 27–30, 2003. 11–22.

Khosravi, A., S. Nahavandi, and D. Creighton. "Interpreting and modeling baggage handling system as a system of systems." *IEEE International Conference on Industrial Technology*. Gippsland, Australia: IEEE, February 10–13, 2009. 1–6.

Kim, Y.J., and R. Jain. *Access Control Service Oriented Architecture Security*. 2009.

Madan, B.B., and Y. Lu. "Attack tolerant big data file system." *ACM Sigmetrics Big Data Analytics Workshop*. Pittsburg, 2013.

Madan, B.B., and K.S. Trivedi. "Security modeling and quantification of intrusion tolerant systems using attack-response graph." *Journal of High Speed Networks* 13, no. 4 (October 2004): 297–304.

Madan, B.B., K. Goseva-Popstajanova, K. Vaidyanathan, and K.S. Trivedi. "A method for modeling and quantifying the security attributes of intrusion tolerant system." *Performance Evaluation* 56, no. 1–4 (March 2004): 167–186.

Madan, B.B., B.C. Wu, S. Phoha, and D. Bein. "Modeling and simulation of failure tolerance in scale free networks." *Winter Simulation Conference*. Baltimore, 2010.

Maier, M.W. "Architecting principles of system-of-systems." *Systems Engineering* 1, no. 4 (1998): 251–313.

Matthews, D.B, and P.A. Collier. "Assessing the value of a C4ISREW system of systems capability." *Proceedings of the 5th International Command and Control Research and Technology Symposium*. Canberra, Australia, October 24–26, 2000.

Miller, S.P., B.C. Neuman, J.C. Schiller, and J.H. Saltzer. "Section E2.1: Kerberos authentication and authorization system." Technical Report, MIT Project Athena, Cambridge, 1987.

Northcutt, S., L. Zeltser, S. Winters, K. Karen Kent, and R.W. Ritchey. *Inside Network Perimeter Security*. 2nd Edition. Indianapolis: Sams Books, 2005.

O'Harrow, R. Zero Day: *The Threat in Cyberspace*. New York: Diversion Books, 2013.

OPNET. *OPNET Modeler Documentation Set-Version: 15.0-System-in-the-Loop (SITL)*, May 2009.

Ortalo, R., Y. Deshwarte, and M. Kaaniche. "Experimenting with quantitative evaluation tools for monitoring operational security." IEEE Transactions on Software Engineering 25, no. 5 (October 1999): 633–650.

Pal, P.P., F. Webber, R.E. Schantz, and J.P. Loyall. "Intrusion tolerant systems." IEEE Information Survivability Workshop (ISW). Boston: IEEE, 2000.

Phillips, C., and L.P. Swiler. "A graph-based system for network vulnerability analysis." *Proceedings of the DARPA Information Survivability Conference and Exposition*. Charlottesville, September 22–26, 1998. 71–79.

Richard Steven, W. *UNIX Network Programming, Volume 2: Interprocess Communications*. Englewood Cliffs: Prentice Hall, 1999.

Roesch, M. "Snort-lightweight intrusion detection for networks." *Proceedings of LISA '99: 13th Systems Administration Conference*. Seattle, November 7–12, 1999. 229–238.

Roy, A., D.S. Kim, and K.S. Trivedi. "Attack countermeasure trees (ACT): towards unifying the constructs of attack and defense trees." *Security and Communication Networks* 5, no. 8 (August 2012): 929–943.

Schneier, B. "Attack trees." Dr Dobbs Journal 24 (December 1999): 21–29.

Sheyner, O., and J. Wing. "Tools for generating and analyzing attack graphs." *2nd International Symposium on Formal Methods for Components and Objects*. Leiden, the Netherlands, 2003. 344–371.

Simanta, S., E. Morris, G.A. Lewis, and D.B. Smith. "Engineering lessons for systems of systems learned from service-oriented systems." *4th Annual IEEE Systems Conference*. San Diego: IEEE, April 5–8, 2010. 634–639.

Srihari, S.N. "Reliability analysis of majority vote systems." *Information Sciences* 26, no. 3 (April 1982): 243–256.

Tolk, A., C. Turnista, and S. Diallo. "Model based alignment and orchestration of heterogeneous homeland security applications enabling composition of system of systems." *Winter Simulation Conference*. Washington, DC, December 9–12, 2007. 842–850.

Urias, V., B. Van Leeuwen, and B. Richardson. "Supervisory Command and Data Acquisition (SCADA) system cyber security analysis using a live, virtual, and constructive testbed." *IEEE MILCOM*. Orlando, October 29 – November 1, 2012. 1–8.

Van Leeuwen, B., D, Burton, U. Onunkwo, and M. McDonald. "Simulated, Emulated, and Physical Investigative Analysis (SEPIA) of networked systems." *IEEE MILCOM*. Boston, October 18–21, 2009.

Wang, F., and R. Uppalli. "SITAR Scalable Intrusion-Tolerant Architecture for Distributed Services-a technology summary." DARPA *Information Survivability Conference and Exposition*. Washington, DC, April 22–24, 2003. 153–155.

Zhou, B., A. Arabo, O. Drew, D. Llewellyn-Jones, M. Merabtui, Q. Shi, A. Waller, R. Cradock, G. Jones, and A. Yau. "Data flow security analysis for systems of systems in public security incedents." *3rd Conference on Advances in Computer Security and Forensics, Liverpool John-Moores University*. Liverpool, July 10–11, 2008.

Part IV
Conclusions

Chapter 22

Toward a Research Agenda for M&S Support of System of Systems Engineering

Andreas Tolk[1] and Larry B. Rainey[2]
[1] *SimIS Inc., Portsmouth, VA, USA*
[2] *Integrity Systems and Solutions, LLC, Colorado Springs, CO, USA*

22.1 RELEVANT EFFORTS TOWARD A RESEARCH AGENDA

As with many other fields, System of Systems Engineering (SoSE) started in many different fields and application domains. Prior to the formalization of the discipline, there was no common body of knowledge (BoK). Because academic documentation in the form of curricula and documentation of professional domains of interests in the form of a BoK must go hand in hand, a solid body of research and education of the workforce was not possible in years past without first an understanding of and consensus on a common foundation. This challenge was not new. In October 2006, a workshop on establishing a research agenda on system of systems architecting was conducted (Valerdi et al., 2008). Thirteen total participants representing commercial, defense, and academia were present. From the commercial sector, companies such as Bosch, Intelligent Systems Technology, and Motorola were present. From the defense sector, companies such as Lockheed Martin, Northrop Grumman, and Boeing were in attendance. And from academia, the University of South California and the Massachusetts Institute of Technology were represented. All 13 participants convened for a daylong workshop of brainstorming and consensus building. At the end of the workshop, the group concluded that there was a need to

---

Modeling and Simulation Support for System of Systems Engineering Applications, First Edition.
Edited by Larry B. Rainey and Andreas Tolk.
© 2015 John Wiley & Sons, Inc. Published 2015 by John Wiley & Sons, Inc.

better understand the following list of ten topics, many of which are topics addressed in this book:

1. *Resilience*: A resilient system is less likely to break down due to failures and can recover even from major disruptions. Understanding how to engineer resilient systems and system of systems was most important to the group.
2. *Illustration of success*: Architecting system of systems requires a holistic thinking effort beyond the usual reductionist approaches that govern many traditional Systems Engineering approaches. Proving that new methods and tools are successful and avoiding mistakes was a high priority.
3. *System versus system of systems attributes*: What differentiates systems and system of systems? Maier's work, summarized in Chapter 2, can be considered to be foundational in this research agenda item.
4. *Model-driven architecting*: How can models help to support better Systems Engineering? Since the workshop was conducted, Model-Based Systems Engineering (MBSE) has become ubiquitous in the model-driven architecting field. The use of a common architecture to capture the various systems, their interfaces, and their functionality to better understand their contributions and capabilities in the system of systems is a common theme in the chapters of this book.
5. *Multiple system of systems architectural views*: As system of systems is independently operated, maintained, and governed, and as different phases and stakeholders need to be supported consistently and coherently, multiple views or viewpoints have to be provided. This is a subtopic of MBSE and in particular addressed in the architectural efforts in enterprise organizations, such as described by Maule in Chapter 7.
6. *Human limits to handling complexity*: Holistically approaching the systemic challenges of highly complex system of systems is necessary, but this poses serious problems to individuals and teams. The recommendations of the group focused on controlled reductionism, which may at best be a Band-Aid, but is not a solution. Several examples in this book illustrate this concept as well as how modeling and simulation (M&S) can help to gain a better understanding of a system of systems.
7. *Net-centric vulnerability*: While the interconnection of systems brings potential synergisms into being, this is also the source of serious new threats and vulnerabilities. Today, these are often addressed as cybersecurity issues. Two chapters of this book are of particular interest in this context, namely, Obrien and Sarkesain's work in Chapter 20 and Madan's observations in Chapter 21.
8. *Evolution*: System of systems continues to develop over time, and the individual systems developments are not always aligned with each other. As a result, the overall system of systems may not only expose the desired functionality but also additional capabilities that may expose emergent behavior.
9. *Guided emergence*: One of the more challenging topics is the idea to be able to guide emergence to benefit from these unexpected possibilities. As discussed in several chapters of this book, emergence is exposed as global patterns that result

from local rules. If these global patterns support the objectives of the user of the system of systems, this is good emergence; otherwise, it is bad emergence. The idea is to better understand how to guide emergence to enable good and avoid bad emergence.

10. *No single-owner system of systems*: The absence of a common governing body is the most challenging attribute of system of systems for managers. Self-organization may help, but feedback is still needed, so that the systems that make up the system of systems are not completely independent.

The chapters of this book address all the points enumerated above by focusing on the use of M&S to support the SoSE processes. As such, they are actually establishing a new discipline contributing to the broader field of SoSE by addressing the following questions:

- What are the processes of SoSE, and which ones can be supported by M&S?
- What are the architecture artifacts needed for SoSE support, and which artifacts actually can directly contribute to or utilize M&S methods?
- What M&S methods and paradigms are particularly useful for SoSE support?

22.2 CURRENT M&S SUPPORT

As with other M&S domains, SoSE can be supported in many ways by M&S methods. Rosen (1998) describes modeling as "the essence of science and the habitat of all epistemology." The modeling relation allows one to evaluate congruencies between two models of a system. As such, modeling by itself already contributes significantly to a better understanding of the system of systems that is envisioned or to be analyzed. Adding simulation results in the ability to produce additional numerical insight into the dynamic behavior of the complex system or the system of systems. Within the chapters of this book, several contributions were made to show how to:

- Analyze current capabilities to identify gaps and future needs
- Evaluate alternative systems or systems concepts to address these shortcomings
- Support the procurement of new systems to close gaps
- Provide training for future users before and during the operational phase
- Utilize human-in-the-loop approaches to address human factors better
- Support ongoing planning and conduct operations

The M&S community distinguishes between the following M&S paradigms:

- *Monte Carlo simulations*: These simulations sample probability distributions for each variable producing a broad variety of possible outcomes applying the method of repetitive trials on a large scale. They do not only produce approximations for the mean of unknown statistical variables; they also produce deviations, variances, correlations, etc. (Rubinstein and Kroese, 2011).
- *Systems dynamics/continuous simulations*: These are dynamic simulations of systems in which the system states change continuously with time, usually described by differential equations. When applied to systems or system of systems, they

contribute to understanding the behavior of nonlinear, highly interconnected systems over time by modeling internal feedback loops, flows with time delays, and stocks and piles (Birta and Arbez, 2007).

- *Discrete-event simulations*: These are dynamic simulations of systems in which the system states change instantaneously when defined events occur. The nature of the state change and the time at which the change occurs mandate precise description (Banks et al., 2009).
- *Agent-based simulations*: Agents are "intelligent software objects" that perceive their situated environment, try to reach their objectives, communicate with other agents, act in their situated environment, observe the results, and learn accordingly. Their interaction is governed by rules. They build the system bottom-up (Yilmaz and Ören, 2009).

Mittal and colleagues (2008) were among the first who systematically looked at the opportunity to utilize M&S methods to support SoSE better by focusing and the discrete-event simulation formalism (DEVS) that was originally directly derived from Systems Engineering ideas and applying them to create better simulation systems. Mittal's contribution to this book also shows the need for close alignment of M&S and Systems Engineering when supporting SoSE. Another piece of pioneering work is described by Kewley and Wood (2012) who use simulation federations to support SoSE for the Armed Forces.

Within this book, Bruzzone, Massei, Garro, and Longo make a strong case for the use of agent-based simulation in Chapter 8. The argument to utilize agents that use domain-specific ontologies for the system of systems communication has also been articulated by Tolk et al. (2011). The reason becomes apparent when the agent-based simulation characteristics are mapped to the characteristics of system of systems as defined by Maier in Chapter 2:

1. *Operational independence of the individual systems*: Agent can be used to implement each system that belongs to the system of systems. This ensures that the operations remain independent, as they are encapsulated in the implementing agent. The rules of the agent guide the execution.
2. *Managerial independence of the systems*: Each agent can contain its own set of rules, objectives, and even value systems and desires. They can communicate with other agents representing other systems to find their optimal strategy based on their guidelines and objectives.
3. *Geographic distribution*: As agents can be mobile, they can literally be geographically dispersed. Alternatively, the distribution can be simulated, including clear guidelines which resources can be shared.
4. *Emergent behavior*: Agents are the software solution most often associated with emerging behavior. They are often even used as examples for emergence of global patterns that result from rules on the local level of the agents.
5. *Evolutionary development*: As each agent is independent, it is easy to evolve them individually as well.

Therefore, when it comes to supporting SoSE, it makes sense to focus on agent-based simulations. This does not exclude the other paradigms enumerated in this section. Agents use discrete-event simulation methods as well as Monte Carlo methods. They can either embed systems dynamics modules or be embedded into them. Simulation methods are becoming more and more hybrid, but the agent-based metaphor remains nearly ideal for SoSE efforts.

22.3 ORGANIZATIONAL CHALLENGES OF SYSTEM OF SYSTEMS

A major challenge of system of systems from the management view is the *managerial independence of the systems*. There is no overarching authority that guides the overall processes. Very often, there is not even a way to influence these system-specific processes, as discussed in several chapters. Self-organization can be observed nonetheless.

The North Atlantic Treaty Organization (NATO) has often found themselves in system of systems in recent efforts, such as the humanitarian relief operations after the tsunami in 2004 (Huber et al., 2008). Various nongovernmental organizations, nonprofit help organizations, local groups, and military support came together to reach the common objective to help the tsunami victims, but no central command was in charge. A research group on command and control challenges in such environments (Alberts et al., 2010) developed a maturity model to measure the ability of independent organizations to support a common goal. This model has high potential to be extended and become applicable to address the issue of missing common governance in system of systems. The levels of collaboration maturity are defined as follows:

- *Conflicting*: While systems are managed and governed independently, the plans are often conflicting. Each system exclusively focuses on its own resources and capabilities to reach their own objectives as if no other participant were present. Therefore, conflicts between participants are the rule. Their objectives are potentially mutually exclusive. Plans and execution will compete with each other.

- *Deconflicted*: Limiting the responsibility of systems or organizations to sectors of responsibility results in deconflicted operations. It ensures that organizations avoid interfering with one another in their sectors. This partition is often artificial and likely suboptimal, as it excludes synergism. The orchestration of activities in each sector is supported by special nodes in intersecting domains specifically designed for this task. Nonetheless, orchestration is limited to minimally synchronizing the execution of independent operations in independent sectors.

- *Coordinated*: The next level of maturity of operations requires joint planning driven by a shared intent or common goal. The synchronized plan allows decision making on lower levels. The execution of the plan remains the responsibility of the local system. Coordinated operations require a common intent and a common awareness, supported by broader access to shared sensors for the sake of gathering and fusing information into a common picture. This is where sociability starts to make a real difference.

- *Collaboration*: This type of operation requires not only collaboration in the planning process but also active orchestration of the execution process. Shared situational awareness supported by joint common operational pictures requires a unifying approach that integrates the heterogeneous contributions of all systems.

Information is shared seamlessly. Execution is not only synchronized but orchestrated, which means that the best system available for each job is selected.

- *Coherent*: This highest level of maturity is characterized by rapid and agile decision-making processes based on seamless and transparent information sharing. Smart and social components of each system will have access to all information they need to make the decision, regardless of where they are, which components they use to gain access to the information, or where the information came from.

Although the NATO Net-enabled Capability Command and Control Maturity Model (N2C2M2) focuses on military organizations, it provides a good foundation to be extended to a more general interoperation maturity model for SoSE. The identified levels—*conflicting*, *deconflicted*, *coordinated*, *collaboration*, and *coherent*—can be generalized from military organizations to systems in the context of SoSE.

22.4 AGENT-DIRECTED DECISION SUPPORT SIMULATION SYSTEMS

As previously discussed, M&S methods are also applied in various domains to support real-life operations. The application of M&S for optimization in industrial applications is an often featured topic in conference proceedings (Carson and Maria, 1997). One of the challenges requiring additional research, however, is the *human limit to handling complexity*. The application of decision support systems as evaluated in Yilmaz and Ören (2009) is of particular interest.

Decision support systems support business and organizational tasks of a decision maker or manager. They help him conduct his operational tasks by compiling useful information from often distributed data, documents, personal or educational knowledge, and business models and strategies. In other words, they obtain, display, and evaluate operationally relevant data.

Using simulation systems in this context introduces new capabilities to managers. They can use the traditional capability of decision support system to obtain data but in addition use the M&S methods exploiting the full potential by producing new numerical insight into the behavior of complex systems. If these simulation systems are agent based, the characteristics enumerated previously further increase the flexibility.

Agent-based decision support simulation systems have the potential to reduce the complexity human decision makers and managers have to handle. They can be used to provide more insight, enable the better evaluation of alternatives, and can help to observe the systems. Eventually, they may even help to guide emergence. However, while first applications are described in some book chapters, more research is needed.

22.5 TOPICS OF A RESEARCH AGENDA TO IMPROVE SoSE

This book, in general, and this chapter, in particular, address several of the original research agenda items identified in the 2006 workshop. While some problems have been solved by recent research results, other challenges are still obviously unresolved, but the application of M&S methods promises to offer an improvement. We make, therefore, the following recommendations regarding topics for the research agenda. This list is neither considered to be complete nor exclusive and is meant to be modified and extended by the community of scholars and practitioners to fit their needs:

1. *Taxonomy for SoSE*: A literature research on SoSE methods immediately shows that the community is not speaking a common language. Publications are full of synonyms and homonyms, which make it hard to communicate and reuse results from other relevant domains. Mapping results to a common taxonomy will support a better understanding of the common concepts. DeLaurentis' (2005) work is already a good starting point. This common understanding of concepts, relationships, and processes is also critical for the efficient use of M&S methods to improve SoSE. This is because successful M&S methods can be associated with taxonomical concepts, ensuring their reuse; otherwise, experts from other SoSE domains may not recognize application-specific elements as being relevant for them.

2. *Theoretic foundations for SoSE*: Several authors have already requested an SoSE methodology, that is, a rigorous engineering analysis that invests heavily in the understanding and framing of the problem under study (Tolk et al., 2011). In particular, modeling has been identified as a successful M&S method to support a better understanding (Zhou et al., 2011). Although these efforts are a good starting point, capturing the ideas in unambiguous and rigorous formal methods is needed as well. The authors are convinced that the field of cybernetics has the potential to better support these efforts. An example is Ashby's (1958) law of requisite variety for the control of complex systems. The application of this law will provide insight into what a common operating picture would look like for a system of systems where coherence, as addressed previously, is the objective. Beer's (1979) viable systems model has the potential to facilitate a better theoretic understanding of resilience. Finally, systems thinking (Boardman and Sauser, 2008) and its formalisms need to be reevaluated in order to provide solid theoretic foundations. The theoretic foundations for SoSE seem to be present, but they may have to be compiled from the variety of related contributing domains into a more coherent BoK.

3. *Organizational and human factors engineering*: The human limit to handling complexity and the organizational constraints for systems with operational and managerial independence continue to be unsolved challenges. Currently proposed solutions often focus on technical proposals, but technical efforts cannot solve conceptual problems. A consolidated effort that brings together management expertise, an educated workforce, and supporting technical solutions is needed. Engineering management for SoSE needs to play a pivotal role to ensure that (i) the existing and new technical solutions are recognized by academicians and management professionals and (ii) professional education of the workforce is provided.

4. *Engineering emergence*: The emergent behavior of system of systems is recognized. The topic of guided emergence was recognized already in the 2006 workshop and evolved into a broader task, namely, to actively pursue positive emergence and avoid negative emergence. Positive emergence should not be a

welcomed coincident but rather the product of engineering efforts. Now researchers are starting to work on methods to gain a deeper understanding if and how this is possible (Chen et al., 2009). Agent-based simulation is well known for its ability to produce emergence as well. Using sophisticated system of systems models to drive agent-based simulations to gain a better understanding is a logical resulting recommendation. One early application may be the use of such models within serious games to create problem awareness and better training for managers (Tolk, 2014).

5. *Cybersecurity*: Another topic that evolved significantly over the last years is security. The operational and managerial independence creates a significant challenge for secure solutions. Every interface or access point provided by contributing systems within the system of systems federation extends the attack surface exposed to potential threats. The security solutions proposed in the chapters of this book are necessary but not sufficient. Loosely coupling systems offers the rapid accessibility of new functionality, but it also opens the threat of unauthorized access or manipulation of sensitive information. As mentioned before, the development of new security protocols will make the system of systems more secure, but managerial and educational processes to raise the awareness of these problems are also needed. Again, M&S methods can support procurement, testing, and training on multiple levels. Kuhl et al. (2007) give examples of efforts that are under development, and some of them are now in operational use to train cybersecurity personnel.

6. *Model-based SoSE*: The advantages of MBSE are well recognized by the traditional Systems Engineering community by now. The use of a model-based common knowledge repository with a multitude of different facets to support customers, stakeholders, and team members of all life cycles and phases of a system in the form of a consistent systems architecture with multiple views or viewpoints is becoming a common approach. These ideas, methods, and supporting tools need to be adapted and evolved to support SoSE as well. As already discussed by Tolk et al. (2011), it is highly recommended to ensure that all artifacts are machine readable so that intelligent agents and other tools can use them to support users and managers, eventually evolving the state of the art toward M&S-based SoSE.

7. *Academic and professional SoSE education*: Although SoSE gained significant academic attention over the last year, the professional education still needs improvement, as the new knowledge has not been transferred well from academia to the workforce. Specifically, managers and commanders at all levels within the command structure of complex organizations are not aware that the professional environment that they work in day-in and day-out is an SoSE application. This can be observed in government, industry, and supporting Federally Funded Research and Development Centers. The requisite education will significantly enhance their daily situational awareness. One example in the focus of this book is the use of M&S methods. Several complex organizations still have adopted the discrete-event simulation paradigm as a "one-size-fits-all" solution that worked well in the traditional environment, without awareness of the utility of agent-based simulation or hybrid approaches. As a result, decisions are not

based on the latest scientific results and can be improved by providing better academic and professional SoSE education on all levels, including decision makers and managers. What exactly needs to get into curricula and continuous education lessons is open for discussion and needs to be captured as the research agenda progresses.

These topics for a research agenda are compiled based on the chapter contributions to this book. While this book may only serve as a starting point toward developing a common BoK and educating the workforce on the SoSE approach, the authors hope that this book will guide the action toward that goal and contribute to future research continuing this effort.

22.6 CONCLUSIONS

SoSE has reached a level of maturity that requires additional research on how to support the professional analysis, procurement, operational use, and improvements of respective systems.

The chapters in this book present a cross section of the state of the art and show how we can potentially improve our efforts with the use of M&S methods. First results are promising, but more research is needed. We need a comprehensive and concise representation of concepts, terms, and activities that make up a professional M&S domain for SoSE support.

The community of scholars and practitioners interested in SoSE is dispersed, heterogeneous, and not always interconnected. The development of broadly applicable tools and methods and the introduction of common elements in the curricula of academic and continuous professional education must be supported by upper management to enable reaching the critical mass for the next leap in this domain of M&S support of SoSE.

REFERENCES

Alberts, D.S., Huber, R.K., Moffat J. *NATO NEC C2 maturity model*. Command and Control Research Program (CCRP) Press, Washington, DC (2010).

Ashby, W.R. Requisite variety and its implications for the control of complex systems. *Cybernetica*, 1(2), 83–99 (1958).

Banks, J., Carson, J.S., Nelson, B.L., Nicol, D.M. *Discrete-Event System Simulation*, 5th Edition. Prentice - Hall, Inc., Upper Saddle River, NJ (2009).

Beer, S. *The Heart of Enterprise*. John Wiley and Sons Ltd., Chichester, UK (1979).

Birta, L.G., Arbez, G. *Modelling and Simulation: Exploring Dynamic System Behaviour*. Springer, Berlin (2007).

Boardman, J., Sauser, B. *Systems Thinking: Coping with 21st Century Problems*. CRC Press /Taylor & Francis Group, Boca Raton, FL (2008).

Carson, Y., Maria, A. Simulation optimization: methods and applications. *Proceedings of the 29th Winter Simulation Conference*, December 7–10, 1997, Atlanta, GA, pp. 118–126, IEEE Computer Society, San Diego, CA (1997).

Chen, C.C., Nagl, S.B., Clack, C.D. Complexity and emergence in engineering systems. *Complex Systems in Knowledge-based Environments: Theory, Models and Applications*, October 10–12, 2005, Waikoloa, Hawaii, pp. 99–128, Springer, Berlin and Heidelberg (2009).

DeLaurentis, D.A. A taxonomy-based perspective for systems of systems design methods. *Proceedings of the IEEE Conference on Systems, Man and Cybernetics,* Vol. 1, October 10–12, 2005, Waikoloa, Hawaii, pp. 86–91, IEEE Press (2005).

Huber, R., Langsaeter, T., Eggenhofer, P., Freire, F., Grilo, A., Grisogono, A. M., Martins, J., Roemer, J., Spaans, M. Titze, K. *The Indian Ocean Tsunami: A Case Study Investigation by NATO RTO SAS-065.* Command and Control Research Program (CCRP) Press, Washington, DC (2008).

Kewley, R.H., Wood, M. Federated Simulation for System of Systems Engineering. *Engineering Principles of Combat Modeling and Distributed Simulation,* edited by A. Tolk, Wiley, Hoboken, NY (2012).

Kuhl, M.E., Kistner, J., Costantini, K., Sudit, M. Cyber attack modeling and simulation for network security analysis. *Proceedings of the 39th Conference on Winter Simulation,* December 9–12, 2007, Washington, DC, pp. 1180–1188, IEEE Press (2007).

Mittal, S., Zeigler, B.P., Martin, J.L.R., Sahin, F., Jamshidi, M. Modeling and Simulation for System of Systems Engineering. *System of Systems Engineering for 21st Century,* edited by M. Jamshid, Wiley, Hoboken, NY (2008).

Rosen, R. *Essays on Life Itself.* Columbia University Press New York, NY (1998).

Rubinstein, R.Y., Kroese, D.P. *Simulation and the Monte Carlo Method,* 2nd Edition. Probability and Statistics, Vol. 707. Wiley, Hoboken, NJ (2011).

Tolk, A. Infranomics Simulation: Supporting System of Systems Understanding by Gaming. *Infranomics,* pp. 215–221, Springer International Publishing Switzerland (2014).

Tolk, A., Adams, K.M., Keating, C.B. Towards Intelligence-based Systems Engineering and System of Systems Engineering. *Intelligence-based Systems Engineering,* edited by A. Tolk, L. Jain. Intelligent Systems Reference Library Vol. 10, pp. 1–22, Springer, Berlin (2011).

Valerdi, R., Axelband, E., Baehren, T., Boehm, B., Dorenbos, D., Jackson, S., Madni, A., Nadler, G., Robitaille, P., Settles, S. A research agenda for systems of systems architecting. *International Journal System of Systems Engineering,* 1(1/2): 171–188 (2008).

Yilmaz, L., Ören, T. (eds.): *Agent-Directed Simulation and Systems Engineering.* Wiley, Berlin (2009).

Zhou, B., Dvoryanchikova, A., Lobov, A., Lastra, J.L.M. Modeling system of systems: A generic method based on system characteristics and interface. *Proceedings of the IEEE Conference on Industrial Informatics (INDIN),* July 26–29, 2011, Caparica, Lisbon, Portugal, pp.361–368, IEEE Press (2011).

Index

actors, 104, 106, 108–9, 116, 192, 199, 256, 309, 328, 362, 459, 492, 494
agent-based modeling, 154, 189, 215, 257–8, 266–8, 343, 359, 520, 537, 549, 556
agent-based simulation, 4–5, 9, 30–32, 46, 68, 74, 161, 164, 167, 169, 337–8, 342–3, 356, 586, 590–591
agent-oriented perspective, 7, 187, 189, 191, 193, 195, 197, 199, 201, 203, 205, 207, 209, 211, 213, 215, 217
agents, 30, 67–9, 73–4, 197–8, 200, 210, 213–14, 216, 237–8, 268–9, 342–4, 352, 354, 417–18, 431, 433, 436–9, 442, 445–7, 586
aircraft, 27, 52–5, 58, 64, 67, 88–93, 95, 152, 181, 188–9, 203, 223–5, 250, 253–5, 257–63, 265, 267, 271–2, 429
  design, 253, 261–4
  types, 260, 262–3
air transportation systems, 7, 26–7, 222, 225, 249, 251, 253, 255, 257, 259, 261, 263, 265, 267, 269, 271
architects, 20, 33–5, 38–9, 76, 79–82, 282, 335, 380, 387–8, 515–16, 532, 536, 563, 567
architecture, 33–5, 37–9, 76, 172–3, 194, 196–7, 316, 318–20, 328–9, 361, 363, 370–371, 373–4, 376, 388–9, 420, 426–7, 432–3, 486–7, 533–4
  design, 76, 114, 128, 532
  decisions, 37–8, 40
  descriptions, 6, 8, 33, 35–7, 40–41, 148, 161, 173–4, 363, 372, 390, 486, 501, 544
  development, 145, 320–322, 324
  documents, 34, 37–8
  options, 310, 316–17, 333
architecture models, 309, 316, 319–21, 323, 325–6, 329–30, 365–6, 376, 387–9
  executable, 365–6, 386, 388
artificial intelligence (AI), 66, 193, 196, 253, 417, 422, 449–51

attacks, 48, 54, 206, 223–4, 245, 290, 428, 430, 516–17, 526, 532, 541, 546, 549, 551, 554, 562, 568–70, 575–8, 580
attributes, 6, 46, 48, 50–52, 61–2, 67, 69, 76–8, 84–94, 108, 133, 170, 173, 179, 319, 325, 327, 330, 333, 566–7
authority, 17, 19, 27, 33–4, 83, 101, 103, 254–5, 267, 305, 307, 399, 482

Ballistic Missile Defense Architecture, 8, 453, 455, 457, 459, 461, 463, 465, 467, 469, 471, 473, 475, 477
Ballistic Missile Defense System, 8, 142, 453–63, 465, 467, 469, 471–7, 536–7, 540, 560, 562–4, 566, 572
battle management (BM), 9, 73, 454, 456–61, 473, 523, 531–2, 545, 548
behavior, 20–23, 24–25, 31–2, 46, 48, 133, 135, 187–9, 198–9, 266–8, 342–3, 361–73, 375–82, 386–91, 417, 419, 436–7, 446–7, 510, 518–19

capabilities, 47, 57–8, 67–9, 132, 156, 170–173, 178–81, 225–6, 230–231, 238–9, 241–3, 255–7, 321–4, 326–7, 342–3, 422, 436–8, 444–7, 474–5, 540–541
cells, dynamic, 549–50, 552, 555–7
circuit designs, 484, 500, 505, 507–8, 510
collaborative system, 13, 26–7, 221, 299
combat systems, 254, 444, 447–8
competency-based training, 423–6, 444
components, 13–17, 19–22, 24–5, 29–32, 67–8, 134, 159, 192, 211, 326–7, 342–3, 362–3, 365–8, 372, 376–7, 380–381, 418–19, 482–4, 486–7, 490
  behaviors, 362–3, 366, 380–381, 389
  interactions, 363–4, 380–381, 389, 455
  systems, 4, 13, 16–17, 19, 21, 29, 101, 221–2, 256, 265, 338–42, 366, 395, 401, 442, 482, 566

Modeling and Simulation Support for System of Systems Engineering Applications, First Edition.
Edited by Larry B. Rainey and Andreas Tolk.
© 2015 John Wiley & Sons, Inc. Published 2015 by John Wiley & Sons, Inc.

593

composability, 6, 45–9, 51, 53–7, 59, 61, 63, 65–7, 69–71, 73–4
composition of behavior models, 8, 361, 363, 365, 367, 369, 371, 373, 375, 377, 381, 387, 389, 391
composition operations, 367–70, 376, 382, 386–7
computational resources, 340–342, 346–7, 350
concept development, 404–7, 409–11
conceptual interoperability model, 6, 46, 56, 73–4, 420, 451
conceptual modeling, 6, 46, 49–53, 57, 66, 70–74, 102, 104, 109–12, 196, 206, 320, 433, 560
constituent systems, 14, 45–9, 75–6, 79–84, 86–92, 95, 97, 139–42, 251, 255, 263, 275, 376, 473–5, 482–5, 487, 553, 565–8, 570–575, 577
contracts, 164, 175–6, 178–9, 181, 306
control systems, 215, 339, 341, 468, 529, 533, 535
cyberattacks, 472, 516, 525, 532, 534, 539–40, 544–5, 557, 560, 562–3, 576
Cyber C2/BM, 533–37, 539, 541–2, 544–5, 547–9, 552–7, 559–60, 562–4
CyberOps, 532, 535–6, 542, 544–9, 551–2, 558, 563
cybersecurity, 515, 518, 532, 538–40, 551–2, 563, 575, 577–8, 590
CyberSensorNet, 557–9

data policy, 100, 102–8, 110–114, 117–19, 121, 124, 128–9, 305, 310, 314, 334
  definition, 100–101, 103, 105–9, 111, 113, 115–17, 119, 121, 129
  methodology, 6, 100, 103–4, 107, 119, 121, 128
  requirements, 103, 107, 113, 116, 118, 120, 124, 128
design
  functional, 104, 107, 116, 118–20, 124
  models, 103, 113–14, 138
  phases, 365, 388, 405–6
  range, 262–3, 265
  space, 76, 88, 94, 254, 365, 388, 465, 468, 475
  verification, 114–15, 117
discrete event simulation, 137, 151, 203, 451, 577, 579, 586, 590
distributed simulation, 51, 62, 64, 72–3, 413, 420, 591
distributed systems, 160, 515, 566, 568, 572, 574
DoD architecture design, 145, 147, 149, 151, 153, 155, 157, 159, 161, 163, 165, 167, 169, 171, 173, 175, 177, 179, 181, 183
DoD Architecture Framework (DoDAF), 6, 34, 36–7, 50, 112, 145–50, 152–4, 156–7, 161, 173–5, 180–183, 246, 365–6, 376–7, 422, 486, 488, 526, 535, 544

dynamic data-driven application system (DDDAS) framework, 7, 337–8, 340–342, 344–5, 356, 358

emergence, 12, 20–25, 32, 40, 48, 50, 63, 141–2, 189, 251, 256, 281, 400, 448–9, 584, 586, 590–591
  spooky, 22–4, 142
  strong, 13–14, 22–5, 436–7, 440, 444–6, 484–5
  weak, 22, 436–7, 440, 444, 449, 484
emergent behavior, 4–5, 13–14, 46, 48, 221–2, 251, 256, 258, 265, 267, 417–18, 436–7, 440–441, 444, 446, 448, 482–4, 584, 586, 589
emergent properties, 4, 11, 13, 20–22, 24–6, 32, 48–9, 274, 417, 518
  strong, 21, 23–4, 32
  weak, 21–2, 24, 436–7, 440, 444, 449, 484
engineered systems, 132, 136, 139, 567
environment, domestic, 219–22, 225, 244
enterprise architecture (EA) system, 437–44
European Space Agency (ESA), 6–7, 119, 129, 303–6, 308–9, 314, 316, 318, 320, 335
events, 63–5, 105, 133–5, 166–7, 172–3, 179, 181, 219–20, 339, 347, 352, 354–5, 364–70, 372–3, 381–2, 386–9, 408, 410–411, 440, 554
extracorporeal membrane oxygenation (ECMO), 8, 479–90, 492–5, 504, 506, 508, 510, 512–13
  circuit, 479, 482, 484, 488–92, 494, 496, 498–500, 502, 505, 508, 510–511
  SoS, 482–4, 487–8, 494, 496, 502, 510
  specialists, 485, 488–9, 491–6, 500–501
  therapy, 483–4, 487–8, 490, 501–2, 505

failures, 64, 131–8, 142, 145, 189, 206, 210–212, 273, 275, 297, 300, 329, 485, 510, 532, 537, 540, 571, 575–6, 584
fidelity, 53, 78–9, 87, 268, 289, 340, 346, 350–351, 356, 411, 465, 496
functional architecture, 116–17, 124, 126, 390
functional development, 403–4, 406–11

gaming, 73, 235–6, 240–244, 393, 591

high level architecture (HLA), 46, 57, 59–64, 66, 69, 193–4, 241, 247, 416, 420, 442, 450
homeland security, 7, 219–23, 225–7, 229, 231, 233–5, 237, 239, 241, 243–7, 302, 564, 574

information exchanges, 4, 14, 25, 27, 54, 124, 149, 164, 255, 321, 324–6, 328, 421, 455
information superiority, 521, 525, 528–9, 540, 551–2

Index

integratability, 6, 46, 53–6, 73
integration, 16, 55–6, 81–2, 124, 159–60, 191, 234–5, 241–2, 270–271, 305, 401–4, 406–11, 420–421, 432–3, 483–4, 535, 538–9, 548–9, 560–561, 572–3
intelligent behavior, 417, 436–7, 440, 444, 446, 449
interoperability, 46, 49, 53–8, 60–61, 63–7, 69, 72–4, 193, 195, 227, 229, 309–10, 420–423, 432, 438, 440, 446, 450, 454, 574–5

joint training, 8, 235, 242, 393–7, 399, 401, 403, 405, 407, 409, 411, 413, 420

knowledge, 4–5, 7, 9, 13, 66–7, 75–6, 134–5, 195, 198, 221, 225–7, 234–5, 238, 244–5, 418, 421, 423–4, 431, 436–8, 499–500

levels of conceptual interoperability model (LCIM), 46, 56–7, 420, 423, 451
Live Virtual Constructive (LVC), 8, 63, 194, 396, 399–400, 415–16, 419, 422–3, 426, 432, 438, 447, 450, 454, 475, 580
loop in system of systems, 8, 415, 417, 419, 421, 423, 425, 427, 429, 431, 433, 437, 439, 441, 445, 447, 449, 451

managerial control, 81–4, 86, 92, 97, 482
metamodel, 150, 215, 316, 329, 331, 333–5, 441
microgrid, 7, 337–47, 350, 352, 355–6, 358
Ministry of Defence Architecture Framework (MODAF), 37, 112, 129, 150, 316, 324
model-based approaches, 76, 83, 113, 115
Model-Based Systems Engineering, 8, 418, 423, 425, 427, 451, 479, 485–7, 499, 512, 584, 590
model design, 149, 174, 364
model development, 146–7, 152, 174, 227, 318
model elements, 173, 318, 324, 328–9, 331
modeling and simulation (M&S), 8, 247, 249, 415, 417, 419, 421, 423, 425, 427, 429, 431, 433, 437, 439, 441, 445, 447, 449–51, 453
  capabilities, 220, 225, 228, 244–5, 394, 405
  environment, 457–8, 466, 471–2
  methods, 257, 457, 585–6, 588–91
  tools, 150, 258, 270, 400, 464, 466, 471–2
modeling architectures, 388, 416, 428, 432–3, 436–7, 443, 447
modeling tools, 24, 152–3, 155–6, 330–331, 333
model methodology, 145, 147, 149, 151, 153, 155, 157, 159, 161, 163, 165, 167, 169, 171, 173, 175, 177, 179, 181, 183
monolithic systems, 12, 16–17, 27, 29–30, 55, 251, 253
Monte Carlo simulations, 60, 460–461, 585

network boundaries, 535–6, 548–9, 552, 557
networked systems, 14, 228, 571, 577–8, 580
neural networks, 23, 464–5

objectives, 18–19, 28, 30, 33, 47–8, 51, 55, 66, 178, 198–9, 228, 262, 394–5, 416, 421, 429, 540, 542, 545, 585–7
ontology, 46, 50, 66–74, 171, 418, 421–2, 436, 438–40, 442–3, 446, 449, 451, 586
operational
  activities, 117–18, 164, 166, 171, 180, 319, 322–3, 538, 561
  architecture, 9, 116, 534–7, 545, 548–50, 552, 556
  independence, 4, 13, 28, 46–7, 63, 75, 79, 81, 87, 221, 254–5, 304, 338, 356, 482, 518, 586
  nodes, 117–18, 166, 324–6, 328

problem situation, 226, 228, 277–9, 281, 289–91

radar systems, 21, 64, 442–3
reference architectures, 6, 147–8
reference model, 50–53, 148, 328
requirements, 49, 54, 60–61, 68–9, 103, 106–7, 115–16, 131–2, 190–191, 315–16, 318–21, 397–9, 401–6, 409, 428, 473–4, 486–8, 499–502, 508, 510–511
  definition, 404–7, 409–11
  functional, 229–30, 316, 401–2, 428–9
  new, 69, 119–20, 305, 404–5, 410–411
  original, 404–5, 408–9
  management, 320, 402
research agenda, 583, 585, 587, 589, 591
resiliency, 230, 517, 534, 538, 540, 542, 551
resource flows, 162, 168–9, 172, 176–7

satellites, 14–15, 48, 78, 88–93, 95–6, 121, 152, 188, 206, 305, 308, 328
security, 148, 157, 160, 164, 167–9, 175, 179, 280, 282, 305, 307–8, 329, 346–7, 533, 538, 552–3, 566–7, 569, 571–5, 577–8
  architecture, 538, 568–70, 578
  attributes, 569–70, 576, 578–9
  controls, 532–3, 540
  requirements, 223, 314, 329, 568–9, 571, 575
semantic interoperability, 8, 56–9, 63, 421, 423, 427, 432, 436–9, 446
semantic levels, 57–8, 69, 421–2, 432, 438, 440, 447
sensors, 92, 207, 294, 341, 344–5, 439, 454, 458, 461–3, 467–70, 473, 475, 495–6, 525–6, 547, 560, 563, 566–8
service-oriented rchitecture (SOA), 148–50, 157, 159–60, 162, 166, 170, 175–7, 179, 326, 422, 567, 572, 574–5, 579

596 Index

simulation
  distributed interactive, 46, 70–72, 416, 449
  components, 45, 54–5, 66–9, 155
  federations, 46, 59, 69, 236, 586
  model, 21, 51–3, 55, 150, 193, 196, 201, 206, 213, 228, 243, 338, 340–344, 346, 351
  results, 200, 203, 215, 242, 267, 269, 356, 358, 467, 585
  systems, 45, 49, 54–6, 58–61, 73, 431, 588
software architecture, 363–4, 389–91, 536
surrogate models, 455, 457–9, 461, 463–75, 477
system health management, 131, 133, 135, 137, 139, 141, 143
systemigrams, 7, 273, 275–7, 279–81, 283–91, 293, 296–7, 299, 301–2
system of systems (SoS), 7–8, 187, 189–91, 193, 195, 197, 199, 201, 203, 205, 207, 209, 211, 213, 215–17, 220, 244, 271–2, 299–301, 579
  attributes, 83–92, 97, 565, 574
  governance, 6, 100–102, 104, 128, 316
  problems, 18, 252–4, 256–9, 261, 267, 270, 456
  security, 9, 565–7, 569–75, 577–9
  tradespace exploration, 75, 77, 79, 81, 83, 85, 87, 89, 91, 93, 95, 97
systems
  architectures, 8, 41, 72, 99, 116, 136, 149–50, 153–4, 168, 182, 301, 335, 361, 363, 365–7, 369, 371, 373, 375, 377, 381, 387–9, 391, 455–6, 476, 502, 505, 529, 534–5, 539, 552
  behaviors, 134–5, 343, 362–3, 367–9, 376–7, 381, 386, 388, 417, 502, 518
  capabilities, 147, 170, 520, 557, 580
  collection of, 19–20, 251, 255, 338
  design, 86–7, 142, 285, 291, 314, 327, 361, 364, 473
  engineer, 45, 405–7, 409–11, 473
  family of, 13, 19–20, 32
  governance, 4, 99–101, 103, 105, 107, 109, 111, 113, 115, 117, 119, 121, 129–30, 159–60, 305, 310, 321, 391
  human-in-the-loop, 418–19, 425, 446
  integration, 74, 138, 145, 157, 161, 421, 450, 540
  life cycle, 49, 188, 190
  model, 11, 46, 365, 381, 486, 496, 589
  portfolio of, 13, 19–20, 29, 33
  theory, 252–3, 297, 417–18
  thinking, 274–7, 281, 293, 297, 299–301, 589, 591

Systems Engineering, 9, 41, 70–72, 98, 129–31, 142–3, 216–17, 271–2, 276, 296–7, 299–302, 304, 335–6, 449–50, 454, 456, 474, 476–7, 579–80, 591
System of Systems Engineering, 4–5, 9, 41, 45–6, 69, 71, 75, 98, 130, 191, 200–201, 216–17, 300–301, 336, 450, 583, 585, 588–91
Systems Modeling Language (SysML), 51, 152, 316, 323, 330, 366, 376–7, 380, 390–391, 486–7, 492, 494, 496, 499–500, 502, 505, 508, 510, 512–13

taxonomy, 5, 73, 161, 250, 252–3, 256, 259, 265–6, 270–271, 440, 443, 475, 532, 589
test and training enabling architecture (TENA), 46, 57, 63–7, 69, 71–2, 420, 442
test, 49, 54, 65–6, 69, 71–2, 138, 156–7, 190, 195, 206, 211, 296, 298, 300, 391, 401–4, 406–11, 413, 425, 433
test events, 63–4, 406, 408
test phase, 399, 408–9
The Open Group Architecture Framework (TOGAF), 150, 316–17, 319–20, 336
threats, 224–5, 234, 314, 329, 419, 447, 457, 467, 471, 516–17, 527, 538–9, 545, 556, 563, 571, 573, 580, 590
training, 53, 58–60, 63, 67, 69, 71–2, 74, 191, 228, 291, 294, 297–8, 393–4, 399–400, 403–4, 415–16, 418, 420, 425, 577
  environment, 393–4, 396–7, 447
  events, 420, 423, 425, 488
  systems, 52, 400–401, 423, 445

Unified Modeling Language (UML), 6, 51, 65, 102, 104, 109, 149–50, 153–4, 157, 161–2, 170, 176, 179, 183, 325, 364, 366, 376–7, 486
user requirements, 3, 116, 124, 171, 324, 405–7
users, 18–20, 22, 32, 48–9, 64–5, 157–9, 164, 191, 213–14, 226, 234, 240, 304, 307–8, 319, 331, 377–85, 398–400, 402–10, 465–6

validation, 51–2, 191, 196, 213, 258, 282, 321, 404–11, 413, 427, 433, 447, 454, 465, 520
verification, 6, 99, 101, 103, 105, 107, 109, 111, 113, 115, 117, 119, 121, 128–9, 196, 365–6, 433, 447, 454, 486
viewpoints, operational, 161, 486–7